Lecture Notes in Artificial Intelligence   9285

Subseries of Lecture Notes in Computer Science

LNAI Series Editors

Randy Goebel
  University of Alberta, Edmonton, Canada
Yuzuru Tanaka
  Hokkaido University, Sapporo, Japan
Wolfgang Wahlster
  DFKI and Saarland University, Saarbrücken, Germany

LNAI Founding Series Editor

Joerg Siekmann
  DFKI and Saarland University, Saarbrücken, Germany

More information about this series at http://www.springer.com/series/1244

Annalisa Appice · Pedro Pereira Rodrigues
Vitor Santos Costa · João Gama
Alípio Jorge · Carlos Soares (Eds.)

Machine Learning and Knowledge Discovery in Databases

European Conference, ECML PKDD 2015
Porto, Portugal, September 7–11, 2015
Proceedings, Part II

 Springer

Editors

Annalisa Appice
University of Bari Aldo Moro
Bari
Italy

Pedro Pereira Rodrigues
University of Porto
Porto
Portugal

Vitor Santos Costa
University of Porto - CRACS/INESC TEC
Porto
Portugal

João Gama
University of Porto - INESC TEC
Porto
Portugal

Alípio Jorge
University of Porto - INESC TEC
Porto
Portugal

Carlos Soares
University of Porto - INESC TEC
Porto
Portugal

ISSN 0302-9743   ISSN 1611-3349 (electronic)
Lecture Notes in Artificial Intelligence
ISBN 978-3-319-23524-0   ISBN 978-3-319-23525-7 (eBook)
DOI 10.1007/978-3-319-23525-7

Library of Congress Control Number: 2015947118

LNCS Sublibrary: SL7 – Artificial Intelligence

Printed on acid-free paper

Springer International Publishing AG Switzerland is part of Springer Science+Business Media
(www.springer.com)

Preface

We are delighted to introduce the proceedings of the 2015 edition of the European Conference on Machine Learning and Principles and Practice of Knowledge Discovery in Databases, or ECML PKDD for short. This conference stems from the former ECML and PKDD conferences, the two premier European conferences on, respectively, Machine Learning and Knowledge Discovery in Databases. Originally independent events, the two conferences were organized jointly for the first time in 2001. The sinergy between the two led to increasing integration, and eventually the two merged in 2008. Today, ECML PKDD is a world-wide leading scientific event that aims at exploiting the synergies between Machine Learning and Data Mining, focusing on the development and application of methods and tools capable of solving real-life problems.

ECML PKDD 2015 was held in Porto, Portugal, during September 7–11. This was the third time Porto hosted the major European Machine Learning event. In 1991, Porto was host to the fifth EWSL, the precursor of ECML. More recently, in 2005, Porto was host to a very successful ECML PKDD. We were honored that the community chose to again have ECML PKDD 2015 in Porto, just ten years later. The 2015 ECML PKDD was co-located with "Intelligent System Applications to Power Systems", ISAP 2015, a well-established forum for scientific and technical discussion, aiming at fostering the widespread application of intelligent tools and techniques to the power system network and business. Moreover, it was collocated, for the first time, with the Summer School on "Data Sciences for Big Data."

ECML PKDD traditionally combines the research-oriented extensive program of the scientific and journal tracks, which aim at being a forum for high quality, novel research in Machine Learning and Data Mining, with the more focused programs of the demo track, dedicated to presenting real systems to the community, the PhD track, which supports young researchers, and the nectar track, dedicated to bringing relevant work to the community. The program further includes an industrial track, which brings together participants from academia, industry, government, and non-governmental organizations in a venue that highlights practical and real-world studies of machine learning, knowledge discovery, and data mining. The industrial track of ECML PKDD 2015 has a separate Program Committee and separate proceedings volume. Moreover, the conference program included a doctoral consortium, three discovery challenges, and various workshops and tutorials.

The research program included five plenary talks by invited speakers, namely, Hendrik Blockeel (University of Leuven and Leiden University), Pedro Domingos (University of Washington), Jure Leskovec (Stanford University), Nataša Milić-Frayling (Microsoft Research), and Dino Pedreschi (Università di Pisa), as well as one ISAP +ECML PKDD joint plenary talk by Chen-Ching Liu (Washington State University). Three invited speakers contributed to the industrial track: Andreas Antrup (Zalando and

University of Edinburgh), Wei Fan (Baidu Big Data Lab), and Hang Li (Noah's Ark Lab, Huawei Technologies).

Three discovery challenges were announced this year. They focused on "MoRe-BikeS: Model Reuse with Bike rental Station data," "On Learning from Taxi GPS Traces," and "Activity Detection Based on Non-GPS Mobility Data," respectively.

Twelve workshops were held, providing an opportunity to discuss current topics in a small and interactive atmosphere: "MetaSel - Meta-learning and Algorithm Selection," "Parallel and Distributed Computing for Knowledge Discovery in Databases," "Interactions between Data Mining and Natural Language Processing," "New Frontiers in Mining Complex Patterns," "Mining Ubiquitous and Social Environments," "Advanced Analytics and Learning on Temporal Data," "Learning Models over Multiple Contexts," "Linked Data for Knowledge Discovery," "Sports Analytics", "BigTargets: Big Multi-target Prediction," "DARE: Data Analytics for Renewable Energy Integration", and "Machine Learning in Life Sciences."

Ten tutorials were included in the conference program, providing a comprehensive introduction to core techniques and areas of interest for the scientific community: "Similarity and Distance Metric Learning with Applications to Computer Vision," "Scalable Learning of Graphical Models," "Meta-learning and Algorithm Selection," "Machine Reading the Web - Beyond Named Entity Recognition and Relation Extraction," "VC-Dimension and Rademacher Averages: From Statistical Learning Theory to Sampling Algorithms," "Making Sense of (Multi-)Relational Data," "Collaborative Filtering with Binary, Positive-Only Data," "Predictive Maintenance," "Eureka! - How to Build Accurate Predictors for Real-Valued Outputs from Simple Methods," and "The Space of Online Learning Problems."

The main track received 380 paper submissions, of which 89 were accepted. Such a high volume of scientific work required a tremendous effort by the Area Chairs, Program Committee members, and many additional reviewers. We managed to collect three highly qualified independent reviews per paper and one additional overall input from one of the Area Chairs. Papers were evaluated on the basis of significance of contribution, novelty, technical quality, scientific, and technological impact, clarity, repeatability, and scholarship. The industrial, demo, and nectar tracks were equally successful, attracting 42, 32, and 29 paper submissions, respectively.

For the third time, the conference used a double submission model: next to the regular conference tracks, papers submitted to the Springer journals Machine Learning (MACH) and Data Mining and Knowledge Discovery (DAMI) were considered for presentation at the conference. These papers were submitted to the ECML PKDD 2015 special issue of the respective journals, and underwent the normal editorial process of these journals. Those papers accepted for one of these journals were assigned a presentation slot at the ECML PKDD 2015 conference. A total of 191 original manuscripts were submitted to the journal track during this year. Some of these papers are still being refereed. Of the fully refereed papers, 10 were accepted in DAMI and 15 in MACH, together with 4+4 papers from last year's call, which were also scheduled for presentation at this conference. Overall, this resulted in a number of 613 submissions (to the scientific track, industrial track and journal track), of which 126 were selected for presentation at the conference, making an overall acceptance rate of about 21%.

Part I and Part II of the proceedings of the ECML PKDD 2015 conference contain the full papers of the contributions presented in the scientific track, the abstracts of the scientific plenary talks, and the abstract of the ISAP+ECML PKDD joint plenary talk. Part III of the proceedings of the ECML PKDD 2015 conference contains the full papers of the contributions presented in the industrial track, short papers describing the demonstrations, the nectar papers, and the abstracts of the industrial plenary talks.

The scientific track program results from continuous collaboration between the scientific tracks and the general chairs. Throughout we had the unfaltering support of the Local Chairs, Carlos Ferreira, Rita Ribeiro, and João Moreira, who managed this event in a thoroughly competent and professional way. We thank the Social Media Chairs, Dunja Mladenić and Márcia Oliveira, for tweeting the new face of ECML PKDD, and the Publicity Chairs, Ricardo Campos and Carlos Ferreira, for their excellent work in spreading the news. The beautiful design and quick response time of the web site is due to the work of our Web Chairs, Sylwia Bugla, Rita Ribeiro, and João Rodrigues. The beautiful image on all the conference materials is based on the logo designed by Joana Amaral e João Cravo, inspired by Porto landmarks. It has been a pleasure to collaborate with the Journal, Industrial, Demo, Nectar, and PhD Track Chairs. ECML PKDD would not be complete if not for the efforts of the Tutorial Chairs, Fazel Famili, Mykola Pechenizkiy, and Nikolaj Tatti, the Workshop Chairs, Stan Matwin, Bernhard Pfahringer, and Luís Torgo, and the Discovery Challenge Chairs, Michel Ferreira, Hillol Kargupta, Luís Moreira-Matias, and João Moreira. We thank the Awards Committee Chairs, Pavel Brazdil, Sašo Džerosky, Hiroshi Motoda, and Michèle Sebag, for their hard work in selecting papers for awards. A special meta thanks to Pavel: ECML PKDD at Porto is only possible thanks to you. We gratefully acknowledge the work of the Sponsorship Chairs, Albert Bifet and André Carvalho, for their key work. Special thanks go to the Proceedings Chairs, Michelangelo Ceci and Paulo Cortez, for the difficult task of putting these proceedings together. We appreciate the support of Artur Aiguzhinov, Catarina Félix Oliveira, and Mohammad Nozari (U. Porto) for helping to check this front matter. We thank the ECML PKDD Steering Committee for kindly sharing their experience, and particularly the General Steering Committe Chair, Fosca Giannotti. The quality of ECML PKDD is only possible due to the tremendous efforts of the Program Committee; our sincere thanks for all the great work in improving the quality of these proceedings. Throughout, we relied on the exceptional quality of the Area Chairs. Our most sincere thanks for their support, with a special thanks to the members who contributed in difficult personal situations, and to Paulo Azevedo for stepping in when the need was there. Last but not least, we would like to sincerely thank all the authors who submitted their work to the conference.

July 2015

Annalisa Appice
Pedro Pereira Rodrigues
Vítor Santos Costa
Carlos Soares
João Gama
Alípio Jorge

Organization

ECML/PKDD 2015 Organization

Conference Co-chairs

João Gama   University of Porto, INESC TEC, Portugal
Alipío Jorge   University of Porto, INESC TEC, Portugal

Program Co-chairs

Annalisa Appice   University of Bari Aldo Moro, Italy
Pedro Pereira Rodrigues   University of Porto, CINTESIS, INESC TEC, Portugal
Vtor SantosCosta   University of Porto, INESC TEC, Portugal
Carlos Soares   University of Porto, INESC TEC, Portugal

Journal Track Chairs

Concha Bielza   Technical University of Madrid, Spain
João Gama   University of Porto, INESC TEC, Portugal
Alipío Jorge   University of Porto, INESC TEC, Portugal
Indré Žllobaité   Aalto University and University of Helsinki, Finland

Industrial Track Chairs

Albert Bifet   Huawei Noah's Ark Lab, China
Michacl May   Siemens, Germany
Bianca Zadrozny   IBM Research, Brazil

Local Organization Chairs

Carlos Ferreira   Oporto Polytechnic Institute, INESC TEC, Portugal
João Moreira   University of Porto, INESC TEC, Portugal
Rita Ribeiro   University of Porto, INESC TEC, Portugal

Tutorial Chairs

Fazel Famili   CNRC, France
Mykola Pechenizkiy   TU Eindhoven, The Netherland
Nikolaj Tatti   Aalto University, Finland

Workshop Chairs

Stan Matwin	Dalhousie University, NS, Canada
Bernhard Pfahringer	University of Waikato, New Zealand
Luís Torgo	University of Porto, INESC TEC, Portugal

Awards Committee Chairs

Pavel Brazdil	INESC TEC, Portugal
Sašo Džeroski	Jožef Stefan Institute, Slovenia
Hiroshi Motoda	Osaka University, Japan
Michèle Sebag	Université Paris Sud, France

Nectar Track Chairs

Ricard Gavaldà	UPC, Spain
Dino Pedreschi	Università di Pisa, Italy

Demo Track Chairs

Francesco Bonchi	Yahoo! Labs, Spain
Jaime Cardoso	University of Porto, INESC TEC, Portugal
Myra Spiliopoulou	Otto-von-Guericke University Magdeburg, Germany

PhD Chairs

Jaakko Hollmén	Aalto University, Finland
Panagiotis Papapetrou	Stockholm University, Sweden

Proceedings Chairs

Michelangelo Ceci	University of Bari, Italy
Paulo Cortez	University of Minho, Portugal

Discovery Challenge Chairs

Michel Ferreira	University of Porto, INESC TEC, Geolink, Portugal
Hillol Kargupta	Agnik, MD, USA
Luís Moreira-Matias	NEC Research Labs, Germany
João Moreira	University of Porto, INESC TEC, Portugal

Sponsorship Chairs

Albert Bifet	Huawei Noah's Ark Lab, China
André Carvlho	University of São Paulo, Brazil
Pedro Pereira Rodrigues	University of Porto, Portugal

Publicity Chairs

Ricardo Campos   Polytechnic Institute of Tomar, INESC TEC, Portugal
Carlos Ferreira   Oporto Polytechnic Institute, INESC TEC, Portugal

Social Media Chairs

Dunja Mladenić   JSI, Slovenia
Márcia Oliveira   University of Porto, INESC TEC, Portugal

Web Chairs

Sylwia Bugla   INESC TEC, Portugal
Rita Ribeiro   University of Porto, INESC TEC, Portugal
João Rodrigues   INESC TEC, Portugal

ECML PKDD Steering Committee

Fosca Giannotti   ISTI-CNR Pisa, Italy
Michèle Sebag   Université Paris Sud, France
Francesco Bonchi   Yahoo! Research, Spain
Hendrik Blockeel   KU Leuven, Belgium and Leiden University,
    The Netherlands
Katharina Morik   University of Dortmund, Germany
Tobias Scheffer   University of Potsdam, Germany
Arno Siebes   Utrecht University, The Netherlands
Peter Flach   University of Bristol, UK
Tijl De Bie   University of Bristol, UK
Nello Cristianini   University of Bristol, Uk
Filip Železný   Czech Technical University in Prague, Czech Republic
Siegfried Nijssen   LIACS, Leiden University, The Netherlands
Kristian Kersting   Technical University of Dortmund, Germany
Rosa Meo   Università di Torino, Italy
Toon Calders   Eindhoven University of Technology, The Netherlands
Chedy Raïssi   INRIA Nancy Grand-Est, France

Area Chairs

Paulo Azevedo   University of Minho
Michael Berthold   Universität Konstanz
Francesco Bonchi   Yahoo Labs Barcelona
Henrik Boström   University of Stockholm
Jean-Françis Boulicaut   Institut National des Sciences Appliquées de Lyon, LIRIS
Pavel Brazdil   University of Porto
André Carvalho   University of São Paulo
Michelangelo Ceci   Università degli Studi di Bari Aldo Moro

Jesse Davis	Katholieke Universiteit Leuven
Luc De Raedt	Katholieke Universiteit Leuven
Peter Flach	University of Bristol
Johannes Fürnkranz	TU Darmstadt
Thomas Gaertner	Fraunhofer IAIS
Bart Goethals	University of Antwerp
Andreas Hotho	University of Kassel
Eyke Hüllermeier	University of Paderborn
George Karypis	University of Minnesota
Kristian Kersting	Technical University of Dortmund
Arno Knobbe	Universiteit Leiden
Pedro Larrañaga	Technical University of Madrid
Peter Lucas	Radboud University Nijmegen
Donato Malerba	Università degli Studi di Bari Aldo Moro
Stan Matwin	Dalhousie University
Katharina Morik	TU Dortmund
Sriraam Natarajan	Indiana University
Eugénio Oliveira	University of Porto
Mykola Pechenizkiy	Eindhoven University of Technology
Bernhard Pfahringer	University of Waikato
Michèle Sebag	CNRS
Myra Spiliopoulou	Otto-von-Guericke University Magdeburg
Jerzy Stefanowski	Poznań University of Technology
Luís Torgo	University of Porto
Stefan Wrobel	Fraunhofer IAIS, Germany
Philip Yu	University of Illinois at Chicago

Program Committee

Leman Akoglu	Narayanaswamy	Jerzy Blaszczynski
Mehmet Sabih Aksoy	Balakrishnan	Konstantinos Blekas
Mohammad Al Hasan	Elena Baralis	Mario Boley
Omar Alonso	Daniel Barbará	Gianluca Bontempi
Aijun An	Gustavo Batista	Christian Borgelt
Aris Anagnostopoulos	Christian Bauckhage	José Luís Borges
Marta Arias	Roberto Bayardo	Marc Boullé
Rubén Armañanzas	Vaishak Belle	Ulf Brefeld
Ira Assent	András Benczúr	Róbert Busa-Fekete
Martin Atzmueller	Bettina Berendt	Toon Calders
Chloé-Agathe Azencott	Michele Berlingerio	Rui Camacho
Paulo Azevedo	Indrajit Bhattacharya	Longbing Cao
Antonio Bahamonde	Marenglen Biba	Henrique Lopes Cardoso
James Bailey	Enrico Blanzieri	Francisco Casacuberta

Gladys Castillo
Loic Cerf
Tania Cerquitelli
Edward Chang
Duen Horng Chau
Sanjay Chawla
Keke Chen
Ling Chen
Weiwei Cheng
Silvia Chiusano
Frans Coenen
Fabrizio Costa
Germán Creamer
Bruno Crémilleux
Marco Cristo
Tom Croonenborghs
Boris Cule
Tomaž Curk
James Cussens
Alfredo Cuzzocrea
Claudia d'Amato
Sašo Džeroski
Maria Damiani
Jeroen De Knijf
Gerard de Melo
Marcílio de Souto
Kurt DeGrave
Juan del Coz
Krzysztof Dembczyński
François Denis
Anne Denton
Mohamed Dermouche
Christian Desrosiers
Luigi Di Caro
Nicola Di Mauro
Jana Diesner
Ivica Dimitrovski
Ying Ding
Stephan Doerfel
Anne Driemel
Chris Drummond
Brett Drury
Devdatt Dubhashi
Wouter Duivesteijn
Bob Durrant
Inês Dutra

Tapio Elomaa
Floriana Esposito
Roberto Esposito
Hadi Fanaee-T
Nicola Fanizzi
Elaine Faria
Fabio Fassetti
Hakan Ferhatosmanoglou
Stefano Ferilli
Carlos Ferreira
Hugo Ferreira
Cèsar Ferri
George Fletcher
Eibe Frank
Élisa Fromont
Fabio Fumarola
Mohamed Medhat Gaber
Fábio Gagliardi Cozman
Patrick Gallinari
José A. Gámez
Jing Gao
Byron Gao
Paolo Garza
Éric Gaussier
Pierre Geurts
Fosca Giannotti
Christophe Giraud-Carrier
Aris Gkoulalas-Divanis
Marco Gori
Pablo Granitto
Michael Granitzer
Maria Halkidi
Jiawei Han
Daniel Hernández Lobato
José Hernández-Orallo
Thanh Lam Hoang
Frank Hoeppner
Geoff Holmes
Arjen Hommersom
Estevam Hruschka
Xiaohua Hu
Minlie Huang
Dino Ienco
Iñaki Inza
Frederik Janssen
Nathalie Japkowicz

Szymon Jaroszewicz
Ulf Johansson
Tobias Jung
Hachem Kadri
Theodore Kalamboukis
Alexandros Kalousis
U. Kang
Andreas Karwath
Hisashi Kashima
Ioannis Katakis
Mehdi Kaytoue
John Keane
Latifur Khan
Dragi Kocev
Levente Kocsis
Alek Kolcz
Irena Koprinska
Jacek Koronacki
Nitish Korula
Petr Kosina
Walter Kosters
Lars Kottof
Georg Krempl
Artus Krohn-Grimberghe
Marzena Kryszkiewicz
Matjaž Kukar
Meelis Kull
Sergei Kuznetsov
Nicolas Lachiche
Helge Langseth
Mark Last
Silvio Lattanzi
Niklas Lavesson
Nada Lavrač
Gianluca Lax
Gregor Leban
Sangkyun Lee
Wang Lee
Florian Lemmerich
Philippe Lenca
Philippe Leray
Carson Leung
Lei Li
Jiuyong Li
Juanzi Li
Edo Liberty

Hsuan-Tien Lin
Shou-de Lin
Yan Liu
Lei Liu
Corrado Loglisci
Eneldo Loza Mencía
Jose A. Lozano
Chang-Tien Lu
Panagis Magdalinos
Giuseppe Manco
Yannis Manolopoulos
Enrique Martinez
Elio Masciari
Florent Masseglia
Luís Matias
Oleksiy Mazhelis
Wannes Meert
Wagner Meira
Ernestina Menasalvas
Corrado Mencar
Rosa Meo
Pauli Miettinen
Dunja Mladenić
Anna Monreale
João Moreira
Emmanuel Müller
Mohamed Nadif
Mirco Nanni
Amedeo Napoli
Houssam Nassif
Benjamin Nguyen
Thomas Niebler
Thomas Nielsen
Siegfried Nijssen
Xia Ning
Niklas Norén
Kjetil Nørvåg
Eirini Ntoutsi
Andreas Nürnberger
Irene Ong
Salvatore Orlando
Gerhard Paaß
David Page
George Paliouras
Panče Panov
Spiros Papadimitriou

Apostolos Papadopoulos
Panagiotis Papapetrou
Ioannis Partalas
Andrea Passerini
Dino Pedreschi
Nikos Pelekis
Jing Peng
Yonghong Peng
Ruggero Pensa
Andrea Pietracaprina
Fabio Pinelli
Marc Plantevit
Pascal Poncelet
Lubos Popelinksky
George Potamias
Ronaldo Prati
Doina Precup
Ricardo Prudêncio
Kai Puolamäki
Buyue Qian
Chedy Raïssi
Liva Ralaivola
Karthik Raman
Jan Ramon
Huzefa Rangwala
Zbigniew Ras
Chotirat Ann
    Ratanamahatana
Jan Rauch
Soumya Ray
Jesse Read
Steffen Rendle
Achim Rettinger
Rita Ribeiro
Fabrizio Riguzzi
Céline Robardet
Marko Robnik-Šikonja
Juan Rodriguez
Irene Rodríguez Luján
André Rossi
Fabrice Rossi
Juho Rousu
Céline Rouveirol
Salvatore Ruggieri
Stefan Rüping
Y. van Saeys

Alan Said
Lorenza Saitta
Ansaf Salleb-Aouissi
Jose S. Sanchez
Raul Santos-Rodriguez
Sam Sarjant
Claudio Sartori
Yücel Saygin
Erik Schmidt
Lars Schmidt-Thieme
Christoph Schommer
Matthias Schubert
Marco Scutari
Thomas Seidl
Nazha Selmaoui
Giovanni Semeraro
Junming Shao
Yun Sing Koh
Andrzej Skowron
Kevin Small
Tomislav Šmuc
Yangqiu Song
Cheng Soon Ong
Arnaud Soulet
Mauro Sozio
Alessandro Sperduti
Eirini Spyropoulou
Steffen Staab
Gregor Stiglic
Markus Strohmaier
Enrique Sucar
Mahito Sugiyama
Johan Suykens
Einoshin Suzuki
Panagiotis Symeonidis
Sándor Szedmák
Andrea Tagarelli
Domenico Talia
Letizia Tanca
Dacheng Tao
Nikolaj Tatti
Maguelonne Teisseire
Alexandre Termier
Evimaria Terzi
Ljupco Todorovski
Vicenç Torra

Roberto Trasarti
Brigitte Trousse
Panayiotis Tsaparas
Vincent Tseng
Grigorios Tsoumakas
Theodoros Tzouramanis
Antti Ukkonen
Takeaki Uno
Athina Vakali
Wil van der Aalst
Guy van der Broeck
Maarten van der Heijden
Peter van der Putten
Matthijs van Leeuwen
    Putten
Martijn van Otterlo
Maarten van Someren
Joaquin Vanschoren
Iraklis Varlamis
Raju Vatsavai
Michalis Vazirgiannis

Julien Velcin
Shankar Vembu
Sicco Verwer
Vassilios Verykios
Herna Viktor
Ricardo Vilalta
Pavlovic Vladimir
Christel Vrain
Jilles Vreeken
Willem Waegeman
Byron Wallace
Fei Wang
Jianyong Wang
Yang Wang
Takashi Washio
Jörg Simon Wicker
Chun-Nam Yu
Jeffrey Yu
Jure Zabkar
Gerson Zaverucha
Demetris Zeinalipour

Filip Železný
Bernard Ženko
Junping Zhang
Kun Zhang
Lei Zhang
Min-Ling Zhang
Nan Zhang
Shichao Zhang
Zhongfei Zhang
Liang Zhao
Ying Zhao
Elena Zheleva
Bin Zhou
Kenny Zhu
Xiaofeng Zhu
Djamel Zighed
Arthur Zimek
Albrecht Zimmermann
Blaž Zupan

Additional Reviewers

Greet Daldewijns
Jessa Bekker
Nuno Castro
Shiyu Chang
Yu Cheng
Paolo Cintia
Heidar Davoudi
Thomas Delacroix
Martin Dimkovski
Michael Färber
Ricky Fok
Emanuele Frandi
Tatiana Gossen
Valerio Grossi
Riccardo Guidotti
Ming Jiang
Nikos Katzouris

Sebastian Kauschke
Jinseok Kim
Jan Kralj
Thomas Low
Stijn Luca
Rafael Mantovani
Pasquale Minervini
Shubhanshu Mishra
Christos Perentis
Fábio Pinto
Dimitrios Rafailidis
Giulio Rossetti
Alexandros Sarafianos
Antonio Vergari
Dimtrios Vogiatzis
Andreas Zioupos

Sponsors

Platinum Sponsors

BNP PARIBAS	http://www.bnpparibas.com/
ONR Global	www.onr.navy.mil/science-technology/onr-global.aspx

Gold Sponsors

Zalando	https://www.zalando.co.uk/
HUAWEI	http://www.huawei.com/en/

Silver Sponsors

Deloitte	http://www2.deloitte.com/
Amazon	http://www.amazon.com/

Bronze Sponsors

Xarevision	http://xarevision.pt/
Farfetch	http://www.farfetch.com/pt/
NOS	http://www.nos.pt/particulares/Pages/home.aspx

Award Sponsor

Machine Learning	http://link.springer.com/journal/10994
Data Mining and Knowledge	http://link.springer.com/journal/10618
Discovery Deloitte	http://www2.deloitte.com/

Lanyard Sponsor

KNIME	http://www.knime.org/

Invited Talk Sponsors

ECCAI	http://www.eccai.org/
Cliqz	https://cliqz.com/
Technicolor	http://www.technicolor.com/
University of Bari Aldo Moro	http://www.uniba.it/english-version

Additional Supporters

INESCTEC	https://www.inesctec.pt/
University of Porto, Faculdade de Economia	http://sigarra.up.pt/fep/pt/web_page.inicial
Springer	http://www.springer.com/
University of Porto	http://www.up.pt/

Official Carrier

TAP	http://www.flytap.com/

Abstracts of Journal Track Articles

A Bayesian Approach for Comparing Cross-Validated Algorithms on Multiple Data Sets

Giorgio Corani and Alessio Benavoli
Machine Learning
DOI: 10.1007/s10994-015-5486-z

We present a Bayesian approach for making statistical inference about the accuracy (or any other score) of two competing algorithms which have been assessed via cross-validation on multiple data sets. The approach is constituted by two pieces. The first is a novel correlated Bayesian t-test for the analysis of the cross-validation results on a single data set which accounts for the correlation due to the overlapping training sets. The second piece merges the posterior probabilities computed by the Bayesian correlated *t*-test on the different data sets to make inference on multiple data sets. It does so by adopting a Poisson-binomial model. The inferences on multiple data sets account for the different uncertainty of the cross-validation results on the different data sets. It is the first test able to achieve this goal. It is generally more powerful than the signed-rank test if ten runs of cross-validation are performed, as it is anyway generally recommended.

A Decomposition of the Outlier Detection Problem into a Set of Supervised Learning Problems

Heiko Paulheim and Robert Meusel
Machine Learning
DOI: 10.1007/s10994-015-5507-y

Outlier detection methods automatically identify instances that deviate from the majority of the data. In this paper, we propose a novel approach for unsupervised outlier detection, which re-formulates the outlier detection problem in numerical data as a set of supervised regression learning problems. For each attribute, we learn a predictive model which predicts the values of that attribute from the values of all other attributes, and compute the deviations between the predictions and the actual values. From those deviations, we derive both a weight for each attribute, and a final outlier score using those weights. The weights help separating the relevant attributes from the irrelevant ones, and thus make the approach well suitable for discovering outliers otherwise masked in high-dimensional data. An empirical evaluation shows that our approach outperforms existing algorithms, and is particularly robust in datasets with many irrelevant attributes. Furthermore, we show that if a symbolic machine learning method is used to solve the individual learning problems, the approach is also capable of generating concise explanations for the detected outliers.

Assessing the Impact of a Health Intervention via User-Generated Internet Content

Vasileios Lampos, Elad Yom-Tov, Richard Pebody, and Ingemar J. Cox
Data Mining and Knowledge Discovery
DOI: 10.1007/s10618-015-0427-9

Assessing the effect of a health-oriented intervention by traditional epidemiological methods is commonly based only on population segments that use healthcare services. Here we introduce a complementary framework for evaluating the impact of a targeted intervention, such as a vaccination campaign against an infectious disease, through a statistical analysis of user-generated content submitted on web platforms. Using supervised learning, we derive a nonlinear regression model for estimating the prevalence of a health event in a population from Internet data. This model is applied to identify control location groups that correlate historically with the areas, where a specific intervention campaign has taken place. We then determine the impact of the intervention by inferring a projection of the disease rates that could have emerged in the absence of a campaign. Our case study focuses on the influenza vaccination program that was launched in England during the 2013/14 season, and our observations consist of millions of geo-located search queries to the Bing search engine and posts on Twitter. The impact estimates derived from the application of the proposed statistical framework support conventional assessments of the campaign.

Beyond Rankings: Comparing Directed Acyclic Graphs

Eric Malmi, Nikolaj Tatti, Aristides Gionis
Data Mining and Knowledge Discovery
DOI: 10.1007/s10618-015-0406-1

Defining appropriate distance measures among rankings is a classic area of study which has led to many useful applications. In this paper, we propose a more general abstraction of preference data, namely directed acyclic graphs (DAGs), and introduce a measure for comparing DAGs, given that a vertex correspondence between the DAGs is known. We study the properties of this measure and use it to aggregate and cluster a set of DAGs. We show that these problems are NP-hard and present efficient methods to obtain solutions with approximation guarantees. In addition to preference data, these methods turn out to have other interesting applications, such as the analysis of a collection of information cascades in a network. We test the methods on synthetic and real-world datasets, showing that the methods can be used to, e.g., find a set of influential individuals related to a set of topics in a network or to discover meaningful and occasionally surprising clustering structure.

Clustering Boolean Tensors

Saskia Metzler and Pauli Miettinen
Data Mining and Knowledge Discovery
DOI: 10.1007/s10618-015-0420-3

Graphs - such as friendship networks - that evolve over time are an example of data that are naturally represented as binary tensors. Similarly to analysing the adjacency matrix of a graph using a matrix factorization, we can analyse the tensor by factorizing it. Unfortunately, tensor factorizations are computationally hard problems, and in particular, are often significantly harder than their matrix counterparts. In case of Boolean tensor factorizations - where the input tensor and all the factors are required to be binary and we use Boolean algebra - much of that hardness comes from the possibility of overlapping components. Yet, in many applications we are perfectly happy to partition at least one of the modes. For instance, in the aforementioned timeevolving friendship networks, groups of friends might be overlapping, but the time points at which the network was captured are always distinct. In this paper we investigate what consequences this partitioning has on the computational complexity of the Boolean tensor factorizations and present a new algorithm for the resulting clustering problem. This algorithm can alternatively be seen as a particularly regularized clustering algorithm that can handle extremely high-dimensional observations. We analyse our algorithm with the goal of maximizing the similarity and argue that this is more meaningful than minimizing the dissimilarity. As a by product we obtain a PTAS and an efficient 0.828-approximation algorithm for rank-1 binary factorizations. Our algorithm for Boolean tensor clustering achieves high scalability, high similarity, and good generalization to unseen data with both synthetic and realworld data sets.

Consensus Hashing

Cong Leng and Jian Cheng
Machine Learning
DOI: 10.1007/s10994-015-5496-x

Hashing techniques have been widely used in many machine learning applications because of their efficiency in both computation and storage. Although a variety of hashing methods have been proposed, most of them make some implicit assumptions about the statistical or geometrical structure of data. In fact, few hashing algorithms can adequately handle all kinds of data with different structures. When considering hybrid structure datasets, different hashing algorithms might produce different and possibly inconsistent binary codes. Inspired by the successes of classifier combination and clustering ensembles, in this paper, we present a novel combination strategy for multiple hashing results, named Consensus Hashing (CH). By defining the measure of consensus of two hashing results, we put forward a simple yet effective model to learn consensus hash functions which generate binary codes consistent with the existing ones. Extensive experiments on several large scale benchmarks demonstrate the overall superiority of the proposed method compared with state-of-the art hashing algorithms.

Convex Relaxations of Penalties for Sparse Correlated Variables With Bounded Total Variation

Eugene Belilovsky, Andreas Argyriou, Gael Varoquaux, Matthew B. Blaschko
Machine Learning
DOI: 10.1007/s10994-015-5511-2

We study the problem of statistical estimation with a signal known to be sparse, spatially contiguous, and containing many highly correlated variables. We take inspiration from the recently introduced k-support norm, which has been successfully applied to sparse prediction problems with correlated features, but lacks any explicit structural constraints commonly found in machine learning and image processing. We address this problem by incorporating a total variation penalty in the k-support framework. We introduce the (k,s) support total variation norm as the tightest convex relaxation of the intersection of a set of sparsity and total variation constraints. We show that this norm leads to an intractable combinatorial graph optimization problem, which we prove to be NP-hard. We then introduce a tractable relaxation with approximation guarantees that scale well for grid structured graphs. We devise several first-order optimization strategies for statistical parameterestimation with the described penalty. We demonstrate the effectiveness of this penalty on classification in the low sample regime, classification with M/EEG neuroimaging data, and image recovery with synthetic and real data background subtracted image recovery tasks. We extensively analyse the application of our penalty on the complex task of identifying predictive regions from low-sample high-dimensional fMRI brain data, we show that our method is particularly useful compared to existing methods in terms of accuracy, interpretability, and stability.

Direct Conditional Probability Density Estimation with Sparse Feature Selection

Motoki Shiga, Voot Tangkaratt, and Masashi Sugiyama
Machine Learning
DOI: 10.1007/s10994-014-5472-x

Regression is a fundamental problem in statistical data analysis, which aims at estimating the conditional mean of output given input. However, regression is not informative enough if the conditional probability density is multi-modal, asymmetric, and heteroscedastic. To overcome this limitation, various estimators of conditional densities themselves have been developed, and a kernel-based approach called leastsquares conditional density estimation (LS-CDE) was demonstrated to be promising. However, LS-CDE still suffers from large estimation error if input contains many irrelevant features. In this paper, we therefore propose an extension of LS-CDE called sparse additive CDE (SA-CDE), which allows automatic feature selection in CDE. SACDE applies kernel LS-CDE to each input feature in an additive manner and penalizes the whole solution by a group-sparse regularizer. We also give a subgradient-based optimization method for SA-CDE training that scales well to high-dimensional large data sets. Through experiments with benchmark and humanoid robot transition datasets, we demonstrate the usefulness of SA-CDE in noisy CDE problems.

DRESS: Dimensionality Reduction for Efficient Sequence Search

Alexios Kotsifakos, Alexandra Stefan, Vassilis Athitsos, Gautam Das, and Panagiotis Papapetrou
Data Mining and Knowledge Discovery
DOI: 10.1007/s10618-015-0413-2

Similarity search in large sequence databases is a problem ubiquitous in a wide range of application domains, including searching biological sequences. In this paper we focus on protein and DNA data, and we propose a novel approximate method method for speeding up range queries under the edit distance. Our method works in a filter-and-refine manner, and its key novelty is a query-sensitive mapping that transforms the original string space to a new string space of reduced dimensionality. Specifically, it first identifies the most frequent codewords in the query, and then uses these codewords to convert both the query and the database to a more compact representation. This is achieved by replacing every occurrence of each codeword with a new letter and by removing the remaining parts of the strings. Using this new representation, our method identifies a set of candidate matches that are likely to satisfy the range query, and finally refines these candidates in the original space. The main advantage of our method, compared to alternative methods for whole sequence matching under the edit distance, is that it does not require any training to create the mapping, and it can handle large query lengths with negligible losses in accuracy. Our experimental evaluation demonstrates that, for higher range values and large query sizes, our method produces significantly lower costs and runtimes compared to two state-of-the-art competitor methods.

Dynamic Inference of Social Roles in Information Cascade

Sarvenaz Choobdar, Pedro Ribeiro, Srinivasan Parthasarathy,
and Fernando Silva

Data Mining and Knowledge Discovery
DOI: 10.1007/s10618-015-0402-5

Nodes in complex networks inherently represent different kinds of functional or organizational roles. In the dynamic process of an information cascade, users play different roles in spreading the information: some act as seeds to initiate the process, some limit the propagation and others are in-between. Understanding the roles of users is crucial in modeling the cascades. Previous research mainly focuses on modeling users behavior based upon the dynamic exchange of information with neighbors. We argue however that the structural patterns in the neighborhood of nodes may already contain enough information to infer users' roles, independently from the information flow in itself. To approach this possibility, we examine how network characteristics of users affect their actions in the cascade. We also advocate that temporal information is very important. With this in mind, we propose an unsupervised methodology based on ensemble clustering to classify users into their social roles in a network, using not only their current topological positions, but also considering their history over time. Our experiments on two social networks, Flickr and Digg, show that topological metrics indeed possess discriminatory power and that different structural patterns correspond to different parts in the process. We observe that user commitment in the neighborhood affects considerably the influence score of users. In addition, we discover that the cohesion of neighborhood is important in the blocking behavior of users. With this we can construct topological fingerprints that can help us in identifying social roles, based solely on structural social ties, and independently from nodes activity and how information flows.

Efficient and Effective Community Search

Nicola Barbieri, Francesco Bonchi, Edoardo Galimberti,
and Francesco Gullo

Data Mining and Knowledge Discovery
DOI: 10.1007/s10618-015-0422-1

Community search is the problem of finding a good community for a given set of query vertices. One of the most studied formulations of community search asks for a connected subgraph that contains all query vertices and maximizes the minimum degree. All existing approaches to min-degree-based community search suffer from limitations concerning efficiency, as they need to visit (large part of) the whole input graph, as well as accuracy, as they output communities quite large and not really cohesive. Moreover, some existing methods lack generality: they handle only single-vertex queries, find communities that are not optimal in terms of minimum degree, and/or require input parameters. In this work we advance the state of the art on community search by proposing a novel method that overcomes all these limitations: it is in general more efficient and effective—one/two orders of magnitude on average, it can handle multiple query vertices, it yields optimal communities, and it is parameter-free. These properties are confirmed by an extensive experimental analysis performed on various real-world graphs.

Finding the Longest Common Sub-Pattern in Sequences of Temporal Intervals

Orestis Kostakis and Panagiotis Papapetrou

Data Mining and Knowledge Discovery
DOI: 10.1007/s10618-015-0404-3

We study the problem of finding the Longest Common Sub-Pattern (LCSP) shared by two sequences of temporal intervals. In particular we are interested in finding the LCSP of the corresponding arrangements. Arrangements of temporal intervals are a powerful way to encode multiple concurrent labeled events that have a time duration. Discovering commonalities among such arrangements is useful for a wide range of scientific fields and applications, as it can be seen by the number and diversity of the datasets we use in our experiments. In this paper, we define the problem of LCSP and prove that it is NP-complete by demonstrating a connection between graphs and arrangements of temporal intervals, which leads to a series of interesting open problems. In addition, we provide an exact algorithm to solve the LCSP problem, and also propose and experiment with three polynomial time and space underapproximation techniques. Finally, we introduce two upper bounds for LCSP and study their suitability for speeding up 1-NN search. Experiments are performed on seven datasets taken from a wide range of real application domains, plus two synthetic datasets.

Generalization Bounds for Learning with Linear, Polygonal, Quadratic and Conic Side Knowledge

Theja Tulabandhula and Cynthia Rudin

Machine Learning
DOI: 10.1007/s10994-014-5478-4

In this paper, we consider a supervised learning setting where side knowledge is provided about the labels of unlabeled examples. The side knowledge has the effect of reducing the hypothesis space, leading to tighter generalization bounds, and thus possibly better generalization. We consider several types of side knowledge, the first leading to linear and polygonal constraints on the hypothesis space, the second leading to quadratic constraints, and the last leading to conic constraints. We show how different types of domain knowledge can lead directly to these kinds of side knowledge.

We prove bounds on complexity measures of the hypothesis space for quadratic and conic side knowledge, and show that these bounds are tight in a specific sense for the quadratic case.

Generalization of Clustering Agreements and Distances for Overlapping Clusters and Network Communities

Reihaneh Rabbany and Osmar R. Zaiane
Data Mining and Knowledge Discovery
DOI: 10.1007/s10618-015-0426-x

A measure of distance between two clusterings has important applications, including clustering validation and ensemble clustering. Generally, such distance measure provides navigation through the space of possible clusterings. Mostly used in cluster validation, a normalized clustering distance, a.k.a. agreement measure, compares a given clustering result against the ground-truth clustering. The two widely-used clustering agreement measures are Adjusted Rand Index (ARI) and Normalized Mutual Information (NMI). In this paper, we present a generalized clustering distance from which these two measures can be derived. We then use this generalization to construct new measures specific for comparing (dis)agreement of clusterings in networks, a.k.a. communities. Further, we discuss the difficulty of extending the current, contingency based, formulations to overlapping cases, and present an alternative algebraic formulation for these (dis)agreement measures. Unlike the original measures, the new co-membership based formulation is easily extendable for different cases, including overlapping clusters and clusters of inter-related data. These two extensions are, in particular, important in the context of finding communities in complex networks.

Generalized Twin Gaussian Processes Using Sharma-Mittal Divergence

Mohamed Elhoseiny and Ahmed Elgammal
Machine Learning
DOI: 10.1007/s10994-015-5497-9

There has been a growing interest in mutual information measures due to its wide range of applications in Machine Learning and Computer Vision. In this manuscript, we present a generalized structured regression framework based on Shama-Mittal divergence, a relative entropy measure, firstly addressed in the Machine Learning community, in this work. Sharma-Mittal (SM) divergence is a generalized mutual information measure for the widely used Rényi, Tsallis, Bhattacharyya, and Kullback-Leibler (KL) relative entropies. Specifically, we study Sharma-Mittal divergence as a cost function in the context of the Twin Gaussian Processes, which generalizes over the KL-divergence without computational penalty. We show interesting properties of Sharma-Mittal TGP (SMTGP) through a theoretical analysis, which covers missing insights in the traditional TGP formulation. However, we generalize this theory based on SM-divergence instead of KL-divergence which is a special case. Experimentally, we evaluated the proposed SMTGP framework on several datasets. The results show that SMTGP reaches better predictions than KL-based TGP (KLTGP), since it offers a bigger class of models through its parameters that we learn from the data.

Half-Space Mass: A Maximally Robust and Efficient Data Depth Method

Bo Chen, Kai Ming Ting, Takashi Washio, and Gholamreza Haffari
Machine Learning
DOI: 10.1007/s10994-015-5524-x

Data depth is a statistical method which models data distribution in terms of centeroutward ranking rather than density or linear ranking. While there are a lot of academic interests, its applications are hampered by the lack of a method which is both robust and efficient. This paper introduces Half-Space Mass which is a significantly improved version of half-space data depth. Half-Space Mass is the only data depth method which is both robust and efficient, as far as we know. We also reveal four theoretical properties of Half-Space Mass: (i) its resultant mass distribution is concave regardless of the underlying density distribution, (ii) its maximum point is unique which can be considered as median, (iii) the median is maximally robust, and (iv) its estimation extends to a higher dimensional space in which the convex hull of the dataset occupies zero volume. We demonstrate the power of Half-Space Mass through its applications in two tasks. In anomaly detection, being a maximally robust location estimator leads directly to a robust anomaly detector that yields a better detection accuracy than halfspace depth; and it runs orders of magnitude faster than L2 depth, an existing maximally robust location estimator. In clustering, the Half-Space Mass version of Kmeans overcomes three weaknesses of K-means.

Improving Classification Performance Through Selective Instance Completion

Amit Dhurandhar and Karthik Sankarnarayanan
Machine Learning
DOI: 10.1007/s10994-015-5500-5

In multiple domains, actively acquiring missing input information at a reasonable cost in order to improve our understanding of the input-output relationships is of increasing importance. This problem has gained prominence in healthcare, public policy making, education, and in the targeted advertising industry which tries to best match people to products. In this paper we tackle an important variant of this problem: Instance Completion, where we want to choose the best k incomplete instances to query from a much larger universe of N($\gg$k) incomplete instances so as to learn the most accurate classifier. We propose a principled framework which motivates a generally applicable yet efficient meta-technique for choosing k such instances. Since we cannot know *a priori* the classifier that will result from the completed dataset, i.e. the final classifier, our method chooses the k instances based on a derived upper bound on the expectation of the distance between the next classifier and the final classifier. We additionally derive a sufficient condition for these two solutions to match. We then empirically evaluate the performance of our method relative to the state-of-the-art methods on 4 UCI datasets as well as 3 proprietary e-commerce datasets used in previous studies. In these experiments, we also demonstrate how close we are likely to be to the optimal solution, by quantifying the extent to which our sufficient condition is satisfied. Lastly, we show that our method is easily extensible to the setting where we have a non uniform cost associated with acquiring the missing information.

Incremental Learning of Event Definitions with Inductive Logic Programming

Nikos Katzouris, Alexander Artikis, and Georgios Paliouras
Machine Learning
DOI: 10.1007/s10994-015-5512-1

Event recognition systems rely on knowledge bases of event definitions to infer occurrences of events in time. Using a logical framework for representing and reasoning about events offers direct connections to machine learning, via Inductive Logic Programming (ILP), thus allowing to avoid the tedious and error-prone task of manual knowledge construction. However, learning temporal logical formalisms, which are typically utilized by logic-based event recognition systems is a challenging task, which most ILP systems cannot fully undertake. In addition, event-based data is usually massive and collected at different times and under various circumstances. Ideally, systems that learn from temporal data should be able to operate in an incremental mode, that is, revise prior constructed knowledge in the face of new evidence. In this work we present an incremental method for learning and revising event-based knowledge, in the form of Event Calculus programs. The proposed algorithmrelies on abductive-inductive learning and comprises a scalable clause refinement methodology, based on a compressive summarization of clause coverage in a stream of examples. We present an empirical evaluation of our approach on real and synthetic data from activity recognition and city transport applications.

Knowledge Base Completion by Learning Pairwise-Interaction Differentiated Embeddings

Yu Zhao, Sheng Gao, Patrick Gallinari, and Jun Guo
Data Mining and Knowledge Discovery
DOI: 10.1007/s10618-015-0430-1

Knowledge base consisting of triple like (subject entity, predicate relation, object entity) is a very important database for knowledge management. It is very useful for humanlike reasoning, query expansion, question answering (Siri) and other related AI tasks. However, knowledge base often suffers from incompleteness due to a large volume of increasing knowledge in the real world and a lack of reasoning capability. In this paper, we propose a Pairwise-interaction Differentiated Embeddings (PIDE) model to embed entities and relations in the knowledge base to low dimensional vector representations and then predict the possible truth of additional facts to extend the knowledge base. In addition, we present a probability-based objective function to improve the model optimization. Finally, we evaluate the model by considering the problem of computing how likely the additional triple is true for the task of knowledge base completion.Experiments on WordNet and Freebase dataset show the excellent performance of our model and algorithm.

Learning from Evolving Video Streams in a Multi-camera Scenario

Samaneh Khoshrou, Jaime dos Santos Cardoso, and Luís Filipe Teixeira
Machine Learning
DOI: 10.1007/s10994-015-5515-y

Nowadays, video surveillance systems are taking the first steps toward automation, in order to ease the burden on human resources as well as to avoid human error. As the underlying data distribution and the number of concepts change over time, the conventional learning algorithms fail to provide reliable solutions for this setting. Herein, we formalize a learning concept suitable for multi-camera video surveillance and propose a learning methodology adapted to that new paradigm. The proposed framework resorts to the universal background model to robustly learn individual object models from small samples and to more effectively detect novel classes. The individual models are incrementally updated in an ensemble based approach, with older models being progressively forgotten. The framework is designed to detect and label new concepts automatically. The system is also designed to exploit active learning strategies, in order to interact wisely with operator, requesting assistance in the most ambiguous to classify observations. The experimental results obtained both on real and synthetic data sets verify the usefulness of the proposed approach.

Learning Relational Dependency Networks in Hybrid Domains

Irma Ravkic, Jan Ramon, and Jesse Davis

Machine Learning
DOI: 10.1007/s10994-015-5483-2

Statistical Relational Learning (SRL) is concerned with developing formalisms for representing and learning from data that exhibit both uncertainty and complex, relational structure. Most of the work in SRL has focused on modeling and learning from data that only contain discrete variables. As many important problems are characterized by the presence of both continuous and discrete variables, there has been a growing interest in developing hybrid SRL formalisms. Most of these formalisms focus on reasoning and representational issues and, in some cases, parameter learning. What has received little attention is learning the structure of a hybrid SRL model from data. In this paper, we fill that gap and make the following contributions. First, we propose Hybrid Relational Dependency Networks (HRDNs), an extension to Relational Dependency Networks that are able to model continuous variables. Second, we propose an algorithm for learning both the structure and parameters of an HRDN from data. Third, we provide an empirical evaluation that demonstrates that explicitly modeling continuous variables results in more accurate learned models than discretizing them prior to learning.

MassExodus: Modeling Evolving Networks in Harsh Environments

Saket Navlakha, Christos Faloutsos, and Ziv Bar-Joseph

Data Mining and Knowledge Discovery
DOI: 10.1007/s10618-014-0399-1

Defining appropriate distance measures among rankings is a classic area of study which has led to many useful applications. In this paper, we propose a more general abstraction of preference data, namely directed acyclic graphs (DAGs), and introduce a measure for comparing DAGs, given that a vertex correspondence between the DAGs is known. We study the properties of this measure and use it to aggregate and cluster a set of DAGs. We show that these problems are NP-hard and present efficient methods to obtain solutions with approximation guarantees. In addition to preference data, these methods turn out to have other interesting applications, such as the analysis of a collection of information cascades in a network. We test the methods on synthetic and real-world datasets, showing that the methods can be used to, e.g., find a set of influential individuals related to a set of topics in a network or to discover meaningful and occasionally surprising clustering structure.

Minimum Message Length Estimation of Mixtures of Multivariate Gaussian and von Mises-Fisher Distribution

Parthan Kasarapu and Lloyd Allison
Machine Learning
DOI: 10.1007/s10994-015-5493-0

Mixture modelling involves explaining some observed evidence using a combination of probability distributions. The crux of the problem is the inference of an optimal number of mixture components and their corresponding parameters. This paper discusses unsupervised learning of mixture models using the Bayesian Minimum Message Length (MML) criterion. To demonstrate the effectiveness of search and inference of mixture parameters using the proposed approach, we select two key probability distributions, each handling fundamentally different types of data: the multivariate Gaussian distribution to address mixture modelling of data distributed in Euclidean space, and the multivariate von Mises-Fisher (vMF) distribution to address mixture modelling of directional data distributed on a unit hypersphere. The key contributions of this paper, in addition to the general search and inference methodology, include the derivation of MML expressions for encoding the data using multivariate Gaussian and von Mises-Fisher distributions, and the analytical derivation of the MML estimates of the parameters of the two distributions. Our approach is tested on simulated and real world data sets. For instance, we infer vMF mixtures that concisely explain experimentally determined three dimensional protein conformations, providing an effective null model description of protein structures that is central to many inference problems in structural bioinformatics. The experimental results demonstrate that the performance of our proposed search and inference method along with the encoding schemes improve on the state of the art mixture modelling techniques.

Mining Outlying Aspects on Numeric Data

Lei Duan, Guanting Tang, Jian Pei, James Bailey,
Akiko Campbell, and Changjie Tang
Data Mining and Knowledge Discovery
DOI: 10.1007/s10618-014-0398-2

When we are investigating an object in a data set, which itself may or may not be an outlier, can we identify unusual (i.e., outlying) aspects of the object? In this paper, we identify the novel problem of mining outlying aspects on numeric data. Given a query object o in a multidimensional numeric data set O, in which subspace is o most outlying? Technically, we use the rank of the probability density of an object in a subspace to measure the outlyingness of the object in the subspace. A minimal subspace where the query object is ranked the best is an outlying aspect. Computing the outlying aspects of a query object is far from trivial. A naïve method has to calculate the probability densities of all objects and rank them in every subspace, which is very costly when the dimensionality is high. We systematically develop a heuristic method that is capable of searching data sets with tens of dimensions efficiently. Our empirical study using both real data and synthetic data demonstrates that our method is effective and efficient.

Multiscale Event Detection in Social Media

Xiaowen Dong, Dimitrios Mavroeidis, Francesco Calabrese,
Pascal Frossard
Data Mining and Knowledge Discovery
DOI: 10.1007/s10618-015-0421-2

Event detection has been one of the most important research topics in social media analysis. Most of the traditional approaches detect events based on fixed temporal and spatial resolutions, while in reality events of different scales usually occur simultaneously, namely, they span different intervals in time and space. In this paper, we propose a novel approach towards multiscale event detection using social media data, which takes into account different temporal and spatial scales of events in the data. Specifically, we explore the properties of the wavelet transform, which is a welldeveloped multiscale transform in signal processing, to enable automatic handling of the interaction between temporal and spatial scales. We then propose a novel algorithm to compute a data similarity graph at appropriate scales and detect events of different scales simultaneously by a single graph-based clustering process. Furthermore, we present spatiotemporal statistical analysis of the noisy information present in the data stream, which allows us to define a novel term-filtering procedure for the proposed event detection algorithm and helps us study its behavior using simulated noisy data. Experimental results on both synthetically generated data and real world data collected from Twitter demonstrate the meaningfulness and effectiveness of the proposed approach. Our framework further extends to numerous application domains that involve multiscale and multiresolution data analysis.

Optimised Probabilistic Active Learning (OPAL) for Fast, Non-Myopic, Cost-Sensitive Active Classification

Georg Krempl, Daniel Kottke, and Vincent Lemaire
Machine Learning
DOI: 10.1007/s10994-015-5504-1

In contrast to ever increasing volumes of automatically generated data, human annotation capacities remain limited. Thus, fast active learning approaches that allow the efficient allocation of annotation efforts gain in importance. Furthermore, cost-sensitive applications such as fraud detection pose the additional challenge of differing misclassification costs between classes. Unfortunately, the few existing cost-sensitive active learning approaches rely on time-consuming steps, such as performing self labelling or tedious evaluations over samples. We propose a fast, non-myopic, and cost-sensitive probabilistic active learning approach for binary classification. Our approach computes the expected reduction in misclassification loss in a labelling candidate's neighbourhood. We derive and use a closed-form solution for this expectation, which considers the possible values of the true posterior of the positive class at the candidate's position, its possible label realisations, and the given labelling budget. The resulting myopic algorithm runs in the same linear asymptotic time as uncertainty sampling, while its non-myopic counterpart requires an additional factor of $O(m \log m)$ in the budget size. The experimental evaluation on several synthetic and real-world data sets shows competitive or better classification performance and runtime, compared to several uncertainty sampling- and error-reduction-based active learning strategies, both in cost-sensitive and cost-insensitive settings.

Poisson Dependency Networks - Gradient Boosted Models for Multivariate Count Data

Fabian Hadiji, Alejandro Molina, Sriraam Natarajan, and Kristian Kersting
Machine Learning
DOI: 10.1007/s10994-015-5506-z

Although count data are increasingly ubiquitous, surprisingly little work has employed probabilistic graphical models for modeling count data. Indeed the univariate case has been well studied, however, in many situations counts influence each other and should not be considered independently. Standard graphical models such as multinomial or Gaussian ones are also often ill-suited, too, since they disregard either the infinite range over the natural numbers or the potentially asymmetric shape of the distribution of count variables. Existing classes of Poisson graphical models can only model negative conditional dependencies or neglect the prediction of counts or do not scale well. To ease the modeling of multivariate count data, we therefore introduce a novel family of Poisson graphical models, called Poisson Dependency Networks (PDNs). A PDN consists of a set of local conditional Poisson distributions, each representing the probability of a single count variable given the others, that naturally facilities a simple Gibbs sampling inference. In contrast to existing Poisson graphical models, PDNs are non-parametric and trained using functional gradient ascent, i.e., boosting. The particularly simple form of the Poisson distribution allows us to develop the first multiplicative boosting approach: starting from an initial constant value, alternatively a log-linear Poisson model, or a Poisson regression tree, a PDN is represented as products of regression models grown in a stage-wise optimization. We demonstrate on several real world datasets that PDNs can model positive and negative dependencies and scale well while often outperforming state-of-the-art, in particular when using multiplicative updates.

Policy Gradient in Lipschitz Markov Decision Processes

Matteo Pirotta, Marcello Restelli, and Luca Bascetta
Machine Learning
DOI: 10.1007/s10994-015-5484-1

This paper is about the exploitation of Lipschitz continuity properties for Markov Decision Processes (MDPs) to safely speed up policy-gradient algorithms.Starting from assumptions about the Lipschitz continuity of the state-transition model, the reward function, and the policies considered in the learning process, we show that both the expected return of a policy and its gradient are Lipschitz continuous w.r.t. policy parameters.By leveraging such properties, we define policy-parameter updates that guarantee a performance improvement at each iteration. The proposed methods are empirically evaluated and compared to other related approaches using different configurations of three popular control scenarios: the linear quadratic regulator, the mass-spring-damper system and the ship-steering control.

Probabilistic Clustering of Time-Evolving Distance Data

Julia Vogt, Marius Kloft, Stefan Stark, Sudhir S. Raman,
Sandhya Prabhakaran, Volker Roth, and Gunnar Rätsch
Machine Learning
DOI: 10.1007/s10994-015-5516-x

We present a novel probabilistic clustering model for objects that are represented via pairwise distances and observed at different time points. The proposed method utilizes the information given by adjacent time points to find the underlying cluster structure and obtain a smooth cluster evolution. This approach allows the number of objects and clusters to differ at every time point, and no identification on the identities of the objects is needed. Further, the model does not require the number of clusters being specified in advance – they are instead determined automatically using a Dirichlet process prior. We validate our model on synthetic data showing that the proposed method is more accurate than state-of-the-art clustering methods. Finally, we use our dynamic clustering model to analyze and illustrate the evolution of brain cancer patients over time.

Ranking Episodes Using a Partition Model

Nikolaj Tatti
Data Mining and Knowledge Discovery
DOI: 10.1007/s10618-015-0419-9

One of the biggest setbacks in traditional frequent pattern mining is that overwhelmingly many of the discovered patterns are redundant. A prototypical example of such redundancy is a freerider pattern where the pattern contains a true pattern and some additional noise events. A technique for filtering freerider patterns that has proved to be efficient in ranking itemsets is to use a partition model where a pattern is divided into two subpatterns and the observed support is compared to the expected support under the assumption that these two subpatterns occur independently. In this paper we develop a partition model for episodes, patterns discovered from sequential data. An episode is essentially a set of events, with possible restrictions on the order of events. Unlike with itemset mining, computing the expected support of an episode requires surprisingly sophisticated methods. In order to construct the model, we partition the episode into two subepisodes. We then model how likely the events in each subepisode occur close to each other. If this probability is high—which is often the case if the subepisode has a high support—then we can expect that when one event from a subepisode occurs, then the remaining events occur also close by. This approach increases the expected support of the episode, and if this increase explains the observed support, then we can deem the episode uninteresting. We demonstrate in our experiments that using the partition model can effectively and efficiently reduce the redundancy in episodes.

Regularized Feature Selection in Reinforcement Learning

Dean Stephen Wookey and George Dimitri Konidaris
Machine Learning
DOI: 10.1007/s10994-015-5518-8

We introduce feature regularization during feature selection for value function approximation. Feature regularization introduces a prior into the selection process, improving function approximation accuracy and reducing overfitting. We show that the smoothness prior is effective in the incremental feature selection setting and present closed-form smoothness regularizers for the Fourier and RBF bases. We present two methods for feature regularization which extend the temporal difference orthogonal matching pursuit (OMP-TD) algorithm and demonstrate the effectiveness of the smoothness prior; smooth Tikhonov OMP-TD and smoothness scaled OMP-TD. We compare these methods against OMP-TD, regularized OMP-TD and least squares TD with random projections, across six benchmark domains using two different types of basis functions.

Soft-max Boosting

Matthieu Geist
Machine Learning
DOI: 10.1007/s10994-015-5491-2

The standard multi-class classification risk, based on the binary loss, is rarely directly minimized. This is due to (i) the lack of convexity and (ii) the lack of smoothness (and even continuity). The classic approach consists in minimizing instead a convex surrogate. In this paper, we propose to replace the usually considered deterministic decision rule by a stochastic one, which allows obtaining a smooth risk (generalizing the expected binary loss, and more generally the cost-sensitive loss). Practically, this (empirical) risk is minimized by performing a gradient descent in the function space linearly spanned by a base learner (a.k.a. boosting). We provide a convergence analysis of the resulting algorithm and experiment it on a bunch of synthetic and real world data sets (with noiseless and noisy domains, compared to convex and non convex boosters).

Tractome: A Visual Data Mining Tool for Brain Connectivity Analysis

Diana Porro-Munoz, Emanuele Olivetti, Nusrat Sharmin,
Thien Bao Nguyen, Eleftherios Garyfallidis, and Paolo Avesani
Data Mining and Knowledge Discovery
DOI: 10.1007/s10618-015-0408-z

Diffusion magnetic resonance imaging data allows reconstructing the neural pathways of the white matter of the brain as a set of 3D polylines. This kind of data sets provides a means of study of the anatomical structures within the white matter, in order to detect neurologic diseases and understand the anatomical connectivity of the brain. To the best of our knowledge, there is still not an effective or satisfactory method for automatic processing of these data. Therefore, a manually guided visual exploration of experts is crucial for the purpose. However, because of the large size of these data sets, visual exploration and analysis has also become intractable. In order to make use of the advantages of both manual and automatic analysis, we have developed a new visual data mining tool for the analysis of human brain anatomical connectivity. With such tool, humans and automatic algorithms capabilities are integrated in an interactive data exploration and analysis process. A very important aspect to take into account when designing this tool, was to provide the user with comfortable interaction. For this purpose, we tackle the scalability issue in the different stages of the system, including the automatic algorithm and the visualization and interaction techniques that are used.

Contents – Part II

Rich Data

Social and Graphs

Research Track

Matrix and Tensor Analysis

BoostMF: Boosted Matrix Factorisation for Collaborative Ranking

Nipa Chowdhury$^{(\boxtimes)}$, Xiongcai Cai, and Cheng Luo

The University of New South Wales, Sydney, NSW 2052, Australia
{nipac,xcai,luoc}@cse.unsw.edu.au

Abstract. Personalised recommender systems are widely used information filtering for information retrieval, where matrix factorisation (MF) has become popular as a model-based approach to personalised recommendation. Classical MF methods, which directly approximate low rank factor matrices by minimising some rating prediction criteria, do not achieve a satisfiable performance for the task of top-N recommendation. In this paper, we propose a novel MF method, namely BoostMF, that formulates factorisation as a learning problem and integrates boosting into factorisation. Rather than using boosting as a wrapper, BoostMF directly learns latent factors that are optimised toward the top-N recommendation. The proposed method is evaluated against a set of state-of-the-art methods on three popular public benchmark datasets. The experimental results demonstrate that the proposed method achieves significant improvement over these baseline methods for the task of top-N recommendation.

Keywords: Recommender system · Collaborative filtering · Matrix factorisation · Learning to rank · Boosting

1 Introduction

Recommender systems (RS) have gained much attention in information retrieval (IR) to guide users when searching information from the information pool. Collaborative filtering (CF) is widely used to build personalised recommender systems such as book recommendation in Amazon [2], movie recommendation in Netflix [2] and friend recommendation in Facebook [2]. It aims to predict the preference of a user on its unseen items by learning the preference from the historic feedback of this user and other like-minded users to provide the user with a list of recommended items or prediction score of items. The personalised prediction problem [1–3,15] in presenting recommendation list can be regarded as estimating the preference function in CF. Usually, this problem can be solved by either i) generating the recommendation list by sorting the predicted ratings in descending order, known as rating-oriented CF or ii) learning the ranking function directly, known as ranking-oriented CF. When the recommendation list itself becomes large, it will be obsolete since people prefer only top listed items

© Springer International Publishing Switzerland 2015
A. Appice et al. (Eds.): ECML PKDD 2015, Part II, LNAI 9285, pp. 3–18, 2015.
DOI: 10.1007/978-3-319-23525-7_1

[2, 15]. So recommender systems should not only be optimised to reflect user tastes and preferences but also rank top items correctly.

Matrix factorisation is a popular model-based CF method, which demonstrates great success in Netflix prize competition [7]. In MF, given N users and M items, the user-item preference matrix $R \in \Re^{N \times M}$ can be approximated by two low rank matrices $P \in \Re^{N \times K}$ and $Q \in \Re^{M \times K}$ as $R \approx P \cdot Q'$ by minimising the sum of squared errors, where $K \ll min(N, M)$ is the dimensionality of latent factors representing user preferences and item characteristics. The major purpose of MF is to obtain some forms of lower-rank approximation to original matrix for understanding the interaction of user preferences and item attractiveness in forms of latent factors [7].

Nevertheless, traditional matrix factorisation algorithms [7, 13] based on rating-oriented CF do not achieve satisfactory ranking performance in the task of top-N recommendation [2, 14, 15, 17]. As users are more concerned about recommended items in the top of the recommendation list, items with higher ratings (i.e., higher possibilities to be preferred by users) should be modelled more correctly than low rating items. Hence, it is important to consider the accuracy of ranked list during learning, and give different emphasises on items with different users' feedback. However, the conventional approach usually does not discriminate the significances of different feedback of items, and the learned latent factors representing user preferences and item characteristics are thus not the optimal ones for generating personalised recommendations. Meanwhile, most of existing methods assume that each latent factor could not contribute differently during the learning of user preferences and item characteristics. This assumption leads to simply update the latent factors as a whole, which may not perform well. In reality, users who originate from different backgrounds are highly proportional to select preferable items based on their different characteristics. These various characteristics are compactly represented by different latent factors. Furthermore, most of existing methods for the top-N recommendation task minimise some error metrics, such as the sum of squared errors, to generate the recommendation list. Unlike optimising against some ranking metrics such as the one used in the paper, this approach is actually an indirect approach that degrades the ranking performance. For example, probabilistic matrix factorisation [13] (PMF), which forms the basis of many model-based recommendation algorithms, adopts even weights on all items and learns all latent factors at a time by minimising the sum of squared errors via stochastic gradient descent.

To improve the accuracy of the top ranked items in the recommendation lists during learning and exploit the contribution of each latent factor separately, we develop a novel method, namely BoostMF, that uses boosting to learn the low rank factor matrices by directly optimising the ranking measure to improve top-N recommendation performance. Specially, rather than treating all observed items with equal importance for each user, BoostMF imposes different emphasises on observed items using a personalised feature selection scheme based on the current estimation of IR evaluation measure. Without computing any structured estimation of ranking loss or continuous approximations of non-smooth IR measure, the proposed method optimises the IR measure directly by integrating boosting into the optimisation, i.e. by gradient descent, of matrix factorisation methods. As the iteration of the optimisation procedure continues, the algorithm is able to place more focus on training examples that have not yet been ranked in top positions correctly. As in real-world deployment, users are more interested in top-N recommended items; this shifting on focus is important and rational for our method to achieve an improved recommendation performance, which will be demonstrated in Section 4. In the end, the learned latent factors representing user preferences and item characteristics are more suitable for generating top-N recommendation. To empirically study the performance of BoostMF, we evaluate our algorithm with some state-of-the-art methods in top-N recommendation and the results demonstrate that our method significantly outperforms these methods for top-N recommendation in terms of recommendation accuracy. Because contextual information is sensitive and expensive to collect, we only focus on user feedback without bothering contextual information. Therefore, we do not compare our method with other rating or ranking-oriented CF methods that use contextual information in addition to user feedback.

The rest of the paper is arranged as follows: in Section 2, we summarise related work and place our work with respect to it. In Section 3, we present the proposed boosted matrix factorisation method. Experimental results are presented in Section 4. Finally, we draw conclusions in Section 5.

2   Related Work

Learning to rank (LTR) is an important research direction in information retrieval where the goal is to present a ranked list of information in response to a query or request [10]. AdaRank [18], MPBoost [19], and RankBoost [4] are well known LTR methods that use boosting to improve ranking performance. If we consider a query as a user and a list of information as items, recommender systems focus on the personalised view of same ranking task as that of LTR. However, incorporating LTR techniques in personalised recommendation is challenging. LTR methods can only handle non-personalised ranking problems rather than personalised ranking and recommendation problems, and also consider that feature vector of items are given and unchanged during learning. But in recommendation settings, user feature and item feature are not explicitly presented during training. The challenge also arises from learning the low rank matrices by optimising the training criterion which is different from the final evaluation criterion that is used to measure the ranking performance. Although different approaches [2,14,15,17] in LTR are adopted to minimise the ambiguity between learning objective criterion and final evaluation measure, these methods either have unsatisfactory performance or incur with computational overhead. The developed BoostMF method in this paper thus aims to simultaneously learn feature vectors and optimise ranking.

Existing methods in ranking-oriented CF can be generally divided into three categories based on the type of issues needed to be addressed. The first class of methods relies on the transformation of ranking measure. CofiRank [17] and CLiMF [14] are the methods that fall into this category. CofiRank uses structured estimation of the ranking loss and CLiMF derives a lower bound of the smooth ranking measure to solve ranking problem in recommendation. However, these transformation results to significantly computational overhead. Our proposed BoostMF algorithm directly accounts the final evaluation criteria into approximating low rank matrices from a high rank matrix. Specifically, based on PMF, we incorporate boosting procedure to learn low rank factor matrices directly for top-N recommendations. To the best of our knowledge, this approach has not been applied before for top-N recommendations. The second class of methods views the recommendation problem as a list-wise ranking problem and uses list-wise loss functions. For example, ListRank-MF [15] uses list-wise loss function based on cross entropy of the top one probability of items. Unified recommendation model (URM) [16] combines both rating-oriented CF, (i.e., PMF) and a ranking-oriented CF, (i.e., ListRank-MF) to improve ranking performance. However, these methods optimise loss function which is not directly related to the final ranking measure, which is not optimal to improve the performance of top-N recommendation. In this regard, BoostMF employs personalised weak ranker at each round to relate the final evaluation measure into the learning process of the model. The third class of methods solves the ranking problem as a regression problem. In collaborative ranking [2], PMF is used to generate feature vectors and regression based LTR algorithm (i.e., point-wise and pairwise) is constructed by these feature vectors to produce the ranking. OrdRec is proposed in [8] as a CF framework following point based approach, and it aims to minimise ordinal regression loss. BPR-MF uses [12] different pair-wise optimisation criterion where pairs are formed by taking one from observed items and the other from unobserved items by assuming a user prefers observed items over unobserved items. But, these methods optimise ranking criterion which is different from the final evaluation measure and hence the final ranking measure is not directly applied to the learning process of the model. The learning model of these methods also imposes equal errors on items misplacement in all positions of the recommendation list. However, in BoostMF, the final ranking measure is directly related to the learning process of the model. BoostMF also uses personalised weight distribution for each user on its rated items to emphasize errors of the learning model on misplacing items in higher positions than lower positions. Thus BoostMF is able to generate better recommendation list which is optimal for the task of top-N recommendation.

3   Boosted Matrix Factorisation (BoostMF)

In this section, we firstly present a key component related to our algorithm probabilistic matrix factorisation (PMF) [13] and then show how to integrate boosting procedure in PMF to learn the best feature vectors for top-N recommendations.

3.1   Probabilistic Matrix Factorisation (PMF)

Assuming there are N users and M items in the data, let matrix $R \in \Re^{N \times M}$ be a user preference matrix. PMF [13] learns two low rank matrices, user factor matrix $P \in \Re^{N \times K}$ and item factor matrix $Q \in \Re^{M \times K}$ to approximate $R \in \Re^{N \times M}$ using probabilistic inference of conditional distributions of observed rating, user priors and item priors, where K is the number of dimensions of latent factors. We use P_u to indicate the latent feature vector of user u, Q_i to indicate the latent feature vector of item i and R_{ui} to indicate the rating that user u gives to item i, respectively. The maximum of the log posterior in PMF can be formulated as

$$P, Q = argmin_{P,Q}\{\frac{1}{2}\sum_{u=1}^{N}\sum_{i=1}^{M} I_{ui}(R_{ui} - P_u Q_i^{'})^2 + \frac{\lambda_p}{2}\|P\|_F^2 + \frac{\lambda_q}{2}\|Q\|_F^2\}, \quad (1)$$

where I_{ui} is an indicator function which equals to 1 for all observed rating, otherwise 0; λ_p and λ_q are regularisation parameters. As the user preference matrix is usually very sparse, $\|P\|_F$ and $\|Q\|_F$ are the Frobenius norms of the matrices P and Q used as regularisation to prevent the learning procedure from overfitting. We use $\lambda_p = \lambda_q = \lambda$ for computational simplicity.

3.2   BoostMF

In matrix factorisation, if one of the factor matrices, say $Q^{'}$ is fixed and only P needs to be learned, then fitting each row of the target matrix R is a linear prediction problem where $Q^{'}$ is the feature vector and each row of P is the model parameter of the linear predictor. The approximation can be formulated as a learning problem for each row $R(u,:) : R(u,:) = P(u,:)*Q^{'}$. Similarly, when P is fixed and $Q^{'}$ needs to be learned, each column of $Q^{'}$ works as the model parameter of the linear prediction model for feature vector P to fit each column of target matrix R. For each column $R(:,i)$, we have: $R(:,i) = P * Q(:,i)^{'}$. In this way, the MF can be thought as a linear regression problem where P and $Q^{'}$ are both unknown and need to be learned. Therefore, an appropriate learning algorithm to solve the linear regression problem is required.

In this work, we use boosting-based techniques to solve the linear regression problem in collaborative learning. Boosting-based techniques come with better convergence properties and stability [5]. We use boosting optimisation technique inside MF to learn low rank factor matrices directly for ranking. We aim at constructing a set of weak learners $\{F^t | t = 1, \ldots, T-1\}$ sequentially to learn user preferences and item characteristics that reside in the data. Based on latent factor selection in the weak learner construction, therefore, the algorithm will be able to stochastically focus on different aspects of user preferences and item characteristics that are modeled by the different selected latent components. By treating each rating as a training instance, a set of training weights $\{W^t | t = 1, \ldots, T-1\}$ is imposed on the ratings. An overall strong learner F is finally assembled by linearly combining weak rankers, which is expected to perform better than any individual learner. The weights of training ratings are updated to reflect the accuracy of the prediction of the weak learner. People usually follow information that appears at the top-N positions in the recommendation list. Therefore, the items that are ranked at the top should be considered more than those at the bottom of the recommendation list. To this end, we dynamically construct personalised weak rankers[1] and modify personalised weights by considering the ranking performance on training items. In next iterations, the learning procedure will give more attention on those items that have not yet been ranked in correct positions. Due to the automatic selection and optimisation of the personalised weak ranker and the dynamic updating of the personalised weights, the learned latent factors for users and items are best suited for top-N recommendation.

The BoostMF method creates weak rankers in the direction that has maximum IR performance improvement over training data. At each boosting round, the method constructs a weak ranker for each user based on IR performance over the items rated by the same user with personalised weight distribution. If user u rates m_k items and the set of items is indicated by $\mathbf{i} = i_1, i_2, \ldots, i_{m_k}$[2] then for round t, BoostMF creates a weak ranker for each user by

$$F_u^{(t)}(l) = \operatorname*{argmax}_{l \in \{1, \ldots, K\}} \left(E \left[\pi_u(W_{u\mathbf{i}}^{(t)} f_{u\mathbf{i}l}^{(t)}), R_{u\mathbf{i}} \right] \right), \tag{2}$$

where $f_{u\mathbf{i}l}^{(t)} = P_{ul}^{(t)} Q_{\mathbf{i}l}^{(t)'}$ is the ranking score according to the l-th dimension of latent factors, $W_{u\mathbf{i}}$ is the weights of user u on its item set $\mathbf{i}$, E represents the IR performance measure and K is the feature dimension of the low rank matrices, respectively. For user u, its permutation list π_u is used to order the items $\mathbf{i}$ by taking as inputs $W_{u\mathbf{i}}^{(t)}$ and $f_{u\mathbf{i}l}^{(t)}$. The design of permutation list π_u is usually correlated with the adoption of E. For simplicity, the weights $W_{u\mathbf{i}}^{(t)}$ is linearly combined with the ranking score $f_{u\mathbf{i}l}^{(t)}$ in the permutation to emphasize its confidence.

The purpose of Equation (2) is to select weak ranker for each user based on the items score $f_{u\mathbf{i}l}^{(t)}$. But in the same time we need to select the weak ranker that will be able to contribute more on items on which previous ranker did not perform well. To provide this information in weak ranker selection as well as to reflect the individual tastes and rating scale of a user, BoostMF uses weight distribution $W_{u\mathbf{i}}^{(t)}$ for each user u on every training item i from item set $\mathbf{i}$. The weight value $W_{u\mathbf{i}}^{(t)}$ is different in each round from user to user and even for the same item belongs to different user. The weight $W_{u\mathbf{i}}^{(t)}$ restricts the factor selection formula in Equation (2) not to select ranker that gives just best E measure, but to select ranker that has the ability to place the items in correct positions on which previous rankers do not perform well. BoostMF increases the values of weights on items that are not ranked well by the dynamically constructed ranking model. So in next iteration, these weight values will make more effect in next ranker selection to improve overall ranking performance.

---

[1] The term weak learner and weak ranker are used interchangeably throughout the paper.

[2] Bold font of i is used to denote set of items and normal to denote single item.

The weight value of an item is calculated based on the performance of the current ranker in placing the item w.r.t. other items in the ranking list. Ideally, we aim for a ranking model that makes no mistake in item placement. But the error in placing two items with rating 5 and 1 has a heavier influence on the IR performance measure than that of placing two items with rating 5 and 4. To reflect this loss, we add the pairwise preference term $(R_{ui} - R_{uj})$ into the weight function to give more penalties for misplacing the items in higher position. Specifically, if the current ranking model for user u is $F_u^{(t)}$ with selected latent factor l, and its updated ranking score on an item i is indicated by $f_{uil}^{(t+1)} = P_{ul}^{(t+1)}Q_{il}^{(t+1)'}$, the weight value of item i for that user on the l-th latent factor at $t+1$ iteration is expressed as,

$$W_{uil}^{(t+1)} = \frac{\sum_{j=1,j\neq i}^{m_k} exp\{-(f_{uil}^{(t+1)} - f_{ujl}^{(t+1)})(R_{ui} - R_{uj})\}}{\max_{i \in m_k} \sum_{j=1,j\neq i}^{m_k} exp\{-(f_{uil}^{(t+1)} - f_{ujl}^{(t+1)})(R_{ui} - R_{uj})\}}. \tag{3}$$

Note that the pairwise preference term in BoostMF is different from the common pairwise preference formulation used in [9, 12]. In these methods, with only implicit feedback, the pair of items consists of one observed item and one unobserved item where the observed item is assumed to have higher preference over the unobserved one. However, the formulation may be inconsistent with the real world scenario because unobserved items could be either unfavoured by the user or simply just unexposed to the user. In contrast, BoostMF uses explicit preferences to construct the personalised pairwise preference term, which is more reliable. Meanwhile, to facilitate the computation, uniform sampling is adopted in almost all of the models with the pairwise preference in order to select the set of pairs of items. However, it is shown [11] that this approach is very inefficient because most of selected items will be correctly ranked after a few of iterations and almost all the gradient magnitude from the selected pairs become less informative. In this regards, BoostMF provides an efficient and informative selection and updating mechanism by constantly focusing on the disordered items for every user across the whole procedure of learning.

At the initialisation, the value of user weight $W_{ui}^{(t)}$ on every item is identical. At the current round t, BoostMF increases the values of weights on items that are not ranked well by the dynamically constructed ranking model. Hence in round $t+1$ these weights will make more contribution to construct the next ranking model that will attempt to rectify the incorrect ranking of these items. The value of $W_{ui}^{(t)}$ yields a clear indication how much the item i is misplaced in the rank list of user u. So in next iteration, this weight will make more effect in next ranker selection to improve the performance.

To model the fact that various users will judge their preferences over different items based on different criteria, BoostMF also selects the direction that has maximum capacity to generate a good ranking list on the training items for each user at every round and performs maximum adjustment in that direction. All other factors for that user are remaining unchanged on the round. Let l denote the dimension of the selected latent factors for the current weak ranker. If the objective function in Equation (1) is denoted by L, then for the ranking

---

Algorithm 1. Boosted Matrix Factorisation

Input: Rating matrix R, no. of iterations T, performance measure E, no. of users N and no. of items M, no. of training items per user m_k, feature dimension K and learning rate η

Output: Low rank factor matrices P and Q

Initialisation: Initialise $P^{(1)}$ and $Q^{(1)}$ randomly, and initialise $W_{ui}^{(1)} = \frac{1}{m_k}$ for each user u on available training items.

   for t=1:T-1 **do**
      for u=1:N **do**
         Select ranking model $F_u^{(t)}(l)$, $l \in \{1, \ldots, K\}$ using Equation (2) for user u on
         its rated item set **i** with weighted distribution of $W_{ui}^{(t)}$.
         Compute $\frac{\delta L}{\delta P_{ul}^{(t)}}$ and $\frac{\delta L}{\delta Q_{il}^{(t)}}$ using Equation (4) and (5).
         Update P_{ul} and Q_{il} by
         $P_{ul}^{(t+1)} = P_{ul}^{(t)} - \eta \frac{\delta L}{\delta P_{ul}^{(t)}}$, $Q_{il}^{(t+1)} = Q_{il}^{(t)} - \eta \frac{\delta L}{\delta Q_{il}^{(t)}}$
         Update $W_{ui}^{(t+1)}$ using Equation (3).
      end for
   end for

Output: $P^{(T)} Q^{(T)'}$

---

model $F_u^{(t)}$, BoostMF updates user and item latent factor by

$$\frac{\delta L}{\delta P_{ul}^{(t)}} = \sum_{i=1}^{m_k} I_{ui}(P_{ul}^{(t)} Q_{il}^{(t)'} - R_{ui})Q_{il}^{(t)} + \lambda P_{ul}^{(t)} \tag{4}$$

$$\frac{\delta L}{\delta Q_{il}^{(t)}} = I_{ui}(P_{ul}^{(t)} Q_{il}^{(t)'} - R_{ui})P_{ul}^{(t)} + \lambda Q_{il}^{(t)}.^3 \tag{5}$$

Compared with the updating stages of latent factors in conventional MF methods, the difference terms in Equation (4) and Equation (5) in BoostMF have also shifted the focus to the contribution of individual latent factor. Instead of combining the weak learner estimation to form final strong learner, BoostMF takes all latent dimensions of $P^{(T)} Q^{(T)'}$ as strong learner after the completion of round $T - 1$, as the weak ranking models are updated during learning. Finally, personalised ranking list is generated by sorting the ratings which are predicted by using all latent dimensions of $P^{(T)}$ and $Q^{(T)'}$. At each boosting round $t = 1, \ldots, T - 1$, BoostMF creates a weak ranker $F_u^{(t)}$ for each user, updates the ranker, modifies weights based on the ranking performance and finally outputs a personalised ensemble model as $F \approx P^{(T)} Q^{(T)'}$. An overview of the algorithm is presented in Algorithm 1.

The complexity of weak ranker selection is in the order of $\mathcal{O}(NK)$, where N is the number of users and K is the feature dimension size. The complexity of the gradient computation in Equation (4) and (5) is in order of $\mathcal{O}(R)$, where R is the number of observed ratings in the given user-item matrix. The computation complexity of the weight updating formula in Equation (3) is $\mathcal{O}(Nm_k^2)$, where m_k denotes the number of training items per user. In collaborative filtering, $R \gg N, M$ and even dominates

---

3 The summation sign is not used on the right hand side of (5), because we formulate the algorithm user-wise and the item set i is rated by one user as shown in Algorithm 1.

the term Nm_k^2. When Nm_k^2 dominates R, BoostMF has complexity in the order of $\mathcal{O}(Nm_k^2)$, otherwise it has linear time complexity in the order of $\mathcal{O}(R)$.

3.3 Theoretical Analysis

In this section, we show theoretical insights by developing an upper error bound of BoostMF following MPBoost [19] and RankBoost [4]. Allowing both rating magnitude in the ranking loss and dynamic changes in the feature vectors for the weak learner model at each boosting round in BoostMF leads to the following theorem:

Theorem 1. *The misplacement loss of the personalised ranking model in BoostMF is bounded by* $\sum_{i=1}^{m_k} \sum_{j=1,j\neq i|R_i>R_j}^{m_k} [\![F_i \leq F_j]\!] + \sum_{i=1}^{m_k} \sum_{j=1,j\neq i|R_i<R_j}^{m_k} [\![F_i \geq F_j]\!] \leq Z_T$, *where* $Z_T = \sum_{i=1}^{m_k} \sum_{j=1,j\neq i}^{m_k} exp\{-[(F_i - F_j)(R_i - R_j)]\}$ *and* $[\![x]\!]$ *is defined to be 1 if predicate x is true and 0 otherwise.*

Proof. The personalised ranking model in BoostMF produces two types of misplacement. The first one is when $F_i \geq F_j$ but $R_i < R_j$ and the second one is when $F_i \leq F_j$ but $R_i > R_j$. Note that $[\![x \geq 0]\!] \leq exp\{\alpha x\}$ and $[\![x \leq 0]\!] \leq exp\{-\alpha x\}$ hold for all $\alpha > 0$ and all real x. We can write the total loss as

$$\sum_{i=1}^{m_k} \sum_{j=1,j\neq i|R_i>R_j}^{m_k} [\![F_i \leq F_j]\!] + \sum_{i=1}^{m_k} \sum_{j=1,j\neq i|R_i<R_j}^{m_k} [\![F_i \geq F_j]\!]$$
$$\leq \sum_{i=1}^{m_k} \sum_{j=1,j\neq i|R_i>R_j}^{m_k} exp\{-[(F_i - F_j)(R_i - R_j)]\}$$
$$+ \sum_{i=1}^{m_k} \sum_{j=1,j\neq i|R_i<R_j}^{m_k} exp\{[-(F_i - F_j)(R_i - R_j)]\}$$
$$= \sum_{i=1}^{m_k} \sum_{j=1,j\neq i}^{m_k} exp\{-[(F_i - F_j)(R_i - R_j)]\} = Z_T \qquad \blacksquare$$

This bound is guaranteed to produce a combined low ranking loss if we choose the weak ranker that minimises $\sum_{i=1}^{m_k} \sum_{j=1,j\neq i}^{m_k} exp\{-[(F_i - F_j)(R_i - R_j)]\}$ on each round t [4]. Minimising the misplacement loss is equivalent to maximising the IR measure [19]. In BoostMF, the weak ranker is set to select the ranking model that maximises the IR measure which is equivalent to minimising the misplacement loss and the weight update formula is set to give more penalties to the ranking model that makes misplacement in higher position. Finally, in terms of misplacement loss, the total error of BoostMF is bounded by Z_T.

4 Experiments

4.1 Datasets and Evaluation Metric

We test the performance of BoostMF on three publicly available datasets for the task of personalised top-N recommendation: MovieLens 100K[4] dataset, Movie-Lens 1M dataset[4] and Netflix[5] dataset. MovieLens 100K dataset consists of 100,000 ratings from 943 users on 1682 movies. MovieLens 1M dataset consists of 1,000,000 ratings from 6040 users and 3900 movies. Ratings are integers and scaled on 1-5. and each user has rated at least 20 movies on both datasets. For Netflix dataset, we use a sampled version, which is extracted from 4% of the

---

[4] http://www.grouplens.org/node/73
[5] B. James and L. Stan, The Netflix prize, (2007).

Netflix dataset with 20% users and 20% movies are randomly selected from the whole pool. The Netflix dataset contains 3,843,340 ratings on scaled 1-5 from 95526 users on 3561 movies.

As our goal is to generate efficient recommendation list that would contain higher rating items in top-N position, we prefer a metric capable of awarding models that correctly rank items in higher positions and penalising models that make more errors in higher positions than in lower positions. Following the standard evaluation metric used in [2,15,17], we use normalised discounted cumulative gain (NDCG) as IR performance measure for testing and evaluation of our algorithm.

4.2 Experimental Setup

We adopt the same experimental protocol from [2,17]. We use 3 different settings of training data based on the number of randomly selected items for each user, namely $S_N=10$, $S_N=20$ and $S_N=50$. The remaining items are used for testing. Users with less than 20, 30 or 60 rated items are removed respectively in each setting to ensure the feasibility to compute NDCG@10. Following the common practice in RS [2,17], items that are not rated by at least 5 users in the dataset are also removed. We also eliminate items from test dataset those are not appeared in training dataset. These settings cause a slightly decrement of user-movie combination than the original dataset. We report the number of users and items available in each setting for all datasets in Table 1. For each setting, we generate 10 versions of the dataset, by randomly sampling items. We report the mean and standard deviation of NDCG@5 and NDCG@10 on those 10 sets over all users. We compare BoostMF with a sets of the state-of-the-art algorithms including PMF [13], and OrdRec [8] which are the state-of-the-art rating-oriented CF methods; ListRank-MF [15], CofiRank [17], and BPR-MF [12], which are the state-of-the-art ranking-oriented CF methods; and URM [16], which combines both rating-oriented and ranking-oriented methods. From the experimental results in [17], CofiRank method that optimises root mean square loss (denote as CofiRank^{Reg}) performs better than CofiRank that optimises NDCG directly (denote as CofiRank^{NDCG}). Therefore, we compare BoostMF with both CofiRank^{Reg} and CofiRank^{NDCG}.

Table 1. No. of users and items for experimental settings $S_N=10$, 20 and 50 on the datasets.

Dataset	No. of users for $S_N=10/20/50$	No. of items for $S_N=10/20/50$
MovieLens 100K	941/743/496	1349/1336/1312
MovieLens 1M	6035/5286/3937	3415/3411/3400
Netflix	45508/35749/20067	3558/3556/3546

4.3   Results

Before comparing the performance of our algorithm with other state-of-the-art approaches, we at first examine whether the weak ranker selection and weight update formula in BoostMF improve the algorithms performance or not. To this end, we create two versions of BoostMF algorithm named (1) RandomBoostMF that selects weak ranker randomly for each user, (2) ModifiedBoostMF that selects ranking model by NDCG but updates item weights without considering the effect of the pairwise preference term $(R_{ui} - R_{uj})$. The comparison of BoostMF with RandomBoostMF indicates whether factor selection by NDCG in BoostMF makes any benefits over random factor selection, and the comparison with ModifiedBoostMF indicates the advantages of using the modified weight update mechanism in Equation (3). We also want to see how these algorithms perform with respect to various feature dimensions. To apply these algorithms, MovieLens 100K dataset with user/item settings S_N=50 is used. We record NDCG@10 for each data fold for factor dimension 5, 10, 15, 25 and 50 and the mean of NDCG over 10 folds for each feature dimension is presented in Fig. 1.

From the results in Fig. 1, we can see that BoostMF performs much better than RandomBoostMF, which verifies that factor selection by NDCG in BoostMF helps improve the performance of top-N recommendation as random selection is not able to generate suitable feature vectors to boost the learning procedure of MF. BoostMF also outperforms ModifiedBoostMF, which shows the success of the developed pairwise preference scheme in the procedure of dynamic weight updating. Most importantly, the performance of BoostMF is stable under different settings of the feature dimension size. It performs the best for feature dimension size of 15 which the NDCG score is 0.7139. The NDCG score slightly decreases for feature dimension size of 50 which is 0.7020 but still much better in comparison to the performance of RandomBoostMF (0.6765) and ModifiedBoostMF (0.6908).

Now we compare our algorithm with PMF, ListRank-MF, URM, OrdRec, BPR-MF and CofiRank. We tune parameters separately on a validation set for all algorithms by cross validation to achieve their best performance on the used datasets. NDCG performance on validation set is used to choose the hyperparameters with the best performance. We implement CofiRank using publicly available software.[6] OrdRec[8] and BPR-MF[12] are implemented by publicly available software Lenskit[7] and Mymedialite[8], respectively. BoostMF uses η=0.01, λ=0.02 and K=5 for MovieLens datasets and η=0.00005, λ=0.000009 and K=10 for Netflix dataset. As each method has different settings of hyperparameters under different settings of experiments, due to the space limitation, we do not state the hyperparameters of other algorithms. We also perform paired t test [6] with significant level of 5%, and all the improvement are statistically significant. The mean and standard deviation over 10 data folds for different approaches with respect to different experimental settings are reported in Table 2-4.

---

[6] http://www.cofirank.org/downloads.
[7] http://lenskit.org/download/
[8] http://mymedialite.net/download/index.html

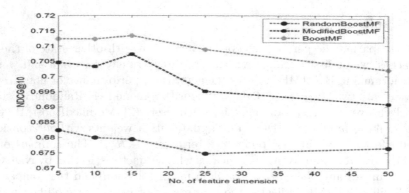

Fig. 1. Performance comparison of RandomBoostMF, ModifiedBoostMF and BoostMF.

According to Table 2, BoostMF significantly outperforms all compared state-of-the-art algorithms in most of the cases. BoostMF achieves 10~12% improvement over CofiRank^{NDCG} for settings S_N=50 and gains 6~8% improvement for settings S_N=10 and S_N=20 on both NDCG@5 and NDCG@10 metrics for MovieLens 100K dataset. It also gains 1.6~4% improvement over PMF, 1~4.7% improvement over URM and 3~5% improvement over CofiRank^{Reg} on both evaluation measures for all experimental settings. BoostMF also shows 0.7~2.3% improvement over ListRank-MF. Although for settings S_N=10, BoostMF performs slightly worse than ListRank-MF on NDCG@5, it performs better than ListRank-MF for all other settings on both metrics. BoostMF outperforms BPR-MF and OrdRec for all experimental settings on both NDCG@5 and NDCG@10 metrics. It achieves 8~11% improvement over BPR-MF and 10~20% improvement over OrdRec.

Results on MovieLens 1M dataset are shown in Table 3. BoostMF achieves significant improvement over CofiRank, BPR-MF and OrdRec for all experimental settings on all the evaluations. It achieves 10-19% improvement over OrdRec, 5-8% improvement over BPR-MF, and 4-9% improvement over CofiRank^{NDCG} and CofiRank^{Reg} respectively on both evaluations for all experimental settings. In comparison with URM, BoostMF achieves 2~2.4% improvement on NDCG@5 metric and 1.1~2.14% improvement on NDCG@10 metric over all experimental settings. BoostMF also outperforms ListRank-MF by 9~10% for settings S_N=10, 6.5~6.8% for setting S_N=20 and also achieves more than 1% improvement for settings S_N=50 on both evaluations. It gains 5~6% improvement for settings S_N=10 and 3~4% improvement for settings S_N=20 and S_N=50 on both metrics when comparison is made with PMF.

From the results in Table 4, it is clear that BoostMF outperforms all other state-of-art approaches on Netflix dataset. It outperforms CofiRank^{NDCG} by 10~12% over all experimental settings on both NDCG computations. It achieves 7~10% improvement over CofiRank^{Reg} for experimental settings S_N=10, S_N=20 and 4~5% improvement for experimental settings S_N=50 on both metrics. Compared to OrdRec and BPR-MF, BoostMF results 8~16% improvement for all settings on both metrics. It also gains 9~14% performance improvement for

Table 2. The NDCG@5 and NDCG@10 accuracy and standard deviation over 10 data folds for PMF, BPR-MF, ListRank-MF, URM, OrdRec, CofiRank and BoostMF on MovieLens 100K dataset. The best performance is in bold.

	S_N=10		S_N=20		S_N=50	
	NDCG@5	NDCG@10	NDCG@5	NDCG@10	NDCG@5	NDCG@10
PMF	0.6330±.009	0.6606±.005	0.6762±.007	0.6864±.007	0.6765±.005	0.6819±.007
BPR-MF	0.5558±.002	0.5942±.003	0.5872±.002	0.6098±.004	0.6292±.001	0.6309±.002
CofiRank^{NDCG}	0.5927±.006	0.6314±.006	0.6098±.005	0.6331±.003	0.5897±.006	0.6096±.005
CofiRank^{Reg}	0.6381±.008	0.6629±.004	0.6398±.003	0.6540±.004	0.6580±.004	0.6708±.002
OrdRec	0.5197±.001	0.5687±.001	0.4852±.003	0.5290±.002	0.58±.002	0.6081±.004
ListRank-MF	**0.6725±.005**	0.6844±.005	0.6834±.004	0.6947±.003	0.6887±.003	0.6982±.004
URM	0.6421±.005	0.6561±.006	0.6778±.004	0.6851±.007	0.6919±.005	0.7034±.004
BoostMF	0.6722±.008	**0.7034±.007**	**0.6921±.005**	**0.7019±.004**	**0.7117±.004**	**0.7135±.004**

Table 3. The NDCG@5 and NDCG@10 accuracy and standard deviation over 10 data folds for PMF, BPR-MF, ListRank-MF, URM, OrdRec, CofiRank and BoostMF on MovieLens 1M dataset. The best performance is in bold.

	S_N=10		S_N=20		S_N=50	
	NDCG@5	NDCG@10	NDCG@5	NDCG@10	NDCG@5	NDCG@10
PMF	0.6814±.007	0.6842±.005	0.7030±.002	0.7043±.003	0.7224±.002	0.7189±.003
BPR-MF	0.6734±.007	0.6873±.006	0.6747±.005	0.6790±.006	0.6711±.007	0.6769±.006
CofiRank^{NDCG}	0.6485±.002	0.6685±.002	0.6587±.005	0.6763±.001	0.6679±.007	0.6812±.007
CofiRank^{Reg}	0.6698±.005	0.6838±.006	0.6728±.005	0.7005±.005	0.6844±.007	0.7049±.006
OrdRec	0.5095±.002	0.5431±.003	0.4948±.002	0.5312±.002	0.6288±.002	0.6485±.001
ListRank-MF	0.6424±.005	0.6423±.004	0.6792±.007	0.6827±.006	0.7406±.004	0.7344±.004
URM	0.7205±.004	0.7222±.002	0.7236±.001	0.7365±.001	0.7328±.003	0.7301±.002
BoostMF	**0.7433±.007**	**0.7389±.007**	**0.7475±.005**	**0.7480±.004**	**0.7528±.004**	**0.7515±.004**

settings S_N=10, 7~8% for settings S_N=20 and more than 4% for settings S_N=50 over PMF on both NDCG evaluations. Over ListRank-MF, BoostMF gains 1.6~3.8% improvement on NDCG@5 metric for settings S_N=20 and S_N=50, and it gains 7~10% improvement on NDCG@10 for experimental settings S_N=10. It also outperforms URM by 2~4.8% on NDCG@5 computation and 1.4~2.8% on NDCG@10 computation for all experimental settings.

To gain a deep understanding of the success of BoostMF, the reasons for the experimental results will be explored as follows. PMF is the rating-oriented collaborative filtering algorithm that minimises sum of squared errors at each step of learning process. Hence, the learning procedure spends its efforts on a criterion that is not directly related to the task of top-N recommendation. OrdRec, which is a regression based rating-oriented CF method, assumes users' feedback as ordinal rather than number. Although it considers users' personalised rating scales, the ranking measure is not directly applied to the learning model. ListRank-MF is the ranking-oriented CF algorithm that aims to present better ranking list, however, unlike BoostMF, the IR evaluation measure of ListRank-MF is not directly related to the learning process of the model. BPR-MF, which is a ranking

Table 4. The NDCG@5 and NDCG@10 accuracy and standard deviation over 10 data folds for PMF, BPR-MF, ListRank-MF, URM, OrdRec, CofiRank and BoostMF on Netflix dataset. The best performance is in bold.

	$S_N=10$		$S_N=20$		$S_N=50$	
	NDCG@5	NDCG@10	NDCG@5	NDCG@10	NDCG@5	NDCG@10
PMF	0.6239±.007	0.5913±.006	0.6481±.004	0.6689±.004	0.6876±.007	0.6980±.005
BPR-MF	0.5724±.003	0.6005±.005	0.5725±.004	0.5982±.005	0.5783±.003	0.6095±.003
CofiRank[NDCG]	0.6203±.001	0.6272±.005	0.6134±.002	0.6139±.003	0.6249±.005	0.6316±.002
CofiRank[Reg]	0.6146±.005	0.6128±.004	0.6252±.003	0.6599±.005	0.6790±.006	0.6931±.005
OrdRec	0.5597±.002	0.6076±.003	0.5908±.005	0.6315±.006	0.6264±.002	0.6556±.003
ListRank-MF	0.6453±.002	0.6359±.001	0.7011±.004	0.7017±.007	0.7156±.007	0.7118±.002
URM	0.6808±.006	0.7217±.002	0.7118±.002	0.7188±.005	0.6831±.002	0.7099±.004
BoostMF	**0.7216±.006**	**0.7364±.002**	**0.7352±.004**	**0.7398±.006**	**0.7317±.003**	**0.7383±.002**

based model, solves personalised ranking problem by optimising area under the curve (AUC). Unlike the optimisation criterion used in BoostMF, AUC imposes equal error on misplacing items irrespective of their positions in the generated recommendation list; thus BPR-MF does not perform well on list based top-N recommendation. URM employs both rating and ranking information together but still the IR evaluation measure is not directly applied in the learning model. Meanwhile, the relative contribution of rating information and ranking information depends on the particular dataset. CofiRank is also a ranking-oriented CF method, but from our experimental results, CofiRank[NDCG] that uses NDCG information directly into learning phases performs worse than CofiRank[Reg] that optimises for regression with root mean square loss. This finding is consistent with experimental results from [2,15,17]. On the other hand, BoostMF is proposed to improve the underlying factor learning in matrix factorisation using boosting with feature selection and to optimise IR measure directly aiming at resolving the mismatch between training objective function and evaluation metric. Without imposing any overhead of IR measure conversion, BoostMF creates ranking model and updates according to the correctness of the ranking list. Thus, BoostMF presents a better ranked recommendation list than the state-of-the-art recommendation approaches by focusing on the factors that are best suited to represent ranking task.

In addition to the comparison, we carry the experiment to see in what extent PMF and BoostMF handle overfitting, which is an important issue when the data is extremely sparse which is common in recommender systems. Note that this is also important for the boosting algorithm whose capability of generalisation is usually considered under non-sparse dataset [4,19]. Specifically, we want to evaluate the IR performance of both algorithms, i.e. PMF and BoostMF, on the test set while the IR performance keeps increasing on training set as the number of training round increases. For experimental settings of $S_N=50$ on MovieLens 100K dataset, we record the performance of the models on both train set and test set for every iteration of PMF and BoostMF. Fig. 2 and 3 show the average NDCG@10 on train set and test set over 10 folds. Learning rate and

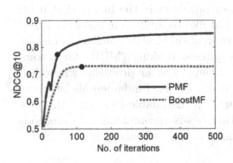

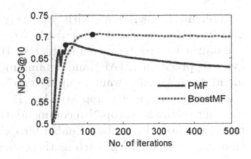

Fig. 2. Average NGCG@10 of PMF and BoostMF over 10 folds on train set

Fig. 3. Average NGCG@10 of PMF and BoostMF over 10 folds on test set

regularisation parameter of PMF and BoostMF are set separately according to their best performance on validation set. The stopping conditions for both algorithms are also set from cross validation and marked as black circle in Fig. 2 and 3.

As shown in the Fig. 2 and 3, PMF suffers from serious overfitting problem whereas BoostMF is very robust to overfitting. This overfitting behaviour of PMF shows its inappropriateness to the ranking problem. As the iteration continues, it ignores the information from IR measure and thus deviates away from improving NDCG. The test performance in PMF is decreasing while the training performance is steadily improving. On the contrary, BoostMF constructs weak learner in the direction that gives maximum ranking accuracy on training data, performs maximum update in that ranking direction and reweights items according to the correctness of the ranking list. All of those three features are the keys for stable ranking as the number of training round increases and thus it avoids overfitting issues.

5   Conclusion

In this paper, we present a novel method, BoostMF, to the problem of matrix factorisation by learning the best feature vectors for ranking and apply it to the task of personalised top-N recommendation. In addition to using latent factors to represent various user preferences and item characteristics, the BoostMF method uses boosting procedure to select best factors to optimise for the ranking task and performs updating only on that factor. In contrast to other ranking-oriented CF methods, the BoostMF method optimises the ranking measure directly by learning low rank factor matrices rather than using the structured estimation of ranking loss or computing continuous approximations of IR measure. To demonstrate the efficiency of BoostMF, we evaluate it against a set of state-of-the-art approaches on three real-world publicly available datasets with different user-item distributions. The experimental results verify that the BoostMF method achieves significant improvement over these baseline methods for the task of top-N recommendation.

Our method will cope with cold start user problems in the future that users have very few ratings (one or two). We would like to apply our boosting-based collaborative filtering model with other IR evaluation metrics, such as minimum average precision (MAP) and minimum reciprocal ranking (MRR) for recommendation. We also want to apply this algorithm to other problems where MF approach is frequently applied. As MF consists of the fundamentals of many existing methods in top-N recommendation, it is reasonable to expect the proposed method also to be valuable for existing top-N recommendation methods that are based on PMF such as these methods shown in Section 2.

References

1. Adomavicius, G., Tuzhilin, A.: Toward the next generation of recommender systems: a survey of the state-of-the-art and possible extensions. IEEE TKDE **17**(6) (2005)
2. Balakrishnan, S., Chopra, S.: Collaborative ranking. In: WSDM (2012)
3. Cai, X., Bain, M., Krzywicki, A., Wobcke, W., Kim, Y., Compton, P., Mahidadia, A.: Learning collaborative filtering and its application on people-to-people recommendation in social networks. In: ICDM (2010)
4. Freund, Y., Iyer, R., Schapire, R., Singer, Y.: An efficient boosting algorithm for combining preferences. J. Mac. Learn. Res. **4** (2003)
5. Friedman, J., Hastie, T., Tibshirani, R.: Additive Logistic Regression: a Statistical View of Boosting. The Annals of Statistics **38**(2) (2000)
6. Goulden, C.: Methods of Statistical Analysis. Wiley (1956)
7. Koren, Y., Bell, R., Volinsky, C.: Matrix factorization techniques for recommender systems. J Comput. **42** (2009)
8. Koren, Y., Sill, J.: Ordrec: an ordinal model for predicting personalized item rating distributions. In: RecSys (2011)
9. Krohn-Grimberghe, A., Drumond, L., Freudenthaler, C., Schmidt-Thieme, L.: Multi-relational matrix factorization using bayesian personalized ranking for social network data. In: WSDM (2012)
10. Liu, T.: Learning to rank for information retrieval. Found. and Trends in Inf. Retr. **3** (2009)
11. Rendle, S., Freudenthaler, C.: Improving pairwise learning for item recommendation from implicit feedback. In: WSDM (2014)
12. Rendle, S., Freudenthaler, C., Gantner, Z., Thieme, L.: BPR: bayesian personalized ranking from implicit feedback. In: UAI (2009)
13. Salakhutdinov, R., Mnih, A.: Probabilistic matrix factorization. In: NIPS (2008)
14. Shi, Y., Karatzoglou, A., Baltrunas, L., Larson, M., Oliver, N., Hanjalic, A.: CLiMF: Learning to maximize reciprocal rank with collaborative less-is-more filtering. In: RecSys (2012)
15. Shi, Y., Larson, M., Hanjalic, A.: List-wise learning to rank with matrix factorization for collaborative filtering. In: RecSys (2010)
16. Shi, Y., Larson, M., Hanjalic, A.: Unifying rating-oriented and ranking-oriented collaborative filtering for improved recommendation. J. of Inf., Sci. (2013)
17. Weimer, M., Karatzoglou, A., Le, Q.V., Smola, A.: CofiRank-maximum margin matrix factorization for collaborative ranking. In: NIPS (2007)
18. Xu, J., Li, H.: AdaRank: a boosting algorithm for information retrieval. In: SIGIR (2007)
19. Zhu, C., Chen, W., Zhu, Z., Gang, W., Wang, D., Chen, Z.: A general magnitude-preserving boosting algorithm for search ranking. In: CIKM (2009)

Convex Factorization Machines

Mathieu Blondel$^{(\boxtimes)}$, Akinori Fujino, and Naonori Ueda

NTT Communication Science Laboratories, Kyoto, Japan
mblondel@gmail.com

Abstract. Factorization machines are a generic framework which allows to mimic many factorization models simply by feature engineering. In this way, they combine the high predictive accuracy of factorization models with the flexibility of feature engineering. Unfortunately, factorization machines involve a non-convex optimization problem and are thus subject to bad local minima. In this paper, we propose a convex formulation of factorization machines based on the nuclear norm. Our formulation imposes fewer restrictions on the learned model and is thus more general than the original formulation. To solve the corresponding optimization problem, we present an efficient globally-convergent two-block coordinate descent algorithm. Empirically, we demonstrate that our approach achieves comparable or better predictive accuracy than the original factorization machines on 4 recommendation tasks and scales to datasets with 10 million samples.

Keywords: Factorization machines · Feature interactions · Recommender systems · Nuclear norm

1 Introduction

Factorization machines [12] [13] are a generic framework which allows to mimic many factorization models simply by feature engineering. Similarly to linear models, factorization machines learn a feature weight vector $w \in \mathbb{R}^d$, where d is the number of features. However, factorization machines also learn a pairwise feature interaction weight matrix $Z \in \mathbb{R}^{d \times d}$. Given a feature vector $x \in \mathbb{R}^d$, factorization machines use w and Z to predict a target $y \in \mathbb{R}$. The main advantage of factorization machines is that they learn the feature interaction weight matrix in factorized form, $Z = VV^{\mathrm{T}}$, where $V \in \mathbb{R}^{d \times k}$ and $k \ll d$ is a rank hyper-parameter. This reduces overfitting, since the number of parameters to estimate is reduced from d^2 to kd, and allows to compute predictions efficiently. Although they can be used for any supervised learning task such as classification and regression, factorization machines are especially useful for recommender systems. As shown in [12][13], factorization machines can mimic many existing factorization models just by choosing an appropriate feature representation for x. Examples include standard matrix factorization, SVD++ [8], timeSVD++[9] and PITF (pairwise interaction tensor factorization) [16]. Moreover, it is easy to incorporate auxiliary features such as

© Springer International Publishing Switzerland 2015
A. Appice et al. (Eds.): ECML PKDD 2015, Part II, LNAI 9285, pp. 19–35, 2015.
DOI: 10.1007/978-3-319-23525-7_2

user and item attributes, contextual information [15] and cross-domain feedback [10]. In [14], it was shown that factorization machines achieve predictive accuracy as good as the best specialized models on the Netflix and KDDcup 2012 challenges. In short, factorization machines are a generic framework which combines the high predictive accuracy of factorization models with the flexibility of feature engineering. Unfortunately, factorization machines have two main drawbacks. First, they involve a non-convex optimization problem. Thus, we can typically only obtain a local solution, the quality of which depends on initialization. Second, factorization machines require the choice of a rank hyper-parameter. In practice, predictive accuracy can be quite sensitive to this choice.

In this paper, we propose a convex formulation of factorization machines based on the nuclear norm. Our formulation is more general than the original one in the sense that it imposes fewer restrictions on the feature interaction weight matrix $\boldsymbol{Z}$. For example, in our formulation, imposing positive semi-definiteness is possible but not necessary. In addition, our formulation does not require choosing any rank hyper-parameter and thus have one less hyper-parameter than the original formulation. For solving the corresponding optimization problem, we propose a globally-convergent two-block coordinate descent algorithm. Our algorithm alternates between estimating the feature weight vector $\boldsymbol{w}$ and a low-rank feature interaction weight matrix $\boldsymbol{Z}$. Estimating $\boldsymbol{w}$ is easy, since the problem reduces to a simple linear model objective. However, estimating $\boldsymbol{Z}$ is challenging, due to the quadratic number of feature interactions. Following a recent line of work [17] [4] [7], we derive a greedy coordinate descent algorithm which breaks down the large problem into smaller sub-problems. By exploiting structure, we can solve these sub-problems efficiently. Furthermore, our algorithm maintains an eigendecomposition of $\boldsymbol{Z}$. Therefore, the entire matrix $\boldsymbol{Z}$ is never materialized and our algorithm can scale to very high-dimensional data. Empirically, we demonstrate that our approach achieves comparable or better predictive accuracy than the original non-convex factorization machines on 4 recommendation tasks and scales to datasets with 10 million samples.

Notation. For arbitrary real matrices, the inner product is defined as $\langle \boldsymbol{A}, \boldsymbol{B} \rangle :=$ $\mathrm{Tr}(\boldsymbol{A}^{\mathrm{T}}\boldsymbol{B})$ and the squared Frobenius matrix norm as $\|\boldsymbol{A}\|_F^2 := \langle \boldsymbol{A}, \boldsymbol{A} \rangle$. We denote the element-wise product between two vectors $\boldsymbol{a} \in \mathbb{R}^d$ and $\boldsymbol{b} \in \mathbb{R}^d$ by $\boldsymbol{a} \circ \boldsymbol{b} := [a_1 b_1, \ldots, a_d b_d]^{\mathrm{T}}$. We denote the Kronecker product between two matrices $\boldsymbol{A} \in \mathbb{R}^{m \times n}$ and $\boldsymbol{B} \in \mathbb{R}^{p \times q}$ by $\boldsymbol{A} \otimes \boldsymbol{B} \in \mathbb{R}^{mp \times nq}$. We denote the set of symmetric $d \times d$ matrices by $\mathbb{S}^{d \times d}$. Given $\boldsymbol{A} \in \mathbb{R}^{m \times n}$, $\mathrm{vec}(\boldsymbol{A}) \in \mathbb{R}^{mn}$ denotes the vector obtained by stacking the columns of $\boldsymbol{A}$. By $[n]$, we denote the set $\{1, \ldots, n\}$. The support of a vector $\boldsymbol{\lambda} \in \mathbb{R}^d$ is defined as $\mathrm{supp}(\boldsymbol{\lambda}) := \{j \in [d] : \lambda_j \neq 0\}$.

2   Factorization Machines

Factorization machines [12][13] predict the output associated with an input $\boldsymbol{x} = [x_1, \ldots, x_d]^{\mathrm{T}} \in \mathbb{R}^d$ using the following simple equation:

$$\tilde{y}(\boldsymbol{x}|\boldsymbol{w}, \boldsymbol{V}) = \boldsymbol{w}^{\mathrm{T}}\boldsymbol{x} + \sum_{j=1}^{d} \sum_{j'=j+1}^{d} (\boldsymbol{V}\boldsymbol{V}^{\mathrm{T}})_{jj'} x_j x_{j'} \qquad (1)$$

where $\boldsymbol{w} \in \mathbb{R}^d$, $\boldsymbol{V} \in \mathbb{R}^{d \times k}$ and $k \ll d$ is a hyper-parameter which defines the rank of the factorization. The vector $\boldsymbol{w}$ contains the weights of individual features for predicting y, while the positive semi-definite matrix $\boldsymbol{Z} = \boldsymbol{V}\boldsymbol{V}^{\mathrm{T}} \in \mathbb{S}^{d \times d}$ contains the weights of pairwise feature interactions. Because factorization machines learn $\boldsymbol{Z}$ in factorized form, the number of parameters to estimate is reduced from d^2 to kd. In addition to helping reduce overfitting, this factorization allows to compute predictions efficiently by using

$$\tilde{y}(\boldsymbol{x}|\boldsymbol{w}, \boldsymbol{V}) = \boldsymbol{w}^{\mathrm{T}}\boldsymbol{x} + \frac{1}{2}\Big(\|\boldsymbol{V}^{\mathrm{T}}\boldsymbol{x}\|^2 - \sum_{s=1}^{k} \|\boldsymbol{v}_s \circ \boldsymbol{x}\|^2\Big),$$

where $\boldsymbol{v}_s \in \mathbb{R}^d$ is the sth column of $\boldsymbol{V}$. Thus, computing predictions costs $O(kd)$, instead of $O(d^2)$ when implemented naively. For sparse $\boldsymbol{x}$, the prediction cost reduces to $O(kN_z(\boldsymbol{x}))$, where $N_z(\boldsymbol{x})$ is the number of non-zero features in $\boldsymbol{x}$.

Although they can be used for any supervised learning task such as classification and regression, factorization machines are especially useful for recommender systems. As shown in [12][13], factorization machines can mimic many existing factorization models just by choosing an appropriate feature representation for $\boldsymbol{x}$. For example, consider a record (u, i, y), where $u \in U$ is a user index, $i \in I$ is an item index and $y \in \mathbb{R}$ is a rating given by u to i. Then factorization machines are exactly equivalent to matrix factorization (c.f., Section A in the supplementary material) simply by converting (u, i, y) to $(\boldsymbol{x}, y)$, where $\boldsymbol{x} \in \mathbb{R}^d$ is expressed in the following binary indicator representation with $d = |U| + |I|$

$$\boldsymbol{x} := [0, \ldots, 0, \overset{u}{1}, 0, \ldots, 0, \underset{|U|}{} 0, \ldots, 0, \overset{|U|+i}{1}, 0, \ldots, 0]^{\mathrm{T}}. \tag{2}$$

Using more elaborated feature representations [12] [13], it is possible to mimic many other factorization models, including SVD++ [8], timeSVD++[9] and PITF (pairwise interaction tensor factorization) [16]. Moreover, it is easy to incorporate auxiliary features such as user and item attributes, contextual information [15] and cross-domain feedback [10]. The ability to quickly try many different features ("feature engineering") is very flexible from a practitioner perspective. In addition, since factorization machines behave much like classifiers or regressors, they are easy to integrate in a consistent manner to a machine learning library (see [3] for a discussion on the merits of library design consistency).

Given a training set consisting of n feature vectors $\boldsymbol{X} = [\boldsymbol{x}_1, \ldots, \boldsymbol{x}_n]^{\mathrm{T}} \in \mathbb{R}^{n \times d}$ and corresponding targets $[y_1, \ldots, y_n]^{\mathrm{T}} \in \mathbb{R}^n$, we can estimate $\boldsymbol{w} \in \mathbb{R}^d$ and $\boldsymbol{V} \in \mathbb{R}^{d \times k}$ using the principle of empirical risk minimization. For example, we can solve the following optimization problem

$$\min_{\boldsymbol{w} \in \mathbb{R}^d, \boldsymbol{V} \in \mathbb{R}^{d \times k}} \sum_{i=1}^{n} \ell\Big(y_i, \tilde{y}(\boldsymbol{x}_i|\boldsymbol{w}, \boldsymbol{V})\Big) + \frac{\alpha}{2}\|\boldsymbol{w}\|^2 + \frac{\beta}{2}\|\boldsymbol{V}\|_F^2, \tag{3}$$

where $\ell(y, \tilde{y})$ is the loss "suffered" when predicting $\tilde{y}$ instead of y. Throughout this paper, we assume ℓ is a twice-differentiable convex function. For instance, for predicting continuous outputs, we can use the squared loss $\ell(y, \tilde{y}) = \frac{1}{2}(\tilde{y} - y)^2$. $\alpha > 0$ and $\beta > 0$ are hyper-parameters which control the trade-off between low loss and low model complexity. In practice, (3) can be solved using the stochastic gradient or coordinate descent methods. Both methods have a runtime complexity of $O(kN_z(\boldsymbol{X}))$ per epoch [13], where $N_z(\boldsymbol{X})$ is the total number of non-zero elements in $\boldsymbol{X}$. Assuming (2) is used, this is the same runtime complexity as for standard matrix factorization. We now state some important properties of the optimization problem (3), which were not mentioned in [12] and [13].

Proposition 1. *The optimization problem* (3) *is i) convex in* $\boldsymbol{w}$, *ii) non-convex in* $\boldsymbol{V}$ *and iii) convex in* v_{js} *(elements of* $\boldsymbol{V}$ *taken separately). If we replace* $\sum_{j'=j+1}^{d}(\boldsymbol{V}\boldsymbol{V}^{\mathrm{T}})_{jj'}x_jx_{j'}$ *by* $\sum_{j'=j}^{d}(\boldsymbol{V}\boldsymbol{V}^{\mathrm{T}})_{jj'}x_jx_{j'}$ *in* (1), *i.e., if we use diagonal elements of* $\boldsymbol{V}\boldsymbol{V}^{\mathrm{T}}$, *then* (3) *is iv) non-convex in both* $\boldsymbol{V}$ *and* v_{js}.

Property ii) means that the stochastic gradient and coordinate descent methods are only guaranteed to reach a local minimum, the quality of which typically depends on the initialization of $\boldsymbol{V}$. Property iii) explains why coordinate descent is a good method for solving (3): it can monotonically decrease the objective (3) until it reaches a local minimum. Property iv) shows that if we use diagonal elements of $\boldsymbol{V}\boldsymbol{V}^{\mathrm{T}}$, (3) becomes a much more challenging optimization problem, possibly subject to more bad local minima. In contrast, our formulation is convex whether or not we use diagonal elements.

3   Convex Formulation

We begin by rewriting the prediction equation (1) as

$$\hat{y}(\boldsymbol{x}|\boldsymbol{w}, \boldsymbol{Z}) = \boldsymbol{w}^{\mathrm{T}}\boldsymbol{x} + \sum_{j=1}^{d}\sum_{j'=1}^{d} z_{jj'}x_jx_{j'} = \boldsymbol{w}^{\mathrm{T}}\boldsymbol{x} + \langle \boldsymbol{Z}, \boldsymbol{x}\boldsymbol{x}^{\mathrm{T}} \rangle,$$

where $z_{jj'}$ denote the entries of the symmetric matrix $\boldsymbol{Z} \in \mathbb{S}^{d \times d}$. Clearly, we need to impose some structure on $\boldsymbol{Z}$ to avoid its $O(d^2)$ memory complexity. We choose to learn a low-rank matrix $\boldsymbol{Z}$, i.e., $\mathrm{rank}(\boldsymbol{Z}) \ll d$. Following recent advances in convex optimization, we can achieve this by regularizing $\boldsymbol{Z}$ with the nuclear norm (a.k.a. trace norm), which is known to be the tightest convex lower bound on matrix rank [11]. Given a symmetric matrix $\boldsymbol{Z} \in \mathbb{S}^{d \times d}$, the nuclear norm is defined as (c.f. supplementary material Section C)

$$\|\boldsymbol{Z}\|_* = \mathrm{Tr}\left(\sqrt{\boldsymbol{Z}^2}\right) = \|\boldsymbol{\lambda}\|_1, \qquad (4)$$

where $\boldsymbol{\lambda}$ is a vector which gathers the eigenvalues of $\boldsymbol{Z}$. We see that regularizing $\boldsymbol{Z}$ with the nuclear norm is equivalent to regularizing its eigenvalues with the ℓ_1 norm, which is known to promote sparsity. Since $\mathrm{rank}(\boldsymbol{Z}) = \|\boldsymbol{\lambda}\|_0 = |\mathrm{supp}(\boldsymbol{\lambda})|$, the nuclear norm thus promotes low-rank solutions. We therefore propose to learn factorization machines by solving the following optimization problem

$$\min_{\boldsymbol{w} \in \mathbb{R}^d, \boldsymbol{Z} \in \mathbb{S}^{d \times d}} \sum_{i=1}^{n} \ell\Big(y_i, \hat{y}(\boldsymbol{x}_i|\boldsymbol{w}, \boldsymbol{Z})\Big) + \frac{\alpha}{2}\|\boldsymbol{w}\|^2 + \beta\|\boldsymbol{Z}\|_*, \qquad (5)$$

where, again, ℓ is a twice-differentiable convex loss function and $\alpha > 0$ and $\beta > 0$ are hyper-parameters. Problem (5) is jointly convex in w and Z. In our formulation, there is no rank hyper-parameter (such as k for V). Instead, the rank of Z is indirectly controlled by β (the larger β, the lower rank(Z)).

Convexity is an important property, since it allows us to derive an efficient algorithm for finding a global solution (i.e., our algorithm is insensitive to initialization). In addition, our convex formulation is more general than the original one in the sense that imposing positive semi-definiteness of Z or ignoring diagonal elements of Z is not necessary (although it is possible, c.f., Section D and Section E in the supplementary material).

Any symmetric matrix $Z \in \mathbb{S}^{d \times d}$ can be written as an eigendecomposition $Z = P\Lambda P^{\mathrm{T}} = \sum_s \lambda_s p_s p_s^{\mathrm{T}}$, where P is an orthogonal matrix with columns $p_s \in \mathbb{R}^d$ and $\Lambda = \mathrm{diag}(\lambda)$ is a diagonal matrix with diagonal entries λ_s. Using this decomposition, we can compute predictions efficiently by

$$\hat{y}(x|w, P\Lambda P^{\mathrm{T}}) = w^{\mathrm{T}}x + \langle P\Lambda P^{\mathrm{T}}, xx^{\mathrm{T}} \rangle = w^{\mathrm{T}}x + \sum_{s=1}^{k} \lambda_s (p_s^{\mathrm{T}}x)^2, \qquad (6)$$

where $k = \mathrm{rank}(Z)$. Thus, prediction cost is the same as non-convex factorization machines, i.e., $O(kN_z(x))$. The algorithm we present in Section 4 always maintains such a decomposition. Therefore, Z is never materialized in memory and we can scale to high-dimensional data. Equation (6) also suggests an interesting interpretation of convex factorization machines. Let $\kappa(p, x) = (p^{\mathrm{T}}x)^2$, i.e., κ is a homogeneous polynomial kernel of degree 2. Then, (6) can be written as $\hat{y}(x|w, P\Lambda P^{\mathrm{T}}) = w^{\mathrm{T}}x + \sum_{s=1}^{k} \lambda_s \kappa(p_s, x)$. Thus, convex factorization machines evaluate the homogeneous polynomial kernel between orthonormal basis vectors $p_1, \ldots, p_k$ and x. In contrast, kernel ridge regression and other kernel machines compute predictions using $\sum_{i=1}^{n} a_i \kappa(x_i, x)$, i.e., the kernel is evaluated between *training instances* and x. Thus, the main advantage of convex factorization machines over traditional kernel machines is that the basis vectors are actually *learned* from data.

4   Optimization Algorithm

To solve (5), we propose a two-block coordinate descent algorithm. That is, we alternate between minimizing with respect to w and Z until convergence. When the algorithm terminates, it returns w and $Z = P\Lambda P^{\mathrm{T}}$.

4.1   Minimizing with Respect to w

For minimizing (5) with respect to w, we need to solve

$$\min_{w \in \mathbb{R}^d} \sum_{i=1}^{n} \ell(y_i, w^{\mathrm{T}}x_i + \pi_i) + \frac{\alpha}{2}\|w\|^2, \qquad (7)$$

where $\pi_i = \langle Z, x_i x_i^{\mathrm{T}} \rangle$. This is a standard linear model objective, except that the predictions are shifted by π_i. Thus, we can solve (7) using standard methods.

Algorithm 1. Minimizing (8) w.r.t. $\boldsymbol{Z}$

Input: $\{(\boldsymbol{x}_i, y_i)\}_{i=1}^n$, initial $\boldsymbol{Z} = \boldsymbol{P}\operatorname{diag}(\boldsymbol{\lambda})\boldsymbol{P}^{\mathrm{T}}$, $\beta > 0$
$\boldsymbol{Z}_{\boldsymbol{\lambda}} := \sum_{s \in \operatorname{supp}(\boldsymbol{\lambda})} \lambda_s \boldsymbol{p}_s \boldsymbol{p}_s^{\mathrm{T}}$
repeat
    Compute $\boldsymbol{p} = $ dominant eigenvector of $\nabla L(\boldsymbol{Z}_{\boldsymbol{\lambda}})$
    Find $\lambda = \operatorname{argmin}_{\lambda \in \mathbb{R}} L\left(\boldsymbol{Z}_{\boldsymbol{\lambda}} + \lambda \boldsymbol{p}\boldsymbol{p}^{\mathrm{T}}\right) + \beta|\lambda|$
    $\boldsymbol{P} \leftarrow [\boldsymbol{P}\ \boldsymbol{p}]$ $\boldsymbol{\lambda} \leftarrow [\boldsymbol{\lambda}\ \lambda]$

    Diagonal refitting case
    $\bar{\boldsymbol{\lambda}} \leftarrow \boldsymbol{\lambda}$
    $\boldsymbol{\lambda} \leftarrow \operatorname{argmin}_{\boldsymbol{\lambda} \in \mathbb{R}^{\operatorname{supp}(\bar{\boldsymbol{\lambda}})}} \tilde{L}(\boldsymbol{\lambda}) + \beta\|\boldsymbol{\lambda}\|_1 = \operatorname{argmin}_{\boldsymbol{\lambda} \in \mathbb{R}^{\operatorname{supp}(\bar{\boldsymbol{\lambda}})}} L(\boldsymbol{Z}_{\boldsymbol{\lambda}}) + \beta\|\boldsymbol{\lambda}\|_1$

    Fully-corrective refitting case
    $\boldsymbol{A} = \operatorname{argmin}_{\boldsymbol{A} \in \mathbb{S}^{k \times k}} L(\boldsymbol{P}\boldsymbol{A}\boldsymbol{P}^{\mathrm{T}}) + \beta\|\boldsymbol{A}\|_*$ where $k = \operatorname{rank}(\boldsymbol{Z}_{\boldsymbol{\lambda}})$
    $\boldsymbol{P} \leftarrow \boldsymbol{P}\boldsymbol{Q}$ $\boldsymbol{\lambda} \leftarrow \operatorname{diag}(\boldsymbol{\Sigma})$ where $\boldsymbol{A} = \boldsymbol{Q}\boldsymbol{\Sigma}\boldsymbol{Q}^{\mathrm{T}}$

until convergence
Output: $\boldsymbol{Z} = \boldsymbol{P}\operatorname{diag}(\boldsymbol{\lambda})\boldsymbol{P}^{\mathrm{T}}$

4.2 Minimizing with Respect to $\boldsymbol{Z}$

For minimizing (5) with respect to $\boldsymbol{Z}$, we need to solve

$$\min_{\boldsymbol{Z} \in \mathbb{S}^{d \times d}} \underbrace{\sum_{i=1}^n \ell\left(y_i, \boldsymbol{w}^{\mathrm{T}}\boldsymbol{x}_i + \langle \boldsymbol{Z}, \boldsymbol{x}_i \boldsymbol{x}_i^{\mathrm{T}}\rangle\right)}_{:=L(\boldsymbol{Z})} + \beta\|\boldsymbol{Z}\|_*. \tag{8}$$

Two standard methods for solving nuclear norm regularized problems are proximal gradient and ADMM. For these methods, the key operation is the proximal operator, which requires an SVD and is thus a bottleneck in scaling to large matrix sizes. In order to address this issue, we adapt greedy coordinate descent algorithms [4] [7] designed for general nuclear norm regularized minimization. The main difference of our algorithm is that we learn an eigendecomposition of $\boldsymbol{Z}$ rather than an SVD, in order to take advantage of the symmetry of $\boldsymbol{Z}$.

Outline. To minimize (8), on each iteration we greedily find the rank-one matrix $\boldsymbol{p}\boldsymbol{p}^{\mathrm{T}}$ that most violates the optimality conditions and add it to $\boldsymbol{Z}$ by $\boldsymbol{Z} \leftarrow \boldsymbol{Z} + \lambda \boldsymbol{p}\boldsymbol{p}^{\mathrm{T}}$, where λ is the optimal weight. Thus, the rank of $\boldsymbol{Z}$ increases by at most 1 on each iteration. In practice, however, we never materialize $\boldsymbol{Z}$ and maintain its eigendecomposition $\boldsymbol{Z} = \boldsymbol{P}\operatorname{diag}(\boldsymbol{\lambda})\boldsymbol{P}^{\mathrm{T}}$ instead. To ensure convergence, we refit the eigendecomposition of $\boldsymbol{Z}$ on each iteration using one of two methods: diagonal refitting (update $\boldsymbol{\lambda}$ only) or fully corrective refitting (update both $\boldsymbol{\lambda}$ and $\boldsymbol{P}$). The entire procedure is summarized in Algorithm 1.

Finding λ and p. Using (4) and (6), we obtain that (8) is equivalent to

$$\min_{\boldsymbol{\lambda} \in \Theta} \underbrace{\sum_{i=1}^{n} \ell\Big(y_i, \boldsymbol{w}^{\mathrm{T}}\boldsymbol{x}_i + \sum_{s \in \mathcal{S}} \lambda_s(\boldsymbol{p}_s^{\mathrm{T}}\boldsymbol{x}_i)^2\Big)}_{:=\tilde{L}(\boldsymbol{\lambda})} + \beta\|\boldsymbol{\lambda}\|_1, \tag{9}$$

where $\mathcal{S}$ is an index set for the elements of the set $\{\boldsymbol{p}\boldsymbol{p}^{\mathrm{T}} : \boldsymbol{p} \in \mathbb{R}^d, \|\boldsymbol{p}\| = 1\}$ and $\Theta := \{\boldsymbol{\lambda} \in \mathbb{R}^{\mathcal{S}} : \mathrm{supp}(\boldsymbol{\lambda})$ is finite$\}$. Thus, we converted a problem with respect to $\boldsymbol{Z}$ in the space of symmetric matrices to a problem with respect to $\boldsymbol{\lambda}$ in the space of (normalized) rank-one matrices. This space can be arbitrarily large. However, the number of non-zero elements in $\boldsymbol{\lambda}$ is at most d. Moreover, $\boldsymbol{\lambda}$ will be typically sparse thanks to the regularization term $\beta\|\boldsymbol{\lambda}\|_1$, i.e., $|\mathrm{supp}(\boldsymbol{\lambda})| = \mathrm{rank}(\boldsymbol{Z}) \ll d$. A difference between (9) and past works [17] [4] is that we do not constrain $\boldsymbol{\lambda}$ to be non-negative, since eigenvalues can be negative, unlike singular values. Constraining $\boldsymbol{\lambda}$ to be non-negative corresponds to a positive semi-definite constraint on $\boldsymbol{Z}$, which we cover in Section D of the supplementary material.

According to the Karush-Kuhn-Tucker (KKT) conditions, for any $s \in \mathcal{S}$, the optimality violation of λ_s at $\boldsymbol{\lambda}$ is given by

$$\nu_s = \begin{cases} |\nabla_s\tilde{L}(\boldsymbol{\lambda}) + \beta|, & \text{if } \lambda_s > 0 \\ |\nabla_s\tilde{L}(\boldsymbol{\lambda}) - \beta|, & \text{if } \lambda_s < 0 \\ \max\left(|\nabla_s\tilde{L}(\boldsymbol{\lambda})| - \beta, 0\right), & \text{if } \lambda_s = 0, \end{cases}$$

where $\nabla_s\tilde{L}(\boldsymbol{\lambda}) = \frac{\partial \tilde{L}(\boldsymbol{\lambda})}{\partial \lambda_s}$. Using the chain rule, we obtain

$$\nabla_s\tilde{L}(\boldsymbol{\lambda}) = \langle \nabla L(\boldsymbol{Z}_\lambda), \boldsymbol{p}_s\boldsymbol{p}_s^{\mathrm{T}}\rangle = \boldsymbol{p}_s^{\mathrm{T}}\nabla L(\boldsymbol{Z}_\lambda)\boldsymbol{p}_s,$$

where $\boldsymbol{Z}_\lambda := \sum_{s \in \mathrm{supp}(\boldsymbol{\lambda})} \lambda_s\boldsymbol{p}_s\boldsymbol{p}_s^{\mathrm{T}}$ and $\nabla L(\boldsymbol{Z}) \in \mathbb{S}^{d \times d}$ is the gradient of L at $\boldsymbol{Z}$. Intuitively, we would like to find the eigenvector $\boldsymbol{p}_s$ which maximizes ν_s:

$$\underset{s \notin \mathrm{supp}(\boldsymbol{\lambda})}{\mathrm{argmax}}\; \nu_s = \underset{s \in \mathcal{S}}{\mathrm{argmax}}\; |\nabla_s\tilde{L}(\boldsymbol{\lambda})| = \underset{s \in \mathcal{S}}{\mathrm{argmax}}\; |\boldsymbol{p}_s^{\mathrm{T}}\nabla L(\boldsymbol{Z}_\lambda)\boldsymbol{p}_s|$$

Thus, $\boldsymbol{p}_s$ corresponds to the dominant eigenvector of $\nabla L(\boldsymbol{Z}_\lambda)$ (eigenvector corresponding to the greatest eigenvalue in absolute value). We can find $\boldsymbol{p}_s$ efficiently using the power iteration method. Since $\nabla L(\boldsymbol{Z}_\lambda)$ is a $d \times d$ matrix, we cannot afford to store it in memory when d is large. Fortunately, the power iteration method only accesses $\nabla L(\boldsymbol{Z}_\lambda)$ through matrix-vector products $\nabla L(\boldsymbol{Z}_\lambda)\boldsymbol{p}$ for some vector $\boldsymbol{p} \in \mathbb{R}^d$. By exploiting the structure of $\nabla L(\boldsymbol{Z}_\lambda)$, we can compute this product efficiently (c.f., Section 4.3 for the squared loss).

Let $\bar{\boldsymbol{\lambda}}$ be the current iterate of $\boldsymbol{\lambda}$. Once we found $\boldsymbol{p}_s$, we can find λ_s by

$$\lambda_s = \underset{\lambda \in \mathbb{R}}{\mathrm{argmin}}\; \tilde{L}\Big(\bar{\boldsymbol{\lambda}} + (\lambda - \bar{\lambda}_s)\boldsymbol{e}_s\Big) + \beta|\lambda| = \underset{\lambda \in \mathbb{R}}{\mathrm{argmin}}\; L\Big(\boldsymbol{Z}_{\bar{\lambda}} + (\lambda - \bar{\lambda}_s)\boldsymbol{p}_s\boldsymbol{p}_s^{\mathrm{T}}\Big) + \beta|\lambda|, \tag{10}$$

where $e_s = [\underbrace{0,\ldots,0},1,0,\ldots,0]^{\mathrm{T}}$. For the squared loss, this problem can be
$\quad\quad\quad\quad s-1$
solved in closed form (c.f., Section 4.3). For other loss functions, we can solve
the problem iteratively.

Diagonal Refitting. Similarly to [4], we can refit $\boldsymbol{\lambda}$ restricted to its current
support. Let $\bar{\boldsymbol{\lambda}}$ be the current iterate of $\boldsymbol{\lambda}$. Then, we solve

$$\min_{\boldsymbol{\lambda}\in\mathbb{R}^{\mathrm{supp}(\bar{\boldsymbol{\lambda}})}} \tilde{L}(\boldsymbol{\lambda}) + \beta\|\boldsymbol{\lambda}\|_1.$$

This can easily be solved by iteratively using (10) for all $s \in \mathrm{supp}(\bar{\boldsymbol{\lambda}})$ until
the sum of violations $\sum_{s\in\mathrm{supp}(\bar{\boldsymbol{\lambda}})} \nu_s$ converges. We call this method "diagonal
refitting", since the matrix $\boldsymbol{\Lambda} = \mathrm{diag}(\boldsymbol{\lambda})$ in $\boldsymbol{Z} = \boldsymbol{P}\boldsymbol{\Lambda}\boldsymbol{P}^{\mathrm{T}}$ is diagonal.

Fully-corrective Refitting. Any matrix $\boldsymbol{Z} \in \mathbb{S}^{d\times d}$ can be written as $\boldsymbol{P}\boldsymbol{A}\boldsymbol{P}^{\mathrm{T}}$,
where $\boldsymbol{P} \in \mathbb{R}^{d\times k}$, $\boldsymbol{A} \in \mathbb{S}^{k\times k}$ ($\boldsymbol{A}$ not necessarily diagonal) and $k = \mathrm{rank}(\boldsymbol{Z})$.
Following a similar idea to [17] and [7], injecting $\boldsymbol{Z} = \boldsymbol{P}\boldsymbol{A}\boldsymbol{P}^{\mathrm{T}}$ in (8), we can
solve

$$\min_{\boldsymbol{A}\in\mathbb{S}^{k\times k}} L(\boldsymbol{P}\boldsymbol{A}\boldsymbol{P}^{\mathrm{T}}) + \beta\|\boldsymbol{A}\|_*, \tag{11}$$

where we used $\|\boldsymbol{P}\boldsymbol{A}\boldsymbol{P}^{\mathrm{T}}\|_* = \|\boldsymbol{A}\|_*$. This problem is similar to (8); only this
time, it is $k \times k$ dimensional instead of $d \times d$ dimensional. Once we obtained $\boldsymbol{A}$,
we can update $\boldsymbol{P}$ and $\boldsymbol{\lambda}$ by $\boldsymbol{P} \leftarrow \boldsymbol{P}\boldsymbol{Q}$ and $\boldsymbol{\lambda} \leftarrow \mathrm{diag}(\boldsymbol{\Sigma})$, where $\boldsymbol{Q}\boldsymbol{\Sigma}\boldsymbol{Q}^{\mathrm{T}}$ is an
eigendecomposition of $\boldsymbol{A}$ (cheap to compute since $\boldsymbol{A}$ is $k \times k$).

We propose to solve (11) by the alternating direction method of multipliers
(ADMM). To do so, we consider the following augmented Lagrangian

$$\min_{\boldsymbol{A}\in\mathbb{S}^{k\times k},\boldsymbol{B}\in\mathbb{S}^{k\times k}} L(\boldsymbol{P}\boldsymbol{A}\boldsymbol{P}^{\mathrm{T}}) + \beta\|\boldsymbol{B}\|_* \text{ s.t. } \boldsymbol{A} - \boldsymbol{B} = 0. \tag{12}$$

ADMM solves (12) using the following iterative procedure:

$$\boldsymbol{A}^{\tau+1} = \underset{\boldsymbol{A}\in\mathbb{S}^{k\times k}}{\mathrm{argmin}} \underbrace{L(\boldsymbol{P}\boldsymbol{A}\boldsymbol{P}^{\mathrm{T}}) + \frac{\rho}{2}\|\boldsymbol{A} - \boldsymbol{B}^{\tau} + \boldsymbol{M}^{\tau}\|^2}_{:=\hat{L}(\boldsymbol{A})} \tag{13}$$

$$\boldsymbol{B}^{\tau+1} = S_{\beta/\rho}\left(\boldsymbol{A}^{\tau+1} + \boldsymbol{M}^{\tau}\right) \tag{14}$$

$$\boldsymbol{M}^{\tau+1} = \boldsymbol{M}^{\tau} + \boldsymbol{A}^{\tau+1} - \boldsymbol{B}^{\tau+1},$$

where ρ is a parameter and S_c is the proximal operator (here, shrinkage operator). In practice, a common choice is $\rho = 1$. The procedure converges when
$\|\boldsymbol{A}^{\tau} - \boldsymbol{B}^{\tau}\|_F^2 \leq \epsilon$. We now explain how to solve (14) and (13).

Given an eigendecomposition $\boldsymbol{A} = \boldsymbol{Q}\boldsymbol{\Sigma}\boldsymbol{Q}^{\mathrm{T}}$, where $\boldsymbol{\Sigma} = \mathrm{diag}(\sigma_1,\ldots,\sigma_k)$, the
shrinkage operator is defined as

$$S_c(\boldsymbol{A}) = \underset{\boldsymbol{B}}{\mathrm{argmin}} \frac{1}{2}\|\boldsymbol{A} - \boldsymbol{B}\|_F^2 + c\|\boldsymbol{B}\|_* = \boldsymbol{Q}\,\mathrm{diag}(\hat{\sigma}_1,\ldots,\hat{\sigma}_k)\boldsymbol{Q}^{\mathrm{T}},$$

where $\hat{\sigma}_s = \text{sign}(\sigma_s) \max(|\sigma_s| - c, 0)$. In other words, we apply the soft-thresholding operator to the eigenvalues of A. For solving the sub-problem (13), we can afford to use the Newton method, since $k \ll d$. Let $\nabla \hat{L}(A) \in \mathbb{S}^{k \times k}$ and $\nabla^2 \hat{L}(A) \in \mathbb{S}^{k^2 \times k^2}$ be the gradient and Hessian of $\hat{L}$ at A. On each iteration, the Newton method updates A by

$$A \leftarrow A - \gamma D$$

where $D \in \mathbb{R}^{k \times k}$ is the solution of the system of linear equations

$$\nabla^2 \hat{L}(A) \text{vec}(D) = \text{vec}\left(\nabla \hat{L}(A)\right) \tag{15}$$

and γ is adjusted by line search (typically, using the Wolfe conditions). Using the chain rule, we can compute $\nabla \hat{L}(A)$ and $\nabla^2 \hat{L}(A)$ by

$$\nabla \hat{L}(A) = P^{\mathrm{T}}\left(\nabla L(Z)|_{Z=PAP^{\mathrm{T}}}\right)P + \rho(A - B^{\tau} + M^{\tau})$$

$$\nabla^2 \hat{L}(A) = P^{\mathrm{T}} \otimes P^{\mathrm{T}}\left(\nabla^2 L(Z)|_{Z=PAP^{\mathrm{T}}}\right)P \otimes P + \rho I$$

To compute $\nabla L(Z)|_{Z=PAP^{\mathrm{T}}}$ and $\nabla^2 L(Z)|_{Z=PAP^{\mathrm{T}}}$, we need to compute the predictions at $Z = PAP^{\mathrm{T}}$. This can be done efficiently by $\hat{y}(x|w, PAP^{\mathrm{T}}) = w^{\mathrm{T}}x + x^{\mathrm{T}}(PA)(P^{\mathrm{T}}x)$.

To solve (15), we can use the conjugate gradient method. This method only accesses the Hessian through Hessian-vector products, i.e., $\nabla^2 \hat{L}(A) \text{vec}(D)$. By using the problem structure together with the property $(A \otimes B)\text{vec}(D) = \text{vec}(BDA^{\mathrm{T}})$, we can usually compute these products efficiently.

4.3 Squared Loss Case

For the case of the squared loss, we obtain very simple expressions and closed-form solutions.

Minimizing with Respect to w. For the squared loss, (7) becomes

$$\min_{w \in \mathbb{R}^d} \frac{1}{2}\|Xw - \tau\|^2 + \frac{\alpha}{2}\|w\|^2,$$

where $\tau \in \mathbb{R}^n$ is a vector with elements $\tau_i = y_i - \langle Z, x_i x_i^{\mathrm{T}}\rangle$. This is a standard ridge regression problem. A closed-form solution can be computed by $w = X^{\mathrm{T}}(XX^{\mathrm{T}} + \alpha I)^{-1}\tau$ in $O(n^3)$ or by $w = (X^{\mathrm{T}}X + \alpha I)^{-1}X^{\mathrm{T}}\tau$ in $O(d^3)$. When n and d are both large, we can use an iterative method (e.g., conjugate gradient) instead.

Finding the Dominant Eigenvector. For finding the dominant eigenvector of $\nabla L(Z_\lambda)$, we use the power iteration method, which needs to compute matrix-vector products $\nabla L(Z_\lambda)p$. For the squared loss, the gradient is given by:

$$\nabla L(Z) = \sum_{i=1}^{n} r_i x_i x_i^{\mathrm{T}} = X^{\mathrm{T}}RX, \tag{16}$$

where $R = \mathrm{diag}(r_1, \ldots, r_n)$ and $r_i = \hat{y}_i - y_i$ is the residual of x_i at (w, Z). Clearly, we can compute $\nabla L(Z_\lambda)p$ efficiently without ever materializing $\nabla L(Z_\lambda)$.

Minimizing with Respect to λ. For the squared loss, we obtain that (10) is equivalent to

$$\lambda_s = \underset{\lambda \in \mathbb{R}}{\mathrm{argmin}}\, \nabla_s \tilde{L}(\bar{\lambda})(\lambda - \bar{\lambda}_s) + \frac{1}{2}\nabla_{ss}^2 \tilde{L}(\bar{\lambda})(\lambda - \bar{\lambda}_s)^2 + \beta|\lambda| = \underset{\lambda \in \mathbb{R}}{\mathrm{argmin}}\, \frac{1}{2}\left(\lambda - \tilde{\lambda}_s\right)^2 + c_s|\lambda|$$

where $\tilde{\lambda}_s := \bar{\lambda}_s - \frac{\nabla_s \tilde{L}(\bar{\lambda})}{\nabla_{ss}^2 \tilde{L}(\bar{\lambda})}$ and $c_s := \frac{\beta}{\nabla_{ss}^2 \tilde{L}(\bar{\lambda})}$. This is the well-known soft-thresholding operator, whose closed-form solution is given by

$$\lambda_s = \mathrm{sign}(\tilde{\lambda}_s) \max(|\tilde{\lambda}_s| - c_s, 0).$$

The first and second derivatives of $\tilde{L}$ with respect to λ_s can be computed efficiently by

$$\nabla_s \tilde{L}(\lambda) = \sum_{i=1}^{n} r_i \langle p_s p_s^{\mathrm{T}}, x_i x_i^{\mathrm{T}} \rangle = \sum_{i=1}^{n} r_i (p_s^{\mathrm{T}} x_i)^2 \tag{17}$$

$$\nabla_{ss}^2 \tilde{L}(\lambda) = \sum_{i=1}^{n} \langle p_s p_s^{\mathrm{T}}, x_i x_i^{\mathrm{T}} \rangle^2 = \sum_{i=1}^{n} (p_s^{\mathrm{T}} x_i)^4, \tag{18}$$

where, again, $r_i = \hat{y}_i - y_i$ is the residual of x_i at (w, Z_λ).

Fully-corrective Refitting. For the squared loss, the Newton method gives the exact solution of (13) in one iteration and γ can be set to 1 (i.e., no line search needed). Given an initial guess $\bar{A}$, if we solve the system

$$\nabla^2 \hat{L}(\bar{A}) \mathrm{vec}(D) = \mathrm{vec}\left(\nabla \hat{L}(\bar{A})\right) \tag{19}$$

w.r.t. $\mathrm{vec}(D)$, then the optimal solution of (13) is $A = \bar{A} - D$. To solve (19), we use the conjugate gradient method, which accesses the Hessian only through Hessian-vector products. Thus, we never need to materialize the Hessian matrix. The gradient and Hessian-vector product expressions are given by

$$\nabla \hat{L}(A) = P^{\mathrm{T}} X^{\mathrm{T}} R X P + \rho(A - B + M) \tag{20}$$

$$\nabla^2 \hat{L}(A) \mathrm{vec}(D) = \mathrm{vec}(P^{\mathrm{T}} X^{\mathrm{T}} \Pi X P) + \rho\, \mathrm{vec}(D), \tag{21}$$

where $R = \mathrm{diag}(r_1, \ldots, r_n)$, $r_i = \hat{y}_i - y_i$ is the residual of x_i at (w, PAP^{T}), $\Pi = \mathrm{diag}(\pi_1, \ldots, \pi_n)$ and $\pi_i = \langle PDP^{\mathrm{T}}, x_i x_i^{\mathrm{T}} \rangle = x_i^{\mathrm{T}}(PDP^{\mathrm{T}})x_i$. Note that the Hessian-vector product is independent of A.

4.4   Computational Complexity

We focus our discussion on minimizing w.r.t. Z when using the squared loss (we assume the implementation techniques described in Section G of the supplementary material are used). For power iteration, the main cost is computing the matrix-vector product $\nabla L(\boldsymbol{Z}_{\boldsymbol{\lambda}})\boldsymbol{p}$. From (16), this costs $O(N_z(\boldsymbol{X}))$. For minimizing with respect to λ_s, the main task consists in computing the first and second derivatives (17) and (18), which costs $O(n)$. For the fully corrective refitting, ADMM alternates between (13) and (14). For (13), the main cost stems from computing the gradient and Hessian-vector product (20) and (21), which takes $O(kN_z(\boldsymbol{X}) + dk^2)$. For (14), the main cost stems from computing the eigendecomposition of a $k \times k$ matrix, which takes $O(k^3)$, where $k \ll d$. If we use the binary indicator representation (2), then convex factorization machines have the same overall runtime cost as convex matrix factorization [7].

4.5   Convergence Guarantees

Our method is an instance of block coordinate descent with two blocks, $\boldsymbol{w}$ and $\boldsymbol{Z}$. Past convergence analysis of block coordinate descent typically requires subproblems to have unique solutions [2, Proposition 2.7.1]. However, (5) is convex in $\boldsymbol{Z}$ but not strictly convex. Hence minimization with respect to $\boldsymbol{Z}$ may have multiple optimal solutions. Fortunately, for the case of two blocks, the uniqueness condition is not needed [6]. For minimization with respect to $\boldsymbol{Z}$, our greedy coordinate descent algorithm is an instance of [4] when using diagonal refitting and of [7] when using fully corrective refitting. Both methods asymptotically converge to an optimal solution, even if we find the dominant eigenvector only approximately. Thus, our two-block coordinate descent method asymptotically converges to a global minimum.

5   Experimental Results

5.1   Synthetic Experiments

We conducted experiments on synthetic data in order to compare the predictive power of different models:

- Convex FM (use diag): $\hat{y} = \boldsymbol{w}^{\mathrm{T}}\boldsymbol{x} + \langle \boldsymbol{Z}, \boldsymbol{x}\boldsymbol{x}^{\mathrm{T}} \rangle$
- Convex FM (ignore diag): $\hat{y} = \boldsymbol{w}^{\mathrm{T}}\boldsymbol{x} + \langle \boldsymbol{Z}, \boldsymbol{x}\boldsymbol{x}^{\mathrm{T}} - \mathrm{diag}(\boldsymbol{x})^2 \rangle$
- Original FM: $\hat{y} = \boldsymbol{w}^{\mathrm{T}}\boldsymbol{x} + \sum_{j=1}^{d}\sum_{j'=j+1}^{d}(\boldsymbol{V}\boldsymbol{V}^{\mathrm{T}})_{jj'}x_j x_{j'}$
- Ridge regression: $\hat{y} = \boldsymbol{w}^{\mathrm{T}}\boldsymbol{x}$
- Kernel ridge regression: $\hat{y} = \sum_{i=1}^{n} a_i \kappa(\boldsymbol{x}_i, \boldsymbol{x})$

For kernel ridge regression, the kernel used was the polynomial kernel of degree 2: $\kappa(\boldsymbol{x}_i, \boldsymbol{x}_j) = (\gamma + \boldsymbol{x}_i^{\mathrm{T}}\boldsymbol{x}_j)^2$. Due to lack of space, the parameter estimation procedure for Convex FM (ignore diag) is explained in the supplementary material. We compared the above models under various generative assumptions.

Data generation. We generated $\boldsymbol{y} = [y_1, \ldots, y_n]^{\mathrm{T}}$ by $y_i = \boldsymbol{w}^{\mathrm{T}}\boldsymbol{x}_i + \langle \boldsymbol{Z}, \boldsymbol{x}_i\boldsymbol{x}_i^{\mathrm{T}} \rangle$ (use diagonal case) or by $y_i = \boldsymbol{w}^{\mathrm{T}}\boldsymbol{x}_i + \langle \boldsymbol{Z}, \boldsymbol{x}_i\boldsymbol{x}_i^{\mathrm{T}} - \mathrm{diag}(\boldsymbol{x}_i)^2 \rangle$ (ignore diagonal case). To generate $\boldsymbol{w} = [w_1, \ldots, w_d]^{\mathrm{T}}$, we used $w_j \sim \mathcal{N}(0,1)\ \forall j \in [d]$ where $\mathcal{N}(0,1)$ is the standard normal distribution. To generate $\boldsymbol{Z} = \boldsymbol{P}\,\mathrm{diag}(\boldsymbol{\lambda})\boldsymbol{P}^{\mathrm{T}}$, we

Table 1. Test RMSE of different methods on synthetic data.

Generative process	Convex FM (use diag)	Convex FM (ignore diag)	Original FM	Ridge	Kernel ridge (polynomial kernel)
dense, PSD, use diag	**68.35**	110.18	104.39	104.67	76.77
dense, PSD, ignore diag	27.45	**5.93**	5.97	56.91	31.74
dense, not PSD, use diag	**92.31**	159.47	165.90	223.76	154.12
dense, not PSD, ignore diag	60.74	**21.17**	139.66	208.55	138.17
sparse, PSD, use diag	**23.12**	25.23	23.82	25.45	25.10
sparse, PSD, ignore diag	8.93	**5.10**	5.92	21.41	14.39
sparse, not PSD, use diag	**12.75**	23.13	30.60	36.43	25.17
sparse, not PSD, ignore diag	11.66	**7.91**	27.46	34.62	21.75

used $p_{js} \sim \mathcal{N}(0,1)\ \forall j \in [d]\ \forall s \in [k]$ and $\lambda_s \in \mathcal{N}(0,1)\ \forall s \in [d]$ (not positive semi-definite [PSD] case) or $\lambda_s \sim \mathcal{U}(0,1)\ \forall s \in [d]$ (positive semi-definite case), where $\mathcal{U}(0,1)$ is the uniform distribution between 0 and 1. For generating $\boldsymbol{X} \in \mathbb{R}^{n \times d}$, we compared two cases. In the dense case, we used $x_{ij} \sim \mathcal{N}(0,1)\ \forall i \in [n]\ \forall j \in [d]$. In the sparse case, we sampled $\bar{d}$ features from a multinomial distribution whose parameters are set uniformly at random. We chose $n = 1000$, $d = 50$, $k = 5$ and $\bar{d} = 5$. We split the data into 75% training and 25% testing and added 1% Gaussian noise to the training targets.

Results. Results (RMSE on test data) are indicated in Table 1. Hyperparameters of the respective methods were optimized by 5-fold cross-validation. The setting which is most favorable to Original FM is when the matrix $\boldsymbol{Z}$ used for generating synthetic data is PSD and diagonal elements of $\boldsymbol{Z}$ are ignored (2nd and 6th rows in Table 1). In this case, Original FM performed well, although worse than Convex FM (ignore diag). However, in other settings, especially when $\boldsymbol{Z}$ is not PSD, convex FM outperformed the Original FM. For example, for dense data, when $\boldsymbol{Z}$ is not PSD and diagonal elements of $\boldsymbol{Z}$ are ignored, Convex FM (use diag) achieved a test RMSE of 60.74, Convex FM (ignore diag) 21.17 and Original FM 139.66. Ridge regression was the worst method in all settings. This is not surprising since it does not use feature interactions. Kernel ridge regression with a polynomial kernel of degree 2 outperformed ridge regression but was worse than convex FM on all datasets.

5.2   Recommender System Experiments

We also conducted experiments on 4 standard recommendation tasks. Datasets used in our experiments are summarized below.

Dataset	n	$d = |U| + |I|$
Movielens 100k	100,000 (ratings)	2,625 = 943 (users) + 1,682 (movies)
Movielens 1m	1,000,209 (ratings)	9,940 = 6,040 (users) + 3,900 (movies)
Movielens 10m	10,000,054 (ratings)	82,248 = 71,567 (users) + 10,681 (movies)
Last.fm	108,437 (tag counts)	24,078 = 12,133 (artists) + 11,945 (tags)

For simplicity, we used the binary indicator representation (2), which results in a design matrix $\boldsymbol{X}$ of size $n \times d$. We split samples uniformly at random between 75% for training and 25% for testing. For Movielens datasets, the task is to predict ratings between 1 and 5 given by users to movies, i.e., $y \in \{1, \ldots, 5\}$.

For Last.fm, the task is to predict the number of times a tag was assigned to an artist, i.e., $y \in \mathbb{N}$. In all experiments, we set $\alpha = 10^{-9}$ for convex and original factorization machines, as well as ridge regression. Because we used the binary indicator representation (2), w plays the same role as unpenalized bias terms (c.f., Section A in the supplementary material).

Solver Comparison. For minimizing our objective function with respect to Z, we compared greedy coordinate descent (GCD) with diagonal refitting and with fully-corrective refitting, the proximal gradient method and ADMM. Minimization with respect to w was carried out using the conjugate gradient method. Results when setting $\beta = 10$ are given in Figure 1. We were only able to run ADMM on Movielens 100K because it needs to materialize Z in memory. Experiments were run on a machine with Intel Xeon X5677 CPU (3.47GHz) and 48 GB memory.

Results. GCD with fully-corrective refitting was consistently the best solver both with respect to objective value and test RMSE. GCD with diagonal refitting converged slower with respect to objective value but was similar with respect to test RMSE, except on Last.fm. The proximal gradient and ADMM methods were an order of magnitude slower than GCD.

Model Comparison. We used the same setup as in Section 5.1 except that we replaced kernel ridge regression with support vector regression (we used the implementation in libsvm, which has a kernel cache and scales better than kernel ridge regression w.r.t. n). For hyper-parameter tuning, we used 3-fold cross-validation (CV). For convex and original factorization machines, we chose β from 10 log-spaced values between 10^{-1} and 10^{2}. For original factorization machines, we also chose k from $\{10, 20, 30, 40, 50\}$. For Movielens 10M, we only chose β from 5 log-spaced values and we set $k = 20$ in order to reduce the search space. For SVR, we chose the regularization parameter C from 10 log-spaced values between 10^{-5} and 10^{5}. For convex factorization machines, we made use of warm-start when computing the regularization path in order to accelerate training. For practical reasons, we used early stopping in order to keep rank(Z) under 50.

Results. Test RMSE, training time (including hyper-parameter tuning using 3-fold CV) and the rank obtained (when applicable) are indicated in Table 2. Except on Movielens 100k, Convex FM (ignore diag) obtained lower RMSE, was faster to converge and obtained lower rank than Convex FM (use diag). This comes however at the cost of more complicated gradient and Hessian expressions (c.f., Section E in the supplementary material for details). Except on Movielens 10M, Convex FM (ignore diag) obtained lower RMSE than Original FM. Training time was also lower thanks to the reduced number of hyper-parameters to search. Ridge regression (RR) was a surprisingly strong baseline, SVR was worse than RR. This is due to the extreme sparsity of the design matrix when using the binary indicator representation (2). Since features co-occur exactly only once, SVR cannot exploit the feature interactions despite the use of polynomial kernel. In contrast, factorization machines are able to exploit feature interactions

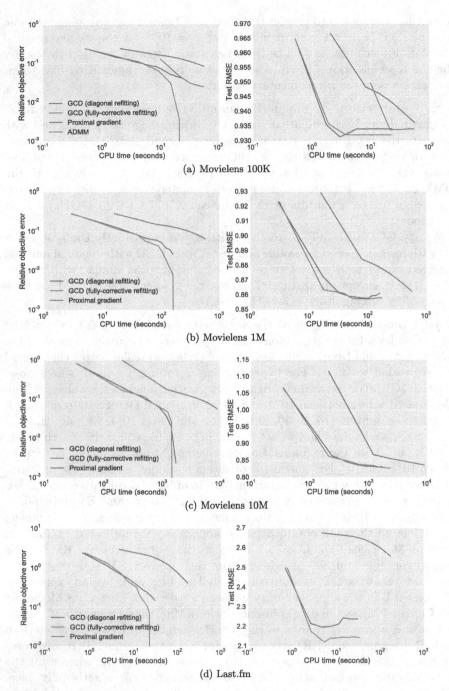

(a) Movielens 100K

(b) Movielens 1M

(c) Movielens 10M

(d) Last.fm

Fig. 1. Solver comparison when using $\alpha = 10^{-9}$ and $\beta = 10$. Left: relative objective error $|(f^t - f^*)/f^*|$, where f^t is the objective value measured on time t and f^* is the optimal objective value. Right: RMSE on test data.

Table 2. Test RMSE, training time (including hyper-parameter tuning using 3-fold cross-validation) and rank of different models on real data. Results are averaged over 3 runs using different train / test splits (rank uses the median).

Dataset		Convex FM (use diag)	Convex FM (ignore diag)	Original FM	Ridge	SVR (polynomial kernel)
Movielens 100k	RMSE	**0.93**	**0.93**	**0.93**	0.95	1.20
	Time	7.09 min	6.72 min	10.05 min	0.28 s	35.30 s
	Rank	23	20	20		
Movielens 1m	RMSE	0.87	**0.85**	0.86	0.91	1.24
	Time	1.07 h	38.74 min	3.93 h	3.14 s	3.68 min
	Rank	27	20	20		
Movielens 10m	RMSE	0.84	0.82	**0.81**	0.87	N/A
	Time	5.02 h	4.29 h	5.84 h	59.35 s	N/A
	Rank	34	17	20		
Last.fm	RMSE	2.21	**2.05**	2.13	2.60	3.24
	Time	7.77 min	6.91 min	14.17 min	0.63 s	36.70 s
	Rank	50	48	40		

despite high sparsity thanks to the parameter sharing induced by the factorization $Z = P \Lambda P^{\mathrm{T}}$.

6   Related Work

Recently, convex formulations for the estimation of a low-rank matrix have been extensively studied. The key idea [5] is to replace the rank of a matrix, which is non-convex, by the nuclear norm (a.k.a. trace norm), which is known to be the tightest convex lower bound on matrix rank [11]. Nuclear norm regularization has been applied to numerous applications, including multi-task learning and matrix completion [18]. The latter is typically formulated as the following optimization problem. Given a matrix $X \in \mathbb{R}^{|U| \times |I|}$ containing missing values, we solve

$$\min_{M \in \mathbb{R}^{|U| \times |I|}} \frac{1}{2} \| \mathcal{P}_\Omega(X) - \mathcal{P}_\Omega(M) \|_F^2 + \lambda \| M \|_*, \qquad (22)$$

where Ω is the set of observed values in X and $(\mathcal{P}_\Omega(M))_{i,j} = (M)_{i,j}$ if $(i,j) \in \Omega$, 0 otherwise. Extensions to tensor factorization have also been proposed for data with more than two modes (e.g., user, item and time) [19]. However, in (22) and tensor extensions, it is not trivial to incorporate auxiliary features such as user (age, gender, ...) and item (release date, director's name, ...) attributes. The most related work to convex factorization machines is [1], in which a collaborative filtering method which can incorporate additional attributes is proposed. However, their method can only handle two modes (e.g., user and item) and no scalable learning algorithm is proposed. The advantage of convex factorization machines is that it is very easy to engineer features, even for more than two modes (e.g., user, item and context).

7   Conclusion

Factorization machines are a powerful framework that can exploit feature interactions even when features co-occur very rarely. In this paper, we proposed a convex formulation of factorization machines. Our formulation imposes fewer restrictions on the feature interaction weight matrix and is thus more general than the original one. For solving the corresponding optimization problem, we presented an efficient globally-convergent two-block coordinate descent algorithm. Our formulation achieves comparable or lower predictive error on several synthetic and real-world benchmarks. It can also overall be faster to train since it has one less hyper-parameter than the original formulation. As a side contribution, we also clarified the convexity properties (or lack thereof) of the original factorization machine's objective function. Future work includes trying (convex) factorization machines on more data (e.g., genomic data, where feature interactions should be useful) and developing algorithms for out-of-core learning.

References

1. Abernethy, J., Bach, F., Evgeniou, T., Vert, J.P.: A new approach to collaborative filtering: Operator estimation with spectral regularization. J. Mach. Learn. Res. **10**, 803–826 (2009)
2. Bertsekas, D.P.: Nonlinear programming. Athena scientific Belmont (1999)
3. Buitinck, L., Louppe, G., Blondel, M., Pedregosa, F., Mueller, A., Grisel, O., Niculae, V., Prettenhofer, P., Gramfort, A., Grobler, J., Layton, R., VanderPlas, J., Joly, A., Holt, B., Varoquaux, G.: API design for machine learning software: experiences from the scikit-learn project. In: ECML PKDD Workshop: Languages for Data Mining and Machine Learning, pp. 108–122 (2013)
4. Dudik, M., Harchaoui, Z., Malick, J.: Lifted coordinate descent for learning with trace-norm regularization. In: AISTATS, vol. 22, pp. 327–336 (2012)
5. Fazel, M., Hindi, H., Boyd, S.P.: A rank minimization heuristic with application to minimum order system approximation. American Control Conference **6**, 4734–4739 (2001)
6. Grippo, L., Sciandrone, M.: On the convergence of the block nonlinear gauss-seidel method under convex constraints. Operations Research Letters **26**(3), 127–136 (2000)
7. Hsieh, C.J., Olsen, P.: Nuclear norm minimization via active subspace selection. In: ICML, pp. 575–583 (2014)
8. Koren, Y.: Factorization meets the neighborhood: a multifaceted collaborative filtering model. In: KDD, pp. 426–434 (2008)
9. Koren, Y.: Collaborative filtering with temporal dynamics. Communications of the ACM **53**(4), 89–97 (2010)
10. Loni, B., Shi, Y., Larson, M., Hanjalic, A.: Cross-domain collaborative filtering with factorization machines. In: de Rijke, M., Kenter, T., de Vries, A.P., Zhai, C.X., de Jong, F., Radinsky, K., Hofmann, K. (eds.) ECIR 2014. LNCS, vol. 8416, pp. 656–661. Springer, Heidelberg (2014)
11. Recht, B., Fazel, M., Parrilo, P.A.: Guaranteed minimum-rank solutions of linear matrix equations via nuclear norm minimization. SIAM Review **52**(3), 471–501 (2010)

12. Rendle, S.: Factorization machines. In: ICDM, pp. 995–1000. IEEE (2010)
13. Rendle, S.: Factorization machines with libfm. ACM Transactions on Intelligent Systems and Technology (TIST) **3**(3), 57–78 (2012)
14. Rendle, S.: Scaling factorization machines to relational data. In: VLDB, vol. 6, pp. 337–348 (2013)
15. Rendle, S., Gantner, Z., Freudenthaler, C., Schmidt-Thieme, L.: Fast context-aware recommendations with factorization machines. In: SIGIR, pp. 635–644 (2011)
16. Rendle, S., Schmidt-Thieme, L.: Pairwise interaction tensor factorization for personalized tag recommendation. In: WSDM, pp. 81–90. ACM (2010)
17. Shalev-Shwartz, S., Gonen, A., Shamir, O.: Large-scale convex minimization with a low-rank constraint. In: ICML, pp. 329–336 (2011)
18. Srebro, N., Rennie, J., Jaakkola, T.S.: Maximum-margin matrix factorization. In: Advances in Neural Information Processing Systems, pp. 1329–1336 (2004)
19. Tomioka, R., Hayashi, K., Kashima, H.: Estimation of low-rank tensors via convex optimization. arXiv preprint arXiv:1010.0789 (2010)

Generalized Matrix Factorizations as a Unifying Framework for Pattern Set Mining: Complexity Beyond Blocks

Pauli Miettinen[✉]

Max-Planck-Institut Für Informatik, Saarbrücken, Germany
pauli.miettinen@mpi-inf.mpg.de

Abstract. Matrix factorizations are a popular tool to mine regularities from data. There are many ways to interpret the factorizations, but one particularly suited for data mining utilizes the fact that a matrix product can be interpreted as a sum of rank-1 matrices. Then the factorization of a matrix becomes the task of finding a small number of rank-1 matrices, sum of which is a good representation of the original matrix. Seen this way, it becomes obvious that many problems in data mining can be expressed as matrix factorizations with correct definitions of what a rank-1 matrix and a sum of rank-1 matrices mean. This paper develops a unified theory, based on generalized outer product operators, that encompasses many pattern set mining tasks. The focus is on the computational aspects of the theory and studying the computational complexity and approximability of many problems related to generalized matrix factorizations. The results immediately apply to a large number of data mining problems, and hopefully allow generalizing future results and algorithms, as well.

1 Introduction

One of the most fundamental tasks in data mining is to explain (or summarize) a data set using a collection of simple and easy-to-understand structures (commonly referred to as *regularities* or *patterns*). A *block* of some kind has been the predominant type of patterns sought by many data mining algorithms; a clique (or quasi-clique) in a network or a frequent itemset or a tile in transaction data are all blocks, or more precisely, (binary) rank-1 (sub-)matrices. Consequently, a collection of these blocks summarizing the data can be seen as a matrix factorization of the data matrix [27].

But in recent years, increased emphasis has been put to patterns that go 'beyond blocks', for example, nested submatrices [17], taxonomies [18], stars, biclique cores, or chains [19], or hyperbolic subgraphs [3] (see Figure 1 for examples of some of these patterns). At first, it might seem like these patterns share very little, if anything, with simple blocks; they are, for example, not rank-1. It seems, then, that we have to re-do much of the work we have already done in analysing the computational aspects of mining collections of blocks.

© Springer International Publishing Switzerland 2015
A. Appice et al. (Eds.): ECML PKDD 2015, Part II, LNAI 9285, pp. 36–52, 2015.
DOI: 10.1007/978-3-319-23525-7_3

More in-depth study, however, starts revealing various similarities between these patterns that are 'beyond blocks' and the simple blocks. The intuition of expressing a data sets using a collection of simple patterns, for example, is the same. Consequently, we can *still* consider the patterns as 'rank-1' sub-matrices, and the whole summarization as a form of matrix factorization – we only need new definitions of ranks and matrix factorizations.

In this paper I propose *generalized outer products* as a unifying framework to express different kinds of patterns. As the name implies, it generalizes the outer product of two vectors (i.e. the operator used to express cliques and other block patterns). The new generalized definitions of matrix rank and factorization follow from the outer product, as we will see in Section 3.

The framework works over any semi-ring, but for the sake of concreteness, the examples are presented using a fixed set of binary patters, introduced in Example 1 and Figure 1. After we have seen the definition of the framework (Section 3), we study how some common concepts, such as the matrix rank, behave under it (Section 3.2).

The framework alone is not useful for data mining researchers, though. To that end, it must help the researchers to obtain interesting results. To demonstrate the proposed framework's capability to do that, we see a series of general results regarding the computational complexity (Section 4) and approximability (Section 5) of some fundamental problems related to the generalized decompositions in binary matrices. These results will immediately yield corresponding results for any pattern fitting to the framework.

The purpose of this paper is to develop the framework and to demonstrate its usefulness via number of general results. As a consequence, some of the results we will see are already known in the literature in some specific cases; indeed, some of the results have been presented for multiple special cases – an unnecessary repetition that can be avoided with the proposed framework. The goal of this paper is *not* to present novel algorithms for mining the patterns. While it is expected that the framework facilitates developing of general algorithms, more in-depth studies to that end are left for future work. That said, we do see number of existing algorithms (mostly in Section 5) that can be used to solve certain general problems in the proposed framework with provable performance guarantees.

Before moving on, let us briefly discuss some related work.

2   Related Work

Finding patterns from data is in the core of data mining, with frequent itemset mining being an early and prominent example. Much of the research focus has nowadays sifted from finding all the patterns to finding an interesting subset of them, with the interestingness being defined either in combinatorial [15,34], information-theoretic [10,33], or other means. Similar problems were also studied, for example, in formal concept analysis [6] and role mining [12].

What all of the aforementioned work shares is that they aim at describing a binary matrix using a set of rank-1 binary sub-matrices. The requirement for sub-matrices is strict: all these methods are restricted to rank-1 matrices that appear in the input data as such (known as from-below or dominated decompositions). When this requirement is lifted, the problem is usually referred to as Boolean matrix factorization [22,25], although especially earlier other names were also used [28].

A seemingly unrelated line of research grew from the problem of finding communities from graphs. Traditionally, communities were considered exclusively as cliques (or bicliques, in case of bipartite graphs), corresponding again to rank-1 sub-matrices of the adjacency matrix. While this connection is often not made explicitly, it can be – and has been [35] – used to design community-detection algorithms, especially for the overlapping case.

Mere cliques (or bicliques), however, might not be enough to properly explain the interesting communities in the graphs [19], and recently many real-world communities were found out to be more hyperbolic than clique-like [3]. The requirement for communities more complex than simply cliques has been encapsulated to the "beyond blocks" slogan.

Communities or itemsets or rank-1 matrices are not the only kind of patterns data miners are interested about, of course. Patterns such as nested or banded submatrices [17] or taxonomies [18], among many others, are equally well expressed in the generalized framework of this paper.

3   Definitions

Throughout this paper, upper-case bold symbols (A) will be used to denote matrices, lower-case bold symbols (a) denote vectors, and lower-case normal symbols (a) denote scalars. If n is an integer, the shorthand notation $[n]$ is used for set $\{1, 2, \ldots, n\}$.

For any matrix A (binary or not), $|A|$ denotes the number of non-zero elements in it.

We work over algebraic structure $\mathbb{T} = (T, \boxplus, \boxtimes, 0, 1)$. The binary operator $\boxplus$ is called addition and the binary operator $\boxtimes$ is called multiplication, with $0 \in T$ and $1 \in T$ being their respective identity elements. $\mathbb{T}$ is required to be at least a semiring, that is, $\boxplus$ is commutative and $\boxtimes$ distributes over $\boxplus$.

Before going forward to the definitions, let us see different types of patterns that will be used in examples throughout the paper.

Example 1. A *biclique*, a *binary rank-1 matrix*, and a *(combinatorial) tile* all refer to the same kind of pattern: a submatrix full of 1s, that is, a block. A *star* (Figure 1, left) and a *biclique core* (Figure 1, middle-left) are forms of patterns in (undirected) graphs: star represents a collection of vertices that are connected to each other only via a single hub vertex, while a biclique core represents a set of vertices that form a complete bipartite graph. A *chain* (Figure 1, middle-right) is a set of vertices where each vertex is connected only to the next one, while *nested matrix* (Figure 1, right) is a bipartite graph where each subsequent vertice's neighbors are a subset of the previous one's neighbors.

Note that in Figure 1, all matrices are permuted for maximum readability; in general, no particular ordering of the rows or columns is required.   ◇

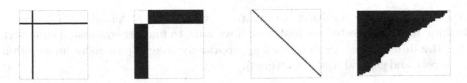

Fig. 1. Different types of patters, from left to right: star, biclique core, chain, and nested. The first three are symmetric square matrices; nested is asymmetric general matrix.

3.1 Generalized Outer Product

The definition of the generalized outer product is the core of the proposed framework. Similar definitions have appeared earlier, but to the best of the author's knowledge, this exact definition of generalized outer product has not been proposed earlier.

Definition 1. *The* generalized outer product operator *of two vectors* $x \in \mathbb{T}^n$ *and* $y \in \mathbb{T}^m$ *with parameters* $\theta \in \Theta$ *is a function* $o : \mathbb{T}^n \times \mathbb{T}^m \times \Theta \to \mathbb{T}^{n \times m}$ *such that for all* $(i, j) \in [n] \times [m]$, *if* $x_i = 0$ *or* $y_j = 0$, *then* $o(x, y, \theta)_{ij} = 0$.

It is helpful to consider the outer products of the four patterns in Figure 1:

Example 2. The star pattern can be generated using generalized outer product $o_s(x, x, k) = (a_{ij})$, where

$$a_{ij} = \begin{cases} 1 & \text{if } x_i = 1 \text{ and } x_j = 1 \text{ and either } i = k \text{ or } j = k \\ 0 & \text{otherwise} . \end{cases}$$

This can be interpreted as follows: binary vector x selects the rows and columns (e.g. vertices) that participate in the pattern, while parameter k chooses the vertex that is the centre of the star. Naturally, if $x_k = 0$ this yields empty pattern. We could require that $x_k = 1$ without significant changes to anything that follows. Similar requirements could be applied to many of the following patterns, as well, but they are not stated for the sake of brevity.

The biclique core pattern can be generated using generalized outer product $o_{bc}(x, x, I \subset [n]) = (a_{ij})$, where

$$a_{ij} = \begin{cases} 1 & \text{if } x_i = 1 \text{ and } x_j = 1 \text{ and exactly one of } i \in I \text{ or } j \in I \text{ holds} \\ 0 & \text{otherwise} . \end{cases}$$

The outer product generating the chain pattern is $o_c(x, y) = (a_{ij})$, where $a_{ij} = 1$ if $x_i = 1$, $y_j = 1$, and $j = i + 1$, and $a_{ij} = 0$ otherwise.

The outer product for nested pattern is $o_n(x, y, s) = (a_{ij})$, where $s \in [m]^n$ defines a *step function* and

$$a_{ij} = \begin{cases} 1 & \text{if } x_i = 1 \text{ and } y_j = 1 \text{ and } j \leq s_i \\ 0 & \text{otherwise} . \end{cases}$$

Notice that for the resulting pattern to be valid, $s_i \leq s_j$ for all $i > j$ if we are looking for *direct* nested submatrices; if we want to find general nested submatrices, the indices must be permuted appropriately (see [17] for more information on direct and general nested patterns).   ◇

In the above examples, all outer products were defined element-wise. The generalized outer products that can be defined element-wise form an important sub-class of generalized outer products, called *decomposable* outer products:

Definition 2. *A generalized outer product operator o is* decomposable *if we have that $o(\boldsymbol{x}, \boldsymbol{y}, \theta)_{ij} = f(x_i, y_j, i, j, \theta)$ for all i and j. We say that o is decomposable to f.*

The indices i and j can be considered as two additional parameters in θ and will often be omitted for brevity's sake.

Another common feature shared with most of the above examples is that the outer product is with vector $\boldsymbol{x}$ itself. This ensures that the product is *symmetric*.

Definition 3. *An outer product $o(\boldsymbol{x}, \boldsymbol{y}, \theta)$ is* symmetric *if $\boldsymbol{x} = \boldsymbol{y}$ and $o(\boldsymbol{x}, \boldsymbol{y}, \theta) = o(\boldsymbol{x}, \boldsymbol{y}, \theta)^T$.*

For decomposable outer products, it is enough to require that $\boldsymbol{x} = \boldsymbol{y}$ as the other constraint follows automatically.

3.2   Generalized Rank

A common way to measure the complexity of the structure of a matrix is its *rank*. Multiple equivalent definitions of a matrix rank exist under normal linear algebra, but most of them do not generalize well to our case. We shall now generalize the so-called *Schein rank* and study some properties of the resulting general rank.

Definition 4. *Matrix $\boldsymbol{M} \in \mathbb{T}^{n \times m}$ has outer product operator o induced rank, $\mathrm{rank}_o(\boldsymbol{M}) = 1$ if there exists vectors $\boldsymbol{x} \in \mathbb{T}^n$ and $\boldsymbol{y} \in \mathbb{T}^m$ and parameters $\theta \in \Theta$ such that $\boldsymbol{M} = o(\boldsymbol{x}, \boldsymbol{y}, \theta)$.*

For brevity, the outer product will be omitted from the rank when it is clear from the context. Hence, if matrix $\boldsymbol{M}$ is rank-1, it implicitly means that its o-induced rank is 1.

Definition 5. *The (generalized)* sum *of two matrices $\boldsymbol{A} \in \mathbb{T}^{n \times m}$ and $\boldsymbol{B} \in \mathbb{T}^{n \times m}$ is the element-wise sum $\boldsymbol{A} \boxplus \boldsymbol{B}$ using the summation $\boxplus$ of $\mathbb{T}$. That is, $(\boldsymbol{A} \boxplus \boldsymbol{B})_{ij} = a_{ij} \boxplus b_{ij}$.*

Definition 6. *A set* $D = \{\boldsymbol{F}_i \in \mathbb{T}^{n \times m} : \mathrm{rank}_o(\boldsymbol{F}_i) = 1\}$ *is the* decomposition *(under o and* $\boxplus$*) of a matrix* $\boldsymbol{M} \in \mathbb{T}^{n \times m}$ *if*

$$\boldsymbol{M} = \boldsymbol{F}_1 \boxplus \boldsymbol{F}_2 \boxplus \cdots \boxplus \boldsymbol{F}_{|D|} . \tag{1}$$

The size *of the decomposition* D *is* $|D|$*.*

If D is a decomposition of $\boldsymbol{A}$ and o is decomposable to f, we get the familiar element-wise matrix product form

$$a_{ij} = \boxplus_{k=1}^{|D|} f(x_{ik}, y_{jk}, \theta) , \tag{2}$$

where $o(\boldsymbol{x}_k, \boldsymbol{y}_k, \theta) = \boldsymbol{F}_k \in D$ for all $i \in [|D|]$.

Hence, we can define the matrix product for decomposable outer products.

Definition 7. *If o is decomposable to* f*, the* matrix product *of* $\boldsymbol{A} \in \mathbb{T}^{n \times k}$ *and* $\boldsymbol{B} \in \mathbb{T}^{k \times m}$ *(under o,* θ*, and* $\boxplus$*) is defined element-wise as*

$$(\boldsymbol{A} \boxtimes_\theta \boldsymbol{B})_{ij} = \boxplus_{l=1}^{k} f(a_{il}, b_{lj}, \theta) . \tag{3}$$

Definition 8. *Let o be decomposable to* f*. The operator* $\langle \cdot, \cdot \rangle : \mathbb{T}^n \times \mathbb{T}^n \to \mathbb{T}$ *is* inner product *if it satisfies the following rules for all* $\boldsymbol{x}, \boldsymbol{y}, \boldsymbol{z} \in \mathbb{T}^n$ *and* $\alpha \in \mathbb{T}$*:*

$$\langle \boldsymbol{x}, \boldsymbol{y} \rangle = \langle \boldsymbol{y}, \boldsymbol{x} \rangle \tag{4a}$$
$$\langle \alpha \boxtimes \boldsymbol{x}, \boldsymbol{y} \rangle = \alpha \boxtimes \langle \boldsymbol{x}, \boldsymbol{y} \rangle \tag{4b}$$
$$\langle \boldsymbol{x} \boxplus \boldsymbol{y}, \boldsymbol{z} \rangle = \langle \boldsymbol{x}, \boldsymbol{z} \rangle \boxplus \langle \boldsymbol{y}, \boldsymbol{z} \rangle \tag{4c}$$
$$\langle \boldsymbol{x}, \boldsymbol{x} \rangle \geq 0 \tag{4d}$$
$$\langle \boldsymbol{x}, \boldsymbol{x} \rangle = 0 \Rightarrow \boldsymbol{x} = 0 . \tag{4e}$$

If decomposable o induces an inner product, we say o is inner-product compatible*.*

Notice that the decomposability is the crucial element here.

Definition 9. *The* rank *(over o) of* $\boldsymbol{A} \in \mathbb{T}^{n \times m}$*,* $\mathrm{rank}_o(\mathbb{T})$*, is the smallest integer* k *such that there exists a decomposition* D *of* $\boldsymbol{A}$ *with size* k*. If no decomposition exist,* $\mathrm{rank}_o(\mathbb{T}) = \infty$*.*

A *singleton* matrix has exactly one non-zero value, and is characterized by triple (i, j, α), corresponding to a matrix $\boldsymbol{A}$ for which $a_{pq} = \alpha$ if $i = p$ and $j = q$ and $a_{pq} = 0$ otherwise. If an outer product operator o can generate all n-by-m singleton matrices $(i, j, \alpha) \in [n] \times [m] \times \mathbb{T}$ (it is *singleton-generating*), the rank of any $\boldsymbol{A} \in \mathbb{T}^{n \times m}$ induced by o is bounded by $\mathrm{rank}_o(\boldsymbol{A}) \leq |\boldsymbol{A}| \leq nm$.

Example 3. The star outer product o_s is singleton-generating: we can set x to have exactly one non-zero and set k to the index of that non-zero. These singleton-stars can be used to represent any symmetric binary matrix A. The biclique core outer product o_{bc} is not singleton-generating, however, and it can induce infinite ranks. Consider, for example, the 2-by-2 identity matrix I_2:

$$I_2 = \begin{pmatrix} 1 & 0 \\ 0 & 1 \end{pmatrix} .$$

As every vertex is connected to itself, and only to itself, there are no bipartite graphs and I_2 cannot be expressed with a set of biclique cores, that is, $\text{rank}_{o_{bc}}(I_2) = \infty$.   ◊

4   Computational Complexity

We will now move to the applications of the proposed framework. Specifically, we will consider some results regarding the computational complexity and approximability of problems related to generalized decompositions in binary matrices.

4.1   Rank-1 Submatrices

Let us start by studying problems regarding finding the largest rank-1 matrix. It is a very common sub-problem in almost all algorithms that need to find a set of patterns. For the results, we need the definition of *hereditary* outer product.

Definition 10. *Binary outer product operator o is* hereditary *if the class of rank-1 matrices induced by it is closed under permutation and deletion of rows and columns.*

Example 4. Stars, biclique cores, and nested matrices are closed under permutations and deletions of rows and columns, and consequently their outer products are hereditary. Chains are not closed under deletion of rows and columns, and the outer product is not hereditary.

The exact definition of the problems depends on how we define "large". We start by studying the case where large pattern is one with large circumference.

Problem 1. In the *binary maximum-circumference o-induced rank-1 submatrix problem* we are given a binary matrix $A \in \{0,1\}^{n \times m}$ and our task is to find vectors $x \in \{0,1\}^n$ and $y \in \{0,1\}^m$ and parameters θ such that $o(x, y, \theta)$ is dominated by A and we maximize the (half-) *circumference* $|x| + |y|$.

Proposition 1. *Let $\mathcal{M}_o$ be the family of all o-induced rank-1 matrices (of any size), where o is hereditary. If the number of distinct rows (or columns) in matrices of $\mathcal{M}_o$ is unbounded, the binary maximum-circumference o-induced rank-1 submatrix problem is NP-hard; if the number of distinct rows (or columns) is bounded, the problem can be solved in polynomial time.*

The result follows straight forwardly from Corollary 4 of Yannakakis [36]. Many interesting patterns have bounded number of distinct rows, for example, bicliques have exactly two kinds of rows: the rows corresponding to the nodes that are in the biclique, and the rows corresponding to the nodes that are not in it. On the other hand, for example nested matrices have unbounded number of distinct rows. Notice also that one cannot have bounded number of distinct rows but unbounded number of distinct columns (or vice versa) in a binary matrix: for k distinct rows one cannot have more than 2^k distinct columns.

The *symmetric* maximum-circumference rank-1 submatrix problem is like its asymmetric case, but we require the input matrix $A \in \{0,1\}^{n \times n}$ and the outer product o be symmetric (and hence the outer product is of type $o(x, x, \theta)$). Adding the symmetry requirement makes the problem harder, as now it is enough that there are infinitely-many rank-1 submatrices and infinitely many matrices with higher ranks.

Proposition 2. *Let S_o be the family of all binary matrices generated by hereditary binary symmetric outer product $o(x, x, \theta)$ and let S_o^c be the family of all symmetric binary matrices not in S_o. If $|S_o| = |S_o^c| = \infty$, then finding the maximum-circumference o-induced symmetric rank-1 submatrix is NP-hard.*

The result follows from Lewis and Yannakakis [21].

Another possible definition for "large" is to study the area: the maximum-area rank-1 problems ask to maximize the area instead of the circumference.

Problem 2. In the *binary maximum-area o-induced rank-1 submatrix problem* we are given a binary matrix $A \in \{0,1\}^{n \times m}$ and our task is to find vectors $x \in \{0,1\}^n$ and $y \in \{0,1\}^m$ and parameters θ such that $o(x, y, \theta)$ is dominated by A and we maximize the area $|x| \, |y|$.

The symmetric binary maximum-area rank-1 submatrix problem is defined analogously. As the area of a symmetric rank-1 matrix is simply $|x|^2$, the hardness of the symmetric problems is a straight forward corollary of Proposition 2.

Proposition 3. *The binary symmetric maximum-area rank-1 submatrix problem is NP-hard exactly when the binary symmetric maximum-circumference rank-1 problem is.*

Proof. The maximum-circumference problem asks us to decide if the given matrix has a symmetric rank-1 submatrix such that $2|x| \geq t$, while in the maximum-area problem the condition is that $|x|^2 \geq t'$. Thus, we can solve the maximum-circumference problem by solving the maximum-area problem with $t' = t^2/4$. $\square$

For the asymmetric case this trivial correspondence does not necessarily hold, as can be readily witnessed by noticing that the binary maximum-area rank-1 submatrix problem under standard algebra corresponds to the maximum-edge biclique, and hence is NP-hard [29].

The third variant of binary maximum rank-1 submatrices are the *maximum-content* submatrices, that is, the submatrices with maximum number of non-zeros (in symmetric matrices, this corresponds to maximum-edge subgraphs). The complexity of these problems can often be reduced to the complexity of the maximum-area or maximum-circumference problems.

Proposition 4. *If there exists a set of parameters θ such that the number of non-zeros in $o(\boldsymbol{x}, \boldsymbol{y}, \theta)$ depends only on circumference $|\boldsymbol{x}| + |\boldsymbol{y}|$ (resp. area $|\boldsymbol{x}| \, |\boldsymbol{y}|$), then finding the maximum-content rank-1 submatrix is NP-hard if finding the maximum-circumference (resp. maximum-area) rank-1 submatrix is NP-hard.*

4.2   Selecting Some Rank-1 Submatrices

We now turn to problems where we are given a set of rank-1 matrices, and our task is to select some of them (e.g. to be presented to the user). This is also a common subproblem in many pattern set mining algorithms, where first a set of candidate solutions is generated, and then a final selection is performed from that set.

Problem 3. In the *smallest binary sub-decomposition problem* we are given a matrix $\boldsymbol{A} \in \{0,1\}^{n \times m}$ and its decomposition $D = \{\boldsymbol{F}_i \in \{0,1\}^{n \times m} : \mathrm{rank}_o(\boldsymbol{F}_i) = 1\}$ and our task is to find the smallest subset $C \subseteq D$ that is still a valid decomposition, i.e., $\boxplus_{F \in C} \boldsymbol{F} = \boldsymbol{A}$.

Proposition 5. *If $\boxplus$ is either logical OR, logical AND, or logical XOR, the smallest binary sub-decomposition problem is NP-hard.*

Proof. We study the cases separately.

OR: This case is very similar to the Tiling databases [15], and we present the reduction only for the sake of completeness. The reduction is from the *minimum set cover* problem [14]. Let $(U, \mathcal{S} \subset 2^U)$ be a set system. Let $\boldsymbol{a} \in \{0,1\}^n$ be an all-1s vector where $|U| = n$, and for each $S \in \mathcal{S}$, define vector $\boldsymbol{f}^{(S)} \in D$ to be the *characteristic vector* of S, that is $f_i^{(S)} = 1$ if $u_i \in S$ and $f_i^{(S)} = 0$ otherwise. It is trivial to see that the smallest sub-decomposition $C \subseteq D$ for which $\bigvee_{\boldsymbol{f}^{(S)} \in C} \boldsymbol{f}^{(S)}$ is equivalent to the minimum set cover.

AND: This is similar to above, except that the reduction constructs the complements of $\boldsymbol{a}$ (which is all-0s) and $\boldsymbol{f}^{(S)}$s and the proof follows from De Morgan's laws.

XOR: In the *decoding of linear codes* problem [14], we are given a binary matrix $\boldsymbol{B} \in \{0,1\}^{n \times k}$ and a binary vector $\boldsymbol{a} \in \{0,1\}^k$ and we need to find a binary vector $\boldsymbol{x} \in \{0,1\}^k$ minimizing $|\boldsymbol{x}|$ such that $\bigoplus_{j=1}^{k} x_j b_{ij} = a_i$ for all $i \in [n]$, where $\oplus$ is the logical XOR operatior. To reduce this to the sub-decomposition problem, it is enough to notice that if we take the column vectors $\boldsymbol{b}_j$ of matrix $\boldsymbol{B}$ as the factors in D we have the smallest binary sub-decomposition problem.[1]   □

---

[1]   A minor technicality is that $\boldsymbol{B}$ for which $\bigoplus_j \boldsymbol{b}_j \neq \boldsymbol{a}$ does not yield to a valid input to the smallest sub-decomposition problem. This can be solved by adding a new column $\boldsymbol{c} = \bigoplus_j \boldsymbol{b}_j$ to $\boldsymbol{B}$, and by adding one row to $\boldsymbol{B}$ and $\boldsymbol{a}$; this row has value 1 in $\boldsymbol{a}$ and $\boldsymbol{c}$, and is 0 in other columns of $\boldsymbol{B}$.

These problems can also be seen as generalizations of a test for linear independency: in some sense, what we need to remove are the factors that are not independent from the others.

4.3   Minimum-Error Sub-Decompositions

In many applications we can assume the input data contains noise, and has high (or even infinite) o-induced rank. In these situations, we might be more interested on approximate decompositions, and instead of finding the smallest exact sub-decomposition, we want to find a sub-decomposition that induces the minimum error.

Problem 4. In the *minimum-error binary sub-decomposition problem* we are given a matrix $A \in \{0,1\}^{n \times m}$ and a set $D = \{F_i \in \{0,1\}^{n \times m} : \mathrm{rank}_o(F_i) = 1\}$ and our task is to find subset $C \subseteq D$ of size k that minimizes

$$\left| \boxplus_{F \in C} F - A \right| . \tag{5}$$

Using the Hamming distance, as in (5), is natural in case of binary decompositions; other error measures are of course possible, especially for matrices taking non-binary values. These problems are no easier than the smallest sub-decomposition problems.

Proposition 6. *If $\boxplus$ is either logical OR, logical AND, or logical XOR, the minimum-error binary decomposition problem is NP-hard.*

Proof. We again work case-by-case:

OR: In the *basis usage* problem we are given a binary vector $a \in \{0,1\}^n$ and a binary matrix $B \in \{0,1\}^{n \times k}$, and our task is to find a binary vector $x \in \{0,1\}^k$ such that $\left| a - \bigvee_{i=1}^{k} x_i b_i \right|$ is minimized, where $x_i b_i$ yields all-0s vector if $x_i = 0$, and b if $x_i = 1$. This clearly a special case of minimum-error binary sub-decomposition, and the claim follows as basis usage problem is NP-hard [25].

AND: This case again follows from De Morgan's laws by taking complements and using the above reduction.

XOR: In the *nearest codeword* problem we are given a binary vector $a \in \{0,1\}^n$ and a binary matrix $B \in \{0,1\}^{n \times k}$, and our task is to find a binary vector $x \in \{0,1\}^k$ such that $\left| a - \oplus_{i=1}^{k} x_i b_i \right|$ is minimized. This, again, is a special case of minimum-error binary sub-decomposition, and the claim follows as the nearest codeword problem is NP-hard [4].   □

4.4   Deciding the Rank

Deciding the (generalized) rank of a matrix is a fundamental question. Unfortunately, the complexity of deciding the rank depends on the interplay between the underlying algebraic structure $\mathbb{T}$ and the generalized outer product o, as the following examples illustrate.

Example 5. Consider binary n-by-m matrices and the normal outer product $o(\boldsymbol{x}, \boldsymbol{y}, \theta) = \boldsymbol{x}\boldsymbol{y}^T$. If the summation $\boxplus$ is the logical OR operator, the rank is the *Boolean rank* of the matrix, and consequently NP-hard (see, e.g. [28]). If, however, the summation $\boxplus$ is the logical XOR operator, finding the rank can be done in polynomial time [31]. $\Diamond$

We say binary outer product $o(\boldsymbol{x}, \boldsymbol{y}, \theta)$ *subsumes* bicliques if there exists parameters θ such that $o(\boldsymbol{x}, \boldsymbol{y}, \theta) = \boldsymbol{x}\boldsymbol{y}^T$ for all $\boldsymbol{x}$ and $\boldsymbol{y}$.

Corollary 1. *Let* $\mathbb{T} = (\{0, 1\}, \vee, \boxtimes, 0, 1)$. *Finding the o-induced rank of* $\boldsymbol{A} \in \mathbb{T}^{n \times m}$ *is* NP-*hard if* o *subsumes the bicliques.*

4.5   Minimum-Error Approximate Decompositions

The *minimum-error fixed-rank decompositions* are to the rank-decision problems what the minimum-error fixed-size sub-decompositions are to the smallest sub-decompositions.

Problem 5. In the *minimum-error fixed-rank binary decomposition* we are given a matrix $\boldsymbol{A} \in \{0, 1\}^{n \times m}$ and an integer k, and our task is to find set $D = \{\boldsymbol{F}_i \in \{0, 1\}^{n \times m} : \mathrm{rank}_o(\boldsymbol{F}_i) = 1, i \in [k]\}$ that minimizes

$$\left| \boxplus_{\boldsymbol{F} \in D} \boldsymbol{F} - \boldsymbol{A} \right| . \tag{6}$$

In the decision version of Problem 5, the input is prepended with parameter $t \in \mathbb{R}$ and instead of minimizing (6), the task is to decide if there exists D such that

$$\left| \boxplus_{\boldsymbol{F} \in D} \boldsymbol{F} - \boldsymbol{A} \right| \leq t . \tag{7}$$

It is easy to see that the complexity of these problems is no easier than that of the related rank-decision problems:

Proposition 7. *If computing the o-rank is* NP-*hard, so is computing the minimum-error approximate decomposition.*

Proof. If we set $t = 0$ in (7), the o-rank of $\boldsymbol{A}$ is the least k for which we have positive answer. $\qquad\square$

5   Approximability

As we saw, most problems related to binary generalized outer products are NP-hard. Approximation algorithms are a common recourse to the NP-completeness, and in this section we will study the approximability of some of the problems studied above. As we shall see, many – but not all – problems are hard even to approximate, giving a post hoc justification to the various heuristics employed in the prior work.

5.1   Approximating Smallest Sub-Decompositions

We start with the problem that is easiest to approximate:

Proposition 8. *If $\boxplus = \vee$, the smallest binary sub-decomposition can be approximated to within $\ln n$ (and no better) in polynomial time.*

Proof. We reduce the problem to the minimum set cover in an approximation-preserving way. Let $U = \{(i,j) : a_{ij} = 1\}$ be the set of all locations of A that are 1. For every matrix $F_k \in D$, define set $F_k = \{(i,j) : (F_k)_{ij} = 1\}$ as the locations of ones in F_k and let the collection $\mathcal{D} = \{F_k : k = 1, \ldots, |D|\}$. As D is a decomposition of A, it is guaranteed that $\bigcup_{k=1}^{|D|} F_k = U$. The task of finding the smallest sub-decomposition of D is equivalent to finding the smallest sub-collection $\mathcal{C} \subset \mathcal{D}$ such that $\bigcup_{F_k \in \mathcal{C}} F_k = U$, that is, the minimum set cover problem. As the reduction preserves the value of the optimization target, it preserves the approximability and hence we can use, for example, the famous greedy $\Theta(\ln n)$-approximation algorithm [16]. On the other hand, the reduction in the proof of Proposition 5 is also approximation-preserving, and consequently, the $\ln n$ bound is tight unless P = NP [13]. $\square$

The approximation-preserving reduction to and from set cover, together with the result of Simon [32], gives us the following interesting corollary:

Corollary 2. *Given an algorithm that, in every iteration of the greedy algorithm, selects the factor $F_i \in D$ that approximates the best choice by a factor of $O(h(n))$, we can approximate the smallest binary sub-decomposition by a factor of $O(h(n) \ln n)$.*

This corollary can be very useful when the decomposition D is given only implicitly, and cannot be exhaustively searched for the best solution in every iteration. For example, the tiling algorithm of [15] needs – in principle – to search every closed itemset of the input data in every iteration. To avoid that, the authors resolve to clever heuristics, but with the cost of approximation guarantees. If the heuristic algorithm could be replaced with one with provable approximation guarantees, Corollary 2 would give us an overall approximation guarantee.

When $\boxplus = \oplus$, the problem becomes harder to approximate. The *minimum weight codeword* problem is similar to the aforementioned decoding of linear codes problem. In the former, we are given a matrix $B \in \{0,1\}^{n \times k}$ and an all-zeros vector $a \in \{0,1\}^n$, and our task is to find a *non-empty* vector $x \in \{0,1\}^n$ such that $Bx = a$ and x has as few 1s as possible. This problem is as hard to approximate as the smallest binary sub-decomposition when $\boxplus = \oplus$.

Proposition 9. *If $\boxplus = \oplus$, the smallest binary sub-decomposition problem is as hard to approximate as the minimum weight codeword problem.*

Proof. Notice first that we can replace the all-zeros vector a in the minimum weight codeword problem by an arbitrary binary vector c by adding that vector also as a column to B and by adding a new row to B and c to enforce that c must be part of the solution, similarly to the proof of Proposition 5.

To see that the smallest binary sub-decomposition is at least as hard to approximate as the minimum weight codeword, notice that the reduction in Proposition 5 is approximation-preserving.

To see that the smallest binary sub-decomposition is no harder to approximate than the minimum weight codeword, consider an instance $(A, D = \{F_1, \ldots, F_k\})$ of the sub-decomposition problem. Re-shape matrix A into a nm-dimensional binary column vector a. Reshape all matrices F_k similarly into binary vectors f_k, and collect them as columns of nm-by-$|D|$ binary matrix B. Our task now is to select the least number of columns of B such that their sum modulo-2 is a, that is, to find the vector x in the minimum-weight codeword problem. This reduction is also approximation-preserving, concluding the proof. $\square$

For the following results, we need the concept of *(randomized) quasi*-NP-*hardness.*

Definition 11. *We say a problem Π is* quasi-NP-hard *if Π cannot be solved in polynomial time unless* $\mathrm{NP} \subseteq \mathrm{DTIME}(n^{\mathrm{polylog}(n)})$. *We say Π is randomized* quasi-NP-hard, *if it cannot be solved in polynomial time unless* $\mathrm{NP} \subseteq \mathrm{RTIME}(n^{\mathrm{polylog}(n)})$, *where* $\mathrm{RTIME}(n^{\mathrm{polylog}(n)})$ *is the set of all languages recognizable by Monte Carlo algorithms with probability exceeding $1/2$ in time* $O(n^{\mathrm{polylog}(n)})$.

Corollary 3. *If $\boxplus = \oplus$, the smallest binary sub-decomposition is randomized* quasi-NP-*hard to approximate to within $2^{\log^{1-\varepsilon} k}$ for any $\varepsilon > 0$. It can, however, be approximated in polynomial time to within εk for any fixed ε.*

Proof. The negative result follows from a result of Dumer et al. [11] and Proposition 9. The positive result comes from Berman and Karpinski [8] and Proposition 9. $\square$

5.2   Approximating Minimum-Error Sub-Decompositions

We now turn our attention to the minimum-error sub-decompositions.

Proposition 10. *Let $(A, D = \{F_i\})$, with $|D| = k$, be an input for the minimum-error binary sub-decomposition problem. If $\boxplus = \vee$, it is quasi-NP-hard to approximate the minimum-error binary sub-decomposition problem to within a factor of $\Omega(2^{(4 \log k)^{1-\varepsilon}})$ and NP-hard to approximate it within $\Omega(2^{\log^{1-\varepsilon} |A|})$ for any $\varepsilon > 0$. The problem can be approximated to within a factor of $2\sqrt{(k + |A|) \log |A|}$ in polynomial time.*

Proof. The hardness-of-approximation result follows directly from the proof of Proposition 6 in case of $\boxplus = \vee$: the reduction from the basis usage problem is approximation-preserving, and corresponding lower bounds are known for the basis usage problem [24,25].

For the positive result, we need the reduction to go the other way. Let $(A, D = \{F_i\})$ be an input for the minimum-error sub-decomposition problem. Re-shape A to an nm-dimensional column vector a, and re-shape factor matrices F_i similarly and collect them into an nm-by-k binary matrix B. This is a valid input for the basis usage problem, and the solution can be approximated using Peleg's algorithm [30]. The result can be mapped back to minimum-error sub-decomposition problem without any change in the target value, and hence the reduction is approximation-preserving. The claim follows from known upper bounds for Peleg's algorithm for basis usage problem [25].   □

It is interesting to notice that the size of A plays no role in Proposition 10, only the number of its non-zeros. Also, quasi-NP-hardness is slightly stronger assumption than randomized quasi-NP-hardness that was assumed in Proposition 9.

The claim (and proof) for $\boxplus = \oplus$ is similar to the above.

Proposition 11. *Let $(A \in \{0,1\}^{n \times m}, D = \{F_i\})$, with $|D| = k$, be an input for the minimum-error binary sub-decomposition problem. If $\boxplus = \oplus$, it is quasi-NP-hard to approximate the minimum-error binary sub-decomposition problem to within a factor of $\Omega(2^{\log^{0.8-\varepsilon} n})$ for any $\varepsilon > 0$ and NP-hard to approximate it to within any constant factor. The problem can be approximated to within a factor of $O(k/\log(nm))$ in randomized polynomial time and to the same factor deterministically in time $(nm)^{O(\log^* nm)}$.*

Proof. This proof is similar to the above ones, and only a sketch of the proof is presented. The reduction from the nearest codeword problem in the proof of Proposition 6 is approximation-preserving, giving the negative result when paired with the results from [4]. The positive results require similar re-writing as above, after which we can use the results from [8] and [1].   □

6   Conclusions and Future Work

This paper presents an approach to unify pattern set mining using generalized outer products. Not every type of pattern can be expressed as a generalized outer product – not, at least, without making the outer products so general that we lose any reasonable way to study them as a group. Yet, as the above discussion has demonstrated, many interesting types of patterns – stars, biclique covers, nested matrices, and others – can easily be expressed in the framework, making it probable that also many yet-to-be-invented types of patterns will fit into it.

When a new type of pattern set mining problem can be expressed within the proposed framework, the researcher gains many benefits: the connections to other, sometimes seemingly unrelated work became more clear, many existing results might already apply, saving the researcher from tedious proofs, and the new results and techniques could generalize as well, immediately benefitting the whole field.

The work presented in this paper is only the beginning. The results concentrate on binary matrices, but recent work has generalized the binary setting to ordered lattices [5], ternary values [23], and rank matrices [20]. The general framework presented here could be extended to these situations, as well.

Another line of research extending the framework is to move from matrices to tensors (i.e. multi-way arrays). Again, there exists precedence in the pattern set mining, where mining higher-order (binary) data has gathered significant research interest [7,9,26].

Instead of finding the smallest or minimum-error pattern set, one can seek for a *planted* pattern, that is, a pattern we know the data contains, but that has been perturbed by noise. Recent research has shown that we can find individual planted patterns relatively well, even under strong noise assumptions [2,31].

There is, then, a lot to do to before the proposed framework can gain its full power. Yet, even this short preliminary work should be enough to demonstrate the potential the framework has, and – hopefully – convince researchers to frame their research within it.

References

1. Alon, N., Panigrahy, R., Yekhanin, S.: Deterministic approximation algorithms for the nearest codeword problem. In: Dinur, I., Jansen, K., Naor, J., Rolim, J. (eds.) PPROX and RANDOM 2009. LNCS, vol. 5687, pp. 339–351. Springer, Heidelberg (2009)
2. Ames, B.P.W., Vavasis, S.A.: Nuclear norm minimization for the planted clique and biclique problems. Math. Program. B **129**(1), 69–89 (2011)
3. Araujo, M., Günnemann, S., Mateos, G., Faloutsos, C.: Beyond blocks: hyperbolic community detection. In: Calders, T., Esposito, F., Hüllermeier, E., Meo, R. (eds.) ECML PKDD 2014, Part I. LNCS, vol. 8724, pp. 50–65. Springer, Heidelberg (2014)
4. Arora, S., Babai, L., Stern, J., Sweedyk, Z.: The hardness of approximate optima in lattices, codes, and systems of linear equations. In: FOCS 1993, pp. 724–733 (1993)
5. Bělohlávek, R., Krmelova, M.: Beyond boolean matrix decompositions: toward factor analysis and dimensionality reduction of ordinal data. In: ICDM 2013, pp. 961–966 (2013)
6. Bělohlávek, R., Vychodil, V.: Discovery of optimal factors in binary data via a novel method of matrix decomposition. J. Comput. Syst. Sci. **76**(1), 3–20 (2010)
7. Belohlavek, R., Vychodil, V.: Factorizing three-way binary data with triadic formal concepts. In: Setchi, R., Jordanov, I., Howlett, R.J., Jain, L.C. (eds.) KES 2010, Part I. LNCS, vol. 6276, pp. 471–480. Springer, Heidelberg (2010)
8. Berman, P., Karpinski, M.: Approximating minimum unsatisfiability of linear equations. In: SODA 2002, pp. 514–516 (2002)
9. Cerf, L., Besson, J., Nguyen, K.N.T., Boulicaut, J.F.: Closed and noise-tolerant patterns in n-ary relations. Data Min. Knowl. Discov. **26**(3), 574–619 (2013)
10. De Bie, T.: Maximum entropy models and subjective interestingness: an application to tiles in binary databases. Data Min. Knowl. Discov. **23**(3), 407–446 (2011)

11. Dumer, I., Micciancio, D., Sudan, M.: Hardness of approximating the minimum distance of a linear code. IEEE Trans. Inform. Theory **49**(1), 22–37 (2003)
12. Ene, A., Horne, W., Milosavljevic, N., Rao, P., Schreiber, R., Tarjan, R.E.: Fast exact and heuristic methods for role minimization problems. In: SACMAT 2008, pp. 1–10 (2008)
13. Feige, U.: A threshold of ln n for Approximating Set Cover. J. ACM **45**(4), 634–652 (1998)
14. Garey, M.R., Johnson, D.S.: Computers and intractability: A guide to the theory of NP-Completeness. W. H. Freeman, New York (1979)
15. Geerts, F., Goethals, B., Mielikäinen, T.: Tiling databases. In: Suzuki, E., Arikawa, S. (eds.) DS 2004. LNCS (LNAI), vol. 3245, pp. 278–289. Springer, Heidelberg (2004)
16. Johnson, D.S.: Approximation Algorithms for Combinatorial Problems. J. Comput. Syst. Sci. **9**, 256–278 (1974)
17. Junttila, E.: Patterns in permuted binary matrices. Ph.D. thesis, Helsinki University Press, Helsinki, August 2011
18. Kötter, T., Günnemann, S., Berthold, M., Faloutsos, C.: Extracting taxonomies from bipartite graphs. In: WWW 2015 Companion, pp. 51–52 (2015)
19. Koutra, D., Kang, U., Vreeken, J., Faloutsos, C.: VoG: summarizing and understanding large graphs. In: SDM 2014, pp. 91–99 (2014)
20. Le Van, T., van Leeuwen, M., Nijssen, S., Fierro, A.C., Marchal, K., De Raedt, L.: Ranked tiling. In: Calders, T., Esposito, F., Hüllermeier, E., Meo, R. (eds.) ECML PKDD 2014, Part II. LNCS, vol. 8725, pp. 98–113. Springer, Heidelberg (2014)
21. Lewis, J.M., Yannakakis, M.: The node-deletion problem for hereditary properties is NP-complete. J. Comput. Syst. Sci. **20**(2), 219–230 (1980)
22. Lucchese, C., Orlando, S., Perego, R.: A Unifying Framework for Mining Approximate Top-k Binary Patterns. IEEE Trans. Knowl. Data Eng. **26**(12), 2900–2913 (2013)
23. Maurus, S., Plant, C.: Ternary matrix factorization. In: ICDM 2014, pp. 400–409 (2014)
24. Miettinen, P.: On the positive-negative partial set cover problem. Inform. Process. Lett. **108**(4), 219–221 (2008)
25. Miettinen, P.: Matrix Decomposition Methods for Data Mining: Computational Complexity and Algorithms. Ph.D. thesis, Department of Computer Science, University of Helsinki (2009)
26. Miettinen, P.: Boolean tensor factorizations. In: ICDM 2011, pp. 447–456 (2011)
27. Miettinen, P.: Fully dynamic quasi-biclique edge covers via Boolean matrix factorizations. In: DyNetMM 2013, pp. 17–24 (2013)
28. Miettinen, P., Mielikäinen, T., Gionis, A., Das, G., Mannila, H.: The Discrete Basis Problem. IEEE Trans. Knowl. Data Eng. **20**(10), 1348–1362 (2008)
29. Peeters, R.: The maximum edge biclique problem is NP-complete. Discrete Appl. Math. **131**(3), 651–654 (2003)
30. Peleg, D.: Approximation algorithms for the Label-Cover$_{MAX}$ and Red-Blue Set Cover problems. J. Discrete Alg. **5**(1), 55–64 (2007)
31. Ramon, J., Miettinen, P., Vreeken, J.: Detecting bicliques in GF[q]. In: Blockeel, H., Kersting, K., Nijssen, S., Železný, F. (eds.) ECML PKDD 2013, Part I. LNCS, vol. 8188, pp. 509–524. Springer, Heidelberg (2013)
32. Simon, H.U.: On approximate solutions for combinatorial optimization problems. SIAM J. Discrete Math. **3**(2), 294–310 (1990)

33. Vreeken, J., van Leeuwen, M., Siebes, A.: Krimp: mining itemsets that compress. Data Min. Knowl. Discov. **23**(1), 169–214 (2011)
34. Xiang, Y., Jin, R., Fuhry, D., Dragan, F.F.: Summarizing transactional databases with overlapped hyperrectangles. Data Min. Knowl. Discov. **23**(2), 215–251 (2011)
35. Yang, J., Leskovec, J.: Overlapping community detection at scale: a nonnegative matrix factorization approach. In: WSDM 2013 (2013)
36. Yannakakis, M.: Node-Deletion Problems on Bipartite Graphs. SIAM J. Comput. **10**(2), 310–327 (1981)

Scalable Bayesian Non-negative Tensor Factorization for Massive Count Data

Changwei Hu[1], Piyush Rai[1(✉)], Changyou Chen[1], Matthew Harding[2], and Lawrence Carin[1]

[1] Department of Electrical and Computer Engineering,
Duke University, Durham, USA
{ch237,piyush.rai,cc448,lcarin}@duke.edu
[2] Sanford School of Public Policy and Department of Economics,
Duke University, Durham, USA
matthew.harding@duke.edu

Abstract. We present a Bayesian non-negative tensor factorization model for count-valued tensor data, and develop scalable inference algorithms (both batch and online) for dealing with massive tensors. Our generative model can handle overdispersed counts as well as infer the rank of the decomposition. Moreover, leveraging a reparameterization of the Poisson distribution as a multinomial facilitates conjugacy in the model and enables simple and efficient Gibbs sampling and variational Bayes (VB) inference updates, with a computational cost that only depends on the number of nonzeros in the tensor. The model also provides a nice interpretability for the factors; in our model, each factor corresponds to a "topic". We develop a set of online inference algorithms that allow further scaling up the model to massive tensors, for which batch inference methods may be infeasible. We apply our framework on diverse real-world applications, such as *multiway* topic modeling on a scientific publications database, analyzing a political science data set, and analyzing a massive household transactions data set.

Keywords: Tensor factorization · Bayesian learning · Latent factor models · Count data · Online bayesian inference

1 Introduction

Discovering interpretable latent structures in complex multiway (tensor) data is an important problem when learning from polyadic relationships among multiple sets of objects. Tensor factorization [5,14] offers a promising way of extracting such latent structures. The inferred factors can be used to analyze objects in each mode of the tensor (e.g., via classification or clustering using the factors), or to do tensor completion.

Of particular interest, in the context of such data, are sparsely-observed *count-valued* tensors. Tensors are routinely encountered in many applications. For example, in analyzing a database of scientific publications, the data may be in form of a sparse four-way count-valued tensor (authors × words × journals

© Springer International Publishing Switzerland 2015
A. Appice et al. (Eds.): ECML PKDD 2015, Part II, LNAI 9285, pp. 53–70, 2015.
DOI: 10.1007/978-3-319-23525-7_4

× years). Another application where multiway count data is routinely encountered is the analysis of contingency tables [11] which represent the co-occurrence statistics of multiple sets of objects.

We present a scalable Bayesian model for analyzing such sparsely-observed tensor data. Our framework is based on a beta-negative binomial construction, which provides a principled generative model for tensors with sparse and potentially overdispersed count data, and produces a non-negative tensor factorization. In addition to performing non-negative tensor factorization and tensor completion for count-valued tensors, our model has the property that each latent factor inferred for a tensor mode also represents a *distribution* (or "topic", as in topic models) over the objects of that tensor mode; our model naturally accomplishes this by placing a Dirichlet prior over the columns of the factor matrix of each tensor mode. In addition to providing an expressive and interpretable model for analyzing sparse count-valued tensors, the model automatically infers the rank of the decomposition, which side-steps the crucial issue of pre-specifying the rank of the decomposition [14, 18, 22].

Our framework also consists of a set of batch and scalable online inference methods. Using a reparameterization of the Poisson distribution as a multinomial allows us to achieve conjugacy, which facilitates closed-form Gibbs sampling as well as variational Bayes (VB) inference. Moreover, we also develop two *online* inference algorithms - one based on online MCMC [7] and the other based on stochastic variational inference [9]. These inference algorithms enable scaling up the model to massive-sized tensor data.

One of the motivations behind our work is analyzing massive multiway data for tasks such as understanding thematic structures in scholarly databases (e.g., to design better recommender systems for scholars), understanding consumer behavior from shopping patterns of large demographies (e.g., to design better marketing and supply strategies), and understanding international relations in political science studies. In our experiments, we provide qualitative analyses for such applications on large-scale real-world data sets, and the scalability behavior of our model.

2   Canonical PARAFAC Decomposition

Given a tensor $\mathcal{Y}$ of size $n_1 \times n_2 \times \cdots \times n_K$, with n_k denoting the size of $\mathcal{Y}$ along the kth mode (or "way") of the tensor, the goal in a Canonical PARAFAC (CP) decomposition [14] is to decompose $\mathcal{Y}$ into a set of K factor matrices $\mathbf{U}^{(1)}, \ldots, \mathbf{U}^{(K)}$ where $\mathbf{U}^{(k)} = [\boldsymbol{u}_1^{(k)}, \ldots, \boldsymbol{u}_R^{(k)}]$, $k = \{1, \ldots, K\}$, denotes the $n_k \times R$ factor matrix associated with mode k. In its most general form, CP decomposition expresses the tensor $\mathcal{Y}$ via a weighted sum of R rank-1 tensors as $\mathcal{Y} \sim f(\sum_{r=1}^{R} \lambda_r . \boldsymbol{u}_r^{(1)} \odot \ldots \odot \boldsymbol{u}_r^{(K)})$. The form of f depends on the type of data being modeled (e.g., f can be Gaussian for real-valued, Bernoulli-logistic for binary-valued, Poisson for count-valued tensors). Here λ_r is the weight associated with the rth rank-1 component, the $n_k \times 1$ column vector $\boldsymbol{u}_r^{(k)}$ represents the rth latent factor of mode k, and $\odot$ denotes vector outer product.

3   Beta-Negative Binomial CP Decomposition

We focus on modeling count-valued tensor data [4] and assume the following generative model for the tensor $\mathcal{Y}$

$$\mathcal{Y} \sim \text{Pois}(\sum_{r=1}^{R} \lambda_r . \boldsymbol{u}_r^{(1)} \odot \ldots \odot \boldsymbol{u}_r^{(K)}) \tag{1}$$

$$\boldsymbol{u}_r^{(k)} \sim \text{Dir}(a^{(k)}, \ldots, a^{(k)}) \tag{2}$$

$$\lambda_r \sim \text{Gamma}(g_r, \frac{p_r}{1 - p_r}) \tag{3}$$

$$p_r \sim \text{Beta}(c\epsilon, c(1 - \epsilon)) \tag{4}$$

We use subscript $\boldsymbol{i} = \{i_1, \ldots, i_K\}$ to denote the index of the $\boldsymbol{i}$-th entry in $\mathcal{Y}$. Using this notation, the $\boldsymbol{i}$ th entry of the tensor can be written as $y_{\boldsymbol{i}} \sim$ $\text{Pois}(\sum_{r=1}^{R} \lambda_r \prod_{k=1}^{K} u_{i_k r}^{(k)})$. We assume that we are given N observations $\{y_{\boldsymbol{i}}\}_{i=1}^{N}$ from the tensor $\mathcal{Y}$.

Since the gamma-Poisson mixture distribution is equivalent to a negative binomial distribution [15], (1) and (3), coupled with the beta prior (Eq 4) on p_r, lead to what we will call the beta-negative binomial CP (BNBCP) decomposition model. A few things worth noting about our model are

- The Dirichlet prior on the factors $\boldsymbol{u}_r^{(k)}$ naturally imposes non-negativity constraints [4] on the factor matrices $\mathbf{U}^{(1)}, \ldots, \mathbf{U}^{(K)}$. Moreover, since each column $\boldsymbol{u}_r^{(k)}$ of these factor matrices sums to 1, $\boldsymbol{u}_r^{(k)}$ can also be thought of a distribution (e.g., a "topic") over the n_k entities in mode k.
- The gamma-beta hierarchical construction of λ_r (Eq 3 and 4) allows inferring the rank of the tensor by setting an upper bound R on the number of factors and letting the inference procedure infer the appropriate number of factors by shrinking the coefficients λ_r's to close to zero for the irrelevant factors.
- The resulting negative binomial model is useful for modeling *overdispersed* count data in cases where the Poisson likelihood may not be suitable.
- Using alternate parameterizations (Section 3.1) of the Poisson distribution in (1) leads to a fully conjugate model and facilitates efficient Gibbs sampling and variational Bayes (VB) inference, in both batch as well as online settings.

3.1   Reparametrizing the Poisson Distribution

The generative model described in Eq (1)-(4) is not conjugate. We now describe two equivalent parametrizations [6,24] of (1), which transform (1)-(4) into a fully conjugate model and facilitate easy-to-derive and scalable inference procedures. These parameterizations are based on a data augmentation scheme described below.

The first parametrization expresses the $\boldsymbol{i}$-th count-valued entry $y_{\boldsymbol{i}}$ of the tensor $\mathcal{Y}$ as a sum of R *latent* counts $\{\tilde{y}_{\boldsymbol{i}r}\}_{r=1}^{R}$

$$y_{\boldsymbol{i}} = \sum_{r=1}^{R} \tilde{y}_{\boldsymbol{i}r}, \quad \tilde{y}_{\boldsymbol{i}r} \sim \text{Pois}(\lambda_r \prod_{k=1}^{K} u_{i_k r}^{(k)}) \tag{5}$$

The second parametrization assumes the vector $\{\tilde{y}_{ir}\}_{r=1}^{R}$ of latent counts is drawn from a multinomial as

$$\tilde{y}_{i1}, \ldots, \tilde{y}_{iR} \sim \text{Mult}(y_i; \zeta_{i1}, \ldots, \zeta_{iR})$$

$$\zeta_{ir} = \frac{\lambda_r \prod_{k=1}^{K} u_{i_k r}^{(k)}}{\sum_{r=1}^{R} \lambda_r \prod_{k=1}^{K} u_{i_k r}^{(k)}} \tag{6}$$

The above parameterizations follows from the following lemma [6,24]:

Lemma 1. *Suppose that $x_1, \ldots, x_R$ are independent random variables with $x_r \sim Pois(\theta_r)$ and $x = \sum_{r=1}^{R} x_r$. Set $\theta = \sum_{r=1}^{R} \theta_r$; let $(z, z_1, \ldots, z_R)$ be another set of random variables such that $z \sim Pois(\theta)$, and $(z_1, \ldots, z_R)|z \sim Mult(z; \frac{\theta_1}{\theta}, \ldots, \frac{\theta_R}{\theta})$. Then the distribution of $\boldsymbol{x} = (x, x_1, \ldots, x_R)$ is the same as the distribution of $\boldsymbol{z} = (z, z_1, \ldots, z_R)$.*

These parameterizations, along with the fact that the columns $\boldsymbol{u}_r^{(k)}$ of each factor matrix are drawn from a Dirichlet, allows us to leverage the Dirichlet-multinomial conjugacy and derive simple Gibbs sampling and variational Bayes (VB) inference update equations, as described in Section 4.

4   Inference

We first present the update equations for batch Gibbs sampling (Section 4.1) and batch VB inference (Section 4.2). We then present two online inference algorithms, based on: (i) conditional density filtering [7], which provides an efficient way to perform online MCMC sampling using conditional sufficient statistics of the model parameters; and (ii) stochastic variational inference [9], which will allow scaling up VB inference by processing data in small minibatches.

We also define two quantities $s_{j,r}^{(k)} = \sum_{i:i_k=j} \tilde{y}_{ir}$ and $s_r = \sum_i \tilde{y}_{i,r}$ which denote aggregates (sufficient statistics) computed using the latent counts $\tilde{y}_{ir}$. These quantities appear at various places in the description of the inference algorithms we develop.

4.1   Gibbs Sampling

- **Sampling $\tilde{y}_{ir}$:** The latent counts $\{\tilde{y}_{ir}\}_{r=1}^{R}$ are sampled from a multinomial (6).

- **Sampling $\boldsymbol{u}_r^{(k)}$:** Due to the Dirichlet-multinomial conjugacy, the columns of each factor matrix have Dirichlet posterior and are sampled as

$$\boldsymbol{u}_r^{(k)} \sim \text{Dir}(a^{(k)} + s_{1,r}^{(k)}, a^{(k)} + s_{2,r}^{(k)}, \ldots, a^{(k)} + s_{n_k,r}^{(k)}) \tag{7}$$

- **Sampling p_r:** Using the fact that $s_r = \sum_i \tilde{y}_{i,r}$ and marginalizing over the $u_{i_k r}^{(k)}$'s in (5), we have $s_r \sim \text{Pois}(\lambda_r)$. Using this, along with (3), we can express

s_r using a negative binomial distribution, i.e., $s_r \sim \mathrm{NB}(g_r, p_r)$. Then, due to the conjugacy between negative binomial and beta, we can sample p_r as

$$p_r \sim \mathrm{Beta}(c\epsilon + s_r, c(1 - \epsilon) + g_r) \tag{8}$$

– **Sampling λ_r:** Again using the fact that $s_r \sim \mathrm{Pois}(\lambda_r)$, and due to the gamma-Poisson conjugacy, we have

$$\lambda_r \sim \mathrm{Gamma}(g_r + s_r, p_r) \tag{9}$$

Computational Complexity: Sampling the latent counts $\{\tilde{y}_{ir}\}_{r=1}^R$ for each nonzero observation y_i (note that for $y_i = 0$, the latent counts are trivially zero) requires computing $\{\zeta_{ir}\}_{r=1}^R$, and computing each ζ_{ir} requires $O(K)$ time (Eq 6). Therefore, sampling all the latent counts $\{\tilde{y}_{ir}\}_{r=1}^R$ requires $O(NRK)$ time. Sampling the latent factors $\{\boldsymbol{u}_r^{(k)}\}_{r=1}^R$ for the K tensor modes requires $O(RK)$ time. Sampling $\{p_r\}_{r=1}^R$ and $\{\lambda_r\}_{r=1}^R$ requires $O(R)$ time each. Of all these steps, sampling the latent counts $\{\tilde{y}_{ir}\}_{r=1}^R$ (which are also used to compute the sufficient statistics $s_{j,r}^{(k)}$ and s_r) is the most dominant step, leading to an overall time-complexity of $O(NRK)$ for the Gibbs sampling procedure.

The linear dependence on N (number of nonzeros) is especially appealing because most real-world count-valued tensors are extremely sparse (have much less than even 1% nonzeros). In contrast to the standard negative-binomial models for count data, for which the inference complexity also depends on the zeros whose number may be massive (and therefore heuristics, such as subsampling the zeros, are needed), the reparametrizations (Section 3.1) used by our model allow us to ignore the zeros in the multinomial sampling step (the sufficient statistics do not depend on the zero entries in the tensor), thereby significantly speeding up the inference.

4.2 Variational Bayes Inference

Using the mean-field assumption [12], we approximate the target posterior distribution by $Q = \prod_{i,r} q(\tilde{y}_{ir}) \prod_{k,r} q(\boldsymbol{u}_r^{(k)}) \prod_r q(\lambda_r) \prod_r q(p_r)$. Our fully conjugate model enables closed-form variational Bayes (VB) inference updates, with the distribution $q(\tilde{y}_{ir})$, $q(\boldsymbol{u}_r^{(k)})$, $q(\lambda_r)$, and $q(p_r)$ being multinomial, Dirichlet, beta, and gamma, respectively. We summarize the update equations for the variational parameters of each of these distributions, below:

– **Updating $\tilde{y}_{ir}$:** Using (6), the updates for y_{ir} are given by $\mathbb{E}[y_{ir}] = y_i \zeta_{ir}$ where ζ_{ir} is defined as $\zeta_{ir} = \frac{\tilde{\zeta}_{ir}}{\sum_{r=1}^R \tilde{\zeta}_{ir}}$ and $\tilde{\zeta}_{ir}$ can be computed as

$$\tilde{\zeta}_{ir} = \exp\{\Psi(s_r + g_r) + \ln(p_r) + \sum_{k=1}^K \Psi(s_{i_k,r}^{(k)} + a^{(k)}) - \Psi[\sum_{k=1}^K (s_{i_k,r}^{(k)} + a^{(k)})]\} \tag{10}$$

where $\Psi(.)$ is the digamma function, which is the first derivative of the logarithm of the gamma function.

– **Updating $u_{i_k r}^{(k)}$:** The mean-field posterior $q(u_r^{(k)})$ is Dirichlet with each of the component means given by $\mathbb{E}[u_{i_k r}^{(k)}] = \frac{\rho_{i_k r}^{(k)}}{\sum_{i_k=1}^{n_k} \rho_{i_k r}^{(k)}}$ where $\rho_{i_k r}^{(k)} = a^{(k)} + s_{i_k, r}^{(k)}$.

– **Updating p_r:** The mean-field posterior $q(p_r)$ is beta with mean given by $\mathbb{E}[p_r] = \frac{p_{ra}}{p_{ra} + p_{rb}}$ where $p_{ra} = c\epsilon + s_r$, $p_{rb} = c(1 - \epsilon) + g_r$.

– **Updating λ_r:** The mean-field posterior $q(\lambda_r)$ is gamma with mean given by $\mathbb{E}[\lambda_r] = \lambda_{ra} \lambda_{rb}$, where $\lambda_{ra} = (g_r + s_r)$ and $\lambda_{rb} = p_r$.

A note on Gibbs vs VB: The per-iteration time-complexity of the VB inference procedure is also $O(NRK)$. It is to be noted however that, in practice, one iteration of VB in this model is a bit more expensive than one iteration of Gibbs, due to the digamma function evaluation for the $\tilde{\zeta}_{ir}$ which is needed in VB when updating the $\tilde{y}_{ir}$'s. Prior works on Bayesian inference for topic models [8] also support this observation.

4.3 Online Inference

Batch Gibbs (Section 4.1) and VB (Section 4.2) inference algorithms are simple to implement and efficient to run on moderately large-sized problems. These algorithms can however be slow to run for massive data sets (e.g., where the number of tensor entries N and/or the dimension of the tensor is massive). The Gibbs sampler may exhibit slow mixing and the batch VB may be slow to converge. To handle such massive tensor data, we develop two online inference algorithms. The first is online MCMC based conditional density filtering [7], while the second is based on stochastic variational inference [9]. Both these inference algorithms allow processing data in small minibatches and enable our model to analyze massive and/or streaming tensor data.

Conditional Density Filtering: The conditional density filtering (CDF) algorithm [7] for our model selects a minibatch of tensor entries at each iteration, samples the latent counts $\{\tilde{y}_{ir}\}_{r=1}^R$ for these entries conditiond on the previous estimates of the model parameters, updates the sufficient statistics $s_{j,r}^{(k)}$ and s_r using these latent counts (as described below), and resamples the model parameters conditioned on these sufficient statistics. Denoting I_t as data indices in minibatch at round t, the algorithm proceeds as

– **Sampling $\tilde{y}_{ir}$:** For all $i \in I_t$, sample the latent counts $\tilde{y}_{ir(i \in I_t)}$ using (6).

– **Updating the conditional sufficient statistics:** Using data from the current minibatch, update the conditional sufficient statistics as:

$$s_{j,r}^{(k,t)} = (1 - \gamma_t) s_{j,r}^{(k,t-1)} + \gamma_t \frac{N}{B} \sum_{i \in I_t : i_k = j} \tilde{y}_{ir} \qquad (11)$$

$$s_r^{(t)} = (1 - \gamma_t) s_r^{(t-1)} + \gamma_t \frac{N}{B} \sum_{i \in I_t} \tilde{y}_{i,r} \qquad (12)$$

Note that the updated conditional sufficient statistics (CSS), indexed by superscript t, is a weighted average of the old CSS, indexed by superscript $t-1$, and of that computing only using the current minibatch (of size B). In addition, the latter term is further weighted by N/B so as to represent the *average* CSS over the *entire* data. In the above, γ_t is defined as $\gamma_t = (t_0+t)^{-\kappa}$, $t_0 \geq 0$, and $\kappa \in (0.5, 1]$ is needed to guarantee convergence [3].

- **Updating $u_r^{(k)}, p_r, \lambda_r$:** Using the updated CSS, draw M samples for each of the model parameters $\{u_r^{(k,m)}, p_r^{(m)}, \lambda_r^{(m)}\}_{m=1}^M$, from the following conditionals:

$$u_r^{(k)} \sim \text{Dir}(a^{(k)} + s_{1,r}^{(k,t)}, \ldots, a^{(k)} + s_{n_k,r}^{(k,t)}) \tag{13}$$

$$p_r \sim \text{Beta}(c\epsilon + s_r^{(t)}, c(1-\epsilon) + g_r) \tag{14}$$

$$\lambda_r \sim \text{Gamma}(g_r + s_r^{(t)}, p_r) \tag{15}$$

and either store the sample averages of $u_r^{(k)}, p_r$, and λ_r, or their analytic means to use for the next CDF iteration [7]. Since the analytic means of the model parameters are available in closed-form in this case, we use the latter option, which obviates the need to draw M samples, thereby also speeding up the inference significantly.

We next describe the stochastic (online) VB inference for our model.

Stochastic Variational Inference: The batch VB inference (Section 4.2) requires using the entire data for the parameter updates in each iteration, which can be computationally expensive and can also result in slow convergence. Stochastic variational inference (SVI), on the other hand, leverages ideas from stochastic optimization [9] and, in each iteration, uses a small randomly chosen minibatch of the data to updates the parameters. Data from the current minibatch is used to compute stochastic gradients of the variational objective w.r.t. each of the parameters and these gradients are subsequently used in the parameter updates. For our model, the stochastic gradients depend on the sufficient statistics computed using the current minibatch I_t: $s_{j,r}^{(k,t)} = \sum_{i \in I_t : i_k = j} \tilde{y}_{ir}$ and $s_r^{(t)} = \sum_{i \in I_t} \tilde{y}_{i,r}$, where $\tilde{y}_{ir}$ is computed using Eq 10. Denoting B as the minibatch size, we reweight these statistics by N/B to compute the *average* sufficient statistics over the entire data [9] and update the other variational parameters as follows:

$$\rho_{i_k r}^{(k,t)} = (1-\gamma_t)\rho_{i_k r}^{(k,t-1)} + \gamma_t(a^{(k)} + (N/B)s_{i_k,r}^{(k,t)}) \tag{16}$$

$$p_{ra}^{(t)} = (1-\gamma_t)p_{ra}^{(t-1)} + \gamma_t(c\epsilon + (N/B)s_r^{(t)}) \tag{17}$$

$$p_{rb}^{(t)} = (1-\gamma_t)p_{rb}^{(t-1)} + \gamma_t(c(1-\epsilon) + g_r) \tag{18}$$

$$\lambda_{ra}^{(t)} = (1-\gamma_t)\lambda_{ra}^{(t-1)} + \gamma_t(g_r + (N/B)s_r^{(t)}) \tag{19}$$

$$\lambda_{rb}^{(t)} = (1-\gamma_t)\lambda_{ra}^{(t-1)} + \gamma_t p_r \tag{20}$$

where γ_t is defined as $\gamma_t = (t_0 + t)^{-\kappa}$, $t_0 \geq 0$, and $\kappa \in (0.5, 1]$ is needed to guarantee convergence [9].

Computational Complexity: In contrast to the batch Gibbs and batch VB, both of which have $O(NRK)$ cost per-iteration, the per-iteration cost of the online inference algorithms (CDF and SVI) is $O(|I_t|RK)$ where $|I_t|$ is the mini-batch size at round t. We use a fixed minibatch size B for each minibatch, so the per-iteration cost is $O(BRK)$.

5 Related Work

Although tensor factorization methods have received considerable attention recently, relatively little work exists on scalable analysis of massive count-valued tensor data. Most of the recently proposed methods for scalable tensor decomposition [2,10,13,17] are based on minimizing the Frobenious norm of the tensor reconstruction error, which may not be suitable for count or overdispersed count data. The rank of decomposition also needs to be pre-specified, or chosen via cross-validation. Moreover, these methods assume the tensor to be fully observed and thus cannot be used for tensor completion tasks. Another key difference between these methods and ours is that scaling up these methods requires parallel or distributed computing infrastructure, whereas our fully Bayesian method exhibits excellent scalability on a single machine. At the same time, the simplicity of the inference update equations would allow our model to be easily parallelized or distributed. We leave this possibility to future work.

One of the first attempts to explicitly handle count data in the context of non-negative tensor factorization includes the work of [4], which is now part of the Tensor Toolbox [1]. This method optimizes the Poisson likelihood, using an alternating Poisson regression sub-routine, with non-negative constraints on the factor matrices. However, this method requires the rank of the decomposition to be specified, and cannot handle missing data. Due to its inability in handling missing data, for our experiments (Section 6), as a baseline, we implement and use a Bayesian version of this model which *can* handle missing data.

Among other works of tensor factorization for count data, the method in [1] can deal with missing values, though the rank still needs to be specified, and moreover the factor matrices are assumed to be real-valued, which makes it unsuitable for interpretability of the inferred factor matrices.

In addition to the Poisson non-negative tensor factorization method of [4], some other non-negative tensor factorization methods [5,20,21] also provide interpretability for the factor matrices. However, these methods usually have one or more of the following limitations: (1) there is no explicit generative model for the count data, (2) the rank needs to be specified, and (3) the methods do not scale to the massive tensor data sets of scales considered in this work.

Methods that facilitate a full Bayesian analysis for massive count-valued tensors, which are becoming increasingly prevalent nowadays, are even fewer. A recent attempt on Bayesian analysis of count data using Poisson likelihood is considered in [19]; however, unlike our model, their method cannot infer the rank and relies on batch VB inference, limiting its scaling behavior. Moreover, the Poisson likelihood may not be suitable for overdispersed counts.

---

[1] http://www.sandia.gov/~tgkolda/TensorToolbox/index-2.6.html

Finally, inferring the rank of the tensor, which is NP-complete in general [14], is another problem for which relatively little work exists. Recent attempts at inferring the rank of the tensor in the context of CP decomposition include [18,22]; however (1) these methods are not applicable for count data, and (2) the inferred factor matrices are real-valued, lacking the type of interpretability needed in many applications.

Our framework is similar in spirit to the matrix factorization setting proposed in [24] which turns out to be a special case of our framework. In addition, while [24] only developed (batch) Gibbs sampling based inference, we present both Gibbs sampling as well as variational Bayesian inference, and design efficient *online* Bayesian inference methods to scale up our framework for handling massive real-world tensor data.

To summarize, in contrast to the existing methods for analyzing tensors, our fully Bayesian framework, based on a proper generative model, provides a flexible method for analyzing massive count-valued tensors, side-stepping crucial issues such as rank-specification, providing good interpretability of the latent factors, while still being scalable for analyzing massive real-world tensors via online Bayesian inference.

6   Experiments

We apply the proposed model on a synthetic and three real-world data sets that range in their sizes from moderate to medium to massive. The real-world tensor data sets we use in our experiments are from diverse application domains, such as analyzing country-country interaction data in political science, topic modeling on *multiway* publications data (with entities being authors, words, and publication venues), and analysis of massive household transactions data. These data sets include:

- **Synthetic Data:** This is a tensor of size $300 \times 300 \times 300$ generated using our model by setting an upper bound $R = 50$ over the number of factors; only 20 factors were significant (based on the values of λ_r), resulting in an effective rank 20.

- **Political Science Data (GDELT):** This is a real-world four-way tensor data of country-country interactions. The data consists of 220 countries, 20 action types, and the interactions date back to 1979 [16]. We focus on a subset of this data collected during the year 2011, resulting in a tensor of size $220 \times 220 \times 20 \times 52$. Section 6.4 provides further details.

- **Publications Data:** This is a $2425 \times 9088 \times 4068$ count-valued tensor, constructed from a database of research papers published by researchers at Duke University[2]; the three tensor modes correspond to authors, words, and venues. Section 6.3 provides further details.

---

[2] Obtained from https://scholars.duke.edu/

- **Transactions (Food) Data:** This is a $117054 \times 438 \times 67095$ count-valued tensor, constructed from a database of transactions data of food item purchases at various stores in the US [3]; the three tensor modes correspond to households, stores, and items. Section 6.5 provides further details.

We compared our model with the following baselines: (*i*) Bayesian Poisson Tensor Factorization (BAYESPTF), which is fully Bayesian version of the Poisson Tensor Factorization model proposed in [4], and (*ii*) Non-negative Tensor Decomposition based on Low-rank Approximation (LRANTD) proposed in [23]. All experiments are done on a standard desktop computer with Intel i7 3.4GHz processor and 24GB RAM.

6.1   Inferring the Rank

To begin with, as a sanity check for our model, we first perform an experiment on the synthetic data described above to see how well the model can recover the true rank (tensor completion results are presented separately in Section 6.2). For this experiment, we run the batch Gibbs sampler (the other inference methods also yield similar results) with 1000 burn-ins, and 1000 collection samples. We experiment with three settings: using 20%, 50% and 80% data for training. The empirical distribution (estimated using the collected MCMC samples) of the effective inferred rank for each of these settings is shown in Figure 1 (left). In each collection iteration, the effective rank is computed after a simple thresholding on the λ_r's where components with very small λ_r are not counted (also see Figure 1

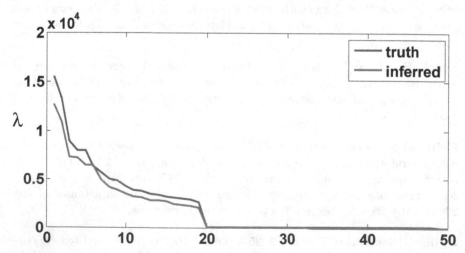

Fig. 1. Distribution over inferred ranks for syntheric data (left), and λ inferred using 80% training data (right).

---

[3] Data provided by United States Department of Agriculture (USDA) under a Third Party Agreement with Information Resources, Inc. (IRI).

(right)). With 80% training data, the distribution shows a distinct peak at 20 and even with smaller amounts of training data (20% and 50%), the inferred rank is fairly close to the ground truth of 20. In Figure 1 (right), we show the spectrum of all the λ_r's comparing the ground truth vs the inferred values;

6.2 Tensor Completion Results

We next experiment on the task of tensor completion, where for each method 95% of the data are used for training and the remaining 5% data is used as the heldout set (note that the data sets we use are extremely sparse in nature, with considerably less than 1% entries of the tensor being actually observed). The results are reported in Table 1 where we show the log likelihood and the mean-absolute error (MAE) in predicting the heldout data. Timing-comparison for the various batch and online inference methods is presented separately in Section 6.6.

For this experiment, we compare our BNBCP model (using the various inference methods) with (1) BAYESPTF - a fully Bayesian variant (we implented it ourselves) of a state-of-the-art Poisson Tensor Factorization model originally proposed in [4] (which cannot however handle missing data), and (2) LRANTD [23] which is an optimization based non-negative tensor decomposition method. As Table 1 shows, our methods achieve better log-likelihood and MAE as compared to these baselines. Moreover, among our batch and online Bayesian inference methods, the online inference methods give competitive or better results as compared to their batch counterparts. In particular, the online MCMC method based on conditional density filtering (BNBCP-CDF) works the best across all the methods (please see Section 6.6 for a timing comparison).

Table 1. Loglikelihood and MAE comparison for different methods (the two baselines, our model with batch inference, and our model with online inference) on four datasets. Note: LRANTD gave out-of-memory error on publications and food transactions data sets so we are unable to report its results on these data sets. We also only report the MAE for LRANTD, and not the log-likelihood, because it uses a Gaussian likelihood model for the data.

DATASETS	TOY DATA	GDELT	PUBLICATION	FOOD	TOY DATA	GDELT	PUBLICATION	FOOD
BAYESPTF	-107563	-4425695	-860808	-2425433	1.012	55.478	1.636	1.468
LRANTD	N/A	N/A	N/A	N/A	1.019	65.049	N/A	N/A
BNBCP-GIBBS	-97580	-3079883	-619258	-2512112	0.989	45.436	1.565	1.459
BNBCP-VB	-99381	-2971769	-632224	-2533086	0.993	**43.485**	1.574	1.472
BNBCP-CDF	-95472	**-2947309**	**-597817**	**-2403094**	0.985	44.243	**1.555**	**1.423**
BNBCP-ONLINEVB	-98446	-3169335	-660068	-2518996	0.989	46.188	1.601	1.461

6.3 Analyzing Publications Database

The next experiment is on a three-way tensor constructed from a scientific publications database. The data consist of abstracts from papers published by various researchers at Duke University [4]. In addition to the paper abstract, the venue

---

[4] Data crawled from https://scholars.duke.edu/

information for each paper is also available. The data collection contains 2425 authors, 9088 words (after removing stop-words), and 4068 venues which results in a 2425 × 9088 × 4068 word-counts tensor, on which we run our model. As the output of the tensor decomposition, we get three factor matrices. Since the latent factors in our model are non-negative and sum to one, each latent factor can also be interpreted as a *distribution* over authors/words/venues, and consequently represents a "topic". Therefore the three factor matrices inferred by our model for this data correspond to authors × topics, words × topics, and venue × topics, which we use to further analyze the data.

We apply the model BNBCP-CDF on this data (with $R = 200$) and using the inferred words × topics matrix, in Table 2 (left) we show the list of 10 most probable words in four factors/topics that seem to correspond to **optics**, **genomics**, **machine learning & signal processing**, and **statistics**. To show the topic representation across different departments, we present a histogram of *departmental affiliations* for 20 authors with highest probabilities in these four factors. We find that, for the genomics factor, the top authors (based on their topic scores) have affiliations related to biology which makes intuitive sense. Likewise, for the statistics factor, most of the top authors are from statistics and biostatistics departments. The top 20 authors in factors that correspond to optics and machine learning & signal processing, on the other hand, are from departments of electrical and computer engineering and/or computer science, etc.

Table 2. Most probable words in topics related to optics, genomics, machine learning/signal processing(ML/SP) and statistics (Stats), and top ranked venues in ML/SP community.

OPTICS	GENOMICS	ML/SP	STATS	TOP VENUES IN ML/SP
GIGAPIXEL	GENE	DICTIONARY	MODEL	ICASSP
MICROCAMERA	CHROMATIN	SPARSITY	PRIORS	IEEE TRANS. SIG. PROC.
CAMERAS	OCCUPANCY	MODEL	BAYESIAN	ICML
APERTURE	CENTROMERE	BAYESIAN	LASSO	SIAM J. IMG. SCI.
LENS	TRANSCRIPTION	COMPRESSED	LATENT	IEEE TRANS. IMG. PROC.
MULTISCALE	GENOME	COMPRESSIVE	INFERENCE	IEEE INT. SYMP. BIOMED. IMG.
OPTICAL	SITES	MATRIX	REGRESSION	NIPS
SYSTEM	EXPRESSION	DENOISING	SAMPLER	IEEE TRANS. WIRELESS COMM.
NANOPROBES	SEQUENCE	GIBBS	SEMIPARAMETRIC	IEEE WORKSHOP STAT. SIG. PROC.
METAMATERIAL	VEGFA	NOISE	NONPARAMETRIC	IEEE TRANS. INF. THEORY

Similarly, using the inferred venues × topics matrix, we list the most likely venues for each topic. Due to space-limitations, here we only present the most likely venues in machine learning & signal processing factor/topic; the result is shown in Table 2 (right-most column). The result shows that venues like ICASSP, IEEE Trans. Signal Proc., ICML, and NIPS all rank at the top in the machine learning & signal processing factor, which again makes intuitive sense.

6.4 Analyzing Political Science Data

We use the model to analyze a real-world political science data set consisting of country-country interactions. Such analyses are typically done by political

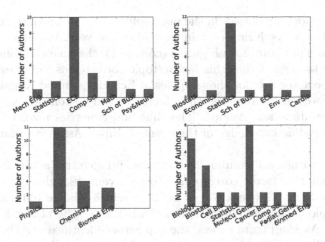

Fig. 2. Histogram of affiliations for top 20 authors in factors related to machine learning/signal processing (top left) and statistics (top right), optics (bottom left), and genomics(bottom right)

scientists to study, analyze and understand complex international multilateral relations among countries. The data set is from the Global Database of Events, Location, and Tone (GDELT) [16]. GDELT records the dyadic interactions between countries in the form of "Country A did something to Country B". In our experiments, we consider 220 countries ("actors") and 20 unique high-level action types in 52 weeks of year 2012. After preprocessing, we have a four-way (country-country-action-time) action counts tensor of size $220 \times 220 \times 20 \times 52$. Note that both first and second tensor mode represents countries; first mode as "sender" and the second mode as "receiver" of a particular action. In this analysis, we set R to be large enough (200) and the model discovered roughly about 120 active components (i.e., components with significant value of λ_r).

We apply the model (BNBCP-CDF; other methods yield similar results) and examine each of the time dimension factors, specifically looking for the

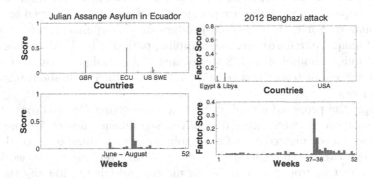

Fig. 3. Country factors (top row) and time factors (bottom row) for Julian Assange asylum in Ecuador (left column) and 2012 Benghazi attack (right column).

significant components (based on the magnitude of λ_r) in which the time dimension factor also peaks during certain time(s) of the year. We show results with two such factors in Figure 3. In Figure 3 (column 1), the time and country (actor) factors seems to suggest that this factor/topic corresponds to the event "Julian Assange". The actor subplot shows spikes at Ecuador, United Kingdom, United States, and Sweden whereas the time factor in the bottom left subplot shows spikes between June and August. The time and countries involved are consistent with the public knowledge of the event of Julian Assange seeking refuge in Ecuador.

Likewise, in Figure 3 (column 2), the time and country (actor) factors seems to suggest that this factor corresponds to the event "Benghazi Attack" which took place on Sept. 12 (week 37) of 2012, in which Islamic militants attacked American diplomatic compound in Benghazi, Libya. The attack killed an US Ambassador. As the Figure shows, the top actors identified are US, Libya and Egypt, and spikes are found at around week 37 and 38, which are consistent with the public knowledge of this event.

The results of these analyses demonstrate that the interpretability of our model can be useful for identifying events or topics in such multiway interaction data.

6.5 Analyzing Transactions Data

We next apply our model (BNBCP-CDF; other methods yield similar results) for analyzing transactions data for food item purchases made at stores. Our data is collected for a demographically representative sample of US consumers who reside in large urban and suburban areas and purchase food in supermarkets and grocery stores. The data were provided by the USDA under a Third Party Agreement with IRI. Each transaction is identified by a unique Universal Product Code (UPC) barcode and the store where the transaction occurred. Some items such as fresh produce do not have UPCs and are identified separately. The households are observed over a four year period, during which they are provided with a technology that allows them to scan each purchase and record additional information such as the store where the purchase was made (and other economic data). Participating households are provided with incentives designed to encourage compliance. For each household-product-store combination we record the number of unique purchases over the sampling period. The database has a total of 117,054 unique households, 438 stores, and 67,095 unique items and we construct a 3-way count tensor of size $117054 \times 438 \times 67095$ with about 6.2 million nonzero entries.

We apply the proposed model on this data by setting $R = 100$ (out of which about 60 components were inferred to have a significant value of λ_r) and looked at the stores factor matrix. Since each column (which sums to 1) of the store factor matrix can be thought of as a *distribution* over the stores, we look at three of the factors from the store factor matrix and tried to identify the stores that rank at the top in that factor. In Table 3, we show results from each of these factors. Factor 1 seems to suggest that it is about the most popular stores (included Walmart, for example), Factor 2 has stores that primarily deal in wholesale (e.g., Costco, Sam's Wholesale Club), and Factor 3 contains stores that sell none or very few food items (e.g., Mobil, Petco). Note that the Walmart Super Center figures prominently in both Factor 1 and Factor 2.

Table 3. Three of the store factors inferred from the transaction data (top-5 stores shown for each)

FACTOR 1	FACTOR 2	FACTOR 3
WALMART SUP. CENTER	SAM'S CLUB	DICK'S SPORTING
WALMART TRADERS	MEIJER	MOBIL
WALMART NEIGHB.	COSTCO	PETCO
WALMART	B J'S WHOLESALE	SALLY BEAUTY
KROGER	WALMART SUP. CENTER	GNC ALL

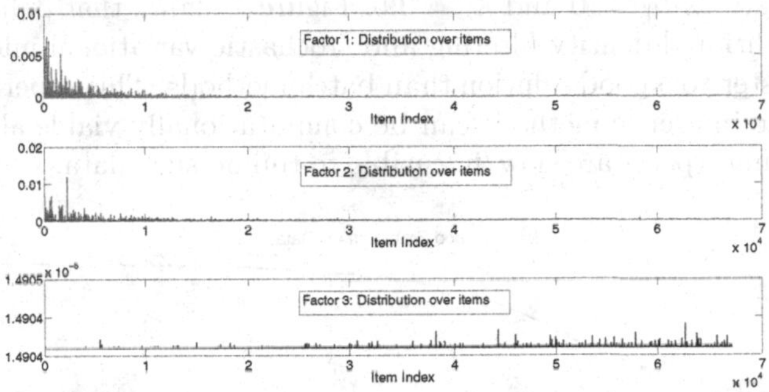

Fig. 4. Distributions over items for three factors (each factor corresponds to a cluster).

We next look at the items factor matrix. In Figure 2, we plot the inferred distribution over items in each of the three clusters described above. For factors 1 and 2 (which correspond to the most popular stores and wholesale stores respectively), the distribution over the items (top and bottom panel in Figure 2) have a reasonably significant mass over a certain range of items (for the items indexed towards the left side in the plots of factors 1 and 2). On the other hand, for factor 3 which corresponds to stores that sell no or very few types of food items, the distribution over the items is rather flat and diffuse with very weak intensities (looking at the scale on the y axis). From the Figure 2, it is also interesting to observe that the set of active items in factors (1 & 2) vs factor 3 seem to be mostly disjoint.

This analysis provides a first attempt to analyze food shopping patterns for American consumers on a large scale. As the world, at large, struggles with a combination of increasing obesity rates and food insecurity, this analysis shows that consumer preferences are densely clustered across both stores and items. This indicates that household tend to have fairly rigid preferences over the stores where they shop. Furthermore, they tend to consume a relatively small number of products from the universe of available products. The concentration in both stores and products is indicative of limited search behavior and substantial behavioral rigidity which may be associated with suboptimal outcomes in terms of nutrition and health.

6.6 Scalability

We now perform an experiment comparing the proposed inference methods (batch and online) to assess their scalability (Figure 5). We first use the Transactions data $(117054 \times 438 \times 67095)$ for this experiment. We would like to note that the state-of-the-art methods for count-valued tensor, such as the Poisson Tensor Factorization (PTF) method from the Tensor Toolbox [4], are simply infeasible to run on this data because of storage explosion issue (the method requires expensive flattening operations of the tensor). The other baseline LRANTD [23] we used in our experiments was also infeasible to run on this data. We set $R = 100$ for each method (about 60 factors were found to be significant, based on the inferred values of the λ_r's) and use a minibatch size of 100000 for all the online inference methods. For the conditional density filtering as well as stochastic variational inference, we set the learning rate as $t_0 = 0$ and $\kappa = 0.5$. Figure 5 shows that online inference methods (conditional density filtering and stochastic variational inference) converge much faster to a good solution than batch methods. This experiment shows that our online inference methods can be computationally viable alternatives if their batch counterparts are slow/infeasible to run on such data.

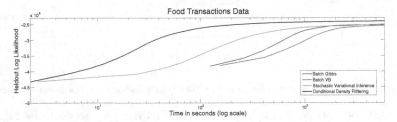

Fig. 5. Time vs heldout log likelihoods with various methods on transactions data

We then perform another experiment on the Scholars data, on which the PTF method of [4] was feasible to run and compare its per-iteration running time with our model (using both batch as well as online inference). Since PTF cannot handle missing data, for this experiment, each method was run with all the data. As Fig 6 shows, our methods have running times that are considerably smaller than that of PTF.

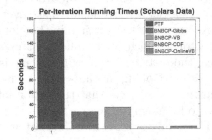

Fig. 6. Timing comparison of various methods on Scholars data

7 Conclusion

We have presented a fully Bayesian framework for analyzing massive tensors with count data, and have designed a suite of scalable inference algorithms for handling massive tensor data. In addition to giving interpretable results and inferring the rank from the data, the proposed model can infer the distribution over objects in each of the tensor modes which can be useful for understanding groups of similar objects, and also for doing other types of qualitative analyses on such data, as shown by our various experiments on real-world data sets. Simplicity of the inference procedure also makes the proposed model amenable for parallel and distributed implementations. e.g., using MapReduce or Hadoop. The model can be a useful tool for analyzing data from diverse applications and scalability of the model opens door to the application of scalable Bayesian methods for analyzing massive multiway count data.

Acknowledgments. The research reported here was supported in part by ARO, DARPA, DOE, NGA and ONR. Any opinions, findings, recommendations, or conclusions are those of the authors and do not necessarily reflect the views of the Economic Research Service, U.S. Department of Agriculture. The analysis, findings, and conclusions expressed in this paper also should not be attributed to either Nielsen or Information Resources, Inc. (IRI). This research was conducted in collaboration with USDA under a Third Party Agreement with IRI.

References

1. Bazerque, J.A., Mateos, G., Giannakis, G.B.: Inference of poisson count processes using low-rank tensor data. In: ICASSP (2013)
2. Beutel, A., Kumar, A., Papalexakis, E.E., Talukdar, P.P., Faloutsos, C., Xing, E.P.: Flexifact: Scalable flexible factorization of coupled tensors on hadoop. In: SDM (2014)
3. Cappé, O., Moulines, E.: On-line expectation-maximization algorithm for latent data models. Journal of the Royal Statistical Society: Series B (Statistical Methodology) **71**(3), 593–613 (2009)
4. Chi, E.C., Kolda, T.G.: On tensors, sparsity, and nonnegative factorizations. SIAM Journal on Matrix Analysis and Applications **33**(4), 1272–1299 (2012)
5. Cichocki, A., Zdunek, R., Phan, A.H., Amari, S.: Nonnegative matrix and tensor factorizations: applications to exploratory multi-way data analysis and blind source separation. John Wiley & Sons (2009)
6. Dunson, D.B., Herring, A.H.: Bayesian latent variable models for mixed discrete outcomes. Biostatistics **6**(1), 11–25 (2005)
7. Guhaniyogi, R., Qamar, S., Dunson, D.B.: Bayesian conditional density filtering. arXiv preprint arXiv:1401.3632 (2014)
8. Heinrich, G., Goesele, M.: Variational Bayes for Generic Topic Models. In: Mertsching, B., Hund, M., Aziz, Z. (eds.) KI 2009. LNCS, vol. 5803, pp. 161–168. Springer, Heidelberg (2009)
9. Hoffman, M.D., Blei, D.M., Wang, C., Paisley, J.: Stochastic variational inference. The Journal of Machine Learning Research **14**(1), 1303–1347 (2013)
10. Inah, J., Papalexakis, E.E., Kang, U., Faloutsos, C.: Haten2: Billion-scale tensor decompositions. In: ICDE (2015)
11. Johndrow, J.E., Battacharya, A., Dunson, D.B.: Tensor decompositions and sparse log-linear models. arXiv preprint arXiv:1404.0396 (2014)
12. Jordan, M.I., Ghahramani, Z., Jaakkola, T.S., Saul, L.K.: An introduction to variational methods for graphical models. Machine Learning **37**(2), 183–233 (1999)

13. Kang, U., Papalexakis, E., Harpale, A., Faloutsos, C.: Gigatensor: scaling tensor analysis up by 100 times-algorithms and discoveries. In: KDD (2012)
14. Kolda, T.G., Bader, B.W.: Tensor decompositions and applications. SIAM Review **51**(3), 455–500 (2009)
15. Kozubowski, T.J., Podgórski, K.: Distributional properties of the negative binomial Lévy process. Centre for Mathematical Sciences, Faculty of Engineering, Lund University, Mathematical Statistics (2008)
16. Leetaru, K., Schrodt, P.A.: Gdelt: Global data on events, location, and tone, 1979–2012. ISA Annual Convention **2**, 4 (2013)
17. Papalexakis, E., Faloutsos, C., Sidiropoulos, N.: Parcube: Sparse parallelizable candecomp-parafac tensor decompositions. ACM Transactions on Knowledge Discovery from Data (2015)
18. Rai, P., Wang, Y., Guo, S., Chen, G., Dunson, D., Carin, L.: Scalable bayesian low-rank decomposition of incomplete multiway tensors. In: ICML (2014)
19. Schein, A., Paisley, J., Blei, D.M., Wallach, H.: Inferring polyadic events with poisson tensor factorization. In: NIPS Workshop (2014)
20. Schmidt, M., Mohamed, S.: Probabilistic non-negative tensor factorisation using markov chain monte carlo. In: 17th European Signal Processing Conference (2009)
21. Shashua, A., Hazan, T.: Non-negative tensor factorization with applications to statistics and computer vision. In: ICML (2005)
22. Zhao, Q., Zhang, L., Cichocki, A.: Bayesian cp factorization of incomplete tensors with automatic rank determination
23. Zhou, G., Cichocki, A., Xie, S.: Fast nonnegative matrix/tensor factorization based on low-rank approximation. IEEE Transactions on Signal Processing 60(6), 2928–2940 (2012)
24. Zhou, M., Hannah, L.A., Dunson, D., Carin, L.: Beta-negative binomial process and poisson factor analysis. In: AISTATS (2012)

A Practical Approach to Reduce
the Learning Bias Under Covariate Shift

Van-Tinh Tran[✉] and Alex Aussem

LIRIS, UMR 5205, University of Lyon 1, 69622 Lyon, France
{van-tinh.tran,aaussem}@univ-lyon1.fr

Abstract. Covariate shift is a specific class of selection bias that arises when the marginal distributions of the input features X are different in the source and the target domains while the conditional distributions of the target Y given X are the same. A common technique to deal with this problem, called importance weighting, amounts to reweighting the training instances in order to make them resemble the test distribution. However this usually comes at the expense of a reduction of the effective sample size. In this paper, we show analytically that, while the unweighted model is globally more biased than the weighted one, it may locally be less biased on low importance instances. In view of this result, we then discuss a manner to optimally combine the weighted and the unweighted models in order to improve the predictive performance in the target domain. We conduct a series of experiments on synthetic and real-world data to demonstrate the efficiency of this approach.

1 Introduction

Selection bias, also termed dataset shift or domain adaptation in the literature [8], occurs when the training distribution $P(x, y)$ and the test distribution $P'(x, y)$ are different. It is pervasive in almost all empirical studies, including Machine Learning, Statistics, Social Sciences, Economics, Bioinformatics, Biostatistics, Epidemiology, Medicine, etc. Selection bias is prevalent in many real-world machine learning problems because the common assumption in machine learning is that the training and the test data are drawn independently and identically from the same distribution. The term "domain adaptation" is used when one builds a model from some fixed source domain, but wishes to deploy it across one or more different target domains. The term "selection bias" is slightly more specific as it assumes implicitly that there exists a binary variable S that controls the selection of examples in the training set, in other words we only have access to the examples that have $S = 1$. For instance, case-control studies in Epidemiology are particularly susceptible to selection bias, including bias resulting from inappropriate selection of controls in case-control studies, bias resulting from differential loss-to-follow-up, incidence-prevalence bias, volunteer bias, healthy-worker bias, and nonresponse bias [4].

It is well known that one may account for the difference between $P(x, y)$ and $P'(x, y)$ by re-weighting the training points using the so-called importance

© Springer International Publishing Switzerland 2015
A. Appice et al. (Eds.): ECML PKDD 2015, Part II, LNAI 9285, pp. 71–86, 2015.
DOI: 10.1007/978-3-319-23525-7_5

weight, denoted as $\beta(x, y) = P'(x, y)/P(x, y)$. Formally, let $\{h_{\theta*}\}_{\theta \in \Theta}$ be a model family from which we want to select an optimal model $h_{\theta*}(x) = h(x, \theta^*)$ for our learning task and let $l(y, h(x, \theta))$ be the loss function we would like to minimize, the optimal model we are searching for is the one that minimizes the expected loss over the test (or target) distribution:

$$\theta^* = \underset{\theta \in \Theta}{\operatorname{argmin}} \sum_{(x, y) \sim P} \beta(x, y)P(x, y)l(y, h(\theta, x))$$

So in practice, weighting the empirical loss of the training instances by $\beta(x, y)$ provides a well-justified solution to the selection bias problem.

In general, the estimation of $\beta(x, y)$ with two different distributions $P(x, y)$ and $P'(x, y)$ is unsolvable, as the two terms could be arbitrarily far apart. One simple assumption we can make about the connection between the distributions of the source and the target domains is that $P(x, y)$ and $P'(x, y)$ differ only in $P(x)$ and $P'(x)$ while their conditional distribution $P(y|x)$ remains unchanged. This specific selection bias is known as *covariate shift* in the literature [10]. In this case, the weighting term reduces to $\beta(x) = P'(x)/P(x)$ and effective adaptation is possible. At first glance, it may appear that covariate shift is not a problem because, for classification, we are only interested in $P(Y|X)$ which remains unchanged. In fact, Shimodaira [10] showed that there are circumstances under which the predictive performance is jeopardized by covariate shift. This happens typically when the parametric model family $\{P(Y|X, \theta)\}_{\theta \in \Theta}$ is misspecified, that is, there does not exist any $\theta \in \Theta$ such that $P(Y|X = x, \theta) = P(Y|X = x)$ for all $x \in \mathcal{X}$, so none of the models in the model family can exactly match the true relation between X and Y.

The intuitive reason why covariate shift under model misspecification is a problem is that the optimal (misspecified) model performs better in dense regions of the input space than in sparse regions, because the dense regions dominate the average classification error, which is what we want to minimize. If the dense regions of X are different in the training and test sets, the optimal model on the training set will no longer be optimal on the test set. In other words, the optimal model depends on $P(x)$, and if $P'(x) \neq P(x)$, then the optimal model for the target domain differs from that for the source domain. It was proven that, if the support of $P'(x)$ (the set of x for which $P'(x) > 0$) is contained in the support of $P(x)$, then the optimal model that maximizes this re-weighted log likelihood function asymptotically converges to the optimal model for the target domain [10] and a large body of research has been devoted to the estimation of $P'(x)/P(x)$ e.g. [13], [5], [11], [2], [1], [6], [7], [9]. However, reweighting methods do not necessarily improve the prediction accuracy as they also dependent on the extent to which the model is misspecified [12].

In this paper, we show analytically that, despite the fact that the unweighted model is globally more biased than the weighted one, the former may locally be less biased on low importance instances. In view of this result, we design a simple algorithm that combines the weighted and the unweighted models in order to improve the predictive performance in the target domain. More specifically, we prove that an optimal $B^\star$ always exists such that, in the region where $\beta(x) \le B^\star$, the biased model trained on the unweighted sample should be preferred to the unbiased one, and vice-versa. We propose a practical procedure to estimate this threshold value from training data.

The remainder of this paper is structured as follows. In Section 2, we define some key concepts used along the paper and state some results that will support our analysis. Then in Section 3, we conduct a theoretical analysis to prove that an optimal (but not necessarily unique) $B^\star$ always exists and discuss a manner to optimally combine the weighted and the unweighted models in order to improve the predictive performance in the target domain. In section 4, a series of experiments are carried out on toy problems and real-world data sets to assess the effectiveness of this approach.

2 Preliminaries

In this section, we define some key concepts used along the paper and state some results that will support our analysis. Consider the supervised learning problem where we observed n training samples, denoted by $((x_t; y_t) : t = 1, ..., n)$, where $x_t \in \mathcal{X} \subset \mathcal{R}^d$ are i.i.d training input points drawn from some probability distribution $p(x)$ and $y_t \in \mathcal{Y} \subset \mathcal{R}$ are the corresponding training output values drawn from a conditional probability distribution $p(y|x)$. We are interested in predicting the output value y at an input point x using a model $h_\theta(x) = h(x, \theta)$ parameterized by $\theta \in \Theta \subset \mathcal{R}^m$. Under covariate shift assumption, the test inputs follow a different probability distribution $p'(x)$ while the conditional probability distribution of test output $p(y|x)$ remains unchanged. The ratio $\beta(x) = \frac{p'(x)}{p(x)}$ is called the *importance* of x. Given a loss function $l(y, h(x, \theta)) : \mathcal{X} \times \mathcal{Y} \times \mathcal{Y} \to [0, \infty)$, we shall consider throughout this paper, the following loss functions:

- **EL-Tr**: Expectation of loss over training distribution $p(x, y) = p(x)p(y|x)$

$$Loss_0(h_\theta) = E_{x,y \sim p}[l(y, h(x, \theta))] = \int p(x) \int p(y|x)l(y, h(x, \theta))dydx$$

- **EL-Te**: Expectation of loss over test distribution $p'(x, y) = p'(x)p(y|x)$

$$Loss_1(h_\theta) = E_{x,y \sim p'}[l(y, h(x, \theta))] = \int p'(x) \int p(y|x)l(y, h(x, \theta))dydx$$

- **EL-IWTr**: Expectation of Importance-weighted loss over training distribution

$$Loss_\beta(h_\theta) = E_{x,y \sim p}[\beta(x)l(y, h(x, \theta))]$$

- **B-LEL-Te**: We then define Local Expectation of loss over test distribution given $\beta(x) \le B$ of any given hypothesis h_θ:

$$loss(h_\theta, \beta(x) \le B) = \int_{\beta(x) \le B} p'(x) \int_{\mathcal{Y}} p(y|x)l(y, h(x, \theta))dydx$$

We also define the optimal parameters of EL-Tr, EL-Te and EL-IWTr:

$$\begin{cases} \theta_0 & = \text{argmin}_\theta \, Loss_0(h_\theta) \\ \theta_1 & = \text{argmin}_\theta \, Loss_1(h_\theta) \\ \theta_\beta & = \text{argmin}_\theta \, Loss_\beta(h_\theta). \end{cases}$$

It may easily be shown that EL-IWTr is equal to EL-Te,

$$E_{x,y\sim p}[\beta(x)l(y, h(x, \theta))] = \int p(x) \int p(y|x)\frac{p'(x)}{p(x)}l(y, h(x, \theta))dydx$$

$$= \int p'(x) \int p(y|x)l(y, h(x, \theta))dydx$$

Therefore, minimizing EL-IWTr is equivalent to minimizing EL-Te. Nonetheless, while h_{θ_β} is globally less biased than h_{θ_0}, we will show next that it is more biased than h_{θ_0} on low-importance instances. Note that B-LEL-Te can be rewritten as:

$$loss(h_\theta, \beta(x) \leq B) = \int_{\beta(x)\leq B} \beta(x) \int_{\mathcal{Y}} p(x)p(y|x)l(y, h(x, \theta))dydx$$

Suppose $\beta(x)$ takes on continuous value in $[b_0, b_M]$ where $b_0 > 0$, we may rewrite B-LEL-Te as following:

$$loss(h_\theta, \beta(x) \leq B) = \int_{b_0}^{B} b \int_{\beta(x)=b} \int_{\mathcal{Y}} p(x)p(y|x)l(y, h(x, \theta))dydxdb$$

Let $\mathcal{L}(h_\theta, \beta(x) = b) = \int_{\beta(x)=b} \int_{\mathcal{Y}} p(x)p(y|x)l(y, h(x, \theta))dydx$, then:

$$loss(h_\theta, \beta(x) \leq B) = \int_{b_0}^{B} b\mathcal{L}(h_\theta, \beta(x) = b)db$$

Similarly, if $\beta(x)$ takes on discrete values in $\{b_i\}_{i=0}^{M}$ such that $b_0 < b_1 < ... < b_M$, we rewrite B-LEL-IWTr as:

$$loss(h_\theta, \beta(x) \leq B) = \sum_{i=0}^{k(B)} b_i\mathcal{L}(h_\theta, \beta(x) = b_i)$$

where $k(B)$ is the largest integer such that $b_{k(B)} \leq B$. From the definitions above, we may write

$$\begin{cases} Loss_1(h_\theta) & = loss(h_\theta, \beta(x) \leq b_M), \\ Loss_0(h_\theta) & = \int_{b_0}^{\infty} \mathcal{L}(h_\theta, \beta(x) = b)db, \quad \text{for continuous } \beta(x), \\ Loss_0(h_\theta) & = \sum_{i=0}^{M} \mathcal{L}(h_\theta, \beta(x) = b_i), \quad \text{for discrete } \beta(x). \end{cases}$$

As aforementioned, a model $h(x, \theta)$ is said to be *correctly specified* if there exist parameter $\theta^* \in \Theta$ such that $h(x, \theta^*) = f(x)$, otherwise it is said to be *misspecified*. It is obvious that if a model is correctly specified, the optimal parameter θ of EL-Tr, EL-Te, and any B-LEL-Te coincide. Therefore, the model that minimizes EL-Tr will perform well on the test data globally (i.e., minimizing EL-Te) as well as locally (i.e., B-LEL-Te) in any region of the form $\beta(x) < B$. Yet, in practice, almost all models are more or less misspecified. So minimizing EL-Tr θ_0 is not necessarily equivalent minimizing EL-Te. Since EL-Te is equal to EL-IWTr, the parameter minimizing of EL-IWTr θ_β, which can be estimated from data, will also minimize EL-Te as shown in [10], [13]. However, due to the model misspecification, θ_β does not necessarily minimize B-LEL-Te. In fact, we will prove that there exist some $B^*(h_{\theta_\beta}) \in [b_0, b_M]$ such that B-LEL-Te of θ_β exceeds that of θ_0 by proving a stronger conclusion that for all model h_θ, with $\theta \in \Theta$, there exist some $B^*(h_\theta) \in [b_0, b_M]$ such that B*-LEL-Te of h_θ exceeds that of h_{θ_0}, in other words any h_θ is **locally more biased** than h_{θ_0} when predicting instance with $\beta(x) \leq B^*$.

In addition, the estimation of θ_β may subject to high variance since it involves instance weighting, which is known to reduce the effective samples size [2], [3]. Hence the idea to use h_{θ_0} of instead of h_{θ_β} to predict the test instances with $\beta(x) \leq B^*$.

3 Problem Analysis

In this section, we conduct theoretical analyses for a simple and then a more general selection bias mechanism. Those analyses will be used to derive a practical procedure aiming at reducing the bias due to covariate shift with misspecified regression or classification learning models.

We first show how EL-Tr is related to B-LEL-Te,

Lemma 1. *Suppose $\beta(x)$ takes on continuous value in $[b_0, b_M]$ with $b_M > b_0 > 0$, then:*

$$Loss_0(h_\theta) = \frac{1}{b_M} loss(h_\theta, \beta(x) \leq b_M) + \int_{b_0}^{b_M} \frac{1}{B^2} loss(h_\theta, \beta(x) \leq B) dB$$

Proof. For continuous $\beta(x)$:

$$\int_{b_0}^{b_M} \frac{1}{B^2} loss(h_\theta, \beta(x) \leq B) dB = \int_{b_0}^{b_M} loss(h_\theta, \beta(x) \leq B) d\left(\frac{-1}{B}\right)$$

$$= loss(h_\theta, \beta(x) \leq B) \left(\frac{-1}{B}\right) |_{b_0}^{b_M} - \int_{b_0}^{b_M} \frac{-1}{B} d(loss(h_\theta, \beta(x) \leq B))$$

By definition, $loss(h_\theta, \beta(x) \leq B) = \int_{b_0}^{B} b\mathcal{L}(b, h_\theta)db$, so $loss(h_\theta, \beta(x) \leq b_0) = 0$ and $d(loss(h_\theta, \beta(x) \leq B)) = B\mathcal{L}(h_\theta, \beta(x) = B)dB$. Thus:

$$\int_{b_0}^{b_M} \frac{1}{B^2} loss(h_\theta, \beta(x) \leq B)dB = \frac{-1}{b_M} loss(h_\theta, \beta(x) \leq b_M)$$

$$+ \int_{b_0}^{b_M} \frac{1}{B}(B\mathcal{L}(h_\theta, \beta(x) = B)dB)$$

By definition, we have $Loss_0(h_\theta) = \int_{b_0}^{b_M} \mathcal{L}(h_\theta, \beta(x) = B)dB$, so:

$$\int_{b_0}^{b_M} \frac{1}{B^2} loss(h_\theta, \beta(x) \leq B)dB = -\frac{1}{b_M} loss(h_\theta, b_M) + Loss_0(h_\theta)$$

which concludes the proof   □

A similar results holds in the discrete case.

Corollary 1. *Suppose $\beta(x)$ takes on discrete values $\{b_i\}_{i=0}^{M}$ such that $b_0 < b_1 < ... < b_M$, then:*

$$Loss_0(h_\theta) = \frac{1}{b_M} loss(h_\theta, \beta(x) \leq b_M) + \sum_{k=0}^{M-1} \left(\frac{1}{b_k} - \frac{1}{b_{k+1}} \right) loss(h_\theta, \beta(x) \leq b_k)$$

Proof.

$$\sum_{k=0}^{M-1} \left(\frac{1}{b_k} - \frac{1}{b_k + 1} \right) loss(h_\theta, \beta(x) \leq b_k) + \frac{1}{b_M} loss(h_\theta, \beta(x) \leq b_M)$$

$$= \left(\frac{1}{b_0} - \frac{1}{b_1} \right) [b_0\mathcal{L}(h_\theta, \beta(x) = b_0)]$$

$$+ \left(\frac{1}{b_1} - \frac{1}{b_2} \right) [b_0\mathcal{L}(h_\theta, \beta(x) = b_0) + b_1\mathcal{L}(h_\theta, \beta(x) = b_1)]$$

$$+ ...$$

$$+ \left(\frac{1}{b_{M-1}} - \frac{1}{b_M} \right) [b_0\mathcal{L}(h_\theta, \beta(x) = b_0) + ... + b_{M-1}\mathcal{L}(h_\theta, \beta(x) = b_{M-1})]$$

$$+ \frac{1}{b_M} [b_0\mathcal{L}(h_\theta, \beta(x) = b_0) + b_1\mathcal{L}(h_\theta, \beta(x) = b_1) + .. + b_M\mathcal{L}(h_\theta, \beta(x) = b_M)]$$

$$= b_0 \mathcal{L}(h_\theta, \beta(x) = b_0) \left[\left(\frac{1}{b_0} - \frac{1}{b_1} \right) + \left(\frac{1}{b_1} - \frac{1}{b_2} \right) + \ldots + \left(\frac{1}{b_{M-1}} - \frac{1}{b_M} \right) + \frac{1}{b_M} \right]$$

$$+ \ldots$$

$$+ b_{M-1} \mathcal{L}(h_\theta, \beta(x) = b_{M-1}) \left[\left(\frac{1}{b_{M-1}} - \frac{1}{b_M} \right) + \frac{1}{b_M} \right]$$

$$+ b_M \mathcal{L}(h_\theta, \beta(x) = b_M) \left[\frac{1}{b_M} \right]$$

$$= \sum_{i=0}^{M} \mathcal{L}(h_\theta, \beta(x) = b_i) = Loss_0(h_\theta) \qquad \square$$

In view of Corollary 1, we may now state the following theorem,

Theorem 1. *Suppose there exists two real values, b_0 and b_1, such that $b_0 < 1 < b_1$ and a subset $X_0 \subset \mathcal{X}$ such that*

$$\beta(x) = \begin{cases} b_0 & \text{if } x \in X_0 \\ b_1 & \text{if } x \notin X_0, \end{cases}$$

then there exists a threshold B^* such that:

$$loss(h_{\theta_1}, \beta(x) \leq B^*) \geq loss_1(h_{\theta_0}, \beta(x) \leq B^*).$$

In fact, B^* can take any value in $[b_0, b_1)$.

Proof. By definition, $Loss_0(h_{\theta_0}) \leq Loss_0(h_{\theta_1})$, using Lemma 1, we may write:

$$Loss_0(h_{\theta_0}) = \frac{1}{b_1} loss(h_{\theta_0}, \beta(x) \leq b_1) + \left(\frac{1}{b_0} - \frac{1}{b_1} \right) loss(h_{\theta_0}, \beta(x) \leq b_0)$$

$$= \frac{1}{b_1} Loss_1(h_{\theta_0}) + \left(\frac{1}{b_0} - \frac{1}{b_1} \right) loss(h_{\theta_0}, \beta(x) \leq b_0)$$

Similarly,

$$Loss_0(h_{\theta_1}) = \frac{1}{b_1} Loss_1(h_{\theta_1}) + \left(\frac{1}{b_0} - \frac{1}{b_1} \right) loss(h_{\theta_1}, \beta(x) \leq b_0)$$

Thus,

$$\frac{1}{b_1} Loss_1(h_{\theta_0}) + \left(\frac{1}{b_0} - \frac{1}{b_1} \right) loss(h_{\theta_0}, \beta(x) \leq b_0) \leq \frac{1}{b_1} Loss_1(h_{\theta_1})$$

$$+ \left(\frac{1}{b_0} - \frac{1}{b_1} \right) loss(h_{\theta_1}, \beta(x) \leq b_0)$$

Finally,

$$loss(h_{\theta_1}, \beta(x) \leq b_0) - loss(h_{\theta_0}, \beta(x) \leq b_0) = \frac{b_0}{b_1 - b_0} [Loss_1(h_{\theta_0}) - Loss_1(h_{\theta_1})]$$

It is easily shown that the right hand side of inequality above is non-negative due to the definition of θ_1. It follows that

$$loss(h_{\theta_1}, \beta(x) \leq b_0) - loss(h_{\theta_0}, \beta(x) \leq b_0) \leq 0$$

which, given the assumption about $\beta(x)$, is equivalent to,

$$loss(h_{\theta_1}, \beta(x) = b_0) - loss(h_{\theta_0}, \beta(x) = b_0) \leq 0$$

Thus the Theorem is true when $B^* = b_0$. It is also true for any other $B^* \in [b_0, b_1)$ as a consequence.   □

When the assumptions of Theorem 1 holds, we say that the covariate shift scheme follows a simple step distribution. The equality in Theorem 1 only occurs when θ_0 minimizes EL-Te and θ_1 minimizes EL-Tr. Such condition indicates that covariate shift does not have an effect on searching for optimal θ, which is a rare case as shown by other studies. Theorem 1 shows that for *simple step distribution* where inclusion in the training sample is either proportional to b_0^{-1} (over-sampled instances), or to b_1^{-1} (under-sampled instances), h_{θ_0} exhibits a lower bias compared to h_{θ_1} on the low importance test instances. This type of selection bias mechanism is actually quite common. For instance, prospective cohort studies in epidemiology are by design prone to covariate shift because selection criteria are associated with the exposure to potential risk factors.

Theorem 2. *For all $\theta \in \Theta$, there exists a threshold $B^*(h_\theta)$ such that*

$$loss(h_\theta, \beta(x) \leq B^*(h_\theta)) \geq loss(h_{\theta_0}, \beta(x) \leq B^*(h_\theta)) \tag{1}$$

$B^*(h_\theta)$ *could take any value in the set below:*

$$B^*(h_\theta) = \underset{B}{\mathrm{argmax}}(loss(h_\theta, \beta(x) \leq B) - loss(h_{\theta_0}, \beta(x) \leq B))$$

The equality occurs whenever θ_1 is also a minimum for EL-Tr.

Proof. We prove by contradiction that Theorem 2 holds. Assume that inequality 1 does not hold for $B^*(h_\theta)$ defined above:

$$loss(h_\theta, \beta(x) \leq B^*(h_\theta)) - loss(h_{\theta_0}, \beta(x) \leq B^*(h_\theta)) < 0 \tag{2}$$

By definition of $B^*(h_\theta)$, we may show that, for all $B \in [b_0, b_M]$,

$$loss(h_\theta, \beta(x) \leq B) - loss(h_{\theta_0}, \beta(x) \leq B) < 0$$

Thus, for all $B \in [b_0, b_M]$

$$loss(h_{\theta_0}, \beta(x) \leq B) > loss(h_\theta, \beta(x) \leq B)$$

Now, using Lemma 1 for continuous $\beta(x)$, we have:

$$Loss_0(h_{\theta_0}) = \frac{1}{b_M} loss(h_{\theta_0}, \beta(x) \leq b_M) + \int_{b_0}^{b_M} \frac{1}{B^2} loss(h_{\theta_0}, \beta(x) \leq B) dB$$

$$> \frac{1}{b_M} loss(h_\theta, \beta(x) \leq b_M) + \int_{b_0}^{b_M} \frac{1}{B^2} loss(h_\theta, \beta(x) \leq B) dB = Loss_0(h_\theta)$$

Hence, $Loss_0(h_{\theta_0}) > Loss_0(h_\theta)$, contradicts the fact that $\theta_0 = \mathrm{argmin}_\theta$ $Loss_0(h_\theta)$ is the optimal hypothesis under the unweighting scheme and $\theta \neq$ $\mathrm{argmin}_\theta Loss_0(h_\theta)$.

If the two terms in inequality 1 are equal, then we can prove similarly that $Loss_0(h_{\theta_0}) = Loss_0(h_\theta)$, which implies that θ_1 is also a minimal solution of EL-Tr. The demonstration for discrete $\beta(x)$ values follows similarly.   □

Theorem 2 states that any model h_θ with $\theta \in \Theta$ is outperformed by h_{θ_0} learned from the unweighted training samples in terms of bias when predicting examples with $\beta(x) \leq B^*(h_\theta)$. This is also applied to model h_{θ_β} which minimizes EL-IWTr. In addition, the estimation of θ_β may exhibit a higher variance due to the effective sample size reduction as discussed in [2,3]. These results altogether suggest that h_{θ_0} should be preferred to h_{θ_β} for predicting the instance's outputs in the region $\beta(x) \leq B^*(h_\theta)$, termed **low-importance region**. Therefore, for any learning task with covariate shift, we shall train two distinct models, one with and the other without the importance weighting scheme. Then, we shall use the latter to predict instances satisfying $\beta(x) \leq B^*(h_\theta)$ and use the former to predict the remaining instances. The optimal value for $B^*(h_\theta)$ may be estimated from the training data. The set of all possible empirical threshold $\hat{B}^*(h_{\theta_\beta})$ can be obtained empirically by solving the following problem :

$$\hat{B}^*(h_\theta) = \mathrm{argmax}_B \frac{1}{n} \sum_{\substack{i \in \{1,..,n\} \\ \beta(x_i) \leq B}} \beta(x_i)[l(y_i, h(x_i, \theta_\beta)) - l(y_i, h(y_i, \theta_0))] \qquad (3)$$

As n grows to infinity, it follows from the law of large numbers that,

$$\hat{B}^*(h_\theta) \to B^*(h_\theta)$$

Therefore, $B^*(h_{\theta_\beta})$ could be estimated empirically either from training data or by cross validation. In this study, we use a 5-fold importance weighted cross validation to estimate $B^*(h_{\theta_\beta})$ as suggested in [11]. It should be emphasized that $B^*(h_{\theta_\beta})$ is not necessarily unique. For instance, any value between b_0 and b_1 in Theorem 1 is admissible as mentioned earlier.

4   Experiments

In this section, we assess the ability of our "hybrid approach" to reduce the learning bias under covariate shift based on Theorem 2. We first discuss the strategies employed to estimate the importance weights: one is based explicitly on the true bias mechanism, the other is based on linear density-ratio model. We emphasize that the latter does not require any prior knowledge of the true sampling probabilities to estimate the $\beta(x)$ values, and uses the test input features instead. In fact, the estimation of distribution is a hard problem, thus it is more appealing to directly estimate $\beta(x)$. Indeed, a large body of work has been devoted to this line of research e.g. [13], [5], [11], [9], [2], [1], [6]. From the many references, we choose the Unconstrained Least-Square Importance Fitting (uLSIF) estimator for $\beta(x)$ that was proved to be successful with covariate shift. We then study a toy regression problem to show if covariate shift corrections based on our method reduces prediction error on the test set when the learning model is misspecified. We then test our approach on real world benchmark data sets, from which the training examples are selected according to various biased sampling schemes as suggested in [6].

4.1   Importance Ratio Estimation

As aforementioned, we use two weighting schemes in ours experiments, one is derived from the true selection bias mechanism an one is Unconstrained Least-Square Importance Fitting (uLSIF), a method based on linear density-ratio models [6]. Formally, it assumes that the density ratio $\beta(x)$ can be approximated by a linear model $\hat{\beta}(x) = \sum_{i=1}^{M} \alpha_i h_i(x)$ where the basis functions h_i, $i = 1, ..., M$ are chosen so that $h_i(x) \geq 0$ for all input value x. The coefficients $\alpha_1, ..., \alpha_M$ are parameters of the linear model and are estimated from data by minimizing the empirical square error between weighted biased distribution (from training data) and the bias-free distribution of x:

$$\min_{\alpha} \frac{1}{2n} \sum_{i=1}^{n} (\hat{\beta}(x_i))^2 - \frac{1}{n'} \sum_{i=1}^{n'} \hat{\beta}(x'_i) + \lambda.Reg(\alpha)$$

where $\{x_i\}_{i=1}^{n}$ and $\{x'_i\}_{i=1}^{n'}$, are the training and test inputs, $Reg(\alpha)$ is the regularization term, introduced to avoid overfitting. A heuristic choice of $h_i(x)$ proposed in [6] is a Gaussian kernel centered at the test points $\{x_i\}_{i=1}^{n'}$ when the number of test points is small (less than 100) or at *template* points $\{x'_i\}_{i=1}^{100}$, which is a random subset of test set when the number of test points is large for computation advantage. The kernel width and the regularization term $Reg(\alpha)$ are optimized by cross-validation with grid search.

4.2   Toy Regression Problem

Consider the following training data generating process: $x \sim N(\mu_0, \sigma_0)$ and $y = f(x) + \epsilon$, where $\mu_0 = 0$, $\sigma_0 = 0.5$, $f(x) = -x + x^3$, and $\epsilon \sim N(0, 0.3)$. In the test data, we have the same relationship between x and y but the distribution of the covariate x is shifted to $x \sim N(\mu_1, \sigma_1)$, where $\mu_1 = 0$, $\sigma_1 = 1$. The training and test distributions, along with their ratio are depicted in Fig. 1a and 1b. The minimization of EL-Tr is obtained using the unweighted Least Square Regression (uLSR) method for the normal regression while minimization of EL-Te is performed by the weighted Least Square Regression (wLSR). As shown in [10], wLSR is unbiased thus it should perform better than uLSR, which is biased, on test data. However, as can be seen in Fig. 1c, uLSR (red dashed line) seems to better approximate the $y = f(x)$ curve (in blue) than wLSR (black dashed line) on instances in the interval $(-1, 1)$. As may be seen in Fig.1d, the hybrid model that optimally combines wLSR and uLSR, based

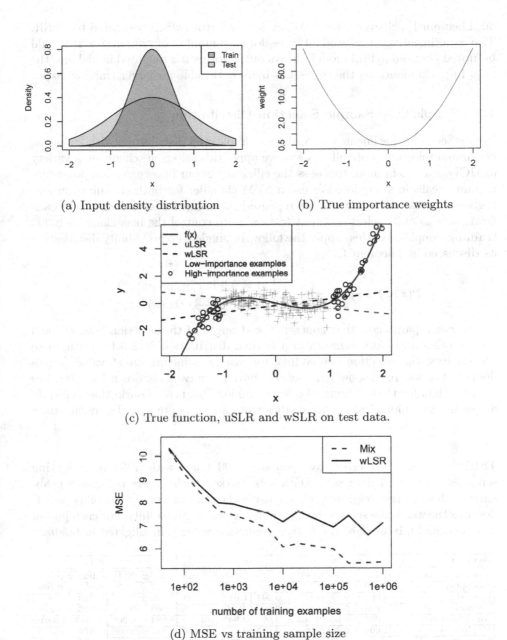

(a) Input density distribution

(b) True importance weights

(c) True function, uSLR and wSLR on test data.

(d) MSE vs training sample size

Fig. 1. An illustrative example of fitting a function f(x) using a linear model with/without the weight importance scheme (wLSR/uLSR) and a combination of both (termed "Mix").

on Theorem 1, achieves a lower Mean Square Error (MSE) compared to wLSR. The experiment was repeated 30 times for each number of sample size. It should be noted that the hybrid model always outperforms the weighted model and the gain in performance on the test set is more noticeable for larger training sizes.

4.3   Simple Step Sample Selection Distribution

In this second experiment, we consider a simple step distribution with known or estimated selection probabilities and we apply this selection scheme on a variety of UCI data sets in order to assess the efficiency of our bias correction procedure in more realistic scenarios. We use a SVM classifier for both classification and regression tasks. Experiments are repeated 100 times for each data set. In each trial, we randomly select an input feature x^c to control the bias along with 300 training samples. We then apply the following single step probability distribution as discussed in Theorem 1,

$$P(s = 1|x = x_i^c) = p_s = \begin{cases} p1 = 0.9 \text{ if } x_i^c \leq mean(x^c) \\ p2 = \frac{0.9}{1+exp(r)} \text{otherwise} \end{cases}$$

where r is a parameter that controls the strength of the selection bias. In each trial r takes a random value from a normal distribution $N(2, 0.1)$. With these parameters, the selection probability for instances having an x^c value (e.g. a degree of exposure to some risk factor) above the mean is between 7 to 10 times smaller than for those having of a lower value. This is a scenario that typically arises in epidemiological cohort studies when subjects are includes in the study

Table 1. Mean test error averaged over 200 trials with different weighting schemes on 15 UCI data sets. Data sets marked with '*' are regression problems. P denotes the weighting scheme using the true selection probability and $\hat{P}$ denotes the weighting scheme using a noisy selection probability. For each pair of weighted and mix models, the better prediction value is highlighted in boldface

Data set	No weighting	P	P mix	$\hat{P}$	$\hat{P}$ mix
India diabetes	1.000 ± 0.020	0.966 ± 0.019	**0.960 ± 0.018**	0.968 ± 0.019	**0.962 ± 0.018**
Ionosphere	1.000 ± 0.128	0.915 ± 0.105	**0.902 ± 0.107**	0.911 ± 0.104	**0.897 ± 0.106**
BreastCancer	1.000 ± 0.039	1.020 ± 0.044	**1.013 ± 0.044**	1.020 ± 0.044	**1.013 ± 0.043**
GermanCredit	1.000 ± 0.008	1.000 ± 0.007	**0.996 ± 0.008**	1.000 ± 0.008	**0.996 ± 0.008**
Australian credit	1.000 ± 0.006	0.963 ± 0.008	**0.947 ± 0.010**	0.964 ± 0.008	**0.947 ± 0.010**
Mushroom	1.000 ± 0.068	0.090 ± 0.057	**0.872 ± 0.060**	0.888 ± 0.058	**0.874 ± 0.056**
Congressional Voting	1.000 ± 0.033	1.026 ± 0.039	**0.993 ± 0.038**	1.030 ± 0.038	**1.000 ± 0.037**
Banknote	1.000 ± 0.040	**0.970 ± 0.043**	0.978 ± 0.038	**0.969 ± 0.042**	0.975 ± 0.039
Airfoil self noise*	1.000 ± 0.023	0.997 ± 0.015	**0.961 ± 0.012**	0.993 ± 0.015	**0.958 ± 0.012**
Abanlone*	1.000 ± 0.032	0.984 ± 0.020	**0.960 ± 0.020**	0.985 ± 0.021	**0.961 ± 0.020**
Auto MGP*	1.000 ± 0.084	0.939 ± 0.066	**0.933 ± 0.067**	0.939 ± 0.066	**0.930 ± 0.067**
Boston Housing*	1.000 ± 0.057	1.037 ± 0.053	**0.994 ± 0.050**	1.037 ± 0.053	**0.994 ± 0.050**
Space GA*	1.000 ± 0.009	1.021 ± 0.007	**0.962 ± 0.008**	1.018 ± 0.008	**0.961 ± 0.008**
Cadata*	1.000 ± 0.013	1.038 ± 0.022	**1.029 ± 0.017**	1.037 ± 0.022	**1.029 ± 0.017**

according to some exposure factor. Consider the two following weighting schemes. The first one: $\beta = p'(x)/p(x) = p(s = 1)/p(s = 1|x) \sim 1/p_s$ assumes that the bias mechanism is known exactly.

$$\beta(x) \sim p_s^{-1} \sim \begin{cases} b1 = 1 \text{ if } x_i^c \leq mean(x^c) \\ b2 = 1 + exp(r) \text{otherwise} \end{cases}$$

In practice, however, the selection probability is rarely known exactly. So let us assume that the estimation of β is subject to some error and let us consider the following approximate weighting scheme:

$$\hat{\beta}(x) \sim p_s^{-1} \sim \begin{cases} b1 = 1 \text{ if } x_i^c \leq mean(x^c) \\ b2 = 1 + exp(\hat{r}) \text{ if otherwise} \end{cases}$$

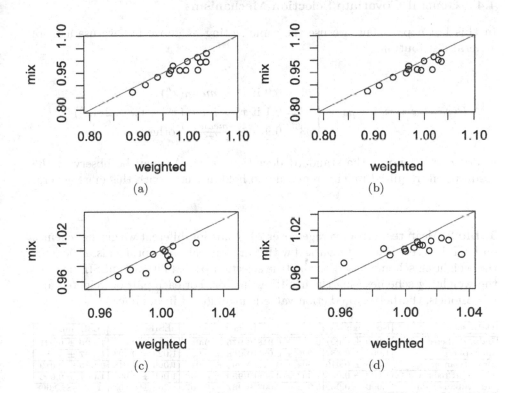

Fig. 2. MSE gain of the mix model vs. MSE gain of the weighted model. Points below the diagonal line indicate that the mix model outperforms the weighted model. Figures (a) and (b): simple step distribution covariate shift used in the first experiment with the weighted model based on (a) the true selection probability and (b) based on the estimated selection probability. Figures (c) and (d): covariate shift in used in the second experiment when the weighted model based was based on: (c) true selection probability; (d) on uLSIF.

where $\hat{r} = r + \mathcal{N}(0, 0.1)$ is our noisy estimate of r. For each weighting scheme, we fit a true weighted model (denoted as P in Table 1) and an approximated weighted model (denoted as $\hat{P}$ in Table 1). As $p_1 < 1$ and $p_2 > 1$, our weighting mechanism satisfies the assumptions of Theorem 1, so we set $B^* = 1$. We report the mean square errors (MSE) in Table 1. All values are normalized by the MSE of the unweighted model (our gold standard). As may be seen from the plots in Fig. 2a and 2b, the combined models outperform the weighted ones. That is, when using either exact probability ratio, the results obtained with P_{mix} are better than that of P. The same observation can be made when the estimated probability ratios are used instead (i.e., $\hat{P}_{mix}$ versus $\hat{P}$) and except on the Banknote data set. The gain is significant at the significance level 5% using the Wilcoxon signed rank test.

4.4   General Covariate Selection Mechanisms

In this last experiment, we use the same setting as above but we use a more general distribution:

$$P(s = 1 | x = x_i^c) = ps = \begin{cases} p1 = 0.9 \text{ if } x_i^c \leq mean(x^c) \\ p2 = 0.1 \text{ if } x_i^c > mean(x^c) + 0.8 \times 2\sigma(x^c) \\ p3 = 0.9 - \frac{x_i^c - mean(x^c)}{2\sigma(x^c)} \text{ otherwise.} \end{cases}$$

where $\sigma(x^c)$ denotes the standard deviation of x^c. As may be observed, the assumptions required in Theorem 1 do no hold anymore with this more general

Table 2. Mean test error averaged over 200 trials for different weighting schemes on UCI data set. Data sets marked with * are for regression problems. P denotes the weighting scheme based on the true selection probability and uLSIF denotes the weighting scheme using the uLSIF estimator. For each pair of weighted and mix models, the better prediction value is highlighted in boldface.

Data set	No weighting	P	P mix	uLSIF	uLSIF mix
India diabetes	1.000 ± 0.021	0.980 ± 0.018	$\mathbf{0.975 \pm 0.018}$	1.016 ± 0.021	$\mathbf{1.006 \pm 0.021}$
Ionosphere	1.000 ± 0.087	1.006 ± 0.087	$\mathbf{0.988 \pm 0.085}$	1.028 ± 0.093	$\mathbf{1.007 \pm 0.087}$
BreastCancer	1.000 ± 0.019	1.004 ± 0.018	$\mathbf{0.993 \pm 0.019}$	$1.000 + 0.018$	$\mathbf{0.993 \pm 0.019}$
GermanCredit	1.000 ± 0.008	1.003 ± 0.008	$\mathbf{0.999 \pm 0.008}$	1.009 ± 0.008	$\mathbf{1.001 \pm 0.008}$
Australian credit	1.000 ± 0.009	0.972 ± 0.007	$\mathbf{0.967 \pm 0.007}$	1.007 ± 0.008	$\mathbf{1.005 \pm 0.008}$
Mushroom	1.000 ± 0.558	1.011 ± 0.054	$\mathbf{0.963 \pm 0.051}$	0.991 ± 0.054	$\mathbf{0.989 \pm 0.054}$
Congressional Voting	1.000 ± 0.037	1.023 ± 0.036	$\mathbf{1.010 \pm 0.037}$	$\mathbf{0.987 \pm 0.036}$	0.997 ± 0.036
Banknote	1.000 ± 0.060	1.083 ± 0.057	$\mathbf{0.962 \pm 0.062}$	$\mathbf{0.962 \pm 0.061}$	0.979 ± 0.058
Airfoil self noise*	1.000 ± 0.007	0.995 ± 0.007	$\mathbf{0.995 \pm 0.007}$	1.011 ± 0.008	$\mathbf{1.001 \pm 0.008}$
Abanlone*	1.000 ± 0.007	1.001 ± 0.008	1.001 ± 0.007	1.005 ± 0.007	$\mathbf{0.998 \pm 0.006}$
Auto MGP*	1.000 ± 0.026	0.990 ± 0.025	$\mathbf{0.970 \pm 0.025}$	1.015 ± 0.027	$\mathbf{0.994 \pm 0.026}$
Boston Housing*	1.000 ± 0.043	0.984 ± 0.031	$\mathbf{0.940 \pm 0.032}$	1.036 ± 0.040	$\mathbf{0.989 \pm 0.042}$
Space GA*	1.000 ± 0.006	1.005 ± 0.005	$\mathbf{0.980 \pm 0.006}$	1.000 ± 0.005	$\mathbf{0.996 \pm 0.005}$
Cadata*	1.000 ± 0.012	1.008 ± 0.013	$\mathbf{1.006 \pm 0.012}$	1.023 ± 0.013	$\mathbf{1.010 \pm 0.012}$

sample selection distribution. According to Eq. 3, we need to estimate $\hat{B}^*(h_\theta)$ empirically from data. We consider again two importance weighting schemes: one is based on the true underlying probability and is referred to as P, while the other is based on the uLSIF estimator. As may be observed from Table 2 and Figures 2c and 2d that performances of the hybrid models are significantly improved with respect to the weighted models, except with the Congressional Voting and Banknote data sets.

5   Conclusions

In this paper, we showed that the standard importance weighting approach used to reduce the bias due to covariate shift can easily be improved when misspecified training models are used. Considering a simple class of selection bias mechanisms, we proved analytically that the unweighted model exhibits a lower prediction bias compared to the globally unbiased model in the low importance input subspace. Even for more general covariate shift scenarios, we proved that there always exist a threshold for the importance weight below which the test instances should be predicted by the globally biased model. In view of this result, we proposed a practical procedure to estimate this threshold and we discussed a simple procedure to combine the weighted and unweighted prediction models. The method was shown to be effective in reducing the bias on several UCI data sets.

Acknowledgments. This work was partially supported by a grant from the European ENIAC Joint Undertaking (INTEGRATE project).

References

1. Bickel, S., Brückner, M., Scheffer, T.: Discriminative learning under covariate shift. The Journal of Machine Learning Research **10**, 2137–2155 (2009)
2. Cortes, C., Mansour, Y., Mohri, M.: Learning bounds for importance weighting. In: Lafferty, J., Williams, C., Shawe-Taylor, J., Zemel, R., Culotta, A. (eds.) Advances in Neural Information Processing Systems, vol. 23, pp. 442–450. Curran Associates, Inc. (2010)
3. Gretton, A., Smola, A., Huang, J., Schmittfull, M., Borgwardt, K., Schölkopf, B.: Covariate shift by kernel mean matching (2009)
4. Hernán, M.A., Hernández-Díaz, S., Robins, J.M.: A structural approach to selection bias. Epidemiology **15**(5), 615–625 (2004)
5. Huang, J., Smola, A.J., Gretton, A., Borgwardt, K.M., Schölkopf, B.: Correcting sample selection bias by unlabeled data. In: Schölkopf, B., Platt, J., Hoffman, T. (eds.) NIPS, pp. 601–608. MIT Press (2006)
6. Kanamori, T., Hido, S., Sugiyama, M.: A least-squares approach to direct importance estimation. J. Mach. Learn. Res. **10**, 1391–1445 (2009)
7. Kanamori, T., Suzuki, T., Sugiyama, M.: Statistical analysis of kernel-based least-squares density-ratio estimation. Machine Learning **86**(3), 335–367 (2012)
8. Moreno-Torres, J.G., Raeder, T., Alaiz-Rodríguez, R., Chawla, N.V., Herrera, F.: A unifying view on dataset shift in classification. Pattern Recognition **45**(1), 521–530 (2012)

9. Nguyen, X., Wainwright, M.J., Jordan, M.I.: Estimating divergence functionals and the likelihood ratio by convex risk minimization. IEEE Transactions on Information Theory **56**(11), 5847–5861 (2010)
10. Shimodaira, H.: Improving predictive inference under covariate shift by weighting the log-likelihood function. Journal of Statistical Planning and Inference **90**(2), 227–244 (2000)
11. Sugiyama, M., Nakajima, S., Kashima, H., von Bünau, P., Kawanabe, M.: Direct importance estimation with model selection and its application to covariate shift adaptation. In: Platt, J., Koller, D., Singer, Y., Roweis, S. (eds.) NIPS (2007)
12. Wen, J., Yu, C.-N., Greiner, R.: Robust learning under uncertain test distributions: relating covariate shift to model misspecification. In: Jebara, T., Xing, E.P. (eds.) Proceedings of the 31st International Conference on Machine Learning (ICML 2014), pp. 631–639 (2014)
13. Zadrozny, B.: Learning and evaluating classifiers under sample selection bias. In: Greiner, R., Schuurmans, D. (eds.) Proceedings of the 21st International Conference on Machine Learning (ICML 2004) (2004)

Hyperparameter Optimization with Factorized Multilayer Perceptrons

Nicolas Schilling[✉], Martin Wistuba, Lucas Drumond,
and Lars Schmidt-Thieme

Information Systems and Machine Learning Lab,
University of Hildesheim, 31141 Hildesheim, Germany
{schilling,wistuba,ldrumond,schmidt-thieme}@ismll.uni-hildesheim.de

Abstract. In machine learning, hyperparameter optimization is a challenging task that is usually approached by experienced practitioners or in a computationally expensive brute-force manner such as grid-search. Therefore, recent research proposes to use observed hyperparameter performance on already solved problems (i.e. data sets) in order to speed up the search for promising hyperparameter configurations in the sequential model based optimization framework.

In this paper, we propose multilayer perceptrons as surrogate models as they are able to model highly nonlinear hyperparameter response surfaces. However, since interactions of hyperparameters, data sets and metafeatures are only implicitly learned in the subsequent layers, we improve the performance of multilayer perceptrons by means of an explicit factorization of the interaction weights and call the resulting model a factorized multilayer perceptron. Additionally, we evaluate different ways of obtaining predictive uncertainty, which is a key ingredient for a decent tradeoff between exploration and exploitation. Our experimental results on two public meta data sets demonstrate the efficiency of our approach compared to a variety of published baselines. For reproduction purposes, we make our data sets and all the program code publicly available on our supplementary webpage.

Keywords: Hyperparameter optimization · Sequential model-based optimization

1 Introduction

Unfortunately, machine learning models are very rarely parameter-free, as they usually contain a set of hyperparameters which have to be chosen appropriately on validation data. As a simple example, the number of latent variables in a matrix factorization cannot be determined using gradient descent as firstly, it is not explicitly given in the objective function and secondly is not a continuous but a discrete parameter. Additionally, the choice of kernel function for an SVM can also be understood as hyperparameter, where gradient descent approaches fail. Besides being a parameter of learned model, hyperparameters can also be part of the objective function, such as regularization constants. Moreover, they can also be part of

© Springer International Publishing Switzerland 2015
A. Appice et al. (Eds.): ECML PKDD 2015, Part II, LNAI 9285, pp. 87–103, 2015.
DOI: 10.1007/978-3-319-23525-7_6

the learning algorithm that is used to optimize the model for the objective function, for example the steplength of a gradient based technique or the threshold of a stopping criterion. Finally, even the choice of preprocessing can be viewed as a hyperparameter. Some of these hyperparameters are continuous, some are categorical, but what they all have in common is that there is no efficient learning algorithm for them. Therefore many researchers rely on searching them on a grid, which is computationally very expensive, as with growing data and growing complexity of models the optimization part usually requires a lot of time.

The performance of a model on test data trained with specific hyperparameters depends on the data set where the machine learning model should be learned, and therefore hyperparameter optimization is usually started from the scratch for each new data set. Thus, possibly valuable information of past hyperparameter performance on other data sets is ignored. Recent work proposes to use this information to be able to perform a more efficient and faster hyperparameter optimization than before [2]. To accomplish this, the sequential model-based optimization framework is applied, where a surrogate model is learned to predict hyperparameter performances in a first step. Then an acquisition function is queried to choose the next hyperparameter to test while maintaining a reasonable tradeoff between exploration and exploitation. As the prediction of the surrogate model can be done in constant time, hyperparameters can be optimized in a controlled way, resulting in less runs of the actual learning algorithm until a promising configuration is found.

This paper targets the problem of hyperparameter learning and more generally model selection across different data sets. We propose to use a multilayer perceptron as surrogate model and show how it can be learned to also include hyperparameter performances of data sets observed in the past. Additionally, we propose a factorized multilayer perceptron that contains a factorization part in the first layer of the network to directly model interactions of hyperparameters and datasets. For both of these surrogates, we propose different ways of assessing their uncertainty which is a key ingredient for hyperparameter optimization in the SMBO framework. Finally, we conduct three different experiments, where the first shows the capability of a surrogate model to predict the response surface. The second experiment compares different ways of estimating prediction uncertainty, and the last demonstrates surrogate performance in a standard SMBO setting against a variety of published baselines.

2   Related Work

In the recent years, the field of hyperparameter optimization has attracted more and more interest from the research community. The current state-of-the-art can be roughly classified into four different method categories.

At first, there are *exhaustive* methods that search the hyperparameter space exhaustively and therefore are usually conducted on a compute cluster as they are computationally expensive. The most simple and most widely used method is a grid search. Another exhaustive method was proposed by [3], where hyperparameters are not sampled on a grid but using probability distributions and work well in cases of low effective dimensionality, i.e. the case where one hyper-parameter does not affect the final performance as much as others.

Secondly, there are the *model-specific* methods that optimize hyperparam-eters for a specific model choice, such as [1] and [7], which is tailored to least squares SVM. For a regression with small sample size, the work of [5] can be applied. Furthermore, [10] deal with hyperparameter optimization in the case of semi-supervised learning. There is a plethora of model-specific methods, but their common downside is that they are tailored to a chosen model class and therefore cannot be applied in general.

A third class of methods to optimize hyperparameter is based on evolutionary algorithms, for instance [11] optimizes kernel hyperparameters of an SVM and therefore can be seen as also a *model-specific* method.

Lastly, a more recent class of hyperparameter optimization methods is based on the *sequential model-based optimization* (SMBO) [9] framework which stems from black-box optimization. The choice of this framework is quite reasonable, as the function that maps hyperparameters for a given model on a given data set to the final validation performance is certainly a black-box. All SMBO methods learn a surrogate model on given hyperparameter choices to infer the perfor-mance of unknown hyperparameters, where the next hyperparameter to test is chosen based on the prediction of the surrogate model and its uncertainty. Gaus-sian processes are used as surrogate model in [19], but are not used to include hyperparameter performances on other data sets. Moreover, SMAC [8] employs random forests as surrogate model, but also does not learn across data sets. The first paper that proposed to include past hyperparameter performances for SMBO-based hyperparameter optimization is [2], their method SCoT employs an SVM Rank as surrogate and uses a second stage Gaussian process that is learned on the output of SVM Rank to allow for uncertainty in the prediction. Another work uses past hyperparameter performances to come up with a good initialization for Bayesian hyperparameter optimization [6]. The work in [12] chooses hyperparameters and models by using active testing on the past obser-vations, which can be seen as SMBO with a very specific choice of surrogate and acquisition function.

Finally, [21] uses a Gaussian process with a more sophisticated choice of kernel function, which is able to generalize over past performances on other data sets, which is very close to the multi-task Gaussian process approach used by [20].

Compared to exhaustive methods, SMBO algorithms are more efficient in the overall number of hyperparameters that have to be evaluated; compared to model-specific methods, they may be applied for every model choice. Moreover, SMBO algorithms learn a model for the hyperparameter space, which itself is very inter-esting as it gives a deeper understanding of hyperparameter interactions.

3   Background

In this section, we will first introduce the problem setting, to then discuss impor-tant properties of surrogate functions. Afterwards, we propose three new surro-gates and finally show how to assess their prediction uncertainty.

3.1 Problem Setting

Let us define by $\mathcal{D}$ the space of all data sets. Furthermore, for a fixed model class $\mathcal{M}$, let us denote by $\mathcal{A}_\lambda$ a machine learning algorithm as a mapping $\mathcal{A}_\lambda :$ $\mathcal{D} \longrightarrow \mathcal{M}$ that maps training data $D^{\text{train}} \in \mathcal{D}$ to a learned model $M_\lambda \in \mathcal{M}$ for a given hyperparameter configuration $\lambda \in \Lambda$ by searching through $\mathcal{M}$ and finding a model that minimizes:

$$\mathcal{A}_\lambda(D^{\text{train}}) := \arg \min_{M_\lambda \in \mathcal{M}} \mathcal{L}(M_\lambda, D^{\text{train}}) . \tag{1}$$

Usually, $\Lambda = \Lambda_1 \times \ldots \times \Lambda_p$, where Λ_i may be a continuous or discrete space. Having learned a model for a given hyperparameter configuration λ, the *hyperparameter optimization problem* can be stated as choosing the $\lambda^\star$, for which the associated model $M_{\lambda^\star}$ has a minimal error on a validation set

$$\lambda^\star := \arg \min_{\lambda \in \Lambda} \mathcal{L}(\mathcal{A}_\lambda(D^{\text{train}}), D^{\text{val}}) := \arg \min_{\lambda \in \Lambda} f(\lambda) . \tag{2}$$

Thus, the problem of hyperparameter optimization can be stated as minimizing computationally expensive black-box function f over Λ. As discussed earlier, these hyperparameters cannot be optimized using standard means, as there is no knowledge of f, and therefore exhaustive search methods such as grid search partition Λ into a discrete subset $G \subset \Lambda$ and optimize f over G, which takes a lot of time as many hyperparameter configurations have to be tested.

A more recent class of hyperparameter optimization methods follows the SMBO framework, where on known hyperparameter responses of f on a discrete subset G, a surrogate model $\Psi(\lambda)$ is learned to most accurately predict f. Once this is accomplished, Ψ is then used to predict promising hyperparameter configurations to choose next, while maintaining a tradeoff between exploration and exploitation. Exploration drives the choice of choosing *distant* hyperparameter configurations, where the surrogate model Ψ is very uncertain. Exploitation chooses hyperparameters in well-known regions of f, which might find local but not necessarily global optima. Therefore, a decent tradeoff between exploration and exploitation is desired.

As [2] proposed, this procedure is not limited to only one data set and can therefore be expanded in a way that Ψ learns the response for given hyperparameters across many data sets $D \in \{D_1, \ldots, D_m\}$ where the response surface has already been observed, to then use the gained knowledge to optimize hyperparameters for a new data set D^{new}. In order to learn such a surrogate, we now denote the input of Ψ and f by x, which also contains dataset information.

$$x = (\lambda, d, m) , \tag{3}$$

where d is a binary dataset indicator and for a given data set D_j defined as

$$d(D_j) = (d_1, ..., d_m) \quad d_i = \delta(i = j) , \tag{4}$$

for δ being the indicator function. By m or more formally $m(D_j)$, we denote descriptive features for data set D_j. They are usually called meta features and can be simple statistics, such as number of attributes, number of instances [2] [21] or more complex features such as the classification accuracy of a decision tree or a linear SVM [15]. Finally, an observation history $\mathcal{H}$ is built to contain all hyperparameter responses for $\lambda \in G$ for all data sets D where hyperparameter optimization has already been accomplished.

The resulting procedure can be seen in Algorithm 1. At the beginning of one trial, we fit the surrogate model Ψ to the given observation history. Then we query an acquisition and choose its maximum to be the next hyperparameter configuration to test. The most widely used acquisition function is the expected improvement (EI) [9], which given a currently best hyperparameter configuration x^{best} is defined as

$$EI(x) := \int_0^\infty I \cdot p(I \mid \Psi, x^{\text{best}}) \, \mathrm{d}I \ . \tag{5}$$

Afterwards, f is evaluated for the proposed hyperparameter configuration and the tuple $(x, f(x))$ is then added into the observation history $\mathcal{H}$.

Algorithm 1. Sequential Model-based Optimization Across Data Sets

---

Input: Hyperparameter space Λ, observation history $\mathcal{H}$, target data set D^{new}, number of iterations T, acquisition function a, surrogate model Ψ.
Output: Best hyperparameter configuration x^{best} for D^{new}
1: **for** $t = 1$ to T **do**
2:     Fit Ψ to $\mathcal{H}$
3:     $x^{\text{new}} = \arg \max_x a\left(x, \Psi(x)\right)$
4:     Evaluate $f\left(x^{\text{new}}\right)$
5:     **if** $f(x^{\text{new}}) > f(x^{\text{best}})$ **then**
6:         $x^{\text{best}} = x^{\text{new}}$
7:     $\mathcal{H} = \mathcal{H} \cup (x^{\text{new}}, f(x^{\text{new}}))$
8: **return** x^{best}

3.2 Requirements for a Surrogate Model

We have identified three main ingredients for a surrogate model to be able to accurately predict hyperparameter responses across data sets.

Nonlinearity. Usually, the hyperparameter response f is highly nonlinear and therefore dictates a surrogate model to also adapt this property. We will see later in our experiments, that even nonlinear models can fail to reproduce the response surface, if the employed basis functions are not well chosen and thus the model does not offer enough complexity.

Prediction Uncertainty. If we fully trust the surrogate model Ψ in its predictions, i.e. use the identity as acquisition function and therefore always query the hyperparameter configuration with the best predicted performance, we are doomed to fail because only exploitation of the model is done, meaning that we always stay in a region of the hyperparameter space Λ where we have started. This is due to the fact that the surrogate model is learned on a few observations of f and therefore will not accurately predict every hyperparameter performance. To circumvent this issue, acquisition functions such as the EI are employed, that try to balance exploration and exploitation. In order for EI to work, the surrogate model needs a predictive posterior, i.e. a probability distribution on $\Psi(x)$ that can be queried for how uncertain the prediction is, thus forming the second key ingredient for a decent surrogate model.

Shared and Data Set Specific Parameters. To successfully learn surrogate model across different problem aspects (i.e. data sets), it should be able to distinguish between these to learn specific data set characteristics. A natural way is to add binary dataset indicators as it was done above. However, to be able to learn more than only a data set bias with these features, we aim to learn factorization models that can also model the interactions of hyperparameters with datasets, hyperparameters with model choices and so on. In this way, we automatically learn latent characteristics of a data set.

Another way to let the surrogate learn across problems is to add meta features that describe the problems, where for data sets, many meta features have already been proposed. If we think one step further and want to generalize over other problem aspects such as preprocessing, choice of model, etc. we have to come up with meta features describing these problem aspects, which does not seem reasonable to us anymore.

3.3 Proposed Models

Factorization Machines. The first surrogate model we propose is a factorization machine which was introduced in [16]. It works as a generalization of factorization models and can mimic all different kind of models if the features are preprocessed in a certain way. To every given feature, i.e. in our case hyperparameters and binary data set indicators, the model associates a vector of $K \in \mathbb{N}$ latent features. The final prediction is then given through

$$\Psi(x) = w_0 + \sum_{i=1}^{n} w_i x_i + \frac{1}{2} \sum_{i=1}^{n} \sum_{j=i+1}^{n} \langle v_i, v_j \rangle x_i x_j \ . \tag{6}$$

The model is also sometimes called a factorized polynomial regressor, as in its essence it is a polynomial regression of degree two, if one sets $w_{i,j} := \langle v_i, v_j \rangle$, though by factorizing this weight the model can be fitted more effectively in sparse settings as the parameters have more instances to learn from. Moreover, by applying a factorization machine, we are also able to learn interactions of data sets and hyperparameters. Ultimatively, we are even able to use continuous features, such as meta features, which a standard matrix factorization model would not allow us to do.

Multilayer Perceptron. The next model we propose to use as a surrogate is the multilayer perceptron, which may be more commonly named as feedforward neural network. A multilayer perceptron consists of L many layers, where each layer comprises N many nodes and is fully connected to the next layer, forming the structure of a directed acyclic graph. At the beginning, $x = x^0$ is used as input for the first layer. The k-th output of a layer l is then defined as

$$x_k^l = \sigma^{l-1}\left(w_{0,k}^{l-1} + \sum_{i=1}^{n} w_{i,k}^{l-1} x_i^{l-1}\right) = \sigma(s_k^l) , \qquad (7)$$

thus acts as input for the subsequent layer, where σ^{l-1} is a sigmoid function, in our case we used the hyperbolic tangent, and w are the weights, i.e. parameters of the model. In this way, the information is propagated forward until predictions are made in the final layer. As our task is regression, the final prediction will be one-dimensional and σ^{L-1} is defined as the identity function

$$\Psi(x) = w_0^{l-1} + \sum_{i=1}^{n} w_i^{l-1} x_i^{l-1} . \qquad (8)$$

Let us have a closer look at what the model does with binary data set indicators. In the input layer, the multilayer perceptron learns exactly N many weights per each data set, which act as a data set bias, and therefore can be used by the model to generalize across data sets. From the second layer onwards, the model acts independently from the data set as all features then are fitted globally. Nevertheless, interactions can still implicitly be modeled throughout the learning process of the network. The question to answer is whether an explicit modelling of these interactions such as in a factorization machine is better than an implicit one.

Factorized Multilayer Perceptron. Finally, the third surrogate model proposed by us is a mixture of both previous models and is therefore called a factorized multilayer perceptron. Closely related to a multilayer perceptron, it also consists of L many layers, where each layer comprises N many nodes, also the final prediction is the same as given in Equation 8. The only difference is that here we explicitly model feature interactions in the input layer, by using the prediction of a factorization machine instead of a linear model. Thus, the k-th output of the first layer is defined as

$$x_k = \sigma(s_k^1) = \sigma\left(w_{0,k} + \sum_{i=1}^{n} w_{i,k} x_i + \frac{1}{2} \sum_{i=1}^{n} \sum_{j=i+1}^{n} \langle v_{i,k}, v_{j,k}\rangle x_i x_j\right) , \qquad (9)$$

where $v_{i,k} \in \mathbb{R}^K$ are the latent characteristics of feature i for the output k. Note that we only do this in the first layer and therefore dropped the layer dependencies to avoid unnecessary clutter.

In this way, we explicitly model the feature interactions of a factorization machine into the first layer of a multilayer perceptron, as the binary data set indicators are naturally only given in the input layer. This model can be learned straightforward using backpropagation [17], the only difference is that we have to consider the update for the latent feature vectors as well. The resuling procedure can be viewed in Algorithm 2, where the updates are denoted for a stochastic gradient descent approach. We dropped the usual momentum term to avoid clutter, the implementation of such a term is straightforward.

Algorithm 2. SGD-Backpropagation for Factorized Multilayer Perceptron

Input: Data Set D, Loss function $\mathcal{L}$, step length $\eta > 0$.

1: **repeat**
2:     Draw $(x, y) \in D$
3:     Predict $\hat{y}(x)$
4:     Compute $\delta_k^l = \frac{\partial \mathcal{L}(\hat{y}(x),y)}{\partial s_k^l} \cdot \frac{d\sigma}{ds_k^l}$ for all layers l and nodes k
5:     Update $w_{i,k}^l = w_{i,k}^l - \eta \delta_k^l x_i$
6:     Precompute $\mu_j^k = \sum_{i=1}^n v_{i,j}^k x_i$
       Update $v_{i,j}^k = v_{i,j}^k - \eta \delta_k^l (x_i \mu_j^k - v_{i,j}^k x_i^2)$
7: **until** Convergence

3.4 Estimating Prediction Uncertainty

The proposed surrogate models are still lacking the ability to predict under uncertainty, which is a key ingredient for running SMBO with a decent trade-off between exploration and exploitation. SMAC uses a random forest, i.e. a bagged ensemble of decision trees, and is thus able to compute a mean and a standard deviation by assuming that the prediction of the ensemble is Gaussian distributed. Alternatively, SCoT uses a ranking approach and learns a Gaussian process on the ranked output, thus obtaining prediction uncertainty through the Gaussian process.

By treating the abovely proposed surrogate models in a Bayesian setting as it is described in [13], it is possible to deduce prediction uncertainty using a Taylor approximation of the objective function. Let us denote by w a vector of all parameters of Ψ, including biases, weights and possibly latent characteristics. Assuming a Gaussian prior with covariance α^{-1} of the form

$$p(w) = \mathcal{N}(w \,|\, \mathbf{0}, \alpha^{-1}\mathbf{I}) \;, \tag{10}$$

the posterior distribution of the parameters w given the data D, α and data set noise σ^2 can be estimated by using a second-order Taylor decomposition on the objective function. The resulting parameter posterior is approximated as

$$p(w \,|\, D, \alpha, \sigma^2) \approx \mathcal{N}(w \,|\, w^\star, A^{-1}) \;, \tag{11}$$

where $A = \beta H + \alpha I$, and H is the Hessian matrix of the loss on the data set. The densitiy of the predictive posterior can then be written as

$$p(y \,|\, x, D, \alpha, \sigma^2) = \int \mathcal{N}(y \,|\, \Psi(x, w), \sigma^2)\mathcal{N}(w \,|\, w^\star, A^{-1})\mathrm{d}w \;. \tag{12}$$

As [13] argues, this integral is not feasible to compute because of the nonlinearity of Ψ, thus a first order approximation is done around w^* yields

$$\Psi(x, w) \approx \Psi(x, w^*) + g^\top (w - w^*) \quad \text{where} \quad g = \nabla_w \Psi(x, w)|_{w=w^*} . \quad (13)$$

Finally, the predictive posterior can be written as Gaussian

$$p(y \mid x, D, \alpha, \sigma^2) = \mathcal{N}(y \mid \Psi(x, w^*), \sigma^2 + g^\top A^{-1} g) . \quad (14)$$

In conclusion, to predict the uncertainty of Ψ for an instance x, we need to estimate the Hessian of the loss of Ψ on D, and a gradient g depending on x. The latter is easy at it only involves a computation of the gradient, which is for a multilayer perceptron a forward and a backward pass through the network.

To compute the inverse Hessian in an analytic fashion is usually not feasible as computing one entry of the Hessian involves a pass over the whole data and then inverting the resulting matrix has an effort that is cubical in the number of parameters, i.e. the dimensionality of w. Out of this reason, we seek to approximate the inverse of the Hessian directly by using a sum of outer products as it is exposed in [4]. As the target loss is least squares, the Hessian can be written as

$$H = \sum_{(x,y)\in D} \nabla\Psi(x, w)\nabla\Psi(x, w)^\top + \sum_{(x,y)\in D} (y - \Psi(x, w))\nabla\nabla\Psi(x, w) . \quad (15)$$

As [4] outlines, for a carefully learned model the second sum can be neglected as the quantity $(y - \Psi(x, w))$ is close to zero. Thus, H can be approximated using only the first term which is a sum of outer products. The inverse of H

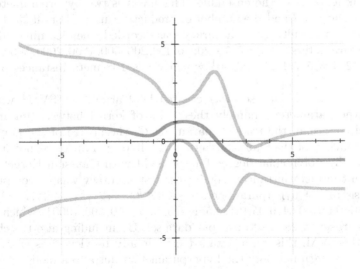

Fig. 1. Predictive Posterior of a multilayer perceptron learned on $(x_1, y_1) = (0, 1)$ and $(x_2, y_2) = (\pi, -1)$. The red line shows the mean, the grey line shows one standard deviation (Best viewed in color)

can directly be computed in an iterative fashion over the data set using the Sherman-Woodbury formula and starting with an initialization of $H^{-1} = \alpha^{-1}I$. In this way, we we effectively compute the inverse of $H + \alpha I$, which is exactly the matrix we seeked to invert. A one dimensional example can be seen in Figure 1, where a multilayer perceptron is learned on two data points.

As an alternative approach, we simply compute an ensemble of surrogates and predict the uncertainty using the estimated mean and standard deviation of all the predictions, as it is also done by SMAC. The resulting variance then stems from differently learned models, which in the case of SMAC results from bagging. As the most simple approach, we propose to learn an average ensemble, where the resulting variance stems only from different initializations of the surrogate model, which is reasonable if the whole optimization problem is not convex and therefore yields different solutions.

4 Experiments

To assess the performance of our proposed surrogate models we will conduct three different experiments on two meta data sets that we have created on our own.

4.1 Meta Data Set Creation

For 25 randomly chosen classification data sets of the UCI repository[1], we merged existing splits into one data set, then shuffled the data set and created one split where 80% of the data was used for training, and the remaining 20% for testing. To create the first data set, we learned AdaBoost[2] by employing decision products as weak learners of the ensemble. This involves two hyperparameters, the number of iterations I and the number of product terms M. For all 25 classification datasets, the resulting test accuracy was recorded when learning AdaBoost with hyperparameters $I \in \{2, 5, 10, 20, 50, 100, 200, 500, 1000, 2000, 5000, 10000\}$ and $M \in \{2, 3, 4, 5, 7, 10, 15, 20, 30\}$ which yields 108 meta instances per data set.

The second meta data set was created by learning an SVM[3] with four involved hyperparameters, namely the choice of kernel between linear, polynomial and Gaussian, the tradeoff parameter C, the degree of the polynomial d and the width γ of the Gaussian kernel. If a hyperparameter is not involved, for example the polynomial degree for the SVM with Gaussian Kernel, we set it to zero in the meta instances. Again, the test accuracy was precomputed on a grid consisting of hyperparameters $C \in \{2^{-5}, \ldots, 2^6\}$, $d \in \{2, \ldots, 10\}$ and $\gamma \in \{0.0001, 0.001, 0.01, 0.05, 0.1, 0.5, 1, 2, 5, 10, 20, 50, 100, 1000\}$, which results in a meta dataset of 288 instances per data set. By including also the choice of kernel for the SVM, this meta data set can already be viewed as cross-model, since we try to learn not only the hyperparameters but also a model choice.

---

[1] http://archive.ics.uci.edu/ml/index.html
[2] http://www.multiboost.org
[3] http://svmlight.joachims.org

As they are an indispensable part of the competing methods, we also added the meta features used by [2] and [21] to our meta data sets. These encompass the number of classes c, the logarithm of the number of predictors $\log(p)$ and finally the logarithm of the quotient of dataset instances and number of predictors $\log(|D|/p)$. Finally, we scaled the meta features to have values in $[0,1]$.

Table 1. Confidence intervals of the resulting RMSE of experiment 1 for all models when reconstructing the response surface

	RF	SVR	FM	MLP	FMLP
SVM	0.0997±0.028	0.1110±0.020	0.1041±0.029	0.0596±0.013	**0.0550** ± 0.016
AdaBoost	0.0462±0.012	0.0840±0.009	0.0579±0.015	0.0380±0.008	**0.0377** ± 0.009

4.2 Experiment 1: Reconstruction of the Response Surface

As a first experiment, we seek to learn models to reconstruct the hyperparameter response surface in order to determine their usefulness for hyperparameter optimization in the sequential model-based optimization framework. The evaluation protocol is designed in a leave-one-out fashion, where we learn a surrogate model on 24 response surfaces plus a few observations of hyperparameter responses of the new dataset to then predict the full response surface. For the test data, we took 4% of the responses as training data, 10% as validation data for hyperparameter optimization of the surrogate model and used the remaining 86% as test data.

As surrogate models, we used a random forest (RF), a support vector regression (SVR), a factorization machine (FM), a multilayer perceptron (MLP) and a factorized multilayer perceptron (FMLP). For the RF, we used the implementation in MLTK[4], for the support vector regression we used the implementation by Joachims [5] . All remaining models were implemented in Java by ourselves and optimized for minimal root mean squared error (RMSE). Hyperparameters of all models have been optimized using grid search, for more detail on the grids, we refer to our supplementary webpage [18]. The resulting 95% confidence intervals of the leave-one-out cross validation are reported in Table 1. Clearly, both neural network models outperform the other models by a considerable margin, where the FMLP tends to achieve the best performance, although the lift to a normal MLP is marginal and not statistically significant. It is observable that results on AdaBoost are much better, therefore indicating that the hyperparameter optimization problem for this specific model is easier than for an SVM. We acknowledge that also the lift of our models compared to the RF is not statistically significant.

---

[4] http://www.cs.cornell.edu/~yinlou/projects/mltk/
[5] http://svmlight.joachims.org/

Unexpectedly, a factorization machine fails to reconstruct the response surface as its RMSE is clearly worse than of the MLP, for instance. This stems from the fact that its expressivity in this setting is rather limited as a standalone model, which can be demonstrated by a small example. If we consider an instance out of the SVM meta data set with RBF kernel, then, leaving out the meta features, the prediction can be shown to have the form:

$$\Psi(x) = w_0 + w_C C + w_\gamma \gamma + w_{C,\gamma} C\gamma , \tag{16}$$

which has the geometrical form of a hyperbolic paraboloid. This clearly fails to reproduce any complex response surface and therefore a plain Factorization Machine is not a good candidate for a surrogate model.

4.3   Experiment 2: Uncertainty Estimation in SMBO

In this experiment we compare the a multilayer perceptron in two different scenarios. Before we proceed, we first introduce the evaluation metrics that we applied in the SMBO setting.

Evaluation Metrics for SMBO. We use two different evaluation metrics, at first the average rank of the individual models, where the second metric is the average hyperparameter rank.

Average Rank. The average rank among different tuning strategies ranks all tuning strategies by the best hyperparameter configuration they have found so far, where ties are solved by granting the average rank. If we for example have four different tuning strategies where one obtains an accuracy of 0.9, two others obtain 0.8 and the third obtains an accuracy of 0.7, we associate the ranks 1, 2.5, 2.5, 4.

Average Hyperparameter Rank. By average hyperparameter rank we do not compare between methods but between hyperparameters found. For a fixed data set D, the hyperparameter responses are ranked according to their performance, then the average hyperparameter rank is simply the average over all folds.

Experiment Setup and Results. We evaluate at first the MLP when computing predictive uncertainty by means of an inverse Hessian matrix (MLPH) as proposed in section 3.4, opposed to the approach where the uncertainty is assessed by using an average ensemble (MLPE). The development of the average hyperparameter rank for both the SVM and the AdaBoost data set is plotted in Figure 2, where results are averaged over 10 runs for both methods. As the figure indicates, the convergence of MLPE is much faster, which is due to several reasons. At first, having an ensemble yields a better approximation of the response surface itself. Secondly, the predicted uncertainty does not seem to help in exploring the hyperparameter space, which then results in an overall small convergence. This is due to the fact, that the inverse Hessian is only approximated in many ways, if we consider Equation 15, which is already a Tailor

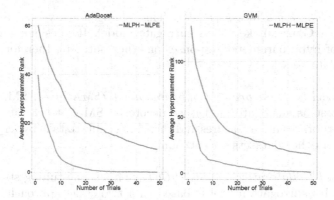

Fig. 2. Development of the average hyperparameter rank with increasing numbers of trials. Clearly, the convergence of the ensemble MLP is much faster than using the inverse Hessian (Best viewed in color)

approximation, the second term was neglected for a carefully trained model. By adding the new target data set to our overall loss, this assumption is likely not valid anymore as the surrogate has almost no knowledge of the new data set and therefore cannot be perfectly trained for it. Moreover, if we consider Figure 1, it still shows quite a bit uncertainty around points which have already been evaluated which might lead to exploitation. The uncertainty only decreases if more points in this region are queried, a luxury that we are not permitted in the SMBO scenario.

Consequently, as it also takes more time to compute the inverse Hessian than learning an ensemble model, we propose to follow the latter strategy. This is also done in the next experiment, where MLP and FMLP are used in an ensemble fashion.

4.4   Experiment 3: Sequential Model Based Optimization

As a final experiment, we test our surrogate models, the MLP and FMLP in the SMBO setting, where we again perform a leave-one-out cross validation over data sets. Hyperparameters of baseline models have been specifically optimized for the average hyperparameter rank, for the models proposed by us we used the optimal hyperparameters of the first experiment. To consider initialization variance, all results are averaged over 10 runs, except for the random search where 1000 runs were executed. We will briefly describe the competing methods in the following.

Tuning Strategies.

Random Search. This is a tuning strategy that neither uses a surrogate model nor uses an acquisition function. It was first proposed in [3], and has proven to work well in scenarios of low effective dimensionality.

Independent Gaussian Process (I-GP). This tuning stategy uses a Gaussian Process with a Gaussian kernel as surrogate model. It does not employ any information of hyperparameter responses on other datasets, therefore does not learn across data sets.

Sequential Model-based Algorithm Configuration++ (SMAC++). SMAC [8] uses a random forest as surrogate model, we denote by SMAC++ a random forest that also incorporates meta features and therefore is able to take hyperparameter performance of other datasets into account.

Surrogate-based Collaborative Tuning (SCoT). This is the tuning strategy proposed by [2]. Its surrogate model is based on a two stage approach, as it first learns a ranking using SVM^{RANK} with an RBF Kernel. Then, a Gaussian Process is learned on the output of the ranking. As indicated by [21], learning an RBF Kernel takes too much time, we followed their suggestion and learned a linear kernel instead.

Gaussian Process with MKL (MKL-GP). As proposed by [21], this tuning strategy is based on a Gaussian Process as surrogate model where the kernel is a mixture of an SE-ARD Kernel combined with a kernel modelling the distances between data sets, which is estimated based on the meta features.

Multilayer Perceptron (MLP). Our tuning strategy based on a multilayer perceptron that associates weights to binary data set indicators. We learn an average-ensemble of 100 models to assess uncertainty in the prediction. The weights of the network are initialized using the Nguyen-Widrow [14] initialization for faster convergence of the model.

Factorized Multilayer Perceptron (FMLP). The final tuning strategy that we propose. It is similar to the multilayer perceptron, but uses an additional factorization part in the first layer to directly model all interactions of hyperparameters, data sets and meta features. As for the MLP, we again learn an ensemble of 100 models to predict uncertainties, the network weights are being initialized as for the MLP, the latent factors are initialized using a Gaussian prior.

Optimal. This is an artificial surrogate model that always predicts the best hyperparameter configurations and is plotted for orientation purposes.

Results. Figure 3 shows the development of the average rank with increasing number of trials. For both meta data sets the first ten trials of SMBO encompass some noise as basically all competing methods start out equally good (or bad). Afterwards, we see that the FMLP performs best in the arguably most interesting region, where there is a proper tradeoff between the optimal hyperparameter found so far and the overall used percentage of the grid, as in the beginning, there is a lot of noise involved and in the end the improvement in hyperparameter

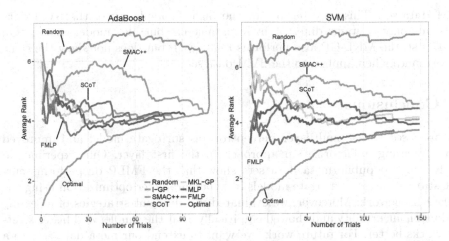

Fig. 3. Development of the average rank with increasing numbers of trials (Best viewed in color or online [18])

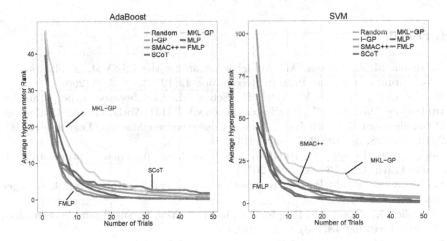

Fig. 4. Development of the average hyperparameter rank with increasing numbers of trials (Best viewed in color or online [18])

performance degrades. Note that the MLP also is very competitive and therefore empirically already a decent tuning strategy.

Figure 4 demonstrates the development of the average hyperparameter rank. This chart gives an impression of how fast the actual performance of proposed hyperparameter configurations converges to the optimal configuration on the grid. Again, we observe that the FMLP works best for both the AdaBoost and the SVM data set, which none of the baselines accomplish, as for example SCoT only works well on the SVM data set. We acknowledge the good results of an independent Gaussian Process on the AdaBoost data, which degrades on the

SVM data set. This may be due to the higher complexity of the SVM data as it not only contains different hyperparameters but also model choices. On AdaBoost, the MKL-GP also performs really well, but does not show the same performance when applied to the SVM data set.

5   Conclusions

We proposed to use multilayer perceptrons as surrogate model and improved them by using a factorization approach in the first layer. Our experimental results on two public meta data sets show that the FMLP outperforms current state of the art surrogate models in hyperparameter optimization using the SMBO framework. Moreover, we evaluated two different strategies of assessing prediction uncertainty and showed empirically, that the simpler and faster strategy works better. For future work, we want to extend our meta data sets to a cross-model problem by using a plethora of base models and then try to learn a common latent feature space for datasets and models. We argue that this is the next step to be made in hyperparameter optimization.

References

1. Adankon, M.M., Cheriet, M.: Model selection for the LS-SVM. Application to handwriting recognition. Pattern Recognition **42**(12), 3264–3270 (2009)
2. Bardenet, R., Brendel, M., Kegl, B., Sebag, M.: Collaborative hyperparameter tuning. In: Dasgupta, S., Mcallester, D. (eds.) JMLR Workshop and Conference Proceedings of the 30th International Conference on Machine Learning (ICML 2013), vol. 28, pp. 199–207, May 2013
3. Bergstra, J., Bengio, Y.: Random search for hyper-parameter optimization. J. Mach. Learn. Res. **13**, 281–305 (2012)
4. Bishop, C.M., et al.: Pattern recognition and machine learning, vol. 4. springer New York (2006)
5. Chapelle, O., Vapnik, V., Bengio, Y.: Model selection for small sample regression. Machine Learning **48**(1–3), 9–23 (2002)
6. Feurer, M., Springenberg, J.T., Hutter, F.: Initializing bayesian hyperparameter optimization via meta-learning. In: Proceedings of the Twenty-Ninth AAAI Conference on Artificial Intelligence (2015)
7. Guo, X.C., Yang, J.H., Wu, C.G., Wang, C.Y., Liang, Y.C.: A novel ls-svms hyperparameter selection based on particle swarm optimization. Neurocomput. **71**(16–18), 3211–3215 (2008)
8. Hutter, F., Hoos, H.H., Leyton-Brown, K.: Sequential model-based optimization for general algorithm configuration. In: Coello, C.A.C. (ed.) LION 2011. LNCS, vol. 6683, pp. 507–523. Springer, Heidelberg (2011)
9. Jones, D.R., Schonlau, M., Welch, W.J.: Efficient global optimization of expensive black-box functions. J. of Global Optimization **13**(4), 455–492 (1998)
10. Kapoor, A., Ahn, H., Qi, Y., Picard, R.W.: Hyperparameter and kernel learning for graph based semi-supervised classification. In: Advances in Neural Information Processing Systems, pp. 627–634 (2005)

11. Koch, P., Bischl, B., Flasch, O., Bartz-Beielstein, T., Weihs, C., Konen, W.: Tuning and evolution of support vector kernels. Evolutionary Intelligence **5**(3), 153–170 (2012)

12. Leite, R., Brazdil, P., Vanschoren, J.: Selecting classification algorithms with active testing. In: Perner, P. (ed.) MLDM 2012. LNCS, vol. 7376, pp. 117–131. Springer, Heidelberg (2012)

13. Murphy, K.P.: Machine learning: a probabilistic perspective. MIT press (2012)

14. Nguyen, D., Widrow, B.: Improving the learning speed of 2-layer neural networks by choosing initial values of the adaptive weights. In: 1990 IJCNN International Joint Conference on Neural Networks, 1990, pp. 21–26. IEEE (1990)

15. Pfahringer, B., Bensusan, H., Giraud-Carrier, C.: Meta-learning by landmarking various learning algorithms. In: Proceedings of the Seventeenth International Conference on Machine Learning, pp. 743–750. Morgan Kaufmann (2000)

16. Rendle, S.: Factorization machines. In: 2010 IEEE 10th International Conference on Data Mining (ICDM), pp. 995–1000. IEEE (2010)

17. Rumelhart, D.E., Hinton, G.E., Williams, R.J.: Learning representations by back-propagating errors. Cognitive modeling 5 (1988)

18. Schilling, N.: Supplementary website. http://hylap.org/publications/hyper-opt-with-factorized-multilayer-perceptrons

19. Snoek, J., Larochelle, H., Adams, R.P.: Practical bayesian optimization of machine learning algorithms. In: Pereira, F., Burges, C., Bottou, L., Weinberger, K. (eds.) Advances in Neural Information Processing Systems, vol. 25, pp. 2951–2959. Curran Associates, Inc. (2012)

20. Swersky, K., Snoek, J., Adams, R.P.: Multi-task bayesian optimization. In: Burges, C., Bottou, L., Welling, M., Ghahramani, Z., Weinberger, K. (eds.) Advances in Neural Information Processing Systems, vol. 26, pp. 2004–2012. Curran Associates, Inc. (2013)

21. Yogatama, D., Mann, G.: Efficient transfer learning method for automatic hyperparameter tuning. In: International Conference on Artificial Intelligence and Statistics (AISTATS 2014) (2014)

Hyperparameter Search Space Pruning – A New Component for Sequential Model-Based Hyperparameter Optimization

Martin Wistuba[✉], Nicolas Schilling, and Lars Schmidt-Thieme

Information Systems and Machine Learning Lab,
University of Hildesheim,
31141 Hildesheim, Germany
{wistuba,schilling,schmidt-thieme}@ismll.uni-hildesheim.de

Abstract. The optimization of hyperparameters is often done manually or exhaustively but recent work has shown that automatic methods can optimize hyperparameters faster and even achieve better final performance. Sequential model-based optimization (SMBO) is the current state of the art framework for automatic hyperparameter optimization. Currently, it consists of three components: a surrogate model, an acquisition function and an initialization technique. We propose to add a fourth component, a way of pruning the hyperparameter search space which is a common way of accelerating the search in many domains but yet has not been applied to hyperparameter optimization. We propose to discard regions of the search space that are unlikely to contain better hyperparameter configurations by transferring knowledge from past experiments on other data sets as well as taking into account the evaluations already done on the current data set.

Pruning as a new component for SMBO is an orthogonal contribution but nevertheless we compare it to surrogate models that learn across data sets and extensively investigate the impact of pruning with and without initialization for various state of the art surrogate models. The experiments are conducted on two newly created meta-data sets which we make publicly available. One of these meta-data sets is created on 59 data sets using 19 different classifiers resulting in a total of about 1.3 million experiments. This is by more than four times larger than all the results collaboratively collected by OpenML.

1 Introduction

Most machine learning algorithms depend on hyperparameters that need to be tuned. In contrast to model parameters, hyperparameters are not estimated during the learning process but have to be set before. Since the hyperparameter tuning often decides whether the performance of an algorithm is state of the art or just moderate, the task of hyperparameter optimization is as important as developing new models [2,7,20,22,25]. Typical hyperparameters are for example the trade-off parameter C of a support vector machine or the

© Springer International Publishing Switzerland 2015
A. Appice et al. (Eds.): ECML PKDD 2015, Part II, LNAI 9285, pp. 104–119, 2015.
DOI: 10.1007/978-3-319-23525-7_7

regularization constant of a Tikhonov-regularized model. Taking a step further, the chosen model as well as preprocessing steps can be considered as hyperparameters [25]. Then, hyperparameter optimization not only involves model selection but also model class selection, choice of learning algorithms and preprocessing.

The conventional way of hyperparameter optimization is a combination of manual search with a grid search. This is an exhaustive search in the hyperparameter space which involves multiple training of the model. For high-complex hyperparameter spaces or large data sets this becomes infeasible. Therefore, methods to accelerate the process of hyperparameter optimization are currently an interesting topic for researchers [3,22,25]. Sequential model-based optimization (SMBO) [15] is a black-box optimization process and has proven to be effective in accelerating the hyperparameter optimization process. SMBO is based on a surrogate model that approximates the response function of a data set for given hyperparameters such that sequentially possibly interesting hyperparameter configurations can be evaluated.

Recent work tries to transfer knowledge about the hyperparameter space from past experiments to a new data set [1,24,29]. They motivate this idea by assuming that regions of the hyperparameter space that perform well for few data sets likely contain promising hyperparameter configurations for new data sets.

1.1 Our Contributions

The SMBO framework currently has at most three components. First, the surrogate model that predicts the performance for each possible hyperparameter configuration. Secondly, the acquisition function which uses the surrogate model to propose the next hyperparameter configuration to evaluate. These are the two mandatory components. The third optional component is some initialization technique which usually starts which a hyperparameter configuration that has proven to be good on many data sets [9,11]. We propose to add a fourth component which is orthogonal to all the others. Our idea is to reduce the hyperparameter search space by using knowledge from past experiments to discard regions that are very likely not interesting. This avoids that the acquisition function chooses hyperparameter configurations in these regions because of high uncertainty and therefore avoids unnecessary function evaluations.

Additionally, we created two meta-data sets and make them publicly available. One is a meta-data set created by running a kernel support vector machine on 50 different data sets with 288 different hyperparameter configurations resulting into 14,000 meta-instances. The second is a large scale meta-data set created by using 19 different classifiers provided by Weka [13] on 59 data sets. In total 1,290,389 meta-instances were created such that the number of runs is by more than 4 times larger than the number of runs collaboratively collected by OpenML [26].

2 Related Work

Pruning is a well known technique to accelerate the search in several domains. Thus, for example, various pruning techniques are applied to the minimax algorithm such as the killer heuristic or null move pruning [8]. Branch-and-Bound [18] is a pruning technique that is applied in the domain of operations research for discrete and combinatorial optimization problems and is very common for NP-hard optimization problems [17]. Nevertheless, we are not aware of any published work that is trying to prune the search space in the SMBO framework for hyperparameter optimization.

Since pruning as proposed by us is some way of transferring knowledge from past experiments to a new experiment, other techniques that try exactly the same are the closest related work but as we will see, orthogonal to our contribution. One common and easy way to use experience in the hyperparameter optimization domain is to define an initialization, a sequence of hyperparameter configurations that are chosen first. These are usually those hyperparameter configurations that performed best on average across data sets [9,11]. The second and last method to do so is by using the surrogate model. Instead of learning the surrogate model only on the new data set, the surrogate model is learned across all data sets [1,24,29]. We want to highlight that all these three possibilities are not mutually exclusive and can be combined and thus these ideas are orthogonal to each other.

Leite et al. [19] propose a similar distance function between data sets as we use. But they propose a hyperparameter selection strategy that is limited to the hyperparameter configurations that have been seen on the meta-training data.

Furthermore, there also exist strategies to optimize hyperparameters that are based on optimization techniques from artificial intelligence such as tabu search [4], particle swarm optimization [12] and evolutionary algorithms [10] as well as gradient-based optimization techniques [6] designed for SVMs.

3 Background

3.1 The Formal Setup

A machine learning algorithm $\mathcal{A}_\lambda$ is a mapping $\mathcal{A}_\lambda : \mathcal{D} \to \mathcal{M}$ where $\mathcal{D}$ is the set of all data sets, $\mathcal{M}$ is the space of all models and $\lambda \in \Lambda$ is the chosen hyperparameter configuration with $\Lambda = \Lambda_1 \times \ldots \times \Lambda_p$ being the p-dimensional hyperparameter space. The learning algorithm estimates a model $M_\lambda \in \mathcal{M}$ that minimizes a regularized loss function $\mathcal{L}$ (e.g. misclassification rate):

$$\mathcal{A}_\lambda \left(D^{(train)} \right) := \arg \min_{M_\lambda \in \mathcal{M}} \mathcal{L} \left(M_\lambda, D^{(train)} \right) + \mathcal{R} \left(M_\lambda \right) . \tag{1}$$

Then, the task of *hyperparameter optimization* is finding the optimal hyperparameter configuration λ^* using a validation set i.e.

$$\lambda^* := \arg \min_{\lambda \in \Lambda} \mathcal{L} \left(\mathcal{A}_\lambda \left(D^{(train)} \right), D^{(valid)} \right) := \arg \min_{\lambda \in \Lambda} f_D \left(\lambda \right) . \tag{2}$$

3.2   Sequential Model-Based Optimization

Exhaustive hyperparameter search methods such as grid search are becoming more and more expensive. Data sets are growing, models are getting more complex and have high-dimensional hyperparameter spaces. Sequential model-based optimization (SMBO) [15] is a black-box optimization framework that replaces the time-consuming function f to evaluate with a cheap-to-evaluate surrogate function Ψ that approximates f. With the help of an acquisition function such as expected improvement [15] it sequentially chooses new points such that a balance between exploitation and exploration is found and f is optimized. In our scenario evaluating f is equivalent to learning a model on some training data for a given hyperparameter configuration and estimating the performance of this model on a hold-out data set.

Algorithm 1 outlines the SMBO framework. It starts with an observation history $\mathcal{H}$ that equals the empty set in cases where no knowledge from past experiments is used [2,14,22] or is non-empty in cases where past experiments are used [1,24,29]. First, the optimization process can be initialized. Then, the surrogate model Ψ is fitted to $\mathcal{H}$ where Ψ can be any regression model. Since the acquisition function usually needs to assess prediction uncertainty of the surrogate, common choices are Gaussian processes [1,22,24,29] or ensembles such as random forests [14]. The acquisition function chooses the next candidate to evaluate. A common choice for the acquisition function is expected improvement [15] but further acquisition functions exist such as probability of improvement [15], the conditional entropy of the minimizer [27] or a criterion based on multi-armed bandits [23]. The evaluated candidate is finally added to the set of observations. After T-many SMBO iterations, the best currently found hyperparameter configuration is returned.

Line 6 is our proposed addition to the SMBO framework. Selecting the identity function as *prune* results in the typical SMBO framework. In the next section we propose a more suitable pruning function.

Algorithm 1. Sequential Model-based Optimization

Input: Hyperparameter space Λ, observation history $\mathcal{H}$, number of iterations T, acquisition function a, surrogate model Ψ, initial hyperparameter configurations $\Lambda^{(\text{init})}$.
Output: Best hyperparameter configuration found.
1: **for** $\lambda \in \Lambda^{(\text{init})}$ **do**
2:     Evaluate $f(\lambda)$
3:     $\mathcal{H} \leftarrow \mathcal{H} \cup \{(\lambda, f(\lambda))\}$
4: **for** $t = \left|\Lambda^{(\text{init})}\right| + 1$ to T **do**
5:     Fit Ψ to $\mathcal{H}$
6:     $\Lambda^{(\text{pruned})} \leftarrow$ prune (Λ)
7:     $\lambda \leftarrow \arg\max_{\lambda \in \Lambda^{(\text{pruned})}} a(\lambda, \Psi)$
8:     Evaluate $f(\lambda)$
9:     $\mathcal{H} \leftarrow \mathcal{H} \cup \{(\lambda, f(\lambda))\}$
10: **return** $\arg\max_{(\lambda, f(\lambda)) \in \mathcal{H}} f(\lambda)$

---

4   Pruning the Search Space

The idea of pruning is to consider only a subset of the hyperparameter configuration space Λ to avoid unnecessary function evaluations in regions where we do not expect any improvements. It is obvious that if it is possible to identify regions that are for sure not of interest without evaluating any point in this region highly accelerates the hyperparameter optimization. We propose to predict the potential of regions by transferring knowledge from past experiments. The key idea is that similar data sets to the new data set have similar or even the same regions that are not interesting and therefore not worth investigating.

4.1   Formal Description

We define a region R by its center $\lambda \in \Lambda$ and diameter $\delta \in \mathbb{R}^p$, $\delta > 0$. The potential of this region after t trials on the new data set $D^{(new)}$ is defined by

$$\text{potential}\left(R = (\lambda, \delta), \Lambda_t\right) := \sum_{D' \in \mathcal{N}\left(D^{(test)}\right)} \tilde{f}_{D'}(\lambda) - \max_{\lambda' \in \Lambda_t} \tilde{f}_{D'}(\lambda') \qquad (3)$$

where Λ_t is the set of already evaluated hyperparameter configurations on $D^{(new)}$ and $\mathcal{N}\left(D^{(new)}\right)$ is the set of data sets that are closest to the new data set. $\tilde{f}_D$ is the normalized version of the response function f_D of data set D. $\tilde{f}_D$ is scaled to the interval $[0, 1]$ such that each data set has the same influence on the potential. Thus, the potential is the predicted improvement when choosing λ over the hyperparameter configurations already evaluated. Since f_D is not fully observed for $D \in \mathcal{D}$, where $\mathcal{D}$ is the meta-training set, we approximate $\tilde{f}_D$ with a plug-in estimator $\hat{y}_D$. We use a Gaussian process [21] that is trained on all normalized meta-instances of a data set such that we get for each training data set a plug-in estimator

$$\tilde{f}_D(\lambda) \sim \hat{y}_D(\lambda) := \mathcal{GP}\left(m_D(\lambda), k_D(\lambda, \lambda')\right) \qquad (4)$$

where we define m_D as the mean function and k_D as the covariance function of $\tilde{f}_D$. As a kernel function we are using the squared exponential kernel

$$k(\lambda, \lambda') := \exp\left(-\frac{\|\lambda - \lambda'\|_2^2}{2\sigma^2}\right) . \qquad (5)$$

This allows to estimate $\tilde{f}_D$ for arbitrary hyperparameter configurations. Then, we replace the definition from Equation 3 with

$$\text{potential}\left(R = (\lambda, \delta), \Lambda_t\right) := \sum_{D' \in \mathcal{N}\left(D^{(new)}\right)} \hat{y}_{D'} - \max_{\lambda' \in \Lambda_t} \hat{y}_{D'}(\lambda') . \qquad (6)$$

To estimate the nearest neighbors of the new data set $D^{(new)}$ we have to define a distance function between data sets. A common choice for this is the

Euclidean distance with respect to the meta-features [1, 29]. Since we experienced better results with a distance function based on rank correlation metrics such as the Kendall tau rank correlation coefficient [16], we are using following distance function

$$\text{KTRC}\left(D_1, D_2, \Lambda_t\right) := \frac{\sum_{\lambda_1, \lambda_2 \in \Lambda_t} \mathbb{I}\left(\hat{y}_{D_1}(\lambda_1) > \hat{y}_{D_1}(\lambda_2) \oplus \hat{y}_{D_2}(\lambda_1) > \hat{y}_{D_2}(\lambda_2)\right)}{(|\Lambda_t| - 1)|\Lambda_t|} \tag{7}$$

where $\oplus$ is the symbol for an exclusive or.

Algorithm 2. Prune

Input: Hyperparameter space Λ, observation history $\mathcal{H}$, region radius δ, fraction of the pruned space ν.
Output: Pruned hyperparameter space $\Lambda^{\text{pruned}} \subseteq \Lambda$.
  1: Estimate the most similar data sets of the new data set $\mathcal{N}\left(D^{new}\right)$ using Equation 7.
  2: Estimate the set Λ' containing the $\nu |G|$ hyperparameter configurations $\lambda' \in G \subset \Lambda$ with little potential using Equation 6.
  3: $\Lambda^{(pruned)} := \{\lambda \in \Lambda \mid \text{dist}(\lambda, \lambda') > \delta, \lambda' \in \Lambda'\}$.
  4: **return** $\Lambda^{(pruned)} \cup \{\lambda \in \Lambda \mid \text{dist}(\lambda, \lambda') \leq \delta, \lambda' \in \Lambda_t\}$

Algorithm 2 summarizes the pruning function. Line 1 estimates the k most similar data sets which we know from past experiments using the KTRC distance function defined in Equation 7. In Line 2 the potential of hyperparameter configurations are estimated using the plug-in estimators (Equation 6) on a fine grid $G \subset \Lambda$. The $\nu |G|$ hyperparameter configurations with little potential define regions where no improvement is predicted. Hence, the pruned hyperparameter space is defined as the set of hyperparameter configurations that are not within an δ-region of these low-potential hyperparameter configurations (Line 3). Additionally, the hyperparameter configurations that are within a δ-region of already evaluated hyperparameter configurations are added (Line 4). The intuition here is that since we have already observed an evaluation in this region, the acquisition function will not choose a hyperparameter combination close to these points for exploration but only for exploitation. Hence, no evaluations will be done by the standard SMBO framework without a very likely improvement. For the distance function between hyperparameter configurations we need to consider one that does not take discrete variables into account. Obviously, the loss does not change smoothly when changing a categorical variable that e.g. indicates which algorithm was chosen. Therefore, we define the distance function in Algorithm 2 as

$$\text{dist}(\lambda, \lambda') := \begin{cases} \infty & \text{if } \lambda \text{ and } \lambda' \text{ differ in a categorical variable} \\ \|\lambda - \lambda'\| & \text{otherwise} \end{cases} \tag{8}$$

5   Experimental Evaluation

First, we will introduce the reader to the state of the art tuning strategies which are used to evaluate pruning. Then, the evaluation metrics are defined and the meta-data sets are introduced. Finally, the results are presented.

5.1   Tuning Strategies

We want to give a short introduction to all the tuning strategies we will consider in our experiments. We are considering both strategies that are using no knowledge from previous experiments and those that do.

Random Search. This is the only strategy that is not using any surrogate model. Hyperparameter configurations are sampled uniformly at random. This is a common strategy in cases where a grid search is not possible. Bergstra and Bengio [3] have shown that this is very effective for hyperparameters with low effective dimensionality.

Independent Gaussian Process (I-GP). This tuning strategy uses a Gaussian process [22] with squared-exponential kernel as a surrogate model. It only uses knowledge from the current data set and is not using any knowledge from previous experiments.

Independent Random Forest (I-RF). Next to Gaussian processes, random forests are the most widely used surrogate models [14] and hence we are using them in our experiments. Like the independent Gaussian process, the I-RF does not use any knowledge from previous experiments.

Sequential Model-based Algorithm Configuration++ (SMAC++). SMAC [14] is a tuning strategy that is based on a random forest as a surrogate model without background knowledge of previous experiments. SMAC++ is our extension to SMAC. SMAC++ is using the typical SMBO framework but the random forest is also trained on the meta-training data.

Surrogate Collaborative Tuning (SCoT). SCoT [1] uses a Gaussian process with squared-exponential kernel with automatic relevance determination and is trained on hyperparameter observations of previous experiments evaluated on other data sets and the few knowledge achieved on the new data set. An SVMRank is learned on the data set and its predictions are used instead of the hyperparameter performances. Bardenet et al. [1] argue that this overcomes the problem of having data sets with different scales of hyperparameter performances. In the original work it was proposed to use an RBF kernel for SVMRank. For reasons of computational complexity we follow the lead of Yogatama and Mann [29] and use a linear kernel instead.

Gaussian Process with MKL (MKL-GP). Similarly to Bardenet et al. [1], Yogatama and Mann [29] propose to use a Gaussian process as a surrogate model for the SMBO framework. Instead of using SVMRank to deal with the different scales, they are adapting the mean of the Gaussian process, accordingly. Additionally, they are using a specific kernel, a linear combination of an SE-ARD kernel with a kernel modelling the distance between data sets.

Optimal. This is an artificial tuning strategy that always evaluates the best hyperparameter configuration and is added to plots for orientation purposes.

Kernel parameters are learned by maximizing the marginal likelihood on the meta-training set [21]. Hyperparameters of the tuning strategies are optimized in a leave-one-out cross-validation on the meta-training set.

The results reported are the average of at least ten repetitions. For the strategies with random initialization (Random, I-GP, I-RF), the mean of 1000 repetitions is reported.

5.2   Evaluation Metrics

In our experiments we are using three different evaluation metrics which we will explain here in detail.

Average Rank. The *average rank among different hyperparameter tuning strategies* or for short simply *average rank* is a relative metric between different tuning strategies. The tuning strategies are ranked by the best hyperparameter configuration that they have found so far, ties are solved by granting them the average rank. If we have for example four different tuning strategies that have found hyperparameter configurations that achieve an accuracy of 0.78, 0.77, 0.77 and 0.76, respectively, then the ranking is 1, 2.5, 2.5 and 4.

Normalized Average Loss. The disadvantage of the average rank is that it gives no information about by which margin the found hyperparameters of one tuning strategy are better than another and it will vary when strategies are added or are removed. One metric that overcomes this disadvantage is the *normalized average loss*. In our experiments we will consider only classification problems such that $f_D(\lambda)$ is the accuracy on data set D using hyperparameter configuration λ. Since the scale of f_D varies for different D we normalize f_D between 0 and 1 such that every data set has the same impact on the evaluation metric. Thus, the normalized average loss at iteration t is defined as

$$\text{NAL}\left(\mathcal{D}, \Lambda_t\right) := \frac{1}{|\mathcal{D}|} \sum_{D \in \mathcal{D}} 1 - \frac{\max_{\lambda \in \Lambda_t} f_D\left(\lambda\right) - \min_{\lambda \in \Lambda} f_D\left(\lambda\right)}{\max_{\lambda \in \Lambda} f_D\left(\lambda\right) - \min_{\lambda \in \Lambda} f_{D(\lambda)}} . \tag{9}$$

Average Hyperparameter Rank. The average hyperparameter rank is another way to overcome the disadvantages of the average rank. Compared to the average rank it is not ranking the tuning strategies but ranking the hyperparameter configurations. Let $r_D(\lambda)$ be the rank of the hyperparameter configuration λ on data set D, then the average hyperparameter rank is defined as

$$\text{AHR}(\mathcal{D}, \Lambda_t) := \frac{1}{|\mathcal{D}|} \sum_{D \in \mathcal{D}} \min_{\lambda \in \Lambda_t} r_D(\lambda) - 1 \ . \tag{10}$$

5.3   Meta-Data Sets

The SVM meta-data set was created by using 50 classification data sets chosen at random. All instances were merged in cases where splits were already given, shuffled and split into 80% train and 20% test. We then used a support vector machine (SVM) [5] to create the meta-instances. We trained the SVM using three different kernels (linear, polynomial and Gaussian) and estimated the labels of the meta-instances by evaluating the trained model on the test split. The hyperparameter space dimension is six, three dimensions for binary features that indicate which kernel was chosen, one for the trade-off parameter C, one for the degree of the polynomial kernel d and the width γ of the Gaussian kernel. If the hyperparameter is not involved, e.g. the degree if we are using the linear kernel, it was set to 0. The test accuracy was precomputed on a grid $C \in \{2^{-5}, \dots, 2^6\}$, $d \in \{2, \dots, 10\}$, $\gamma \in \{10^{-4}, 10^{-3}, 10^{-2}, 0.05, 0.1, 0.5, 1, 2, 5, 10, 20, 50, 10^2, 10^3\}$ resulting into 288 meta-instances per data set. Since meta-features are a vital part for many surrogate models and mandatory for SCoT and MKL-GP, we added the meta-features that were used by [1, 29] to our meta-data. First , we extracted the number of training instances n, the number of classes c and the number of predictors m. The final meta-features are c, $\log(m)$ and $\log(n/m)$ scaled to [0.1].

The Weka meta-data set was created using 59 classification data sets which were preprocessed like the classification data sets used for the SVM meta-data set. We used 19 different Weka classifiers [13] and produced 21,871 hyperparameter configurations per data set. The dimension of the hyperparameter space is 102 including the indicator variables for the classifier. Thus, this meta-data set focuses stronger on the model class selection. Overall, this meta-data set contains 1,290,389 instances. In comparison, OpenML [26] has collaboratively collected 344,472 runs.[1]

The meta-data sets are available on our supplementary website together with a visualization of the meta-data as well as more details about how the meta-data sets were created and a detailed list which data sets were used [28].

5.4   Hyperparameter Optimization for SVMs

To show that the proposed plug-in estimators work (Equation 4), we did not use all 288 hyperparameter configurations for training but only 50 per data set. The

---

[1] Status 2015/03/27 by http://openml.org

evaluation is nevertheless done on all 288 of the new data set. We choose G to contain these 288 configurations and fixed $|\mathcal{N}(D^{new})| = 2$, $\nu = 1 - |G|^{-1}$ and δ such that the two closest neighbored hyperparameter configurations of the test region are within δ-distance.

We want to conduct two different experiments. First, we want to compare a surrogate model with pruning to current state of the art tuning strategies. We once again want to stress that pruning in the SMBO framework is an orthogonal contribution such that these results are actually of minor interest. Second, we want to compare different surrogate models with and without pruning or initialization. Pruning is a useful contribution as long as it does not worsen the optimization speed in general and accelerates it in some cases.

Figure 1 shows the results of the comparison of pruning to the current state of the art method. As a surrogate model we decided to choose the Gaussian process that is *not* learned across data sets since it is the most common and simple surrogate model. Surprisingly, the pruning alone with the Gaussian process is able to outperform all the competitor strategies with respect to all three evaluation metrics.

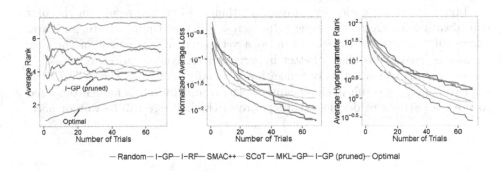

Fig. 1. Pruning is an orthogonal contribution to the SMBO framework. Nevertheless, we compare a pruned independent Gaussian process to many current state of the art tuning strategies without pruning.

Figures 2 to 6 show the results of different surrogate models. We distinguish four different cases: i) only the surrogate model, ii) the surrogate model with pruning, iii) the surrogate model with three steps of initialization and iv) the surrogate model with three steps of initialization and pruning. Figures 2 and 3 show the results for the surrogate models that do not learn across data sets and the remaining three Figures show the results for the surrogate models that learn across data sets. Our expectation before the experiments were that the lift is higher i) for the experiments without initialization and ii) for the experiments with the surrogate models that do not learn across data sets. The reason for this is simple. An initialization is a fixed policy that proposes hyperparameter configurations that has been good on average while pruning discards regions that were

not useful. Thus, pruning will also have an effect of initialization. The difference between initialization and pruning is that initialization proposes a specific
hyperparameter while pruning reduces the full hyperparameter space to a set
of good hyperparameter configurations and pruning is applied at each iteration
and not just for the initial iterations. Additionally, pruning is a way to transfer knowledge between data sets such that those strategies that do not use this
knowledge at all benefit more and are prevented from conducting unnecessary
exploration queries.

This is exactly what the results of the experiments show. The SMBO experiments with pruning have comparable good starting points like those with initialization. If we compare the results of the independent Gaussian process and
random forest for the setting with only initialization with the one with only
pruning, we clearly see the unnecessary exploration queries after a good start.
The setting with both initialization and pruning does not suffer from this problem and thus is clearly the best strategy. This effect is weaker for the surrogate
models that are learned across data sets in Figures 4 and 6. Only for SCoT
(Figure 5) pruning does not accelerate the hyperparameter optimization on *this*
meta-data set but it also does not worsen it. Table 1 shows the results for all
evaluation metrics and surrogate models.

The reader may notice two important things. First, the results in the plot
will always converge to the same value across different tuning strategies if you
allow only enough trials. Second, even a very small improvement of the performance *just* by choosing a better hyperparameter configurations is already a
success especially since this optimization is usually limited in time. This little
improvement may result in significantly better results for a new model compared
to the competitors or decides whether a research challenge will be won or not.

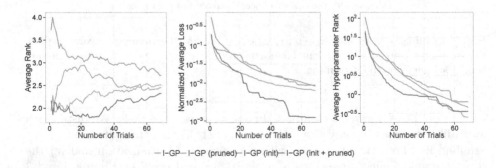

Fig. 2. Average rank, normalized average loss and average hyperparameter rank for
I-GP on the SVM meta-data set.

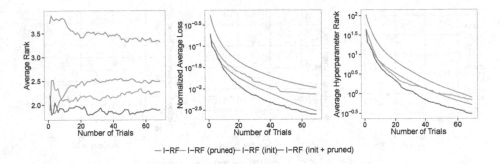

Fig. 3. Average rank, normalized average loss and average hyperparameter rank for
I-RF on the SVM meta-data set.

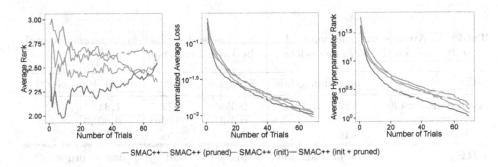

Fig. 4. Average rank, normalized average loss and average hyperparameter rank for
SMAC++ on the SVM meta-data set.

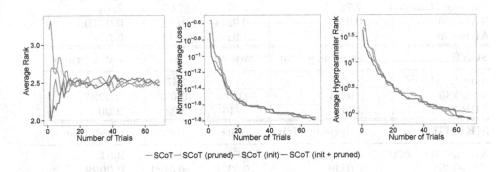

Fig. 5. Average rank, normalized average loss and average hyperparameter rank for
SCoT on the SVM meta-data set.

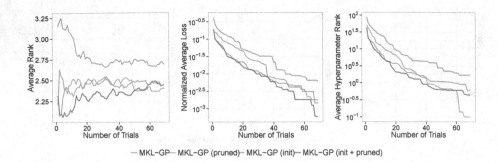

— MKL–GP— MKL–GP (pruned)— MKL–GP (init)— MKL–GP (init + pruned)

Fig. 6. Average rank, normalized average loss and average hyperparameter rank for MKL-GP on the SVM meta-data set.

Table 1. Average rank, normalized average loss and average hyperparameter rank after 30 trials on the SVM meta-data set. Best results are bold.

I-GP	no pruning/init	pruned	init	init + pruned
Average Rank@30	3.12	2.35	2.72	**1.81**
NAL@30	0.0224	0.0131	0.0291	**0.0055**
AHR@30	3.48	2.60	3.98	**1.97**
I-RF	no pruning/init	pruned	init	init + pruned
Average Rank@30	3.51	2.11	2.51	**1.87**
NAL@30	0.0281	0.0149	0.0116	**0.0070**
AHR@30	4.75	2.64	2.98	**2.14**
SMAC++	no pruning/init	pruned	init	init + pruned
Average Rank@30	2.72	2.45	2.65	**2.18**
NAL@30	0.0251	0.0228	0.0256	**0.0210**
AHR@30	5.42	4.92	4.52	**3.53**
SCoT	no pruning/init	pruned	init	init + pruned
Average Rank@30	2.55	2.55	2.47	**2.43**
NAL@30	0.0244	0.0244	**0.0237**	**0.0237**
AHR@30	3.44	3.44	3.02	**2.90**
MKL-GP	no pruning/init	pruned	init	init + pruned
Average Rank@30	2.68	2.52	2.49	**2.31**
NAL@30	0.0349	0.0232	0.0120	**0.0099**
AHR@30	6.30	3.48	3.00	**2.40**

5.5   Hyperparameter Optimization for Weka

In the last chapter, we have seen little improvement in cases where an initialization is combined with surrogate models that are learning across data sets. We expect pruning to be useful in two scenarios: if i) the dimensionality of the hyperparameter space is very high and ii) the meta-data set is too large such that surrogate models that are learning across data sets are no longer a cost-efficient alternative to evaluating the true function. Since most surrogate models are based on Gaussian processes, a further problem is storing the kernel matrix. In our next meta-data set we are using more than a million meta-instances which result into a kernel matrix of dimensions $10^6 \times 10^6$ which needs 8 TB of memory for storing it.

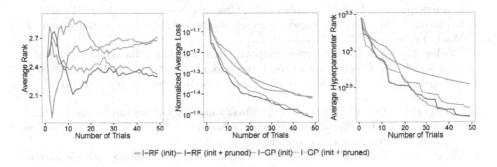

Fig. 7. Average rank, normalized average loss and average hyperparameter rank for I-RF and I-GP on the Weka meta-data set.

For the Weka meta-data set we conducted a similar experiment as for the SVM meta-data set. Due to the size we restricted ourselves to the tuning strategies that do not learn across data sets. Previously, we have seen that a tuning strategy without initialization and pruning is outperformed by a large margin by the same strategy only using pruning. Hence, we show here only the comparison between the strategy i) only using an initialization step and ii) using both initialization and pruning. Figure 7 concludes our experiments. As we have seen on the SVM meta-data set, pruning again indicates that it is a useful addition to the SMBO framework by further accelerating the hyperparameter optimization.

6   Conclusion and Future Work

We propose pruning as an orthogonal contribution the the SMBO framework and show in elaborated experiments on two different data set that it accelerates the hyperparameter optimization in most cases and in the worst case does not worsen it. It can be especially considered for tuning strategies that do not use information from the past for the surrogate model. Additionally, we created a new meta-data set which is the largest to the best of our knowledge with about four times more experiments than OpenML and make it publicly available.

Acknowledgments. The authors gratefully acknowledge the co-funding of their work by the German Research Foundation (DFG) under grant SCHM 2583/6-1.

References

1. Bardenet, R., Brendel, M., Kégl, B., Sebag, M.: Collaborative hyperparameter tuning. In: Proceedings of the 30th International Conference on Machine Learning, ICML 2013, Atlanta, GA, USA, 16–21 June 2013, pp. 199–207 (2013)
2. Bergstra, J., Bardenet, R., Bengio, Y., Kégl, B.: Algorithms for hyper-parameter optimization. In: Advances in Neural Information Processing Systems 24: 25th Annual Conference on Neural Information Processing Systems 2011. Proceedings of a meeting held 12–14 December 2011, Granada, Spain, pp. 2546–2554 (2011)
3. Bergstra, J., Bengio, Y.: Random search for hyper-parameter optimization. J. Mach. Learn. Res. **13**, 281–305 (2012)
4. Cawley, G.: Model selection for support vector machines via adaptive step-size tabu search. In: Proceedings of the International Conference on Artificial Neural Networks and Genetic Algorithms, Prague, Czech Republic, pp. 434–437, April 2001
5. Chang, C.C., Lin, C.J.: LIBSVM: A library for support vector machines. ACM Transactions on Intelligent Systems and Technology **2**, 27:1–27:27 (2011). software available at http://www.csie.ntu.edu.tw/ cjlin/libsvm
6. Chapelle, O., Vapnik, V., Bousquet, O., Mukherjee, S.: Choosing multiple parameters for support vector machines. Machine Learning **46**(1–3), 131–159 (2002)
7. Coates, A., Ng, A.Y., Lee, H.: An analysis of single-layer networks in unsupervised feature learning. In: Proceedings of the Fourteenth International Conference on Artificial Intelligence and Statistics, AISTATS 2011, Fort Lauderdale, USA, April 11–13, 2011, pp. 215–223 (2011)
8. David-Tabibi, O., Netanyahu, N.S.: Verified null-move pruning. ICGA Journal **25**(3), 153–161 (2002)
9. Feurer, M., Springenberg, J.T., Hutter, F.: Using meta-learning to initialize bayesian optimization of hyperparameters. In: ECAI workshop on Metalearning and Algorithm Selection (MetaSel), pp. 3–10 (2014)
10. Friedrichs, F., Igel, C.: Evolutionary tuning of multiple svm parameters. Neurocomput. **64**, 107–117 (2005)
11. Gomes, T.A.F., Prudêncio, R.B.C., Soares, C., Rossi, A.L.D., Carvalho, A.C.P.L.F.: Combining meta-learning and search techniques to select parameters for support vector machines. Neurocomputing **75**(1), 3–13 (2012)
12. Guo, X.C., Yang, J.H., Wu, C.G., Wang, C.Y., Liang, Y.C.: A novel ls-svms hyper-parameter selection based on particle swarm optimization. Neurocomput. **71**(16–18), 3211–3215 (2008)
13. Hall, M., Frank, E., Holmes, G., Pfahringer, B., Reutemann, P., Witten, I.H.: The weka data mining software: An update. SIGKDD Explor. Newsl. **11**(1), 10–18 (2009)
14. Hutter, F., Hoos, H.H., Leyton-Brown, K.: Sequential model-based optimization for general algorithm configuration. In: Coello, C.A.C. (ed.) LION 2011. LNCS, vol. 6683, pp. 507–523. Springer, Heidelberg (2011)

15. Jones, D.R., Schonlau, M., Welch, W.J.: Efficient global optimization of expensive black-box functions. J. of Global Optimization **13**(4), 455–492 (1998)
16. Kendall, M.G.: A New Measure of Rank Correlation. Biometrika **30**(1/2), 81–93 (1938)
17. Land, A.H., Doig, A.G.: An Automatic Method for Solving Discrete Programming Problems. Econometrica **28**, 497–520 (1960)
18. Lawler, E.L., Wood, D.E.: Branch-And-Bound Methods: A Survey. Operations Research **14**(4), 699–719 (1966)
19. Leite, R., Brazdil, P., Vanschoren, J.: Selecting classification algorithms with active testing. In: Perner, P. (ed.) MLDM 2012. LNCS, vol. 7376, pp. 117–131. Springer, Heidelberg (2012)
20. Pinto, N., Doukhan, D., DiCarlo, J.J., Cox, D.D.: A high-throughput screening approach to discovering good forms of biologically inspired visual representation. PLoS Computational Biology **5**(11), e1000579 (2009). PMID: 19956750
21. Rasmussen, C.E., Williams, C.K.I.: Gaussian Processes for Machine Learning. Adaptive Computation and Machine Learning. The MIT Press (2005)
22. Snoek, J., Larochelle, H., Adams, R.P.: Practical bayesian optimization of machine learning algorithms. In: Advances in Neural Information Processing Systems 25: 26th Annual Conference on Neural Information Processing Systems 2012. Proceedings of a meeting held December 3–6, 2012, Lake Tahoe, Nevada, United States, pp. 2960–2968 (2012)
23. Srinivas, N., Krause, A., Seeger, M., Kakade, S.M.: Gaussian process optimization in the bandit setting: no regret and experimental design. In: Fürnkranz, J., Joachims, T. (eds.) Proceedings of the 27th International Conference on Machine Learning (ICML 2010), pp. 1015–1022. Omnipress (2010)
24. Swersky, K., Snoek, J., Adams, R.P.: Multi-task bayesian optimization. In: Advances in Neural Information Processing Systems 26: 27th Annual Conference on Neural Information Processing Systems 2013. Proceedings of a meeting held December 5–8, 2013, Lake Tahoe, Nevada, United States, pp. 2004–2012 (2013)
25. Thornton, C., Hutter, F., Hoos, H.H., Leyton Brown, K.: Auto-weka: combined selection and hyperparameter optimization of classification algorithms. In: Proceedings of the 19th ACM SIGKDD International Conference on Knowledge Discovery and Data Mining, KDD 2013, pp. 847–855. ACM, New York (2013)
26. Vanschoren, J., van Rijn, J.N., Bischl, B., Torgo, L.: Openml: Networked science in machine learning. SIGKDD Explorations **15**(2), 49–60 (2013)
27. Villemonteix, J., Vazquez, E., Walter, E.: An informational approach to the global optimization of expensive-to-evaluate functions. Journal of Global Optimization **44**(4), 509–534 (2009)
28. Wistuba, M.: Supplementary website, June 2015. http://hylap.org/publications/Hyperparameter-Search-Space-Pruning
29. Yogatama, D., Mann, G.: Efficient transfer learning method for automatic hyperparameter tuning. In: International Conference on Artificial Intelligence and Statistics (AISTATS 2014) (2014)

Multi-Task Learning with Group-Specific Feature Space Sharing

Niloofar Yousefi[1][(✉)], Michael Georgiopoulos[1],
and Georgios C. Anagnostopoulos[2]

[1] Department of Electrical Engineering and Computer Science,
University of Central Florida, 4000 Central Florida Blvd., Orlando, FL 32816, USA
{niloofar.yousefi,michaelg}@ucf.edu
[2] Department of Electrical and Computer Engineering,
Florida Institute of Technology, 150 W. University Blvd., Melbourne, FL 32901, USA
georgio@fit.edu

Abstract. When faced with learning a set of inter-related tasks from a limited amount of usable data, learning each task independently may lead to poor generalization performance. (MTL) exploits the latent relations between tasks and overcomes data scarcity limitations by co-learning all these tasks simultaneously to offer improved performance. We propose a novel Multi-Task Multiple Kernel Learning framework based on Support Vector Machines for binary classification tasks. By considering pair-wise task affinity in terms of similarity between a pair's respective feature spaces, the new framework, compared to other similar MTL approaches, offers a high degree of flexibility in determining how similar feature spaces should be, as well as which pairs of tasks should share a common feature space in order to benefit overall performance. The associated optimization problem is solved via a block coordinate descent, which employs a consensus-form Alternating Direction Method of Multipliers algorithm to optimize the Multiple Kernel Learning weights and, hence, to determine task affinities. Empirical evaluation on seven data sets exhibits a statistically significant improvement of our framework's results compared to the ones of several other Clustered Multi-Task Learning methods.

1 Introduction

Multi-Task Learning (MTL) is a machine learning paradigm, where several related task are learnt simultaneously with the hope that, by sharing information among tasks, the generalization performance of each task will be improved. The underlying assumption behind this paradigm is that the tasks are related to each other. Thus, it is crucial how to capture task relatedness and incorporate it into an MTL framework. Although, many different MTL methods [1,7,12,15,18,27] have been proposed, which differ in how the relatedness across multiple tasks is modeled, they all utilize the parameter or structure sharing strategy to capture the task relatedness.

However, the previous methods are restricted in the sense that they assume all tasks are similarly related to each other and can equally contribute to the joint

© Springer International Publishing Switzerland 2015
A. Appice et al. (Eds.): ECML PKDD 2015, Part II, LNAI 9285, pp. 120–136, 2015.
DOI: 10.1007/978-3-319-23525-7_8

learning process. This assumption can be violated in many practical applications as "outlier" tasks often exist. In this case, the effect of "negative transfer", *i.e.*, sharing information between irrelevant tasks, can lead to a degraded generalization performance.

To address this issue, several methods, along different directions, have been proposed to discover the inherent relationship among tasks. For example, some methods [3,26–28], use a regularized probabilistic setting, where sharing among tasks is done based on a common prior. These approaches are usually computationally expensive. Another family of approaches, known as the Clustered Multi-Task Learning (CMTL), assumes that tasks can be clustered into groups such that the tasks within each group are close to each other according to a notion of similarity. Based on the current literature, clustering strategies can be broadly classified into two categories: task-level CMTL and feature-level CMTL.

The first one, task-level CMTL, assumes that the model parameters used by all tasks within a group are close to each other. For example, in [2,13,17], the weight vectors of the tasks belonging to the same group are assumed to be similar to each other. However, the major limitations for these methods are: (i) that such an assumption might be too risky, as similarity among models does not imply that meaningful sharing of information can occur between tasks, and (ii) for these methods, the group structure (number of groups or basis tasks) is required to be known a priori.

The other strategy for task clustering, referred to as feature-level CMTL, is based on the assumption that task relatedness can be modeled as learning shared features among the tasks within each group. For example, in [19] the tasks are clustered into different groups and it is assumed that tasks within the same group can jointly learn a shared feature representation. The resulting formulation leads to a non-convex objective, which is optimized using an alternating optimization algorithm converging to local optima, and suffers potentially from slow convergence. Another similar approach has been proposed in [25], which assumes that tasks should be related in terms of feature subsets. This study also leads to a non-convex co-clustering structure that captures task-feature relationship. These methods are restricted in the sense that they assume that tasks from different groups have nothing in common with each other. However, this assumption is not always realistic, as tasks in disjoint groups might still be inter-related, albeit weekly. Hence, assigning tasks into different groups may not take full advantage of MTL. Another feature-level clustering model has been proposed in [29], in which the cluster structure can vary from feature to feature. While, this model is more flexible compared to other CMTL methods, it is, however, more complicated and also less general compared to our framework, as it tries to find a shared feature representation for tasks by decomposing each task parameter into two parts: one to capture the shared structure between tasks and another to capture the variations specific to each task. This model is further extended in [16], where a multi-level structure has been introduced to learn task groups in the context of MTL. Interestingly, it has been shown that there is an equivalent relationship between CMTL and alternating structure optimization [30], wherein the basic idea is to identify a shared low-dimensional predictive structure for all tasks.

In this paper, we develop a new MTL model capable of modeling a more general type of task relationship, where the tasks are implicitly grouped according to a notion of feature similarity. In our framework, the tasks are not forced to have a common feature space; instead, the data automatically suggests a flexible group structure, in which a common, similar or even distinct feature spaces can be determined between different pairs of tasks. Additionally, our MTL framework is kernel-based and, thus, may take advantage of the non-linearity introduced by the feature mapping of the associated Reproducing Kernel Hilbert Space (RKHS) $\mathcal{H}$. Also, to avoid a degradation in generalization performance due to choosing an inappropriate kernel function, our framework employs a Multiple Kernel Learning (MKL) strategy [21], hence, rendering it a Multi-Task Multiple Kernel Learning (MT-MKL) approach.

It is worth mentioning that a widely adopted practice for combining kernels is to place an L_p-norm constraint on the combination coefficients $\boldsymbol{\theta} = [\theta_1, \ldots, \theta_M]$, which are learned during training. For example, a conically combination of task objectives with an L_p-norm feasible region is introduced in [23] and further extended in [22]. Also, another method introduced in [24] proposes a partially shared kernel function $k_t \triangleq \sum_{m=1}^{M} (\mu^m + \lambda_t^m) k_m$, along with L_1-norm constraints on $\boldsymbol{\mu}$ and $\boldsymbol{\lambda}$. The main advantage of such a method over the traditional MT-MKL methods, which consider a common kernel function for all tasks (by letting $\lambda_t^m = 0, \forall t, m$), is that it allows tasks to have their own task-specific feature spaces and, potentially, alleviate the effect of negative transfer. However, popular MKL formulations in the context of MTL, such as this one, are capable of modeling two types of tasks: those that share a global, common feature space and those that employ their own, task-specific feature space. In this work we propose a more flexible framework, which, in addition to allowing some tasks to use their own specific feature spaces (to avoid negative transfer learning), it permits forming arbitrary groups of tasks sharing the same, group-specific (instead of a single, global), common feature space, whenever warranted by the data. This is accomplished by considering a group lasso regularizer applied to the set of all pair-wise differences of task-specific MKL weights. For no regularization penalty, each task is learned independently of each other and will utilize its own feature space. As the regularization penalty increases, pairs of MKL weights are forced to equal each other leading the corresponding pairs of tasks to share a common feature space. We demonstrate that the resulting optimization problem can be solved by employing a 2-block coordinate descent approach, whose first block consists of the Support Vector Machine (SVM) weights for each task and which can be optimized efficiently using existing solvers, while its second block comprises the MKL weights from all tasks and is optimized via a consensus-form, Alternating Direction Method of Multipliers (ADMM)-based step.

The rest of the paper is organized as follows: In Sect. 2 we describe our formulation for jointly learning the optimal feature spaces and the parameters of all the tasks. Sect. 3 provides an optimization technique to solve our non-smooth convex optimization problem derived in Sect. 2. Sect. 4 presents a Rademacher complexity-based generalization bound for the hypothesis space corresponding to our model. Experiments are provided in Sect. 5, which demonstrate the effectiveness of our proposed model compared to several MTL methods. Finally, in Sect. 6 we conclude our work and briefly summarize our findings.

Notation: In what follows, we use the following notational conventions: vectors and matrices are depicted in bold face. A prime $'$ denotes vector/matrix transposition. The ordering symbols $\succeq$ and $\preceq$ when applied to vectors stand for the corresponding component-wise relations. If $\mathbb{Z}_+$ is the set of postivie integers, for a given $S \in \mathbb{Z}_+$, we define $\mathbb{N}_S \triangleq \{1, \ldots, S\}$. Additional notation is defined in the text as needed.

2 Formulation

Assume T supervised learning tasks, each with a training set $\{(x_t^n, y_t^n)\}_{n=1}^{n_t}, t \in \mathbb{N}_T$, which is sampled from an unknown distribution $P_t(x, y)$ on $\mathcal{X} \times \{-1, 1\}$. Here, $\mathcal{X}$ denotes the native space of samples for all tasks and ± 1 are the associated labels. Without loss of generality, we will assume an equal number n of training samples per task. The objective is to learn T binary classification tasks using discriminative functions $f_t(x) \triangleq \langle w_t, \phi_t(x) \rangle_{\mathcal{H}_{t,\theta}} + b_t$ for $t \in \mathbb{N}_T$, where w_t is the weight vector associated to task t. Moreover, the feature space of task t is served by $\mathcal{H}_{t,\theta} = \bigoplus_{m=1}^{M} \sqrt{\theta_t^m} \mathcal{H}_m$ with induced feature mapping $\phi_t \triangleq [\sqrt{\theta_t^1} \phi_1' \cdots \sqrt{\theta_t^M} \phi_M']'$ and endowed with the inner product $\langle \cdot, \cdot \rangle_{\mathcal{H}_{t,\theta}} = \sum_{m=1}^{M} \theta_t^m \langle \cdot, \cdot \rangle_{\mathcal{H}_m}$. The reproducing kernel function for this feature space is given as $k_t(x_t^i, x_t^j) = \sum_{m=1}^{M} \theta_t^m k_m(x_t^i, x_t^j)$ for all $x_t^i, x_t^j \in \mathcal{X}$. In our framework, we attempt to learn the w_t's and b_t's jointly with the θ_t's via the following regularized risk minimization problem:

$$
\min_{w \in \Omega(w), \theta \in \Omega(\theta), b} \sum_{t=1}^{T} \frac{\|w_t\|^2}{2} + C \sum_{t=1}^{T} \sum_{i=1}^{n} [1 - y_t^i f_t(x_t^i)]_+ + \lambda \sum_{t=1}^{T-1} \sum_{s>t}^{T} \|\theta_t - \theta_s\|_2
$$

$$
\Omega(w) \triangleq \{w = (w_1, \cdots, w_T) : w_t \in \mathcal{H}_{t,\theta}, \theta \in \Omega(\theta)\}
$$

$$
\Omega(\theta) \triangleq \{\theta - (\theta_t, \cdots, \theta_T) : \theta_t \succeq 0, \|\theta_t\|_1 \leq 1, \forall t \in \mathbb{N}_T\} \tag{1}
$$

where $w \triangleq (w_t, \cdots, w_T)$ and $\theta \triangleq (\theta_t, \cdots, \theta_T)$, $\Omega(w)$ and $\Omega(\theta)$ are the corresponding feasible sets for w and θ respectively, and $[u]_+ = \max\{u, 0\}$, $u \in \mathbb{R}$ denotes the hinge function. Finally, C and λ are non-negative regularization parameters.

The last term in Problem 1 is the sum of pairwise differences between the tasks' feature weight vectors. For each pair of (θ_t, θ_s), the pairwise penalty $\|\theta_t - \theta_s\|_2$ may favor a small number of non-identical θ_t. Therefore, it ensures that a flexible (common, similar or distinct) feature space, will be selected between tasks t and s. In this manner, a flexible group structure of shared features across multiple tasks can be achieved by this framework. It is also worth mentioning that two special cases are covered by the proposed model: (i) if $\lambda \to \infty$ (λ is only required to be sufficiently large), for all task pairs $\|\theta_t - \theta_s\|_2 \to 0$ and, thus, all tasks share a single common feature space. (ii) As $\lambda \to 0$, the proposed model reduces to T independent classification tasks.

It is easy to verify that Problem 1 is a convex minimization problem, which can be solved using a block coordinate descent method alternating between the minimization with respect to $\boldsymbol{\theta}$ and the $(\boldsymbol{w}, \boldsymbol{b})$ pair. Motivated by the non-smooth nature of the last regularization term, in Sect. 3 we develop a consensus version of the ADMM to solve the minimization problem with respect to $\boldsymbol{\theta}$.

3   The Proposed Consensus Optimization Algorithm

Problem 1 can be formulated as the following equivalent problem, which entails T inter-related SVM training problems:

$$
\min_{\boldsymbol{\theta}, \boldsymbol{w}, \boldsymbol{b}, \boldsymbol{\xi}} \sum_{t=1}^{T} \sum_{m=1}^{M} \frac{\|\boldsymbol{w}_t^m\|_{\mathcal{H}_m}^2}{2\theta_t^m} + C \sum_{t=1}^{T} \sum_{i=1}^{n} \xi_t^i + \lambda \sum_{t=1}^{T-1} \sum_{s>t}^{T} \|\boldsymbol{\theta}_t - \boldsymbol{\theta}_s\|_2
$$

$$
s.t.\ y_t^i \left(\langle \boldsymbol{w}_t, \phi(x_t^i) \rangle_{\mathcal{H}_t} + b_t \right) \geq 1 - \xi_t^i,\ \xi_t^i \geq 0,\ \forall\, t \in \mathbb{N}_T, i \in \mathbb{N}_n
$$

$$
\boldsymbol{\theta}_t \succeq \mathbf{0}, \|\boldsymbol{\theta}_t\|_1 \leq 1, \forall\, t \in \mathbb{N}_T \tag{2}
$$

It can be shown that the primal-dual form of Problem 2 with respect to $\boldsymbol{\theta}$ and $\{\boldsymbol{w}, \boldsymbol{b}, \boldsymbol{\xi}\}$ is given by

$$
\min_{\boldsymbol{\theta}_t \in \Omega(\boldsymbol{\theta})} \max_{\boldsymbol{\alpha}_t \in \Omega(\boldsymbol{\alpha})} \sum_{t=1}^{T} \boldsymbol{\alpha}_t' \mathbf{1}_n - \frac{1}{2} \sum_{t=1}^{T} \sum_{m=1}^{M} \theta_t^m (\boldsymbol{\alpha}_t' Y_t K_t^m Y_t \boldsymbol{\alpha}_t) + \lambda \sum_{t=1}^{T-1} \sum_{s>t}^{T} \|\boldsymbol{\theta}_t - \boldsymbol{\theta}_s\|_2
$$

$$
\Omega(\boldsymbol{\alpha}) \triangleq \{\boldsymbol{\alpha} = (\boldsymbol{\alpha}_t, \cdots, \boldsymbol{\alpha}_T) : 0 \preceq \boldsymbol{\alpha}_t \preceq C\mathbf{1}_n,\ \boldsymbol{\alpha}_t' \boldsymbol{y}_t = 0,\ \forall\, t \in \mathbb{N}_T\}
$$

$$
\Omega(\boldsymbol{\theta}) \triangleq \{\boldsymbol{\theta} = (\boldsymbol{\theta}_t, \cdots, \boldsymbol{\theta}_T) : \boldsymbol{\theta}_t \succeq \mathbf{0}, \|\boldsymbol{\theta}_t\|_1 \leq 1, \forall\, t \in \mathbb{N}_T\} \tag{3}
$$

where $\mathbf{1}_n$ is a vector containing n 1's, $Y_t \triangleq diag(\boldsymbol{y}_t)$, $K_t^m \in \mathbb{R}^{n \times n}$ is the kernel matrix, whose (i, j) entry is given as $k_m(x_t^i, x_t^j)$, $\boldsymbol{\theta}_t \triangleq [\theta_t^1, \ldots, \theta_t^M]'$, and $\boldsymbol{\alpha}_t$ is the Lagrangian dual variable for the minimization problem w.r.t.$\{\boldsymbol{w}_t, b_t, \boldsymbol{\xi}_t\}$.

It is not hard to verify that the optimal objective value of the dual problem is equal to the optimal objective value of the primal one, as the strong duality holds for the primal-dual optimization problems w.r.t.$\{\boldsymbol{w}, \boldsymbol{b}, \boldsymbol{\xi}\}$ and $\boldsymbol{\alpha}$ respectively. Therefore, a block coordinate descent framework[1] can be applied to decompose Problem 3 into two subproblems. The first subproblem, which is the maximization problem with respect to $\boldsymbol{\alpha}$, can be efficiently solved via LIBSVM [8], and the second subproblem, which is the minimization problem with respect to $\boldsymbol{\theta}$, takes the form

$$
\min_{\boldsymbol{\theta}_t} \lambda \sum_{t=1}^{T-1} \sum_{s>t}^{T} \|\boldsymbol{\theta}_t - \boldsymbol{\theta}_s\|_2 + \sum_{t=1}^{T} \boldsymbol{\theta}_t' \boldsymbol{q}_t
$$

$$
s.t.\ \boldsymbol{\theta}_t \succeq \mathbf{0}, \|\boldsymbol{\theta}_t\|_1 \leq 1,\ \forall\, t \in \mathbb{N}_T \tag{4}
$$

---

[1] A MATLAB® implementation of our framework is available at https://github.com/niloofaryousefi/ECML2015

where we defined $q_t^m \triangleq -\frac{1}{2}\alpha_t' Y_t K_t^m Y_t \alpha_t$ and $q_t \triangleq [q_t^1, \ldots, q_t^M]'$. Due to the non-smooth nature of Problem 4, we derive a consensus ADMM-based optimization algorithm to solve it efficiently. Based on the exposition provided in Sections 5 and 7 of [6], it is straightforward to verify that Problem 4 can be written in ADMM form as

$$\min_{s,\theta,z} \lambda \sum_{i=1}^{N} h_i(s_i) + g(\theta) + I_{\Omega(\theta)}(z)$$

$$s.t. \; s_i - \tilde{\theta}_i = 0, \; i \in \mathbb{N}_N$$

$$z - \theta = 0 \tag{5}$$

where $N \triangleq \frac{T(T-1)}{2}$, and the local variable $s_i \in \mathbb{R}^{2M}$ consists of two vector variables $(s_i)_j$ and $(s_i)_{j'}$, where $(s_i)_j = \theta_{\mathcal{M}(i,j)}$. Note that the index mapping $t = \mathcal{M}(i,j)$ maps the jth component of the local variable s_i to the tth component of the global variable θ. Also, $\tilde{\theta}_i$ can be considered as the global variable's idea of what the local variable s_i should be. Moreover, for each i, the function $h_i(s_i)$ is defined as $\|(s_i)_j - (s_i)_{j'}\|_2$, and the objective term $g(\theta)$ is given as $\sum_{t=1}^{T} \theta_t' q_t$. Finally, $I_{\Omega(\theta)}(z)$ is the indicator function for the constraint set θ (i.e., $I_{\Omega(\theta)}(z) = 0$ for $z \in \Omega(\theta)$, and $I_{\Omega(\theta)}(z) = \infty$ for $z \notin \Omega(\theta)$).

The augmented Lagrangian (using scaled dual variables) for Problem 5 is

$$L_\rho(s,\theta,z,u,v) = \lambda \sum_{i=1}^{N} h_i(s_i) + g(\theta) + I_{\Omega(\theta)}(z) + (\rho/2) \sum_{i-1}^{N} \|s_i - \tilde{\theta}_i + u_i\|_2^2$$

$$+ (\rho/2)\|z - \theta + v\|_2^2, \tag{6}$$

where u_i and v are the dual variables for the constraints $s_i = \tilde{\theta}_i$ and $z = \theta$ respectively. Applying ADMM on the Lagrangian function given in (6), the following steps are carried out in the kth iteration

$$s_i^{k+1} = \arg\min_{s_i}\{\lambda h_i(s_i) + (\rho/2)\|s_i - \tilde{\theta}_i^k + u_i^k\|_2^2\} \tag{7}$$

$$\theta^{k+1} = \arg\min_{\theta}\{g(\theta) + (\rho/2)\sum_{i=1}^{N}\|s_i^{k+1} - \tilde{\theta}_i + u_i^k\|_2^2 + (\rho/2)\|z^k - \theta + v^k\|_2^2\} \tag{8}$$

$$z^{k+1} = \arg\min_{z}\{I_{\Omega(\theta)}(z) + (\rho/2)\|z - \theta^{k+1} + v^k\|_2^2\} \tag{9}$$

$$u_i^{k+1} = u_i^k + s_i^{k+1} - \tilde{\theta}_i^{k+1} \tag{10}$$

$$v^{k+1} = v^k + z^{k+1} - \theta^{k+1} \tag{11}$$

where, for each $i \in \mathbb{N}_N$, the s- and u-updates can be carried out independently and in parallel. It is also worth mentioning that the s-update is a proximal operator evaluation for $\|.\|_2$ which can be simplified to

$$s_i^{k+1} = \mathcal{S}_{\lambda/\rho}(\tilde{\theta}_i^k + u_i^k), \; \forall \, i \in \mathbb{N}_N \tag{12}$$

where $\mathcal{S}_\kappa$ is the vector-valued soft thresholding (or shrinkage) operator and which is defined as

$$\mathcal{S}_\kappa(\boldsymbol{a}) \triangleq (1 - \kappa/\|\boldsymbol{a}\|_2)_+ \boldsymbol{a}, \quad \mathcal{S}_\kappa(0) \triangleq 0. \tag{13}$$

Furthermore, as the objective term g is separable in $\boldsymbol{\theta}_t$, the $\boldsymbol{\theta}$-update can be decomposed into T independent minimization problems, for which a closed from solution exists

$$\boldsymbol{\theta}_t^{k+1} = \frac{1}{T-1} \left[\sum_{\mathcal{M}(i,j)=t} \left((\boldsymbol{s}_i)_j^{k+1} + (\boldsymbol{u}_i)_j^k \right) + \left(\boldsymbol{z}_t^k + \boldsymbol{v}_t^k \right) - (1/\rho)\boldsymbol{q}_t \right], \ \forall\, t \in \mathbb{N}_T \tag{14}$$

Algorithm 1. Algorithm for solving Problem 3.

Require: $\boldsymbol{X}_1, \ldots, \boldsymbol{X}_T, \boldsymbol{Y}_1, \ldots, \boldsymbol{Y}_T, C, \lambda$
Ensure: $\boldsymbol{\theta}_1, \ldots, \boldsymbol{\theta}_T, \boldsymbol{\alpha}_1, \ldots, \boldsymbol{\alpha}_T$
1: **Initialize:** $\boldsymbol{\theta}_1^{(0)}, \ldots, \boldsymbol{\theta}_T^{(0)}, r = 1$
2: **Calculate:** Base kernel matrices K_t^m using $\boldsymbol{X}_t$'s for the T tasks and the M kernels.

3: **while** not converged **do**
4:     $\boldsymbol{\alpha}^{(r)} \leftarrow \arg\max_{\boldsymbol{\alpha} \in \Omega(\boldsymbol{\alpha})} \sum_{t=1}^T \boldsymbol{\alpha}_t' \boldsymbol{e} - \frac{1}{2} \sum_{t=1}^T \sum_{m=1}^M (\theta_t^m)^{(r-1)} (\boldsymbol{\alpha}_t' Y_t K_t^m Y_t \boldsymbol{\alpha}_t)$
5:     $(q_t^m)^{(r)} \leftarrow -\frac{1}{2}(\boldsymbol{\alpha}_t')^{(r)} Y_t K_t^m Y_t (\boldsymbol{\alpha}_t)^{(r)}, \ \forall t, m$
6:     $\boldsymbol{\theta}^{(r)} \leftarrow \arg\min_{\boldsymbol{\theta} \in \Omega(\boldsymbol{\theta})} \lambda \sum_{t=1}^{T-1} \sum_{s>t}^T \|\boldsymbol{\theta}_t - \boldsymbol{\theta}_s\|_2 + \sum_{t=1}^T \boldsymbol{\theta}_t' \boldsymbol{q}_t^{(r)}$ using Algorithm 2
7: **end while**
8: $\boldsymbol{\alpha}^* = \boldsymbol{\alpha}^{(r)}$
9: $\boldsymbol{\theta}^* = \boldsymbol{\theta}^{(r)}$

In the third step of the ADMM, we project $(\boldsymbol{\theta}^{k+1} - \boldsymbol{v}^k)$ onto the constraint set $\Omega(\boldsymbol{\theta})$. Note that, this set is separable in $\boldsymbol{\theta}$, so the projection step can also be performed independently and in parallel for each variable $\boldsymbol{z}_t$, *i.e.*,

$$\boldsymbol{z}_t^{k+1} = \boldsymbol{\Pi}_{\Omega(\boldsymbol{\theta})}(\boldsymbol{\theta}_t^{k+1} + \boldsymbol{v}_t^k), \ \forall\, t \in \mathbb{N}_T. \tag{15}$$

The $\boldsymbol{z}_t$-update can also be seen as the problem of finding the intersection between two closed convex sets $\Omega_1(\boldsymbol{\theta}) = \{\boldsymbol{\theta}_t \succeq 0, \ \forall\, t \in \mathbb{N}_T\}$ and $\Omega_2(\boldsymbol{\theta}) = \{\|\boldsymbol{\theta}_t\|_1 \le 1, \ \forall\, t \in \mathbb{N}_T\}$, which can be handled using Dykstra's alternating projections method [5,11] as follows

$$\boldsymbol{y}_t^{k+1} = \boldsymbol{\Pi}_{\Omega_1(\boldsymbol{\theta})}(\boldsymbol{\theta}_t^{k+1} + \boldsymbol{v}_t^k - \boldsymbol{\beta}_t^k) = \frac{1}{2} \left[\boldsymbol{\theta}_t^{k+1} + \boldsymbol{v}_t^k - \boldsymbol{\beta}_t^k \right]_+, \ \forall\, t \in \mathbb{N}_T \tag{16}$$

$$\boldsymbol{z}_t^{k+1} = \boldsymbol{\Pi}_{\Omega_2(\boldsymbol{\theta})}(\boldsymbol{y}_t^{k+1} + \boldsymbol{\beta}_t^k) = \mathbf{P}_M(\boldsymbol{y}_t^{k+1} + \boldsymbol{\beta}_t^k) + \frac{1}{M}\mathbf{1}_M, \ \forall\, t \in \mathbb{N}_T \tag{17}$$

$$\boldsymbol{\beta}_t^{k+1} = \boldsymbol{\beta}_t^k + \boldsymbol{y}_t^{k+1} - \boldsymbol{z}_t^{k+1}, \ \forall\, t \in \mathbb{N}_T \tag{18}$$

where $\mathbf{P}_M \triangleq \left(\mathbf{I}_M - \frac{\mathbf{1}_M \mathbf{1}_M'}{M}\right)$ is the centering matrix. Furthermore, the $\mathbf{y}_t$- and $\mathbf{z}_t$ updates are the Euclidean projections onto $\Omega_1(\boldsymbol{\theta})$ and $\Omega_2(\boldsymbol{\theta})$ respectively with dual variables $\boldsymbol{\beta}_t \in \mathbb{R}^{M \times 1}$, $t = 1, \ldots, T$. Finally, we update the dual variables $\mathbf{u}_i$ and $\mathbf{v}$ using the equations given in (10) and (11).

Algorithm 2. Consensus ADMM algorithm to solve optimization Problem 4

Require: $\mathbf{q}_1^{(r)}, \ldots, \mathbf{q}_T^{(r)}, \rho$
Ensure: $\boldsymbol{\theta}_1^{(r)}, \ldots, \boldsymbol{\theta}_T^{(r)}$
1: **Initialize:** $\hat{\boldsymbol{\theta}}_1^{(0)}, \ldots, \hat{\boldsymbol{\theta}}_T^{(0)}, k = 0$
2: **while** not converged **do**
3:     **for** $i \in \mathbb{N}_N, t \in \mathbb{N}_T$ **do**
4:         $\mathbf{s}_i^{k+1} \leftarrow \mathcal{S}_{\lambda/\rho}(\tilde{\boldsymbol{\theta}}_i^k + \mathbf{u}_i^k)$
5:         $\hat{\boldsymbol{\theta}}_t^{k+1} \leftarrow \frac{1}{T-1}\left[\sum_{\mathcal{M}(i,j)=t}\left((\mathbf{s}_i)_j^{k+1} + (\mathbf{u}_i)_j^k\right) + \left(\mathbf{z}_t^k + \mathbf{v}_t^k\right) - (1/\rho)\mathbf{q}_t\right]$
6:         $\mathbf{y}_t^{k+1} \leftarrow \frac{1}{2}\left[\hat{\boldsymbol{\theta}}_t^{k+1} + \mathbf{v}_t^k - \boldsymbol{\beta}_t^k\right]_+$
7:         $\mathbf{z}_t^{k+1} \leftarrow \mathbf{P}_M(\mathbf{y}_t^{k+1} + \boldsymbol{\beta}_t^k) + \frac{1}{M}\mathbf{1}_M$
8:         $\boldsymbol{\beta}_t^{k+1} \leftarrow \boldsymbol{\beta}_t^k + \mathbf{y}_t^{k+1} - \mathbf{z}_t^{k+1}$
9:         $\mathbf{u}_i^{k+1} \leftarrow \mathbf{u}_i^k + \mathbf{s}_i^{k+1} - \tilde{\boldsymbol{\theta}}_i^{k+1}$
10:         $\mathbf{v}_t^{k+1} \leftarrow \mathbf{v}_t^k + \mathbf{z}_t^{k+1} - \hat{\boldsymbol{\theta}}_t^{k+1}$
11:     **end for**
12: **end while**
13: $\boldsymbol{\theta}^{(r)} \leftarrow \hat{\boldsymbol{\theta}}^{(k+1)}$

3.1 Convergence Analysis and Stopping Criteria

Convergence of Algorithm 2 can be derived based on two mild assumptions similar to the standard convergence theory of the ADMM method discussed in [6]; (i) the objective functions $h(\mathbf{s}) = \sum_{i=1}^N \|(\mathbf{s}_i)_j - (\mathbf{s}_i)_{j'}\|_2$ and $g(\boldsymbol{\theta}) = \sum_{t=1}^T \boldsymbol{\theta}_t' \mathbf{q}_t$ are closed, proper and convex, which implies that the subproblems arising in the $\mathbf{s}$-update (7) and $\boldsymbol{\theta}$-update (8) are solvable, and (ii) the augmented Lagrangian (0) for $\rho = 0$ has a saddle point. Under these two assumptions, it can be shown that our ADMM-based algorithm satisfies the following

- Convergence of residuals : $\mathbf{s}_i^k - \tilde{\boldsymbol{\theta}}_i^k \to \mathbf{0}$, $\forall\, i \in \mathbb{N}_N$, and $\mathbf{z}^k - \boldsymbol{\theta}^k \to \mathbf{0}$ as $k \to \infty$.
- Convergence of dual variables: $\mathbf{u}_i^k \to \mathbf{u}_i^*, \forall i \in \mathbb{N}_N$, and $\mathbf{v}^k \to \mathbf{v}^*$ as $k \to \infty$, where $\mathbf{u}^*$ and $\mathbf{v}^*$ are the dual optimal points.
- Convergence of the objective : $h(\mathbf{s}^k) + g(\mathbf{z}^k) \to p^*$ as $k \to \infty$, which means the objective function (4) converges to its optimal value as the algorithm proceeds.

Also, the algorithm is terminated, when the primal and dual residuals satisfy the following stopping criteria

$$\|e_{p_1}^k\|_2 \le \epsilon_1^{pri}, \qquad \|e_{p_2}^k\|_2 \le \epsilon_2^{pri}, \qquad \|e_{p_3}^k\|_2 \le \epsilon_3^{pri}$$

$$\|e_{d_1}^k\|_2 \leq \epsilon_1^{dual}, \quad \|e_{d_2}^k\|_2 \leq \epsilon_2^{dual}, \quad \|e_{d_3}^k\|_2 \leq \epsilon_3^{dual} \tag{19}$$

where the primal residuals of the kth iteration are given as $e_{p_1}^k = s^k - \theta^k$, $e_{p_2}^k = z^k - \theta^k$ and $e_{p_3}^k = y^k - z^k$. Similarly $e_{d_1}^k = \rho(\theta^{k+1} - \theta^k)$, $e_{d_2}^k = \rho(z^k - z^{k+1})$ and $e_{d_3}^k = \rho(y^k - y^{k+1})$ are dual residuals at iteration k. Also, the tolerances $\epsilon^{pri} > 0$, and $\epsilon^{dual} > 0$ can be chosen appropriately using the method described in Chapter 3 of [6].

3.2 Computational Complexity

Algorithm 1 needs to compute and cache TM kernel matrices; however, they are computed only once in $\mathcal{O}(TMn^2)$ time. Also, as long as the number of tasks T is not excessive, all the matrices can be computed and stored on a single machine, since (i) the number M of kernels, is typically chosen small (*e.g.*, we chose $M = 10$), and (ii) the number n of training samples per task is not usually large; if it were large, MTL would probably not be able to offer any advantages over training each task independently. For each iteration of Algorithm 1, T independent SVM problems are solved at a time cost of $\mathcal{O}(n^3)$ per task. Therefore, if Algorithm 2 converges in K iterations, the runtime complexity of Algorithm 1 becomes $\mathcal{O}(Tn^3 + KMT^2)$ per iteration. Note, though, that K is not usually more than a few tens of iterations [6].

On the other hand, if the number of tasks T is large, the nature of our problem allows our algorithm to be implemented in parallel. The α-update can be handled as T independent optimization problems, which can be easily distributed to T subsystems. Each subsystem N needs to compute once and cache M kernel matrices for each task. Then, for each iteration, one SVM problem is required to be solved by each subsystem, which takes $\mathcal{O}(n^3)$ time. Moreover, our ADMM-based algorithm updating the θ parameters can also be implemented in parallel over $i \in \mathbb{N}_N$. Assuming that exchanging data and updates between subsystems consumes negligible time, the ADMM only requires $\mathcal{O}(KM)$ time. Therefore, taking advantage of a distributed implementation, the complexity of Algorithm 1 is only $\mathcal{O}(n^3 + KM)$ per iteration.

4   Generalization Bound Based on Rademacher Complexity

In this section, we provide a Rademacher complexity-based generalization bound for the Hypothesis Space (HS) considered in Problem 1, which can be identified with the help of the following Proposition [1].

---

[1] Note that Proposition 1 here utilizes the first part of Proposition 12 in [20] and does not require the strong duality assumption, which is necessary for the second part of Proposition 12 in [20].

Proposition 1. *(Proposition 12 in [20], part (a)) Let $C \subseteq X$ and let $f, g : C \mapsto \mathbb{R}$ be two functions. For any $\nu > 0$, there must exist a $\eta > 0$, such that the optimal solution of (20) is also optimal in (21)*

$$\min_{x \in C} f(x) + \nu g(x) \tag{20}$$

$$\min_{x \in C, g(x) \leq \eta} f(x) \tag{21}$$

Using Proposition 1, one can show that Problem 1 is equivalent to the following problem

$$\min_{w \in \Omega'(w)} C \sum_{t=1}^{T} \sum_{i=1}^{n} l\left(w_t, \phi_t\left(x_t^i\right), y_t^i\right)$$

$$\Omega'(w) \triangleq \{w = (w_1, \cdots, w_T) : w_t \in \mathcal{H}_{t,\theta}, \theta \in \Omega'(\theta), \|w_t\|^2 \leq R_t, t \in \mathbb{N}_T\} \tag{22}$$

where

$$\Omega'(\theta) \triangleq \Omega(\theta) \cap \left\{ \theta = (\theta_t, \cdots, \theta_T) : \sum_{t=1}^{T-1} \sum_{s>t}^{T} \|\theta_t - \theta_s\|_2 \leq \gamma \right\}$$

The goal here is to choose the w and θ from their relevant feasible sets, such that the objective function of (22) is minimized. Therefore, the relevant hypothesis space for Problem 22 becomes

$$\mathcal{F} \triangleq \left\{ x \mapsto [\langle w_1, \phi_1 \rangle, \ldots, \langle w_T, \phi_T \rangle]' : \forall t\, w_t \in \mathcal{H}_{t,\theta}, \|w_t\|^2 \leq R_t, \theta \in \Omega'(\theta) \right\} \tag{23}$$

Note that finding the Empirical Rademacher Complexity (ERC) of $\mathcal{F}$ is complicated due to the non-smooth nature of the constraint $\sum_{t=1}^{T-1} \sum_{s>t}^{T} \|\theta_t - \theta_s\|_2 \leq \gamma$. Instead, we will find the ERC of the HS $\mathcal{H}$ defined in (24); notice that $\mathcal{F} \subseteq \mathcal{H}$.

$$\mathcal{H} \triangleq \left\{ x \mapsto [\langle w_1, \phi_1 \rangle, \ldots, \langle w_T, \phi_T \rangle]' : \forall t\, w_t \in \mathcal{H}_{t,\theta}, \|w_t\|^2 \leq R_t, \theta \in \Omega''(\theta) \right\} \tag{24}$$

where

$$\Omega''(\theta) \triangleq \Omega(\theta) \cap \left\{ \theta = (\theta_t, \cdots, \theta_T) : \sum_{t=1}^{T-1} \sum_{s>t}^{T} \|\theta_t - \theta_s\|_2^2 \leq \gamma^2 \right\} \tag{25}$$

Using the first part of Theorem (12) in [4], it can be shown that the ERC of $\mathcal{H}$ upper bounds the ERC of function class $\mathcal{F}$. Thus, the bound derived for $\mathcal{H}$ is also valid for $\mathcal{F}$. The following theorem provides the generalization bound for $\mathcal{H}$.

Theorem 1. *Let $\mathcal{H}$ defined in (24) be the multi-task HS for a class of functions $f = (f_1, \ldots, f_T) : \mathcal{X} \mapsto \mathbb{R}^T$. Then for all $f \in \mathcal{H}$, for $\delta > 0$ and for fixed $\rho > 0$, with probability at least $1 - \delta$ it holds that*

$$R(f) \leq \hat{R}_\rho(f) + \frac{2}{\rho}\hat{\mathfrak{R}}_S(\mathcal{H}) + 3\sqrt{\frac{\log \frac{1}{\delta}}{2Tn}} \tag{26}$$

where

$$\hat{\mathfrak{R}}_S(\mathcal{H}) \leq \hat{\mathfrak{R}}_{ub}(\mathcal{H}) = \sqrt{\frac{\sqrt{3}\gamma RM}{nT}} \tag{27}$$

where $\hat{\mathfrak{R}}_S(\mathcal{H})$, the ERC of $\mathcal{H}$, is given as

$$\hat{\mathfrak{R}}_S(\mathcal{H}) = \frac{1}{nT}\mathrm{E}_\sigma\left\{\sup_{f=(f_1,\ldots,f_T)\in\mathcal{F}}\sum_{t=1}^{T}\sum_{i=1}^{n}\sigma_t^i f_t(x_t^i)\middle|\{x_t^i\}_{t\in\mathbb{N}_T,i\in\mathbb{N}_n}\right\} \tag{28}$$

the ρ-empirical large margin error $\hat{R}_\rho(f)$, for the training sample $S = \left\{(x_t^i, y_t^i)\right\}_{i,t=1}^{n,T}$ is defined as

$$\hat{R}_\rho(f) = \frac{1}{nT}\sum_{t=1}^{T}\sum_{i=1}^{n}\min\left(1, [1 - y_t^i f_t(x_t^i)/\rho]_+\right)$$

Also, $R(f) = \Pr[yf(x) < 0]$ is the expected risk w.r.t. 0-1 loss, n is the number of training samples for each task, T is the number of tasks to be trained, and M is the number of kernel functions utilized for MKL.

The proof of this theorem is omitted due to space constraints. Based on Theorem 1, the second term in (26), the upper bound for ERC of $\mathcal{H}$, decreases as the number of tasks increases. Therefore, it is reasonable to expect that the generalization performance to improve, when the number T of tasks or the number n of training samples increase. Also, due to the formulation's group lasso (L_1/L_2-norm) regularizer on the pair-wise MKL weight differences, the ERC in (27) depends on M as $\mathcal{O}\sqrt{M}$. It is worth mentioning, that, while this could be improved to $\mathcal{O}\sqrt{\log M}$ as in [9], if one considers instead a L_p/L_q-norm regularizer, we won't pursue this avenue here. Let us finally note, that (26) allows one to construct data-dependent confidence intervals for the true, pooled (averaged over tasks) misclassification rate of the MTL problem under consideration.

5   Experiments

In this section, we demonstrate the merit of the proposed model via a series of comparative experiments. For reference, we consider two baseline methods referred to as **STL** and **MTL**, which present the two extreme cases discussed in Sect. 2. We also compare our method with five state-of-the-art methods which, like ours, fall under the CMTL family of approaches. These methods are briefly described below.

- **STL:** single-task learning approach used as a baseline, according to which each task is individually trained via a traditional single-task MKL strategy.
- **MTL:** a typical MTL approach, for which all tasks share a common feature space. An SVM-based formulation with multiple kernel functions was utilized and the common MKL parameters for all tasks were learned during training.
- **CMTL** [17]: in this work, the tasks are grouped into disjoint clusters, such that the model parameters of the tasks belonging to the same group are close to each other.
- **Whom** [19]: clusters the task, into disjoint groups and assumes that tasks of the same group can jointly learn a shared feature representation.
- **FlexClus** [29]: a flexible clustering structure of tasks is assumed, which can vary from feature to feature.
- **CoClus** [25]: a co-clustering structure is assumed aiming to capture both the feature and task relationship between tasks.
- **MeTaG** [16]: a multi-level grouping structure is constructed by decomposing the matrix of tasks' parameters into a sum of components, each of which corresponds to one level and is regularized with a L_2-norm on the pairwise difference between parameters of all the tasks.

5.1 Experimental Settings

For all experiments, all kernel-based methods (including **STL**, **MTL** and our method) utilized 1 Linear, 1 Polynomial with degree 2, and 8 Gaussian kernels with spread parameters $\{2^0, \ldots, 2^7\}$ for MKL. All kernel functions were normalized as $k(\boldsymbol{x}, \boldsymbol{y}) \leftarrow k(\boldsymbol{x}, \boldsymbol{y})/\sqrt{k(\boldsymbol{x}, \boldsymbol{x})k(\boldsymbol{y}, \boldsymbol{y})}$. Moreover, for **CMTL**, **Whom** and **CoClus** methods, which require the number of task clusters to be pre-specified, cross-validation over the set $\{1, \ldots, T/2\}$ was used to select the optimal number of clusters. Also, the regularization parameters of all methods were chosen via cross-validation over the set $\{2^{-10}, \ldots, 2^{10}\}$.

5.2 Experimental Results

We assess the performance of our proposed method compared to the other methods on 7 widely-used data sets including 3 real-world data sets: Wall-Following Robot Navigation (*Robot*), Statlog Vehicle Silhouettes (*Vehicle*) and Statlog Image Segmentation (*Image*) from the UCI repository [14], 2 handwritten digit data sets, namely MNIST Handwritten Digit (*MNIST*) and Pen-Based Recognition of Handwritten Digits (*Pen*), as well as *Letter* and *Landmine*.

The data sets from the UCI repository correspond to three multi-class problems. In the *Robot* data set, each sample is labeled as: "Move-Forward, "SlightRight-Turn", "Sharp-Right-Turn" and "Slight-Left-Turn". These classes are designed to navigate a robot through a room following the wall in a clockwise direction. The *Vehicle* data set describes four different types of vehicles as "4 Opel", "SAAB", "Bus" and "Van". On the other hand, the instances of the *Image* data set were drawn randomly from a database of 7 outdoor images which are labeled as "Sky", "Foliage", "Cement", "Window", "Path" and "Grass".

Also, two multi-class handwritten digit data sets, namely *MNIST* and *Pen*, consist of samples of handwritten digits from 0 to 9. Each example is labeled as one of ten classes. A one-versus-one strategy was adopted to cast all multi-class learning problems into MTL problems, and the average classification accuracy across tasks was calculated for each data set. Moreover, an equal number of samples from each class was chosen for training for all five multi-class problems.

We also compare our method on two widely-used multi-task data sets, namely the *Letter* and *Landmine* data sets. The former one is a collection of handwritten words collected by Rob Kassel of MIT's spoken Language System Group, and involves eight tasks: 'C' vs. 'E', 'G' vs. 'Y', 'M' vs. 'N', 'A' vs. 'G', 'I' vs. 'J', 'A' vs. 'O', 'F' vs. 'T' and 'H' vs. 'N'. Each letter is represented by a 8 by 16 pixel image, which forms a 128 dimensional feature vector per sample. We randomly chose 200 samples for each letter. An exception is letter J, for which only 189 samples were available. The *Landmine* data set consists of 29 binary classification tasks collected from various landmine fields. The objective is to recognize whether there is a landmine or not based on a region's characteristics, which are described by four moment-based features, three correlation-based features, one energy ratio feature, and one spatial variance feature.

In all our experiments, for all methods, we considered training set sizes of 10%, 20% and 50% of the original data set to investigate the influence of the data set size on generalization performance. An exception was the *Landmine* data set, for which we used 20% and 50% of the data set for training purposes due to its small size. The rest of data were split into equal sizes for validation and testing.

In Table 1, we report the average classification accuracy over 20 runs of randomly sampled training sets for each experiment. Note that we utilized the method proposed in [10] for our statistical analysis. More specifically, Friedman's and Holm's post-hoc tests at significance level $\alpha = 0.05$ were employed to compare our proposed method with the other methods.

As shown in Table 1, for each data set, Friedman's test ranks the best performing model as first, the second best as second and so on. The superscript next to each value in Table 1 indicates the rank of the corresponding model on the relevant data set, while the superscript next to each model reflects its average rank over all data sets for the corresponding training set size. Note that methods depicted in boldface are deemed statistically similar to our model, since their corresponding p-values are not smaller than the adjusted α values obtained by Holm's post-hoc test. Overall, it can be observed that our method dominates three, six and five out of seven methods, when trained with 10%, 20% and 50% training set sizes respectively.

Table 1. Experimental comparison between our method and seven benchmark methods

10%	STL$^{(7)}$	MTL$^{(5.42)}$	CMTL$^{(6.33)}$	Whom$^{(3.25)}$	FlexClus$^{(4.33)}$	Coclus$^{(4)}$	MetaG$^{(5)}$	Our Method$^{(1.67)}$
Robot	84.51$^{(7)}$	84.82$^{(6)}$	84.15$^{(8)}$	**88.90**$^{(1)}$	88.34$^{(4)}$	87.83$^{(5)}$	88.77$^{(2)}$	88.67$^{(3)}$
Vehicle	79.73$^{(8)}$	80.38$^{(6)}$	80.23$^{(7)}$	83.14$^{(4)}$	82.45$^{(5)}$	**86.79**$^{(1)}$	83.53$^{(3)}$	84.51$^{(2)}$
Image	97.08$^{(7)}$	97.43$^{(3)}$	97.09$^{(6)}$	97.27$^{(4)}$	98.05$^{(2)}$	97.24$^{(5)}$	97.05$^{(8)}$	**98.19**$^{(1)}$
Pen	98.16$^{(7)}$	98.28$^{(5.5)}$	95.78$^{(8)}$	98.28$^{(5.5)}$	98.67$^{(3)}$	**99.26**$^{(1)}$	98.57$^{(4)}$	99.12$^{(2)}$
MNIST	94.09$^{(7)}$	94.87$^{(4)}$	94.49$^{(6)}$	95.56$^{(3)}$	94.59$^{(5)}$	93.09$^{(8)}$	96.13$^{(2)}$	**96.70**$^{(1)}$
Letter	84.12$^{(6)}$	83.12$^{(8)}$	85.62$^{(3)}$	86.82$^{(2)}$	83.72$^{(7)}$	85.46$^{(4)}$	85.41$^{(5)}$	**87.41**$^{(1)}$

20%	STL$^{(6)}$	MTL$^{(4.43)}$	CMTL$^{(6.14)}$	Whom$^{(3.29)}$	FlexClus$^{(5.57)}$	Coclus$^{(4.57)}$	MetaG$^{(4.71)}$	Our Method$^{(1.14)}$
Robot	87.67$^{(7)}$	88.23$^{(6)}$	85.08$^{(8)}$	**90.76**$^{(1)}$	90.15$^{(3)}$	88.43$^{(5)}$	89.12$^{(4)}$	90.34$^{(2)}$
Vehicle	85.88$^{(4)}$	86.16$^{(3)}$	82.29$^{(8)}$	85.67$^{(6)}$	85.29$^{(7)}$	87.15$^{(2)}$	85.78$^{(5)}$	**87.76**$^{(1)}$
Image	97.41$^{(6)}$	98.02$^{(3)}$	97.32$^{(7)}$	98.46$^{(2)}$	97.44$^{(5)}$	97.50$^{(4)}$	97.29$^{(8)}$	**98.54**$^{(1)}$
Pen	98.57$^{(7)}$	99.01$^{(6)}$	96.06$^{(8)}$	99.14$^{(3)}$	99.13$^{(4)}$	99.30$^{(2)}$	99.02$^{(4)}$	**99.63**$^{(1)}$
MNIST	96.13$^{(6)}$	96.71$^{(4)}$	96.56$^{(5)}$	96.76$^{(3)}$	95.04$^{(7)}$	94.09$^{(8)}$	96.84$^{(2)}$	**97.86**$^{(1)}$
Landmine	58.76$^{(7)}$	61.89$^{(7)}$	65.28$^{(2)}$	62.53$^{(5)}$	62.46$^{(6)}$	63.52$^{(3)}$	62.59$^{(4)}$	**65.82**$^{(1)}$
Letter	88.75$^{(4)}$	89.98$^{(2)}$	88.24$^{(5)}$	88.88$^{(1)}$	83.79$^{(7)}$	82.20$^{(8)}$	87.99$^{(6)}$	**90.72**$^{(1)}$

50%	STL$^{(5.64)}$	MTL$^{(3.85)}$	CMTL$^{(6.29)}$	Whom$^{(3.29)}$	FlexClus$^{(6.21)}$	Coclus$^{(5.29)}$	MetaG$^{(4.42)}$	Our Method$^{(1)}$
Robot	91.26$^{(5.5)}$	91.49$^{(3)}$	86.26$^{(8)}$	91.70$^{(2)}$	91.26$^{(5.5)}$	89.04$^{(7)}$	91.27$^{(4)}$	**92.41**$^{(1)}$
Vehicle	88.33$^{(3)}$	88.71$^{(2)}$	83.91$^{(8)}$	87.3$^{(5)}$	86.72$^{(7)}$	87.55$^{(4)}$	86.81$^{(6)}$	**89.83**$^{(1)}$
Image	98.40$^{(6)}$	98.43$^{(5)}$	97.56$^{(8)}$	98.58$^{(2)}$	98.04$^{(7)}$	98.52$^{(3)}$	98.49$^{(4)}$	**99.07**$^{(1)}$
Pen	98.77$^{(7)}$	99.23$^{(5)}$	96.17$^{(8)}$	99.32$^{(4)}$	99.33$^{(3)}$	99.34$^{(2)}$	99.21$^{(6)}$	**99.77**$^{(1)}$
MNIST	97.20$^{(6)}$	97.37$^{(4)}$	97.31$^{(5)}$	97.78$^{(3)}$	96.60$^{(7)}$	95.87$^{(8)}$	98.46$^{(2)}$	**98.64**$^{(1)}$
Landmine	63.76$^{(8)}$	64.98$^{(6)}$	66.76$^{(2)}$	65.57$^{(4)}$	64.87$^{(7)}$	65.15$^{(5)}$	66.24$^{(3)}$	**67.15**$^{(1)}$
Letter	91.18$^{(4)}$	91.62$^{(2)}$	90.97$^{(5)}$	91.25$^{(3)}$	86.47$^{(7)}$	86.27$^{(8)}$	90.66$^{(6)}$	**92.49**$^{(1)}$

Table 2. Comparison of our method against the other methods with the Holm test

10%	STL	MTL	CMTL	**Whom**	**FlexClus**	**Coclus**	**MeTaG**
Test statistic	3.93	2.13	3.49	1.25	2.40	2.62	2.29
p value	0.0005	0.0138	0.0022	**0.2869**	**0.0777**	**0.1214**	**0.1214**
Adjusted α	0.0071	0.0083	0.0100	0.0125	0.01667	0.0250	0.0500

20%	STL	MTL	CMTL	**Whom**	FlexClus	Coclus	MeTaG
Test statistic	3.71	2.51	3.82	1.64	3.38	2.62	2.73
p value	0.00021	0.0121	0.0001	**0.1017**	0.0007	0.0088	0.0064
Adjusted α	0.0083	0.0250	0.0071	0.0500	0.0100	0.01667	0.0125

50%	STL	**MTL**	CMTL	**Whom**	FlexClus	Coclus	MeTaG
Test statistic	3.55	2.18	4.04	1.75	3.98	3.27	2.61
p value	0.0004	**0.0291**	0.0001	**0.0809**	0.0001	0.0011	0.0089
Adjusted α	0.0100	0.0250	0.0071	0.0500	0.0083	0.0125	0.01667

Also, in Figure 1, we provide better insight of how the grouping of task feature spaces might be determined in our framework. For the purpose of visualization, we applied two Gaussian kernel functions with spread parameters 2 and 2^8 and used the *Letter* multi-task data set.

In this figure, the x and y axes represent the weights of these two kernel functions for each task. From Figure 1(a), when a small training size (10%) is chosen, it can be seen that our framework yields a cluster of 3 tasks, namely { "A" vs "G", "A" vs "O", "G" vs "Y"} that share a common feature space to benefit

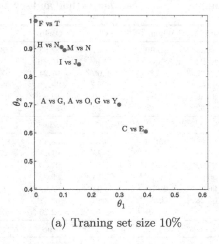

(a) Traning set size 10%

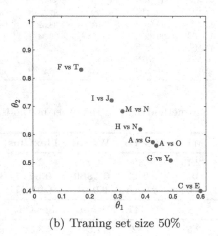

(b) Traning set size 50%

Fig. 1. Feature space parameters for *Letter* multi-task data set

from each other's data. However, as the number n of training samples per task increases, every task is allowed to employ its own feature space to guarantee good performance. This is shown in Figure 1 (b), which displays the results obtained for a 50% training set size. Note, that the displayed MKL weights lie on the $\theta_1 + \theta_2 = 1$ line due to the framework's L_1 MKL weight constraint.

6   Conclusions

In this work, we proposed a novel MT-MKL framework for SVM-based binary classification, where a flexible group structure is determined between each pair of tasks. In this framework, tasks are allowed to have a common, similar, or distinct feature spaces. Recently, some MTL frameworks have been proposed, which also consider clustering strategies to capture task relatedness. However, our method is capable of modeling a more general type of task relationship, where tasks may be implicitly grouped according to a notion of feature space similarity. Also, our proposed optimization algorithm allows for a distributed implementation, which can be significantly advantageous for MTL settings involving large number of tasks. The performance advantages reported on 7 multi-task SVM-based classification problems largely seem to justify our arguments in favor of our framework.

Acknowledgments. N. Yousefi acknowledges support from National Science Foundation (NSF) grants No. 0806931 and No. 1161228. Moreover, M. Georgiopoulos acknowledges partial support from NSF grants No. 0806931, No. 0963146, No. 1200566, No. 1161228, and No. 1356233. Finally, G. C. Anagnostopoulos acknowledges partial support from NSF grant No. 1263011. Any opinions, findings, and conclusions or recommendations expressed in this material are those of the authors and do not necessarily reflect the views of the NSF.

References

1. Argyriou, A., Clémençon, S., Zhang, R.: Learning the graph of relations among multiple tasks. In: ICML 2014 workshop on New Learning Frameworks and Models for Big Data (2013)
2. Argyriou, A., Evgeniou, T., Pontil, M.: Convex multi-task feature learning. Machine Learning **73**(3), 243–272 (2008)
3. Bakker, B., Heskes, T.: Task clustering and gating for bayesian multitask learning. The Journal of Machine Learning Research **4**, 83–99 (2003)
4. Bartlett, P.L., Mendelson, S.: Rademacher and gaussian complexities: Risk bounds and structural results. The Journal of Machine Learning Research **3**, 463–482 (2003)
5. Bauschke, H., Borwein, J.M.: Dykstra's alternating projection algorithm for two sets. Journal of Approximation Theory **79**(3), 418–443 (1994)
6. Boyd, S., Parikh, N., Chu, E., Peleato, B., Eckstein, J.: Distributed optimization and statistical learning via the alternating direction method of multipliers. Foundations and Trends in Machine Learning **3**(1), 1–122 (2011)
7. Caruana, R.: Multitask learning. Machine Learning **28**(1), 41–75 (1997)
8. Chang, C.C., Lin, C.J.: LIBSVM: A library for support vector machines. ACM Transactions on Intelligent Systems and Technology **2**, 27:1–27:27 (2011). http://www.csie.ntu.edu.tw/~cjlin/libsvm
9. Cortes, C., Mohri, M., Rostamizadeh, A.: Generalization bounds for learning kernels. In: Proceedings of the 27th International Conference on Machine Learning (ICML 2010), pp. 247–254 (2010)
10. Demšar, J.: Statistical comparisons of classifiers over multiple data sets. The Journal of Machine Learning Research **7**, 1–30 (2006)
11. Dykstra, R.L.: An algorithm for restricted least squares regression. Journal of the American Statistical Association **78**(384), 837–842 (1983)
12. Evgeniou, A., Pontil, M.: Multi-task feature learning. Advances in Neural Information Processing Systems **19**, 41 (2007)
13. Evgeniou, T., Micchelli, C.A., Pontil, M.: Learning multiple tasks with kernel methods. Journal of Machine Learning Research, 615–637 (2005)
14. Frank, A., Asuncion, A.: UCI machine learning repository (2010). http://archive.ics.uci.edu/ml

15. Gu, Q., Li, Z., Han, J.: Joint feature selection and subspace learning. In: IJCAI Proceedings-International Joint Conference on Artificial Intelligence, vol. 22, p. 1294 (2011)
16. Han, L., Zhang, Y.: Learning multi-level task groups in multi-task learning. In: Proceedings of the 29th AAAI Conference on Artificial Intelligence (AAAI) (2015)
17. Jacob, L., Vert, J.p., Bach, F.R.: Clustered multi-task learning: a convex formulation. In: Advances in Neural Information Processing Systems, pp. 745–752 (2009)
18. Jalali, A., Sanghavi, S., Ruan, C., Ravikumar, P.K.: A dirty model for multi-task learning. In: Advances in Neural Information Processing Systems, pp. 964–972 (2010)
19. Kang, Z., Grauman, K., Sha, F.: Learning with whom to share in multi-task feature learning. In: Proceedings of the 28th International Conference on Machine Learning (ICML 2011), pp. 521–528 (2011)
20. Kloft, M., Brefeld, U., Sonnenburg, S., Zien, A.: Lp-norm multiple kernel learning. The Journal of Machine Learning Research **12**, 953–997 (2011)
21. Lanckriet, G.R., Cristianini, N., Bartlett, P., Ghaoui, L.E., Jordan, M.I.: Learning the kernel matrix with semidefinite programming. The Journal of Machine Learning Research **5**, 27–72 (2004)
22. Li, C., Georgiopoulos, M., Anagnostopoulos, G.C.: Conic multi-task classification. In: Calders, T., Esposito, F., Hüllermeier, E., Meo, R. (eds.) ECML PKDD 2014, Part II. LNCS, vol. 8725, pp. 193–208. Springer, Heidelberg (2014)
23. Li, C., Georgiopoulos, M., Anagnostopoulos, G.C.: Pareto-path multitask multiple kernel learning. IEEE Transactions on Neural Networks and Learning Systems **26**(1), 51–61 (2015)
24. Tang, L., Chen, J., Ye, J.: On multiple kernel learning with multiple labels. In: IJCAI, pp. 1255–1260 (2009)
25. Xu, L., Huang, A., Chen, J., Chen, E.: Exploiting task-feature co-clusters in multi-task learning. In: Proceedings of the Twenty-Ninth AAAI Conference on Artificial Intelligence (AAAI 2015) (2015)
26. Xue, Y., Liao, X., Carin, L., Krishnapuram, B.: Multi-task learning for classification with dirichlet process priors. The Journal of Machine Learning Research **8**, 35–63 (2007)
27. Zhang, Y., Yeung, D.Y.: A convex formulation for learning task relationships in multi-task learning. arXiv preprint arXiv:1203.3536 (2012)
28. Zhang, Y., Yeung, D.Y.: A regularization approach to learning task relationships in multitask learning. ACM Transactions on Knowledge Discovery from Data (TKDD) **8**(3), 12 (2014)
29. Zhong, W., Kwok, J.: Convex multitask learning with flexible task clusters. arXiv preprint arXiv:1206.4601 (2012)
30. Zhou, J., Chen, J., Ye, J.: Clustered multi-task learning via alternating structure optimization. In: Advances in Neural Information Processing Systems, pp. 702–710 (2011)

Opening the Black Box: Revealing Interpretable Sequence Motifs in Kernel-Based Learning Algorithms

Marina M.-C. Vidovic[1]([⊠]), Nico Görnitz[1], Klaus-Robert Müller[1,2]([⊠]),
Gunnar Rätsch[3]([⊠]), and Marius Kloft[4]([⊠])

[1] Berlin Institute of Technology, 10587 Berlin, Germany
marina.vidovic@ml.tu-berlin.de,
{nico.goernitz,klaus-robert.mueller}@tu-berlin.de
[2] Department of Brain and Cognitive Engineering, Korea University, Anam-dong,
Seongbuk-gu, Seoul 136-713, Republic of Korea
[3] Memorial Sloan-Kettering Cancer Center, New York, NY 10065, USA
raetsch@mskcc.org
[4] Humboldt University of Berlin, 10099 Berlin, Germany
kloft@hu-berlin.de

Abstract. This work is in the context of kernel-based learning algorithms for sequence data. We present a probabilistic approach to automatically extract, from the output of such string-kernel-based learning algorithms, the subsequences—or *motifs*—truly underlying the machine's predictions. The proposed framework views motifs as free parameters in a probabilistic model, which is solved through a global optimization approach. In contrast to prevalent approaches, the proposed method can discover even difficult, long motifs, and could be combined with any kernel-based learning algorithm that is based on an adequate sequence kernel. We show that, by using a discriminate kernel machine such as a support vector machine, the approach can reveal discriminative motifs underlying the kernel predictor. We demonstrate the efficacy of our approach through a series of experiments on synthetic and real data, including problems from handwritten digit recognition and a large-scale *human* splice site data set from the domain of computational biology.

1 Introduction

In the view of the rapidly increasing amount of data collected in science and technology, effective automation of decisions is necessary. To this end, kernel-based methods [13,17,19,26,31,32] such as support vector machines (SVM) [5,7] have found diverse applications due to their distinct merits such as the descent computational complexity, high usability, and the solid mathematical foundation [24]. Kernel-based learning allows us to obtain more complex non-linear learning machines from simple linear ones in a canonical way, since the learning and data representation processes are decoupled in a modular fashion. Yet, after more than a decade of research, kernel methods are widely considered as black boxes, and it remains an unsolved problem to make their decisions

© Springer International Publishing Switzerland 2015
A. Appice et al. (Eds.): ECML PKDD 2015, Part II, LNAI 9285, pp. 137–153, 2015.
DOI: 10.1007/978-3-319-23525-7_9

accessible or interpretable to domain experts. This is especially pressing in natural and life sciences, where not maximum prediction accuracy but unveiling the underlying natural principles is the foremost aim.

In several important application fields, the data exhibits an inherent sequence structure. This includes DNA sequences in genomics, text data in natural language processing, and speech data in speech recognition. A state-of-the-art approach to learn from such sequence data consists in the weighted-degree (WD) kernel [4,27,28,31] in combination with a kernel-based learning machine such as an SVM. Given two discrete sequences $x = (x_1, \ldots, x_L)$, $x' = (x'_1, \ldots, x'_L) \in \mathcal{A}^L$ of length L over the alphabet $\mathcal{A}$ with $|\mathcal{A}| < \infty$, the weighted-degree kernel is defined by

$$\kappa(x, x') = \sum_{\ell=1}^{\ell_{\max}} \sum_{j=1}^{L-\ell+1} \mathbb{I}\{x[j]^\ell = x'[j]^\ell\}, \tag{1}$$

where $x[j]^\ell$ denotes the length-ℓ subsequence of x starting at position j and terminating at position $j + \ell - 1$. In a nutshell, it breaks x and x' into all possible subsequences up to a maximum length $\ell_{\max} \leq L$ and computes the number of matching subsequences. The WD-kernel SVM has been shown to achieve state-of-the-art prediction accuracies in many genomic discrimination tasks, including the detection of transcription start sites [38] and splice sites [37]—achieving the winning entry in the international comparison by [1] of 19 leading gene finders and remains still unbeaten. Efficient implementations such as the one contained in the SHOGUN machine-learning toolbox [33], which employs effective feature hashing techniques [36], have been applied to problems where millions of sequences, each with more than thousand positions, are processed at the same time [34].

Like many other kernels, the WD kernel is a black-box that hinders direct interpretation and analysis of the classifier that is output by the kernel-based learning algorithm (for other approaches for interpreting non-linear classification see e.g. [2,3,14,25,41]). It is an aim of this paper to work toward unveiling the function of such a classifier by computing the most important subsequences that determine the classifier's decision—the so-called *motifs*. A motif is a widespread and typical pattern in the input data that has, or is conjectured to have, a significance or

Fig. 1. Example of a *motif*, that is, an "interesting" subsequence in a sequence learning task that has a significance or impact on the label. The task here was gene detection and the motif has been generated using the *WebLogo 3* software [8]. The motif is illustrated as a *positional weight matrix* (PWM), where the size of a letter indicates the probability of its occurrence at a certain position in the motif. The likeliest entries are arranged top down.

impact on the associated label. For instance in the detection of gene starts, a motif is a nucleotide sequence (i.e., a string over the alphabet $\mathcal{A} = \{A, C, G, T\}$), which frequently appears at the start positions of genes in the DNA. For instance in Figure 1, we give an illustration of the motif TACTGTATATATATACAGTA.

The main contributions of this work can be summarized as follows:

1. Putting forward the work of [35] on positional oligomer importance matrices (POIMs), we propose a novel probabilistic framework to finally go the full way from the output of a WD-kernel SVM to the relevant motifs truly underlying the kernel machine's predictions.
2. To deal with the sheer exponentially large size of the feature space associated with the WD kernel, we propose a very efficient optimization framework based on advanced sequence decomposition techniques.
3. Our approach is able to even find multiple motifs consisting of hundreds of positions, while previous approaches are limited to either comparably short or contiguous motifs.
4. We demonstrate the efficiency and efficacy of our approach on synthetic data sets, on the USPS hand-written digits dataset, as well as on a *human* splice data set, where we achieve near-perfect motif reconstruction quality when evaluated by means of the JASPAR database [29].

2   Preliminaries

A first step towards the identification of motifs from the WD-kernel classifiers is achieved in [35], where the concept of positional oligomer importance matrices (POIMs) is introduced, which we review below, after giving more details on the concept of the WD kernel.

2.1   Weighted-Degree (WD) Kernel

The weighted-degree kernel is formally defined in (1). It is important to note, however, that we may equivalently represent the WD kernel by the corresponding binary feature embedding Φ, with $\kappa(x, x') = \langle \Phi(x), \Phi(x') \rangle$, where each entry of $\Phi(x)$ represents a valid positional subsequence y of length $\ell \in \{1, \ldots, \ell_{\max}\}$ starting at position $j \in \{1, \ldots, L - \ell + 1\}$. A WD-kernel SVM then simply fits the parameter w of the linear model $s(x) := \langle w, \Phi(x) \rangle$, which can, more concisely, be expressed as

$$s(x) = \sum_{\ell=1}^{\ell_{\max}} \sum_{i=1}^{L-\ell+1} w_{(x[i]^\ell, i)} \tag{2}$$

since $\Phi(x)$ is inherently *sparse* (only the entries in $\Phi(x)$ corresponding to the subsequences $y = x[i]^\ell$ with $\ell \in \{1, \ldots, \ell_{\max}\}$ and $i \in \{1, \ldots, L - \ell + 1\}$ are non-zero).

2.2   Positional Oligomer Importance Matrices (POIMs)

Given the base sequence length L, a *positional k-gram* is a subsequence $(y, j) \in \Sigma^k \times \{1, \ldots, L - k + 1\}$ of length k starting at a position j. *Positional oligomer importance matrices* (POIMs) assign each positional k-gram with an importance score. This allows us to visualize the significance of the various positional k-grams as illustrated in Fig. 2. To formally introduce the POIM approach, let Σ be a discrete alphabet, let $\mathcal{X} \sim \mathcal{U}(\Sigma^L)$ be a random variable that uniformly takes values in Σ^L, and let $x \in \Sigma^L$ be a realization thereof. For any positional k-gram (y, j) starting at position j, denote as

$$Q_{k,y,j} := \mathbb{E}[s(\mathcal{X})|\mathcal{X}[j]^k = y] - \mathbb{E}[s(\mathcal{X})], \tag{3}$$

the *POIM of order* k is defined as the tupel

$$Q \equiv Q_k := \left(Q_{k,y,j}\right)_{(y,j) \in \Sigma^k \times \{1,\dots,L-k+1\}}.$$

We may interpret (3) as a measure for the contribution of the positional k-gram (y, j) to the SVM prediction function s as follows: a high value of $w_{(y,j)}$, by (2), implies a strong contribution to the prediction score $s(x)$ if and only if $y = x[j]^k$. We can very well visualize POIMs in terms of heatmaps as illustrated in Fig. 2, from which we may obtain the most discriminative features by manual inspection. As a first step towards a more automatic analysis of POIMs, [40] propose an extension of the POIM method, the so-called *differential POIM*, which aims to identify the most relevant motif lengths as well as the corresponding starting positions. Formally, the differential POIM Ω is defined as a $\ell_{\max} \times L$ matrix $\Omega := \left(\Omega_{\ell,j}\right)$ with entries

$$\Omega_{\ell,j} := \begin{cases} q_{\max}^{\ell,j} - \max\{q_{\max}^{\ell-1,j}, q_{\max}^{\ell-1,j+1}\} & \text{if } \ell \in \{2,\dots,L\} \\ 0 & \text{elsewise}, \end{cases}$$

where $q_{\max}^{\ell,j} := \max\limits_{y \in \Sigma^\ell} |Q_{\ell,y,j}|$. We can interpret $\Omega_{\ell,j}$ as an overall score for the general importance of the subsequence of length ℓ at position j.

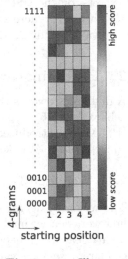

Fig. 2.   Illustration of a POIM of k-grams (k = 4) over the binary alphabet $\mathcal{A}$ = $\{0,1\}$ and sequence length L = 5 for a trained kernel predictor. Each positional 4-gram corresponds to a cell, where the color indicates the significance of the positional 4-gram to the kernel predictor.

2.3   Shortcomings of POIMs

Although being a major step towards the explanation of trained WD kernel models, POIMs suffer from the fact that their size grows exponentially with the length of the motif, which renders their computation feasible only for rather small motif sizes, typically $k \leq 12$. It also hampers manual inspection (in order to determine candidate motifs) already for rather small motif sizes such as $k \approx 5$ and is prohibitive for $k \geq 10$. For example, a POIM of order $k = 5$ contains, at each position, already $4^5 \approx 1,000$ oligomers that a domain expert would have to manually inspect. Slightly increasing the motif length to $k = 10$ leads to an unfeasible amount of $4^{10} \approx 1,000,000$ subsequences per position in the POIM.

2.4   What is Coming Up: The Proposed Approach in a Nutshell

In this paper, we tackle obtaining motifs from a trained kernel machine via the use of POIMs from a different perspective. In a nutshell, our approach is the other way round (!): we propose a probabilistic framework to reconstruct, from a given motif, the POIM that is the most likely to be generated by the motif. By subsequently minimizing the reconstruction error with respect to the truly given POIM, we can in fact optimize over the motif in order to find the one that is the most likely to have generated the POIM at hand. The latter poses a substantial numerical challenge due to the extremely high dimensionality of the feature space. Figure 3 illustrates our approach.

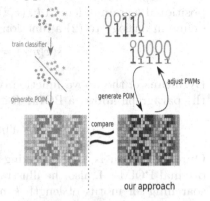

Fig. 3.   Illustration of the proposed approach: extracting a motif (top right) from a trained kernel machine (top left) by approximating the corresponding POIM (bottom left) by another POIM (bottom right) that is derived from a set of candidate motifs, over which we optimize (top right).

3   Methodology for Revealing Discriminative Motifs by Mimicking POIMs

In this section, we introduce the proposed motifPOIM methodology for extraction of motifs from POIMs, state the optimization problem, and derive an efficient optimization procedure. In a nutshell, our motifPOIM methodololgy (illustrated in Figure 3) is based on associating each candidate motif by a probability of occurrence at a certain location—which we call *probabilistic positional motif* (PPM)—and then (re-)construct from each PPM the POIM that is the most likely to be generated from the candidate PPM, which we call motifPOIM. The final motif is obtained by optimizing over the candidate motifs such that the reconstruction error of the motifPOIM with respect to the truly given POIM is minimized.

To this end, let us formally define the PPM as a tuple $m_k := (r, \mu, \sigma)$, where $r \in \mathbb{R}^{|\Sigma| \times k}$ and $\mu, \sigma \in \mathbb{R}$. We think of m_k as a candidate motif with PWM r and estimated starting position μ of which the variable σ encodes the uncertainty in the location of the motif. For this PPM we define a probabilistic model, with a probability of the starting position given by a Gaussian function with parameters μ and σ

$$P^1_{(z,i)}(m_k) := \frac{1}{\sqrt{2\pi}\sigma} exp\left(-\frac{(i-\mu)^2}{2\sigma^2} \right),$$

and a probability for the motif sequence itself, given by the product of its PWM entries

$$P^2_{(z,i)}(m_k) := \prod_{\ell=1}^{k} r_{z_\ell, \ell}.$$

Under this probabilistic model, we define, in analogy to the SVM weight vector w occurring in (2), a motif weight vector $v \equiv v(m_k)$ with entries

$\left(v(m_k)\right)_{z,i} = v_{(z,i)}(m_k)$ defined as $v_{(z,i)}(m_k) := P^1_{(z,i)}(m_k)P^2_{(z,i)}(m_k)$, for any positional k-gram of length k, $(z,i) \in \Sigma^k \times \{1, \ldots, L-k+1\}$. Consequently, we define in analogy to (2) a function

$$\bar{s}(x|m_k) := \sum_{i=1}^{L-k+1} v_{(x[i]^k,i)}(m_k). \tag{4}$$

By means of the above function, we can construct, from a PPM as defined in the paragraph above, a POIM $R \equiv R(m_k)$ with entries

$$R_{y,j}(m_k) := \mathbb{E}[\bar{s}(\mathcal{X}|m_k)|\mathcal{X}[j]^k = y] - \mathbb{E}[\bar{s}(\mathcal{X}|m_k)]. \tag{5}$$

Our overall aim is, by optimizing over the motifPOIM R, to approximate the original POIM (cf. also the illustration given by Figure 3). Due to the fact that searching for motifs of length k means computing POIMs of degree k, which is for longer PPMs ($k \geq 5$) computationally expensive, we have modified our optimization problem in a way that finding long PPMs can be accomplished using POIMs of lower degrees $\tilde{k} \in \{2,3\}$. The basic idea is to split longer PPMs of length k into shorter overlapping PPMs of length $\tilde{k} \leq k$ and use only the small POIM of degree $\tilde{k}$ for our optimization approach. First we define a set of smaller overlapping motifs, the SubPPMs, which should be devoted to the large PPM: A PPM of length k is modeled as a set of D SubPPMs, $D := k - \tilde{k} + 1$ with length $\tilde{k} \leq k$. The SubPPMs are defined by:

$$\tilde{m}_d(m_k, \tilde{k}) := (\tilde{r}, \tilde{\mu}, \sigma), \ \forall \ d = 0, \ldots, D-1$$

with $\tilde{\mu} := \mu + d$ and $\tilde{r} := r[d, d + \tilde{k}]$, where $r[d, d + \tilde{k}]$ is the d-th until the $(d + \tilde{k})$-th column of the PPMs PWM r.

3.1   Optimization Problem

We now derive the optimization problem for the extraction of motifs from POIMs. The core idea is to determine a motif m_k with an corresponding motif-POIM $R(m_k)$ that approximates the original POIM Q_k. To this end, let us introduce some notation. Let $\mathcal{K} \subset \mathbb{N}$ be the set of all motif lengths to be considered and $k_{\max} = \max_{k \in \mathcal{K}} k$ the maximum length. The vector $T \in \mathbb{N}_0^{k_{\max}}$ contains the number of PPMs for each motif length, where T_k is the given number of PPMs of length k for all $k \in \mathcal{K}$. For example, when $\mathcal{K} = \{2, 4, 10\}$ and $T = (0, 6, 0, 3, 0, 0, 0, 0, 0, 2)$, then the goal is to find 6 PPMs of length 2, 3 PPMs of length 4, and 2 PPMs of length 10. Our optimization method is as follows: given the set $\mathcal{K}$ and the vector T, we randomly initialize the PPMs $m_{k,t}$ $t = 1, \ldots, T_k$, $k \in \mathcal{K}$ and generate a set of motifPOIMs for the SubPPMs $\tilde{m}_d(m_k, \tilde{k})$, $d = 0, \ldots, D-1$. The optimization variables are the T_k many PPMs for all $k \in \mathcal{K}$. For obtaining the priorities of the PPMs we weight the PPMs by $\lambda_{k,t}$, $t = 1, \ldots, T_k$, $k \in \mathcal{K}$ and additionally optimize over the weights. Hence, the optimization variables are:

- PPM $m_{k,t} = (r_{k,t}, \mu_{k,t}, \sigma_{k,t})$, $t = 1, \ldots, T_k$, $k \in \mathcal{K}$,

   where $\mu_{k,t} \in \mathbb{R}, \sigma_{k,t} \in \mathbb{R}, r_{k,t} \in \mathbb{R}^{|\Sigma| \times k}$, $t = 1, \ldots, T_k$, $k \in \mathcal{K}$

– weight of $m_{k,t}$ $\qquad$ $\lambda_{k,t} \in \mathbb{R},$ $\qquad$ $t = 1, \ldots, T_k$, $k \in \mathcal{K}$.

A PPM generates a motifPOIM, which is given by the sum of D motif-POIMs generated by its SubPPMs. The sum of the weighted motifPOIMs, $\lambda_{k,t} R(m_{k,t})$, $t = 1, \ldots, T_k$, should estimate the POIM $Q_{\tilde{k}}$ for each $k \in \mathcal{K}$. The optimization problem is now that of minimizing the distance between the sum of the motifPOIMs and the original POIM, which leads to a non-convex optimization problem with the following objective function:

$$f(\eta) = \frac{1}{2} \sum_{k \in \mathcal{K}} \sum_{y \in \Sigma^{\tilde{k}}} \sum_{j=1}^{T_k} \left(\sum_{t=1}^{D-1} \lambda_{k,t} \sum_{d=0} R_{y,j}(\tilde{m}_d(m_{k,t}, \tilde{k})) - Q_{\tilde{k},y,j} \right)^2, \quad (6)$$

where $\eta = (m_{k,t}, \lambda_{k,t}, \tilde{k})_{t=1,\ldots,T_k, k \in \mathcal{K}}$. The associated constrained non-linear optimization problem is thus as follows:

$$\min_{(m_{k,t}, \lambda_{k,t})_{t=1,\ldots,T_k, k \in \mathcal{K}}} f(\eta) \quad (7)$$

$$\text{subject to} \quad \epsilon \leq \sigma_{k,t} \leq k, \qquad t = 1, \ldots, T_k, \, k \in \mathcal{K}$$

$$1 \leq \mu_{k,t} \leq L - k + 1, \qquad t = 1, \ldots, T_k, \, k \in \mathcal{K}$$

$$0 \leq \lambda_{k,t} \leq W, \qquad t = 1, \ldots, T_k, \, k \in \mathcal{K}$$

$$\epsilon \leq r_{k,t,o,s} \leq 1, \qquad t = 1, \ldots, T_k, \, k \in \mathcal{K}$$

$$o = 1, \ldots, |\Sigma|, s = 1, \ldots, k, \sum_{o=1}^{|\Sigma|} r_{k,t,o,s} - 1$$

where $W \in \mathbb{R}^+$. The objective function $f(\eta)$ is defined on compact set U, since all parameters are defined in a closed and bounded, convex space. Consequently, if U is not empty, $f(\eta)$ is a continuously differentiable function, since its conforming parts, that is, the Gaussian function and the product of the PWM entries, all are continuously differentiable. Thus the global minimum of the optimization problem (7) is guaranteed to exist. Due to the non-convex nature of (7), however, there may exist multiple local minima.

3.2 Efficient Computation of motifPOIM

To allow for numerical optimization of (7), we need an efficient way of computing (5). To this end, note that (5) consists of two summands. The right-hand summand can be computed as follows:

$$\mathbb{E}[\bar{s}(\mathcal{X}|m_k)] = \frac{1}{|\Sigma^L|} \sum_{x \in \Sigma^L} \bar{s}(x; m_k) = \frac{1}{|\Sigma^L|} \sum_{x \in \Sigma^L} \sum_{\ell=1}^{k} \sum_{i=1}^{L-\ell+1} v_{(x[i]^\ell, i)}(m_k)$$

$$= \sum_{\ell=1}^{k} \sum_{i=1}^{L-\ell+1} \frac{1}{|\Sigma^L|} \sum_{x \in \Sigma^L} v_{(x[i]^\ell, i)}(m_k) = \sum_{\ell=1}^{k} \sum_{i=1}^{L-\ell+1} \frac{1}{|\Sigma^\ell|} \sum_{z \in \Sigma^\ell} v_{(z,i)}(m_k)$$

$$= \sum_{\ell=1}^{k} \sum_{z \in \Sigma^\ell} \sum_{i=1}^{L-\ell+1} v_{(z,i)}(m_k) \mathbb{P}(\mathcal{X}[i]^\ell = z). \quad (8)$$

Furthermore, by an analogous computation, we compute the left-hand summand in (5) and obtain

$$\mathbb{E}[\bar{s}(\mathcal{X}|m_k)|\mathcal{X}[j]^k = y] = \sum_{\ell=1}^{k} \sum_{z \in \Sigma^\ell} \sum_{i=1}^{L-\ell+1} v_{(z,i)}(m_k)\mathbb{P}(\mathcal{X}[i]^\ell = z|\mathcal{X}[j]^k = y). \quad (9)$$

We now consider this probability term and its influence on the summation in (5). To this end, we introduce the following notation as in [37].

Definition 1. *Two positional subsequences (z, i) and (y, j) of length ℓ and k are independent if and only if they do not share any position; in this case we write $(y, j) \not\prec (z, i)$ and $(y, j) \prec (z, i)$ otherwise (i.e., when they are dependent). If they are dependent and also agree on all shared positions we say they are compatible and we write $(y, j) \precsim (z, i)$ (and $(y, j) \not\precsim (z, i)$ if they are not compatible).*

According to the cases discussed in the above definition, the conditioned probability term can take the following values:

$$\mathbb{P}(\mathcal{X}[i]^\ell = z|\mathcal{X}[j]^k = y) = \begin{cases} \frac{1}{|\Sigma^\ell|} & \text{if } (y, j) \not\prec (z, i) \\ 0 & \text{if } (y, j) \not\precsim (z, i) \\ \frac{|\Sigma^c|}{|\Sigma^\ell|} & \text{if } (y, j) \precsim (z, i) \end{cases}, \quad (10)$$

where c is the number of shared and compatible positions of two positional subsequences:

$$c((y, j), (z, i)) = \begin{cases} \ell - |i - j| & \text{if } i < j \text{ and } (y, j) \precsim (z, i) \\ \ell & \text{if } i = j \text{ and } (y, j) \precsim (z, i) \\ k - |i - j| & \text{if } i > j \text{ and } (y, j) \precsim (z, i) \\ 0 & \text{else.} \end{cases}$$

Taken the case $(y, j) \not\prec (z, i)$, the probability terms in the motifPOIM formula (5) subtract to zero, so that the positional subsequence (z, i) is not considered in the sum $R_{y,j}(m_k)$. Hence, in order to compute $R_{y,j}(m_k)$, it is sufficient to sum over two positional subsequence sets, where one contains all (z, i) with $(y, j) \precsim (z, i)$, $\mathcal{I}_{(y,j)}^{\precsim}$, and the others contains all (z, i) with $(y, j) \not\precsim (z, i)$, $\mathcal{I}_{(y,j)}^{\not\precsim}$:

$$R_{y,j}(m_k) = \sum_{(z,i) \in \mathcal{I}_{(y,j)}^{\precsim}} v_{(z,i)}(m_k)\Big(\frac{|\Sigma^c|}{|\Sigma^k|} - \frac{1}{|\Sigma^k|}\Big) + \sum_{(z,i) \in \mathcal{I}_{(y,j)}^{\not\precsim}} v_{(z,i)}(m_k)\Big(-\frac{1}{|\Sigma^k|}\Big)), \quad (11)$$

where $\mathcal{I}_{(y,j)}^{\circ} := \Big\{(z, i) \in \Sigma^{|y|} \times \{1, \ldots, L-|y|+1\} | (y, j) \circ (z, i)\Big\}$ and $\circ \in \{\precsim, \not\precsim\}$.

4   Empirical Analysis

In this section, we analyze our proposed mathematical model (7) empirically. After introducing the experimental setup, we evaluate our approach on the USPS data set, containing grayscale handwritten digit images. Afterwards, we conduct a biology experiment with a synthetic data set where we fully control the underlying ground truth. Finally, we investigate our model on a real *human* splice data set and compare our results to motifs contained in the JASPAR database [29]. As kernel-based learning algorithm, we use a support vector machine in all experiments.

4.1 Experimental Setup

For SVM training, we use the SHOGUN machine-learning toolbox [33]. The regularization constant C of the SVM and the degree d of the weighted-degree kernel are set to $C = 1$ and $d = 20$ for the biological experiments, which are proven default values. For the experiments on the USPS data, we set $d = 8$ and select C through model selection.

After SVM training, the POIM Q is generated through the Python script COMPUTE_POIMS.PY included in the SHOGUN toolbox. The Python framework obtains the trained SVM and the POIM of order k as parameters and returns the differential POIM and the regular POIMs $Q_l, l = 1, \ldots, k_{poim}$. We set $k = 7$ because of memory requirements (storing all POIMs up to a degree of 10 requires about 4 gigabytes of space). Note that this is no restriction as our modified optimization problem (7) requires POIMs of degree two or three only. Nevertheless, POIMS of higher degree than three can provide additional useful information since they contain prior information about the optimization variables.

We then compute the differential POIM using the Python scripts included in the SHOGUN toolbox, where we search for points of accumulation of high scoring entries, from which we estimate the number of motifs as well as their length and starting position. Throughout the experiments, we use a greedy approach for estimating the initial values of PWMs given a POIM. Once the motif interval is estimated, we select the leading nucleotide from the highest scoring column entry within the interval from the corresponding POIM and initialize the respective PWM entry with a value of 0.7 and 0.1 for non-matches. Indeed, we found that this approach is more stable and reliable than using random initializations. These parameters serve as initialization for our non-convex optimization problem (7). To compute a PWM from the computed POIMs, we employ the L-BFGS-B Algorithm [23], where the parameters λ and σ are initialized as 1 and 0.01, respectively.

As a measure of the motif reconstruction quality (MRQ), we employ in the biological experiments the same score as in the established JASPAR SPLICE database [30]. Given a ground truth sequence motif t we test the reconstruction quality of an equally-sized, revealed motif r according to the following formula:

$$\text{MRQ} = \sum_{p=1}^{k} \left[\frac{1}{k} - \frac{1}{2k} \sum_{c \in \{A,C,G,T\}} (t_{cp} - r_{cp})^2 \right]$$ We also introduce a second measure, the maximal-value MRQ (mvMRQ), which is defined in exactly the same way as the MRQ but uses the *maximum posteriori* motif $\hat{r} \in \{0,1\}^{4 \times k}$, that

is, it considers only the most likely sequence in the motif, which can have the advantage of discarding potential noise in the data and motif.

4.2   Experimental Results for USPS Dataset

We first evaluate the proposed methodology on the USPS data set [15,16], which includes 9298 images of handwritten digits, encoded through gray scale values ranging in $[-1, 1]$. For pre-processing, the data was converted to a binary format by setting a threshold at -0.2 for the gray scale values. To preserve locality in the vectorial image representation, we further preprocessed the data by scanning the image using a Hilbert curve of order 4, which is a proven method for mapping images to sequences [6,9]. Fig. 4 (a) shows the path of the Hilbert-curve scan for the handwritten images of the digit three. To determine the justification of the use of a high-dimensional weighted degree kernel, we compare it with a linear kernel on the gray scale values as well as with the weighted degree kernel of degree one only. The results in terms of multi-class classification accuracy are shown in Fig 4 (b), where the SVM was trained in one-vs.-all scheme. We observe that a weighted degree kernel of degree 8 (dimensionality: $2^8 * 256 = 65536$) performs best in our experiments.

For the remaining experiments, we focus on the binary classification tasks of the handwritten digits three vs. eight and two vs. nine, respectively. These

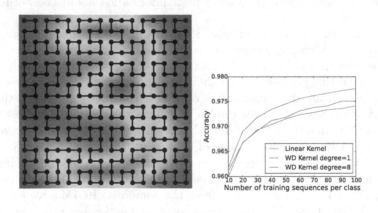

(a) Hilbert Curve   (b) SVM performance

Fig. 4. (a) Foreground: illustration of Hilbert-curve scanning (of order 4) of an image depicting of the handwritten digit three. The image is converted into a sequence through a curve that traverses the image in a way that mimics a fractal structure. It has been shown in [6,9] that this strategy is able to well capture the image's locality structure. The heatmap in the background shows the average feature values for the images of the digit three.
(b) The one-vs.-all SVM prediction accuracy is shown as a function of the number of training sequences per class for various WD sequence kernels over the Hilbert-scanned sequences and for a linear kernel on the gray-scale pixel values. The WD kernel of degree 8 performers best, even for only a small number of training sequences.

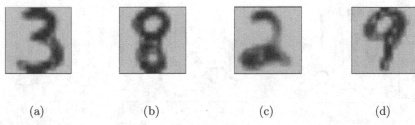

(a)   (b)   (c)   (d)

Fig. 5. Illustration of the results found by our proposed framework, when training a WD-kernel SVM of degree 8 for the handwritten digits three vs. eight and two vs. nine, respectively. The highest scoring positions in the motif are highlighted in red. Note that these are very characteristic positions for the dissimilarities between both digits. The background the average feature values for the images of the respective digit.

respective digit pairs are considered to be especially difficult to discriminate. For both digit pairs we train a WD-kernel SVM of degree 8 on the Hilbert-scanned sequences. Afterward, we compute the POIM as described in Section 4.1 and use our presented methodology to find a motif that incorporates the discriminative positions of the SVM decision for both classes. In this experiment, we simply fix the length of the motif to 256, which thus coincides with the sequence length. The step of intializing the POIM parameter through analyzing the differential POIM is thus omitted in this experiment. The results, illustrated in Figure 5, show the precise coherence between the discriminative motifs found and the obvious individually characteristic differences of the two digits, respectively. For instance in the discriminative task three-vs.-eight, we can observe that the most distinctive positions in the motif of the digit-eight (highlighted in red in Figure 5 (b)) are exactly the parts that are missing in the digit-three image.

4.3   Results for Synthetic Splice Site Experiments

Next, we evaluate the proposed methodology for biology DNA sequence data, by generating a synthetic data set, where we have full access to the underlying ground truth. This experiment aims at demonstrating the ability of our method in reconstructing the truly underlying motifs.

To this end, we generate the following sample sets: the sample set S_1 consists of 10,000 DNA sequences of length 30 over the alphabet $\{A, C, G, T\}^{30}$, randomly drawn from a uniform distribution $\mathcal{U}(\Sigma^L)$ over Σ^L. We subsequently modify 25% of the sequences by replacing the positions 6 to 11 by the synthetic target sequence CCTATA. These modified sequences form the positively labeled examples, while the remaining 75% of sequences are assigned with a negative label. The sample set S_2 includes the motif GATACATTAGGC of length 12 starting at position 16 in the positively labeled sequences. In the third sample set S_3 we insert both motifs at the same time.

The result of the realization of this synthetic experiments using the base sample S_1 and S_2 are shown in Figure 6. The corresponding motif/PWM computed by our approach correctly identifies the true underlying motif sequence as the most likely path in the PWM. More detailed results are shown in Table 1,

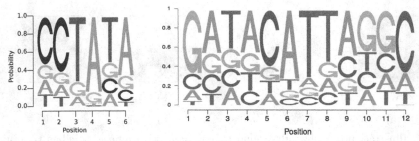

Fig. 6. Illustration of the motifs computed by our approach in the synthetic experiment: the size of a letter indicates the probability of occurrence of the corresponding nucleotide at a certain position in the motif. The left- and right-hand figures show the results for the synthetic data sets S_1 and S_2, respectively. Note that the truly underlying motifs where CCTATA for S_1 and GATACATTAGGC for S_2.

Table 1. Experimental results for the synthetic experiments on the three different sample sets S_1, S_2, and S_3.

sample set	SVM acc	#iter	time (s)	fevals	λ_{opt}	MRQ	mvMRQ
S_1	0.9987	157	13.2	116	1.0	0.93	1.0
S_2	1.0	31	19.7	64	1.0	0.65	1.0
S_3	1.0	31	25.87	64	0.42	0.85	1.0 (motif 1)
					0.58	0.84	1.0 (motif 2)

where, besides the MRQ and the mvMRQ value, we report also on the runtime of our approach, as well as the number of function evaluations, the optimal parameters for λ, the number of iterations needed, and the achieved SVM accuracy. Inspecting the mvMRQ, one can observe that even for the difficult dataset S_3, where we implanted both motives into the training sequences, we reconstruct both truly underlying motifs with 100% accuracy. The runtime of our approach ranges between 13 and 26 seconds.

4.4   Real-World Experiments on Human Splice Data

In this section, we evaluate our methodology on a *human* splice data set, which we downloaded from http://www.fml.tuebingen.mpg.de/raetsch/projects/lsmkl. For verifying our results we use the JASPAR database [29] (Available from http://jaspar.genereg.net), which provides us with a collection of important DNA motifs and also contains a splice site database. Note that real DNA sequences may contain non-polymorphic loci, which is why such a motif is not discriminative and we may thus not expect the SVM to identify this locus. We thus catch this special case and place this positional oligomer in the solution sequence. We apply the full experimental pipeline described in Section 4.1 to this data set.

Table 2. Execution times and optimal parameters for the *human* splice data set.

σ fixed	λ_{opt}	f_{opt}	time (s)	f evals	MRQ
0.01	0.005	78.97	37.91	24	90.1
0.1	0.84	59.48	28.3	20	97.58
1	1.67	57.18	33.53	17	97.03

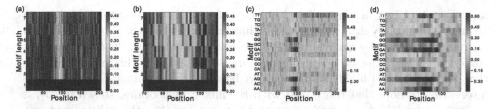

Fig. 7. Results of the real-world human splice experiment: Figures (a) and (c) show the differential POIM and the POIM of degree 2, respectively, for the entire sequence length of 200, while Figures (b) and (d) zoom into the "interesting" positions 70–110 only.

Figure 7 shows the preliminary results in terms of the differential POIM and the corresponding POIM of degree 2, shown for the entire sequence (see Figures 7 (a) and (c), respectively) as well as zoomed in for the "interesting" positions 70–110 of the sequence (see Figures 7 (b) and (d)). According to Figure 7 (b) the largest entry corresponds to a 7-mer that is found at position 95; furthermore, we observe high scoring entries for 7-,6- and 5-mers at position 85, from which we conclude that the discriminative motif starts at position 85 and ends at position 102. Thus, the motif we are searching is expected to have a length of 18 nucleotides, which we use as an initialization for our motifPOIM approach. We also account for non-polymorphic loci and find that the nucleotides A and G appear in all DNA sequences of the data set, always at the positions 100 and 101, respectively. We thus place them in the final PWM with a probability of 10%. The JASPAR splice database provides us with splice site motifs of length 20 only, which is why we search for motifs of the same size instead of the expected motif length 18.

The final results are shown in Figure 8, where the true underlying motif taken from the JASPAR splice database is shown in Figure (a), while the motif computed by our approach is shown in Figures (b)–(d). We observe a striking accordance with the true motif as evidenced by a high consensus score of 98.39

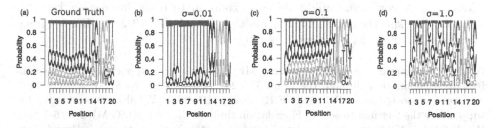

Fig. 8. Further results of the real-world human splice experiment: Figure (a) shows the (normalized) real splice sequence as taken from JASPAR. Figure (b)–(d) show the (normalized) computed PWMs for different values of the parameter σ. The best JASPAR score of 97.58, is achieved with $\sigma = 0.1$. This is, interestingly, followed by $\sigma = 1$ with a JASPAR score of 97.03 although the reconstructed motif of b) with $\sigma = 0.01$ and a score of 90.1 appears much more similar to the true motif in a).

for $\sigma = 0.1$, shown in Figure (c). Note that, for example, a completely random sequence (uniformly drawn nucleotides) has an average consensus of 89.31, which is greatly exceeded by our result. It is interesting to note that the function value corresponding to the best consensus score is suboptimal; this might indicate that the function is highly nonconvex with many local minima. Moreover, it is interesting to note that the PWM with the mixed nucleotides, shown in (d), is assigned a much higher accordance with the true motif than the well ordered one, shown in (b), which is more similar to the original JASPAR PWM. Furthermore, from Table 2, we observe moderate execution times of up to 32 seconds.

5   Conclusion and Discussion

Putting forward the work of [35] on positional oligomer importance matrices (POIMs), we have developed a new probabilistic methodology to automatically extract discriminative motifs from trained weighted-degree kernel machines such as support vector machines. To deal with the exponentially large size of the feature space associated with the SVM weight vector and the corresponding POIM ("[..] we realize that the list of POs can be prohibitively large for manual inspection." [35], page 8), we proposed an efficient optimization framework.

The results clearly illustrate the power of our approach in discovering discriminative motifs. For the experiment on handwritten digits, the proposed approach excels in finding intuitive motifs, as can be seen in Figure 5. In the synthetic experiments, the hidden motifs could be found and almost perfectly reconstructed. For the human splice site experiments, we recovered known motifs up to a very high precision of 98.39% as compared to the Jaspar splice data base.

We will provide the core algorithms as an add-on to the Python interface of the SHOGUN Machine Learning Toolbox. It is not only an established machine-learning framework, moreover, it already incorporates the possibility to extract positional-oligomer importance matrices (POIMs) of trained support vector machines using a WD-kernel. Ultimately, the usage by experimentalists will determine the utility of this approach and govern the direction of further extensions. A core issue might be the extension to other interesting kernels, such as, e.g., spectrum kernels [22], multiple kernels [17–19,21], other learning methods [11,12], or learning settings [10,20,39].

Acknowledgments. Acknowledgments MK acknowledges support by the German Research Foundation through the grant KL 2698/1-1 and KL 2698/2-1. MMCV and NG were supported by BMBF ALICE II grant 01IB15001B. We also acknowledges the support by the German Research Foundation through the grant DFG MU 987/6-1 and RA 1894/1-1. KRM thanks for partial funding by the National Research Foundation of Korea funded by the Ministry of Education, Science, and Technology in the BK21 program and the German Ministry for Education and Research as Berlin Big Data Center BBDC, funding mark 01IS14013A.

References

1. Abeel, T., de Peer, Y.V., Saeys, Y.: Towards a gold standard for promoter prediction evaluation. Bioinformatics (2009)
2. Bach, S., Binder, A., Montavon, G., Klauschen, F., Müller, K.R., Samek, W.: On pixel-wise explanations for non-linear classifier decisions by layer-wise relevance propagation. PLOS ONE (2015)
3. Baehrens, D., Schroeter, T., Harmeling, S., Kawanabe, M., Hansen, K., Müller, K.R.: How to explain individual classification decisions. JMLR **11**, 1803–1831 (2010)
4. Ben-Hur, A., Ong, C.S., Sonnenburg, S., Schölkopf, B., Rätsch, G.: Support vector machines and kernels for computational biology. PLoS Comput Biology **4**(10), e1000173 (2008). http://www.ploscompbiol.org/article/info:doi/10.1371/journal.pcbi.1000173
5. Boser, B., Guyon, I., Vapnik, V.: A training algorithm for optimal margin classifiers. In: Haussler, D. (ed.) COLT. pp. 144–152. ACM (1992)
6. Chung, K.L., Huang, Y.L., Liu, Y.W.: Efficient algorithms for coding hilbert curve of arbitrary-sized image and application to window query. Information Sciences **177**(10), 2130–2151 (2007)
7. Cortes, C., Vapnik, V.: Support vector networks. Machine Learning **20**, 273–297 (1995)
8. Crooks, G., Hon, G., Chandonia, J., Brenner, S.: Weblogo: A sequence logo generator. Genome Research **14**, 1188–1190 (2004)
9. Dafner, R., Cohen-Or, D., Matias, Y.: Context-based space filling curves. In: Computer Graphics Forum, vol. 19, pp. 209–218. Wiley Online Library (2000)
10. Goernitz, N., Braun, M., Kloft, M.: Hidden markov anomaly detection. In: Proceedings of The 32nd International Conference on Machine Learning, pp. 1833–1842 (2015)
11. Görnitz, N., Kloft, M., Brefeld, U.: Active and semi-supervised data domain description. In: Buntine, W., Grobelnik, M., Mladenić, D., Shawe-Taylor, J. (eds.) ECML PKDD 2009, Part I. LNCS, vol. 5781, pp. 407–422. Springer, Heidelberg (2009)
12. Görnitz, N., Kloft, M., Rieck, K., Brefeld, U.: Active learning for network intrusion detection. In: AISEC, p. 47. ACM Press (2009)
13. Görnitz, N., Kloft, M.M., Rieck, K., Brefeld, U.: Toward supervised anomaly detection. Journal of Artificial Intelligence Research (2013)
14. Hansen, K., Baehrens, D., Schroeter, T., Rupp, M., Müller, K.R.: Visual interpretation of kernel-based prediction models. Molecular Informatics **30**(9), September 2011. WILEY-VCH Verlag
15. Hastie, T., Tibshirani, R., Friedman, J., Hastie, T., Friedman, J., Tibshirani, R.: The elements of statistical learning, vol. 2. Springer (2009)
16. Hull, J.J.: A database for handwritten text recognition research. IEEE Transactions on Pattern Analysis and Machine Intelligence **16**(5), 550–554 (1994)
17. Kloft, M., Brefeld, U., Sonnenburg, S., Zien, A.: lp-Norm Multiple Kernel Learning. JMLR **12**, 953–997 (2011)
18. Kloft, M., Brefeld, U., Düessel, P., Gehl, C., Laskov, P.: Automatic feature selection for anomaly detection. In: Proceedings of the 1st ACM Workshop on AISec, pp. 71–76. ACM (2008)
19. Kloft, M., Brefeld, U., Sonnenburg, S., Laskov, P., Müller, K.R., Zien, A.: Efficient and accurate lp-norm multiple kernel learning. Advances in Neural Information Processing Systems **22**(22), 997–1005 (2009)

20. Kloft, M., Laskov, P.: Online anomaly detection under adversarial impact. In: AISTATS, pp. 405–412 (2010)

21. Kloft, M., Rückert, U., Bartlett, P.: A unifying view of multiple kernel learning. Machine Learning and Knowledge Discovery in Databases pp. 66–81 (2010)

22. Leslie, C.S., Eskin, E., Noble, W.S.: The spectrum kernel: A string kernel for svm protein classification. In: Pacific Symposium on Biocomputing, pp. 566–575 (2002)

23. Liu, D.C., Nocedal, J.: On the limited memory BFGS method for large scale optimization. Math. Program. **45**(3), 503–528 (1989). http://dx.doi.org/10.1007/BF01589116

24. Mohri, M., Rostamizadeh, A., Talwalkar, A.: Foundations of machine learning. MIT press (2012)

25. Montavon, G., Braun, M.L., Krueger, T., Müller, K.R.: Analyzing local structure in kernel-based learning: Explanation, complexity and reliability assessment. Signal Processing Magazine, IEEE **30**(4), 62–74 (2013)

26. Müller, K.R., Mika, S., Rätsch, G., Tsuda, K., Schölkopf, B.: An introduction to kernel-based learning algorithms. IEEE Transactions on Neural Networks **12**(2), 181–201 (2001). http://dx.doi.org/10.1109/72.914517

27. Rätsch, G., Sonnenburg, S., Srinivasan, J., Witte, H., Müller, K.R., Sommer, R.J., Schölkopf, B.: Improving the caenorhabditis elegans genome annotation using machine learning. PLoS Comput. Biol. **3**(2), e20 (2007)

28. Rätsch, G., Sonnenburg, S.: Accurate splice site prediction for caenorhabditis elegans. Kernel Methods in Computational Biology, 277–298 (2004). MIT Press series on Computational Molecular Biology, MIT Press

29. Sandelin, A., Alkema, W., Engström, P., Wasserman, W.W., Lenhard, B.: Jaspar: an open-access database for eukaryotic transcription factor binding profiles. Nucleic Acids Research **32**(Database–Issue), 91–94 (2004)

30. Sandelin, A., Höglund, A., Lenhardd, B., Wasserman, W.W.: Integrated analysis of yeast regulatory sequences for biologically linked clusters of genes. Functional & Integrative Genomics **3**(3), 125–134 (2003)

31. Schölkopf, B., Smola, A.: Learning with Kernels. MIT Press, Cambridge (2002)

32. Schölkopf, B., Smola, A., Müller, K.R.: Nonlinear component analysis as a kernel eigenvalue problem. Neural Computation **10**(5), 1299–1319 (1998)

33. Sonnenburg, S., Rätsch, G., Henschel, S., Widmer, C., Behr, J., Zien, A., Bona, F.D., Binder, A., Gehl, C., Franc, V.: The SHOGUN machine learning toolbox. Journal of Machine Learning Research **11**, 1799–1802 (2010)

34. Sonnenburg, S., Rätsch, G., Schäfer, C., Schölkopf, B.: Large scale multiple kernel learning. Journal of Machine Learning Research **7**, 1531–1565 (2006)

35. Sonnenburg, S., Zien, A., Philips, P., Rätsch, G.: POIMs: positional oligomer importance matrices – understanding support vector machine based signal detectors. Bioinformatics (2008). (received the Outstanding Student Paper Award at ISMB 2008)

36. Sonnenburg, S., Franc, V.: Coffin: a computational framework for linear SVMs. In: ICML, pp. 999–1006 (2010)

37. Sonnenburg, S., Schweikert, G., Philips, P., Behr, J., Rätsch, G.: Accurate Splice Site Prediction. BMC Bioinformatics, Special Issue from NIPS workshop on New Problems and Methods in Computational Biology Whistler, Canada, December 18, 2006, vol. 8(Suppl. 10), p. S7, December 2007

38. Sonnenburg, S., Zien, A., Rätsch, G.: ARTS: Accurate Recognition of Transcription Starts in Human. Bioinformatics **22**(14), e472–480 (2006)

39. Zeller, G., Goernitz, N., Kahles, A., Behr, J., Mudrakarta, P., Sonnenburg, S., Raetsch, G.: mtim: rapid and accurate transcript reconstruction from rna-seq data. arXiv preprint arXiv:1309.5211 (2013)
40. Zien, A., Philips, P., Sonnenburg, S.: Computing Positional Oligomer Importance Matrices (POIMs). Research Report; Electronic Publication 2, Fraunhofer Institute FIRST, December 2007
41. Zien, A., Rätsch, G., Mika, S., Schölkopf, B., Lengauer, T., Müller, K.R.: Engineering support vector machine kernels that recognize translation initiation sites in DNA. BioInformatics **16**(9), 799–807 (2000)

Pattern and Sequence Mining

Fast Generation of Best Interval Patterns for Nonmonotonic Constraints

Aleksey Buzmakov[1,2]($\boxtimes$), Sergei O. Kuznetsov[2], and Amedeo Napoli[1]

[1] LORIA (CNRS – Inria NGE – University de Lorraine),
Vandœuvre-lès-Nancy, France
`aleksey.buzmakov@inria.fr, amedeo.napoli@loria.fr`
[2] National Research University Higher School of Economics, Moscow, Russia
`skuznetsov@hse.ru`

Abstract. In pattern mining, the main challenge is the exponential explosion of the set of patterns. Typically, to solve this problem, a constraint for pattern selection is introduced. One of the first constraints proposed in pattern mining is support (frequency) of a pattern in a dataset. Frequency is an anti-monotonic function, i.e., given an infrequent pattern, all its superpatterns are not frequent. However, many other constraints for pattern selection are not (anti-)monotonic, which makes it difficult to generate patterns satisfying these constraints. In this paper we introduce the notion of projection-antimonotonicity and $\vartheta - \sum o\psi\iota\alpha$ algorithm that allows efficient generation of the best patterns for some nonmonotonic constraints. In this paper we consider stability and Δ-measure, which are nonmonotonic constraints, and apply them to interval tuple datasets. In the experiments, we compute best interval tuple patterns w.r.t. these measures and show the advantage of our approach over postfiltering approaches.

Keywords: Pattern mining · Nonmonotonic constraints · Interval tuple data

1 Introduction

Interestingness measures were proposed to overcome the problem of combinatorial explosion of the number of valid patterns that can be discovered in a dataset [1]. For example, pattern support, i.e., the number of objects covered by the pattern, is one of the most famous measures of pattern quality. In particular, support satisfies the property of anti-monotonicity (aka "a priori principle"), i.e., the larger the pattern is the smaller the support is [2,3]. Many other measures can be mentioned such as utility constraint [4], pattern stability [5,6], pattern leverage [7], margin closeness [8], MCCS [9], cosine interest [10], pattern robustness [11], etc.

Some of these measures (e.g., support, robustness for generators [11], or upper bound constraint of MCCS [9]) are "globally anti-monotonic", i.e., for any two patterns $X \sqsubseteq Y$ we have $\mathcal{M}(X) \geq \mathcal{M}(Y)$, where $\mathcal{M}$ is a measure and $\sqsubseteq$ denotes

© Springer International Publishing Switzerland 2015
A. Appice et al. (Eds.): ECML PKDD 2015, Part II, LNAI 9285, pp. 157–172, 2015.
DOI: 10.1007/978-3-319-23525-7_10

the (subsumption) order relation on patterns. When a measure is anti-monotonic, it is relatively easy to find patterns whose measure is higher than a certain threshold (e.g., patterns with a support higher than a threshold). In contrast some other measures are called "locally anti-monotonic", i.e., for any pattern X there is an immediate subpattern $Y \prec X$ such that $\mathcal{M}(Y) \geq \mathcal{M}(X)$. Then the right strategy should be selected for traversing the search space, e.g., a pattern Y should be extended only to patterns X such that $\mathcal{M}(Y) \geq \mathcal{M}(X)$. For example, for "locally anti-monotonic" cosine interest [10], the extension of a pattern Y consists in adding only attributes with a smaller support than any attribute from Y. The most difficult case for selecting valid patterns occurs when a measure is not locally anti-monotonic. Then, valid patterns can be retained by postfiltering, i.e., finding a (large set of) patterns satisfying an antimonotone constraint and filtering them w.r.t. the chosen nonmonotonic measure (i.e., neither monotonic nor anti-monotonic) [6,8,11], or using heuristics such as leap search [12] or low probability of finding interesting patterns in the current branch [7].

Most of the measures are only applicable to one type of patterns, e.g., pattern leverage or cosine interest can be applied only to binary data since their definitions involve single attributes. "Pattern independent measures" usually relies on support of the pattern and/or on support of other patterns from the search space. In particular, support, stability [5], margin-closeness [8] and robustness [11] are pattern independent measures. In this paper we work with interval tuple data, where only pattern independent measures as well as specific measures for interval tuples can be applied. In addition, given a measure, it can be difficult to define a good threshold. Thus various approaches for finding top-K patterns were introduced [13–15], with the basic idea to automatically adjust the threshold for a measure $\mathcal{M}$.

In this paper we introduce a new algorithm $\vartheta - \sum o\varphi\iota\alpha$, i.e., Sofia, for "Searching for Optimal Formal Intents Algorithm" for a interestingness threshold , for extracting the best patterns of a kind, e.g., itemsets, interval tuples, strings, graph patterns, etc. $\vartheta - \sum o\varphi\iota\alpha$ algorithm is applicable to a class of measures called "projection-antimonotonic measures" or more precisely "measures anti-monotonic w.r.t. a chain of projections". This class includes globally anti-monotonic measures such as support, locally anti-monotonic measures such as cosine interest and some of the nonmonotonic measures such as stability or robustness of closed patterns. The main novelty of this paper is $\vartheta - \sum o\varphi\iota\alpha$, a new efficient algorithm for finding best patterns of different kinds w.r.t. projection-antimonotonic measures which constitutes a rather large class of measures.

The remaining of the paper is organized as follows. The formalization of the current approach is based on Formal Concept Analysis (FCA) [16] and pattern structures [17] which are introduced in Section 2. Then, $\vartheta - \sum o\varphi\iota\alpha$ algorithm is detailed in Section 3 first for an arbitrary measure and second for the Δ-measure. Experiments and a discussion are proposed in Section 4, before conclusion.

2 Data Model

2.1 FCA and Pattern Structures

Formal Concept Analysis (FCA) is a formalism for knowledge discovery and data mining thanks to the design of concept lattices [16]. It is also convenient for describing models of itemset mining, and, since [18], lattices of closed itemsets (i.e., concept lattices) and closed descriptions are used for concise representation of association rules. For more complex data such as sequences and graphs one can use an extension of the basic model, called pattern structures [17]. With pattern structures it is possible to define closed descriptions and to give a concise representation of association rules for different descriptions with a natural order (such as subgraph isomorphism order) [19,20].

A *pattern structure* is a triple $(G, (D, \sqcap), \delta)$, where G is a set of objects, $(D, \sqcap)$ is a complete meet-semilattice of descriptions and $\delta : G \to D$ maps an object to a description.

The intersection $\sqcap$ gives the similarity of two descriptions. Standard FCA can be presented in terms of a pattern structure. A *formal context* (G, M, I), where G is a set of objects, M is a set of attributes and $I \subseteq G \times M$ an incidence relation giving information about attributes related to objects, is represented as a pattern structure $(G, (\wp(M), \cap), \delta)$, where $(\wp(M), \cap)$ is a semilattice of subsets of M with $\cap$ being the set-theoretical intersection. If $x = \{a, b, c\}$ and $y = \{a, c, d\}$, then $x \sqcap y = x \cap y = \{a, c\}$. The mapping $\delta : G \to \wp(M)$ is given by $\delta(g) = \{m \in M \mid (g, m) \subset I\}$ and returns the description of a given object as a set of attributes.

The following mappings or diamond operators give a Galois connection between the powerset of objects and descriptions:

$$A^\diamond := \bigsqcap_{g \in A} \delta(g), \qquad\qquad \text{for } A \subseteq G$$

$$d^\diamond := \{g \in G \mid d \sqsubseteq \delta(g)\}, \qquad\qquad \text{for } d \in D$$

Given a subset of objects A, $A^\diamond$ returns the description which is common to all objects in A. Given a description d, $d^\diamond$ is the set of all objects whose description subsumes d. A partial order $\sqsubseteq$ (subsumption) on descriptions from D is defined w.r.t. the similarity operation $\sqcap$: $c \sqsubseteq d \Leftrightarrow c \sqcap d = c$, and c is subsumed by d.

A *pattern concept* of a pattern structure $(G, (D, \sqcap), \delta)$ is a pair (A, d), where $A \subseteq G$, called *pattern extent* and $d \in D$, called *pattern intent*, such that $A^\diamond = d$ and $d^\diamond = A$. A pattern extent is a closed set of objects, and a pattern intent is a closed description, e.g., a closed itemset when descriptions are given as sets of items (attributes). As shown in [19], descriptions closed in terms of counting inference (which is a standard data mining approach), such as closed graphs [21], are elements of pattern intents.

A pattern extent corresponds to the maximal set of objects A whose descriptions subsume the description d, where d is the maximal common description for objects in A. The set of all pattern concepts is partially ordered w.r.t. inclusion on extents, i.e., $(A_1, d_1) \leq (A_2, d_2)$ iff $A_1 \subseteq A_2$ (or, equivalently, $d_2 \sqsubseteq d_1$), making a lattice, called pattern lattice.

2.2   Interval Pattern Structure

A possible instantiation of pattern structures is interval pattern structures introduced to support efficient processing of numerical data without binarization [20]. Given k numerical or interval attributes whose values are of the form $[a, b]$, where $a, b \in \mathbb{R}$, the language of a pattern space is given by tuples of intervals of size k. For simplicity, we denote intervals of the form $[a, a]$ by a.

Figure 1a exemplifies an interval dataset. It contains 6 objects and 2 attributes. An interval as a value of an attribute corresponds to an uncertainty in the value of the attribute. For example, the value of m_1 for g_2 is known exactly, while the value of m_2 is lying in $[1, 2]$. Given this intuition for intervals it is natural to define similarity of two intervals as their convex hull, since by adding new objects one increases the uncertainty. For example, for g_1 the value of m_1 is 0, while for g_6 it is 1, thus given the set $\{g_1, g_6\}$, the uncertainty of m_1 in this set is $[0, 1]$, i.e., the similarity of g_1 and g_6 w.r.t. m_1 is $[0, 1]$. More formally, given two intervals $[a, b]$ and $[c, d]$, the similarity of these two intervals is given by $[a, b] \sqcap [c, d] = [\min(a, c), \max(b, d)]$. Given a tuple of intervals, the similarity is computed component-wise. For example, $g_1^\diamond \sqcap g_6^\diamond = \langle [0, 1]; [0, 2] \rangle$. Reciprocally, $\langle [0, 1]; [0, 2] \rangle = \{g_1, g_2, \cdots, g_6\}$.

The resulting concept lattice is shown in Figure 1b. Concept extents are shown by indices of objects, intents are given in angle brackets, the numbers on edges and on concepts are related to interestingness of concepts and will be described in the next subsection.

2.3   Stability Index of a Concept

For real datasets, the number of patterns can be very large, even computing the number of closed patterns is a #P-complete problem [22]. Different measures were tested for selecting most interesting patterns, such as stability [5]. Stability measures the independence of a concept intent w.r.t. randomness in data.

Given a concept $\mathcal{C}$, *concept stability* $\mathtt{Stab}(\mathcal{C})$ is the relative number of subsets of the concept extent (denoted by $\mathtt{Ext}(\mathcal{C})$), whose descriptions, i.e., the result of $(\cdot)^\diamond$ is equal to the concept intent (denoted by $\mathtt{Int}(\mathcal{C})$).

$$\mathtt{Stab}(\mathcal{C}) := \frac{|\{s \in \wp(\mathtt{Ext}(\mathcal{C})) \mid s^\diamond = \mathtt{Int}(\mathcal{C})\}|}{|\wp(\mathtt{Ext}(\mathcal{C}))|} \tag{1}$$

Here $\wp(P)$ is the powerset of P. The larger the stability, the more objects can be deleted from the context without affecting the intent of the concept, i.e., the intent of the most stable concepts is likely to be a characteristic pattern of a given phenomenon and not an artifact of a dataset.

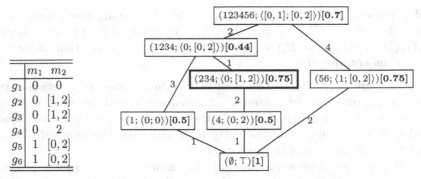

(a) An interval context. (b) An interval concept lattice with corresponding stability indexes. Objects are given by their indices.

Fig. 1. A formal context and the corresponding lattice.

We say that a concept is stable if its stability is higher than a given threshold θ; a pattern p is stable if there is a concept in the lattice with p as the intent and the concept is stable.

Example 1. Figure 1b shows a lattice for the context in Figure 1a. Concept extents are given by their indices, i.e., $\{g_1, g_2\}$ is given by 12. The extent of the highlighted concept C is $\text{Ext}(C) = \{g_2, g_3, g_4\}$, thus, its powerset contains 2^3 elements. Descriptions of 2 subsets of $\text{Ext}(C)$ ($\{g_4\}$ and $\emptyset$) are different from the intent of C, $\text{Int}(C) = \{m_3\}$, while all other subsets of $\text{Ext}(C)$ have a common set of attributes equal to $\langle 0; [1, 2] \rangle$. So, $\text{Stab}(C) = \frac{2^3 - 2}{2^3} = 0.75$. Stability of other concepts is shown in brackets. It should be noticed that stability of all comparable patterns for $\text{Int}(C)$ in the lattice is smaller than the stability of C, which highlights the nonmonotonicity of stability.

Concept stability is closely related to the robustness of a closed pattern [11]. Indeed, robustness is the probability of a closed pattern to be found in a subset of the dataset. To define this probability, the authors define a weight for every subset given as a probability of obtaining this subset by removing objects from the dataset, where every object is removed with probability α, e.g., given a subset of objects $X \subseteq G$, the probability of the induced subset is given by $p(D_\alpha = (X, (D, \sqcap), \delta)) = \alpha^{|X|}(1-\alpha)^{|G \setminus X|}$. Stability in this case is the robustness of closed pattern if the weights of subsets of the dataset are equal to $2^{-|G|}$.

The problem of computing concept stability is #P-complete [5]. A fast computable stability estimate was proposed in [23], where it was shown that this estimate ranks concepts almost in the same way as stability does. In particular, $\text{Stab}(C) \leq 1 - 2^{-\Delta(C)}$, where $\Delta(C) = \min_{\mathcal{D} \leq C} |\text{Ext}(C) \setminus \text{Ext}(\mathcal{D})|$, i.e., the minimal difference in supports between concept C and all its nearest subconcepts. For a threshold θ, patterns p with $\Delta(p) \geq \theta$ are called Δ-stable patterns.

Example 2. Consider the example in Figure 1. Every edge in the figure is labeled with the difference in support between the concepts this edge connects. Thus,

Δ of a pattern is the minimum label of the edges going down from the concept. The value $\Delta((\{g_2, g_3, g_4\}; \langle 0; [1, 2]\rangle))$ is equal to 2. Another example is $\Delta((G; \langle [0, 1]; [0, 2]\rangle)) = 2$. For this example we can also see that Δ-measure is not anti-monotonic either.

Δ-measure is related to the work of margin-closeness of an itemset [8]. In this work, given a set of patterns, e.g., frequent closed patterns, the authors rank them by the minimal distance in their support to the closest superpattern divided over the support of the pattern. In our case, the minimal distance is exactly the Δ-measure of the pattern.

Stability and Δ-measure are not anti-monotonic but rather projection-antimonotonic. Patterns w.r.t. such kind of measures can be mined by a specialized algorithm introduced in Section 3. But before we should introduce projections of pattern structures in order to properly define projection-antimonotonicity and the algorithm.

2.4 Projections of Pattern Structures

The approach proposed in this paper is based on projections introduced for reducing complexity of computing pattern lattices [17].

A *projection* $\psi : D \to D$ is an "interior operator", i.e., it is (1) monotonic ($x \sqsubseteq y \Rightarrow \psi(x) \sqsubseteq \psi(y)$), (2) contractive ($\psi(x) \sqsubseteq x$) and (3) idempotent ($\psi(\psi(x)) = \psi(x)$). A *projected pattern structure* $\psi((G, (D, \sqcap), \delta))$ is a pattern structure $(G, (D_\psi, \sqcap_\psi), \psi \circ \delta)$, where $D_\psi = \psi(D) = \{d \in D \mid \exists d^* \in D : \psi(d^*) = d\}$ and $\forall x, y \in D, x \sqcap_\psi y := \psi(x \sqcap y)$.

Example 3. Consider the example in Figure 1. If we remove a column corresponding to an attribute, e.g., the attribute m_2, from the context in Figure 1a, we define a projection, given by $\psi(\langle [a, b]; [c, d]\rangle) = \langle [a, b]; [-\infty, +\infty]\rangle$, meaning that no value of m_2 is taken into account.

Given a projection ψ we call $\psi(D) = \{d \in D \mid \psi(d) = d\}$ the *fixed set of* ψ. Note that, if $\psi(d) \neq d$, then there is no other $\tilde{d}$ such that $\psi(\tilde{d}) = d$ because of idempotency of projections. Hence, any element outside the fixed set of the projection ψ is pruned from the description space. Given the notion of a fixed set we can define a partial order on projections.

Definition 1. *Given a pattern structure* $\mathbb{P} = (G, (D, \sqcap), \delta)$ *and two projections* ψ_1 *and* ψ_2, *we say that* ψ_1 *is simpler than* ψ_2 (ψ_2 *is more detailed than* ψ_1), *denoted by* $\psi_1 < \psi_2$, *if* $\psi_1(D) \subset \psi_2(D)$, *i.e.,* ψ_1 *prunes more descriptions than* ψ_2.

Our algorithm is based on this order on projections. The simpler a projection ψ is, the less patterns we can find in $\psi(\mathbb{P})$, and the less computational efforts one should take. Thus, we compute a set of patterns for a simpler projection, then we remove unpromising patterns and extend our pattern structure and the found patterns to a more detailed projection. This allows us to reduce the size of patterns within a simpler projection in order to reduce the computational complexity of more detailed projection.

2.5 Projections of Interval Pattern Structures

Let us first consider interval pattern structures with only one attribute m. Let us denote by $W = \{w_1, \cdots, w_{|W|}\}$ all possible values of the left and right endpoints of the intervals corresponding to the attribute in a dataset, so that $w_1 < w_2 < \cdots < w_{|W|}$. By reducing the set W of possible values for the left or the right end of the interval we define a projection. For example, if $\{w_1\}$ is the only possible value for the left endpoint of an interval and $\{w_{|W|}\}$ is the only possible value of the right endpoint of an interval, then all interval patterns are projected to $[w_1, w_{|W|}]$. Let us consider this in more detail.

Let two sets $L, R \subset W$ such that $w_1 \in L$ and $w_{|W|} \in R$ be constraints on possible values on the left and right endpoints of an interval, respectively. Then a projection is defined as follows:

$$\psi_{m[L,R]}([a,b]) = [\max\{l \in L | l \le a\}, \min\{r \in R | r \ge b\}]. \tag{2}$$

Requiring that $w_1 \in L$ and $w_{|W|} \in R$ we ensure that the sets used for minimal and maximal functions are not empty. It is not hard to see that (2) is a projection. The projections given by (2) are ordered w.r.t. simplicity (Definition 1). Indeed, given $L_1 \subseteq L$ and $R_1 \subseteq R$, we have $\psi_{m[L_1,R_1]} < \psi_{m[L,R]}$, because of inclusion of fixed sets. Let us notice that a projection $\psi_{m[W,W]}$ does not modify the lattice of concepts for the current dataset, since any interval for the value set W is possible. We also notice that a projection $\psi_{m[L,R]}$ is defined for one interval, while we can combine the projections for different attributes in a tuple to a single projection for the whole tuple $\psi_{m_1[L_1,R_1]m_2[L_2,R_2]\ldots}$.

Example 4. Consider example in Figure 1. Let us consider a projection

$$\psi_{m_1[\{0,1\},\{1\}]m_2[\{0,2\},\{0,2\}]}.$$

The fixed set of this projection consists of $\{[0,1], 1\} \times \{0, 2, [0,2]\}$, i.e., 6 intervals. Let us find the projection of $(g_2)^\diamond = \langle 0; [1,2] \rangle$ in a component-wise way: $\psi_{m_1[\{0,1\},\{1\}]}(0) = [0,1]$, since 0 is allowed on the left endpoint of an interval but not allowed to be on the right endpoint of an interval; $\psi_{m_2[\{0,2\},\{0,2\}]}([1,2]) = [0,2]$ since 1 is not allowed on the left endpoint of an interval. Thus,

$$\psi_{m_1[\{0,1\},\{1\}]m_2[\{0,2\},\{0,2\}]}(\langle 0; [1,2]\rangle) = \langle [0,1]; [0,2]\rangle.$$

The lattice corresponding to this projection is shown in Figure 2.

3 $\vartheta - \sum o\varphi\iota\alpha$ Algorithm

3.1 Anti-monotonicity w.r.t. a Projection

Our algorithm is based on the projection-antimonotonicity, a new idea introduced in this paper. Many interestingness measures for patterns, e.g., stability, are not (anti-)monotonic w.r.t. subsumption order on patterns. A measure $\mathcal{M}$

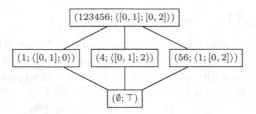

Fig. 2. Projected lattice from example in Figure 1 by projection $\psi_{m_1[\{0,1\},\{1\}]m_2[\{0,2\},\{0,2\}]}$. See Example 4.

is called *anti-monotonic*, if for two patterns $q \sqsubseteq p$, $\mathcal{M}(q) \geq \mathcal{M}(p)$. For instance, support is a anti-monotonic measure w.r.t. pattern order and it allows for efficient generation of patterns with support larger than a threshold [2,3,18]. The projection-antimonotonicity is a generalization of standard anti-monotonicity and allows for efficient work with a larger set of interestingness measures.

Definition 2. *Given a pattern structure* $\mathbb{P}$ *and a projection* ψ, *a measure* $\mathcal{M}$ *is called* anti-monotonic w.r.t. the projection ψ, *if*

$$(\forall p \in \psi(\mathbb{P}))(\forall q \in \mathbb{P}, \psi(q) = p)\ \mathcal{M}_\psi(p) \geq \mathcal{M}(q), \tag{3}$$

where $\mathcal{M}_\psi(p)$ *is the measure* $\mathcal{M}$ *of a pattern* p *computed in* $\psi(\mathbb{P})$.

Here, for any pattern p of a projected pattern structure we check that a preimage q of p for ψ has a measure smaller than the measure of p. It should be noticed that a measure $\mathcal{M}$ for a pattern p can yield different values if $\mathcal{M}$ is computed in $\mathbb{P}$ or in $\psi(\mathbb{P})$. Thus we use the notation $\mathcal{M}_\psi$ for the measure $\mathcal{M}$ computed in $\psi(\mathbb{P})$. The property of a measure given in Definition 2 is called projection-antimonotonicity.

It should be noticed that classical anti-monotonic measures are projection-antimonotonic for any projection. Indeed, because of contractivity of ψ ($\psi(p) \sqsubseteq p$), for any anti-monotonic measure one has $\mathcal{M}(\psi(p)) \geq \mathcal{M}(p)$. This definition covers also the cases where a measure $\mathcal{M}$ is only locally anti-monotonic, i.e., given a pattern p there is an immediate subpattern $q \prec p$ such that $\mathcal{M}(q) \geq \mathcal{M}(p)$, see e.g., the cosine interest of an itemset, which is only locally anti-monotonic [10]. Moreover, this definition covers also some measures that are not locally anti-monotonic. As we mentioned in Examples 1 and 2 stability and Δ-measure are not locally anti-monotonic. However, it can be shown that they are anti-monotonic w.r.t. any projection [24]. Moreover, following the same strategy one can prove that robustness of closed patterns from [11] is also anti-monotonic w.r.t. any projection. In particular, the robustness of closed patterns defines a anti-monotonic constraint w.r.t. any projection.

Thus, given a measure $\mathcal{M}$ anti-monotonic w.r.t. a projection ψ, if p is a pattern such that $\mathcal{M}_\psi(p) < \theta$, then $\mathcal{M}(q) < \theta$ for any preimage q of p for ψ. Hence, if, given a pattern p of $\psi(\mathbb{P})$, one can find all patterns q of $\mathbb{P}$ such that $\psi(q) = p$, it is possible to first find all patterns of $\psi(\mathbb{P})$ and then to filter them w.r.t. $\mathcal{M}_\psi$ and a threshold, and finally to compute the preimages of filtered patterns. It allows one to cut earlier unpromising branches of the search space or adjust a threshold for finding only a limited number of best patterns.

3.2 Anti-monotonicity w.r.t. a Chain of Projections

However, given just one projection, it can be hard to efficiently discover the patterns, because the projection is either hard to compute or the number of unpromising patterns that can be pruned is not high. Hence we introduce *a chain of projections* $\psi_0 < \psi_1 < \cdots < \psi_k = \mathbb{1}$, where a pattern lattice for $\psi_0(\mathbb{P})$ can be easily computed and $\mathbb{1}$ is the identity projection, i.e., $(\forall x)\mathbb{1}(x) = x$. For example, to find frequent itemsets, we typically search for small frequent itemsets and then extend them to larger ones. This corresponds to extension to a more detailed projection.

Definition 3. *Given a pattern structure $\mathbb{P}$ and a chain of projections $\psi_0 < \psi_1 < \cdots < \psi_k = \mathbb{1}$, a measure $\mathcal{M}$ is called* anti-monotonic w.r.t. the chain of projections *if $\mathcal{M}$ is anti-monotonic w.r.t. all ψ_i for $0 \leq i \leq k$.*

Example 5. Let us construct a chain of projections satisfying (2) for the example in Figure 1. The value set for the first attribute is $W_1 = \{0, 1\}$ and the value set for the second is $W_2 = \{0, 1, 2\}$. Let us start the chain from a projection $\psi_0 = \psi_{m_1[\{0\},\{1\}]m_2[\{0\},\{2\}]}$. This projection allows only for one pattern $\langle [0, 1]; [0, 2] \rangle$, i.e., the concept lattice is easily found. Then we increase the complexity of a projection by allowing more patterns. For example, we can enrich the first component of a tuple without affecting the second one, i.e., a projection $\psi_1 = \psi_{m_1[\{0,1\},\{0,1\}]m_2[\{0\},\{2\}]}$. This projection allows for 3 patterns, i.e., any possible interval of the first component and only one interval [0,2] for the second component. Let us notice that it is not hard to find preimages for ψ_0 in $\psi_1(D)$. Indeed, for any pattern p from $\psi_0(D)$ one should just modify either the left side of the first interval of p by one value, or the right side of the first interval of p.

Then we can introduce a projection that slightly enrich the second component of a tuple, e.g., $\psi_2 = \psi_{m_1[\{0,1\},\{0,1\}]m_2[\{0,1\},\{1,2\}]}$ and finally we have $\psi_3 = \psi_{m_1[W_1,W_1]m_2[W_2,W_2]}$. Finding preimages in this chain is not a hard problem, since on every set we can only slightly change left and/or right side of the second interval in a tuple. Thus, starting from a simple projection and making transitions from one projection to another, we can cut unpromising branches and efficiently find the set of interesting patterns.

3.3 Algorithms

Given a measure anti-monotonic w.r.t. a chain of projections, if we are able to find all preimages of any element in the fixed set of ψ_i that belong to a fixed set of ψ_{i+1}, then we can find all patterns of $\mathbb{P}$ with a value of $\mathcal{M}$ higher than a given threshold θ. We call this algorithm $\vartheta - \sum o\varphi\iota\alpha$ and its pseudocode is given in Algorithm 1. In lines 11-12 we find all patterns for $\psi_0(\mathbb{P})$ satisfying the

---

Data: A pattern structure $\mathbb{P}$, a chain of projections $\Psi = \{\psi_0, \psi_1, \cdots, \psi_k\}$, a measure $\mathcal{M}$ anti-monotonic for the chain Ψ, and a threshold θ for $\mathcal{M}$.

1 **Function** ExtendProjection(i, θ, $\mathcal{P}_{i-1}$)

    Data: i is the projection number to which we should extend ($0 < i \le k$), θ is a threshold value for $\mathcal{M}$, and $\mathcal{P}_{i-1}$ is the set of patterns for the projection ψ_{i-1}.

    Result: The set $\mathcal{P}_i$ of all patterns with the value of measure $\mathcal{M}$ higher than the threshold θ for ψ_i.

2     $\mathcal{P}_i \longleftarrow \emptyset$;

3   /* Put all preimages in $\psi_i(\mathbb{P})$ for any pattern p */

4   **foreach** $p \in \mathcal{P}_{i-1}$ **do**

5         $\mathcal{P}_i \longleftarrow \mathcal{P}_i \cup \text{Preimages}(i, p)$

6   /* Filter patterns in $\mathcal{P}_i$ to have a value of $\mathcal{M}$ higher than θ */

7   **foreach** $p \in \mathcal{P}_i$ **do**

8   **if** $\mathcal{M}_{\psi_i}(p) \le \theta$ **then**

9             $\mathcal{P}_i \longleftarrow \mathcal{P}_i \setminus \{p\}$

10 **Function** Algorithm_$\vartheta - \sum o\varphi\iota\alpha$

    Result: The set $\mathcal{P}$ of all patterns with a value of $\mathcal{M}$ higher than the threshold θ for $\mathbb{P}$.

11   /* Find all patterns in $\psi_0(\mathbb{P})$ with a value of $\mathcal{M}$ higher than θ */

12     $\mathcal{P} \longleftarrow \text{FindPatterns}(\theta, \psi_0)$;

13   /* Run through out the chain Ψ and find the patterns for $\psi_i(\mathbb{P})$ */

14   **foreach** $0 < i \le k$ **do**

15         $\mathcal{P} \longleftarrow \text{ExtendProjection}(i, \theta, \mathcal{P})$;

Algorithm 1. The $\vartheta - \sum o\varphi\iota\alpha$ algorithm for finding patterns in $\mathbb{P}$ with a value of a measure $\mathcal{M}$ higher than a threshold θ.

constraint that a value of $\mathcal{M}$ is higher than a threshold. Then in lines 13-15 we iteratively extend projections from simpler to more detailed ones. The extension is done by constructing the set $\mathcal{P}_i$ of preimages of the set $\mathcal{P}_{i-1}$ (lines 2-5) and then by removing the patterns that do not satisfy the constraint from $\mathcal{P}_i$ (lines 6-9).

The algorithm is sound and complete, since first, a pattern p is included into the set of preimages of p ($\psi(p) = p$) and second, if we remove a pattern p from the set $\mathcal{P}$, then the value $\mathcal{M}(p) < \theta$ and, hence, the measure value of any preimage of p is less than θ by the projection-antimonotonicity of $\mathcal{M}$. The worst case time complexity of $\vartheta - \sum o\varphi\iota\alpha$ algorithm is

$$\mathbb{T}(\vartheta - \sum o\varphi\iota\alpha) = \mathbb{T}(FindPatterns(\psi_0)) +$$
$$+ k \cdot \max_{0 < i \le k} |\mathcal{P}_i| \cdot (\mathbb{T}(Preimages) + \mathbb{T}(\mathcal{M})), \tag{4}$$

where k is the number of projections in the chain, $\mathbb{T}(\mathcal{X})$ is time for computing operation $\mathcal{X}$. Since projection ψ_0 can be chosen to be very simple, in a typical case the complexity of $FindPatterns(\theta, \psi_0)$ can be low or even constant. The complexities of $Preimages$ and $\mathcal{M}$ depend on the measure, the chain of projections, and the kind of patterns. In many cases $\max_{0 < i \le k} |\mathcal{P}_i|$ can be exponential in the size of the input, because the number of patterns can be exponential. It can be a difficult task to define the threshold θ such that the maximal cardinality of $\mathcal{P}_i$ is not larger than a given number. This can be solved by an automatically adjustment of the threshold θ, which is not discussed here.

3.4 $\vartheta - \sum o \varphi \iota \alpha$ Algorithm for Interval Tuple Data

In this subsection we consider a pattern structure $\mathbb{K} = (G, (D_I, \sqcap), \delta)$, where D_I is a semilattice of interval tuple descriptions. We say that every component of a tuple p corresponds to an attribute $m \in M$, where M is the set of interval attributes. Thus, the size of any tuple in D_I is $|M|$, and for any attribute $m \in M$ we can denote the corresponding interval by $m(p)$. We also denote the value set of m by W_m. Since the set W_m is totally ordered we also denote by $W_m^{(j)}$ and $W_m^{(-j)}$ the sets containing the first j (smallest) elements and the last j (largest) elements from W_m, respectively.

A projection chain for interval tuple data is formed in the same way as discussed in Example 5. We start from the projection containing only one pattern corresponding to the largest interval in each component, i.e., for an attribute m the projection is of the form $\psi_m[W_m^{(1)}, W_m^{(-1)}]$. Then to pass to a next projection, we select the attribute m, and for this attribute we extend the projection from $\psi_m[W_m^{(j)}, W_m^{(-j)}]$ to $\psi_m[W_m^{(j+1)}, W_m^{(-j-1)}]$. Thus, there are $k = \max\limits_{m \in M} |W_m| \cdot |M|$ projections.

Finding preimages in this case is not hard, since to make a projection more detailed one should just extend the corresponding interval in left and/or on right end of the interval, i.e., there are only 4 possible preimages for a pattern when passing from one projection to another in this chain. Thus, we have proved the following

Proposition 1. *The worst case complexity for $\vartheta - \sum o \varphi \iota \alpha$ algorithm for interval tuple data is*

$$\mathbb{T}(\vartheta - \sum o \varphi \iota \alpha_{intervals}) = \max_{m \in M} |W_m| \cdot |M| \cdot \max_{0 < i \leq k} |\mathcal{P}_i| \cdot \mathbb{T}(\mathcal{M}). \quad (5)$$

3.5 $\vartheta - \sum o \varphi \iota \alpha$ Algorithm for Closed Patterns

Closed frequent itemsets are widely used as a condensed representation of all frequent itemsets since [18]. Here we show how we can adapt the algorithm for closed patterns. A closed pattern in $\psi_{i-1}(\mathbb{P})$ is not necessarily closed in $\psi_i(\mathbb{P})$. However, the extents of $\psi(\mathbb{P})$ are extents of $\mathbb{P}$ [17]. Thus, we associate the closed patterns with extents and then work with extents instead of patterns, i.e., a pattern structure $\mathbb{P} = (G, (D, \sqcap), \delta)$ is transformed into $\mathbb{P}_C = (G, (D_C, \sqcap_C), \delta_C)$, where $D_C = 2^G$. Moreover, for all $x, y \in D_C$ we have $x \sqcap_C y = (x^\diamond \sqcap y^\diamond)^\diamond$, where diamond operator is computed in $\mathbb{P}$ and $\delta_C(g \in G) = \{g\}$. Hence, every pattern p in D_C corresponds to a closed pattern $p^\diamond$ in D. A projection ψ of $\mathbb{P}$ induces a projection ψ_C of $\mathbb{P}_C$, given by $\psi_C(X \subseteq G) = \psi(X^\diamond)^\diamond$ with $(\cdot)^\diamond$ for $\mathbb{P}$.

Table 1. Patterns found for every projection in a chain for the example in Figure 1. Patterns are grey if they are removed for the corresponding projetion and they are labeled with "–" if they have not yet been found.

#	Pattern Ext.	Δ-measure ψ_0 ψ_1 ψ_2 ψ_3			
1	$\{g_1, g_2, g_3, g_4, g_5, g_6\}$	6	2	2	2
2	$\{g_1, g_2, g_3, g_4\}$	–	4	1	1
3	$\{g_5, g_6\}$	–	2	2	2
4	$\{g_1\}$	–	–	1	1
5	$\{g_2, g_3, g_4\}$	–	–	3	2
6	$\{g_4\}$	–	–	–	1

3.6 Δ-measure and $\vartheta - \sum o\varphi\iota\alpha$ Algorithm

In this subsection we show that Δ-measure is anti-monotonic for any projection; it is a stronger condition than the one required by Definition 3. Δ-measure works for closed patterns, and, hence, we identify every description by its extent (Subsection 3.5).

Proposition 2. Δ *is anti-monotonic for any projection* ψ.

Proof. By properties of a projection, an extent of $\psi(\mathbb{P})$ is an extent of $\mathbb{P}$ [17]. Let us consider an extent E and an extent of its descendant in $\psi(\mathbb{P})$. Let us suppose that E_p is a preimage of E for the projection ψ. Since E_c and E_p are extents in $\mathbb{P}$, the set $E_{cp} = E_c \cap E_p$ is an extent in $\mathbb{P}$ (the intersection of two closed sets is a closed set). Since E_p is a preimage of E, then $E_p \nleq E_c$ (otherwise, E_p is a preimage of E_c and not of E). Then, $E_{cp} \neq E_p$ and $E_{cp} \leq E_p$. Hence, $\Delta(E_p) \leq |E_p \setminus E_{cp}| \leq |E \setminus E_c|$. So, given a preimage E_p of E, $(\forall E_c \subseteq E)\Delta(E_p) \leq |E \setminus E_c|$, i.e., $\Delta(E_p) \leq \Delta(E)$. Thus, we can use Δ-measure in combination with $\vartheta - \sum o\varphi\iota\alpha$.

3.7 Example of Δ-Stable Patterns in Interval Tuple Data

Let us consider the example in Figure 1 and show how we can find all Δ-stable patterns with a threshold $\theta = 2$. The chain of projections for this example is given in Example 5, it contains 4 projections:

$$\psi_0 = \psi_{m_1[\{0\},\{1\}]m_2[\{0\},\{2\}]} \qquad\qquad \psi_1 = \psi_{m_1[\{0,1\},\{0,1\}]m_2[\{0\},\{2\}]}$$
$$\psi_2 = \psi_{m_1[\{0,1\},\{0,1\}]m_2[\{0,1\},\{1,2\}]} \qquad \psi_3 = \psi_{m_1[\{0,1\},\{0,1\}]m_2[\{0,1,2\},\{0,1,2\}]}$$

Since we are looking for closed patterns, every pattern can be identified by its extent. In Table 1 all patterns are given by their extents, i.e., by elements of D_C. For every pattern Δ-measure is shown for every ψ_i. A cell is shown in grey if the pattern is no more considered (the value of Δ less than 2). A cell has a dash "–", if a pattern in the row has not been generated for this projection.

For the example in Figure 1 the global process is as follows. At the beginning $\psi_0(D_I)$ contains only one element corresponding to pattern extent 123456 (a short cut for $\{g_1, g_2, g_3, g_4, g_5, g_6\}$) with a description $\langle[0,1];[0,2]\rangle$. Then, in $\psi_1(G,(D_I,\sqcap),\delta)$ possible preimages of 123456 are patterns with descriptions $\langle0;[0,2]\rangle$ and $\langle1;[0,2]\rangle$ given by pattern extents 1234 and 56, respectively. Then we continue with these three patterns which are all Δ-stable for the moment. The pattern extents 123456 and 56 have no preimages for the transition $\psi_1 \rightarrow \psi_2$, while the pattern extent 1234 has two preimages with descriptions $\langle0;[0,1]\rangle$ and $\langle0;[1,2]\rangle$ for this projection, which correspond to pattern extents 1 and 234. The first one is not Δ-stable and thus is no more considered. Moreover, the pattern extent 1234 is not Δ-stable (because of 234) and should also be removed. Finally, in transition $\psi_2 \rightarrow \psi_3$ only extent-pattern 234 has a preimage, a pattern extent 4, which is not Δ-stable. In such a way, we have started from a very simple projection ψ_0 and achieved the projection ψ_3 that gives us the Δ-stable patterns of the target pattern structure.

4   Experiments and Discussion

In this section we compare our approach to approaches based on postfiltering. Indeed, there is no approach that can directly mine stable-like pattern, e.g., stable, Δ-stable or robust patterns. The known approaches use postfiltering to mine such kind of patterns [6,8,11,24]. Recently it was also shown that it is more efficient to mine interval tuple data without binarization [20]. In their paper the authors introduce algorithm MinIntChange for working directly with interval tuple data. Thus we compare $\vartheta - \sum o\varphi\iota\alpha$ and MinIntChange for finding Δ-stable patterns. We find Δ-stable concepts with $\vartheta - \sum o\varphi\iota\alpha$ and then adjust frequency threshold θ such that all Δ-stable patterns are among the frequent ones.

The experiments are carried out on an "Intel(R) Core(TM) i7-2600 CPU @ 3.40GHz" computer with 8Gb of memory under Ubuntu 14.04 operating system. The algorithms are not parallelized and are coded in C++.

4.1   Dataset Simplification

For interval tuple data stable patterns can be very deep in the search space, such that neither of the algorithms can find them quickly. Thus, we join some similar values for every attribute in an interval in the following way. Given a threshold $0 < \beta$, two consequent numbers w_i and w_{i+1} from a value set W are joined in the same interval if $w_{i+1} - w_i < \beta$. In order to properly set the threshold β, we use another threshold $0 < \gamma < 1$, which is much easier to set.

If we assume that the values of the attribute m are distributed around several states with centers $\tilde{w}^1, \cdots, \tilde{w}^l$, then it is natural to think that the difference between the closest centers $\mathtt{abs}(\tilde{w}^i - \tilde{w}^{i\pm1})$ are much larger than the difference between the closest values. Ordering all values in the increasing order and finding the maximal difference $\delta_{\max}$ can give us an idea of typical distance between the states in the data. Thus, γ is defined as a proportion of this distance that should be considered as a distance between states, i.e., we put $\beta = \gamma \cdot \delta_{\max}$. If the

Table 2. Runtime in seconds of   and `MinIntChange` for different datasets.

DS	# Objs	# Attrs	γ	Δ	# Ptrns	θ	t	t_{MIC}
EM	61	9	0.3	3	3	21	< 0.1	57
BK	96	4	0.3	4	50	46	< 0.1	11
CN	105	20	0.8	2	5362	30	2.4	28
CU	108	5	0.3	5	4	27	< 0.1	1.5
FF	125	3	0.3	6	3	48	< 0.1	1
AP	135	4	0.01	5	1	19	< 0.1	34
EL	211	12	0.3	6	33	83	< 0.1	34
BA	337	16	0.5	4	736	91	1.5	32
AU	398	7	0.3	7	17	234	0.7	73
HO	506	13	0.8	10	1	340	0.7	57
QU	2178	25	0.3	40	1	659	1.3	28
AB	4177	8	0.3	46	3	1400	11	86
CA	8192	21	0.3	85	6	2568	112	24
PT	9065	48	0.3	2	1	2	45	14

distance between closest values in W are always the same, then even $\gamma = 0.99$ does not join values in intervals. However, if there are two states and the values are distributed very closely to one of these two states, then even $\gamma = 0.01$ can join values into one of two intervals corresponding to the states.

4.2 Datasets

We take several datasets from the Bilkent University database[1]. The datasets are summarized in Table 2. The names of datasets are given by standard abbreviations used in the database of Bilkent University. For every dataset we provide the number of objects and attributes and the threshold γ for which the experiments are carried out. For example, database EM has 61 objects, 9 numeric attributes, and the threshold γ is set to 0.3. Categorical attributes and rows with missing values, if any, are removed from the datasets.

4.3 Experiments

In Table 2 we show the computation time for finding the best Δ-stable pattern (or patterns if they have the same value for Δ-measure) for $\vartheta - \sum o\varphi\iota\alpha$ and for `MinIntChange`. The last algorithm is abbreviated as `MIC`. Since `MinIntChange` algorithm sometimes produces too many patterns, i.e., we do not have enough memory in our computer to check all of them, we interrupt the procedure and show the corresponding time in grey. We also show the number of the best patterns and the corresponding threshold Δ. The support threshold θ for finding the best Δ-stable patterns is also shown. For example, dataset CN contains 5362 best Δ-stable patterns, all having a Δ of 2. To find all these patterns with a post-filtering, we should mine frequent patterns with a support threshold lower than 30 or $\frac{30}{105} = 30\%$. $\vartheta - \sum o\varphi\iota\alpha$ computes all these patterns in 2.4 seconds, while `MinIntChange` requires at least 28 seconds and the procedure was interrupted without continuation.

---

[1] http://funapp.cs.bilkent.edu.tr/DataSets/

As we can see, $\vartheta - \sum o\varphi\iota\alpha$ is significantly faster than `MinIntChange` in all datasets. In the two datasets `CA` and `PT`, `MinIntChange` was stopped before computing all patterns and the runtime did not exceed the runtime of $\vartheta - \sum o\varphi\iota\alpha$. However, in both cases, `MinIntChange` achieved less than 10% of the required operations.

5 Conclusion

In this paper we have introduced a new class of interestingness measures that are anti-monotonic w.r.t. a chain of projections. We have designed a new algorithm, called $\vartheta - \sum o\varphi\iota\alpha$, which is able to efficiently find the best patterns w.r.t. such interestingness measures for interval tuple data. The experiments reported in the paper are the witness of the efficiency of the $\vartheta - \sum o\varphi\iota\alpha$ algorithms compared to indirect approaches based on postfiltering. Many future research directions are possible. Different measures should be studied in combination with $\vartheta - \sum o\varphi\iota\alpha$. One of them is robustness, which is very close to stability and can be applied to nonbinary data. Moreover, the choice of a projection chain is not a simple one and can affect the algorithm efficiency. Thus, a deep study of suitable projection chains should be carried out.

Acknowledgments. this research was supported by the Basic Research Program at the National Research University Higher School of Economics (Moscow, Russia) and by the BioIntelligence project (France).

References

1. Vreeken, J., Tatti, N.: Interesting patterns. In: Aggarwal, C.C., Han, J. (eds.) Freq. Pattern Min., pp. 105–134. Springer International Publishing, Heildelberg (2014)
2. Mannila, H., Toivonen, H., Verkamo, A.I.: Efficient algorithms for discovering association rules. In: Knowl. Discov. Data Min., pp. 181–192 (1994)
3. Agrawal, R., Srikant, R., et al.: Fast algorithms for mining association rules. In: Proc. 20th Int. Conf. Very Large Data Bases, VLDB, Vol. 1215, pp. 487–499 (1994)
4. Yao, H., Hamilton, H.J.: Mining itemset utilities from transaction databases. Data Knowl. Eng. **59**(3), 603–626 (2006)
5. Kuznetsov, S.O.: On stability of a formal concept. Ann. Math. Artif. Intell. **49**(1–4), 101–115 (2007)
6. Roth, C., Obiedkov, S.A., Kourie, D.G.: On succinct representation of knowledge community taxonomies with formal concept analysis. Int. J. Found. Comput. Sci. **19**(02), 383–404 (2008)
7. Webb, G.I.: Self-sufficient itemsets. ACM Trans. Knowl. Discov. Data **4**(1), 1–20 (2010)
8. Moerchen, F., Thies, M., Ultsch, A.: Efficient mining of all margin-closed itemsets with applications in temporal knowledge discovery and classification by compression. Knowl. Inf. Syst. **29**(1), 55–80 (2011)
9. Spyropoulou, E., De Bie, T., Boley, M.: Interesting pattern mining in multi-relational data. Data Min. Knowl. Discov., 1–42 (April 2013)

10. Cao, J., Wu, Z., Wu, J.: Scaling up cosine interesting pattern discovery: A depth-first method. Inf. Sci. (Ny) **266**, 31–46 (2014)
11. Tatti, N., Moerchen, F., Calders, T.: Finding Robust Itemsets under Subsampling. ACM Trans. Database Syst. **39**(3), 1–27 (2014)
12. Yan, X., Cheng, H., Han, J., Yu, P.S.: Mining significant graph patterns by leap search. In: Proc. 2008 ACM SIGMOD Int. Conf. Manag. Data - SIGMOD 2008, pp. 433–444. ACM Press, New York, June 2008
13. Han, J., Wang, J., Lu, Y., Tzvetkov, P.: Mining top-k frequent closed patterns without minimum support. In: Proceedings. 2002 IEEE Int. Conf. Data Mining, ICDM 2003, pp. 211–218 (2002)
14. Xin, D., Cheng, H., Yan, X., Han, J.: Extracting redundancy-aware top-k patterns. In: Proc. 12th ACM SIGKDD Int. Conf. Knowl. Discov. Data Min. - KDD 2006, p. 444. ACM Press, New York, August 2006
15. Webb, G.I.: Filtered-top-k association discovery. Wiley Interdiscip. Rev. Data Min. Knowl. Discov. **1**(3), 183–192 (2011)
16. Ganter, B., Wille, R.: Formal Concept Analysis: Mathematical Foundations, 1st edn. Springer, Heildelberg (1999)
17. Ganter, B., Kuznetsov, S.O.: Pattern structures and their projections. In: Delugach, H.S., Stumme, G. (eds.) ICCS 2001. LNCS (LNAI), vol. 2120, pp. 129–142. Springer, Heidelberg (2001)
18. Pasquier, N., Bastide, Y., Taouil, R., Lakhal, L.: Efficient Mining of Association Rules Using Closed Itemset Lattices. Inf. Syst. **24**(1), 25–46 (1999)
19. Kuznetsov, S.O., Samokhin, M.V.: Learning closed sets of labeled graphs for chemical applications. In: Kramer, S., Pfahringer, B. (eds.) ILP 2005. LNCS (LNAI), vol. 3625, pp. 190–208. Springer, Heidelberg (2005)
20. Kaytoue, M., Kuznetsov, S.O., Napoli, A.: Revisiting numerical pattern mining with formal concept analysis. In: Proc. 22nd Int. Jt. Conf. Artif. Intell. Barcelona, IJCAI 2011, Catalonia, Spain, July 16–22, 2011, pp. 1342–1347 (2011)
21. Yan, X., Han, J., Afshar, R.: CloSpan: mining closed sequential patterns in large databases. In: Proc. SIAM Int'l Conf. Data Min., pp. 166–177 (2003)
22. Kuznetsov, S.O.: On Computing the Size of a Lattice and Related Decision Problems. Order **18**(4), 313–321 (2001)
23. Buzmakov, A., Kuznetsov, S.O., Napoli, A.: Scalable estimates of concept stability. In: Glodeanu, C.V., Kaytoue, M., Sacarea, C. (eds.) ICFCA 2014. LNCS, vol. 8478, pp. 157–172. Springer, Heidelberg (2014)
24. Buzmakov, A., Egho, E., Jay, N., Kuznetsov, S.O., Napoli, A., Raïssi, C.: On projections of sequential pattern structures (with an application on care trajectories). In: Proc. 10th Int. Conf. Concept Lattices Their Appl., pp. 199–208 (2013)

Non-parametric Jensen-Shannon Divergence

Hoang-Vu Nguyen and Jilles Vreeken[✉]

Max Planck Institute for Informatics and Saarland University, Saarbrücken, Germany
{hnguyen,jilles}@mpi-inf.mpg.de

Abstract. Quantifying the difference between two distributions is a common problem in many machine learning and data mining tasks. What is also common in many tasks is that we only have empirical data. That is, we do not know the true distributions nor their form, and hence, before we can measure their divergence we first need to assume a distribution or perform estimation. For exploratory purposes this is unsatisfactory, as we want to explore the data, not our expectations. In this paper we study how to non-parametrically measure the divergence between two distributions. More in particular, we formalise the well-known Jensen-Shannon divergence using cumulative distribution functions. This allows us to calculate divergences directly and efficiently from data without the need for estimation. Moreover, empirical evaluation shows that our method performs very well in detecting differences between distributions, outperforming the state of the art in both statistical power and efficiency for a wide range of tasks.

1 Introduction

Measuring the difference between two distributions – their divergence – is a key element of many data analysis tasks. Let us consider a few examples. In time series analysis, for instance, to detect either changes or anomalies we need to quantify how different the data in two windows is distributed [18,23]. In discretisation, if we want to maintain interactions, we should only merge bins when their multivariate distributions are similar [13]. In subgroup discovery, the quality of a subgroup depends on how much the distribution of its targets deviates from that of its complement data set [3,6].

To optimally quantify the divergence of two distributions we need the actual distributions. Particularly for exploratory tasks, however, we typically only have access to empirical data. That is, we do not know the actual distribution, nor even its *form*. This is especially true for real-valued data. Although we can always make assumptions (parametric) or estimating them by kernel density estimation (KDE), these are not quite ideal in practice. For example, both parametric and KDE methods are prone to the curse of dimensionality [22]. More importantly, they restrict our analysis to the specific types of distributions or kernels used. That is, if we are not careful we are exploring our expectations about the data, not the data itself. To stay as close to the data as possible, we hence study a non-parametric divergence measure.

© Springer International Publishing Switzerland 2015
A. Appice et al. (Eds.): ECML PKDD 2015, Part II, LNAI 9285, pp. 173–189, 2015.
DOI: 10.1007/978-3-319-23525-7_11

In particular, we propose CJS, an information-theoretic divergence measure for numerical data. We build it upon the well-known Jensen-Shannon (JS) divergence. Yet, while the latter works with probability distribution functions (pdfs), which need to be estimated, we consider *cumulative* distribution functions (cdfs) which can be obtained directly from data. CJS has many appealing properties. In a nutshell, it does not make assumptions on the distributions or their relation, it permits non-parametric computation on empirical data, and is robust against the curse of dimensionality.

Empirical evaluation on both synthetic and real-world data for a wide range of exploratory data analysis tasks including change detection, anomaly detection, discretisation, and subgroup discovery shows that CJS consistently outperforms the state of the art in both quality and efficiency.

Overall, the main contributions of this paper are as follows:

(a) a new information-theoretic divergence measure CJS,
(b) a non-parametric method for computing CJS on empirical data, and
(c) a wide range of experiments on various tasks that validate the measure.

The road map of this paper is as follows. In Section 2, we introduce the theory of CJS. In Section 3, we review related work. In Section 4 we evaluate CJS empirically. We round up with discussion in Section 5 and finally conclude in Section 6. For readability and succinctness, we postpone the proofs for the theorems to the online Appendix.[1]

2   Theory

We consider numerical data. Let X be a univariate random variable with $dom(X) \subseteq \mathbb{R}$, and let $\mathbf{X}$ be a multivariate random variable $\mathbf{X} = \{X_1, \ldots, X_m\}$, with $\mathbf{X} \subseteq \mathbb{R}^m$. Our goal is to measure the difference between two distributions $p(\mathbf{X})$ and $q(\mathbf{X})$ over the same random variable, where we have n_p and n_q data samples, respectively. We will write p and q to denote the pdfs, and say P and Q for the respective cdfs. All logarithms are to base 2, and by convention we use $0 \log 0 = 0$.

Ideally, a divergence measure gives a zero score iff $p(\mathbf{x}) = q(\mathbf{x})$ for every $\mathbf{x} \in dom(\mathbf{X})$. That is, $p(\mathbf{X}) = q(\mathbf{X})$. Second, it is often convenient if the score is symmetric. Third, it should be well-defined without any assumption on the values of $p(\mathbf{x})$ and $q(\mathbf{x})$ for $\mathbf{x} \in dom(\mathbf{X})$. That is, no assumption the relation between p and q needs to be made. Fourth, to explore the data instead of exploring our expectations, the measure should permit non-parametric computation on empirical data. Finally, as real-world data often has high dimensionality and limited observations, the measure should be robust to the curse of dimensionality.

To address each of these desired properties, we propose CJS, a new information-theoretic divergence measure. In short, CJS embraces the spirit of Kullback-Leibler (KL) and Jensen-Shannon (JS) divergences, two well-known information-theoretic divergence measures. They both have been employed

---

[1] http://eda.mmci.uni-saarland.de/cjs/

widely in data mining [8,12]. As we will show, however, in their traditional form both suffer from some drawbacks w.r.t. exploratory analysis. We will alleviate these issues with CJS.

2.1 Univariate Case

To ease presentation, let us discuss the univariate case; when $\mathbf{X}$ is a single variable.

On univariate distributions, we consider a single univariate random variable X. We start with Kullback-Leibler divergence – one of the first information-theoretic divergences proposed in statistics [9]. Conventionally, it is defined as follows.

$$\mathrm{KL}(p(X) \parallel q(X)) = \int p(x) \log \frac{p(x)}{q(x)} dx \quad .$$

Importantly, it holds that $\mathrm{KL}(p(X) \parallel q(X)) = 0$ iff $p(X) = q(X)$. Although KL is assymetic itself, we can easily achieve symmetry by using $\mathrm{KL}(p(X) \parallel q(X)) + \mathrm{KL}(q(X) \parallel p(X))$. In addition, KL does suffer from two issues, however. First, it is undefined if $q(x) = 0$ and $p(x) \neq 0$, or vice versa, for some $x \in dom(X)$. Thus, p and q have to be *absolutely continuous* w.r.t. each other for their KL score to be defined [11]. As a result, KL requires an assumption on the relationship between p and q. Second, KL works with pdfs which need parametric or KDE estimation.

Another popular information-theoretic divergence measure is the Jensen-Shannon divergence [11]. It is defined as

$$\mathrm{JS}(p(X) \parallel q(X)) = \int p(x) \log \frac{p(x)}{\frac{1}{2}p(x) + \frac{1}{2}q(x)} dx \quad .$$

As for KL, for JS we also have that $\mathrm{JS}(p(X) \parallel q(X)) = 0$ iff $p(X) = q(X)$, and we can again obtain symmetry by considering $\mathrm{JS}(p(X) \parallel q(X)) + \mathrm{JS}(q(X) \parallel p(X))$. In contrast, JS is well defined independent of the values of $p(x)$ and $q(x)$ with $x \in dom(X)$. However, it still requires us to know or estimate the pdfs.

To address this, that is, to address the computability of JS on empirical data, we propose to redefine it by replacing pdfs with cdfs. This gives us a new divergence measure, CJS, for cumulative JS divergence.

Definition 1 (Univariate CJS). *The cumulative JS divergence of $p(X)$ and $q(X)$, denoted* $\mathrm{CJS}(p(X) \parallel q(X))$, *is*

$$\int P(x) \log \frac{P(x)}{\frac{1}{2}P(x) + \frac{1}{2}Q(x)} dx + \frac{1}{2\ln 2} \int (Q(x) - P(x)) \, dx \quad .$$

As we will explain shortly below, the second integral is required to make the score non-negative. Similar to KL and JS, we address symmetry by considering $\mathrm{CJS}(p(X) \parallel q(X)) + \mathrm{CJS}(q(X) \parallel p(X))$. Similar to JS, our measure does not make any assumption on the relation of p and q. With the following theorem we proof that CJS is indeed a divergence measure.

Theorem 1. $\mathrm{CJS}(p(X) \parallel q(X)) \geq 0$ *with equality iff* $p(X) = q(X)$.

Proof. Applying in sequence the log-sum inequality, and the fact that $\alpha \log \frac{\alpha}{\beta} \geq \frac{1}{\ln 2}(\alpha - \beta)$ for any $\alpha, \beta > 0$, we obtain

$$\int P(x) \log \frac{P(x)}{\frac{1}{2}P(x) + \frac{1}{2}Q(x)} dx \geq \int P(x) dx \log \frac{\int P(x) dx}{\int (\frac{1}{2}P(x) + \frac{1}{2}Q(x)) dx}$$

$$\geq \frac{1}{2\ln 2} \int (P(x) - Q(x))\, dx \quad .$$

For the log-sum inequality, equality holds if and only if $\frac{P(x)}{\frac{1}{2}P(x) + \frac{1}{2}Q(x)} = \delta$ for every $x \in dom(X)$ with δ being a constant. Further, equality of the second inequality holds if and only if $\int P(x) dx = \int (\frac{1}{2}P(x) + \frac{1}{2}Q(x)) dx$. Combining the two, we arrive at $\delta = 1$, i.e. $P(x) = Q(x)$ for every $x \in dom(X)$. Taking the derivatives of the two sides, we obtain the result. □

In Sec. 2.3, we will show in more detail that by considering cdfs, CJS permits non-parametric computation on empirical data. Let us now consider multivariate variables.

2.2 Multivariate Case

We now consider multivariate $\mathbf{X}$. In principle, the multivariate versions of KL and JS are obtained by replacing X with $\mathbf{X}$. We could arrive at a multivariate version of CJS in a similar way. However, if we were to do so, we would have to work with the joint distribution over *all* dimensions in $\mathbf{X}$, which would make our score prone to the curse of dimensionality. To overcome this, we build upon a factorised form of KL, as follows.

Theorem 2. $\mathrm{KL}\left(p(\mathbf{X}) \parallel q(\mathbf{X})\right) =$

$$\mathrm{KL}(p(X_1) \parallel q(X_1)) + \mathrm{KL}(p(X_2 \mid X_1) \parallel q(X_2 \mid X_1))$$
$$+ \ldots +$$
$$\mathrm{KL}(p(X_m \mid \mathbf{X} \setminus \{X_m\}) \parallel q(X_m \mid \mathbf{X} \setminus \{X_m\}))$$

where

$$\mathrm{KL}(p(X_i \mid X_1, \ldots, X_{i-1}) \parallel q(X_i \mid X_1, \ldots, X_{i-1}))$$
$$= \int \mathrm{KL}(p(X_i \mid x_1, \ldots, x_{i-1}) \parallel q(X_i \mid x_1, \ldots, x_{i-1}))$$
$$\times\, p(x_1, \ldots, x_{i-1}) \times dx_1 \times \ldots \times dx_{i-1}$$

is named an $(i - 1)$-*order conditional* KL *divergence.*

Proof. We extend the proof of Theorem 2.5.3 in [5] to the multivariate case. □

Theorem 2 states that $\textsc{kl}(p(\mathbf{X}) \,\|\, q(\mathbf{X}))$ is the summation of the difference between univariate (conditional) pdfs. This form of $\textsc{kl}$ is less prone the curse of dimensionality thanks to the low-order conditional divergence terms. We design the multivariate version of $\textsc{cjs}$ along the same lines. In particular, directly following Theorem 2 multivariate $\textsc{cjs}$ is defined as.

Definition 2 (Fixed-Order CJS). $\textsc{cjs}(p(X_1,\ldots,X_d) \,\|\, q(X_1,\ldots,X_d))$ *is*

$$\textsc{cjs}(p(X_1) \,\|\, q(X_1)) + \textsc{cjs}(p(X_2 \mid X_1) \,\|\, q(X_2 \mid X_1))$$
$$+ \ldots +$$
$$\textsc{cjs}(p(X_d \mid \mathbf{X} \setminus \{X_d\}) \,\|\, q(X_d \mid \mathbf{X} \setminus \{X_d\}))$$

where

$$\textsc{cjs}(p(X_i \mid X_1,\ldots,X_{i-1}) \,\|\, q(X_i \mid X_1,\ldots,X_{i-1}))$$
$$= \int \textsc{cjs}(p(X_i \mid x_1,\ldots,x_{i-1}) \,\|\, q(X_i \mid x_1,\ldots,x_{i-1}))$$
$$\times \, p(x_1,\ldots,x_{i-1}) \times dx_1 \times \ldots \times dx_{i-1}$$

is named an $(i-1)$-order conditional $\textsc{cjs}$ *divergence.*

From Definition 2, one can see the analogy between multivariate $\textsc{cjs}$ and the factorised form of $\textsc{kl}$. However, unlike $\textsc{kl}$, when defined as in Definition 2 $\textsc{cjs}$ may be variant to how we factorise the distribution, that is, the permutation of dimensions. To circumvent this we derive a permutation-free version of $\textsc{cjs}$ as follows. Let $\mathcal{F}$ be the set of bijective functions $\sigma : \{1,\ldots,m\} \to \{1,\ldots,m\}$.

Definition 3 (Order-Independent CJS). $\textsc{cjs}(p(\mathbf{X}) \,\|\, q(\mathbf{X}))$ *is*

$$\max_{\sigma \in \mathcal{F}} \sum_{i=2}^{d} \textsc{cjs}\left(p(X_{\sigma(1)},\ldots,X_{\sigma(m)}) \,\|\, q(X_{\sigma(1)},\ldots,X_{\sigma(m)})\right) \quad .$$

Definition 3 eliminates the dependence on any specific permutation by taking the maximum score over all permutations. Now we need to show that multivariate $\textsc{cjs}$ is indeed a divergence.

Theorem 3. $\textsc{cjs}(p(\mathbf{X}) \,\|\, q(\mathbf{X})) \geq 0$ *with equality iff* $p(\mathbf{X}) = q(\mathbf{X})$.

Proof. For readability, we postpone the proof to the online Appendix.   □

We now know that $\textsc{cjs}$ is a suitable divergence measure for multivariate distributions. To compute $\textsc{cjs}$, however, we would have to search for the optimal permutation among $m!$ permutations. When m is large, this is prohibitively costly. We tackle this by proposing $\textsc{cjs}_{pr}$, a practical version of $\textsc{cjs}$.

Definition 4 (Practical CJS). $\textsc{cjs}_{pr}(p(\mathbf{X}) \,\|\, q(\mathbf{X}))$ *is*

$$\textsc{cjs}\left(p(X_{\sigma(1)},\ldots,X_{\sigma(m)}) \,\|\, q(X_{\sigma(1)},\ldots,X_{\sigma(m)})\right)$$

where $\sigma \in \mathcal{F}$ *is a permutation such that* $\textsc{cjs}(X_{\sigma(1)}) \geq \ldots \geq \textsc{cjs}(X_{\sigma(m)})$.

In other words, CJS_{pr} chooses the permutation corresponding to the sorting of dimensions in descending order of CJS values. The intuition behind this choice is that the difference between $p(X_i \mid \ldots)$ and $q(X_i \mid \ldots)$ is likely reflected through the difference between $p(X_i)$ and $q(X_i)$. Thus, by ordering dimensions in terms of their CJS values, we can approximate the optimal permutation. Although a greedy heuristic, our experiments reveal that CJS_{pr} works well in practice. For exposition, from now on we simply assume that σ is the *identity mapping function*, i.e. the permutation of dimensions is $X_1, \ldots, X_m$. Following the proof of Theorem 3, we also have that CJS_{pr} is a divergence measure.

Theorem 4. $\text{CJS}_{pr}(p(\mathbf{X}) \parallel q(\mathbf{X})) \geq 0$ *with equality iff* $p(\mathbf{X}) = q(\mathbf{X})$.

In the remainder of the paper we will consider CJS_{pr} and for readability simply refer to it as CJS.

2.3   Computing CJS

To compute $\text{CJS}(p(\mathbf{X}) \parallel q(\mathbf{X}))$, we need to compute unconditional and conditional CJS. For the former, suppose that we want to compute $\text{CJS}(p(X) \parallel q(X))$ for $X \in \mathbf{X}$. Let $v \leq X[1] \leq \ldots \leq X[n_p] \leq V$ be realisations of X drawn from $p(X)$. Further, let $P_{n_p}(x) = \frac{1}{n_p} \sum_{j=1}^{n_p} I(X[j] \leq x)$. Following [15], we have

$$\int P(x)dx = \sum_{j=1}^{n_p-1} (X[j+1] - X[j]) \frac{j}{n_p} + (V - X[n_p]) \quad .$$

The other terms required for calculating $\text{CJS}(p(X) \parallel q(X))$ (cf., Definition 1), e.g. $\int Q(x)dx$, are similarly computed. More details can be found in [15].

Computing conditional CJS terms, however, requires pdfs – which are unknown. We resolve this in a non-parametric way using *optimal* discretisation. That is, we first compute $\text{CJS}(p(X_1) \parallel q(X_1))$. Next, we calculate $\text{CJS}(p(X_2 \mid X_1) \parallel q(X_2 \mid X_1))$ by searching for the discretisation of X_1 that maximises this term. At step $k \geq 3$, we compute $\text{CJS}(p(X_k \mid X_1, \ldots, X_{k-1}) \parallel q(X_k \mid X_1, \ldots, X_{k-1}))$ by searching for the discretisation of X_{k-1} that maximises this term. Thus, we only discretise the dimension picked in the previous step and do not re-discretise any earlier chosen dimensions. First and foremost, this increases the efficiency of our algorithm. Second, and more importantly, it facilitates interpretability as we only have to consider one discretisation per dimension.

Next, we show that the discretisation at a step can be done efficiently and optimally by dynamic programming. For simplicity, let $\mathbf{X}' \subset \mathbf{X}$ be the set of dimensions already picked *and* discretised. We denote X as the dimension selected in the previous step but not yet discretised. Let X_c be the dimension selected in this step. Our goal is to find the discretisation of X maximising $\text{CJS}(p(X_c \mid \mathbf{X}', X) \parallel q(X_c \mid \mathbf{X}', X))$.

To accomplish this, let $X[1] \leq \ldots \leq X[n_p]$ be realisations of X drawn from *the samples of* $p(X)$. We write $X[j, u]$ for $\{X[j], X[j+1], \ldots, X[u]\}$ where $j \leq u$.

Note that $X[1, n_p]$ is in fact X. We use

$$\text{CJS}(p(X_c \mid \mathbf{X}', \langle X[j, u] \rangle) \parallel q(X_c \mid \mathbf{X}', \langle X[j, u] \rangle))$$

to denote $\text{CJS}(p(X_c \mid \mathbf{X}') \parallel q(X_c \mid \mathbf{X}'))$ computed using the $(u - j + 1)$ samples of $p(X)$ corresponding to $X[j]$ to $X[u]$, projected onto X. For $1 \leq l \leq u \leq n_p$, we write

$$f(u, l) = \max_{dsc:|dsc|=l} \text{CJS}\left(p(X_c \mid \mathbf{X}', X^{dsc}[1, u]) \parallel q(X_c \mid \mathbf{X}', X^{dsc}[1, u])\right)$$

where dsc is a discretisation of $X[1, u]$, $|dsc|$ is its number of bins, and $X^{dsc}[1, u]$ is the discretised version of $X[1, u]$ produced by dsc. For $1 < l \leq u \leq n_p$, we have

Theorem 5. $f(u, l) = \max\limits_{j \in [l-1, u)} \mathcal{A}_j$ where

$$\mathcal{A}_j = \frac{j}{u} f(j, l - 1) + \frac{u - j}{u} \text{CJS}\left(p(X_c \mid \mathbf{X}', \langle X[j+1, u] \rangle) \parallel q(X_c \mid \mathbf{X}', \langle X[j + 1, u] \rangle)\right)$$

Proof. For readability, we postpone the proof to the online Appendix. □

Theorem 5 shows that the optimal discretisation of $X[1, u]$ can be derived from that of $X[1, j]$ with $j < u$. This allows us to design a dynamic programming algorithm to find the discretisation of X maximising $\text{CJS}(p(X_c \mid \mathbf{X}', X) \parallel q(X_c \mid \mathbf{X}', X))$.

2.4 Complexity Analysis

We now discuss the time complexity of computing $\text{CJS}(p(\mathbf{X}) \parallel q(\mathbf{X}))$. When discretising a dimension $X \in \mathbf{X}$, if we use its original set of data samples as cut points, the time complexity of solving dynamic programming is $O(n_p^2)$, rather restrictive for large data. Most cut points, however, will not be used in the optimal discretisation. To gain efficiency, we can hence impose a maximum grid size $max\_grid = n_p^\epsilon$ and limit the number of cut points to $c \times max\_grid$ with $c > 1$. To find these candidate cut points, we follow Reshef et al. [20] and apply equal-frequency binning on X with the number of bins equal to $(c \times max\_grid + 1)$. Note that this pre-processing trades off accuracy for efficiency. Other types of pre-processing are left for future work.

Regarding ϵ and c, the larger they are, the more candidate discretisations we consider, and hence, at a higher the computational cost, the better the result. Our empirical results show that $\epsilon = 0.5$ and $c = 2$ offers a good balance between quality and efficiency, and we will use these values in the experiments. The cost of discretising each dimension X then is $O(n_p)$. The overall complexity of computing $\text{CJS}(p(\mathbf{X}) \parallel q(\mathbf{X}))$ is therefore $O(m \times n_p)$. Similarly, the complexity of computing $\text{CJS}(q(\mathbf{X}) \parallel p(\mathbf{X}))$ is $O(m \times n_q)$.

2.5   Summing Up

We note that CJS is asymmetric. To have a symmetric distance, we use

$$\text{CJS}_{sym}(p(\mathbf{X}) \parallel q(\mathbf{X})) = \text{CJS}(p(\mathbf{X}) \parallel q(\mathbf{X})) + \text{CJS}(q(\mathbf{X}) \parallel p(\mathbf{X})) \quad .$$

In addition, we present two important properties pertaining specifically to univariate CJS_{sym}. Although in the interest of space we will not explore these properties empirically, but they may be important to know for other applications of our measure.

Theorem 6. $\text{CJS}_{sym}(p(X) \parallel q(X)) \leq \int (P(x) + Q(x))\, dx.$

Proof. For readability, we postpone the proof to the online Appendix.   □

Theorem 7. *Univariate* $\sqrt{\text{CJS}_{sym}}$ *is a metric.*

Proof. We follow the proof of Theorem 1 in [7].   □

Theorem 6 tells us that the value of univariate CJS_{sym} is bounded above, which facilitates interpretation [11]. Theorem 7 on the other hand says that the square root of univariate CJS_{sym} is a metric distance. This is beneficial for, e.g. query optimisation in multimedia databases.

3   Related Work

Many divergence measures have been proposed in the literature. Besides Kullback-Leibler and Jensen-Shannon, other well-known divergence measures include the Kolmogorov-Smirnov test (KS), the Cramér-von Mises criterion (CM), Earth Mover's Distance (EMD), and the quadratic measure of divergence (QR) [13]. Each has its own strengths and weaknesses – most particularly w.r.t. exploratory analysis. For example, multivariate KS, CM, EMD, and QR all operate on the joint distributions over *all* dimensions. Thus, they inherently suffer from the curse of dimensionality, which reduces their statistical power when applied on non-trivial numbers of dimensions. In addition, EMD needs probability mass functions (pmfs). While readily available for discrete data, real-valued data first needs to be discretised. There currently exists no discretisation method that directly optimises EMD, however, by which the results may turn out ad hoc. Recently, Perez-Cruz [17] studied how to estimate KL using cdfs. Park et al. [16] proposed CKL, redefining KL by replacing pdfs with cdfs. While computable on empirical data, as for regular KL it may be undefined when $p(\mathbf{x}) = 0$ or $q(\mathbf{x}) = 0$ for some $\mathbf{x} \in dom(\mathbf{X})$. Further, CKL was originally proposed as a univariate measure. Wang et al. [24] are the first to formulate JS using cdfs. However, their CJS relies on joint cdfs and hence suffers from the curse of dimensionality.

Many data mining tasks require divergence measures. For instance, for change detection on time-series, it is necessary to test whether two windows of data are sampled from the same underlying distribution. Song et al. [23] proposed such a test, using Gaussian kernels to approximate the data distribution – including the joint distribution over all dimensions. Generalisations of KL computed using Gaussian kernels have shown to be powerful alternatives [8, 12]. KL is also used for anomaly detection in time series, where we can compute an anomaly score for a window against the reference data set [18]. In interaction-preserving discretisation we need to assess how different (multivariate) distributions are between two consecutive bins. This can be done through contrast mining [3], or by using QR [13]. In multi-target subgroup discovery, also known as exceptional model mining [10], we need to compare the distributions of subgroup against that of its complement data set. Leman et al. use a quadratic measure of divergence [10], whereas Duivesteijn et al. consider the edit distance between Bayesian networks [6]. In Section 4, we will consider the efficacy of CJS for each of these areas.

Nguyen et al. [15] proposed a correlation measure inspired by factorised KL using cumulative entropy [19]. Although it permits reliable non-parametric computation on empirical data, it uses ad hoc clustering to compute conditional entropies. Nguyen et al. [14] showed that these are inferior to optimal discretisation, in their case for total correlation. In CJS we use the same general idea of optimal discretisation, yet the specifics for measuring divergence are nontrivial and had to be developed from scratch.

4 Experiments

Next, we empirically evaluate CJS. In particular, we will evaluate the statistical power at which it quantifies differences between data distributions, and its scalability to data size and dimensionality. In addition, we evaluate its performance in four exploratory data mining tasks. We implemented CJS in Java, and make our code available for research purposes.[2] All experiments were performed single-threaded on an Intel(R) Core(TM) i7-4600U CPU with 16GB RAM. We report wall-clock running times.

We compare CJS to MG [23] and RSIF [12], two measures of distribution difference recently proposed for change detection on time series. In short, to compare two samples S_p and S_q, MG randomly splits S_p into S_p^1 and S_p^2. Next, it uses S_p^1 to model the distribution of data. Then it fits S_p^2 and S_q into the model. The difference in their fitness scores is regarded as the difference between S_p and S_q. RSIF on the other hand uses a non-factorised variant of KL divergence. To compute this divergence, it estimates the ratio $\frac{p(\mathbf{X})}{q(\mathbf{X})}$. As third baseline, we consider QR, a quadratic measure of distribution difference recently proposed by Nguyen et al. [13]. It works on $P(\mathbf{X})$ and $Q(\mathbf{X})$, i.e. the cdfs of all dimensions. Note that by their definition these three competitors are prone to the curse of dimensionality. Finally, we include CKL, extended to the multivariate setting similarly to CJS.

---

[2] http://eda.mmci.uni-saarland.de/cjs/

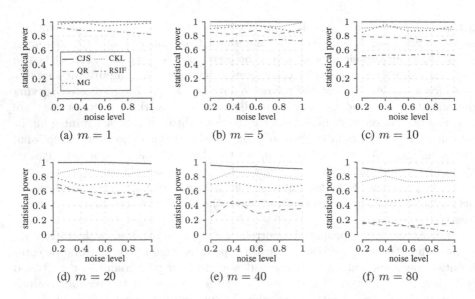

Fig. 1. [Higher is better] Statistical power vs. dimensionality of CJS, CKL, QR, RSIF, and MG on synthetic data sets. Overall, CJS achieves the best statistical power across different dimensionality and noise levels.

4.1 Statistical Power

Our aim here is to examine if our measure is really suitable for quantifying the difference between two data distributions. For this purpose, we perform statistical tests using synthetic data. To this end, the null hypothesis is that the two distributions are similar. To determine the cutoff for testing the null hypothesis, we first generate 100 pairs of data sets of the same size (n) and dimensionality (m), and having the same distribution f_1. Next, we compute the divergence score for each pair. Subsequently, we set the cutoff according to the significance level $\alpha = 0.05$. We then generate 100 pairs of data sets, again with the same n and m. However, two data sets in such a pair have different distributions. One follows distribution f_1 while the other follows distribution f_2. The power of the measure is the proportion of the 100 new pairs of data sets whose divergence scores exceed the cutoff. We simulate a noisy setting by adding Gaussian noise to the data. We show the results in Fig. 1 for $n = 1\,000$ and varying over m with f_1 and f_2 two Gaussian distributions with different mean vectors and covariance matrices. For other data sizes and distributions we observe the same trend.

Inspecting these results, we find that CJS obtains higher statistical power than other measures. Moreover, it is very stable across dimensionality and noise. Other measures, especially QR and RSIF, deteriorate with high dimensionality. Overall, we find that CJS reliably measures the divergence of distributions, regardless of dimensionality or noise.

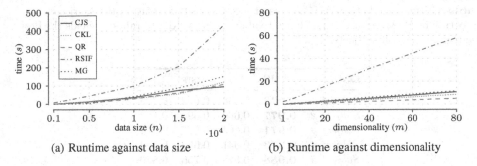

(a) Runtime against data size   (b) Runtime against dimensionality

Fig. 2. [Lower is better] Runtime scalability of CJS, CKL, QR, RSIF, and MG on synthetic data sets. Overall, CJS scales similarly to CKL, QR, and MG and much better than RSIF.

4.2 Scalability

Next, we study the scalability of our measures with respect to the data size n and dimensionality m. For scalability to n, we generate data sets with $m = 10$ and n varied from 1 000 to 20 000. For scalability to m, we generate data sets with $n = 1\,000$ and m varied from 1 to 80. We present the results in Fig. 2. We observe that our measure is efficient. It scales similarly as CKL, QR, and MG, and much better than RSIF. Combining this with our results regarding statistical power, we conclude that CJS yields the best balance between quality and efficiency.

The results show that CJS outperforms CKL while the two have similar runtime. We therefore exclude CKL in the remainder.

4.3 Change Detection on Time Series

Divergence measures are widely used for change detection on time series [4,12,21]. The main idea is that given a window size W, at each time instant t, we measure the difference between the distribution of data over the interval $[t - W, t)$ to that over the interval $[t, t + W)$. A large difference is an indicator that a change may have occurred. The quality of change detection is thus dependent on the quality of the measure. In this experiment, we apply CJS in change detection. In particular, we use it in the retrospective change detection model proposed by Liu et al. [12].

As data, we use the PAMAP data set,[3] which contains human activity monitoring data. Essentially, it consists of data recorded from sensors attached to 9 human subjects. Each subject performs different types of activities, e.g. *standing*, *walking*, *running*, and each activity is represented by 51 sensor readings recorded per second. Since each subject has different physical characteristics, we consider his/her data to be a separate data set. One data set is very small, so we discard it. We hence consider 8 time series over 51 dimensions with in the order of 100 000 time points. In each time series, the time instants when the

---

[3] http://www.pamap.org/demo.html

Table 1. [Higher is better] AUC scores of CJS, QR, RSIF, and MG in time-series change detection on PAMAP data sets. Highest values are in **bold**. Overall, CJS yields the best accuracy across all subjects.

Data	CJS	QR	RSIF	MG
Subject 1	**0.972**	0.658	0.662	0.775
Subject 2	**0.977**	0.669	0.694	0.782
Subject 3	**0.971**	0.663	0.857	0.954
Subject 4	**0.973**	0.641	0.662	0.642
Subject 5	**0.988**	0.678	0.756	0.850
Subject 6	**0.977**	0.662	0.497	0.550
Subject 7	**0.978**	0.646	0.782	0.705
Subject 8	**0.973**	0.741	0.552	0.424
Average	**0.976**	0.670	0.683	0.710

respective subject changes his/her activities are regarded as change points. As the change points are known, we evaluate how well each measure tested retrieves these cut points. It is expected that each measure should assign higher difference scores at the change points in comparison to other normal time instants. As performance metric we construct Receiver Operating Characteristic (ROC) curves and consider the Area Under the ROC curve (AUC) [8,12,23].

Table 1 gives the results. We see that CJS consistently achieves the best AUC over all subjects. Moreover, it outperforms its competitors with relatively large margins.

4.4   Anomaly Detection on Time Series

Closely related to change detection is anomaly detection [2,18]. The core idea is that a reference data set is available as training data. For example, obtained for instance from historical records. It is used for building a statistical model capturing the generation process of normal data. Then, a window is slid along the test time series to compute the anomaly score for each time instant, using the model constructed. With CJS, we can perform the same task by simply comparing the distribution over a window against that of the reference set. That is, no model construction is required. In contrast to GGM [18] – a state of the art method for anomaly detection in time series – CJS can be considered as a 'lazy' detector. We will assess how CJS performs against GGM. For this, we use the TEP data set, as it was used by Qiu et al. [18]. It contains information on an industrial production process. The data has 52 dimensions. Following their setup, we set the window size to 10. We vary the size of the training set to assess stability.

Fig. 3 presents the results. We see that CJS outperfoms GGM at its own game. In particular, we see that CJS is less sensitive to the size of the training set than GGM, which could be attributed to its 'lazy' approach. Overall, the conclusion is that CJS reliably measures the difference of multivariate distributions.

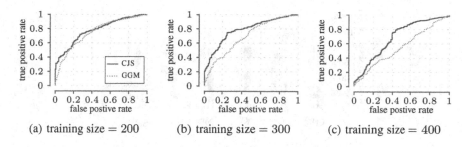

(a) training size = 200   (b) training size = 300   (c) training size = 400

Fig. 3. [Higher is better] ROC curves of CJS and GGM regarding time-series anomaly detection on the TEP data set. AUC scores of CJS are respectively 0.780, 0.783, and 0.689. AUC scores of GGM are respectively 0.753, 0.691, and 0.552. Overall, CJS outperforms GGM.

4.5 Multivariate Discretisation

When discretising multivariate data the key goal is to discretise the data such that the output data preserves the most important multivariate interactions in the input data [3,13]. Only when we do so it will be possible to use techniques that require discrete data – such as pattern mining – to pick up on truly interesting correlations. One of the major components of interaction-preserving discretisation is to measure the difference of data distributions in different bins. The difference scores are then used to decide if bin merge takes place or not.

In principle, the better such measure, the better correlations can be maintained. For example, the better pattern-based compressors such as COMPREX [1] can compress it. In this experiment, we apply CJS in IPD [13] – a state of the art technique for interaction-preserving discretisation. To evaluate, we apply COMPREX to the discretised data and compare the total encoded size. We compare against original IPD, which uses QR. For testing purposes, we use 6 public data sets available in the UCI Repository.

We display the results in Fig. 4. The plot shows the relative compression rates with IPD as the bases, per data set. Please note that lower compression costs are better. Going over the results, we can see that CJS improves the performance IPD in 4 out of 6 data sets. This implies that CJS reliably assesses the difference of multivariate distributions in different bins [13].

4.6 Multi-Target Subgroup Discovery

In subgroup discovery we are after finding queries – patterns – that identify subgroups of data points for which the distribution of some target attribute varies strongly compare to either the complement, or the whole data. As the name implies, in multi-target subgroup discovery we do not consider a univariate targets, but multivariate ones.

Formally, let us consider a data set $\mathbf{D}$ with attributes $A_1, \dots, A_k$ and targets $T_1, \dots, T_l$. A subgroup S on $\mathbf{D}$ is characterized by condition(s) imposed on some attribute(s). A condition on an attribute A has the form of an interval.

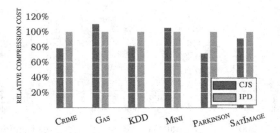

Fig. 4. [Lower is better] Relative compression costs of CJS and IPD in interaction-preserving discretisation. COMPREX [1] is the compressor. The compression costs of IPD are the bases. Overall, CJS outperforms IPD.

The subset of $\mathbf{D}$ corresponding to $\mathcal{S}$ is denoted as $\mathbf{D}_{\mathcal{S}}$. The set of remaining data points, the complement set, is $\overline{\mathbf{D}}_{\mathcal{S}} = \mathbf{D} \setminus \mathbf{D}_{\mathcal{S}}$. Within subgroup discovery, exceptional model mining is concerned with detecting $\mathcal{S}$ such that $p(T_1, \ldots, T_l \mid \mathbf{D}_{\mathcal{S}})$ is different from $p(T_1, \ldots, T_l \mid \overline{\mathbf{D}}_{\mathcal{S}})$ [6,10]. The higher the difference, the better.

In this experiment, we use CJS for quantifying the distribution divergence non-parametrically. Apart from that, we apply as-is the search algorithm proposed in [6] for discovering high quality subgroups. As data sets, we use 3 public ones. Two from the UCI Repository, namely, the *Bike* dataset of 731 data points over 6 attributes with 2 targets, and the *Energy* dataset of 768 rows over 8 attributes also with 2 targets. Third, we consider the *Chemnitz* dataset of 1440 rows over 3 attributes and with 7 targets.[4] Our objective here is to see if CJS can assist in discovering interesting subgroups on these data sets. The representative subgroups on three data sets are in Table 2 (all subgroups are significant at significance level $\alpha = 0.05$).

Going over the results, we see CJS to detect subgroups having different distribution in targets compared to that of their respective complement set. For instance, on *Bike* we discover the subgroup *temperature* $\geq 6.5 \land$ *temperature* < 10.7. In this subgroup, we find that its numbers of registered and non-registered bikers are significantly lower than those of its complement set. This is intuitively understandable, as at these low temperatures one expects to see only a few bikers, and especially few casual ones. In contrast, for the subgroup *temperature* $\geq 27.1 \land$ *temperature* < 31.2, the numbers of bikers in both targets are very high. This again is intuitively understandable.

From the *Energy* data, we find that the two subgroups *surface area* $\geq 624.8 \land$ *surface area* < 661.5 and *roof area* < 124.0 have much higher heating and cooling loads compared to their complement sets.

The previous two data sets contain 2 targets only. In contrast, Chemnitz data set has 7 targets, which poses a more challenging task. Nevertheless, with CJS we can detect informative subgroups as it can capture divergences between distributions that are involved in different numbers of targets – not all target attributes have to be 'divergent' at the same time, after all. In particular,

---

[4] http://www.mathe.tu-freiberg.de/Stoyan/umwdat.html

Table 2. Representative subgroups discovered by CJS on Bike, Energy, and Chemnitz data sets. On Chemnitz, only targets where the divergence is large are shown. Overall, CJS helps detect high quality and informative subgroups on all three data sets.

Data	Target	Mean subgroup (D_S)	complement ($\overline{D}_S$)
Bike	$6.5 \leq temperature < 10.7$ (support $= 63$)		
	registered bikers	166	913
	non-registered bikers	1 889	3 840
	$27.1 \leq temperature < 31.2$ (support $= 127$)		
	registered bikers	1 347	743
	non-registered bikers	4 406	3 499
Energy	$624.8 \leq surface\ area < 661.5$ (support $= 128$)		
	heating	38.6	20.8
	cooling	40.2	23.1
	$roof\ area < 124.0$ (support $= 192$)		
	heating	31.6	19.1
	cooling	33.1	21.7
Chemnitz	$4.25 \leq temperature < 7.5$ (support $= 370$)		
	dust	53.5	109.7
	SO_2	80.6	184.4
	NO_2	20.4	41.4
	NO_x	50.2	94.3
	$wind < -0.75$ (support $= 395$)		
	NO	69.0	39.0
	NO_x	106.4	74.1

the subgroup $temperature \geq 4.25 \wedge temperature < 7.5$ has its divergence traced back to five targets. On the other hand, there are only two targets responsible for the divergence of the subgroup $wind < -0.75$.

Overall, we find that CJS can be successfully applied to non-parametrically discover subgroups in real-world data with multiple targets.

5   Discussion

The experiments show that CJS is efficient and obtains high statistical power in detecting divergence for varying dimensionality and noise levels. Further, we demonstrated that CJS is well-suited for a wide range of exploratory tasks, namely time-series change detection and anomaly detection, interaction-preserving discretisation, and multi-target subgroup discovery. The improvement in performance of CJS over existing measures can be traced back to its three main properties: (a) it does not make any assumption on the relation between two distributions, (b) it allows non-parametric computation on empirical data, and (c) it is less sensitive to the curse of dimensionality.

Yet, there is room for alternative methods as well as further improvements. For instance, in this paper, we pursue the non-parametric setting. As long as the knowledge on data distributions is known, one can resort to parametric methods to compute other divergence measures, e.g. KL and JS. A promising direction is to extend CJS to heterogeneous data types. That is, in addition to numerical data, we can consider categorical data as well. A possible solution to this end is to combine JS and CJS. More in particular, JS is used to handle categorical data; CJS is used for numerical data; and discretisation can be used to bridge both worlds. The details, however, are beyond the scope of this work. As future work, we also plan to develop new subgroup discovery methods that integrate CJS more deeply into the mining process. This will help us to better exploit the capability of CJS in this interesting branch of exploratory analysis.

6   Conclusion

In this paper, we proposed CJS, an information-theoretic divergence measure to quantify the difference of two distributions. In short, CJS requires neither assumptions on the forms of distributions nor their relation. Further, it permits efficient non-parametric computation on empirical data. Extensive experiments on both synthetic and real-world data showed that our measure outperforms the state of the art in both statistical power and efficiency in a wide range of exploratory tasks.

Acknowledgments. The authors thank the anonymous reviewers for insightful comments. Hoang-Vu Nguyen and Jilles Vreeken are supported by the Cluster of Excellence "Multimodal Computing and Interaction" within the Excellence Initiative of the German Federal Government.

References

1. Akoglu, L., Tong, H., Vreeken, J., Faloutsos, C.: Comprex: compression based anomaly detection. In: CIKM. ACM (2012)
2. Arnold, A., Liu, Y., Abe, N.: Temporal causal modeling with graphical granger methods. In: KDD, pp. 66–75 (2007)
3. Bay, S.D.: Multivariate discretization for set mining. Knowledge and Information Systems 3(4), 491–512 (2001)
4. Chandola, V., Vatsavai, R.R.: A gaussian process based online change detection algorithm for monitoring periodic time series. In: SDM, pp. 95–106 (2011)
5. Cover, T.M., Thomas, J.A.: Elements of Information Theory. Wiley-Interscience, New York (2006)
6. Duivesteijn, W., Knobbe, A.J., Feelders, A., van Leeuwen, M.: Subgroup discovery meets bayesian networks - an exceptional model mining approach. In: ICDM, pp. 158–167 (2010)

7. Endres, D.M., Schindelin, J.E.: A new metric for probability distributions. IEEE Transactions on Information Theory **49**(7), 1858–1860 (2003)
8. Kawahara, Y., Sugiyama, M.: Change-point detection in time-series data by direct density-ratio estimation. In: SDM, pp. 389–400 (2009)
9. Kullback, S., Leibler, R.: On information and sufficiency. Annals of Mathematical Statistics **22**(1), 79–86 (1951)
10. Leman, D., Feelders, A., Knobbe, A.J.: Exceptional model mining. In: Daelemans, W., Goethals, B., Morik, K. (eds.) ECML PKDD 2008, Part II. LNCS (LNAI), vol. 5212, pp. 1–16. Springer, Heidelberg (2008)
11. Lin, J.: Divergence measures based on the Shannon entropy. IEEE Transactions on Information Theory **37**(1), 145–151 (1991)
12. Liu, S., Yamada, M., Collier, N., Sugiyama, M.: Change-point detection in time-series data by relative density-ratio estimation. Neural Networks **43**, 72–83 (2013)
13. Nguyen, H.V., Müuller, E., Vreeken, J., Efros, P., Böhm, K.: Unsupervised interaction-preserving discretization of multivariate data. Data Min. Knowl. Discov. **28**(5–6), 1366–1397 (2014)
14. Nguyen, H.V., Müller, E., Vreeken, J., Efros, P., Böhm, K.: Multivariate maximal correlation analysis. In: ICML, pp. 775–783 (2014)
15. Nguyen, H.V., Müller, E., Vreeken, J., Keller, F., Böhm, K.: CMI: an information-theoretic contrast measure for enhancing subspace cluster and outlier detection. In: SDM, pp. 198–206 (2013)
16. Park, S., Rao, M., Shin, D.W.: On cumulative residual Kullback-Leibler information. Statistics and Probability Letters **82**, 2025–2032 (2012)
17. Perez-Cruz, F.: Kullback-Leibler divergence estimation of continuous distributions. In: ISIT, pp. 1666 1670. IEEE (2008)
18. Qiu, H., Liu, Y., Subrahmanya, N.A., Li, W.: Granger causality for time-series anomaly detection. In: ICDM, pp. 1074–1079 (2012)
19. Rao, M., Chen, Y., Vemuri, B.C., Wang, F.: Cumulative residual entropy: A new measure of information. IEEE Transactions on Information Theory **50**(6), 1220–1228 (2004)
20. Reshef, D.N., Reshef, Y.A., Finucane, H.K., Grossman, S.R., McVean, G., Turnbaugh, P.J., Lander, E.S., Mitzenmacher, M., Sabeti, P.C.: Detecting novel associations in large data sets. Science **334**(6062), 1518–1524 (2011)
21. Saatci, Y., Turner, R.D., Rasmussen, C.E.: Gaussian process change point models. In: ICML, pp. 927–934 (2010)
22. Scott, D.W.: Multivariate Density Estimation: Theory, Practice, and Visualization. John Wiley & Sons Inc, New York (1992)
23. Song, X., Wu, M., Jermaine, C.M., Ranka, S.: Statistical change detection for multi-dimensional data. In: KDD, pp. 667–676 (2007)
24. Wang, F., Vemuri, B.C., Rangarajan, A.: Groupwise point pattern registration using a novel cdf-based Jensen-Shannon divergence. In: CVPR, pp. 1283–1288 (2006)

Swap Randomization of Bases of Sequences for Mining Satellite Image Times Series

Nicolas Méger[1]([⊠]), Christophe Rigotti[2], and Catherine Pothier[3]

[1] LISTIC Laboratory, Université Savoie Mont Blanc, Polytech Annecy-Chambéry,
B.P. 80439, 74944 Annecy-le-vieux Cedex, France
nicolas.meger@univ-smb.fr
[2] LIRIS Laboratory (UMR 5205), Université de Lyon, CNRS, INRIA, INSA-Lyon,
20 Avenue A. Einstein, 69621 Villeurbanne Cedex, France
christophe.rigotti@insa-lyon.fr
[3] LGCIE Laboratory, Université de Lyon, INSA-Lyon, 20 Av. A. Einstein,
69621 Villeurbanne Cedex, France
catherine.pothier@insa-lyon.fr

Abstract. Swap randomization has been shown to be an effective technique for assessing the significance of data mining results such as Boolean matrices, frequent itemsets, correlations or clusterings. Basically, instead of applying statistical tests on selected attributes, the global structure of the actual dataset is taken into account by checking whether obtained results are likely or not to occur in randomized datasets whose column and row margins are equal to the ones of the actual dataset. In this paper, a swap randomization approach for bases of sequences is proposed with the aim of assessing sequential patterns extracted from Satellite Image Time Series (SITS). This assessment relies on the spatiotemporal locations of the extracted patterns. Using an entropy-based measure, the locations obtained on the actual dataset and a single swap randomized dataset are compared. The potential and generality of the proposed approach is evidenced by experiments on both optical and radar SITS.

1 Introduction

Earth observation satellite technology is continuously being enhanced, providing end users with ever ever-growing data volumes. Improvements relate to the number of acquisition channels, the spatial resolution and the revisit frequency. The revisit capability makes possible to gather acquisitions of a same geographical zone through time and form *Satellite Image Time Series (SITS)*. SITS are large datasets containing complex spatiotemporal information that can be affected both by atmospheric perturbations and sensor problems. In order to fully exploit such SITS, information retrieval and data mining techniques are being developed. Among them, unsupervised data mining techniques demonstrate their potential when it comes to describe and discover spatiotemporal phenomena. They rely either on global models such as clusterings (e.g., [13] or [21]) or on local patterns such as sequential patterns (e.g., [16] or [14]). In particular, a SITS can be considered as a special kind of base of sequences, as first introduced in [1]. In that

A. Appice et al. (Eds.): ECML PKDD 2015, Part II, LNAI 9285, pp. 190–205, 2015.
DOI: 10.1007/978-3-319-23525-7_12

initial context, each sequence gives the transactions of a customer whereas, in the case of a SITS, each sequence contains the descriptions of the values of a pixel through time and is thus located spatially. As proposed in [17], *Grouped Fequent Sequential patterns (GFS-patterns)* can be extracted from such a base of sequences. Besides expressing pixel temporal evolutions, these sequential patterns also take into account the spatial information brought by SITS: each GFS-pattern is required to affect a group of pixels that are sufficiently numerous and connected to each other. Reciprocally, each pixel can be affected by different GFS-patterns. As a consequence, pixel groups corresponding to extracted GFS-patterns can partially or fully overlap each other: they can refine each other. Extracting GFS-patterns thus differs from segmenting or clustering a SITS. Experiments reported in [17] or [22] show that GFS-patterns can be used both on radar and optical data, for various applications ranging from agricultural to crustal deformation monitoring. Despite their ability to address various types of datasets and applications, these patterns can be numerous, even if maximal ones are focused on. How to select the most significant ones without making any assumption? We aim to answer that question by adapting swap randomization to the SITS mining context.

In statistics, the significance of a result (e.g., the number of correlations found in a dataset) can be assessed via randomization testing methods [12]. Basically, they check whether the result observed on the actual dataset is likely to be obtained or not on randomized datasets. These datasets are meant to sufficiently differ from the actual one while sharing some of its structural properties such as the number of 0's and 1's in the case of a Boolean matrix. With this aim in view, randomized datasets are built by shuffling the actual dataset. Considering randomized datasets avoids generating random ones by sampling a distribution law that has to be defined a priori. Swap randomization follows these guidelines and focuses on more fined-grained structural properties such as the column and row margins of a Boolean matrix [5]. In data mining, as evidenced in [9], [10] or [15], swap randomization can be exploited to assess the significance of global models characterizing the whole actual dataset. These models can be clusterings, sets of frequent itemsets, sets of correlations or singular values. Even if they do not describe the entire dataset, local patterns such as frequent itemsets can also be evaluated individually (e.g., [10] or [15]).

To our knowledge, no swap randomization techniques handling bases of sequences or SITS have been proposed so far. In this paper, such a proposal is made with the aim of evaluating GFS-patterns [17] individually. While being dedicated to GFS-patterns, the presented approach could also be used for any kind of sequential patterns or episodes. Assessing GFS-patterns is not a trivial task. Their spatiotemporal nature must be taken into consideration and the following questions must be answered: which fine-grained structure should be maintained when randomizing the base of sequences representing a SITS? Which GFS-pattern-related information should be considered for their individual assessment? How to compare the information observed on the actual dataset with the one obtained for the randomized datasets? How to be efficient when considering a SITS containing millions of pixel values? Our answers are as follows: with regards to the structure to be maintained while randomizing, the distributions of the values of each image and each pixel sequence are preserved. The assessment of a GFS-pattern is then performed by comparing its spatiotemporal locations on the actual dataset with the ones on the randomized datasets. This comparison relies on the *Normalised Mutual Information (NMI)* [6], an entropy-based measure. Efficiency is achieved by performing the comparison using a single randomized dataset, as opposed to hundreds of randomized datasets when considering the standard swap randomization approach. This paper is organized as follows: Section 2 gives some preliminary definitions regarding SITS and GFS-patterns. The swap randomization approach proposed to shuffle bases of sequences representing a SITS is detailed in Sect. 3. Section 4 explicates GFS-pattern assessment and its use for SITS summarization. Experiments are presented in Sect. 5. They show that the proposed approach is general enough to mine either radar or optical SITS, yields relevant patterns on real datasets and can support different applications such as land cover or crustal deformation monitoring. Section 6 concludes this paper and gives future work directions.

2   Grouped Frequent Sequential Patterns

In this section, the definition of *Grouped Frequent Sequential Patterns (GFS-patterns)*, as first introduced in [17], is recalled. Let us consider a SITS, i.e., a satellite image time series covering the same area at n different dates. Within each image, each pixel is associated with a value, e.g., the reflectance intensity of the geographical zone it represents. These values are discretized to get *event types* (symbols) encoding *events* under the form of a pair (t, e) with e an event type and t its occurrence date (here the date will be the index of the image in the series). Event types can correspond to ranges obtained by image quantization or to pixel clusters. A *symbolic SITS* is a set of *pixel evolution sequences*, each one containing the coordinates (x, y) of a pixel and its corresponding event sequence, i.e., a tuple of events $\langle(t_1, e_1), (t_2, e_2), ..., (t_n, e_n)\rangle$. In pattern mining, a typical base of sequences is a set of sequences of discrete events, in which each sequence has a unique sequence identifier. Each location (x, y) being unique, a symbolic SITS is a base of sequences and the standard notions of sequential patterns, support and frequent sequential patterns introduced in [1] can be easily reused as follows[1]. A *sequential pattern* α is a tuple of m event types $\langle\alpha_1, \alpha_2, ..., \alpha_m\rangle$. The *support* of α in a SITS, denoted by $support(\alpha)$, is the number of pixel evolution sequences in which α occurs at least once. Note that the event types do not need to occur contiguously. Sequential pattern α is a *frequent sequential pattern* if $support(\alpha) \geq \sigma$ with σ a support threshold. Reusing the definitions of sequential patterns permits to take advantage of the efficient extraction techniques developed in this domain (e.g., [1], [25] or [20]). The pixels where a pattern α occurs are said to be *covered* by α. For a SITS, the notion of support can be interpreted very naturally as an area. In order to obtain pixels forming regions in space, an average connectivity measure is also used. It is based on the *8-nearest neighbors*

---

[1] Sequences are simpler here since there is a single event type for each timestamp.

(8-NN) convention [8]. For α, the *connectivity* of a pixel (x, y) is the number of pixels covered by α among the 8 nearest neighbors of (x, y) (i.e., the pixels surrounding (x, y)). The *average connectivity* of α, denoted $AC(\alpha)$, is simply the average of the connectivity over all pixels covered by α. Finally, a *Grouped Frequent Sequential pattern (GFS-pattern)* α is a frequent sequential pattern such that $AC(\alpha) \geq \kappa$ with κ a positive real number termed *average connectivity threshold*. Depending on the parameter settings and the dataset, numerous GFS-patterns can be produced. In order to reduce the redundancy among the patterns, a standard method is to retain only the maximal ones (e.g., [19]). This approach is also used here. The *maximal* GFS-patterns of a collection of GFS-patterns $\mathcal{C}$ are the elements in $\mathcal{C}$ that are not subpattern of any other pattern in $\mathcal{C}$. In other words, the GFS-patterns focusing on the most specific evolutions are retained. Though the number of GFS-patterns can be drastically reduced by adopting such a strategy, it can still be large. How to select the most significant ones without making any additional assumption with respect to covered pixels (e.g., assumptions about the shape or the texture of pixel groups)? We propose to answer that question by adapting the swap randomization of Boolean matrices to the SITS mining context.

3 Swap Randomization of Base of Sequences Representing SITS

Swap randomization is aimed at generating Boolean matrices having the same row and column margins without assuming any underlying distribution law. To this end, the elements of the matrices are swapped. A swap is defined as follows [23]: let B be a $m \times n$ Boolean matrix. Let u and v be two rows. Let i and j be two columns. If $B_{u,i} = B_{v,j} = 0$ and $B_{u,j} = B_{v,i} = 1$ then rows (or columns) are changed so that $B_{u,i} = B_{v,j} = 1$ and $B_{u,j} = B_{v,i} = 0$: values 0 and 1 are swapped. By construction, such a swap does not modify column and row margins. These margins give the number of occurrences of symbol '1' (or symbol '0', its dual symbol) for each column and each row. An example is given in Fig. 1. Boolean matrix B' is obtained from matrix B via a single swap such that $u = 2$, $v = 4$, $i = 1$ and $j = 3$. Swapped 0's and 1's are underlined.

In [23], Ryser shows that it is possible, starting from a given Boolean matrix, to generate all possible Boolean matrices having the same row and column margins by applying a series of swaps, each swap being applied to the latest matrix that had been obtained. In [5], on the basis of this result, the authors show that it is possible to randomly generate equiprobable matrices having the same row and column margins. More precisely, starting from a given Boolean matrix, a series of swap is performed by choosing rows and columns at random. Rows and columns can be chose more than once. As a consequence, swaps can be undone. Each swap can be seen as a random step from a vertex to another one in a graph whose vertices represent all possible matrices and whose edges represent transitions that can be performed by swapping 0' s and 1's. The series of swaps can thus be interpreted as a random walk on a graph that, in turn, can be formalized as a Markov

$$B = \begin{pmatrix} 0\ 1\ 0\ 0 \\ \underline{1}\ 0\ \underline{0}\ 1 \\ 1\ 0\ 1\ 1 \\ \underline{0}\ 0\ \underline{1}\ 0 \end{pmatrix}, B' = \begin{pmatrix} 0\ 1\ 0\ 0 \\ \underline{0}\ 0\ \underline{1}\ 1 \\ 1\ 0\ 1\ 1 \\ \underline{1}\ 0\ \underline{0}\ 0 \end{pmatrix}$$

Fig. 1. Boolean matrix B' is obtained from B by swapping underlined values. Both matrices have the same row and column margins.

chain. In such a chain, the authors explain that the probability of state (i.e., each vertex/matrix when considering the graph) to be reached by a sequence of transitions can differ from one state to another. The proposed solution consists in adding self-loops to have all vertices being reached by the same amount of edges, which guarantees that vertices, and thus Boolean matrices are equiprobable [5]. One important question remains: how many random walk steps are needed to get a Boolean matrix that is sufficiently randomized, i.e., that sufficiently differs from the actual dataset? This is still an open research question. See [3] and [2] for discussions regarding the obtention of p-values using a Markov chain. Nevertheless, empirical results are available (e.g., [10]). Holding in place with self-loops is not efficient when trying to get data sets that are sufficiently randomized. An optimization can be achieved by relying on the Metropolis-Hastings algorithm (e.g., [5] or [10]). Another simpler and efficient optimization is proposed in [10]. It is based on the same approach than [5] but requires less self-loops. It relies on a set P containing all pairs (u, i) such that $B_{u,i} = 1$. This structure is made available throughout the whole algorithm. The swapping procedure differs from the standard one: u, v, i and j are not fully chosen at random. They are chosen by randomly selecting two pairs (u, i) and (v, j) in P. If pairs (u, j) and (v, i) are not in P, then $B_{u,j} = B_{v,i} = 0$ and the swap is made effective. Otherwise, the swap attempt is counted as a self-loop. By avoiding a full random walk, the convergence is accelerated and the overhead induced by the management of P is absorbed. In [10], using this algorithm, it is empirically estimated that the number of random walk steps should be in order of the number of 1's of the matrix to converge to a sufficiently randomized Boolean matrix.

Swap randomization is basically applied to Boolean matrices to assess data mining results using p-values. The bottom line is to define a null hypothesis stating that the result observed for the actual dataset is likely to be observed on randomized datasets having the same structure, i.e., the same column and row margins. If the null hypothesis is rejected then the result is considered to be significant. In order to run such a test, a metric of interest has to be chosen. With regards to correlations, it is proposed in [10] to compute the number of correlations or the maximum and the minimum correlation values. The same kind of strategy is also used to analyze sets of frequent itemsets by considering the number of extracted frequent itemsets, the fraction of frequent itemsets that are preserved and the fraction of frequent itemsets that disappear. For this latter case, the analysis is run by directly comparing these numbers and fractions, without using p-values. Still, if required, it would be possible to compute them. Finally, clusterings are studied through clustering errors. Besides global models, local patterns such as frequent itemsets can also be evaluated individually through their support measure directly or via p-values (e.g., [10]) or [15]). The ratio between the support observed on the actual dataset and the mean support observed for randomized datasets is also mentionned as an interesting alternative. The experiments reported in [5], [9], [10] or [15] all demonstrate the potential of the swap randomization approach in the case of Boolean matrices.

With regard to a $m \times n$ non-Boolean symbolic matrix S, i.e., a matrix containing elements defined with more than two distinct symbols such as '0' and '1', the standard Boolean swap defined in [23] can be extended as follows : let u and v be two rows, and let i and j be two columns. If $S_{u,i} = S_{v,j} = \alpha$ and $S_{u,j} = S_{v,i} = \beta$ with α and β two distinct symbols, then rows (or columns) are changed so that $S_{u,i} = S_{v,j} = \beta$ and $S_{u,j} = S_{v,i} = \alpha$: symbols α and β are swapped. This *symbolic swap* preserves row and column margins. For each symbol used to define S, these margins give the number of its occurrences for each row and each column. A symbolic swap is illustrated in Fig. 2. Non-boolean symbolic matrix C' is obtained from C via a single swap such that $u = 1$, $v = 3$, $i = 1$ and $j = 2$. Swapped symbols '2' and '3' are underlined. Both matrices share the same row and column margins. Sadly, it is not possible to generate all non-Boolean symbolic matrices having the same row and column margins by swapping data. Fig. 2 gives an example: no swap series can be found to transform D into D' though both matrices have the same row and column margins. Consequently, if swap randomization is performed on such matrices, then swap randomized datasets must be compared with the actual dataset to check whether they sufficiently differ from each other.

Following the principles of swap randomization as defined for Boolean matrices, we aim to assess GFS-patterns by randomizing bases of sequences, and more specifically symbolic SITS. This randomization is thus required to maintain a fine-grained structure of the dataset while breaking event connectivity within each image and event ordering within each pixel evolution sequence. This raises the following question: which structure can be preserved? In order to break event connectivity and ordering only, we propose to maintain event type frequencies within each image and each pixel evolution sequence. This can be achieved by considering spatiotemporal swaps, i.e symbolic swaps. Indeed, as long as more than two even types are considered, a symbolic SITS representing n acquisitions of m pixels can be transformed into a $m \times n$ non-Boolean symbolic matrix (and

$$C = \begin{pmatrix} 3\ 2 \\ 1\ 1 \\ 2\ 3 \end{pmatrix}, C' = \begin{pmatrix} 2\ 3 \\ 1\ 1 \\ 3\ 2 \end{pmatrix}, D = \begin{pmatrix} 1\ 2 \\ 2\ 3 \\ 3\ 1 \end{pmatrix}, D' = \begin{pmatrix} 2\ 1 \\ 3\ 2 \\ 1\ 3 \end{pmatrix}$$

Fig. 2. Non-boolean symbolic matrix C' is obtained from C by swapping underlined values. C and C' have the same row and column margins. D' can not obtained from D by swapping data though they share the same column and row margins.

vice versa). In such a matrix, an element located at row k and column l gives the event type describing a pixel whose coordinates are mapped bijectively to k in the lth image. Consequently, in order to swap randomize a symbolic SITS, we propose to adapt the algorithm described in [10] by performing a series of *swap attempts* which are defined as follows:

Definition 1. *(swap attempt)* Let S be a $m \times n$ non-Boolean symbolic matrix representing a symbolic SITS defined over E, the set of event types. Let $P = \{\{(u,i),(v,j)\}|S_{u,i} = S_{v,j} = \alpha, \forall \alpha \in E\}$. A *swap attempt* is selecting $p = \{(u,i),(v,j)\} = \alpha \in P$ randomly. If $\exists\, p' \in P$ such that $p' = \{(u,j),(v,i)\} = \beta \mid \beta \neq \alpha$, then a symbolic swap is performed so that $S_{u,i} = S_{v,j} = \beta$ and $S_{u,j} = S_{v,i} = \alpha$. Otherwise no swap is performed but it is still counted as a self-loop.

By performing such spatiotemporal swaps, a first structure level of SITS is maintained, i.e., event type frequencies. Maintaining event type frequencies in images is equivalent to preserving their histograms which are standard first level image descriptors [11]. With respect to pixel evolution sequences, their first structure level can also be given by event type frequencies. From the application point of view, this makes sense. At the image level, an image affected by clouds should not be converted into an image expressing the presence of vegetation (and vice versa). Similarly, vegetation should not be transformed into a glacier. At the pixel evolution sequence level, since each sequence relates to a specific location, if the presence of water is expressed through a sequence, then there is no reason to change it to a sequence relating to bare soils. The same holds for a pixel whose sequence is giving variations between snow and rocks with little vegetation: swap randomization should not transform it into a sequence of permanent vegetation. Maintaining the spatiotemporal structure of a SITS is a strategy similar to the one adopted in [24] to randomize time series collections. In that case, a time series collection is represented by J real-valued matrices, where J is the number of wavelet coefficients used to describe the series, i.e., the maximum detail level. An element located at position (i,j) of the fth matrix gives the value of the fth wavelet coefficient for series i at time point j. These matrices are independently randomized by approximately preserving the temporal distributions (row distributions) and the series domain distributions (column distributions) of the wavelet coefficients. Hence, this approach could be adapted to SITS randomization. Nevertheless, in addition to performing a discrete wavelet transform of the original time series and randomizing several matrices (one per coefficient), an inverse discrete wavelet transform is required to transform each randomized dataset back to the original representation. Finally, if one were to assess GFS-patterns using this approach, then every randomized dataset should also be quantized. Back to our approach, even if the SITS first structure level is preserved, the connectivity and the order of the event types forming GFS-patterns is affected. This allows to detect the GFS-patterns that are due or not to such a structure. As for the algorithm of [10], convergence is accelerated through the use of set P and self-loops allow to generate equiprobable datasets. However, in practice, as already stated previously in this section, it is not possible to generate all $m \times n$ non Boolean symbolic matrices (and thus symbolic SITS) having the same row and column margins. Still, as shown empirically in Sect. 5, it is possible to generate and explore randomized datasets that differ from each other and that also differ from the actual SITS sufficiently. As long as it makes sense to preserve row and column margins, this kind of technique can also be applied to other types of bases of sequences.

4   GFS-Pattern Assessment and SITS Summarization

Using the SITS swap randomization approach proposed in Sect. 3, we aim to assess GFS-patterns individually. As explained in Sect. 3, when considering the swap randomization of Booleean matrices, frequent itemsets can be assessed through their support measures directly, support ratios or p-values (e.g., [10]) or [15]). With regard to GFS-patterns, considering their support measure only is not sufficient since their spatiotemporal nature is not taken into account fully. The coordinates and the temporal locations (starting dates, ending dates, timespans, etc.) of the pixels affected by a GFS-pattern must also be considered. Therefore, we propose to focus on pixel coordinates and ending dates by relying on *Spatio Temporal Localization Map (STL-maps)*. An STL-map is an image generated for each GFS-pattern given a symbolic SITS (randomized or not). In such an image, if a pixel is covered by the GFS-pattern for which the image was generated, then its value gives the ending date of the earliest occurrence available for the corresponding coordinates. Otherwise, no ending date is stored (a *black pixel value* is used). By construction, STL-maps also include the information related to the support of GFS-pattern. As shown in Sect. 5, and though other types of temporal locations are also interesting, considering ending dates only allows to perform an efficient and reliable GFS-pattern assessment. Efficiency is also achieved by considering a single swap randomized symbolic SITS only: this avoid generating lots of STL-maps and running numerous comparisons.

How to compare the STL-map M, obtained on the actual SITS for a pattern α, with M', the STL-map obtained for α on a single swap-randomized SITS? How to compare them without having to make any assumption about their relation? At this stage, we are interested by the following two settings:

- M and M' are dissimilar: M is singular as it can not be obtained for a randomized dataset with the same structure in terms of event type frequencies,
- M and M' are similar: the swap-randomization does not destroy the occurrences of α and thus C expresses a prominent phenomena explained by the margins.

The first setting is in line with the standard swap randomization approach while the second one is usually not considered since one-tailed tests are focused on. Still, the second setting is of primary interest. Geographical zones affected by few changes are expressed through event types that are sowewhat always the same. Hence, the corresponding events are hardly randomized. If we were to reject them, the SITS exploration would be biased towards GFS-patterns expressing changes and interesting areas such as deserts, lakes or cities would disappear from extracted descriptions. How to assess and distinguish the latter two settings using a single measure? Let Ω be the sample space containing all ending dates. Let us consider each ending date x of M as the realization of a discrete random variable X and each ending date y of M' as the realization of a discrete random variable Y. We propose to rely on the *Normalized Mutual Information (NMI)* as presented in details in [6].

$$NMI(X;Y) = \frac{\sum_{x,y\in\Omega^2} P(x,y) \log \frac{P(x,y)}{P(x)P(y)}}{min(H(X),H(Y))} \tag{1}$$

where $H(X) = -\sum_{x\in\Omega} P(x) \log P(x)$ and $P(x,y)$ represents the probability of co-occurrence of the two ending dates x and y at the same pixel position, in M and M'. The NMI quantifies the information content shared by two random variables. In other words, knowing the realizations of two random variables X and Y, it measures the extent to which the realizations of variable X can be deduced from the ones of Y, and vice versa. It can been therefore seen as a measure of the mutual dependence between X and Y. The more X and Y are independent (respectively dependent), the more the NMI tends to 0 (respectively 1) since no bit is shared between the two variables. A particular case must be handled: the black pixels. These pixels show no realizations, no ending dates. Since we extract GFS-patterns that may only cover little fractions of the observed zone, black pixels can be numerous with respect to non-black ones. If these numerous black pixels were to be considered as showing another *special* ending date, a lot of black pixels in M could be associated to other black pixels in M': their joint probability would be high, raising the NMI measure artificially and masking the other, but more important, joint probabilities. Consequently, the joint probability of black pixels is not considered. Nevertheless, because of the swap-randomization, black pixels can differ from M to M': these other cases are taken into account thanks to joint probabilities having one of the two values set to a black pixel value.

Once the NMI is computed for each STL-map/GFS-pattern, then STL-maps/GFS-patterns are ranked accordingly. The NMI-based ranking that is obtained can be easily browsed to build a SITS summary by focusing on both ends of the ranking. Phenomena that can not be obtained on a swap-randomized SITS have low NMI scores and prominent phenomena that are still present in a swap-randomized STIS have high NMI scores. As shown in Sect. 5, if several swap randomized SITS are computed, then rankings are stable for high and low NMI GFS-patterns: a single swap randomized SITS can thus be considered. By relying on the NMI, no assumption about the relation between the ending dates is done. Beside extracting GFS-patterns, this allows us to produce summaries which are as unsupervised as possible.

5   Experiments

The swap randomization approach presented in this paper was assessed by conducting experiments on two different SITS, a radar one and an optical one. Their characteristics are given by Table 1. For each SITS, raw data are transformed into a single synthesized channel dedicated to the application domain. Regarding Etna, phase delays were computed [7] by Marie-Pierre Doin (ISTerre laboratory, CNRS). These floats express vertical and/or lateral displacements w.r.t. a master acquisition. An example is given by Fig. 9 where Mount Etna is revealed in the upper part of the image. For NC, the *Normalized Difference Vegetation Index (NDVI)* [4] was generated by Rémi Andréoli (Bluecham S.A.S. *www.bluecham.net*). It expresses the presence of biomass. An example is shown in Fig. 10: the ocean (resp. land) is mainly located in the lower right part (resp. upper left part) of the image. Radar shadows, atmospheric perturbations, clouds and sensor defaults are still present in these synthesized channels. Preprocessing details are available in Table 1.

The experiments were run on a standard computing platform (a single core on a 2.7 GHz Intel Core i7) using our own prototype *SITS-miner* implemented in C and Python. On the side of parameter settings, average connectivity threshold κ is set to 5 neighbors to extract zones making sense spatially. This is a standard setting [17]. In order to assess reasonable amounts of GFS-patterns, we focus on maximal ones, as explained in Sect. 2. With regard to minimum support threshold σ, it is set such that the *richest/most diverse* description is obtained. This achieved by finding the lowest value of σ such that the number of maximal GFS-patterns is maximum: the widest possible range of surfaces, from σ to the surface of the image itself, is considered. Following this strategy, minimum support threshold σ was found to be 7000 for both SITS (covering about 2.11% of an image in Etna and 2.66% in NC). By consuming no more than 655 MB of RAM and in less than one minute, 508 maximal GFS-patterns are extracted from Etna and 297 maximal GFS-patterns are mined in NC^2.

These patterns were assessed using the swap randomization approach and the NMI ranking procedure described in this paper. Regarding swap randomization, the parameter to be set is N_s, the number of swap attempts to be performed. In [10], it is empirically estimated that N_s should be in order of the number 1's of the matrix to converge to a sufficiently randomized Boolean matrix. In our case, we will consider the number of events multiplied by about 20 to adopt a very conservative setting: $N_s = 100.000.000$. This setting makes sense since it can be empirically shown that the two SITS are sufficiently randomized to get stable NMI values for the patterns we are interested in, i.e., those located at both ends of the NMI rankings (see Sect. 4). Let us the consider the 20 highest and the 20 lowest NMI patterns obtained for 100M swaps. Their respective NMI values were also computed for $N_s = 20M, 40M, \ldots, 140M$, and are reported as randomizations labelled 0 to 6 in the figures 3, 4, 5 and 6. As it can be observed, they rapidly converge to levels that are quite stable, especially around 100M of swaps. With regard to swapped

---

^{2} The reader is referred to [17] and [18] for discussions regarding the impact of σ and of the number of event types on the number of extracted patterns.

Table 1. SITS properties, preprocessing and extraction settings.

SITS name	Etna	NC
provider/credit	ESA	USGS/NASA Landsat
satellite	ENVISAT	LANDSAT 7
SITS type	Synthetic Aperture Radar	Multispectral
time period	16 images 2003-2010	16 images 2000-2011
site	Geohazards Supersite: Mount Etna	UNESCO World Heritage Site: lagoons of New Caledonia
application	crustal deformation monitoring	soil erosion monitoring
data quality	pixel values are not always available (radar shadows), atmospheric perturbations	a lot of clouds, sensor defaults
image size	598×553	513×513
resolution	$160\ m$	$30\ m$
synthesized channel	phase delays	NDVI
discretization	quantization/all images (33rd and 66th centiles)	quantization/each image (33rd and 66th centiles)
event types	'1': motion towards satellite (satellite on the left) '2': stable '3': motion away from satellite	'1': few biomass '2': average biomass '3': lot of biomass
parameters	$\sigma = 7000, \kappa = 5$	$\sigma = 7000, \kappa = 5$

randomized SITS themselves, we generated 1000 swap randomized datasets for each SITS to evaluate them. Though 73.9% of the Etna events and 16.2% of the NC events can not be swapped, in average, 6.5% of the Etna events and 32.9% of the NC events were swapped. The standard deviation of these swapped event rates tends to 0, which shows the stability of our swap randomization process. Finally, if we consider a single randomization and focus on effective swaps (self-loops are not counted), it should be mentioned that 1.070.219 different swap randomized datasets are explored when randomizing Etna. Among them, one dataset is generated 8 times and others are obtained only once. In the case of SITS NC, 8.911.591 different datasets are generated once, one dataset is obtained 4189 times and another one is reach 44 times. Consequently, and though no all SITS having the same column and row margins can be reached (see Sect. 3), the proposed swap randomization approach does explore a lot of different SITS having the same structure. As proposed in Sect. 4, for efficiency reasons, rankings are established using a single swap randomized dataset. This makes sense for both SITS since rankings are stable for high and low NMI GFS-patterns. As shown by Fig. 7 and Fig. 8, the rank standard deviation is less than 1 for both ends of the ranking. It was computed using the rankings obtained for the 1000 swap randomized datasets we generated for both SITS. Similar results are obtained when plotting the rank standard deviation against the rank mode or a reference ranking. For both SITS, memory consumption and execution times do no not exceed 1.66 GB of RAM and 700 seconds to perform the pattern extraction, the STL-map computation for the maximal patterns, a single randomization and the final ranking of STL-maps.

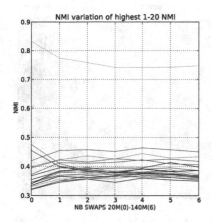

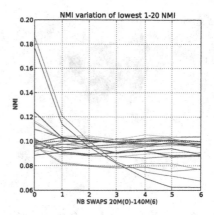

Fig. 3. 20 highest NMI values vs. N_s, Etna.

Fig. 4. 20 lowest NMI values vs. N_s, Etna.

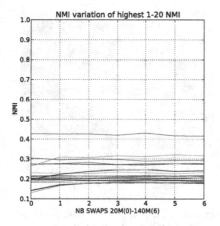

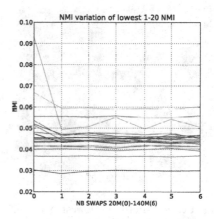

Fig. 5. 20 highest NMI values vs. N_s, NC.

Fig. 6. 20 lowest NMI values vs. N_s, NC.

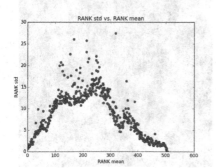

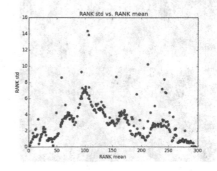

Fig. 7. Rank std. vs. rank mean, Etna.

Fig. 8. Rank std. vs. rank mean, NC.

Fig. 9. Total phase delays, from negative values (black) to positive values (white), 2003/01/22, Marie-Pierre Doin, Etna.

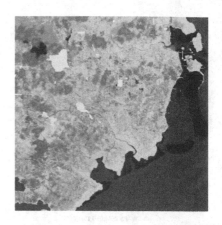

Fig. 10. NDVI, from low values (black) to high values (white), 2004/01/13, Bluecham S.A.S., NC.

Fig. 11. STL-map: 1st lowest NMI pattern $\langle 1,1,2,1,1,1,1,3 \rangle$, Etna.

Fig. 12. STL-map: 1st highest NMI pattern $\langle 1,2,3,3,3,3,3,3,3,3,3,3,3,3,3 \rangle$, Etna.

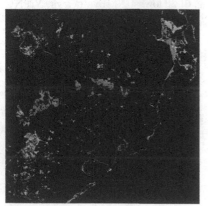

Fig. 13. STL-map: 6th lowest NMI pattern $\langle 2,2,1,1,1,2 \rangle$, NC.

Fig. 14. STL-map: 2nd highest NMI pattern $\langle 3,3,3,3,3,3,3,3,3,3,3,3,3,3 \rangle$, NC.

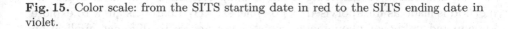

Fig. 15. Color scale: from the SITS starting date in red to the SITS ending date in violet.

Regarding qualitative results, it is possible to extract known and unknown meaningful phenomena, at both ends of the NMI rankings and for both datasets. Different STL-maps, representative of the well ranked ones, are shown in figures 11-12 (for the Etna SITS) and in figures 13-14 (for the NC SITS). Pixels where there is no occurence of the pattern are represented in black for NC and in gray for Etna (depicting a digital elevation model available for the area). The color scale used to represent the occurrence dates is given Fig. 15. In Fig. 11, pattern $\langle 1,1,2,1,1,1,1,3 \rangle$ (1st lowest NMI pattern) shows, at the foot of the volcano, a zone moving towards the satellite before going away from the satellite (for this SITS the location of the satellite is on the left side of the image). It matches a sedimentary zone that is affected by movements due to subduction plates. In Fig. 12, pattern $\langle 1,2,3,3,3,3,3,3,3,3,3,3,3,3,3 \rangle$ (1st highest NMI pattern) denotes a short motion towards the satellite and then a very long motion away from the satellite. It covers a part of the east flank of the volcano, called the *Valle del Bove*, which is known to be slipping into the sea. In Fig. 13, pattern $\langle 2,2,1,1,1,2 \rangle$ (6th lowest NMI pattern) traces losses of vegetation due to anthropic activities (mining area at center and middle-left, mining facilities bottom-left). It also uncovers the impact of drought on a lakeshore (top-left) and exhibits sediment deposition (top-right). Notice that the color scale shows clear differences among the dates of occurrence of the phenomena. In Fig. 14, the simple pattern $\langle 3,3,3,3,3,3,3,3,3,3,3,3,3,3 \rangle$ (2nd highest NMI pattern) locates dense vegetation along the coastline. The STL-maps obtained on the NC SITS are commercialized through the web-based decision support system operated by Bluecham S.A.S (Qëhnelö plateform www.yate.nc). Finally, the fact that encouraging results are obtained for very different datasets (radar or optical, different spatiotemporal resolutions and different rates of swappable events) shows the general nature of the approach.

6 Conclusion

This paper extends the swap randomization of Boolean matrices to the swap randomization of a base of sequences representing a Satellite Image Time Series (SITS). The proposed approach is aimed at assessing spatiotemporal patterns extracted from SITS. It preserves event frequencies, spatially and temporally, while breaking event connectivity and ordering. Once swap randomized datasets are generated, patterns are ranked using the Normalized Mutual Information (NMI). Low NMI patterns underline singular phenomena that are unlikely in randomized datasets while high NMI patterns express prominent phenomena that cannot be destroyed via swap randomization. Experiments on an optical and a radar SITS evidence the stability of the swap randomization approach and its ability to explore a lot of different datasets. They also confirm that efficiency can be achieved by considering a single swap randomized dataset. Since the method is made as unsupervised as possible, extracted patterns allow to explore known and unknown phenomena, which gives access to different application domains ranging from agricultural monitoring to crustal deformation monitoring. Results regarding soil erosion monitoring are already commercialized. Future work include handling multispectral SITS, building clustering on top of extracted patterns and pushing NMI constraints within the extraction process.

References

1. Agrawal, R., Srikant, R.: Mining sequential patterns. In: Proceedings of the Eleventh International Conference on Data Engineering, Taipei, Taiwan, pp. 3–14, March 1995
2. Besag, J.: Markov chain monte carlo methods for statistical inference (2004). http://www-users.mat.umk.pl/~wniem/SemMgr/besag_MCMC.pdf
3. Besag, J., Clifford, P.: Generalized monte carlo significance tests. Biometrika **76**(4), 633–642 (1989)
4. Chuvieco, E., Huete, A.: Fundamentals of Satellite Remote Sensing. CRC Press, Boca Raton (2009)
5. Cobb, G.W., Chen, Y.: An application of markov chain monte carlo to community ecology. The American Mathematical Monthly **110**(4), 265–288 (2003)
6. Cover, T.M., Thomas, J.A.: Elements of Information Theory (Wiley Series in Telecommunications and Signal Processing). Wiley-Interscience (2006)
7. Doin, M., Lodge, F., Guillaso, S., Jolivet, R., Lasserre, C., Ducret, G., Grandin, R., Pathier, E., Pinel, V.: Presentation of the small baseline nsbas processing chain on a case example: the etna deformation monitoring from 2003 to 2010 using envisat data. In: Proceedings of the Fringe Symposium, pp. 3434–3437. ESA SP-697, ESA Communications, Frascati, September 2011
8. Fisher, R., Dawson-Howe, K., Fitzgibbon, A., Robertson, C., Trucco, E.: Dictionary of Computer Vision and Image Processing. John Wiley and Sons, New York (2005)
9. Gionis, A., Mannila, H., Mielikäinen, T., Tsaparas, P.: Assessing data mining results via swap randomization. In: Proceedings of the Twelfth ACM SIGKDD International Conference on Knowledge Discovery and Data Mining, Philadelphia, PA, USA, pp. 167–176, August 2006
10. Gionis, A., Mannila, H., Mielikäinen, T., Tsaparas, P.: Assessing data mining results via swap randomization. TKDD **1**(3) (2007)
11. Gonzalez, R.C., Woods, R.E.: Digital Image Processing, 3rd edn. Prentice-Hall Inc., Upper Saddle River (2006)
12. Good, P.: Permutation tests : a practical guide to resampling methods for testing hypotheses. Springer series in statistics. Springer, New York (2000)
13. Gueguen, L., Datcu, M.: Image time-series data mining based on the information-bottleneck principle. IEEE Trans. Geoscience and Remote Sensing **45**(4), 827–838 (2007)
14. Guttler, F., Ienco, D., Teisseire, M., Nin, J., Poncelet, P.: Towards the use of sequential patterns for detection and characterization of natural and agricultural areas. In: Laurent, A., Strauss, O., Bouchon-Meunier, B., Yager, R.R. (eds.) IPMU 2014, Part I. CCIS, vol. 442, pp. 97–106. Springer, Heidelberg (2014)

15. Hanhijärvi, S., Ojala, M., Vuokko, N., Puolamäki, K., Tatti, N., Mannila, H.: Tell me something I don't know: randomization strategies for iterative data mining. In: Proceedings of the 15th ACM SIGKDD International Conference on Knowledge Discovery and Data Mining, Paris, France, pp. 379–388, June-July 2009

16. Honda, R., Konishi, O.: Temporal rule discovery for time-series satellite images and integration with RDB. In: Siebes, A., De Raedt, L. (eds.) PKDD 2001. LNCS (LNAI), vol. 2168, p. 204. Springer, Heidelberg (2001)

17. Julea, A., Méger, N., Bolon, P., Rigotti, C., Doin, M., Lasserre, C., Trouvé, E., Lazarescu, V.: Unsupervised spatiotemporal mining of satellite image time series using grouped frequent sequential patterns. IEEE Trans. Geoscience and Remote Sensing 49(4), 1417–1430 (2011)

18. Julea, A., Méger, N., Rigotti, C., Trouvé, E., Jolivet, R., Bolon, P.: Efficient spatio-temporal mining of satellite image time series for agricultural monitoring. Trans. MLDM 5(1), 23–44 (2012)

19. Luo, C., Chung, S.M.: Efficient mining of maximal sequential patterns using multiple samples. In: Proceedings of the 2005 SIAM International Conference on Data Mining, SDM 2005, Newport Beach, CA, USA, pp. 415–426. SIAM, April 2005

20. Pei, J., Han, J., Wang, W.: Constraint-based sequential pattern mining: the pattern-growth methods. J. Intell. Inf. Syst. 28(2), 133–160 (2007)

21. Petitjean, F., Inglada, J., Gançarski, P.: Satellite image time series analysis under time warping. IEEE Trans. Geoscience and Remote Sensing 50(8), 3081–3095 (2012)

22. Rigotti, C., Lodge, F., Méger, N., Pothier, C., Jolivet, R., Lasserre, C.: Monitoring of tectonic deformation by mining satellite image time series. In: Reconnaissance de Formes et Intelligence Artificielle (RFIA), Rouen, France, June 2014

23. Ryser, H.J.: Combinatorial properties of matrices of zeros and ones. Canadian Journal of Mathematics 9, 371–377 (1957)

24. Vuokko, N., Kaski, P.: Significance of patterns in time series collections. In: Proceedings of the Eleventh SIAM International Conference on Data Mining, SDM 2011, April 28–30, Mesa, Arizona, USA, pp. 676–686 (2011)

25. Zaki, M.J.: Sequence mining in categorical domains: Incorporating constraints. In: Proceedings of the Ninth International Conference on Information and Knowledge Management, CIKM 2000, pp. 422–429. ACM, New York, November 2000

The Difference and the Norm — Characterising Similarities and Differences Between Databases

Kailash Budhathoki and Jilles Vreeken[✉]

Max Planck Institute for Informatics and Saarland University, Saarbrücken, Germany
{kbudhath,jilles}@mpi-inf.mpg.de

Abstract. Suppose we are given a set of databases, such as sales records over different branches. How can we characterise the differences and the norm between these datasets? That is, what are the patterns that characterise the general distribution, and what are those that are important to describe the individual datasets? We study how to discover these pattern sets simultaneously and without redundancy – automatically identifying those patterns that aid describing the overall distribution, as well as those pointing out those that are characteristic for specific databases. We define the problem in terms of the Minimum Description Length principle, and propose the DIFFNORM algorithm to approximate the MDL-optimal summary directly from data. Empirical evaluation on synthetic and real-world data shows that DIFFNORM efficiently discovers descriptions that accurately characterise the difference and the norm in easily understandable terms.

1 Introduction

Suppose we are given a set of databases, such as the sales records over different branches of a chain. How can we characterise the differences and the norm between these datasets? That is, what are the patterns that are common to all databases, and what are those that are important to characterise the individual databases? For example, whereas *bread* and *butter* may be an important pattern in all stores, *pasta* and *ketchup* may only be descriptive for the store on campus. When we mine only the complete data we risk missing the locally important patterns, and when we mine the databases individually we risk missing the bigger picture. We want to discover all important patterns, without redundancy, and such that it is clear which databases they are characteristic for.

More in particular, given a set of databases, we want to discover a *set of patterns* per database or combination of databases that the user is interested in. These pattern sets should only include patterns that are descriptive for the databases associated with the set, and overall these sets should be as non-redundant as possible. That is, together these pattern sets should succinctly summarise the given data.

We formalise this goal in terms of the Minimum Description Length principle [5,13]. That is, we define the best model as the *set of pattern sets* that describes the data most succinctly without loss. By this objective, a pattern

© Springer International Publishing Switzerland 2015
A. Appice et al. (Eds.): ECML PKDD 2015, Part II, LNAI 9285, pp. 206–223, 2015.
DOI: 10.1007/978-3-319-23525-7_13

will only be included in the model if it simplifies the description – if it aids compression. This means our model will not be redundant, nor will it include noise.

To describe an individual database we only have to consider those patterns that are associated with that database. This allows us to associate patterns with the databases they are most characteristic for. As characteristic does not necessarily mean 'same frequency', we do not want to punish patterns for having different frequencies in the different databases they are associated with. To avoid such undue bias we carefully construct a score for this setup using prequential coding, a form of Refined MDL [5].

To discover good model directly from data we introduce the DIFFNORM algorithm. DIFFNORM iteratively searches for that pattern that maximally simplifies the current description. To this end it searches for those itemsets X and Y in its model that are most frequently co-used to describe the same transaction, and considers their union as a candidate pattern. The intuition is that these codes for X and Y are redundant, and that by introducing that $X \cup Y$ to the model the description will become more succinct.

Empirical evaluation on synthetic and real-world data shows that DIFFNORM efficiently discovers descriptions that in easily understandable terms accurately characterise the difference and the norm. On synthetic data it recovers the ground truth of both local and global patterns, without picking up on noise. On real world data it discovers succinct and interpretable pattern sets that characterise the split over the data well.

The remainder of this paper is organised as usual. For readability we postpone details on selected derivations to the online Appendix.[1]

2   Related Work

Comparing two or more transaction databases is a common task, yet there exist surprisingly few techniques that can characterise similarities and differences of databases in easily understandable terms. Traditional frequent pattern [1] as well as supervised pattern mining approaches [10] for example, discover far too many patterns for the result to be interpretable. Pattern set mining circumvents the pattern explosion [17]. Existing unsupervised methods such as TILING [4], SLIM [14], and MTV [8] only characterise one database at a time, while supervised methods only describe what sets databases apart. Running these algorithms on multiple (combinations of) databases and comparing the results does not work in practice – small differences in the data distribution can lead to very different pattern sets which are difficult to compare.

Earlier, Vreeken et al. [16] proposed a dissimilarity measure for transaction data based on KRIMP [17]. The main idea is to infer a pattern set per database, and then measure how many bits more we need to describe the other databases with these patterns – the more similar the data, the small the difference. Here,

---

[1] http://eda.mmci.uni-saarland/diffnorm/

on the other hand, we are interested in characterising all databases at the same time, without redundancy.

In Joint Subspace Matrix Factorization (JSMF) the goal is to discover the common subspace between the two datasets, as well as those that are representative of the specific datasets. Most relevant, as it considers binary databases, is Joint Subspace Boolean Matrix Factorization (JSBMF) [9]. To avoid overfitting, it requires the user to specify the number of patterns per pattern set. Our approach is parameter free. Moreover, as JSMF is defined for pairs of databases, and not trivially extendable to arbitrary combinations of databases, it cannot simultaneously and without redundancy find the patterns over multiple subspaces.

3   Preliminaries

In this section we discuss preliminaries and introduce notation.

3.1   Notation

We consider transaction data. Let $\mathcal{I}$ be a set of items, e.g. products for sale in a store. A transaction $t \in \mathcal{P}(\mathcal{I})$ then corresponds to the set of items a customer bought. A database D over $\mathcal{I}$ is a bag of transactions, e.g. the sales transactions on a given day. We consider bags $\mathcal{D}$ of d such databases, e.g. the sales transactions for different branches of store.

We assume this bag to be indexed such that by $D_i \in \mathcal{D}$ we can access the transactions sold at the i'th branch. Let $\mathcal{J} = \{1, \cdots, d\}$ be the set of indexes. An index set $j \in \mathcal{P}(\mathcal{J})$ then identifies a subset of databases $\{D_i \in \mathcal{D} \mid i \in j\}$. Finally, $U \subseteq \mathcal{P}(\mathcal{J})$ identifies those subsets of databases the user specifies as interesting.

We say that a transaction $t \in D$ supports an itemset $X \subseteq \mathcal{I}$, iff $X \subseteq t$. The support $supp_D(X)$ of X in D is the number of transactions in the database where X occurs. The relative support of X is its frequency, $freq_D(X) = supp_D(X)/|D|$, with $|D|$ for the number of transactions in D. Further, let $||D|| = \sum_{t \in D} |t|$ the total number of items. For $\mathcal{D}$, we write $|\mathcal{D}| = \sum_{D_i \in \mathcal{D}} |D_i|$, and define $||\mathcal{D}||$ analogue.

All logarithms are to base 2, and by convention we use $0 \log 0 = 0$.

3.2   MDL, a Brief Primer

The MDL (Minimum Description Length) [5,12] principle, like its close cousin MML (Minimum Message Length) [18,19], is a practical version of Kolmogorov complexity [6,7]. All three embrace the slogan *Induction by Compression*. For MDL, this principle can be roughly described as follows.

Given a set of models $\mathcal{M}$, the best model $M \in \mathcal{M}$ is the one that minimises

$$L(M) + L(\mathcal{D} \mid M)$$

where $L(M)$ is the length, in bits, of the description of model M, and $L(\mathcal{D} \mid M)$ is the length, in bits, of the description of the data when encoded with M.

This is called two-part MDL, or *crude* MDL. As opposed to *refined* MDL, where model and data are encoded together [5]. We use two-part MDL because we are specifically interested in the model: the pattern sets that yield the best compression. Although refined MDL has stronger theoretical foundations, it can only be computed in special cases. From refined MDL we will use prequential coding to encode the data without bias. Note that MDL requires the compression to be *lossless* in order to allow for fair comparison between different $M \in \mathcal{M}$.

To use MDL in practice we have to define our model class $\mathcal{M}$, how to describe a model $M \in \mathcal{M}$, and how a model M describes the data $\mathcal{D}$. In MDL we are only interested in the length of the description, and never in the encoded data. That is, we are only concerned with the *length* of the encoding, not with materialised codes.

4   MDL for the Difference and the Norm

We first informally introduce our problem, and then formalise our objective.

4.1   The Problem, Informally

Suppose we are given a bag $\mathcal{D}$ of transaction databases. Loosely speaking, by MDL we are after those patterns – itemsets – that together describe these databases best. More in particular, we want to optimally jointly characterise the database subsets U that the user specified as interesting. Our model $\mathcal{S}$ will hence consist of a *set of patterns* S_j for every $j \in U$. Every individual database $D_i \in \mathcal{D}$ will be described – characterised – using the union of all $S_j \in \mathcal{S}$ associated with D_i in the sense that i is an element of j. This allows us to associate patterns with that database subset j they are most characteristic for. Not only does this makes the overall description of the databases more efficient – no duplication is necessary – it also makes the model more insightful – if a pattern is characteristic for all databases, it will be included in the pattern set that is associated with all databases, when it is characteristic only for one database it will only be included in the pattern set associated with that particular database, etc.

We will now formally introduce our objective.

4.2   Our Models

A model $\mathcal{S}$ is a set of pattern sets $\mathcal{S} \subseteq \mathcal{P}(\mathcal{I})$, such that every $S_j \in \mathcal{S}$ is associated with one of the database subsets $j \in U$ the user identifies as interesting. To describe an individual database $D_i \in \mathcal{D}$, we consider the union of all pattern sets in $\mathcal{S}$ that are associated with D_i, and to make sure every database over $\mathcal{I}$ can be encoded without loss, we also add all singletons Formally, we write $\pi_i(\mathcal{S}) = \{S_j \in \mathcal{S} \mid i \in j\}$ for the subset of $\mathcal{S}$ relevant to D_i, and define the coding set C_i for D_i as $C_i = \mathcal{I} \cup \bigcup \pi_i(\mathcal{S})$.

4.3   Encoded Length of the Data

Next, we discuss how we describe data $\mathcal{D}$ given a model $\mathcal{S}$, and in particular how to calculate the encoded length $L(\mathcal{D} \mid \mathcal{S})$. We do so bottom up, starting by how to encode an individual transaction $t \in D$ given an arbitrary coding set C. We do so using a *cover* function $cover(t, C)$ that returns a set of patterns from C such that $\bigcup cover(t, C) = t$.

To encode the patterns in the cover of t, we will use optimal prefix codes. The length of an optimal prefix code is given by Shannon entropy [2], $-\log \Pr(X)$. To compute these lengths, we hence need the probability of a pattern X in the cover of the data. Let $usg_D(X, C) = |\{t \in D \mid X \in cover(t, C)\}|$ be the number of times a pattern $X \in C$ is used in the cover of D. Wherever clear from context we simply write $usg(X)$, and slightly abusing notation, we say $usg(C)$ for the sum of usages of coding set C, i.e. $usg(C) = \sum_{X \in C} usg(X)$. The probability of X is then $\Pr(X) = \frac{usg(X)}{\sum_{Y \in C} usg(Y)}$, and the length of its optimal prefix code $L(code(X) \mid C) = -\log \Pr(X)$.

More in particular, we will use a *prequential* coding scheme [5]. Prequential codes are Universal codes [13], which means they are asymptotically optimal *without* having to know the usages in advance. That is, unlike for KRIMP [17] we do not have to make arbitrary choices for how to encode the usages in the model – choices that may incur undue bias. The idea behind prequential coding is simple: after every received code we re-calculate all probabilities over the data received so far, initialising the usages to ϵ. This means that at any stage we have a valid probability distribution and hence can send optimal prefix codes. Surprisingly, the *order* in which we transmit codes does not affect the encoded length – a sum of logarithms is the logarithm of a product, of which we can move its terms around at will.

For the encoded length of a transaction $t \in D$ we have

$$L(t \mid C) = L_{\mathbb{N}}(|t|) + \sum_{X \in cover(t, C)} L(code(X) \mid C) \quad , \tag{0}$$

where we first encode the cardinality of the transaction, and then the patterns in its cover. For the cardinality, we use $L_{\mathbb{N}}$, the Universal code for integers [13] which for $n \geq 1$ is defined as $L_{\mathbb{N}}(n) = \log^*(n) + \log(c_0)$ with $\log^* = \log(n) + \log\log(n) + \ldots$. To make it a valid code it has to satisfy the Kraft inequality, and hence we set $c_0 = 2.865064$.

For the encoded length of a database D given a coding set C we then have

$$L(D \mid C) = L_{\mathbb{N}}(|D|) + \sum_{t \in D} L(t \mid C) \quad , \tag{0}$$

where we encode the number of transactions in D using $L_{\mathbb{N}}$ and then each of the transactions in turn. Aggregating the lengths of all prequential prefix codes, we have

$$L(D \mid C) = \left[L_{\mathbb{N}}(|D|) + \sum_{t \in D} L_{\mathbb{N}}(|t|) \right] + \left[\log \frac{\prod_{j=0}^{usg(C)-1} (j + \epsilon|C|)}{\prod_{X \in C} \prod_{j=0}^{usg(X)-1} (j + \epsilon)} \right] . \tag{0}$$

Note that the first two terms are constant for all models for the same data, and can hence be ignored during optimisation. The right hand term is the length of the data when encoded using prequential coding. By common convention, for $\epsilon = 0.5$ we have

$$L(D \mid C) = L_{\mathbb{N}}(|D|) + \sum_{t \in D} L_{\mathbb{N}}(|t|) + \log \Gamma(usg(C) + 0.5|C|) -$$

$$\log \Gamma(0.5|C|) - \sum_{X \in C} \log\left((2usg(X) - 1)!! \right) - usg(X) \quad ,$$

where !! denotes the double factorial defined as $(2k - 1)!! = \prod_{i=1}^{k}(2i - 1)$, and Γ is the Gamma function, which is an extension of the factorial function to the complex plane. That is, $\Gamma(x + 1) = x\Gamma(x)$, with relevant base cases $\Gamma(1) = 1$ and $\Gamma(0.5) = \sqrt{\pi}$. We refer the interested reader to the online appendix for more details on prequential coding and its computation. Finally, by encoding the number of databases in $\mathcal{D}$, and then simply encoding every individual database in order, we have

$$L(\mathcal{D} \mid \mathcal{S}) = L_{\mathbb{N}}(|\mathcal{J}|) + \sum_{D_i \in \mathcal{D}} L(D_i \mid C_i) \quad ,$$

for the encoded size of $\mathcal{D}$ given a model $\mathcal{S}$. This leaves discussing the encoding of $\mathcal{S}$.

4.4 Encoded Length of the Model

Let us first discuss $L(S_j)$, the encoded length of a pattern set $S_j \in \mathcal{S}$. We define

$$L(S_j) = L_{\mathbb{N}}(|S_j|) + \sum_{X \in S_j} \left(L_{\mathbb{N}}(|X|) - \sum_{x \in X} \log freq_{\mathcal{D}}(x) \right)$$

in which we first encode the number of patterns, then their cardinalities. Third, we transmit the elements of X using optimal prefix codes – allowing us to reconstruct patterns up to the names of the items – and do so using the marginal item probabilities over $\mathcal{D}$. By this choice a pattern X is equally expensive regardless of the datasets for which S_j is relevant. Note that we do not have to encode the pattern usages as we encode the data prequentially.

Finally, for the encoded length of a model $\mathcal{S}$ we have

$$L(\mathcal{S}) = L_{\mathbb{N}}(|\mathcal{I}|) + L_{\mathbb{N}}(||\mathcal{D}||) + \log \binom{||\mathcal{D}|| - 1}{|\mathcal{I}| - 1} + \sum_{S_j \in \mathcal{S}} L(S_j) \quad ,$$

where we encode the length of the alphabet, the number of items in the data, and then the support per item using an index over a canonically ordered enumeration of all possibilities of distributing $||\mathcal{D}||$ events over $|\mathcal{I}|$ labels. This cost is constant for the same data and can hence be ignored when optimising the model. It is necessary, however, if we want to compare different encodings or model classes.

4.5 The Problem, Formally

Combining the above, the total encoded length of data $\mathcal{D}$ and a model $\mathcal{S}$ is defined as

$$L(\mathcal{D}, \mathcal{S}) = L(\mathcal{S}) + L(\mathcal{D} \mid \mathcal{S}) \quad .$$

By MDL we are after the model that minimises the total encoded length. Formally, our problem definition is as follows.

Minimal Pattern Sets Problem. *Let $\mathcal{I}$ be a set of items, $\mathcal{D}$ a bag of transaction databases over $\mathcal{I}$, U a set of index sets for $\mathcal{D}$, cover a cover algorithm, and $\mathcal{F}$ the space of all admissible models, $\mathcal{F} = \mathcal{P}(\mathcal{P}(\mathcal{I}))^{|U|}$. Find the set of pattern sets $\mathcal{S} \in \mathcal{F}$ with the smallest $\bigcup \mathcal{S}$ such that the corresponding total compressed size $L(\mathcal{D}, \mathcal{S})$ is minimal.*

The search space we have to consider for this problem is rather large – even if we take into account that only patterns that occur in the data can be used to describe the data. Moreover, it does not exhibit structure we can exploit to efficiently find the optimal pattern sets, such as submodularity or (weak) monotonicity.

Hence, we resort to heuristics.

5 Algorithm

To discover good models directly from data, we propose the DIFFNORM algorithm.

5.1 The Cover Algorithm

First, however, we need a cover function $cover(t, C)$ to determine which patterns from C will be used to describe transaction t. Ideally $cover$ minimises $L(\mathcal{D}, \mathcal{S})$. However, as there exists a complex non-linear relation between the total encoded length and the individual usages of patterns, optimising the cover is non-trivial [15]. We therefore adopt the greedy heuristic successfully used in KRIMP [17]. That is, we greedily cover transaction t with non-overlapping patterns from C. We do so in **Standard Cover Order**, i.e. we consider the patterns in C sorted descending on cardinality, on support, and lexicographically. The intuition is that by doing so we need as few as possible, as frequent as possible patterns to cover t. Algorithm 1 gives the pseudo-code.

Algorithm 1. GREEDYCOVER

Input: A transaction t over items $\mathcal{I}$ and a coding set C
Output: A $cover(t, C) \subseteq C$
1 **for** $X \in C$ in **Standard Cover Order do**
2   **if** $X \subseteq t$ **then return**$\{X\} \cup cover(t \setminus X, C)$;

3 **return** $\emptyset$

5.2 The DIFFNORM Algorithm

Next we discuss the DIFFNORM algorithm. We give the pseudo code as Algorithm 2. The main intuition is that we iteratively reduce redundancy in the current description of the data by adding combinations of existing patterns. That is, we take a SLIM-like approach [14]. We start with empty pattern sets (line 1). We iteratively generate candidates in the form of $X \cup Y$ with $X, Y \in \mathcal{S} \cup \mathcal{I}$. We consider these in order of estimated gain (2). (We postpone the details of ΔL to Sec. 5.4.) Note that we can easily impose additional constraints (e.g. minimum support) to accommodate user preferences.

Per candidate, we calculate the difference in bits when adding it to the coding set for each database (line 3–4). We use these gains to determine to which pattern set(s) $S_j \in \mathcal{S}$ we will add the candidate (5–7). We do so greedily (6). We first sort the user specified index sets U descending on gain, cardinality, and last lexicographically. We iteratively pick the top-most index set, and updating the gain scores of the remaining sets by removing the gain for data sets already covered by the chosen index sets, and stop when we cannot select an index set with positive gain.

As the new pattern may have superseded the use of older ones, we have to PRUNE the model [17]. We give the pseudo-code as Algorithm 3. In a nutshell, we simply iteratively re-consider every pattern in $\mathcal{S}$ for which the usage has decreased – as these are now more expensive to encode – ordered by how much the usage has decreased. After pruning we iterate until we cannot find any patterns that improve the total encoded length. Before we return the patterns, we order them by their relative importance – the number of bits we would have to spend extra if the pattern would not be included.

5.3 Candidate Generation and Evaluation

The naive approach to optimising a model is to first mine all frequent patterns $\mathcal{F}$ in $\mathcal{D}$, and then iteratively consider these as candidates. KRAMP [14] is the locally optimal strategy of iteratively adding that $Z \in \mathcal{F}$ to the model that maximises compression. Being quadratic in the size of the candidate set, this approach is prohibitively costly. KRIMP considers these candidates in a fixed order, greedily selecting those that improve compression [17]. Considering every candidate only once and in a static order KRIMP is linear in the number of candidates, but quality suffers and as all candidates need to be pre-mined and ordered materialised the approach remains costly.

Algorithm 2. DIFFNORM

Input: A bag $\mathcal{D}$ of transaction databases over items $\mathcal{I}$, and a database index set U including at least the individual indices over $\mathcal{D}$

Output: An approximation of the MDL-optimal model $\mathcal{S}$ for $\mathcal{D}$

1 $\mathcal{S} \leftarrow \{\emptyset \mid j \in U\}$;
2 **for** $Z \in \{X \cup Y \mid X, Y \in \mathcal{S} \cup \mathcal{I}\}$ descending on $\Delta\widehat{L}(\mathcal{D}, \mathcal{S} \oplus Z)$ **do**
3   **for** $D_i \in \mathcal{D}$ **do**
4          $gain_i \leftarrow \Delta L(D_i \mid C_i \oplus Z)$;
5      $w \leftarrow \{\Delta L(\mathcal{D}, \mathcal{S} \oplus_j Z) \mid j \in U\}$;
6      $U' \leftarrow$ WEIGHTEDGREEDYCOVER$(\mathcal{J}, U, w)$;
7      $\mathcal{S}' \leftarrow \mathcal{S}$ with Z added to every S_j with $j \in U'$;
8      $\mathcal{S} \leftarrow$ PRUNE$(\mathcal{D}, \mathcal{S}, \mathcal{S}')$;
9 Order every $S_j \in \mathcal{S}$ descending on $\Delta L(\mathcal{D}, \mathcal{S} \ominus_j Z)$;
10 **return** $\mathcal{S}$;

Algorithm 3. PRUNE

Input: A bag $\mathcal{D}$ of databases over $\mathcal{I}$, a previous model $\mathcal{S}$ and a current model $\mathcal{T}$

Output: A pruned model $\mathcal{T}$

1 $Cands \leftarrow$ all patterns $X \in \mathcal{T}$ for which $usg(X, \mathcal{T}) < usg(X, \mathcal{S})$;
2 **for** $X \in Cands$ in **Standard Pruning Order do**
3   **if** $L(\mathcal{D}, \mathcal{T} \ominus X) < L(\mathcal{D}, \mathcal{T})$ **then**
4          $\mathcal{T} \leftarrow \mathcal{T} \ominus X$;
5   Add all patterns $Y \in \mathcal{T}$ for which usg reduced to $Cands$;
6 **return** $\mathcal{T}$;

Instead, we can iteratively refine the current model by searching for redundancies. Translated to our setting, SLAM [14] is the locally optimal approach. It iteratively evaluates all pairwise combinations $X, Y \in \mathcal{S} \cup \mathcal{I}$, accepting that $X \cup Y$ which maximises compression. SLIM [14] considers the same candidates, but evaluates these in order of estimated quality, accepting the first that improves compression. This leads to much improved run time and overall description length close to SLAM.

Loosely speaking DIFFNORM follows the same adage as SLIM. However, unlike SLIM, we consider multiple pattern sets – each of which relevant to different set of databases. When we extend SLIM naively, we would generate overly many candidates and evaluate them on by far too many pattern sets and databases. To refine this process we make use of the fact that MDL punishes redundancy – which means that patterns will only be included in pattern sets they are most relevant for.

First we adapt the candidate generation process. We observe that it is very unlikely that $X \cup Y$ will be used much when X and Y are drawn from pattern sets that are not used to describe the same database. This observation allows us to refine the SLIM strategy as follows. Instead of considering all pairs $X, Y \in \mathcal{S} \cup \mathcal{I}$, we consider only $X \cup Y$ if they co-occur in a coding set C for a database D. Formally, we consider only $X \cup Y$ for $X \in S_j$ and $Y \in S_k$ with $j \cap k \neq \emptyset$ as candidates.

Next, we take a closer look at the candidate evaluation process. When we consider a pattern $X \cup Y$ with X from a pattern set S_j that is more 'specific' than the pattern set S_k that we draw Y from, that is, $j \subset k$, it will be very unlikely that $X \cup Y$ will be a good candidate to add to S_k – otherwise, X would have resided in S_k. We use this intuition and in these cases only consider to add this candidate to S_j, not to S_k. More in general, we evaluate the candidate in all $S_l \in S$ with $l \subseteq j$. When j and k overlap, but j is not a strict subset of k, we evaluate the candidate in all $S_l \in S$ with $l \subset j$ or $l \subset k$.

5.4 Estimating Candidate Quality

As we aim to minimise the description length, the quality of a candidate Z is the gain in total compressed size when we would add Z to pattern set $S_j \in S$, i.e. $\Delta L(D, S \oplus_j Z)$. Formally,

$$\Delta L(D, S \oplus_j Z) = L(D, S) - L(D, S \oplus_j Z)$$
$$= \Delta L(S_j \oplus Z) + \sum_{i \in j} \Delta L(D_i \mid C_i \oplus Z))$$
$$= L(S_j) - L(S_j \oplus Z) + \sum_{i \in j} L(D_i \mid C_i) - L(D_i \mid C_i \oplus Z))$$

Note that $\Delta L(D_i \mid C_i \oplus Z)$ is constant regardless to which pattern set $S_j \in S$ we add Z – as long as i is in the index set j. Calculating the actual gain for every candidate is prohibitively costly, however – we need to cover all relevant databases to re-determine the usages. Instead, we therefore estimate the gain in bits when adding a pattern Z to pattern set S_j, i.e. $\Delta\widehat{L}(D, S \oplus_j Z)$. We then use WEIGHTEDGREEDYCOVER to get the total estimated gain, $\Delta\widehat{L}(D, S \oplus Z)$, from $\Delta\widehat{L}(D, S \oplus_j Z) \ \forall \ j \in U$. To this end we assume that as candidate we consider the union of patterns $X, Y \in S \cup I$, and that adding $X \cup Y$ to pattern S_j will affect only the usages of X and Y and not that of other patterns in S. Formally, we have

$$\Delta\widehat{L}(D, S \oplus_j X \cup Y) = \Delta\widehat{L}(S_j \oplus X \cup Y) + \sum_{i \in j} \Delta\widehat{L}(D_i \mid C_i \oplus X \cup Y) \quad ,$$

where for the estimated difference in encoded length of S_j we have

$$\Delta\widehat{L}(S_j \oplus X \cup Y) = L(S_j) - L(S_j \oplus X \cup Y)$$
$$= L_{\mathbb{N}}(|X \cup Y|) - \sum_{x \in XUY} \log \mathit{freq}_D(x) \quad .$$

Somewhat more intimidating, for the estimated encoded length of the data we
have

$$\Delta\widehat{L}(D_i \mid C_i \oplus X \cup Y) = \log(\Gamma(usg(C) + \epsilon|C|)) - \log(\Gamma(\widehat{usg}(C') + \epsilon|C'|)) +$$
$$\log(\Gamma(\widehat{usg}(X, C') + \epsilon)) - \log(\Gamma(usg(X, C) + \epsilon)) +$$
$$\log(\Gamma(\widehat{usg}(Y, C') + \epsilon)) - \log(\Gamma(usg(Y, C) + \epsilon)) +$$
$$\log(\Gamma(\widehat{usg}(X \cup Y, C') + \epsilon)) - \log(\Gamma(\epsilon)) +$$
$$\log(\Gamma(\epsilon|C'|)) - \log(\Gamma(\epsilon|C|))$$

were $C' = C \cup \{X \cup Y\}$, and $\widehat{usg}(Z, C')$ is the estimation of the usage of
pattern Z when covering the data using C'. We estimate the usage of $X \cup Y$
optimistically, assuming it will be used wherever X and Y were co-used. That
is, we say $\widehat{usg}(X \cup Y, C') = |utids(X) \cap utids(Y)|$, where $utids(X) = \{tid(t) \mid t \in D, X \in cover(t, C)\}$ are the ids of the transactions covered using X. Following
the same assumption, we have $\widehat{usg}(X, C') = usg(X, C) - \widehat{usg}(X \cup Y, C')$, and
analogue for Y.

Since we only generate and evaluate the candidates against their relevant coding sets C_i, we do the same when estimating gain. Further, to avoid re-computing
all estimates at every iteration we cache the estimated gains of patterns. However, whenever a candidate Z is added to or pruned from S the usages of other
patterns $X \in S$ may change – and hence so should the estimates of any candidates that use X. We re-estimate the gains of these candidates, and maintain
those for the other candidates.

5.5   Complexity

Finally, we analyse the computational complexity of DiffNorm. In worst
case, a model S contains all the frequent patterns $\mathcal{F}$. Let $|S|$ be the total
number of patterns in model S. At worst, generating the candidates takes
$\mathcal{O}((|S| + |\mathcal{I}|)^2) \subseteq \mathcal{O}(|\mathcal{F}|^2)$ steps. Calculating the gain takes $\mathcal{O}(|S|) \subseteq \mathcal{O}(|\mathcal{F}|)$
steps. WeightedGreedyCover takes $\mathcal{O}(|U| \times \log |U|)$ steps for sorting U and
$\mathcal{O}(|U|^2)$ steps for greedy selection and gain re-computation. Finally, Prune
takes $\mathcal{O}(|S|^2 \times |\mathcal{D}|)$ steps. Altogether, the worst case computational complex-
ity is $\mathcal{O}(|\mathcal{F}|^3 \times |\mathcal{D}|)$. In practice, DiffNorm is fast. First, MDL restricts the
number of patterns in the model, pruning keeps the model non-redundant, and
model changes rarely affect many patterns. Second, we generate candidates not
naively from S but over coding sets C, and evaluate candidates only on the
relevant databases D_i.

6   Experiments

We implemented our algorithm in C++ and provide the source code for the
research purposes, along with the used datasets, and synthetic dataset generator.[2] All experiments were executed single-threaded on Intel Xeon E5-2643 v3

---

[2] http://eda.mmci.uni-saarland/diffnorm/

Table 1. Base statistics of the datasets used in the experiments. We report the number of rows, the size of the alphabet, the total number of items, and the number of databases.

Dataset	$|\mathcal{D}|$	$|\mathcal{I}|$	$\|\mathcal{D}\|$	$|\mathcal{J}|$
Adult	48 842	97	726 165	2
ChessBig	28 056	58	196 392	18
Nursery	12 960	32	116 640	5
Mushroom	8 124	119	186 852	2
PageBlocks	5 473	44	60 203	5
Led7	3 200	24	25 600	6
Chess	3 196	75	118 252	2

machines with 256 GB memory running Linux. We report the wall-clock running times.

We consider both synthetic and real-world data. We give the basic statistics of the real-world datasets in Table 1. For each dataset we give the number of rows, size of the alphabet $|\mathcal{I}|$, total number of items and number of databases. For readability we use the shorthand notation $L\% = \frac{L(\mathcal{D},\mathcal{S})}{L(\mathcal{D},\mathcal{S}_0)}\%$ for the relative compressed size of $\mathcal{D}$ with $\mathcal{S}_0$ the model consisting of only empty pattern sets – lower is better.

In all experiments we consider $U = \{\{1\}, \ldots, \{|\mathcal{D}|\}, \Omega\}$ where $\Omega = \{1, \ldots, |\mathcal{D}|\}$. That is, we want a pattern set S_i per individual $D_i \in \mathcal{D}$, and in addition we want to have a pattern set S_Ω that contains the patterns characteristic to all databases in $\mathcal{D}$.

6.1 Synthetic Data

First, we consider synthetic data to study the behaviour of DIFFNORM on data with known ground truth. We divide the possible data into four categories: data with no patterns included, data where patterns are local to individual databases, data where patterns occur globally in every database, and data where we mix global and local patterns, i.e. data containing both local and global patterns.

For each setup we generate a $\mathcal{D}$ of two databases of 5 000 rows each over 120 items. We randomly plant non-overlapping patterns of cardinality uniformly chosen over the range of 4 to 8, with random frequency over the range 10% to 30%. In addition, we add 5% uniform noise. We run DIFFNORM with a minimum support of 4.5%. Table 2 shows the result of DIFFNORM per synthetic dataset, i.e. number of planted patterns, the total encoded size given the simplest model $\mathcal{S}_0$, relative compressed size $L\%$. Further, following [20], we report the number of exactly recovered patterns, the number of discovered patterns that are unions or subsets of unions of planted patterns, the number of discovered patterns that correspond to intersections between planted patterns, and the number of patterns that are tainted with, or completely due to noise.

Table 2. DIFFNORM **recovers true patterns.** Results on synthetic data. Per dataset we give the number of planted patterns, the baseline description length, and the relative compression $L\%$ we obtain. Further, we report the total number of patterns DIFFNORM discovers, and break this down into the number of exactly recovered patterns ($=$), the number of discovered patterns that are (subsets of) unions of planted patterns ($\cup$), the number of discovered patterns that are intersections of planted patterns ($\cap$), the number of patterns unrelated to planted patterns (?).

Dataset	# planted	$L(\mathcal{D}, \mathcal{S}_0)$	$L\%$	$\lvert\mathcal{S}\rvert$	$=$	$\cup$	$\cap$	?
				Discovered Patterns				
Random	0	387 201	100	0	0	0	0	0
Local	16	587 557	22.1	17	16	1	0	0
Global	10	1 180 530	22.4	17	10	7	0	0
Mixture	19	1 008 546	11.9	29	19	10	0	0

We find that for *Random*, DIFFNORM correctly infers that the data does not contain any patterns. As for the other datasets, DIFFNORM discovers exactly all the planted patterns. In addition, DIFFNORM discovers patterns that are the union, or a subset thereof, of planted patterns X and Y – this is due to generative process. As we allow multiple patterns on the same row, particularly when X and Y are very frequent their combination can also become frequent, and therewith descriptive for the data. Overall, DIFFNORM identifies all the interesting patterns from the synthetic datasets.

6.2   Real World Data

Next, we investigate the performance of DIFFNORM on real-world data. In particular, we first consider seven datasets from Frequent Itemset Mining Implementations (FIMI) repository[3] For these experiments we set a minimum support of 2. The result on FIMI datasets is given in Table 3. We see that DIFFNORM is efficient, requiring only seconds for these datasets. Moreover, we find that it achieves very good compression ratios (lower is better), and returns only modest numbers of patterns.

This leaves us to compare these numbers. This is more difficult than it may seem. For starters, comparing description lengths only makes sense when we consider the same model class and exactly the same data. Comparing on the number of discovered patterns is not trivial either. Comparing to the number of (closed) frequent itemsets [11] is not fair as it is not meant to give a summary of the data. Supervised methods [21] only report patterns that set classes apart, and do not describe the data awhole. Summarisation methods such as SLIM [14] do give succinct description per database, but lack a way to identify patterns common between the databases. Considering the number of patterns discovered summed

---

[3] http://fimi.ua.ac.be/data/

Table 3. DIFFNORM discovers succinct descriptions. Results of DIFFNORM on the real data sets. For DIFFNORM we give the baseline compression cost $L(\mathcal{D}, \mathcal{S}_0)$, the relative compressed size $L\%$ (lower is better), the wall-clock time in seconds, and the total number of discovered patterns. For comparison, in addition we report the number of patterns DIFFNORM and SLIM [14] discover when we concatenate all databases into $\mathcal{D}_\cup = \bigcup \mathcal{D}$.

	DIFFNORM ($\mathcal{D}$)				DN($\mathcal{D}_\cup$)	SLIM ($\mathcal{D}_\cup$)
Dataset	$L(\mathcal{D}, \mathcal{S}_0)$	$L\%$	*time (s)*	$|\mathcal{S}|$	$|\mathcal{S}|$	$|\mathcal{S}|$
Adult	3 237 869	25.6	73.5	757	782	2702
ChessBig	838 358	75.3	11	899	769	1420
Nursery	471 928	58.3	6.5	294	371	308
Mushroom	1 020 100	25.8	17	442	435	1667
PageBlocks	186 765	4.3	0	26	48	105
Led7	68 034	34.2	0	80	56	194
Chess	660 914	20.6	7.5	265	264	653

over all databases would be hugely inflated. Most fair, we find is to compare to the number of patterns discovered over the whole data, i.e. over $\mathcal{D}_\cup = \bigcup \mathcal{D}$. We run both DIFFNORM and SLIM on this database with minsup 2 and report the number of discovered patterns. We see that DIFFNORM discovers roughly the same number of patterns as before, while SLIM on the other hand generally finds many more patterns. This is likely due to overfitting as its encoding scheme does not encode pattern lengths and codes without loss.

Next, we investigate how the patterns that DIFFNORM discovers are distributed over the different databases. That is, in Figure 1 we show the sizes of the discovered pattern sets, starting with the size S_Ω, the pattern set associated with all databases, and then the sizes of each of the S_i corresponding to $D_i \in \mathcal{D}$. We see that the patterns are nicely distributed over $\mathcal{S}$, it is not the case that all patterns are in either the global, or just in the local pattern sets.

We proceed to evaluate how well DIFFNORM optimises our objective. As we do not know the true optimum, we look at how the relative compression $L\%$ converges over the search iterations. An iteration here refers to the event when a pattern is accepted by DIFFNORM. As shown in Figure 2(a), for the *Adult* dataset, the relative compression reduces very sharply in the beginning and after a certain number of iterations it converges more slowly as it then needs to refine the more general patterns discovered in the first iterations. Note that as we PRUNE the number of iterations and the final number of patterns differ.

DIFFNORM relies heavily on the quality of estimating ΔL. We evaluate the quality of our estimate by checking how well $\Delta \widehat{L}(\mathcal{D}, \mathcal{S} \oplus X \cup Y)$ correlates to the actual gain $\Delta L(\mathcal{D}, \mathcal{S} \oplus X \cup Y)$. We consider *Adult* and plot the estimated and actual gain for all the candidates considered by DIFFNORM in Figure 2(b). We see that the two are strongly correlated, as most of the points lie along the

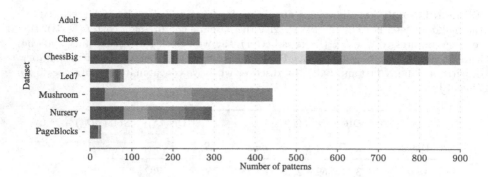

Fig. 1. DIFFNORM **discovers both local and global patterns.** For FIMI datasets, we show the number of patterns in each pattern set discovered by DIFFNORM as indicated by the width of each colored box. The leftmost purple colored box indicates the global pattern set S_Ω.

diagonal, particularly those for high gain candidates. This also explains the shape of the convergence curve in Figure 2(a) – candidates with higher estimated gains are tested during early stages of the algorithm. We find that for lower estimated gains the correlation is weaker. This is explained by our assumption that all usages of all patterns in S remain constant, except for X and Y.

In our final experiment, we evaluate DIFFNORM qualitatively. For this, we consider the *ICDM* dataset.[4] This data consists of the abstracts – stemmed and stop-word removed – of 859 papers published at ICDM. We divide the data into two classes: one of the abstracts that do contain the word *mining* (359 rows) versus the remainder (500 rows). With the minimal support set to 5, DIFFNORM takes 71.5 seconds, and discovers 637 patterns in total, 35 for the first class, 54 for the second, and 548 in S_Ω. As expected, we find that the patterns found in abstracts containing *mining* point more towards exploratory analysis. The patterns discovered from abstracts not containing *mining* point more towards machine learning. On the global patterns, we find commonly used phrases in research papers like "state of the art", "evaluation", etc. We give 5 highly characteristic exemplars drawn from the top-10 of each pattern set in Table 4.

7   Discussion

The experiments show that DIFFNORM works well in practice. On synthetic data we recover all planted patterns exactly, and returns these top-ranked in the output. On the real world data DIFFNORM discovers succinct descriptions, returning on average less than half as many patterns as SLIM [14]. Moreover, the ICDM abstracts data show that the results are clean and easily interpretable. Finally,

---

[4] Available from the authors of [3].

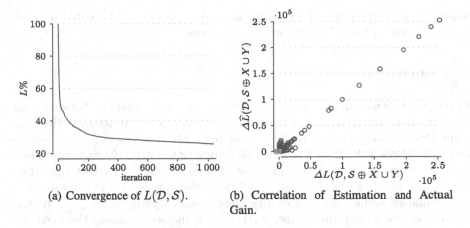

(a) Convergence of $L(\mathcal{D}, \mathcal{S})$.

(b) Correlation of Estimation and Actual Gain.

Fig. 2. DIFFNORM searches efficiently and estimates accurately. For *Adult* we show (left) the convergence of the relative compression $L\%$ per search iteration, and (right) the correlation between the estimated and actual gains of candidates. Candidates marked as circles were accepted whereas those marked as crosses were rejected.

Table 4. DIFFNORM finds meaningful patterns. Results of DIFFNORM on the *ICDM* dataset when split on abstracts including the word 'mining' and those that do not. Shown are five patterns per pattern set, where S_Ω is the pattern set associated with both databases.

S_{mining}	$S_{\neg\text{mining}}$	S_Ω
association rule large database	accuracy learn work	algorithm exp. problem result
fp tree	svm machine	framework general model
prune previous	cluster partition	method large set
strategy freq. pattern discover	classifier train	state [of the] art
support threshold	approach learn	evaluation technique

we showed that DIFFNORM efficiently optimises its objective score thanks to effective quality estimation of candidates.

Although these results are very encouraging, we see many possibilities to further improve DIFFNORM. A particular strong point is that our concept of multiple pattern sets and prequential coding can be extended to other data and pattern types, such as serial episodes [15]. Moreover, it will make for engaging future work to extend DIFFNORM such that it can automatically discover the optimal U for a given set of databases, and/or simultaneously find the optimal partitioning of a single given database.

Last, we have to note that MDL is not a magic wand. That is, even though we use prequential coding our objective function involves choices, and so does the optimisation. Currently we encode patterns in the pattern sets using the global singleton frequencies. In certain settings it may more sense to use the frequencies over the relevant databases instead. Extending DIFFNORM to allow for overlap would likely lead to even more succinct descriptions.

8　Conclusion

We studied how we can characterise the differences and similarities between a set of databases using pattern sets. We formalised the problem in terms of the Minimum Description Length principle [5], defining the best set of pattern sets as the one that gives the most succinct description of the data. To find good models directly from data we introduced the parameter-free DIFFNORM algorithm. Empirical evaluation showed that DIFFNORM discovers easily interpretable and non-redundant summaries that clearly identify which patterns are globally, and which ones are locally important. Future work includes refining the encoding and extending towards other data and pattern types, as well as exploring how well the patterns DIFFNORM selects perform in classification.

Acknowledgments. The authors thank the anonymous reviewers for detailed comments. Kailash Budhathoki and Jilles Vreeken are supported by the Cluster of Excellence "Multimodal Computing and Interaction" within the Excellence Initiative of the German Federal Government.

References

1. Aggarwal, C.C., Han, J. (eds.): Frequent Pattern Mining. Springer (2014)
2. Cover, T.M., Thomas, J.A.: Elements of Information Theory. Wiley-Interscience New York (2006)
3. De Bie, T.: Maximum entropy models and subjective interestingness: an application to tiles in binary databases. Data Min. Knowl. Disc. **23**(3), 407–446 (2011)
4. Geerts, Floris, Goethals, Bart, Mielikäinen, Taneli: Tiling databases. In: Suzuki, Einoshin, Arikawa, Setsuo (eds.) DS 2004. LNCS (LNAI), vol. 3245, pp. 278–289. Springer, Heidelberg (2004)
5. Grünwald, P.: The Minimum Description Length Principle. MIT Press (2007)
6. Kolmogorov, A.N.: Three approaches to the quantitative definition of information. Problemy Peredachi Informatsii **1**(1), 3–11 (1965)
7. Li, M., Vitányi, P.: An Introduction to Kolmogorov Complexity and its Applications. Springer (1993)
8. Mampaey, M., Vreeken, J., Tatti, N.: Summarizing data succinctly with the most informative itemsets. ACM TKDD **6**, 1–44 (2012)
9. Miettinen, P.: On finding joint subspace Boolean matrix factorizations. In: SDM, pp. 954–965. SIAM (2012)
10. Nijssen, P., Guns, T., De Raedt, L.: Correlated itemset mining in ROC space: a constraint programming approach. In: KDD, pp. 647–656. Springer (2009)
11. Pasquier, N., Bastide, Y., Taouil, R., Lakhal, L.: Discovering frequent closed itemsets for association rules. In: Beeri, C., Bruneman, P. (eds.) ICDT 1999. LNCS, vol. 1540, pp. 398–416. Springer, Heidelberg (1998)
12. Rissanen, J.: Modeling by shortest data description. Automatica **14**(1), 465–471 (1978)

13. Rissanen, J.: A universal prior for integers and estimation by minimum description length. Annals Stat. **11**(2), 416–431 (1983)
14. Smets, K., Vreeken, J.: SLIM: Directly mining descriptive patterns. In: SDM, pp. 236–247. SIAM (2012)
15. Nikolaj, T., Jilles, V.: The long and the short of it: Summarizing event sequences with serial episodes. In: KDD. ACM (2012)
16. Vreeken, J., van Leeuwen, M., Siebes, A.: Characterising the difference. In: KDD, pp. 765–774 (2007)
17. Vreeken, J., van Leeuwen, M., Siebes, A.: KRIMP: Mining itemsets that compress. Data Min. Knowl. Disc. **23**(1), 169–214 (2011)
18. Wallace, C.S.: Statistical and inductive inference by minimum message length. Springer (2005)
19. Wallace, C.S., Boulton, D.M.: An information measure for classification. Comput. J. **11**(1), 185–194 (1968)
20. Webb, G., Vreeken, J.: Efficient discovery of the most interesting associations. ACM TKDD **8**(3), 1–31 (2014)
21. Zimmermann, A., Nijssen, S.: Supervised pattern mining and applications to classification. In: Aggarwal, C.C., Han, J. (eds.) Frequent Pattern Mining, pp. 425–442. Springer (2014)

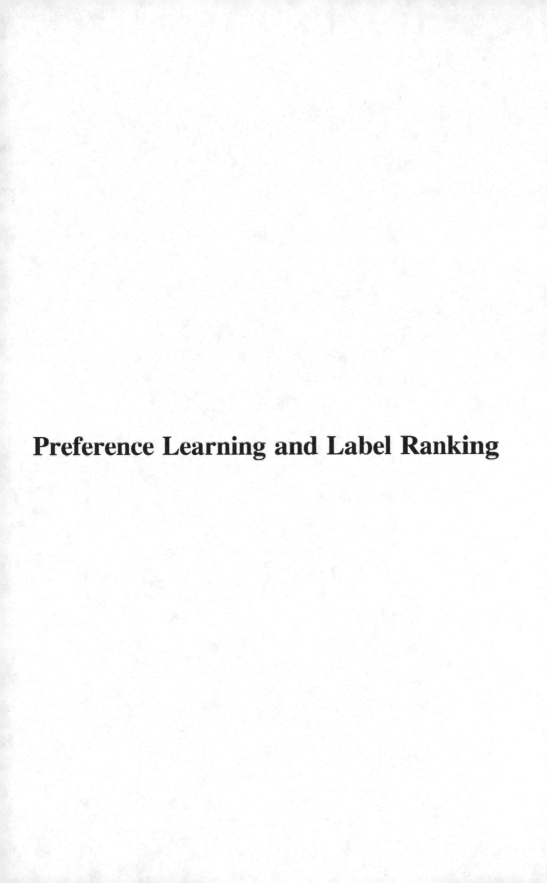

Preference Learning and Label Ranking

Dyad Ranking Using A Bilinear Plackett-Luce Model

Dirk Schäfer[1] and Eyke Hüllermeier[2]([⊠])

[1] University of Marburg, Marburg, Germany
dirk.schaefer@uni-marburg.de
[2] Department of Computer Science, University of Paderborn, Paderborn, Germany
eyke@upb.de

Abstract. Label ranking is a specific type of preference learning problem, namely the problem of learning a model that maps instances to rankings over a finite set of predefined alternatives. These alternatives are identified by their name or *label* while not being characterized in terms of any properties or features that could be potentially useful for learning. In this paper, we consider a generalization of the label ranking problem that we call *dyad ranking*. In dyad ranking, not only the instances but also the alternatives are represented in terms of attributes. For learning in the setting of dyad ranking, we propose an extension of an existing label ranking method based on the Plackett-Luce model, a statistical model for rank data. Moreover, we present first experimental results confirming the usefulness of the additional information provided by the feature description of alternatives.

Keywords: Label ranking · Plackett-Luce model · Meta-learning

1 Introduction

Preference learning is an emerging subfield of machine learning, which deals with the induction of preference models from observed or revealed preference information [7]. Such models are typically used for prediction purposes, for example, to predict context-dependent preferences of individuals on various choice alternatives. Depending on the representation of preferences, individuals, alternatives, and contexts, a large variety of preference models are conceivable, and many such models have already been studied in the literature.

A specific type of preference learning problem is the problem of *label ranking*, namely the problem of learning a model that maps instances to rankings (total orders) over a finite set of predefined alternatives [26]. An instance, which defines the context of the preference relation, is typically characterized in terms of a set of attributes or features; for example, an instance could be a person described by properties such as sex, age, income, etc. As opposed to this, the alternatives to be ranked, e.g., the political parties of a country, are only identified by their name (label), while not being characterized in terms of any properties or features.

© Springer International Publishing Switzerland 2015
A. Appice et al. (Eds.): ECML PKDD 2015, Part II, LNAI 9285, pp. 227–242, 2015.
DOI: 10.1007/978-3-319-23525-7_14

In this paper, we introduce *dyad ranking* as a generalization of the label ranking problem. In dyad ranking, not only the instances but also the alternatives are represented in terms of attributes. For learning in the setting of dyad ranking, we propose an extension of an existing label ranking method based on the Plackett-Luce model, a statistical model for rank data.

The paper is organized as follows. In the next section, we introduce the problem of dyad ranking. Following a discussion of related work in Section 3, we then propose the aforementioned method for dyad ranking in Section 4. In Section 5, we present first experimental results, both for synthetic data and a case study in meta-learning, confirming the usefulness of the additional feature information of alternatives. The paper ends with some concluding remarks in Section 6.

2   Dyad Ranking

As will be explained in more detail later on (cf. Section 3), the learning problem addressed in this paper has connections to several existing problems in the realm of preference learning. In particular, it can be seen as a combination of *dyadic prediction* [19–21] and *label ranking* [26], hence the term "dyad ranking". Since our method for tackling this problem is an extension of a label ranking method, we will introduce dyad ranking here as an extension of label ranking.

2.1   Label Ranking

Let $\mathcal{Y} = \{y_1, \ldots, y_K\}$ be a finite set of (choice) alternatives; adhering to the terminology commonly used in supervised machine learning, and accounting for the fact that label ranking can be seen as an extension of multi-class classification, the y_i are also called *class labels* or simply *labels*. We consider total order relations $\succ$ on $\mathcal{Y}$, that is, complete, transitive, and antisymmetric relations, where $y_i \succ y_j$ indicates that y_i precedes y_j in the order. Since a ranking can be seen as a special type of preference relation, we shall also say that $y_i \succ y_j$ indicates a preference for y_i over y_j. We interpret this order relation in a wide sense, so that $a \succ b$ can mean that the alternative a is more liked that alternative b by a person, but also for example that an algorithm a outperforms algorithm b.

Formally, a total order $\succ$ can be identified with a permutation π of the set $[K] = \{1, \ldots, K\}$, such that $\pi(i)$ is the index of the label on position i. We denote the class of permutations of $[K]$ (the symmetric group of order K) by $\mathbb{S}_K$. By abuse of terminology, though justified in light of the above one-to-one correspondence, we refer to elements $\pi \in \mathbb{S}_K$ as both permutations and rankings.

In the setting of label ranking, preferences on $\mathcal{Y}$ are "contextualized" by instances $\boldsymbol{x} \in \mathbb{X}$, where $\mathbb{X}$ is an underlying instance space. Thus, each instance $\boldsymbol{x}$ is associated with a ranking $\succ_{\boldsymbol{x}}$ of the label set $\mathcal{Y}$ or, equivalently, a permutation $\pi_{\boldsymbol{x}} \in \mathbb{S}_K$. More specifically, since label rankings do not necessarily depend on instances in a deterministic way, each instance $\boldsymbol{x}$ is associated with a probability distribution $\mathbf{P}(\cdot \mid \boldsymbol{x})$ on $\mathbb{S}_K$. Thus, for each $\pi \in \mathbb{S}_K$, $\mathbf{P}(\pi \mid \boldsymbol{x})$ denotes the probability to observe the ranking π in the context specified by $\boldsymbol{x}$.

As an illustration, suppose $\mathbb{X}$ is the set of people characterized by attributes such as sex, age, profession, and marital status, and labels are music genres: $\mathcal{Y} = \{\text{Rock}, \text{Pop}, \text{Classic}, \text{Jazz}\}$. Then, for $x = (m, 30, \text{teacher}, \text{married})$ and $\pi = (2, 1, 3, 4)$, $\mathbf{P}(\pi \,|\, x)$ denotes the probability that a 30 years old married man, who is a teacher, prefers Pop music to Rock to Classic to Jazz.

The goal in label ranking is to learn a "label ranker", that is, a model

$$\mathcal{M} : \mathbb{X} \longrightarrow \mathbb{S}_K$$

that predicts a ranking π for each instance x given as an input. More specifically, seeking a model with optimal prediction performance, the goal is to find a risk (expected loss) minimizer

$$\mathcal{M}^* \in \operatorname*{argmin}_{\mathcal{M} \in \mathbf{M}} \int_{\mathbb{X} \times \mathbb{S}_K} \mathcal{L}(\mathcal{M}(x), \pi) \, d\mathbf{P} \; ,$$

where $\mathbf{M}$ is the underlying model class, $\mathbf{P}$ is the joint measure $\mathbf{P}(x, \pi) = \mathbf{P}(x)\mathbf{P}(\pi \,|\, x)$ on $\mathbb{X} \times \mathbb{S}_K$ and $\mathcal{L}$ is a loss function on $\mathbb{S}_K$.

As training data $\mathcal{D}$, a label ranker uses a set of instances x_n ($n \in [N]$), together with information about the associated rankings π_n. Ideally, complete rankings are given as training information, i.e., a single observation is a tuple of the form $(x_n, \pi_n) \in \mathbb{X} \times \mathbb{S}_K$. From a practical point of view, however, it is important to allow for incomplete information in the form of a ranking of some but not all of the labels in $\mathcal{Y}$:

$$y_{\pi(1)} \succ_x y_{\pi(2)} \succ_x \cdots \succ_x y_{\pi(J)} \; , \tag{1}$$

where $J \leq K$ and $\{\pi(1), \ldots, \pi(J)\} \subset [K]$. For example, for an instance x, it might be known that $y_2 \succ_x y_1 \succ_x y_5$, while no preference information is given about the labels y_3 or y_4.

2.2 Dyad Ranking as an Extension of Label Ranking

In the setting of label ranking as introduced above, instances are supposed to be characterized in terms of properties—typically, an instance is represented as an r-dimensional feature vector $x = (x_1, \ldots, x_r)$. As opposed to this, the alternatives to be ranked, the labels y_i, are only identified by their name, just like categories in classification.

Needless to say, a learner may benefit from knowledge about properties of the alternatives, too. In fact, if the preferences of an instance are somehow connected to such properties, then alternatives with similar properties should also be ranked similarly. In particular, by sharing information via features, it would in principle be possible to rank alternatives that have never be seen in the training process so far.

Returning to our above example of ranking music genres, suppose we know (or at least are quite sure) that Rock $\succ_x$ Classic $\succ_x$ Jazz for a person x. We would then expect that Pop is ranked more likely close to the top than close to the bottom, simply because Pop music is more similar to Rock than to Classic or Jazz. In contrast to a label ranker, for which the music genres are just uninformative names, we are able to make a prediction of that kind thanks to our knowledge about the different types of music.

Given that useful properties of alternatives are indeed often available in practice, we introduce dyad ranking as an extension of label ranking, in which alternatives are elements of a feature space:

$$\boldsymbol{y} = (y_1, y_2, \ldots, y_c) \in \mathbb{Y} = \mathbb{Y}_1 \times \mathbb{Y}_2 \times \cdots \times \mathbb{Y}_c \tag{2}$$

Then, a *dyad* is a pair

$$\boldsymbol{z} = (\boldsymbol{x}, \boldsymbol{y}) \in \mathbb{Z} = \mathbb{X} \times \mathbb{Y} \tag{3}$$

consisting of an instance $\boldsymbol{x}$ and an alternative $\boldsymbol{y}$. We assume training information to be given in the form of rankings

$$\rho_i : \boldsymbol{z}^{(1)} \succ \boldsymbol{z}^{(2)} \succ \ldots \succ \boldsymbol{z}^{(M_i)} \tag{4}$$

of a finite number of dyads, where M_i is the length of the ranking. Typically, though not necessarily, all dyads in (11) share the same context $\boldsymbol{x}$, i.e., they are all of the form $\boldsymbol{z}^{(j)} = (\boldsymbol{x}, \boldsymbol{y}^{(j)})$; in this case, (11) can also be written

$$\rho_i : \boldsymbol{y}^{(1)} \succ_{\boldsymbol{x}} \boldsymbol{y}^{(2)} \succ_{\boldsymbol{x}} \cdots \succ_{\boldsymbol{x}} \boldsymbol{y}^{(M_i)} . \tag{5}$$

Likewise, a prediction problem will typically consist of ranking a subset

$$\left\{ \boldsymbol{y}^{(1)}, \boldsymbol{y}^{(2)}, \ldots, \boldsymbol{y}^{(M)} \right\} \subseteq \mathbb{Y}$$

in a given context $\boldsymbol{x}$. Given a *dyad ranker*, i.e., a model that produces a ranking of dyads as an output, this can be accomplished by applying that ranker to the set of dyads

$$\left(\boldsymbol{x}, \boldsymbol{y}^{(1)} \right), \left(\boldsymbol{x}, \boldsymbol{y}^{(2)} \right), \ldots, \left(\boldsymbol{x}, \boldsymbol{y}^{(M)} \right)$$

and then projecting the result to the alternatives, i.e., transforming a ranking of the form (11) into one of the form (5). This setting, which generalizes label ranking in the sense that additional information in the form of feature vectors is provided for the labels, is the main subject of this paper and will subsequently be referred to as *contextual dyad ranking*.

3   Related Work

As already mentioned earlier, the problem of dyad ranking is not only connected to label ranking, but also to several other types of ranking and preference learning problems that have been discussed in the literature. Although a comprehensive review of related work is beyond the scope of this paper, we shall give a brief overview in this section.

The term "dyad ranking" derives from the framework of *dyadic prediction* as introduced by Menon and Elkan [20]. This framework can be seen as a generalization of the setting of collaborative filtering (CF), in which *row-objects* (e.g., clients) are distinguished from *column-objects* (e.g., products). Moreover, with each combination of such objects, called a dyad by Menon and Elkan, a value (e.g., a rating) is associated. While in CF, row-objects and column-objects are only represented by their name (just like the alternatives in label ranking), they are allowed to have a feature representation (called side-information) in dyadic prediction. Menon and Elkan are trying to exploit this information to improve performance in matrix completion, i.e., predicting the values for those object combinations that have not been observed so far, in very much the same way as we are trying to make use of feature information in the context of label ranking.

Methods for *learning-to-rank* or *object ranking* [6,13] have received a lot of attention in the recent years, especially in the field of information retrieval. In general, the goal is to learn a *ranking function* that accepts a subset $\mathbf{O} \subset \mathbb{O}$ of objects as input, where $\mathbb{O}$ is a reference set of objects (e.g., the set of all books). As output, the function produces a ranking (total order) $\succ$ of the objects $\mathbf{O}$. The ranking function is commonly implemented by means of a scoring function $U : \mathbb{O} \to \mathbb{R}$, i.e., objects are first scored and then ranked according to their scores. In order to induce a function of that kind, the learning algorithm is provided with training information, which typically comes in the form of exemplary pairwise preferences between objects. As opposed to label ranking, the alternatives to be ranked are described in terms of properties (feature vectors), while preferences are not contextualized. In principle, methods for object ranking could be applied in the context of dyad ranking, too, namely by equating the object space $\mathbb{O}$ with the "dyad space" $\mathbb{Z}$ in (3); in fact, dyads can be seen as a specific type of object, i.e., as objects with a specific structure. Especially close in terms of the underlying methodology is the so-called *listwise approach* in learning-to-rank [4].

Close to our setting is also the (kernel-based) framework of *conditional ranking* [24]. Here, relational data is represented in terms of a graph structure, in which nodes correspond to objects and (directed) edges are labeled with associations between these objects. Conditional ranking then refers to the problem of ranking a set of nodes relative to another (target) node, namely, of ranking the former in decreasing order of association with the latter. Associations are modeled in terms of a specific type of kernel function called *preference kernel*, and an SVM-like training procedure (with quadratic instead of hinge loss) is used for model induction. The framework is quite flexible and covers different learning problems as special cases, depending on the type of graph (bipartite or complete), the type of edge labels and the type of training information [23].

4   A Bilinear Plackett-Luce Model

4.1   The Plackett-Luce Model

The Plackett-Luce (PL) model is a parameterized probability distribution on the set of all rankings over a set of alternatives $y_1, \ldots, y_K$. It is specified by a parameter vector $\boldsymbol{v} = (v_1, v_2, \ldots v_K) \in \mathbb{R}_+^K$, in which v_i accounts for the "skill" of the option y_i. The probability assigned by the PL model to a ranking represented by a permutation π is given by

$$\mathbf{P}(\pi \,|\, \boldsymbol{v}) = \prod_{i=1}^{K} \frac{v_{\pi(i)}}{v_{\pi(i)} + v_{\pi(i+1)} + \ldots + v_{\pi(K)}} \tag{6}$$

This model is a generalization of the well-known Bradley-Terry model [18], a model for the pairwise comparison of alternatives, which specifies the probability that "a wins against b" in terms of

$$\mathbf{P}(a \succ b) = \frac{v_a}{v_a + v_b}.$$

Obviously, the larger v_a in comparison to v_b, the higher the probability that a is chosen. Likewise, the larger the parameter v_i in (6) in comparison to the parameters v_j, $j \neq i$, the higher the probability that y_i appears on a top rank.

As a nice feature of Plackett-Luce, we note that marginals (i.e., probabilities of rankings of a subset of the alternatives) can be computed very easily for this model: The probability of an incomplete ranking (1) is given by

$$\mathbf{P}(\pi \,|\, \boldsymbol{v}) = \prod_{i=1}^{J} \frac{v_{\pi(i)}}{v_{\pi(i)} + v_{\pi(i+1)} + \ldots + v_{\pi(J)}} \,,$$

i.e., by an expression of exactly the same form as (6), except that the number of factors is J instead of K.

4.2   Label Ranking Using the PL Model

A method for label ranking based on the PL model was proposed in [5]. The main idea of this approach is to contextualize the skill parameters of the labels y_i by modeling them as functions of the context $\boldsymbol{x}$. More precisely, to guarantee the non-negativity of the parameters, they are modeled as log-linear functions:

$$v_k = v_k(\boldsymbol{x}) = \exp\left(\sum_{d=1}^{r} w_d^{(k)} \cdot x_d \right) = \exp\left(\left\langle \boldsymbol{w}^{(k)}, \boldsymbol{x} \right\rangle \right) . \tag{7}$$

The parameters of the label ranking model, namely the $w_d^{(k)}$ ($1 \leq k \leq K$, $1 \leq d \leq r$), are estimated by maximum likelihood inference.

Given estimates of these parameters, prediction for new query instances $\boldsymbol{x}$ can be done in a straightforward way: $\hat{\boldsymbol{v}} = (\hat{v}_1, \ldots, \hat{v}_K)$ is computed based on (7), and a ranking $\hat{\pi}$ is determined by sorting the labels y_k in decreasing order of their (predicted) skills $\hat{v}_k$. This ranking $\hat{\pi}$ is a reasonable prediction, as it corresponds to the mode of the distribution $\mathbf{P}(\cdot \,|\, \hat{\boldsymbol{v}})$.

4.3   Dyad Ranking Using the PL model

In (7), the skill of the label y_k is modeled as a log-linear function of x, with a label-specific weight vector $w^{(k)}$. In the context of dyad ranking, this approach can be generalized to the modeling of skills for dyads as follows:

$$v(z) = v(x, y) = \exp\left(\langle w, \Phi(x, y) \rangle\right) , \tag{8}$$

where Φ is a joint feature map [25]. A common choice for such a feature map is the Kronecker product:

$$\Phi(x, y) = x \otimes y = \left(x_1 \cdot y_1, x_1 \cdot y_2, \ldots, x_r \cdot y_c\right) = vec\left(xy^\top\right) , \tag{9}$$

which is a vector of length $r \cdot c$ consisting of all pairwise products of the components of x and y. Thus, the inner product $\langle w, \Phi(x, y) \rangle$ can be rewritten as a bilinear form $x^\top W y$ with an $r \times c$ matrix $W = (w_{i,j})$; the entry $w_{i,j}$ can be considered as the weight of the interaction term $x_i y_j$. This choice of the joint-feature map yields a bilinear version of the PL model:

$$v(z) = v(x, y) = \exp\left(x^\top W y\right) \tag{10}$$

Suppose training data $\mathcal{D}$ to be given in the form of a set of rankings (11), i.e., rankings $\rho_1, \ldots, \rho_N$ of the following kind:

$$\rho_n : (x_n^{(1)}, y_n^{(1)}) \succ (x_n^{(2)}, y_n^{(2)}) \succ \ldots \succ (x_n^{(M_n)}, y_n^{(M_n)}) \tag{11}$$

The likelihood of the parameter vector w is then given by

$$L(w) = \mathbf{P}(\mathcal{D} \mid w) = \prod_{n=1}^{N} \prod_{m=1}^{M_n} \frac{\exp\left(w^\top (x_n^{(m)} \otimes y_n^{(m)})\right)}{\sum_{l=m}^{M_n} \exp\left(w^\top (x_n^{(l)} \otimes y_n^{(l)})\right)} ,$$

and the log-likelihood by

$$\ell(w) = \sum_{n=1}^{N} \sum_{m=1}^{M_n} w^\top (x_n^{(m)} \otimes y_n^{(m)}) - \sum_{n=1}^{N} \sum_{m=1}^{M_n} \log\left(\sum_{l=m}^{M_n} \exp\left(w^\top (x_n^{(l)} \otimes y_n^{(l)})\right)\right).$$

Like in the case of the linear PL model, the learning problem can now be formalized as finding the maximum likelihood (ML) estimate, i.e., the parameter w that maximizes the log-likelihood:

$$w_{ML} = \underset{w}{\operatorname{argmax}} \; \ell(w) , \tag{12}$$

To save the costly computations of the Hessian during ML estimation, a quasi-Newton type algorithm (L-BFGS, [17]) is used in our implementation. Further remarks on the identifiability of the model parameters are provided below.

4.4 Identifiability of the Bilinear PL Model

The bilinear PL model introduced above defines a probability distribution on dyad rankings that is parameterized by the weight matrix $\mathbf{W}$. An interesting question concerns the identifiability of this model. Recall that, for a parameterized class of models $\mathcal{M}$, identifiability requires a bijective relationship between models $M_\theta \in \mathcal{M}$ and parameters θ, that is, models are uniquely identified by their parameters. Or, stated differently, parameters $\theta \neq \theta^*$ induce different models $M_\theta \neq M_{\theta^*}$. Identifiability is a prerequisite for a meaningful interpretation of parameters and, perhaps even more importantly, guarantees unique solutions for optimization procedures such as maximum likelihood estimation.

Obviously, the original PL model (6) with constant skill parameters $v = (v_1, \ldots, v_K)$ is not identifiable, since the model is invariant against multiplication of the parameter by a constant factor $c > 0$: The models parameterized by v and $v^* = (cv_1, \ldots, cv_K)$ represent exactly the same probability distribution, i.e., $\mathbf{P}(\pi \,|\, v) = \mathbf{P}(\pi \,|\, v^*)$ for all rankings π. The PL model is, however, indeed identifiable up to this kind of multiplicative scaling. Thus, by fixing one of the weights to the value 1, the remaining $K - 1$ weights can be uniquely identified.

Now, what about the identifiability of our bilinear PL model, i.e., to what extent is such a model uniquely identified by the parameter $\mathbf{W}$? We can show the following result.

Proposition 1: *Suppose the feature representation of labels does not include a constant feature, i.e., $|\mathbb{Y}_i| > 1$ for each of the domains in (2), and that the feature representation of instances includes at most one such feature (accounting for a bias, i.e., an intercept of the bilinear model). Then, the bilinear PL model with skill values defined according to (10) is identifiable.*

Proof (sketch): Recall that the standard PL model is invariant against multiplication with a positive constant, and that this is the only invariance of the model. Since the bilinear PL model defined by (10) is log-linear in $\mathbf{W}$, invariance on the level of this parameter can only be additive. Now, suppose there are two parameters $\mathbf{W} \neq \mathbf{W}^*$ that both induce the same distribution on the set of all potential dyad subsets, which means that

$$x^\top \mathbf{W} y = x^\top \mathbf{W}^* y + \gamma \tag{13}$$

for all dyads (x, y), where γ is a constant that may depend on the parameters $\mathbf{W}$ and $\mathbf{W}^*$ but *not* on the dyads (x, y). More specifically, for the case of contextual dyad ranking, γ is also allowed to depend on x, but again, must not depend on y. Under our assumptions, however, this independence cannot hold. In fact, denoting the elements of $\mathbf{W}$ and $\mathbf{W}^*$ by $w_{i,j}$ and $w_{i,j}^*$, respectively, (13) means that

$$\sum_{i=1}^{r}\sum_{j=1}^{c}(w_{i,j} - w_{i,j}^*)x_i y_j = \sum_{i=1}^{r}\sum_{j=1}^{c}\Delta w_{i,j} x_i y_j = \gamma \ .$$

Then, exploiting the fact that not all $\Delta w_{i,j}$ can vanish at the same time, it is not difficult to show that a variation of some values y_j, which will also have an influence on the difference γ, is always possible.

4.5   Comparison Between the Linear and Bilinear PL Model

It is not difficult to see that the linear model (7), subsequently referred to as LinPL, is indeed a special case of the bilinear model (10), called BilinPL. In fact, the former is recovered from the latter by means of a $(1\text{-of-}K)$ dummy encoding of the alternatives: The label y_k is encoded by a K-dimensional vector with a 1 in position k and 0 in all other positions. The columns of the matrix $\mathbf{W}$ are then given by the weight vectors $\boldsymbol{w}^{(k)}$ in (7).

The other way around, LinPL can also be applied in the setting of dyad ranking, provided the domain $\mathbb{Y}$ of the alternatives is finite. To this end, one would simply introduce one "meta-label" Y_k for each feature combination $(y_1, \ldots, y_c)$ in (2) and apply a standard label ranking method to the set of these meta-labels. Therefore, both approaches are in principle equally expressive. Still, an obvious problem of this transformation is the potential size[1]

$$K = |\mathbb{Y}| = |\mathbb{Y}_1| \times |\mathbb{Y}_2| \times \ldots \times |\mathbb{Y}_c|$$

of the label set thus produced, which might be huge. In fact, the number of parameters that need to be learned for the model (7) is $r \cdot |\mathbb{Y}|$, i.e., $r \cdot a^c$ under the assumptions that each feature has a values. For comparison, the number of parameters is only $r \cdot c$ in the bilinear model. Moreover, all information about relationships between the alternatives (such as shared features or similarities) are lost, since a standard label ranker will only use the name of a meta-label while ignoring its properties.

Against the background of these consideration, one should expect dyad ranking to be advantageous to standard label ranking provided the assumptions underlying the bilinear model (10) are indeed valid, at least approximately. In that case, learning with (meta-)labels and disregarding properties of the alternatives would come with an unnecessary loss of information (that would need to be compensated by additional training data). In particular, using the standard label ranking approach is supposedly problematic in the case of many meta-labels and comparatively small amounts of training data.

Having said that, dyad ranking could be problematic if the model (10) is in fact a misspecification: If the features are not meaningful, or the bilinear model is not properly reflecting their interaction, then learning on the basis of (10) cannot be successful.

In this regard, it is also interesting to mention that both approaches can be combined. To this end, the feature vectors $\boldsymbol{y}$ are extended by a $(1\text{-of-}K)$ dummy-encoding, i.e., dyad ranking is used with feature vectors of the following form:

$$\boldsymbol{y} = (\, y_1, y_2, \ldots, y_c, \underbrace{0, \ldots, 0, 1, 0, \ldots, 0}_{\text{length } K} \,) \tag{14}$$

---

[1] This is an upper bound, since in practice, not all feature combinations are necessarily realized.

Using this representation, subsequently called LinSidePL, the learner is in principle free to exploit the side-information y_i or to ignore it and only use the dummy-labels.

In summary, the main observations can be summarized as follows:

- The linear PL model, like standard label ranking in general, assumes all alternatives to be known beforehand and to be included in the training process. If generalization beyond alternatives encountered in the training process is needed, then BilinPL can be used while LinPL cannot.
- If the assumption (10) of the bilinear model is correct, then BilinPL should learn faster than LinPL, as it needs to estimate fewer parameters. Yet, since LinPL can represent all dependencies that can be represented by BilinPL, the learning curve of the former should reach the one of latter with growing sample size.
- If the bilinear model (10) is actually a misspecification, then LinPL is likely to perform better than BilinPL, at least with enough training data being available (for small training sets, BilinPL could still be better).

5   Experiments

In order to verify the expectations summarized above, we conducted experiments with both synthetic and real data sets. In addition to LinPL (as implemented in [5]), BilinPL and LinSidePL, we included Ranking by Pairwise Comparison (RPC, [12]) and Constrained Classification (CC, [9,10]), which are both state-of-the-art label ranking methods, as additional baselines.[2]

Predictive performance was measured in terms of Kendall's tau coefficient [15], a rank correlation measure commonly used for this purpose in the label ranking literature [26,27]. It is defined as

$$\tau = \frac{C(\pi, \hat{\pi}) - D(\pi, \hat{\pi})}{K(K-1)/2} \ , \tag{15}$$

with C and D the number of concordant (put in the same order) and discordant (put in the reverse order) label pairs, respectively, and K the length of the rankings π and $\hat{\pi}$ (number of labels). Kendall's tau assumes values in $[-1, +1]$, with $\tau = +1$ for the perfect prediction $\hat{\pi} = \pi$ and $\tau = -1$ if $\hat{\pi}$ is the exact reversal of π.

5.1   Synthetic Data

Ideal synthetic ranking data is created by sampling from the Plackett-Luce distribution according to the BilinPL model specification under the setting (5) of contextual dyad ranking. A realistic scenario is simulated in which labels can be missing, i.e., observed rankings are incomplete [11]. To this end, a biased coin is

---

[2] CC was used in its online variant as described in [12].

flipped for every label, and it is decided with probability $p \in [0, 1]$ to keep or to delete it. We choose a missing rate of $p = 0.3$, which means that on average 70% of all labels of the training set are kept while the remaining labels are dismissed. Feature vectors of length $c = 4$ for labels and length $r = 3$ for instances were generated by sampling the elements from a standard normal distribution (except for one instance feature, which is a constant). The weight components were sampled randomly from a normal distribution with mean 1 and standard deviation 9. The predictive performance is then determined on a sufficiently large number of (complete) test examples and averaged over 10 repetitions.

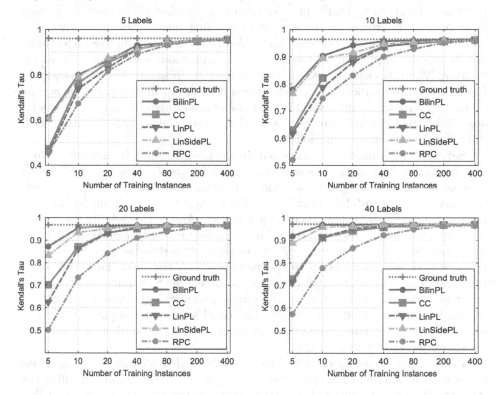

Fig. 1. Learning curves (generalization performance as a function of the number of training examples) of the ranking methods for different numbers of labels.

The learning curves thus produced are shown in Figure 1 for different numbers of labels. Overall, all ranking methods are able to learn and predict correctly if enough training data are available. In the limit, they all reach the performance of the "ground truth": given complete knowledge about the true PL model, the optimal (Bayes) prediction is the mode of that distribution (note that the average performance of that predictor is still not perfect, since sampling from the distribution will not always yield the mode). As expected, BilinPL and LinSidePL both benefit from the additional label description compared to the other label ranking approaches over a wide range of different training set sizes and numbers of labels.

Apart from predictive accuracy, it is worth mentioning that, in comparison with the BilinPL model, standard label ranking methods also exhibit poor runtime characteristics.

5.2   Case Study in Meta-Learning

As conjectured in Section 4.5 and confirmed in Section 5.1, BilinPL is potentially advantageous to LinPL in cases where the number of alternatives (labels) is large in comparison to the amount of training information being available and, moreover, these alternatives can be described in terms of suitable features. An interesting application for which these assumptions seem to hold is meta-learning [2]. In this section, we therefore employ the framework of meta-learning for algorithm recommendation as described in [2,3,14]. In particular, we aim at predicting a ranking over several variants of a class of algorithms such as genetic algorithms (GA), which can be obtained by instantiating the algorithm with different parameter combinations.

Several choices need to be made within the meta-learning framework, including the way of how meta-data is acquired (see Figure 2). The meta-features as part of the meta-data should be able to relate a data set to the relative performance of the candidate algorithms. They are usually made up by a set of numbers acquired by using descriptive statistics. Another possibility consists of probing a few parameter settings of the algorithm under consideration. The performance values of those *landmarkers* can then be used as instance-features for the meta-learner. In addition to the meta-features, the meta-data consists of rankings of the candidate algorithms, i.e., a sorting of the variants in decreasing order of performance. Using the meta-learning terminology, these rankings correspond to the so-called meta-target. The novel aspect in this paper is the use of qualitative performance data in the form of rankings[3] in conjunction with the consideration of side-information.

In analogy to the majority classifier typically used as a baseline in multi-class classification, the meta-learning literature suggests a simple approach called the Average Ranks (AR) method [2]. This approach corresponds to what is called the Borda count in the ranking literature and produces a default prediction by sorting the alternatives according to their average position in the observed rankings.

Learning to Rank Genetic Algorithms. This case study aims at recommending GA parameter settings for instances of the symmetric traveling salesman problem (TSP). The GA performance averages are taken to construct rankings, in which a single performance value corresponds to the distance of the shortest route found by a GA. The GAs share the properties of using the same

---

[3] This also comprises partial rankings and pairwise preferences as special cases.

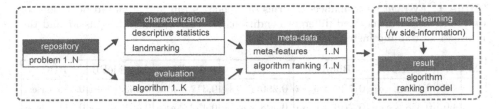

Fig. 2. The components of the "meta-learning for algorithm recommendation" framework shown above are based on [2]. The left box shows the meta-data acquisition process which consists of learning problem (or data set) characterization and the evaluation of the algorithms on the problems (or data sets). The box on the right side, the meta-level learning part, shows the meta-learning process and its outcome. In this case study, the meta-learner must be able to deal with qualitative data in form of rankings and is furthermore allowed to use additional knowledge (side-information) about the algorithms if it is available.

selection criterion, which is "roulette-wheel", the same mutation operator, which is "exchange mutation" and "elitism" of 10 chromosomes [22]. We tested the performance of three groups of GAs on a set of TSP instances. The groups are determined by their choice of the crossover operator, which can be cycle (CX), order (OX) or partially mapped crossover (PMX) [16].

The set of meta-features represent the instance vectors for the ranking models. They are composed of the number of cities and the performances of three landmarkers.

In total, 246 problems are considered, with the number of cities ranging between 10 and 255.[4] For each problem, the city locations (x, y) are drawn randomly from the uniform distribution on $[1, 100]^2$. Moreover, 72 different GAs are considered as alternatives with their parameters as optional label descriptions. They share the number of generations, 500, and the population size of 100. The combinations of all the other parameters, namely, crossover type, crossover rate and mutation rate, are used for characterization:

- Crossover types: {CX, OX, PMX}
- Crossover rates: {0.5, 0.6, 0.7, 0.8, 0.9}
- Mutation rates: {0.08, 0.09, 0.1, 0.11, 0.12}.

The three landmarker GAs have a crossover rate of 0.6 and a mutation rate of 0.12, combined with one of the three crossover types, respectively. They are excluded from the set of alternatives to be ranked. The label and dyad rankers are faced with rankings under different conditions (M, N), with N the number of training instances and M the average length of the rankings (M of the 72 alternatives are chosen at random while the others are discarded).

The results in Table 1 are quite consistent with our first study and again confirm that additional information about labels can principally be exploited by a learner to achieve better predictive performances. In particular, BilinPL is

---

[4] The data set can be obtained from https://www.cs.uni-paderborn.de/fachgebiete/intelligente-systeme/

Table 1. Average performance in terms of Kendall's tau and standard deviations of different meta-learners and different conditions (average rankings lengths M and the numbers of training instances N).

M	N	AR	BilinPL	CC	LinPL	LinSidePL	RPC
5	30	0.192 ± 0.063	$\mathbf{0.727} \pm \mathbf{0.014}$	0.290 ± 0.063	0.317 ± 0.049	0.663 ± 0.031	0.158 ± 0.052
	60	0.358 ± 0.046	$\mathbf{0.766} \pm \mathbf{0.014}$	0.428 ± 0.040	0.452 ± 0.041	0.681 ± 0.026	0.311 ± 0.038
	90	0.404 ± 0.030	$\mathbf{0.770} \pm \mathbf{0.014}$	0.573 ± 0.042	0.575 ± 0.037	0.691 ± 0.018	0.372 ± 0.035
	120	0.430 ± 0.029	$\mathbf{0.777} \pm \mathbf{0.009}$	0.610 ± 0.031	0.619 ± 0.022	0.697 ± 0.015	0.387 ± 0.032
10	30	0.423 ± 0.054	$\mathbf{0.775} \pm \mathbf{0.007}$	0.539 ± 0.054	0.551 ± 0.049	0.696 ± 0.018	0.397 ± 0.043
	60	0.487 ± 0.017	$\mathbf{0.781} \pm \mathbf{0.004}$	0.690 ± 0.021	0.696 ± 0.013	0.718 ± 0.012	0.493 ± 0.037
	90	0.523 ± 0.014	$\mathbf{0.781} \pm \mathbf{0.007}$	0.726 ± 0.015	0.726 ± 0.012	0.727 ± 0.010	0.576 ± 0.018
	120	0.522 ± 0.015	$\mathbf{0.783} \pm \mathbf{0.006}$	0.750 ± 0.014	0.748 ± 0.014	0.735 ± 0.011	0.620 ± 0.020
20	30	0.516 ± 0.037	$\mathbf{0.781} \pm \mathbf{0.005}$	0.722 ± 0.019	0.722 ± 0.015	0.728 ± 0.014	0.622 ± 0.018
	60	0.549 ± 0.014	$\mathbf{0.784} \pm \mathbf{0.005}$	0.763 ± 0.013	0.758 ± 0.014	0.741 ± 0.015	0.714 ± 0.022
	90	0.561 ± 0.014	$\mathbf{0.787} \pm \mathbf{0.006}$	0.779 ± 0.010	0.774 ± 0.013	0.750 ± 0.015	0.751 ± 0.021
	120	0.571 ± 0.022	$\mathbf{0.787} \pm \mathbf{0.008}$	0.786 ± 0.010	0.782 ± 0.010	0.758 ± 0.013	0.772 ± 0.014
30	30	0.554 ± 0.028	$\mathbf{0.782} \pm \mathbf{0.005}$	0.753 ± 0.013	0.746 ± 0.018	0.734 ± 0.015	0.717 ± 0.019
	60	0.567 ± 0.008	$\mathbf{0.785} \pm \mathbf{0.003}$	0.782 ± 0.007	0.775 ± 0.009	0.751 ± 0.010	0.767 ± 0.011
	90	0.578 ± 0.008	0.787 ± 0.004	$\mathbf{0.791} \pm \mathbf{0.005}$	0.786 ± 0.005	0.758 ± 0.010	0.781 ± 0.006
	120	0.580 ± 0.011	0.786 ± 0.006	$\mathbf{0.794} \pm \mathbf{0.005}$	0.789 ± 0.007	0.761 ± 0.011	0.787 ± 0.005

able to take advantage of this information for small values of M and favorably compares to the other label rankers (and, in addition, has of course the advantage of being able to rank GA variants that have not been used in the training phase). As expected, standard label rankers (in this case, CC) surpass BilinPL only for a sufficiently large amount of training data.

6 Summary and Outlook

In this paper, we proposed dyad ranking as an extension of the label ranking problem, a specific type of preference learning problem in which preferences on a finite set of choice alternatives are represented in the form of a contextualized ranking. While the context is described in terms of a feature vector, the alternatives are merely identified by their label.

In practice, however, information about properties of the alternatives is often available, too, and such information could obviously be useful from a learning point of view. In dyad ranking, not only the context but also the alternatives are therefore characterized as feature vectors.

The concrete method we developed, BilinPL, is a generalization of an existing label ranking method based on the Plackett-Luce model. First experimental results using synthetic data as well as a case study in meta-learning confirm that BilinPL tends to be superior to standard label ranking methods if feature information about alternatives is available, at least if training data is scarce in comparison to the number of alternatives to be ranked.

Since the PL approach is only one among several existing label ranking methods, one may wonder to what extent other methods are amenable to the incorporation of label features. This is a question we seek to address in future work. Another interesting idea is to combine label ranking with (unsupervised) representation learning for feature construction [1]: first, labels are embedded in a feature space so as to reflect their similarity in a proper way, and the feature representation thus produced is then used in dyad ranking. Last but not least, there are several interesting applications of dyad ranking, notably those in which standard label ranking has already been used, though without exploiting feature information about choice alternatives. An example of that kind is preference-based reinforcement learning, where label ranking is used to sort actions given states [8]. Since actions do have a natural representation in terms of features or parameters in many reinforcement learning problems, there is obviously scope for enhancement through the incorporation of dyad ranking.

References

1. Bengio, Y., Courville, A.C., Vincent, P.: Representation learning: A review and new perspectives. IEEE Trans. Pattern Anal. Mach. Intell. **35**(8), 1798–1828 (2013)
2. Brazdil, P., Giraud-Carrier, C., Soares, C., Vilalta, R.: Metalearning: Applications to Data Mining, 1st edn. Springer Publishing Company, Incorporated (2008)
3. Brazdil, P., Soares, C., Coasta, J.P.D.: Ranking learning algorithms: Using IBL and meta-learning on accuracy and time results. Machine Learning **50**, 251–277 (2003)
4. Cao, Z., Qin, T., Liu, T.-Y., Tsai, M.-F., Li, H.: Learning to rank: from pairwise approach to listwise approach. In: Proceedings of the 24th International Conference on Machine learning (ICML 2007), pp. 129–136. ACM (2007)
5. Cheng, W., Dembczynski, K., Hüllermeier, E.: Label ranking methods based on the Plackett-Luce model. In: Proceedings of the 27th International Conference on Machine Learning (ICML 2010), pp. 215–222 (2010)
6. Cohen, W.W., Schapire, R.E., Singer, Y.: Learning to order things. Journal of Artificial Intelligence Research **10**(1), 243–270 (1999)
7. Fürnkranz, J., Hüllermeier, E.: Preference learning: An introduction. Preference Learning (2010)
8. Fürnkranz, J., Hüllermeier, E., Cheng, W., Park, S.H.: Preference-based reinforcement learning: A formal framework and a policy iteration algorithm. Machine Learning **89**(1), 123–156 (2012)
9. Har-Peled, S., Roth, D., Zimak, D.: Constraint classification: a new approach to multiclass classification. In: Cesa-Bianchi, N., Numao, M., Reischuk, R. (eds.) ALT 2002. LNCS (LNAI), vol. 2533, pp. 365–379. Springer, Heidelberg (2002)
10. Har-Peled, S., Roth, D., Zimak, D.: Constraint classification for multiclass classification and ranking. In: Becker, S., Thrun, S., Obermayer, K. (eds.) Advances in Neural Information Processing Systems, vol. 15, pp. 809–816. MIT Press (2003)

11. Hüllermeier, E., Cheng, W.: Superset learning based on generalized loss minimization. In: Proc. ECML/PKDD 2015, European Conference on Machine Learning and Knowledge Discovery in Databases, Porto, Portugal (2015)
12. Hüllermeier, E., Fürnkranz, J., Cheng, W., Brinker, K.: Label ranking by learning pairwise preferences. Artificial Intelligence **172**(16), 1897–1916 (2008)
13. Kamishima, T., Kazawa, H., Akaho, S.: A survey and empirical comparison of object ranking methods. In: Fürnkranz, J., Hüllermeier, E. (eds.) Preference Learning, pp. 181–201. Springer (2011)
14. Kanda, J., Soares, C., Hruschka, E., de Carvalho, A.: A meta-learning approach to select meta-heuristics for the traveling salesman problem using MLP-based label ranking. In: Huang, T., Zeng, Z., Li, C., Leung, C.S. (eds.) ICONIP 2012, Part III. LNCS, vol. 7665, pp. 488–495. Springer, Heidelberg (2012)
15. Kendall, M.G.: A new measure of rank correlation. Biometrika **30**(1/2), 81–93 (1938)
16. Larranaga, P., Kuijpers, C.M.H., Murga, R.H., Inza, I., Dizdarevic, S.: Genetic algorithms for the traveling salesman problem: A review of representations and operators. Artificial Intelligence Review **13**, 129–170 (1999)
17. Liu, D.C., Nocedal, J.: On the limited memory BFGS method for large scale optimization. Mathematical Programming **45**(1–3), 503–528 (1989)
18. Marden, J.I.: Analyzing and Modeling Rank Data, 1st edn. Chapman & Hall (1995)
19. Menon, A.K., Elkan, C.: Dyadic prediction using a latent feature log-linear model (2010). arXiv preprint arXiv:1006.2156
20. Menon, A.K., Elkan, C.: A log-linear model with latent features for dyadic prediction. In: Proceedings of the 2010 IEEE International Conference on Data Mining, ICDM 2010, pp. 364–373. IEEE Computer Society (2010)
21. Menon, A.K., Elkan, C.: Predicting labels for dyadic data. Data Mining and Knowledge Discovery **21**(2), 327–343 (2010)
22. Mitchell, M.: An Introduction to Genetic Algorithms. MIT Press, Cambridge (1998)
23. Pahikkala, T., Airola, A., Stock, M., De Baets, B., Waegeman, W.: Efficient regularized least-squares algorithms for conditional ranking on relational data. Machine Learning **93**, 321–356 (2013)
24. Pahikkala, T., Waegeman, W., Airola, A., Salakoski, T., De Baets, B.: Conditional ranking on relational data. In: Balcázar, J.L., Bonchi, F., Gionis, A., Sebag, M. (eds.) ECML PKDD 2010, Part II. LNCS, vol. 6322, pp. 499–514. Springer, Heidelberg (2010)
25. Tsochantaridis, I., Joachims, T., Hofmann, T., Altun, Y.: Large margin methods for structured and interdependent output variables. Journal of Machine Learning Research, 1453–1484 (2005)
26. Vembu, S., Gärtner, T.: Label ranking algorithms: a survey. In: Preference Learning, pp. 45–64. Springer (2011)
27. Zhou, Y., Liu, Y., Yang, J., He, X., Liu, L.: A taxonomy of label ranking algorithms. Journal of Computers **9**(3), 557–565 (2014)

Fast Training of Support Vector Machines for Survival Analysis

Sebastian Pölsterl[1]([✉]), Nassir Navab[1,2], and Amin Katouzian[1]

[1] Chair for Computer Aided Medical Procedures,
Technische Universität München, Munich, Germany
{poelster,navab,katouzian}@in.tum.de
[2] Johns Hopkins University, Baltimore, MD, USA

Abstract. Survival analysis is a commonly used technique to identify important predictors of adverse events and develop guidelines for
patient's treatment in medical research. When applied to large amounts
of patient data, efficient optimization routines become a necessity. We
propose efficient training algorithms for three kinds of linear survival
support vector machines: 1) ranking-based, 2) regression-based, and 3)
combined ranking and regression. We perform optimization in the primal using truncated Newton optimization and use order statistic trees to
lower computational costs of training. We employ the same optimization
technique and extend it for non-linear models too. Our results demonstrate the superiority of our proposed optimization scheme over existing
training algorithms, which fail due to their inherently high time and space
complexities when applied to large datasets. We validate the proposed
survival models on 6 real-world datasets, and show that pure ranking-
based approaches outperform regression and hybrid models.

Keywords: Survival analysis · Support vector machine · Optimization

1 Introduction

Recently, researchers have become interested in studying the effective use of
electronic health records to improve outcomes of medical procedures, reduce
health care costs, evaluate the efficiency of newly developed drugs, and predict
health trends or adverse events (see e.g. [13] for an overview). In the latter case,
survival analysis is employed to examine how a particular set of covariates affects
the time until the occurrence of an event of interest, such as death or reaching
a specific state of disease progression. The objective in survival analysis is to
establish a connection between covariates and the time between the start of
the study and an event. What makes survival analysis differ from traditional
machine learning is the fact that parts of the training data can only be partially
observed – they are *censored*. In a clinical study, patients are often monitored
for a particular time period, and events occurring in this particular period are
recorded. If a patient experiences an event, the exact time of the event can

© Springer International Publishing Switzerland 2015
A. Appice et al. (Eds.): ECML PKDD 2015, Part II, LNAI 9285, pp. 243–259, 2015.
DOI: 10.1007/978-3-319-23525-7_15

be recorded – the patient's record is *uncensored*. In contrast, *right censored* records refer to patients that remained event-free during the study period and it is unknown whether an event has or has not occurred after the study ended. Consequently, survival models demand for proper training algorithms that take this unique characteristic of such a dataset into account.

Cox's proportional hazards model [6] is the standard for analyzing time-to-event data, despite having several shortcomings: 1) it assumes that hazard functions for any two individuals are proportional, i.e., their ratio is constant over time, 2) it is not applicable to data with more features than samples, 3) it fails if features are highly correlated, and 4) its decision function is linear in the covariates. The advantage of large-margin methods for classification and regression has motivated researchers to adapt these models for survival analysis. Authors in [17,22] cast survival analysis as a regression problem and adapted support vector regression, whereas Eleuteri et al. [9] formulated a loss function derived from quantile regression. Steck et al. [23] observed that survival analysis can be expressed as a ranking problem, which led to extensions of Rank Support Vector Machines (RankSVMs) [10,24]. Finally, Van Belle et al. [26] proposed a hybrid solution between the ranking and regression approach.

The main disadvantage of ranking-based techniques is that their objective function depends on a quadratic number of constraints with respect to the number of training samples, which makes training intractable with medium to large sized datasets. By clustering data according to survival times, authors in [25] showed that the computational complexity can be lowered without considerable loss in performance. For regular RankSVMs, which do not account for censoring, authors in [2,19] proposed the use of order statistic trees to alleviate this problem.

In this paper, we extend the work of Lee et al. [19] to efficiently train ranking, and regression-based survival models by re-formulating their approach to be applicable to survival analysis in the presence of right censoring. In [10,24], ranking-based survival support vector machines were based on the hinge loss and optimization was carried out in the dual using a generic quadratic programming solver. In contrast, we use the squared hinge loss and perform truncated Newton optimization, which leads to a more efficient training algorithm. A further improvement is due to order statistic trees to avoid explicitly storing all pairwise comparisons of samples, which requires $O(n^2)$ space, where n is the number of samples. Moreover, we introduce a straightforward training technique for a combined regression and ranking approach. When considering non-linear functions, we demonstrate that training can still be carried out efficiently using the primal formulation. Finally, experimental results of 7 synthetic and 6 real world datasets justify the advantages of our proposed solution.

2   Survival Analysis

The objective in training a survival model is to derive a model's parameters in the presence of censoring. After training, the model can be used to predict the survival time of patients based on a given set of features. For a set of n patients, we know for the i-th patient: 1) the exact time $c_i \geq 0$ of censoring, i.e., the time until which the patient was observed, and 2) the time $t_i \geq 0$ when a patient experienced an event, if any. From these two quantities, we define the survival time y_i as

$$y_i = \min(t_i, c_i) = \begin{cases} t_i & \text{if } \delta_i = 1 \\ c_i & \text{if } \delta_i = 0, \end{cases}$$

where $\delta_i \in \{0, 1\}$ is the event indicator. Thus, training data for a survival model consists of triples $(\mathbf{x}_i, y_i, \delta_i)$, where $\mathbf{x}_i$ is a d-dimensional feature vector.

During training, information about the occurrence of an event is only partially available for censored patients, i.e., those that did not experience an event or dropped out of the study. When training a survival model, one has to consider that two patients i and j are only comparable if both experienced an event or only one of them experienced an event and the time of the event occurred before the time of censoring, formally: $(y_i < y_j \wedge \delta_i = 1) \vee (y_i > y_j \wedge \delta_j = 1)$. If two patients do not satisfy this condition, they are incomparable and their relation cannot be used to deduce a survival model.

Here, we discuss two approaches to survival analysis: the first approach treats survival analysis as a ranking problem, and the second approach as a regression problem. Finally, we present an objective function that combines both ideas. Our implementation of the methods proposed in this paper are publicly available.[1]

3   Survival Analysis as Ranking Problem

In ranking, the goal is to recover the correct order of samples according to their relevance. For survival analysis, relevance corresponds to the survival time. However, not all pairwise comparisons are meaningful in the presence of right censoring. The set $\mathcal{P} = \{(i, j) \mid y_i > y_j \wedge \delta_j = 1\}_{i,j=1,\dots,n}$ defines the pairs of comparable samples that can be used for training and $p = |\mathcal{P}|$ the cardinality of this set, which is bounded by $O(n^2)$. We minimize our objective function similar to the work in [19], but additionally account for right censoring during training.

Definition 1. *The objective function of ranking-based linear survival support vector machine is defined as*

$$f(\boldsymbol{w}) = \frac{1}{2}\boldsymbol{w}^T\boldsymbol{w} + \frac{\gamma}{2}\sum_{i,j\in\mathcal{P}}\max(0, 1 - (\boldsymbol{w}^T\boldsymbol{x}_i - \boldsymbol{w}^T\boldsymbol{x}_j))^2, \qquad (1)$$

where $\boldsymbol{w} \in \mathbb{R}^d$ are the coefficients and $\gamma > 0$ is a regularization parameter. A new set of data points $\boldsymbol{X}_{\text{new}}$, can be ranked with respect to their predicted survival time according to elements of $\boldsymbol{X}_{\text{new}}\boldsymbol{w}$.

---

[1] https://github.com/tum-camp/survival-support-vector-machine

The sum in the second term of (1) has a complexity of $O(n^2)$ and thus training with only a few thousand samples is already intractable. We will first derive a gradient-based minimization of the objective function, based on Newton's method, and then outline a more efficient optimization, which does not depend on the number of comparable pairs, using truncated Newton optimization and order statistic trees.

The objective function (1) can be expressed in matrix form as

$$f(\boldsymbol{w}) = \frac{1}{2}\boldsymbol{w}^T\boldsymbol{w} + \frac{\gamma}{2}\left(\mathbb{1} - \boldsymbol{A}\boldsymbol{X}\boldsymbol{w}\right)^T \boldsymbol{D_w}\left(\mathbb{1} - \boldsymbol{A}\boldsymbol{X}\boldsymbol{w}\right), \tag{2}$$

where $\mathbb{1}$ is a vector of all ones, $\boldsymbol{X} = [\boldsymbol{x}_1, \ldots, \boldsymbol{x}_n]^T$, and $\boldsymbol{A} \in \mathbb{R}^{p \times n}$ a sparse matrix with $A_{ki} = 1$ and $A_{kj} = -1$ if $(i,j) \in \mathcal{P}$ and zero otherwise. $\boldsymbol{D_w}$ is a $p \times p$ diagonale matrix that has an entry for each $(i,j) \in \mathcal{P}$ that indicates whether this pair is a support vector, i.e., $1 - (\boldsymbol{w}^T\boldsymbol{x}_i - \boldsymbol{w}^T\boldsymbol{x}_j) > 0$ [19]. For the k-th item of $\mathcal{P}$, representing the pair (i,j), the corresponding entry in $\boldsymbol{D_w}$ is defined as

$$(\boldsymbol{D_w})_{k,k} = \begin{cases} 1 & \text{if } \boldsymbol{w}^T\boldsymbol{x}_j > \boldsymbol{w}^T\boldsymbol{x}_i - 1 \\ 0 & \text{else} \end{cases}. \tag{3}$$

Thus, we obtain an objective function that is convex in $\boldsymbol{w}$ and can apply Newton's method to minimize it. One update in Newton's method with step size μ becomes

$$\boldsymbol{w}^{\text{new}} = \boldsymbol{w} - \mu\left(\frac{\partial^2 f}{\partial\boldsymbol{w}\partial\boldsymbol{w}^T}\right)^{-1}\frac{\partial f}{\partial\boldsymbol{w}} \tag{4}$$

with partial derivatives

$$\frac{\partial f}{\partial\boldsymbol{w}} = \boldsymbol{w} + \gamma\boldsymbol{X}^T\left(\boldsymbol{A}^T\boldsymbol{D_w}\boldsymbol{A}\boldsymbol{X}\boldsymbol{w} - \boldsymbol{A}^T\boldsymbol{D_w}\mathbb{1}\right) \tag{5}$$

$$\frac{\partial^2 f}{\partial\boldsymbol{w}\partial\boldsymbol{w}^T} = \boldsymbol{I} + \gamma\boldsymbol{X}^T\boldsymbol{A}^T\boldsymbol{D_w}\boldsymbol{A}\boldsymbol{X}. \tag{6}$$

Note that we used the generalized Hessian in the second derivative, because $f(\boldsymbol{w})$ is not twice differentiable at $\boldsymbol{w}$ [16].

Next, we simplify the derivatives by expressing the product $\boldsymbol{A}^T\boldsymbol{D_w}\boldsymbol{A}$ in terms of a new matrix $\boldsymbol{A_w} \in \{-1, 0, 1\}^{p_w, n}$ that is a restricted version of $\boldsymbol{A}$, limited to rows corresponding to support vectors:

$$\boldsymbol{A}^T\boldsymbol{D_w}\boldsymbol{A} = \boldsymbol{A_w}^T\boldsymbol{A_w}, \tag{7}$$

where p_w denotes the number of pairs $(i,j) \in \mathcal{P}$ – rows of $\boldsymbol{A}$ – where $\boldsymbol{w}^T\boldsymbol{x}_j > \boldsymbol{w}^T\boldsymbol{x}_i - 1$.

Algorithm 1. Survival Support Vector Machine Training.

Input: Training data $\mathcal{D} = \{(\boldsymbol{x}_i, y_i, \delta_i)\}_{i=1}^n$, hyper-parameter $\gamma > 0$.
Output: Coefficients $\boldsymbol{w}$.

1 Randomly resolve ties in survival times y_i $\forall i \in \{1, \ldots, n\}$;
2 $\boldsymbol{w}^0 \leftarrow \boldsymbol{0}$;
3 $t \leftarrow 0$;
4 **while** *not converged* **do**
5   Use conjugate gradient to determine search direction $\boldsymbol{u} = \left(\frac{\partial^2 f}{\partial \boldsymbol{w} \partial \boldsymbol{w}^T} \right)^{-1} \frac{\partial f}{\partial \boldsymbol{w}}$
       with $\boldsymbol{w} = \boldsymbol{w}^t$;
6   Choose step size μ by backtracking line search;
7   Update $\boldsymbol{w}^{t+1} \leftarrow \boldsymbol{w}^t + \mu \boldsymbol{u}$;
8      $t \leftarrow t + 1$;
9 **end**
10 $\boldsymbol{w} \leftarrow \boldsymbol{w}^t$;

Definition 2. *Formula* (2) *and its derivatives can be re-formulated using* $\boldsymbol{A_w}$ *to eliminate* $\boldsymbol{D_w}$.

$$f(\boldsymbol{w}) = \frac{1}{2}\boldsymbol{w}^T\boldsymbol{w} + \frac{\gamma}{2}\left(p_w + \boldsymbol{w}^T\boldsymbol{X}^T\left(\boldsymbol{A}_w^T\boldsymbol{A}_w\boldsymbol{X}\boldsymbol{w} - 2\boldsymbol{A}_w^T\mathbb{1}\right)\right) \quad (8)$$

$$\frac{\partial f}{\partial \boldsymbol{w}} = \boldsymbol{w} + \gamma\boldsymbol{X}^T\left(\boldsymbol{A}_w^T\boldsymbol{A}_w\boldsymbol{X}\boldsymbol{w} - \boldsymbol{A}_w^T\mathbb{1}\right) \quad (9)$$

$$\frac{\partial^2 f}{\partial \boldsymbol{w}\partial \boldsymbol{w}^T} = \boldsymbol{I} + \gamma\boldsymbol{X}^T\boldsymbol{A}_w^T\boldsymbol{A}_w\boldsymbol{X} \quad (10)$$

3.1 Truncated Newton Optimization

Medical research is often challenging due to high-dimensional data: a patient's health record comprises several hundred features, and microarray data consists of several thousand measurements. In this applications, explicitly computing and storing the Hessian matrix can be prohibitive, therefore, we use a truncated Newton method that uses a linear conjugate gradient method to compute the search direction [7,16,20]. This only requires the computation of the Hessian-vector product $\boldsymbol{Hv}$, which can be computed by

$$\boldsymbol{Hv} = \boldsymbol{v} + \gamma\boldsymbol{X}^T\boldsymbol{A}_w^T\boldsymbol{A}_w\boldsymbol{X}\boldsymbol{v}. \quad (11)$$

Thus, the complexity of a single conjugate gradient iteration is $O(nd + p + d)$, when multiplying from the right, which is lower than $O(pd^2 + pd + d)$ to obtain the full Hessian matrix. Truncated Newton optimization consists of an outer loop to update the coefficients $\boldsymbol{w}$ and an inner loop to find the search direction via conjugate gradient (see algorithm 1).

3.2 Efficient Calculation of Search Direction

In each iteration of Newton's method, A_w has to be recomputed due to its dependency on w, which requires iterating over all comparable pairs, being of order $\binom{n}{2}$. Therefore, the complexity of learning a new model is still quadratic in the number of samples. Next, we will derive an improved algorithm that avoids constructing A_w explicitly. First, we derive the conditions under which an entry in A_w is non-zero, followed by proposing a compact representation of an entry in $A_w^T A_w$, which finally leads to an efficient optimization scheme that is independent of the size of $\mathcal{P}$.

Proposition 1. *For $k \in \{1, \ldots, p_w\}$ and $q \in \{1, \ldots, n\}$, $(A_w)_{k,q} = 1$ if all of the following conditions are satisfied:*

(a) *survival time of q-th sample is* lower *than survival time of some sample $s \in \{1, \ldots, n\}$ (s outlives q): $y_q < y_s$.*
(b) *the q-th sample is uncensored: $\delta_q = 1$.*
(c) *the pair $(s, q) \in \mathcal{P}$ is a support vector: $w^T x_s < w^T x_q + 1$.*

Proposition 2. *For $k \in \{1, \ldots, p_w\}$ and $q \in \{1, \ldots, n\}$, $(A_w)_{k,q} = -1$ if all of the following conditions are satisfied:*

(a) *survival time of q-th sample is* higher *than survival time of some sample $s \in \{1, \ldots, n\}$ (q outlives s): $y_q > y_s$.*
(b) *the s-th sample is uncensored: $\delta_s = 1$.*
(c) *the pair $(q, s) \in \mathcal{P}$ is a support vector: $w^T x_s > w^T x_q - 1$.*

Proof. Note that the only difference between both propositions is the order of samples s and q with respect to their survival times. Thus, the first proposition can be transformed into the second by swaping s and q, and vice versa. Conditions (a) and (b) are directly derived from the definition of A. Each row of A and A_w contains exactly one element that is 1, one element that is -1, and the rest is all zeros. For each pair of samples (row of A), the sample with the shorter survival time is assigned 1, and the other sample -1, which is reflected by condition (a). In addition, each pair must be comparable, i.e., the sample with the shorter survival time must be uncensored, which leads to condition (b). Finally, condition (c) is due to the multiplication AD_w that restricts rows of A to pairs of samples that are support vectors. □

If proposition 1 or 2 holds, the result of the multiplication $(A_w)_{k,i} \cdot (A_w)_{k,j}$ is either 1 or -1, if $i = j$ or $i \neq j$, respectively, for $k \in \{1, \ldots, p_w\}$ and $i, j \in \{1, \ldots, n\}$. In the latter case, the conditions of propositions 1 and 2 are equal.

Combining all cases, the product $(A_w)_{k,i} \cdot (A_w)_{k,j}$ is defined as

$$(A_w)_{k,i} \cdot (A_w)_{k,j} = \begin{cases} 1 & \text{if } i = j, \ (A_w)_{k,i} = (A_w)_{k,j} = 1, \\ & \text{and proposition 1 holds for } q = i, \\ 1 & \text{if } i = j, \ (A_w)_{k,i} = (A_w)_{k,j} = -1, \\ & \text{and proposition 2 holds for } q = i, \\ -1 & \text{if } i \neq j, \ (A_w)_{k,i} = 1, (A_w)_{k,j} = -1, \\ & \text{and proposition 1 holds for } q = i, s = j \\ & \Leftrightarrow \text{proposition 2 holds for } q = j, s = i, \\ -1 & \text{if } i \neq j, \ (A_w)_{k,i} = -1, (A_w)_{k,j} = 1, \\ & \text{and proposition 1 holds for } q = j, s = i, \\ & \Leftrightarrow \text{proposition 2 holds for } q = i, s = j, \\ 0 & \text{else.} \end{cases} \quad (12)$$

We can compactly express $\left(A_w^T A_w\right)_{i,j} = \sum_{k=1}^{p_w} (A_w)_{k,i} \cdot (A_w)_{k,j}$ using above definitions and by defining the following two sets and their cardinalities.

$$\mathrm{SV}_i^+ = \{s \mid y_s > y_i \wedge w^T x_s < w^T x_i + 1 \wedge \delta_i = 1\} \qquad l_i^+ = |\mathrm{SV}_i^+| \quad (13)$$
$$\mathrm{SV}_i^- = \{s \mid y_s < y_i \wedge w^T x_s > w^T x_i - 1 \wedge \delta_s = 1\} \qquad l_i^- = |\mathrm{SV}_i^-| \quad (14)$$

The set SV_i^+ represents proposition 1, and SV_i^- represents proposition 2. This allows us to compactly express an entry of $A_w^T A_w$ as

$$(A_w^T A_w)_{i,j} = \begin{cases} l_i^+ + l_i^- & \text{if } i = j, \\ -1 & \text{if } i \neq j, \text{ and } j \in \mathrm{SV}_i^+ \text{ or } j \in \mathrm{SV}_i^-, \\ 0 & \text{else,} \end{cases} \quad (15)$$

where the second case is due to only one addend being non-zero, because each pair of samples is compared only once.

The term $A_w^T A_w X v$ is part of the objective function, its gradient, and the Hessian-vector product. Applying the formulation in (15), we obtain

$$(A_w^T A_w X v)_i = (l_i^+ + l_i^-)x_i^T v - \sum_{s \in \mathrm{SV}_i^+} x_s v - \sum_{s \in \mathrm{SV}_i^-} x_s v \quad (16)$$
$$= (l_i^+ + l_i^-)x_i^T v - \sigma_i^+ - \sigma_i^-.$$

and

$$X^T A_w^T A_w X v = X^T \begin{pmatrix} (l_1^+ + l_1^-)x_1^T v - (\sigma_1^+ + \sigma_1^-) \\ \vdots \\ (l_n^+ + l_n^-)x_n^T v - (\sigma_n^+ + \sigma_n^-) \end{pmatrix}. \quad (17)$$

Additionally, the objective function and its gradient contain the term $\boldsymbol{A}_{\boldsymbol{w}}^T \mathbb{1}$, where one component is computed as

$$
\begin{aligned}
(\boldsymbol{A}_{\boldsymbol{w}}^T \mathbb{1})_i &= |\mathrm{SV}_i^+ \cup \mathrm{SV}_i^-| \\
&= |\{(s,t) \mid y_t < y_i < y_s \wedge \delta_t = 1 \wedge \delta_i = 1 \wedge \\
&\qquad \boldsymbol{w}^T \boldsymbol{x}_s - 1 < \boldsymbol{w}^T \boldsymbol{x}_i < \boldsymbol{w}^T \boldsymbol{x}_t + 1\}| \\
&= l_i^- - l_i^+.
\end{aligned}
\tag{18}
$$

By substituting (17) and (18) together with $p_{\boldsymbol{w}} = \sum_{i=1}^n l_i^+ = \sum_{i=1}^n l_i^-$ into (8), (9), and (11), all terms that depend on $\boldsymbol{A}_{\boldsymbol{w}}$ during optimization can be eliminated. Assuming that l_i^+, l_i^-, σ_i^+, and σ_i^- have been computed already, the complexity of evaluating the objective function, gradient, and Hessian-vector product is $O(nd+d)$. Subsequently, we will discuss an efficient method to obtain these values.

3.3   Improving Optimization by Order Statistic Trees

The main difficulty is that the order of actual survival times y_i and predictions $\boldsymbol{w}^T \boldsymbol{x}_i$ have to be considered when constructing the sets SV_i^+ and SV_i^-. Assuming that samples have been sorted in ascending order according to $\boldsymbol{w}^T \boldsymbol{x}_i$, we illustrate how both sets can be constructed by the following example:

i	1	2	3	4	5	6	7	8	9
$\boldsymbol{w}^T \boldsymbol{x}_i$	-0.7	-0.1	0.15	0.2	0.3	0.8	1.6	1.7	2.3
y_i	1	9	6	5	8	2	7	3	4
δ_i	0	0	1	0	1	1	1	0	0

As we can see, the first element for which $\mathrm{SV}_i^+ \neq \varnothing$ occurs at $i = 3$, because both the first and second sample are censored ($\delta_i = 0$), which violates condition (b) of proposition 1. For $i = 3$, we obtain $\mathrm{SV}_3^+ = \{s | y_s > 6 \wedge \boldsymbol{w}^T \boldsymbol{x}_s < 1.15\} = \{2, 5\}$. The next set ($i = 4$) is again empty, because of censoring, and $\mathrm{SV}_5^+ = \{s | y_s > 8 \wedge \boldsymbol{w}^T \boldsymbol{x}_s < 1.3\} = \{2\}$. This example shows, that SV_i^+ is non-empty if and only if the i-th sample is uncensored, and that SV_{i+1}^+ can be constructed incrementally from the set SV_i^+:

$$
\begin{aligned}
&\{s | \boldsymbol{w}^T \boldsymbol{x}_s < \boldsymbol{w}^T \boldsymbol{x}_{i+1} + 1 \wedge \delta_{i+1} = 1\} \\
=&\{s | \boldsymbol{w}^T \boldsymbol{x}_s < \boldsymbol{w}^T \boldsymbol{x}_i + 1\} \cup \{s | \boldsymbol{w}^T \boldsymbol{x}_i + 1 \leq \boldsymbol{w}^T \boldsymbol{x}_s < \boldsymbol{w}^T \boldsymbol{x}_{i+1} + 1 \wedge \delta_{i+1} = 1\}.
\end{aligned}
$$

When constructing the set SV_i^-, we can obtain a similar incremental update rule when iterating the list of samples according to decreasing values of $\boldsymbol{w}^T \boldsymbol{x}_i$. Here, $\mathrm{SV}_9^- = \varnothing$, because no element with $\boldsymbol{w}^T \boldsymbol{x}_s > 1.3$ satisfies conditions (a) and (b) of proposition 2, and $\mathrm{SV}_8^- = \{s | y_s < 3 \wedge \boldsymbol{w}^T \boldsymbol{x}_s > 0.7 \wedge \delta_s = 1\} = \{6\}$. An incremental update when going from i to $i - 1$ is defined as

$$
\begin{aligned}
\{s | \boldsymbol{w}^T \boldsymbol{x}_s > \boldsymbol{w}^T \boldsymbol{x}_{i-1} - 1 \wedge \delta_s = 1\} &= \{s | \boldsymbol{w}^T \boldsymbol{x}_s > \boldsymbol{w}^T \boldsymbol{x}_i - 1 \wedge \delta_s = 1\} \\
&\cup \{s | \boldsymbol{w}^T \boldsymbol{x}_i - 1 \geq \boldsymbol{w}^T \boldsymbol{x}_s > \boldsymbol{w}^T \boldsymbol{x}_{i-1} - 1 \wedge \delta_s = 1\}.
\end{aligned}
$$

To maintain the respective sets of relevant samples for computing SV_i^+ and SV_i^-, we incrementally add elements y_i and $\boldsymbol{x}_i^T \boldsymbol{v}$ to an order statistic tree that allows retrieving $|\{s|y_s > y_i\}|$ and $|\{s|y_s < y_i\}|$ in logarithmic time. Note that both sets in the incremental update of SV_i^- consider censoring, whereas for SV_i^+ censoring is only relevant for the second set, but not the first. For the former, we use an order statistic tree to sort *uncensored* samples according to their survival time y_i, and for the latter we sort *all* samples, disregarding censoring. Formally, an order statistic tree is defined as follows.

Definition 3. *An order statistic tree is a balanced binary search tree that stores key-value pairs and has the following properties.*

1. *For an internal node x with left child $\text{left}(x)$ and right child $\text{right}(x)$:*

$$\text{key}(\text{left}(x)) \leq \text{key}(x) \text{ and } \text{key}(\text{right}(x)) \geq \text{key}(x).$$

2. *For n elements in the tree, the height of the tree is limited by $O(\log n)$.*
3. *Each node x in the tree stores two additional attributes size and sum.*
   (a) size denotes the size of the subtree mounted at x:

$$\text{size}(x) = \begin{cases} 0 & \text{if } x = \varnothing \\ \text{size}(\text{left}(x)) + \text{size}(\text{right}(x)) + 1 & \text{else} \end{cases}$$

   (b) sum denotes the sum of all values in the subtree mounted at x:

$$\text{sum}(x) = \begin{cases} 0 & \text{if } x = \varnothing \\ \text{sum}(\text{left}(x)) + \text{sum}(\text{right}(x)) + \text{value}(x) & \text{else} \end{cases}$$

4. *The correct value for above attributes is maintained after insertion.*

Based on aforementioned definitions, we use algorithm 2 to compute l_i^+, xv_i^+, l_i^- and xv_i^-. The auxiliary function CountSmaller is defined in algorithm 3, and CountLarger works in a similar manner. The complexity of these functions corresponds to the complexity of finding an element in a binary search tree, which is $O(\log n)$. Hence, the overall complexity of algorithm 2 is $O(n \log n)$, and the Hessian-vector product in (11) can be carried out in $O(nd + d + n \log n)$, after sorting according to $\boldsymbol{w}^T \boldsymbol{x}_i$, which costs $O(n \log n)$. Thus, one conjugate gradient iteration does not depend on the number of comparable pairs p anymore, which scales quadratically in the number of samples. Finally, the overall complexity of training a ranking-based survival support vector machine as outlined in algorithm 1 is

$$[O(n \log n) + O(nd + d + n \log n)] \cdot \bar{N}_{\text{CG}} \cdot N_{\text{Newton}}, \tag{19}$$

where $\bar{N}_{\text{CG}}$ and N_{Newton} are the average number of conjugate gradient iterations and the total number of Newton updates, respectively.

Algorithm 2. Efficient computation of l_i^+, l_i^-, σ_i^+, and σ_i^-.

Input: Training data $\mathcal{D} = \{(\boldsymbol{x}_k, y_k, \delta_k)\}_{k=1}^n$, coefficient vectors $\boldsymbol{w}$ and $\boldsymbol{v}$.
Output: l_i^+, l_i^-, σ_i^+, and σ_i^- $\forall i \in \{1, \ldots, n\}$

1   Sort all $\boldsymbol{w}^T \boldsymbol{x}_i$ in ascending order, such that $\boldsymbol{w}^T \boldsymbol{x}_{\pi(1)} \leq \cdots \leq \boldsymbol{w}^T \boldsymbol{x}_{\pi(n)}$;
2   $T \leftarrow$ an empty order statistic tree;
3   $j \leftarrow 1$;
4   **for** $i \leftarrow 1$ **to** n **do**
5   **while** $j \leq n$ *and* $\boldsymbol{w}^T \boldsymbol{x}_{\pi(j)} < \boldsymbol{w}^T \boldsymbol{x}_{\pi(i)} + 1$ **do**
6   Insert $(y_{\pi(j)}, \boldsymbol{x}_{\pi(j)}^T \boldsymbol{v})$ into T;
7         $j \leftarrow j + 1$;
8   **end**
9   **if** $\delta_{\pi(i)} = 1$ **then**
10         $(l_{\pi(i)}^+, xv_{\pi(i)}^+) \leftarrow$ CountLarger($root\ of\ T$, $y_{\pi(i)}$);
11   **else**
12         $(l_{\pi(i)}^+, xv_{\pi(i)}^+) \leftarrow (0, 0)$;
13   **end**
14   **end**
15   $j \leftarrow n$;
16   $T \leftarrow$ an empty order statistic tree;
17   **for** $i \leftarrow n$ **to** 1 **do**
18   **while** $j \geq 1$ *and* $\boldsymbol{w}^T \boldsymbol{x}_{\pi(j)} > \boldsymbol{w}^T \boldsymbol{x}_{\pi(i)} - 1$ **do**
19   **if** $\delta_{\pi(j)} = 1$ **then** Insert $(y_{\pi(j)}, \boldsymbol{x}_{\pi(j)}^T \boldsymbol{v})$ into T;
20         $j \leftarrow j - 1$;
21   **end**
22      $(l_{\pi(i)}^-, xv_{\pi(i)}^-) \leftarrow$ CountSmaller($root\ of\ T$, $y_{\pi(i)}$);
23   **end**

Algorithm 3. CountSmaller

Input: node x in order statistic tree, survival time y_i
Output: l_i^- (number of uncensored samples with $y_s < y_i$), and
     $\sigma_i^- = \sum_{s \in \mathrm{SV}_i^-} \boldsymbol{x}_i^T \boldsymbol{v}$

1   **if** $x = \varnothing$ **then**
2      $l_i^- \leftarrow 0; \sigma_i^- \leftarrow 0$;
3   **else if** key(x) $= y_i$ **then**
4      $l_i^- \leftarrow$ size(left(x));
5      $\sigma_i^- \leftarrow$ sum(left(y));
6   **else if** key(x) $< y_i$ **then**
7      $(l_i^-, \sigma_i^-) \leftarrow$ CountSmaller(right(x), y_i);
8      $l_i^- \leftarrow l_i^- +$ size(x) $-$ size(right(x));
9      $\sigma_i^- \leftarrow \sigma_i^- +$ sum(x) $-$ sum(right(x));
10   **else** // key(x) $> y_i$
11      $(l_i^-, \sigma_i^-) \leftarrow$ CountSmaller(left(x), y_i);
12   **end**

4   Survival Analysis as Regression Problem

Instead of treating survival analysis as a ranking problem, authors have proposed regression-based approaches using an absolute loss as well [17, 22]. In contrast to a ranking-based model, a regression model can predict the exact time of an event. Training algorithms for such a model need to be aware of censored patient record as well. For right censored patients – those who did not experience an event – no information about the correctness of predicted survival times beyond the time of censoring is available. A valid error can only be computed for patients that experienced an event during the study period, or if the predicted survival time is too early, i.e., before the time of censoring. Experiments in [26] revealed that survival models based on ε-insensitive support vector regression worked equally well if the insensitive zone is set to zero. Hence, our regression objective is based on an ordinary least square problem with ℓ_2 penalty and the additional consideration of right censoring.

$$f_{\mathrm{Regr.}}(\boldsymbol{w}, b) = \frac{1}{2}\boldsymbol{w}^T\boldsymbol{w} + \frac{\gamma}{2}\sum_{i=0}^{n}\left(\zeta_{\boldsymbol{w},b}(y_i, x_i, \delta_i)\right)^2 \tag{20}$$

$$\zeta_{\boldsymbol{w},b}(y_i, \boldsymbol{x}_i, \delta_i) = \begin{cases} \max(0, y_i - \boldsymbol{w}^T\boldsymbol{x}_i - b) & \text{if } \delta_i = 0, \\ y_i - \boldsymbol{w}^T\boldsymbol{x}_i - b & \text{if } \delta_i = 1, \end{cases} \tag{21}$$

where $b \in \mathbb{R}$ is the intercept.

By combining all parameters into a single vector $\boldsymbol{\omega} = (b, \boldsymbol{w})^T$, and extending $\boldsymbol{X}$ by a column of all ones to accommodate the intercept, the objective can be expressed in matrix form as follows:

$$f_{\mathrm{Regr.}}(\boldsymbol{\omega}) = \frac{1}{2}\boldsymbol{\omega}^T\boldsymbol{\omega} + \frac{\gamma}{2}(\boldsymbol{y} - \boldsymbol{X}\boldsymbol{\omega})^T \boldsymbol{R}_{\boldsymbol{\omega}}(\boldsymbol{y} - \boldsymbol{X}\boldsymbol{\omega}) \tag{22}$$

where $\boldsymbol{R}_{\boldsymbol{\omega}}$ is a diagonal matrix with the i-th element being 1 if $y_i > \boldsymbol{w}^T\boldsymbol{x}_i + b$ or $\delta_i = 1$, and zero otherwise. Due to $f_{\mathrm{Regr.}}$ being a convex quadratic function, we can use truncated Newton optimization to minimize it, as described in algorithm 1. In addition, we can easily create a hybrid model that addresses the ranking and regression objective concurrently; its objective function is defined as

$$f_{\mathrm{hybrid}}(\boldsymbol{w}, b) = \frac{1}{2}\boldsymbol{w}^T\boldsymbol{w} + \frac{\gamma}{2}\left[\alpha \sum_{i,j\in\mathcal{P}} \max(0, 1 - (\boldsymbol{w}^T\boldsymbol{x}_i - \boldsymbol{w}^T\boldsymbol{x}_j))^2\right.$$
$$\left. + (1 - \alpha)\sum_{i=0}^{n}\left(\zeta_{\boldsymbol{w},b}(y_i, \boldsymbol{x}_i, \delta_i)\right)^2\right]. \tag{23}$$

The hyper-parameter $\alpha \in [0, 1]$ controls the relative weight of the regression and ranking objective. Clearly, if $\alpha = 1$ it reduces to the ranking objective, and if $\alpha = 0$ to the regression objective.

5   Non-linear Extension

So far, we only discussed linear survival support vector machines and their efficient training in the primal. If data are more complex, one might want to model non-linear functions through the use of kernel functions. Commonly, the representer theorem [18] is employed and optimization is carried out in the dual rather than the primal. The weights w are then a linear combination of the training samples. However, if training data is large, the number of support vectors increases as well, resulting in excessive computational costs. Chapelle et al. [5] showed that solving the non-linear problem is equivalent to the combination of Kernel PCA and training in the primal. Thus, efficient training of non-linear survival models is straightforward using the optimization scheme outlined above.

6   Experiments

In our experiments, we first studied the efficiency of our proposed algorithm to minimize the ranking-based objective function and then investigated the predictive performance of ranking, regression, and hybrid approaches. We standardized continuous features to have zero mean and unit standard deviation, and randomly resolved ties in survival times before optimization. For regression, we used the logarithm of survival times y_i as target value.

6.1   Computational Efficiency

In the first set of experiments, we compared the training time of three different formulations of the ranking-based objective function: the simple formulation in (2), the alternative formulation in (8), and our efficient proposed formulation in (17). We generated synthetic survival data of varying size following [4]. Data consisted of 10 normal distributed features and two redundant features, which were linear combinations of a subset of the first ten features. Correlations between the first ten features were defined as follows: $r_{1,3} = 0.03$, $r_{2,5} = 0.42$, $r_{3,5} = 0.08$, $r_{3,9} = 0.03$, $r_{5,8} = -0.55$, $r_{6,9} = 0.32$, and the remainder all zero. Survival times were Gompertz distributed and depended on a linear combination of all features. Finally, half of the samples were randomly censored. Our choice of order statistic trees were red-black trees [3] and AVL trees [1]. To minimize the influence of the operating system's process scheduler in our measurements, we report the lowest training time of ten repetitions in wall time.

Figure 1 shows the lowest training time following algorithm 1. The naive and improved optimization failed with more than 20,000 samples because of excessive memory requirements due to explicitly constructing the sparse matrix A and A_w, respectively. For all datasets, optimization converged after less than 20 iterations. Although A has to be constructed only once for the simple optimization, training time quickly degenerates because it repeatedly has to be multiplied by Xw, which takes $O(pn)$ time. The improved optimization updates A_w after

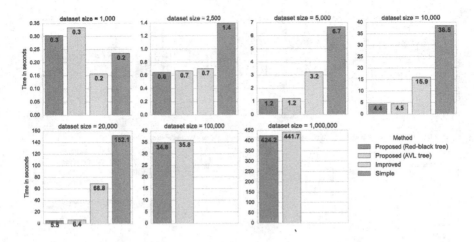

Fig. 1. Training time of survival models using ranking objective with truncated Newton optimization. Simple refers to the objective function in (2) and improved to the one in (8). Our proposed algorithm uses the efficient formulation in (17) with red-black trees or AVL trees.

Table 1. Overview of datasets used in our experiments.

Dataset	n	d	Events	Outcome
AIDS study [12]	1,151	13	96 (8.3%)	AIDS defining event or death
Breast cancer [8]	198	80	62 (31.3%)	Distant metastases
Coronary artery disease [21]	1,204	60	196 (15.9%)	Myocardial infarction or death
Framingham Offspring [15]	4,892	150	1,166 (23.8%)	Coronary vessel disease
Veteran's Lung Cancer [14]	137	6	128 (93.4%)	Death
Worcester Heart Attack Study [12]	500	14	215 (43.0%)	Death

each iteration of Newton's method, but only needs to perform $O(p_w n)$ operations when multiplied by Xw, which results in a lower training time. Using order statistic trees, the training time and memory requirements can be lowered significantly; for very large datasets, red-black trees were superior to AVL trees.

6.2 Prediction Performance

We evaluated the predictive performance of our proposed method for survival analysis on six real-world datasets of varying size, number of features, and amount of censoring (see table 1). In addition to the three models proposed here, we included Cox's proportional hazards model [6] with ℓ_2 (ridge) penalty, and ranking-based survival SVM with hinge loss [10,24]. The regularization parameter γ for survival SVM controls the weight of the (squared) hinge loss, whereas for Cox's proportional hazards model, $\lambda = \gamma^{-1}$ controls the weight of the ℓ_2 penalty.

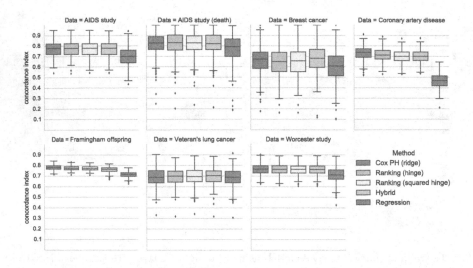

Fig. 2. Concordance index of Cox's proportional hazards model with ℓ_2 (ridge) penalty and four different survival SVM models: ranking objective with hinge loss, ranking objective with squared hinge loss, regression objective, and combined ranking and regression (hybrid).

Optimal performance was determined by a grid search over hyper-parameters. We set γ and λ to 2^i, where we altered i from -12 to 12 in steps of 2. Similar for α, which ranged from 0.05 to 0.95 in steps of 0.05. The maximum number of iterations of Newton's method was one thousand. Performance was measured by Harrell's concordance index (c index) [11], which is the ratio of correctly ordered pairs to comparable pairs. A c index of 0.5 corresponds to a random model and 1.0 to a perfect model. In addition, we measured the root mean squared error (RMSE) on uncensored patients to evaluate regression models. For each parameter setting, we randomly split each dataset into two equally sized parts, one for training and one for testing. Results reported here are with respect to the configuration that performed best on the training portion of 200 different random splits.

Figure 2 summarizes the results of our experiments with respect to c index. We observed that ranking-based approaches to survival analysis, using hinge or squared hinge loss, were comparable to Cox's proportional hazards model with ℓ_2 penalty and superior to a regression-based approach. We believe this is why the combined ranking-regression technique did not exceed the performance of the pure ranking approach. In fact, hyper-parameter search assigned more weight to the ranking objective in all cases but one. The only exception occurred for the breast cancer dataset, where $\alpha = 0.25$ was chosen and the hybrid model performed best. The reason for this becomes obvious when looking at the RMSE shown in figure 3. Predictions of survival time are off by a large extent on all datasets, which renders the regression objective unsuitable. This can be explained by the distribution of survival times, which are – even

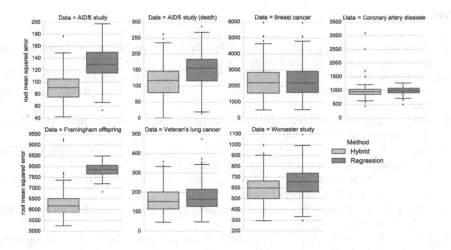

Fig. 3. Root mean squared error (RMSE) of regression-based and hybrid survival support vector machine.

after log-transformation – far from normally distributed, and thus violate a basic assumption of ordinary least squares. In [26] however, regression was based on absolute loss and outperformed ranking. A possible explanation might be the fact that squared loss is more sensitive to outliers than absolute loss. This problem could be alleviated by introducing sample weights to reduce the influence of outliers in the squared loss function. Finally, the performance of all approaches varied to a similar degree among 200 randomly selected train-test splits. We obtained similar results for non-linear survival models.

7 Conclusion

In this paper, we proposed an efficient method for training ranking-based and regression-based survival support vector machines. Our algorithm accounts for right censoring of patient records and avoids explicitly constructing a matrix of pairwise constraints – quadratic in the number of samples – by using order statistic trees. We experimentally showed that the reduced time and space complexity allow efficient training of survival models based on millions of patients, which would otherwise not been possible on commodity hardware. In addition to its high efficiency, the algorithm can be easily adapted for training non-linear as well as hybrid ranking and regression survival models. This opens up the opportunity to build survival models from large sets of medical health records to obtain new insights about the impact of particular factors on a disease.

Acknowledgments. We thank the Leibniz Supercomputing Centre (LRZ, www.lrz.de) for providing the computational resources for our experiments. Data were provided by the Framingham Heart Study of the National Heart Lung and Blood Institute of the National Institutes of Health and Boston University School of Medicine (Contract No.N01-HC-25195).

References

1. Adelson-Velsky, G., Landis, E.: An algorithm for the organization of information. In: Doklady Akademii Nauk SSSR, vol. 146, pp. 263–266 (1962)
2. Airola, A., Pahikkala, T., Salakoski, T.: Training linear ranking SVMs in linearithmic time using red–black trees. Pattern Recogn. Lett. **32**(9), 1328–1336 (2011)
3. Bayer, R.: Symmetric binary B-trees: Data structure and maintenance algorithms. Acta Inform. **1**(4), 290–306 (1972)
4. Bender, R., Augustin, T., Blettner, M.: Generating survival times to simulate Cox proportional hazards models. Stat. Med. **24**(11), 1713–1723 (2005)
5. Chapelle, O., Keerthi, S.S.: Efficient algorithms for ranking with SVMs. Information Retrieval **13**(3), 201–215 (2009)
6. Cox, D.R.: Regression models and life tables (with discussion). J. Roy. Stat. Soc. B **34**, 187–220 (1972)
7. Dembo, R.S., Steihaug, T.: Truncated Newton algorithms for large-scale optimization. Math. Programming **26**(2), 190–212 (1983)
8. Desmedt, C., Piette, F., Loi, S., Wang, Y., Lallemand, F., Haibe-Kains, B., Viale, G., Delorenzi, M., Zhang, Y., d'Assignies, M.S., Bergh, J., Lidereau, R., Ellis, P., Harris, A.L., Klijn, J.G., Foekens, J.A., Cardoso, F., Piccart, M.J., Buyse, M., Sotiriou, C.: Strong Time Dependence of the 76-Gene Prognostic Signature for Node-Negative Breast Cancer Patients in the TRANSBIG Multicenter Independent Validation Series. Clin. Cancer Res. **13**(11), 3207–3214 (2007)
9. Eleuteri, A., Taktak, A.F.G.: Support vector machines for survival regression. In: Biganzoli, E., Vellido, A., Ambrogi, F., Tagliaferri, R. (eds.) CIBB 2011. LNCS, vol. 7548, pp. 176–189. Springer, Heidelberg (2012)
10. Evers, L., Messow, C.M.: Sparse kernel methods for high-dimensional survival data. Bioinformatics **24**(14), 1632–1638 (2008)
11. Harrell, F.E., Califf, R.M., Pryor, D.B., Lee, K.L., Rosati, R.A.: Evaluating the Yield of Medical Tests. J. Am. Med. Assoc. **247**(18), 2543–2546 (1982)
12. Hosmer, D., Lemeshow, S., May, S.: Applied Survival Analysis: Regression Modeling of Time to Event Data. John Wiley & Sons, Inc. (2008)
13. Jensen, P.B., Jensen, L.J., Brunak, S.: Mining electronic health records: towards better research applications and clinical care. Nat. Rev. Genet. **13**(6), 395–405 (2012)
14. Kalbfleisch, J.D., Prentice, R.L.: The Statistical Analysis of Failure Time Data. John Wiley & Sons, Inc. (2002)
15. Kannel, W.B., Feinleib, M., McNamara, P.M., Garrision, R.J., Castelli, W.P.: An Investigation of Coronary Heart Disease in Families: The Framingham Offspring Study. Am. J. Epidemiol. **110**(3), 281–290 (1979)
16. Keerthi, S.S., DeCoste, D.: A Modified Finite Newton Method for Fast Solution of Large Scale Linear SVMs. J. Mach. Learn. Res. **6**, 341–361 (2005)
17. Khan, F.M., Zubek, V.B.: Support vector regression for censored data (SVRc): a novel tool for survival analysis. In: 8th IEEE Int. Conf. on Data Mining, pp. 863–868 (2008)
18. Kimeldorf, G.S., Wahba, G.: A correspondence between bayesian estimation on stochastic processes and smoothing by splines. Ann. Math. Stat. **41**, 495–502 (1970)
19. Lee, C.P., Lin, C.J.: Large-Scale Linear RankSVM. Neural Comput. **26**(4), 781–817 (2014)

20. Mangasarian, O.: A finite newton method for classification. Optimization Methods and Software **17**(5), 913–929 (2002)
21. Ndrepepa, G., Braun, S., Mehilli, J., Birkmeier, K.A., Byrne, R.A., Ott, I., Hösl, K., Schulz, S., Fusaro, M., Pache, J., Hausleiter, J., Laugwitz, K.L., Massberg, S., Seyfarth, M., Schömig, A., Kastrati, A.: Prognostic value of sensitive troponin T in patients with stable and unstable angina and undetectable conventional troponin. Am. Heart J. **161**(1), 68–75 (2011)
22. Shivaswamy, P.K., Chu, W., Jansche, M.: A support vector approach to censored targets. In: 7th IEEE Int. Conf. on Data Mining, pp. 655–660 (2007)
23. Steck, H., Krishnapuram, B., Dehing-oberije, C., Lambin, P., Raykar, V.C.: On ranking in survival analysis: bounds on the concordance index. In: Adv. Neural Inf. Process. Syst., vol. 20, pp. 1209–1216 (2008)
24. Van Belle, V., Pelckmans, K., Suykens, J.A., Van Huffel, S.: Support vector machines for survival analysis. In: Proc. 3rd Int. Conf. Comput. Intell. Med. Healthc, pp. 1–8 (2007)
25. Van Belle, V., Pelckmans, K., Suykens, J.A., Van Huffel, S.: Survival SVM: a practical scalable algorithm. In: Proc. of 16th European Symposium on Artificial Neural Networks, pp. 89–94 (2008)
26. Van Belle, V., Pelckmans, K., Van Huffel, S., Suykens, J.A.K.: Support vector methods for survival analysis: a comparison between ranking and regression approaches. Artif. Intell. Med. **53**(2), 107–118 (2011)

Superset Learning Based on Generalized Loss Minimization

Eyke Hüllermeier[1]([✉]) and Weiwei Cheng[2]

[1] Department of Computer Science, University of Paderborn, Paderborn, Germany
eyke@upb.de
[2] Amazon Inc., Berlin, Germany

Abstract. In standard supervised learning, each training instance is associated with an outcome from a corresponding output space (e.g., a class label in classification or a real number in regression). In the superset learning problem, the outcome is only characterized in terms of a superset—a subset of candidates that covers the true outcome but may also contain additional ones. Thus, superset learning can be seen as a specific type of weakly supervised learning, in which training examples are ambiguous. In this paper, we introduce a generic approach to superset learning, which is motivated by the idea of performing model identification and "data disambiguation" simultaneously. This idea is realized by means of a generalized risk minimization approach, using an extended loss function that compares precise predictions with set-valued observations. As an illustration, we instantiate our meta learning technique for the problem of label ranking, in which the output space consists of all permutations of a fixed set of items. The label ranking method thus obtained is compared to existing approaches tackling the same problem.

1 Introduction

Superset learning is a specific type of learning from weak supervision, in which the outcome (response) associated with a training instance is only characterized in terms of a subset of possible candidates. Thus, superset learning is somehow in-between supervised and semi-supervised learning, with the latter being a special case (in which supersets are singletons for the labeled examples and cover the entire output space for the unlabeled ones). There are numerous applications in which only partial information about outcomes is available [13].

Correspondingly, the superset learning problem has received increasing attention in recent years, and has been studied under various names, such as *learning from ambiguously labeled examples* or *learning from partial labels* [6,11,15,5]. The contributions so far also differ with regard to their assumptions on the incomplete information being provided. In this paper, we only assume the actual outcome to be covered by the subset—hence the name *superset* learning.

We introduce an approach to superset learning based on direct loss minimization with a suitably generalized loss function. While previous work on superset learning has mainly been focused on (multi-class) classification, our approach is

© Springer International Publishing Switzerland 2015
A. Appice et al. (Eds.): ECML PKDD 2015, Part II, LNAI 9285, pp. 260–275, 2015.
DOI: 10.1007/978-3-319-23525-7_16

Fig. 1. Data generating process in the setting of superset learning.

completely generic and does not make any specific assumptions about the output space. In fact, we argue that superset learning is specifically interesting for complex, structured output prediction, because information about such outputs is indeed often incomplete. This is why, in the second part of the paper, we apply our approach to the problem of label ranking, where outputs take the form of rankings. More specifically, by instantiating our approach to superset learning for the case of label ranking, we develop a new method for this problem, which turns out to perform quite strongly in first experimental studies.

The rest of the paper is organized as follows. In the next section, we introduce the basic problem setting and the main notation to be used throughout the paper. Our new approach to superset learning is then introduced in Section 3.[1] In Sections 4 and 5, we recall the label ranking problem and introduce our new method.[2] The paper concludes with a summary and an outlook on future work in Section 6.

2 Setting and Notation

Consider a standard setting of supervised learning with an input (instance) space $\mathcal{X}$ and an output space $\mathcal{Y}$. The goal is to learn a mapping from $\mathcal{X}$ to $\mathcal{Y}$ that captures, in one way or the other, the dependence of outputs (responses) on inputs (predictors). The learning problem essentially consists of choosing an optimal model (hypothesis) M^* from a given model space (hypothesis space) $\mathbf{M}$, based on a set of training data

$$\mathcal{D} = \left\{ (\boldsymbol{x}_n, y_n) \right\}_{n-1}^{N} \in (\mathcal{X} \times \mathcal{Y})^N . \tag{1}$$

More specifically, optimality typically refers to optimal prediction accuracy, i.e., a model is sought whose expected prediction loss or *risk*

$$\mathcal{R}(M) = \int L\big(y, M(\boldsymbol{x})\big) \, d\, \mathbf{P}(\boldsymbol{x}, y) \tag{2}$$

is minimal; here, $L : \mathcal{Y} \times \mathcal{Y} \longrightarrow \mathbb{R}$ is a loss function, and $\mathbf{P}$ is an (unknown) probability measure on $\mathcal{X} \times \mathcal{Y}$ modeling the underlying data generating process.

In this paper, we are interested in the case where output values $y_n \in \mathcal{Y}$ are not necessarily observed precisely; instead, only a superset $Y_n \subseteq \mathcal{Y}$ is observed.

---

[1] This approach is leaned on [8], where a similar problem is studied in the context of learning from "fuzzy data".

[2] A first version of this method has been presented at M-PREF 2013, 7th Multidisciplinary Workshop on Advances in Preference Handling, Beijing, China.

Therefore, the learning algorithm does not have direct access to the (precise) data (1), but only to the (imprecise, ambiguous) observations

$$\mathcal{O} = \{(\boldsymbol{x}_n, Y_n)\}_{n=1}^{N} \in (\mathcal{X} \times 2^{\mathcal{Y}})^N . \tag{3}$$

More specifically, we assume a data generating process as sketched in Figure 1: Given an instance $\boldsymbol{x} \in \mathcal{X}$, an underlying process first generates a precise outcome $y \in \mathcal{Y}$, which is then turned into an imprecise observation in the form of a superset $Y \ni y$. We refer to this process of generating Y as "ambiguation" or "imprecisiation" of y.

In the following, we denote by $\mathbf{Y} = Y_1 \times Y_2 \times \cdots \times Y_N$ the (Cartesian) product of the supersets observed for $\boldsymbol{x}_1, \ldots, \boldsymbol{x}_N$. Moreover, each $\boldsymbol{y} = (y_1, \ldots, y_N) \in \mathbf{Y}$ is called an *instantiation* of the imprecisely observed data. More generally, we call $\mathcal{D}$ in (1) an instantiation of $\mathcal{O}$ if the instances $\boldsymbol{x}_n$ coincide and $y_n \in Y_n$ for all $n \in [N] = \{1, \ldots, N\}$.

Prior to proceeding, let us emphasize that the Y_n are considered as constraints on *actual* outcomes y_n, not on any kind of *ideal* outcomes or predictions for the instance $\boldsymbol{x}_n$. In regression, for example, outcomes y_n could be random variables with expected value $\mu(\boldsymbol{x}_n)$ and standard deviation $\sigma(\boldsymbol{x}_n)$. What we assume, then, is $Y_n \ni y_n$ but not necessarily $Y_n \ni \mu(\boldsymbol{x}_n)$.

3   A Loss Minimization Approach

Given the data generating process as outlined above, the likelihood of a model $M \in \mathbf{M}$ can be defined by the probability of the data given the model, i.e.,

$$\ell(M) = \mathbf{P}(\mathcal{O}, \mathcal{D} \,|\, M) = \mathbf{P}(\mathcal{D} \,|\, M)\mathbf{P}(\mathcal{O} \,|\, \mathcal{D}, M) . \tag{4}$$

A reasonable assumption is that the imprecise observations Y_n only depend on the underlying true outcomes y_n but not on the model M or, in other words, that $\mathcal{O}$ is conditionally independent of M given $\mathcal{D}$. Under this assumption, $\mathbf{P}(\mathcal{O} \,|\, \mathcal{D}, M) = \mathbf{P}(\mathcal{O} \,|\, \mathcal{D})$ and (4) becomes

$$\ell(M) = \mathbf{P}(\mathcal{O}, \mathcal{D} \,|\, M) = \mathbf{P}(\mathcal{D} \,|\, M)\mathbf{P}(\mathcal{O} \,|\, \mathcal{D}) . \tag{5}$$

As can be seen, the likelihood of M under the superset data is a weighted average of standard likelihoods $\mathbf{P}(\mathcal{D} \,|\, M)$, with each precise data sample $\mathcal{D}$ being weighted by the probability $\mathbf{P}(\mathcal{O} \,|\, \mathcal{D})$ of observing $\mathcal{O}$ if the true underlying data were $\mathcal{D}$. In some cases, specific knowledge about these probabilities, i.e., about the process of imprecisiation, is available; for example, in a classification setting, a connection between true labels and observed partial labels is established in terms of a so-called mixing matrix in [17]. However, in lack of any specific knowledge of that kind, the most reasonable assumption we can make is

$$\mathbf{P}(Y \,|\, y) = \begin{cases} \text{const} & \text{if } Y \ni y \\ 0 & \text{if } Y \not\ni y \end{cases} \tag{6}$$

We call this the *superset assumption*, as it does not assume anything else than the observation Y being a superset of y; in fact, the uniform distribution (6) is the "weakest" distribution in accordance with this assumption, namely the one with the highest entropy among all distributions allocating the entire probability mass on supersets of y.

Now, it is easy to see that the likelihood (4) will vanish as soon as $y_n \notin Y_n$ for at least one of the observations, while $\mathbf{P}(\mathcal{O} \mid \mathcal{D})$ is a non-negative constant that does not depend on M if $y_n \in Y_n$ for all $n \in [N]$. Thus, maximizing the likelihood is equivalent to finding

$$M^* \in \operatorname*{argmax}_{M \in \mathbf{M}} \max_{\boldsymbol{y} \in \mathbf{Y}} \prod_{n=1}^{N} \mathbf{P}(y_n \mid M, \boldsymbol{x}_n) \tag{7}$$

or, equivalently,

$$M^* \in \operatorname*{argmin}_{M \in \mathbf{M}} \min_{\boldsymbol{y} \in \mathbf{Y}} \sum_{n=1}^{N} -\log \mathbf{P}(y_n \mid M, \boldsymbol{x}_n) \ . \tag{8}$$

3.1 Generalized Loss Minimization

Recall the principle of *empirical risk minimization* (ERM): A model M^* is sought that minimizes the *empirical risk*

$$\mathcal{R}_{emp}(M) = \frac{1}{N} \sum_{n=1}^{N} L\big(y_n, M(\boldsymbol{x}_n)\big) \ , \tag{9}$$

i.e., the average loss on the training data $\mathcal{D} = \{(\boldsymbol{x}_i, y_i)\}_{i=1}^{N}$. The empirical risk (9) serves as a surrogate of the true risk (2). In order to avoid the problem of possibly *overfitting* the data, not (9) itself is typically minimized but a *regularized* version thereof. This is of minor importance here, however, and the approach outlined in the following can be generalized from standard ERM to regularized risk minimization in a straightforward way.

Now, coming back to our superset learning problem, it is interesting to note that the approach (8) can be seen as a special case of ERM, with the loss function $L(\cdot)$ given by the logistic loss: $L(y, \hat{y}) = L(y, M(\boldsymbol{x}))$ is the (negative) logarithm of the probability of y under the distribution specified by $M(\boldsymbol{x})$. For example, suppose that $\mathbf{M}$ is the class of linear regression models with normally distributed error term, i.e., $y = M_{\boldsymbol{w}}(\boldsymbol{x}) = \boldsymbol{w}^\top \boldsymbol{x} + \epsilon$. Then,

$$M^* \in \operatorname*{argmin}_{M_{\boldsymbol{w}} \in \mathbf{M}} \min_{\boldsymbol{y} \in \mathbf{Y}} \sum_{n=1}^{N} \big(y_n - \boldsymbol{w}^\top \boldsymbol{x}_n\big)^2 \ .$$

As can be seen, each candidate model M is evaluated optimistically according to

$$\overline{\mathcal{R}}_{emp}(M_{\boldsymbol{w}}) = \min_{\boldsymbol{y} \in \mathbf{Y}} \sum_{n=1}^{N} \big(y_n - \boldsymbol{w}^\top \boldsymbol{x}_n\big)^2 \ ,$$

i.e., the standard (squared) loss it makes on the instantiation $\boldsymbol{y}$ that is most favorable for M, and then the model M^* with the best optimistic evaluation is chosen.

Of course, the logistic loss could in principle be replaced by any other loss function $L(\cdot)$ of interest; this is in fact even a prerequisite for working with non-probabilistic models, i.e., if a model M merely produces predictions in $\mathcal{Y}$ but not complete probability distributions. A model M is then evaluated according to

$$\overline{\mathcal{R}}_{emp}(M) = \min_{\boldsymbol{y} \in \mathbf{Y}} \frac{1}{N} \sum_{n=1}^{N} L\big(y_n, M(\boldsymbol{x}_n)\big) \ .$$

Moreover, given a loss that is decomposable (over examples), the "optimism" can be moved into the loss:

$$\min_{\boldsymbol{y} \in \mathbf{Y}} \sum_{n=1}^{N} L\big(y_n, M(\boldsymbol{x}_n)\big) = \sum_{n=1}^{N} \min_{y_n \in Y_n} L\big(y_n, M(\boldsymbol{x}_n)\big)$$
$$= \sum_{n=1}^{N} L^*\big(y_n, M(\boldsymbol{x}_n)\big)$$

with the generalized loss function

$$L^*(Y, \hat{y}) = \min\big\{L(y, \hat{y}) \,|\, y \in Y\big\} \tag{10}$$

that compares (precise) predictions with set-valued observations. We call this loss the *optimistic superset loss* (OSL). Note that this loss covers the *superset error* $[\![\hat{y} \notin Y]\!]$, which is commonly used in superset label learning for classification [14], as a special case.

In summary, our approach to superset learning is based on the minimization of the empirical risk with respect to this generalized loss function. Thus, each candidate model $M \in \mathbf{M}$ is evaluated in terms of

$$\overline{\mathcal{R}}_{emp}(M) = \frac{1}{N} \sum_{n=1}^{N} L^*\big(Y_n, M(\boldsymbol{x}_n)\big) \ , \tag{11}$$

and an optimal model M^* is one that minimizes (11) — or, as mentioned before, a regularized version thereof.

3.2 Data Disambiguation

In the context of learning from data, not only the data is providing information about the (unknown) model, but also the other way around. This view is made explicit in the Bayesian approach to data analysis, where the joint model/data probability $\mathbf{P}(M, \mathcal{D})$ can be written either way, as $\mathbf{P}(M)\mathbf{P}(\mathcal{D}\,|\,M)$ and $\mathbf{P}(\mathcal{D})\mathbf{P}(M\,|\,\mathcal{D})$. From a Bayesian perspective, the superset learning problem could be tackled quite naturally by not only starting with a prior on the

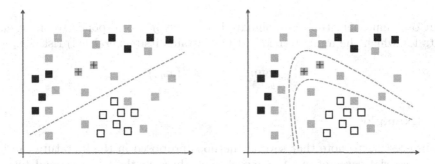

Fig. 2. Model identification and data disambiguation go hand in hand. Left: Assuming a linear model, the two example marked by a cross are most likely positive. Right: Fitting a nonlinear model, disambiguation of these examples is less obvious.

model class $\mathbf{M}$ but also on the data, for example defining a uniform prior on each superset Y_n and zero probability outside. Inference would then come down to attuning these priors, e.g., by turning priors into posteriors on the model space and the data space in an alternating way. Eventually, this will yield a joint model/data (posterior) probability $\mathbf{P}(M, \mathcal{D})$ that will not only inform about a most plausible model M^* but also about a most plausible instantiation $\boldsymbol{y}^*$ of the imprecise data. In other words, it will help *disambiguating* the data.

Our approach supports data disambiguation, too, albeit in a different way. As can be seen from the "double-max" operation in (7), model and data are selected in the most favorable combination. Thus, disambiguation essentially relies on the inductive bias implemented by the model class $\mathbf{M}$ [9]. In fact, against the background of the learning bias, some instantiations of the ambiguous data appear to be more plausible than others. This is illustrated in Figure 2 for a simple scenario of binary classification, in which some instances are known to be positive (marked in black, $y_n = +1$), some are known to be negative (white, $y_n = -1$), whereas some are unlabeled (grey, $Y_n = \{-1, +1\}$). Now, consider the two unlabeled instances marked with a cross, for example. Looking at each example in isolation, nothing can be said about the actual (precise) label. However, when looking at the data as a whole, in conjunction with the assumption of a linear decision boundary between the two classes, the positive class is clearly more plausible than the negative class (left picture). Yet, looking at the data with a slightly less biased view and also allowing for a nonlinear (e.g., quadratic) discriminant, these cases are more difficult to disambiguate: Both the positive and negative class appear to be plausible, since both can be obtained with plausible models $M \in \mathbf{M}$, i.e., models that are in agreement with the rest of the data. This example also shows that the stronger the bias, i.e., the more background knowledge is incorporated in the learning process, the easier disambiguation of the data becomes.

In our approach, the disambiguated outcome $\boldsymbol{y}^*$ corresponds to those elements for which the minimizer M^* of (11) attains its (generalized) risk, i.e.,

$$y_n^* = \operatorname*{argmin}_{y_n \in Y_n} L\big(y_n, M^*(\boldsymbol{x}_n)\big) \ .$$

3.3   Examples

It is interesting to note that several methods proposed in the literature can be seen as special cases of our framework, i.e., these methods correspond to the minimization of the generalized loss (11) following to a suitable imprecisiation of the data. For example, the ϵ-insensitive loss $L(y, \hat{y}) = \max(|y-\hat{y}|-\epsilon, 0)$ used in support vector regression [16] corresponds to the OSL (10) with L the standard L_1 loss $L(y, \hat{y}) = |y - \hat{y}|$ and precise data y_n being replaced by interval-valued data $Y_n = [y_n - \epsilon, y_n + \epsilon]$ (cf. Figure 3).

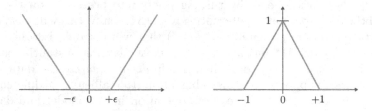

Fig. 3. The ϵ-insensitive loss (left) and the hat loss (right).

Perhaps more interestingly, we obtain semi-supervised learning with support vector machines as a special case by considering unlabeled data as instances labeled with the superset $\{-1, +1\}$ (like in our above example). The generalized loss (10), with L the standard hinge loss, then corresponds to the (non-convex) "hat loss" (cf. Figure 3). More generally, if the loss L is a margin loss of the form $L(y, s) = f(ys)$, comparing a class label $y \in \{-1, +1\}$ with a predicted score $s \in \mathbb{R}$ in terms of a non-increasing function $f : \mathbb{R} \longrightarrow \mathbb{R}$, it is easy to verify that (10) is given by $L^*(Y, S) = f(|ys|)$ for $Y = \{-1, +1\}$ (and, of course, $L^*(Y, S) = L(Y, s) = f(ys)$ for $Y = \{-1\}$ and $Y = \{+1\}$).

3.4   Superset Learning for Structured Output Prediction

Existing work on superset learning has been focused almost exclusively on (multiclass) classification. Obviously, our approach is not restricted to this problem; instead, the output space $\mathcal{Y}$ is completely generic. In fact, one may even argue that superset learning is more interesting for problems with complex, structured outcomes, since outcomes of that kind are often only partially specified in practice. A partial structure is then quite naturally associated with a subset of $\mathcal{Y}$, namely the set of all consistent completions—note that this view is somehow in contrast to the common view of a label set Y_n as a *corruption* of the true label, and of the additional labels as *distractors* [13]. In the following, we shall instantiate our approach for a problem of that kind, namely label ranking [19], where the output space consists of rankings (permutations)

4   Label Ranking

Let $\mathcal{C} = \{c_1, \ldots, c_K\}$ be a finite set of (choice) alternatives, referred to as *labels*. We consider total order relations $\succ$ on $\mathcal{C}$, where $c_i \succ c_j$ indicates that c_i precedes c_j in the order. Since a ranking can be seen as a special type of preference relation, we shall also say that $c_i \succ c_j$ indicates a preference for c_i over c_j. Formally, a total order $\succ$ can be identified with a permutation $\bar{\pi}$ of the set $[K]$, such that $\bar{\pi}(i)$ is the position of c_i in the order. Let the output space $\mathcal{Y}$ be given by the set of permutations of $[K]$ (the symmetric group of order K).

In the setting of label ranking, preferences are "contextualized" by instances $x \in \mathcal{X}$. Thus, each instance x is associated with a ranking $\succ_x$ of the label set $\mathcal{C}$ or, equivalently, a permutation $\bar{\pi}_x \in \mathcal{Y}$. More specifically, since label rankings do not necessarily depend on instances in a deterministic way, each instance x is associated with a probability distribution $\mathbf{P}(\cdot \mid x)$ on $\mathcal{Y}$. Thus, for each $\bar{\pi} \in \mathcal{Y}$, $\mathbf{P}(\bar{\pi} \mid x)$ denotes the probability to observe $\bar{\pi}$ in the context specified by x.

The goal in label ranking is to learn a "label ranker", that is, a model $\mathcal{M} : \mathcal{X} \longrightarrow \mathcal{Y}$ that predicts a ranking $\hat{\pi}$ for each instance x given as an input. As training data $\mathcal{D}$, a label ranker uses a set of instances x_n ($n \in [N]$), together with information about the associated rankings π_n. Ideally, complete rankings are given as training information, i.e., a single observation is a tuple of the form $(x_n, \pi_n) \in \mathcal{X} \times \mathcal{Y}$; we call an observation of that kind a *complete* example. From a practical point of view, however, it is important to allow for incomplete information in the form of a ranking of some but not all of the labels in $\mathcal{C}$:

$$c_{\tau(1)} \succ_x c_{\tau(2)} \succ_x \cdots \succ_x c_{\tau(J)} \;, \tag{12}$$

where $J < K$ and $\{\tau(1), \ldots, \tau(J)\} \subset [K]$. In the following, we will write complete rankings $\bar{\pi}$ with an upper bar (as we already did above). If a ranking π is not complete, then $\pi(j)$ is the position of c_j in the incomplete ranking, provided this label is contained, and $\pi(j) = 0$ otherwise.

Information in the form of an incomplete ranking π is naturally represented in terms of a subset $Y = E(\pi) \subseteq \mathcal{Y}$, namely the set of all of its linear extensions $E(\pi)$ (complete rankings preserving the order of those labels contained in π). Note that, if $\bar{\pi}$ is a completion of π, then $\bar{\pi}(k) \geq \pi(k)$ for all $k \in [K]$.

4.1   Prediction Accuracy

The prediction accuracy of a label ranker is typically assessed by comparing the true ranking $\bar{\pi}$ with the prediction $\hat{\pi}$ in terms of a distance measure on rankings. Among the most commonly used measures is the Kendall distance, which is defined by the number of inversions, that is, index pairs $\{i, j\} \subset [K]$ such that the order of c_i and c_j in $\bar{\pi}$ is inverted in $\hat{\pi}$:

$$D(\bar{\pi}, \hat{\pi}) = \sum_{1 \leq i < j \leq K} [\![\operatorname{sign}(\bar{\pi}(i) - \bar{\pi}(j)) \neq \operatorname{sign}(\hat{\pi}(i) - \hat{\pi}(j))]\!] \tag{13}$$

The well-known Kendall rank correlation measure is an affine transformation of (13) to the range $[-1, +1]$. Besides, the sum of L_1 or L_2 losses on the ranks of the individual labels are often used as an alternative:

$$D_1(\bar{\pi}, \hat{\pi}) = \sum_{i=1}^{K} |\bar{\pi}(i) - \hat{\pi}(i)|, \quad D_2(\bar{\pi}, \hat{\pi}) = \sum_{i=1}^{K} (\bar{\pi}(i) - \hat{\pi}(i))^2 \tag{14}$$

These measures are closely connected with two other well-known rank correlation measures: Spearman's footrule is an affine transformation of D_1 to the interval $[-1, +1]$, and Spearman's rank correlation (Spearman's rho) is such a transformation of D_2.

4.2   Label Ranking Methods

Several methods for label ranking have been proposed that try to exploit, in one way or the other, the complex though highly regular structure of the output space $\mathbb{S}_K$. These include generalizations of standard machine learning methods such as nearest neighbor estimation and decision tree learning [4], as well as statistical inference based on parametrized models of rank data [3]. Moreover, several *reduction techniques* have been proposed, that is, meta-learning techniques that reduce the original label ranking problem into one or several classification problems that are easier to solve [7,10].

Since the (base) learner used to realize label ranking is actually of minor interest for our purpose, we shall stick to a simple nearest neighbor approach in this paper. The most obvious way of exploiting our framework for superset learning to realize such an approach consists of predicting, for a new query instance x_0, the ranking

$$\hat{\pi} \in \operatorname*{argmin}_{\pi \in \mathcal{Y}} \sum_{n=1}^{nn} L^*(E(\pi_n), \pi) , \tag{15}$$

where $\pi_1, \ldots, \pi_{nn}$ are the (incomplete) rankings coming from the nn nearest neighbors of x_0, and L^* is the OSL extension of a loss such as (13) or (14). However, depending on the loss chosen, the problem of finding a minimizer in (15) may become computationally expensive. Therefore, we subsequently introduce a new meta-learning technique for label ranking, which is based on the idea of reducing the original problem to standard classification problems.

5   Label Ranking based on Labelwise Decomposition

Unlike existing reduction techniques, which transform the original label ranking problem to a single large or a quadratic number of small binary classification problems [7,10], our approach is based on a *labelwise* decomposition into K ordinal classification problems. As will be explained in more detail in the following, the basic idea is to train one model per label, namely a model that maps instances to ranks.

5.1 Complete Training Information

If the training data $\mathcal{D}$ is precise, i.e., consists of complete examples $(\boldsymbol{x}_n, \bar{\pi}_n)$, then each such example informs about the rank $\bar{\pi}(k)$ of the label c_k in the ranking associated with $\boldsymbol{x}_n$. Thus, a quite natural idea is to learn a model

$$M_k : \mathcal{X} \longrightarrow [K]$$

that predicts the rank of c_k, given an instance $\boldsymbol{x} \in \mathcal{X}$ as an input. Indeed, such a model can be trained easily on the (label-specific) data

$$\mathcal{D}_k = \Big\{ (\boldsymbol{x}_n, r_n) \,|\, (\boldsymbol{x}_n, \bar{\pi}_n) \in \mathcal{D}, \, r_n = \bar{\pi}_n(k) \Big\}. \tag{16}$$

The classification problems thus produced are multi-class problems with K classes, where each class corresponds to a possible rank. More specifically, since these ranks have a natural order, we are facing an *ordinal classification* problem. Thus, training of the models M_k ($k \in [K]$) can in principle be accomplished by any existing method for ordinal classification.

5.2 Incomplete Training Information

As mentioned before, the original training data is not necessarily precise; instead, for a training instance $\boldsymbol{x}_n$, only an incomplete ranking π_n of a subset of the labels in $\mathcal{C}$ might have been observed, while the complete ranking $\bar{\pi}_n$ is not given. In this case, the above method is not directly applicable: If at least one label is missing, i.e., $|\pi_n| < K$, then none of the true ranks $\bar{\pi}_n(k)$ is precisely known; consequently, the training data (16) cannot be constructed.

Nevertheless, even in the case of incomplete rankings, non-trivial information can be derived about the rank $\bar{\pi}(k)$ for at least some of the labels c_k. In fact, if $|\pi| = J$ and $\pi(k) = r > 0$, then

$$\bar{\pi}(k) \in Y = \big\{ r, r+1, \ldots, r+K-J \big\} \ .$$

Of course, if $\pi(k) = 0$ (i.e., c_k is not present in the ranking), only the trivial information $\bar{\pi}(k) \in [K]$ can be derived. Yet, more precise information can be obtained under additional assumptions on the process of imprecisiation, which in this case is responsible for removing labels from the complete ranking. For example, if π is known to be the top of the ranking $\bar{\pi}$, then

$$\begin{cases} \bar{\pi}(k) = \pi(k) & \text{if } \pi(k) > 0 \\ \bar{\pi}(k) \in \{J+1, \ldots, K\} & \text{if } \pi(k) = 0 \end{cases} . \tag{17}$$

This scenario is practically relevant, since top-ranks are observed in many applications.

In general, the type of training data that can be derived for a label c_k in the case of incomplete rank information is of the form

$$\mathcal{O} = \left\{ (\boldsymbol{x}_n, Y_n) \right\}_{n=1}^{N} \subset \mathcal{X} \times 2^{[K]} , \tag{18}$$

that is, an instance $\boldsymbol{x}_n$ together with a set of possible ranks Y_n. Again, this is exactly the type of data assumed as an input by our approach to superset learning.

5.3   Generalized Nearest Neighbor Estimation

As already mentioned, we use a simple nearest neighbor approach for prediction: Given a new query instance $\boldsymbol{x}_0$, a prediction $\hat{\pi}$ is obtained by combining the (incomplete) rankings $\pi_1, \ldots, \pi_{nn}$ coming from the nn nearest neighbors of $\boldsymbol{x}_0$ in the training data $\mathcal{O}$. Denote by $Y_{k,n}$ ($k \in [K], n \in [nn]$) the (possibly imprecise) rank information for label c_k provided by π_n. Moreover, consider a distance $D(\cdot)$ on $\mathcal{Y}$ that is labelwise decomposable, i.e., which can be written in the form

$$D(\bar{\pi}, \hat{\pi}) = \sum_{k=1}^{K} L(\bar{\pi}(k), \hat{\pi}(k)).$$

Obviously, the L_1 and L_2 loss in (14) are both of this type. Then, the empirical risk of $\hat{\pi}$, i.e., the loss of this prediction in the neighborhood of $\boldsymbol{x}_0$, is given by

$$\sum_{n=1}^{nn} D(\bar{\pi}_n, \hat{\pi}) = \sum_{n=1}^{nn} \sum_{k=1}^{K} L(\bar{\pi}_n(k), \hat{\pi}(k)) \tag{19}$$

$$= \sum_{k=1}^{K} \sum_{n=1}^{nn} L(\bar{\pi}_n(k), \hat{\pi}(k)) \tag{20}$$

$$= \sum_{k=1}^{K} L_k(\hat{\pi}(k)), \tag{21}$$

where $L_k(r)$ is the cost of putting label c_k on position r. Taking into account that in general only incomplete rankings π_n are observed, the loss $L(\cdot)$ should be replaced by its generalization (10) and, therefore, L_k should be defined as

$$L_k(r) = \sum_{n=1}^{nn} L^*(Y_{k,n}, r) .$$

Thus, an optimal solution would consist of assigning c_k the position $\hat{\pi}(k) = r$ for which $L_k(r)$ is minimal. However, noting that each position $r \in [K]$ must be assigned at most once, this approach is obviously not guaranteed to produce a feasible solution. Instead, the minimization of (19) requires the solution of an *optimal assignment problem* [2]:

- labels $c_k \in C$ must be uniquely assigned to ranks $r = \hat{\pi}(k) \in [K]$;
- assigning c_k to rank r causes a cost of $L_k(r)$;
- the goal is to minimize the sum of all assignment costs.

Assignment problems of that kind have been studied extensively in the literature, and efficient algorithms for their solution are available. The well-known Hungarian algorithm [12], for example, solves the above problem in time $O(K^3)$. Such algorithms can be used to produce a prediction $\hat{\pi}$ that minimizes the sum of assignment costs $L_1(\hat{\pi}(1)) + \ldots + L_K(\hat{\pi}(K))$, and therefore to realize our nearest neighbor approach to label ranking. In the next section, we experimentally analyze this approach with L given by D_1 in (14).

5.4   Experiments

In this section, we experimentally compare our new method, referred to as LWD (for Label-Wise Decomposition), with another nearest neighbor approach to label ranking. This approach is based on the (local) estimation of the parameters of a probabilistic model called the Plackett-Luce (PL) model [3]. It is known to achieve state-of-the-art performance, not only among the nearest neighbor approaches but among label ranking methods in general. Apart from that, the comparison with PL is specifically interesting for the following reason: The approach is based on finding the probabilistic model, identified by a parameter vector $v = (v_1, \ldots, v_K)$, for which the likelihood of observing the (neighbor) rankings is maximized:

$$v^* \in \underset{v \in \mathbb{R}_+^K}{\mathrm{argmax}} \prod_{n=1}^{nn} \mathrm{PL}(\pi_n \mid v)$$

Now, with PL being a probability measure on the set of permutations $\mathcal{Y}$, the probability of an incomplete ranking π_n is given by the corresponding marginal, namely

$$\mathbf{P}(\pi_n \mid v) = \sum_{\overline{\pi} \in E(\pi_n)} \mathrm{PL}(\overline{\pi} \mid v) \ .$$

Thus, as can be seen, ambiguous examples are dealt with by *summing* over the corresponding superset, as opposed to *maximizing* as suggested by our approach (7). Since summation is more in line with averaging over all candidates than selecting the most plausible one, this approach is obviously less in the spirit of superset learning through *data disambiguation*.

As data sets, we used several benchmarks for label ranking that have also been used in previous studies [10]; these are semi-synthetic data sets, namely label ranking versions of (real) UCI multi-class data. Moreover, we used two real label ranking data sets: The Sushi data[3] consists of 5000 instances (customers) described by 11 features, each one associated with a ranking of 10 types of sushis. The Students data [1] consists of 404 students (each characterized by 126 attributes) with associated rankings of five goals (want to get along with my

---

[3] http://kamishima.new/sushi/

Table 1. Properties of the data sets.

data set	# inst. (N)	# attr. (d)	# labels (K)
authorship	841	70	4
glass	214	9	6
iris	150	4	3
pendigits	10992	16	10
segment	2310	18	7
vehicle	846	18	4
vowel	528	10	11
wine	178	13	3
sushi	5000	11	10
students	404	126	5

Table 2. Performance in terms of Kendall's tau on synthetic data: missing-at-random (above) and top-rank setting (below).

	complete ranking		30% missing labels		60% missing labels	
	LWD	PL	LWD	PL	LWD	PL
authorship	.933±.016	.936±.015	.925±.018	.833±.030	.891±.021	.601±.054
glass	.840±.075	.841±.067	.819±.078	.669±.064	.721±.072	.395±.068
iris	.960±.036	.960±.036	.932±.051	.896±.069	.876±.068	.787±.111
pendigits	.940±.002	.939±.002	.924±.002	.770±.004	.709±.005	.434±.007
segment	.953±.006	.950±.005	.914±.009	.710±.013	.624±.020	.381±.020
vehicle	.853±.031	.859±.028	.836±.032	.753±.032	.767±.037	.520±.050
vowel	.876±.021	.851±.020	.821±.022	.612±.027	.536±.034	.327±.033
wine	.938±.050	.947±.047	.933±.054	.919±.059	.921±.062	.863±.094
authorship	.933±.016	.936±.015	.932±.017	.927±.017	.923±.015	.886±.022
glass	.840±.075	.841±.067	.838±.074	.809±.066	.815±.075	.675±.069
iris	.960±.036	.960±.036	.956±.036	.926±.051	.932±.048	.868±.070
pendigits	.940±.002	.939±.002	.933±.002	.918±.002	.837±.004	.794±.004
segment	.953±.006	.950±.005	.943±.005	.874±.008	.844±.010	.674±.015
vehicle	.853±.031	.859±.028	.851±.033	.838±.030	.818±.032	.765±.035
vowel	.876±.021	.851±.020	.867±.021	.785±.020	.800±.021	.588±.024
wine	.938±.050	.947±.047	.936±.049	.926±.061	.930±.059	.907±.066

parents, want to feel good about myself, want to have nice things, want to be different from others, want to be better than others). See Table 1 for a summary of the data.

Two missing label scenarios (imprecisiation procedures) were simulated, namely a "missing-at-random" setting and the top-rank setting (17). In the first case, a biased coin is flipped for every label in a ranking to decide whether to keep or delete that label; the probability for a deletion is specified by a parameter $p \in [0, 1]$. Thus, $p \times 100\%$ of the labels will be missing on average. Similarly, in the second case, only the J top-labels in a ranking are kept, where J has a binomial distribution with parameters K and $1 - p$.

The results in Tables 2 and 3 are presented as averages of 5×10-fold cross validation in terms of the Kendall correlation measure; other measures such as

Table 3. Performance in terms of Kendall's tau on real-world data: missing-at-random (above) and top-rank setting (below).

sushi	0%	10%	20%	30%	40%	50%	60%	70%
LWD	.323±.012	.322±.011	.320±.011	.319±.010	.315±.011	.308±.011	.296±.011	.277±.010
PL	.321±.010	.320±.010	.318±.010	.311±.010	.298±.011	.278±.010	.246±.010	.203±.012
LWD	.325±.012	.324±.011	.324±.011	.323±.011	.323±.011	.323±.011	.321±.011	.316±.011
PL	.321±.010	.320±.010	.320±.011	.320±.011	.319±.010	.316±.010	.310±.010	.303±.011

students	0%	10%	20%	30%	40%	50%	60%	70%
LWD	.641±.051	.641±.051	.640±.050	.640±.051	.638±.052	.637±.051	.633±.054	.626±.055
PL	.386±.028	.384±.027	.382±.026	.377±.029	.365±.025	.350±.027	.327±.027	.274±.033
LWD	.641±.051	.641±.051	.640±.051	.641±.051	.640±.051	.640±.052	.638±.050	.628±.052
PL	.386±.028	.385±.028	.386±.028	.385±.027	.383±.029	.379±.026	.377±.026	.371±.028

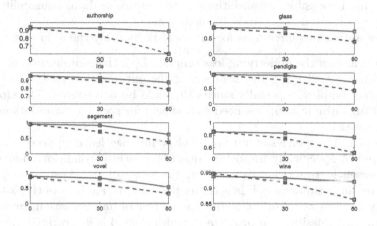

Fig. 4. Performance of LWD (solid lines) and PL (dashed line) in the missing-at-random setting.

(14) led to similar results. The number of nearest neighbors was determined through internal cross-validation. As a distance measure on $\mathcal{X}$, the standard Euclidean distance was used.

These results clearly support the conclusion that, while LWD and PL are quite en par in the complete ranking case, the latter is much more sensitive toward missing label information than the former. In fact, the performance of LWD is comparably stable, and its drop in performance due to missing label information is less pronounced than in the case of PL; this observation is especially clear in the missing-at-random setting (see Figure 4), whereas the differences in performance are less visible in the top-rank setting. In any case, these results are very interesting in light of our previous remarks on the comparison between *averaging* (product-sum inference) and *maximizing* (product-maximum inference) and clearly provide first evidence in favor of the effectiveness of learning through disambiguation in the context of structured output prediction.

6   Summary and Outlook

Our approach to superset learning is based on the idea of simultaneously finding the most plausible combination of model and data. As we explained, this idea could in principle also be realized by means of a probabilistic approach, and indeed, the principle of likelihood maximization was on the origin of our considerations. However, a full-fledged probabilistic approach is quite demanding and requires working with probability distributions both in the model and the data space. While perhaps being less principled, our approach relaxes these requirements: The plausibility of a model is captured in terms of how well it fits the data (according to a given loss); moreover, by merely distinguishing between possible and impossible instantiations of the imprecise data, plausibility in the data space is treated as a purely bivalent notion.

There are various directions for future work, notably the following:

- Depending on the underlying loss function $L(\cdot)$, the computation of the corresponding OSL (10) and solution of the optimization problem (11) may become complex, especially since (10) could be non-convex. Therefore, efficient algorithmic solutions need to be found for specific instantiations of our framework.
- Theoretical properties of our approach to superset learning need to be investigated. A specifically important question concerns conditions under which successful learning, for example in the sense of (stochastic) convergence toward an optimal model, is actually possible. An analysis of this kind obviously requires assumptions about the process of imprecisiation. Imagine, for example, a classification problem in which class A is *deterministically* added to the observed superset whenever the true class is B and vice versa. Learning to distinguish A from B is obviously impossible in that case. See [14] for a first analysis of learnability in the context of superset label learning problem (superset learning for binary classification).
- The idea of tackling structured output prediction by superset learning appears to be interesting, and our results for label ranking are indeed promising. This idea should therefore be realized for other types of structured output prediction, too, for example multi-label classification [18].

References

1. Boekaerts, M., Smit, K., Busing, F.M.T.A.: Salient goals direct and energise students' actions in the classroom. Applied Psychology: An International Review 4(S1), 520–539 (2012)
2. Burkard, R.E., Dell'Amico, M., Martello, S.: Assignment Problems. SIAM (2009)
3. Cheng, W., Dembczynski, K., Hüllermeier, E.: Label ranking based on the Plackett-Luce model. In: Proc. ICML 2010, Int. Conf. on Machine Learning, Haifa, Israel (2010)
4. Cheng, W., Hühn, J., Hüllermeier, E.: Decision tree and instance-based learning for label ranking. In: Proc. ICML 2009, 26th International Conference on Machine Learning, Montreal, Canada (2009)

5. Cour, T., Sapp, B., Taskar, B.: Learning from partial labels. Journal of Machine Learning Research **12**, 1501–1536 (2011)
6. Grandvalet, Y.: Logistic regression for partial labels. In: IPMU 2002, Int. Conf. Information Processing and Management of Uncertainty in Knowledge-Based Systems, pp. 1935–1941, Annecy, France (2002)
7. Har-Peled, S., Roth, D., Zimak, D.: Constraint classification for multiclass classification and ranking. In: Proc. NIPS 2002, pp. 785–792 (2003)
8. Hüllermeier, E.: Learning from imprecise and fuzzy observations: Data disambiguation through generalized loss minimization. International Journal of Approximate Reasoning **55**(7), 1519–1534 (2014)
9. Hüllermeier, E., Beringer, J.: Learning from ambiguously labeled examples. Intelligent Data Analysis **10**(5), 419–440 (2006)
10. Hüllermeier, E., Fürnkranz, J., Cheng, W., Brinker, K.: Label ranking by learning pairwise preferences. Artificial Intelligence **172**, 1897–1917 (2008)
11. Jin, R., Ghahramani, Z.: Learning with multiple labels. In: 16th Annual Conference on Neural Information Processing Systems, Vancouver, Canada (2002)
12. Kuhn, H.W.: The Hungarian method for the assignment problem. Naval Research Logistics Quarterly **2**(1–2), 83–97 (1955)
13. Liu, L.P., Dietterich, T.G.: A conditional multinomial mixture model for superset label learning. In: Proc. NIPS (2012)
14. Liu, L.P., Dietterich, T.G.: Learnability of the superset label learning problem. In: Proc. ICML 2014, Int. Conference on Machine Learning, Beijing, China (2014)
15. Nguyen, N., Caruana, R.: Classification with partial labels. In: Proc. KDD 2008, 14th Int. Conf. on Knowledge Discovery and Data Mining, Las Vegas, USA (2008)
16. Schölkopf, B., Smola, A.J.: Learning with Kernels: Support Vector Machines, Regularization, Optimization, and Beyond. MIT Press (2001)
17. Sid-Sueiro, J.: Proper losses for learning from partial labels. In: Proc. NIPS (2012)
18. Sun, Y.Y., Zhang, Y., Zhou, Z.H.: Multi-label learning with weak label. In: Proc. 24th AAAI Conference on Artificial Intelligence, Atlanta, Georgia, USA (2010)
19. Zhou, Y. Lui, Y., Yang, J., He, X., Liu, L.: A taxonomy of label ranking algorithms. Journal of Computers **9**(3) (2014)

Probabilistic, Statistical,
and Graphical Approaches

Bayesian Modelling of the Temporal Aspects of Smart Home Activity with Circular Statistics

Tom Diethe[✉], Niall Twomey, and Peter Flach

Intelligent Systems Laboratory, University of Bristol, Bristol, UK
{tom.diethe,niall.twomey,peter.flach}@bristol.ac.uk

Abstract. Typically, when analysing patterns of activity in a smart home environment, the daily patterns of activity are either ignored completely or summarised into a high-level "hour-of-day" feature that is then combined with sensor activities. However, when summarising the temporal nature of an activity into a coarse feature such as this, not only is information lost after discretisation, but also the strength of the periodicity of the action is ignored. We propose to model the temporal nature of activities using circular statistics, and in particular by performing Bayesian inference with Wrapped Normal ($\mathcal{WN}$) and $\mathcal{WN}$ Mixture ($\mathcal{WNM}$) models. We firstly demonstrate the accuracy of inference on toy data using both Gibbs sampling and Expectation Propagation (EP), and then show the results of the inference on publicly available smart-home data. Such models can be useful for analysis or prediction in their own right, or can be readily combined with larger models incorporating multiple modalities of sensor activity.

1 Introduction

One of the central hypotheses of a "smart home" is that a number of different sensor technologies may be combined to build accurate models of the Activities of Daily Living (ADL) of its residents. These models can then be used to make informed decisions relating to medical or health-care issues. For example, such models could help by predicting falls, detecting strokes, analysing eating behaviour, tracking whether people are taking prescribed medication, or detecting periods of depression and anxiety. Since 2007, the Centre for Advanced Studies in Adaptive Systems (CASAS) research group has been collecting data from homes with various different sensor layouts and differing numbers of residents (see *e.g.* [2]).

In most of the approaches taken to date [6,7,13], classifiers are learnt which put weights over individual sensors, and then take linear combinations of these weights to produce a decision function for the set of active sensors at any given time. In addition, an extra "hour-of-day" feature is often added, which in some sense attempts to capture the periodic nature of many of the activities under examination. However this can produce undesirable effects, since this is a rather coarse discretisation. This in turn can result in border effects, such as activities that are short-lived but often span an hour boundary.

© Springer International Publishing Switzerland 2015
A. Appice et al. (Eds.): ECML PKDD 2015, Part II, LNAI 9285, pp. 279–294, 2015.
DOI: 10.1007/978-3-319-23525-7_17

We propose instead that it is more satisfactory to take a model-based app-roach, in which the temporally periodic nature of the activities (*i.e.* circadian or diurnal rhythms) is taken directly into account. A natural framework for this is the area of "circular" statistics [5,9,18], where univariate data is defined on an angular scale, typically the (unit) circle.

In addition we suggest that, rather than using frequentist methods to fit cir-cular distributions to the data, a full Bayesian approach would be advantageous in this setting. To begin with, this allows for a principled way of incorporating prior knowledge (or results of a previous round of inference in order to perform on-line learning) if such knowledge exists. However, beyond this, inferring the full distribution over the parameters facilitates model comparison and hypothe-sis testing. Furthermore, if the results of inference are to be used in a decision-making context, such as for the medical application being considered here, the optimal decision is the Bayesian decision [17]. The model-based approach is also appealing as it allows us to consider building larger models, such as hierarchi-cal models that enable us to reason about the differences between individuals and groups of people (using shared hyper-priors), and also to consider transfer learning.

In order to solve the (intractable) inference problems, we will take two approaches. Firstly, we will use Gibbs sampling [3], which is a Markov chain Monte Carlo (MCMC) algorithm for obtaining a sequence of observations which are approximated from a specified multivariate probability distribution. Gibbs sampling has the advantage of being easy to implement, and is particularly well-adapted to sampling the posterior distribution of a Bayesian network, since Bayesian networks are typically specified as a collection of conditional distribu-tions.

We will also consider the deterministic approximation method Expectation Propagation (EP) [11], a generalisation of Belief Propagation (BP) in which the true posterior distribution is approximated with a simpler distribution, which is close in the sense of Kullback-Leibler (KL) divergence. EP approximates the belief states with expectations, such as means and variances, giving it much wider scope than would be possible with BP.

2   Related Work

Many methods and statistical techniques have been developed to analyse and understand circular data, mainly from a frequentist perspective. The popular approaches have been embedding, wrapping and intrinsic approaches (see *e.g.* [5,9]). Here we focus on the wrapping approach, and specifically the Wrapped Normal ($\mathcal{WN}$) distribution [9]. A survey of Bayesian analysis of circular data using the wrapping method was given by [15], and the approaches herein build upon this work.

The use of circular statistics to model circadian or diurnal rhythms was first considered by [9], and also discussed by [16], in which various procedures for the analysis of circadian rhythms at population, organism, cellular and molecular levels were examined, ranging from visual inspection of time plots to several mathematical methods of time series analysis.

A multivariate $\mathcal{WN}$ Mixture ($\mathcal{WNM}$) model was defined used by [1] for the modelling of high-rate quantisation of phase data of speech, in which the authors used the Expectation Maximisation (EM) algorithm to learn the location and covariance parameters. Note however that a maximisation algorithm such as EM is capable of only returning a single point from the distribution, rather than a full distribution over the parameters.

Recently two non-parametric Bayesian models of circular variables based on Dirichlet Process (DP) Mixtures of normal distributions were introduced [14]: the first was a projected DP mixture of of bi-variate normals and the second was based on $\mathcal{WN}$s. Inference was done in this case using Gibbs slice sampling, and has the appeal that in theory it is possible to learn the number of mixture components rather than having to pre-specify or use model comparison. However, inference in this case is extremely expensive, with large numbers of iterations (40,000 were used) required, and large numbers of data points are required to fit the large number of parameters in the model.

3   Methods

Let x be a circular random variable defined on the circumference of a circle. The corresponding circular probability density function (pdf) $f(\cdot)$ is periodic with period γ: $f(x) = f(x + w\gamma), \forall w \in \mathbb{Z}, \gamma \geq 0$. Usually the distributions are defined over the unit circle, in which case $\gamma = 2\pi$, but arbitrary $\gamma \geq 0$ can be considered by a simple rescaling of x. The function $f(\cdot)$ integrates to 1 over $(0, \gamma]$. For notational simplicity, we will assume that all circular variables are constrained to their principal values, obtained by taking the modulo operation $x \leftarrow x \mod \gamma$.

The circular distance between two points x, z for a given period γ is given by [9, eq.2.3.13]:

$$d_\gamma(x, z) = \min\left(x - z, \gamma - (x - z)\right) = \frac{\gamma}{2} - \left|\frac{\gamma}{2} - |x - z|\right|. \tag{1}$$

There exist distributions directly defined on the (unit) circle, such as the von-Mises or Circular Normal distribution (see [9, section2.2.4]), but for reasons given below we will focus on the $\mathcal{WN}$ distribution.

3.1   The Wrapped Normal ($\mathcal{WN}$) Distribution

A "wrapped" distribution is one that results from wrapping the pdf of a linear random variable to the circumference a (unit) circle (infinitely many times). The corresponding distributions are called wrapped distributions, and any continuous pdf can be wrapped in this way. The Wrapped Normal ($\mathcal{WN}$) distribution is the circular analog of the normal distribution, achieved by wrapping in this way. In practice, the von-Mises and the $\mathcal{WN}$ distribution are very similar [9]. However, the wrapped Normal distribution is more convenient for Bayesian inference, as many of the technical details can be brought over from the (well studied) Normal distribution – for example, it is closed under convolution [9]. The probability density function of the wrapped normal distribution is [9]

$$f_{\mathcal{WN}}(x; \mu, \sigma, \gamma) = \frac{1}{\sigma\sqrt{2\pi}} \sum_{k=-\infty}^{\infty} \exp\left[\frac{-(x - \mu + \gamma k)^2}{2\sigma^2}\right], \qquad (2)$$

with $x \in [0, 2\pi)$, location parameter $\mu \in [0, 2\pi)$, and uncertainty parameter $\sigma > 0$. We will use $\tau = \frac{1}{\sigma^2}$ to denote the precision. Because the summands of the series converge to zero, it is natural to approximate the pdf with the finite series:

$$\hat{f}_{\mathcal{WN}}(x; \mu, \sigma, \gamma) = \frac{1}{\sigma\sqrt{2\pi}} \sum_{k=-K}^{K} \exp\left[\frac{-(x - \mu + \gamma k)^2}{2\sigma^2}\right] \approx f_{\mathcal{WN}}(x; \mu, \sigma, \gamma), \quad (3)$$

where only $2K + 1$ summands are considered. However, one can intuitively see that for small values of K, this will only be a good approximation for small values of σ.

The $\mathcal{WN}$ can also be expressed in terms of the Jacobi theta function (see [5, eq.(2.2.15)]), which leads to a second approximation that is more accurate for large values of σ.

$$\tilde{f}_{\mathcal{WN}}(x; \mu, \sigma, \gamma) \approx f_{\mathcal{WN}}(x; \mu, \sigma, \gamma)$$

$$= \frac{1}{\gamma}\left(1 + 2\sum_{k=1}^{K} e^{-\frac{\sigma^2}{2}\left(\frac{2\pi k}{\gamma}\right)^2} \cos\left(\frac{2\pi k}{\gamma}(x - \mu)\right)\right), \qquad (4)$$

where only K summands are considered. Theoretical bounds are given in [8] that show that the errors of both approximations decrease exponentially with the number of summands, and show that the first representation performs well for small σ whereas the other performs well for large σ.

3.2 Bayesian Inference

The $\mathcal{WN}$ distribution possesses the additive property [5], *i.e.* the convolution of two $\mathcal{WN}$ distributed variables is also a an $\mathcal{WN}$ distribution. Hence for the purposes of Bayesian inference, the conjugate prior for the location parameter μ of a $\mathcal{WN}$ distribution is another $\mathcal{WN}$ distribution, which we denote as $\mathcal{WN}_0(\mu; \mu_0, \sigma_0, \gamma)$. The conjugate prior for the precision τ is the Gamma distribution, denoted by $\mathcal{G}a(\tau; \alpha_0, \beta_0)$ for shape and rate parameters α_0 and β_0 respectively, as would be the case for the Normal distribution.

In Figure 1a we show the factor graph for the $\mathcal{WN}$ model, where the shading of the x variable indicates that it is observed, and the box around x and the $\mathcal{WN}$ factor is a plate, indicating that this part of the graph is repeated N times. Inference can be performed in this model using Gibbs sampling, where we use the approximations given in Equation 3 and Equation 4, and where we the former is used if $\sigma^2 < 0.15$ and the latter in the reverse case, as suggested by [8].

3.3 $\mathcal{WN}$ Mixture ($\mathcal{WNM}$) Models

We define the $\mathcal{WN}$ Mixture ($\mathcal{WNM}$) distribution (*i.e.* a mixture of $\mathcal{WN}$ distributions) in the following way,

$$f_{\mathcal{WNM}}(x; \mu, \sigma, \gamma, \phi, M) = \sum_{m=1}^{M} w_m \mathcal{WN}(x; \mu_k, \sigma_k^2, \gamma), \tag{5}$$

where $w \in \mathbb{R}^M : \sum_{m=1}^{M} w_m = 1, w_m \geq 0$, *i.e.* $w \sim \text{Cat}(\phi)$ are the mixing coefficients which are drawn from a categorical (discrete) distribution of dimension $M > 0$ with a probability vector ϕ. The conjugate prior for ϕ is the Dirichlet distribution, with a concentration parameter vector $\alpha_\phi \in \mathbb{R}^M : \alpha > 0$.

In Figure 1b we show the factor graph for the $\mathcal{WNM}$ model using gate notation for representing the mixture model [12]. ϕ is the Dirichlet distributed variable, from which the discrete variable $\Psi \in \mathbb{R}^m$ is drawn, where m is the number of mixture components, representing the gate selector is sampled.

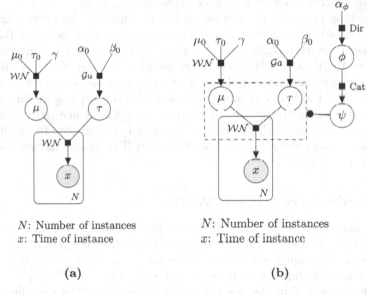

N: Number of instances	N: Number of instances
x: Time of instance	x: Time of instance
(a)	**(b)**

Fig. 1. (a) Wrapped Normal ($\mathcal{WN}$) and (b) $\mathcal{WN}$ Mixture ($\mathcal{WNM}$) models.

As has been noted by [14], the standard $\mathcal{WNM}$ model suffers from issues of identifiability, which we also found when trying to perform inference using the model. There the authors tackle the problem by "unwrapping" the distribution by conditioning on the wrapping number k_i, which results in a complex sampling procedure. Here we will take a simpler approach, that also allows us to use EP as well as Gibbs sampling.

3.4   Approximate $\mathcal{WN}$ ($\mathcal{AWN}$)

We approximate the $\mathcal{WN}$ with mixture of $\tilde{M}$ normal distributions. If we insist that $\tilde{M}$ is odd, and define a vector of offsets, $\boldsymbol{\delta} = (\delta/2)_{\delta=-\tilde{M}-1}^{\tilde{M}-1}$, the $\mathcal{AWN}$ model is defined by

$$f_{\mathcal{AWN}}(x; \mu, \sigma^2, \tilde{M}) = \frac{1}{\tilde{M}} \sum_{\delta \in \boldsymbol{\delta}} \mathcal{N}(x; \mu + \delta\gamma, \sigma^2). \tag{6}$$

This model cascades a series of $\tilde{M}$ Gaussian distributions along the real line where adjacent distributions are a distance of γ apart, all distributions share the same variance, and the mean of the central component is constrained to be found within the periodic range, $[0, \gamma)$ (this is the only component that will fall within this range). The components whose means fall outside the periodic range contribute to modelling by mimicking the wrapped tails of the $\mathcal{WN}$ model. Indeed, as $\tilde{M}$ tends towards infinity the $\mathcal{AWN}$ approximation approaches $\mathcal{WN}$. $\mathcal{AWN}$ models requires specification of three parameters: μ, σ^2 and $\tilde{M}$, and the factor graph for this model is shown in Figure 2a.

By modelling periodic distributions in this manner, we can approximate the posterior distributions of the $\mathcal{WN}$ parameters as one would estimate Bayesian mixture model parameters. We can again use Gibbs sampling to perform inference for the $\mathcal{AWN}$ model. However, since we have replaced the $\mathcal{WN}$ distribution with standard normal distributions, we can also use Expectation Propagation (EP). EP has a major advantage over Gibbs sampling in this setting, which is that it is relatively easy to compute model evidence (see Equation 7 in subsection 3.6) which will allow us to do model comparison. In the first set of experiments (see subsection 4.1 and 5.1) we will compare the two inference methods for this model.

3.5   Approximate $\mathcal{WNM}$ ($\mathcal{AWNM}$)

Generalisation of the $\mathcal{AWN}$ models to an $\mathcal{AWNM}$ is achieved by straightforward application of a standard mixture model gate over the parameter means, variances and approximation factors. The factor graph for this is given in Figure 2b, where mixing factors have been introduced.

As with the $\mathcal{AWN}$ model, we can again use either Gibbs sampling or EP to perform inference for the $\mathcal{AWNM}$ model. The computation of evidence Equation 7 plays an even greater role here, since it gives us a method to select the number of mixture components K (see 5.1). In the first set of experiments (see subsection 4.1 and 5.1) we will compare the two inference methods for this model.

3.6   Model Comparison

We also perform Bayesian model comparison, in which we marginalise over the parameters for the type of model being used, with the remaining variable being

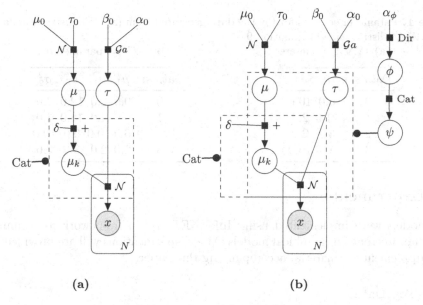

(a)   **(b)**

Fig. 2. (a) $\mathcal{AWN}$ and (b) $\mathcal{AWNM}$ models.

the identity of the model itself. The resulting marginalised likelihood, known as the model evidence, is the probability of the data given the model type, not assuming any particular model parameters. Using D for data, θ to denote model parameters, H as the hypothesis, the marginal likelihood for the model H is

$$p(D|H) = \int p(D|\theta, H)\, p(\theta|H)\ \mathrm{d}\theta \tag{7}$$

This quantity can then be used to compute the "Bayes factor" [4], which is the posterior odds ratio for a model H_1 against another model H_2,

$$\frac{p(H_1|D)}{p(H_2|D)} = \frac{p(H_1)p(D|H_1)}{p(H_2)p(D|H_2)}. \tag{8}$$

3.7 Rose Diagrams

A useful variant of the circular histogram is a "rose diagram", in which the bars of a histogram are replace by segments. The area of each segment is proportional to the frequency of the corresponding group. As such, for groups of equal width, the radius should be proportional to the square root of the relative frequency [9]. We will use these, but with a slight abuse (since the maximum value of a pdf is arbitrary) we will plot the $\mathcal{WN}$ and $\mathcal{WNM}$ pdfs over the rose diagrams with the maximum of the pdf coinciding with the outside of the plot.

Table 1. Parameter settings for toy the data generated from (a) $\mathcal{WN}$ distribution and (b) $\mathcal{WNM}$ distribution of Equation 9

(a) $\mathcal{WN}$ parameters			(b) $\mathcal{WNM}$ parameteres				
Data set	μ	σ^2	Data set	μ_1	σ_1^2	μ_2	σ_2^2
0	0.0	10.0	0	0.0	2.0	12.0	2.0
1	21.0	2.0	1	6.0	4.0	18.0	4.0
2	3.0	2.0	2	6.0	10.0	9.0	10.0
3	10.0	10.0	3	2.0	2.0	3.0	2.0

4   Experiments

All models were implemented using Infer.NET [10], a framework for running Bayesian inference in graphical models. Model specifications will are provided in the supplementary material accompanying this paper.

4.1   Toy Data

In order to evaluate the models, we first created toy datasets where we sampled from $\mathcal{WN}$ and $\mathcal{WNM}$ distributed data. For testing the uni-modal models, data were generated from $\mathcal{WN}$ distribution with the settings for μ and σ^2 given in Table 1a. For testing the mixture models, data were generated from the following mixture model:

$$f(x) = 0.6 \; \mathcal{WN}(x; \mu_1, \sigma_1^2, \gamma) + 0.4 \; \mathcal{WN}(x; \mu_2, \sigma_2^2, \gamma) \tag{9}$$

where $\mu_1, \sigma_1^2, \mu_2, \sigma_2^2$ were set as in Table 1b. The first two are in some sense "easy", since the means are well separated, with the two cases being used to ensure there were no inference pathologies. The third and fourth are harder problems as the variances are large with respect to the difference in means, where in data set 2 the variances are large and in data set 3 the variances are smaller.

We measure the mean difference (MD) for the estimated moments of the $\mathcal{WN}$ components:

$$MD_\mu = \frac{1}{n} \sum_{i=1}^{n} |d_\gamma(\mu_i, \hat{\mu}_i)|, \quad MD_\sigma = \frac{1}{n} \sum_{i=1}^{n} |\sigma_i - \hat{\sigma}_i| \tag{10}$$

where $d_\gamma(x, z)$ is the circular distance defined in Equation 1 and n is the number of random repetitions used.

4.2   The CASAS HH101 Dataset

We next examine some real-world data collected by the CASAS research group [2]. The HH101 data set[1] contains 3 months of single-resident apartment data

---

[1] http://casas.wsu.edu/datasets/hh101.zip

with partial annotations, with 30 different activities appearing in the annotations. The house was equipped with motion sensors, door sensors, temperature sensors, and ambient light sensors, which were recorded asynchronously. We chose this data for the length of recording, and due to the fact that it was from a single resident, to avoid further complications caused by multiple residents. The layout of the house with sensor locations marked with circles can be seen in Figure 3.

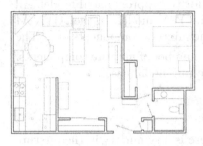

We note that there are sometimes errors in the data, such as ON/OFF events not being paired up correctly. When parsing the data we take a conservative approach, finding only OFF events that follow ON events. As with the sensors, there are sometimes errors in the activity labelling. We use the same conservative method. Note also that there are sometimes activity labels that are orphaned – *i.e.* there is no BEGIN/END trigger but simply a single label next to a sensor activation – these are ignored.

Fig. 3. Floorplan of the CASAS HH101 dataset.

Figure 4 shows the log of the total time spent performing each activity for each of the labelled activities in the CASAS HH101 dataset. It's worth noting that this dataset is dominated by 3 activities (Sleep, Sleep_Out_Of_Bed, and Watch_TV), which is perhaps in part due to the ease of labelling these activities, and in part due the fact that the resident was an elderly person. This will clearly play an important role in the quality of inference, simply due to the number of examples available.

Despite not modelling the sensor activations themselves, our data instances are in fact dependent on the sensor activations, since the dataset only contains annotations where sensor activations exist. In order to provide samples of the times of activity occurrences to our models, we could take the start end times of the activity and then re-sample from within this range (uniformly or otherwise). Here for simplicity we assume that the sensor activations in the period between the start and end annotations themselves provide independent samples of the times of an activity.

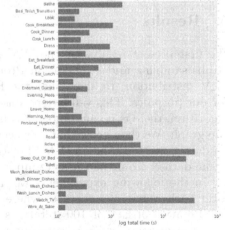

Fig. 4. Log of total time spent performing each activity for each of the labelled activities in the CASAS HH101 dataset.

4.3   Priors

The period γ of all $\mathcal{WN}$ distributions in our experiments were set to 24, representing the 24 hours in the day. In the $\mathcal{WN}$ model we set the location parameter of the prior over the location to 0, and the precision to $\left(\frac{\gamma}{2}\right)^{-2}$, meaning that two σ (wrapped standard deviations) in each direction will reach around the period, which roughly corresponds to a uniform distribution over the circle. We set the Gamma prior hyper-parameters were set to $\alpha_0 = 1$, $\beta_0 = 1$, which simply favours smaller precisions (and therefore larger variances).

In the $\mathcal{AWN}$ model we set the location parameter of the prior over the location to $\frac{\gamma}{2}$, as this is the uninformative prior for the approximated model. All other hyper-parameters were the same as for the $\mathcal{WN}$ model. In the $\mathcal{AWNM}$ model we set the location parameter of each of the mixture components to $\frac{\gamma}{2}$. The prior precision was set to $\left(\frac{\gamma}{2K}\right)^{-2}$ where K is the number of mixture components. The Gamma hyper-parameters were as with the uni-modal case.

4.4 Symmetry Breaking

In a normal mixture model, it is well known there is a symmetry in the mixture component assignments that needs to be broken by randomly initialising each data point to one of the components. In the $\mathcal{AWNM}$ model, this symmetry is also present, but there is an additional symmetry caused by the approximation. Fortunately, both symmetries can be broken using a different method, where the means of the components are initialised to $\frac{m\gamma}{M}, k = 1, \ldots, M$, where M is the number of mixture model components (not approximation components $\tilde{M}$), *i.e.* we distribute the prior means evenly around the circle. Once the means have been initialised in this way, it is no longer necessary to randomly assign the mixture components (and in fact may slow down convergence).

5 Results

We first present results for the $\mathcal{WN}$ model and the $\mathcal{AWN}$ model using both Gibbs sampling and Expectation Propagation (EP) on data generated from a $\mathcal{WN}$ distribution, to show that the EP $\mathcal{AWN}$ model is sufficiently accurate for our purposes. This validation is useful, since although EP is a deterministic algorithm, there is no guarantee of convergence if there are any loops present in the graph. We then show that this accuracy carries over to the $\mathcal{AWNM}$ model on data generated from an $\mathcal{WNM}$ distribution. We then show results on a smart home dataset from the Casas group.

In the following experiments we monitored the convergence of the models after each round of inference, where a round was determined to be a single full iteration of EP, or 100 iterations of Gibbs. The convergence criterion was that the means of each component had not moved by more than 30 seconds ($= \gamma/1800 \approx 0.01$) from one the previous round (other criteria are possible, but this was simple and effective).

5.1   Toy Data

Uni-modal Data. Details of the data generating process are in subsection 4.1 using the parameter settings in Table 1a, where we generated 100 data points and performed 5 repetitions of each data set with different random seeds. The results of learning the $\mathcal{WN}$ model using Gibbs sampling, and $\mathcal{AWN}$ using Gibbs sampling and EP are shown in Figure 5, where performance is measured in terms of MD_μ and MD_σ as defined in Equation 10. We can see that the $\mathcal{AWN}$ model using Gibbs sampling performs almost identically to the $\mathcal{WN}$ model in the estimation of both moments of the distribution. The $\mathcal{AWN}$ model using EP has slightly degraded performance in terms of estimating the location μ, but is able to accurately estimate σ. The average running times were $\mathcal{WN}$: $0.12s$, $\mathcal{AWN}$(Gibbs): $0.40s$, and $\mathcal{AWN}$(EP): $0.33s$.

| Data set | Model | | | | | |
| | WN(Gibbs) | | AWN(Gibbs) | | AWN(EP) | |
	Average of MDμ	Average of MDσ	Average of MDμ	Average of MDσ	Average of MDμ	Average of MDσ
0	0.15	0.29	0.14	0.29	0.30	0.30
1	0.07	0.13	0.06	0.13	0.06	0.13
2	0.06	0.13	0.06	0.13	0.07	0.13
3	0.15	0.29	0.14	0.29	0.15	0.29
Overall Average	0.11	0.21	0.10	0.21	0.15	0.21

Fig. 5. Results on data generated from a uni-modal $\mathcal{WN}$ distribution, comparing the $\mathcal{WN}$ model with $\mathcal{AWN}$ model for both Gibbs and EP.

Mixture Model Data. In the following experiments we generated 100 data points in each data set, and repeated the experiments 5 times with different random seeds. The results in Figure 6 indicate that for fairly small data sets, the EP version of the model is in fact more accurate in terms of MD for the estimated moments. EP and Gibbs required on average over all of the experiments ≈ 20 and 2100 iterations to converge respectively, and EP reached convergence in on average roughly one fifth of the computation time required by Gibbs sampling.

| Data set | Average of MDμ1 | | Average of MDσ1 | | Average of MDμ2 | | Average of MDσ2 | |
	AWNM(EP)	AWNM(Gibbs)	AWNM(EP)	AWNM(Gibbs)	AWNM(EP)	AWNM(Gibbs)	AWNM(EP)	AWNM(Gibbs)
0	0.17	0.07	0.08	0.04	0.15	0.15	0.13	0.73
1	0.08	0.10	0.17	0.17	0.31	0.26	0.20	0.57
2	0.65	1.34	0.45	0.91	1.15	0.94	1.54	1.79
3	0.19	0.98	0.11	0.11	1.15	1.26	0.33	0.27
Overall Average	0.27	0.62	0.21	0.30	0.69	0.65	0.55	0.84

Fig. 6. *Small data set*: Accuracy of inference of the $\mathcal{AWNM}$ model on data generated from an $\mathcal{WNM}$ distribution (details in subsection 4.1 Table 1b).

The results in Figure 7 indicate that for larger data sets, the Gibbs version of the model is more accurate in terms of MD for the estimated moments than the EP version. This is explained by the difficulty of data set "2", which corresponds to the "pathological" case outlined in the model comparison discussion below, and as such the errors that we see EP making here are that it estimates there there is a single mode rather than two, which is not wholly unreasonable. EP and Gibbs required on average over all of the experiments roughly 28 and 1115 iterations to converge respectively, and EP reached convergence in roughly half the computation time than is required by Gibbs sampling.

Data set	Average of MDμ1 AWNM(EP)	AWNM(Gibbs)	Average of MDσ1 AWNM(EP)	AWNM(Gibbs)	Average of MDμ2 AWNM(EP)	AWNM(Gibbs)	Average of MDσ2 AWNM(EP)	AWNM(Gibbs)
0	0.06	0.10	0.01	0.12	0.05	0.62	0.02	1.32
1	0.06	0.14	0.04	0.23	0.07	0.59	0.06	1.11
2	0.77	1.41	0.43	0.87	2.16	1.23	0.38	1.45
3	0.35	0.24	0.20	0.18	1.02	0.46	0.31	0.29
Overall Average	**0.31**	**0.47**	**0.17**	**0.35**	**0.82**	**0.72**	**0.19**	**1.04**

Fig. 7. *Larger data set*: Accuracy of inference of the $\mathcal{AWNM}$ model on data generated from an $\mathcal{WNM}$ distribution (details in subsection 4.1 Table 1b).

Model Selection. We are able to take advantage of the fact that we are using EP in the $\mathcal{AWNM}$ model to perform model selection, since the model evidence computations are more straightforward for EP than for Gibbs sampling, and in fact have already been implemented in Infer.NET. In order to test the ability to use model evidence for model selection purposes, we ran the following experiment. We generated data from a $\mathcal{WNM}$ distribution with $K = \{1, 2, 3, 4\}$ components. We sampled the mixture weights for the components from a symmetric Dirichlet distribution $\text{Dir}(10, 10)$ (which gives roughly equal mass to each of the components) and then sampled 200 data points from the mixture distribution according to those weights. We then computed the model evidence for the $\mathcal{AWNM}$ model with $K = \{1, 2, 3, 4\}$ components, *i.e.* we learnt a model for each possible pair of true K and model K (16 in total). The results are shown in Figure 8. The true K values lie on the diagonal (*i.e.* where the correct K was supplied to the model). As can be seen in bold, the model gives the highest evidence to the those values of K across each row, meaning that by selecting the model with the highest evidence we would indeed choose the correct value for K.

However Figure 9 shows a seemingly pathological case. In this example the true means $\mu_1 = 6, \mu_2 = 9$ are quite close together, with a large $\sigma^2 = 10$ for both components. We can see that the model at first seems to converge to the correct means, but then appears to diverge away. At the end of this inference run, the estimated weights for the components were $w_1 \approx 0.02, w_2 \approx 0.98$, showing that the model had put all of the mass on the second component, with the mean being close to the average of the true mean

True K	Model K 1	2	3	4
1	**-710.3**	-712.6	-718.9	-720.1
2	-763.9	**-742.2**	-746.6	-751.5
3	-819.1	-811.4	**-810.7**	-815.3
4	-858.5	-846.9	-860.5	**-841.7**

Fig. 8. Model evidence computation for the $\mathcal{AWNM}$ model using EP. See text for details.

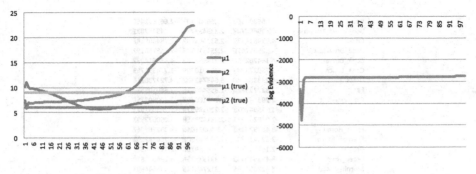

(a) Estimated and true means.   (b) Model evidence.

Fig. 9. A pathological case. The x-axis in both figures shows the EP iteration count. The true means $\mu_1 = 9, \mu_2 = 6$ are quite close, with a large $\sigma^2 = 10$ for both. Note that the model evidence continues to rise, despite the estimated means diverging from the truth. This is because in this case there is insufficient evidence for a bimodal model due to the high variances.

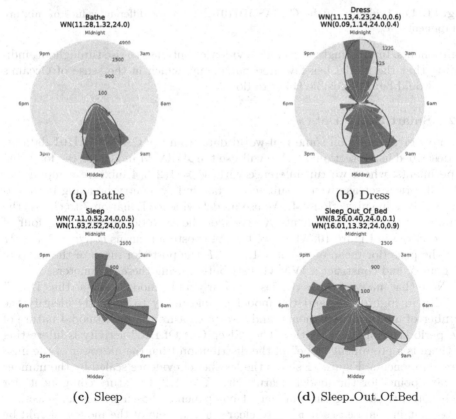

(a) Bathe   (b) Dress

(c) Sleep   (d) Sleep_Out_Of_Bed

Fig. 10. Posterior means of the $\mathcal{AWNM}$ model fitted to activities from the CASAS HH101 dataset. Note that the model correctly captures the multi-modal nature of the activities. The $\mathcal{WN}$ parameters are given in the subtitles of each subplot (mixture weights not shown).

Activity	K 1	2	3
Bathe	-2.65894389	-2.66002403	-2.661022468
Bed_Toilet_Transition	-10.19686747	-2.333433735	-2.349457831
Cook	-4.919811739	-2.632668882	-2.284385382
Cook_Breakfast	-2.25312476	-2.254319898	-2.255758157
Cook_Dinner	-4.144262708	-2.324962255	-4.364614997
Cook_Lunch	-1.750226804	-1.763010309	-1.774391753
Dress	-4.233888048	-3.645742382	-3.727730711
Eat	-2.827253886	-2.855440415	-2.880103627
Eat_Breakfast	-2.293162839	-2.297160752	-2.300803758
Eat_Dinner	-11.75947075	-2.220389972	-2.163091922
Eat_Lunch	-2.010291971	-2.031423358	-2.050072993
Enter_Home	-3.487927461	-3.40022946	-3.352487047
Evening_Meds	-3.223430398	-2.777507102	-2.766882102
Groom	-2.167755102	-1.994489796	-2.321428571
Leave_Home	-3.453540445	-3.391535756	-3.236617819
Morning_Meds	-4.422052195	-2.312746145	-2.316844603
Personal_Hygiene	-4.240142814	-4.241790439	-4.144003307
Phone	-3.553333333	-3.283061224	-3.321496599
Read	-5.354507351	-2.660965548	-2.62588984
Relax	-3.424166667	-3.185714286	-3.153554422
Sleep	-3.569722472	-3.219689469	-3.219689469
Sleep_Out_Of_Bed	-3.955734676	-3.896021116	-3.896580311
Toilet	-2.164069125	-3.981019063	-3.964657046
Wash_Breakfast_Dishes	-2.656940397	-2.661735099	-2.666066225
Wash_Dinner_Dishes	-3.874854167	-2.685444444	-2.520513889
Wash_Dishes	-4.064689332	-3.543100821	-3.334431419
Wash_Lunch_Dishes	-1.510576132	-1.534403292	-1.555349794

Fig. 11. Log evidence on the CASAS HH101 dataset for different values of mixture components M.

components. Interestingly, the model evidence continues to rise throughout, indicating that the model has favoured parsimony, which in the sense of Occam's razor would be the sensible thing to do.

5.2 Smart Home Data

We now give results on some real-world data from the CASAS HH101 dataset, as described in subsection 4.2. We will use the $\mathcal{AWNM}$ model for the following experiments, where we run inference with $M = 1, 2, 3, 4$ mixture components, $\tilde{M} = 3$, priors set as given in subsection 4.3, and symmetry breaking initialisation as given in subsection 4.4. We use model evidence Equation 7 to choose the number of mixture components K. We plot the posterior moments for four of the activities in Figure 10, where we take the posterior means $\mu_m, m = 1, \dots, M$ and the posterior mean of $\tau_m, m = 1, \dots, M$, the posterior mean of the mixture weights ϕ, and construct a $\mathcal{WNM}$ distribution using these parameters.

Note that many of the activities are clearly multi-modal, such as the "Dress" activity in Figure 10b, and the model is able firstly to correctly identify the number of mixture components, and also to capture the multi-modal nature of the periodicity of the activities. The "Sleep_Out_Of_Bed" activity is interesting as there is a prominent "lobe" of the distribution from the narrower of the mixture components. Figure 11 shows the log model evidence scaled by the number of data points for the models learnt with $K = 1, 2, 3$ mixture components for each activity, showing the number of components chosen by model selection. Note that in some cases it is quite clear cut that one of the models should be preferred, but in other cases the choice is more borderline.

6 Discussion

The results indicate that the Approximate $\mathcal{WNM}$ ($\mathcal{AWNM}$) is able to accurately estimate the moments of $\mathcal{WN}$ Mixture ($\mathcal{WNM}$) distributed data, as demonstrated by the experiments on the toy dataset. Furthermore, the Expectation Propagation (EP) implementation is appealing, since it gives comparable results to the Gibbs sampling approach, but is generally faster and also enables model selection through evidence computation.

The inferred posteriors could then be used to (a) generate continuous feature(s) to be used in a classifier (probabilistic or otherwise), *e.g.* by using the log probability of activities given time-stamp. We would expect to see some improvement over a simple "hour-of-day" feature as it is a more refined representation of the distribution over time.

Potentially more interesting, however, is that since we have full distribution over the parameters, we can use these in a larger probabilistic model. For example, we can easily perform a modelling of the periodicity of the sensors activations in the same way, and then learn a mapping from sensor to activity which would in effect be a form of periodic regression.

7 Conclusions

In this paper we have shown that Bayesian inference for the $\mathcal{WN}$ distribution (using Gibbs) is easy to implement and accurate for data generated from a the model. The $\mathcal{WNM}$ suffers from identifiability issues, so we introduced an approximate version $\mathcal{AWN}$ which can be easily implemented using either Gibbs or EP in the modelling framework Infer.NET [10], and that this model accurately approximates the $\mathcal{WN}$ model. We then showed that we could extend this to the mixture modelling $\mathcal{AWNM}$, and demonstrated how model evidence can be used for model selection (choosing the number of mixture components). We then showed some results of preforming inference using the $\mathcal{AWNM}$ model on a real-world smart home data set.

7.1 Further Work

An appealing extension would be to construct a multivariate $\mathcal{WNM}$ model to model all of the sensor activations (from binary sensors) in a smart home together, with the resulting covariance matrix giving a description of the periodic linkage between sensors. Following on from this, it would be interesting to combine such a multivariate model with the univariate model, either by adding an extra dimension for the activities, or by constructing a circular regression task, for example by using the circular regression approach outlined by [15].

Another appealing extension would be to consider a hierarchical model for different residents in a given home, where common hyper-priors are shared between the residents, and individual priors are then inferred for each resident. This would be a natural path to being able to transfer such models to new homes and new residents.

Acknowledgments. This work was performed under the Sensor Platform for HEalthcare in Residential Environment (SPHERE) Interdisciplinary Research Collaboration (IRC) funded by the UK Engineering and Physical Sciences Research Council, Grant EP/K031910/1.

References

1. Agiomyrgiannakis, Y., Stylianou, Y.: Wrapped Gaussian mixture models for modeling and high-rate quantization of phase data of speech. IEEE Transactions on Audio, Speech, and Language Processing **17**(4), 775–786 (2009)
2. Cook, D.J., Schmitter-Edgecombe, M.: Assessing the quality of activities in a smart environment. Methods Inf. Med. **48**(5), 480–485 (2009)
3. Gelman, A., Carlin, J.B., Stern, H.S., Rubin, D.B.: Bayesian data analysis, vol. 2. Taylor & Francis (2014)
4. Goodman, S.N.: Toward evidence-based medical statistics. 2: The Bayes factor. Annals of Internal Medicine **130**(12), 1005–1013 (1999)
5. Jammalamadaka, S.R., Sengupta, A.: Topics in circular statistics, Series on Multivariate Analysis, vol. 5. World Scientific (2001)
6. Kim, E., Helal, S., Cook, D.: Human activity recognition and pattern discovery. IEEE Pervasive Computing **9**(1), 48–53 (2010)
7. Krishnan, N., Cook, D.J., Wemlinger, Z.: Learning a taxonomy of predefined and discovered activity patterns. Journal of Ambient Intelligence and Smart Environments **5**(6), 621–637 (2013)
8. Kurz, G., Gilitschenski, I., Hanebeck, U.D.: Efficient evaluation of the probability density function of a wrapped normal distribution. In: Sensor Data Fusion: Trends, Solutions, Applications (SDF), 2014, pp. 1–5. IEEE (2014)
9. Mardia, K.V., Jupp, P.E.: Directional statistics. John Wiley & Sons, Chichester (2000)
10. Minka, T., Winn, J., Guiver, J., Webster, S., Zaykov, Y., Yangel, B., Spengler, A., Bronskill, J.: Infer.NET 2.6. Microsoft Research Cambridge (2014). http://research.microsoft.com/infernet
11. Minka, T.P.: Expectation propagation for approximate Bayesian inference. In: Proceedings of the 17th Conference on Uncertainty in Artificial Intelligence, pp. 362–369. Morgan Kaufmann Publishers Inc. (2001)
12. Minka, T., Winn, J.: Gates. In: Advances in Neural Information Processing Systems, pp. 1073–1080 (2009)
13. Nazerfard, E., Das, B., Holder, L.B., Cook, D.J.: Conditional random fields for activity recognition in smart environments. In: Proceedings of the 1st ACM International Health Informatics Symposium, pp. 282–286. ACM (2010)
14. Nuñez-Antonio, G., Ausín, M.C., Wiper, M.P.: Bayesian nonparametric models of circular variables based on Dirichlet process mixtures of normal distributions. Journal of Agricultural, Biological, and Environmental Statistics, 1–18 (2014)
15. Ravindran, P., Ghosh, S.K.: Bayesian analysis of circular data using wrapped distributions. Journal of Statistical Theory and Practice **5**(4), 547–561 (2011)
16. Refinetti, R., Cornélissen, G., Halberg, F.: Procedures for numerical analysis of circadian rhythms. Biological Rhythm Research **38**(4), 275–325 (2007)
17. Robert, C.: The Bayesian choice: from decision-theoretic foundations to computational implementation. Springer Science & Business Media (2007)
18. Wilks, D.S.: Statistical methods in the atmospheric sciences, vol. 100. Academic press (2011)

Message Scheduling Methods for Belief Propagation

Christian Knoll[1]([✉]), Michael Rath[1], Sebastian Tschiatschek[2],
and Franz Pernkopf[1]

[1] Signal Processing and Speech Communication Laboratory,
Graz University of Technology, Graz, Austria
christian.knoll@tugraz.at
[2] Learning and Adaptive Systems Group, Department of Computer Science,
ETH Zurich, Zürich, Switzerland
sebastian.tschiatschek@inf.ethz.ch

Abstract. Approximate inference in large and densely connected graphical models is a challenging but highly relevant problem. Belief propagation, as a method for performing approximate inference in loopy graphs, has shown empirical success in many applications. However, convergence of belief propagation can only be guaranteed for simple graphs. Whether belief propagation converges depends strongly on the applied message update scheme, and specialized schemes can be highly beneficial. Yet, residual belief propagation is the only established method utilizing this fact to improve convergence properties. In experiments, we observe that residual belief propagation fails to converge if local oscillations occur and the same sequence of messages is repeatedly updated. To overcome this issue, we propose two novel message update schemes. In the first scheme we add noise to oscillating messages. In the second scheme we apply weight decay to gradually reduce the influence of these messages and consequently enforce convergence. Furthermore, in contrast to previous work, we consider the correctness of the obtained marginals and observe significant performance improvements when applying the proposed message update schemes to various Ising models with binary random variables.

Keywords: Residual belief propagation · Asynchronous message scheduling · Convergence analysis

1 Introduction

Probabilistic reasoning for complex distributions arises in many practical problems including computer vision, medical diagnosis systems, and speech processing [9]. These complex distributions are often modeled as probabilistic graphical models (PGMs). PGMs representing the joint distribution over many random

---

F. Pernkopf—This work was supported by the Austrian Science Fund (FWF) under the project number P25244-N15.

A. Appice et al. (Eds.): ECML PKDD 2015, Part II, LNAI 9285, pp. 295–310, 2015.
DOI: 10.1007/978-3-319-23525-7_18

variables (RVs) of practical problems are often complex and include many loops. Thus performing exact inference is increasingly intricate, in fact exact inference is intractable for general PGMs [1]. Message passing, a powerful method to approximate the marginal distribution, was first introduced to the field of machine learning as Belief Propagation (BP) by Pearl [19]. It is a parallel update scheme where messages are recursively exchanged between RVs until the marginal probabilities converge.

The conjecture that asynchronously updating the messages leads to better convergence performance of BP is widely accepted [2,5,22]. Thus, there was a recent interest in improving the performance of BP by applying dynamic message scheduling. One efficient way for scheduling is residual belief propagation (RBP) [2], where only the message that changes the most is sent. RBP has a provable convergence rate that is at least as good as the convergence rate of BP, while still providing good marginals. The quality of the obtained marginals in [2,24] is comparable to existing methods. Nonetheless, a detailed analysis of the quality of the marginals in comparison to the exact marginals is missing to the best of our knowledge. Dynamic message scheduling increases the number of graphs where BP converges. Yet, on graphs with many loops the occurrence of message oscillation is observed. In this case, a small set of messages is repeatedly selected for update and periodically takes the same values.

Inspired by this observation we introduce and investigate two different methods for dynamic message scheduling. The first method directly improves upon RBP if message oscillations occur. *Noise injection* belief propagation (NIBP) detects message oscillations of RBP. Adding random noise to the message that is propagated prevents these oscillations and improves convergence of BP. The second method is based on the assumption that messages repeatedly taking the same values do not contribute to convergence of the overall PGM. A sequence of oscillating messages does obviously not change the constraints in favor of convergence. We apply weight decay to the residual and consequently, support non oscillating messages to be updated. This way we avoid message oscillations before they even occur. *Weight decay* belief propagation (WDBP) solely changes the scheduling by the damping, whereas directly applying a damping term to the beliefs can also improve the convergence properties [21].

Our proposed methods are evaluated on different realizations of Ising grid graphs. Graphs of such structure have a rich history in statistical physics [8,15], and these models are appealing, as phase transitions can be analytically determined. Phase transitions separate convergent from divergent behavior and can be related to PGMs and the behavior of BP. It is shown in [25,26] how the fixed point solutions of BP correspond to the local minima of the Gibbs free energy.

On difficult Ising graphs we compare the performance of the proposed methods to RBP and asynchronous belief propagation (ABP). The convergence behavior is usually analyzed in terms of the number of times BP converges (i.e converged runs) and the speed of convergence (i.e. convergence rate). In addition, we compare the approximated marginals to the exact marginals, which are obtained by the junction tree algorithm [12].

Our two main findings are: (i) we show empirically that NIBP significantly increases the number of times convergence is achieved and (ii) WDBP accomplishes a quality of marginals superior to the remaining methods, while maintaining good convergence properties.

The rest of this paper is structured as follows. In Section 2 we give a short background on probabilistic graphical models and belief propagation. We introduce our proposed approach to message scheduling in Section 3 and relate it to existing methods. Our experimental results are presented and discussed in Section 4. Related work is deferred to Section 5 for the sake of reading flow. Section 6 summarizes the paper and provides some final conclusions.

2 Preliminaries

In this section we briefly introduce PGMs and the BP algorithm. Some applications and a detailed treatment of PGMs can be found in [11,20]. Let X be a binary random variable (RV) taking values $x \in \mathbb{S} = \{-1, 1\}$. We consider the finite set of random variables $\mathbf{X} = \{X_1, \ldots, X_N\}$.

An undirected graphical model (UGM) consists of an undirected graph $G = (\mathbf{X}, \mathbf{E})$, where $\mathbf{X} = \{X_1, \ldots, X_N\}$ represents the nodes and $\mathbf{E}$ the edges. Two nodes X_i and X_j, $i \neq j$ can be connected by an undirected edge $e_{i,j} \in \mathbf{E}$ that specifies an interaction between these two nodes. Note that we use the same terminology for the nodes as for the RVs since there is a one-to-one relationship. The set of neighbors of X_i is defined by $\Gamma(X_i) = \{X_j \in \mathbf{X} \backslash X_i : e_{i,j} \in \mathbf{E}\}$. We use an UGM to model the joint distribution

$$P(\mathbf{X} = \mathbf{x}) = \frac{1}{Z} \prod_{(i,j) \,:\, e_{i,j} \in \mathbf{E}} \Phi_{X_i, X_j}(x_i, x_j) \prod_{i=1}^{N} \Phi_{X_i}(x_i), \qquad (1)$$

where the first product runs over all edges, and where Φ_{X_i, X_j} are the pairwise potentials and Φ_{X_i} is the local potential.

Our formulation of BP is similar to the one introduced in [25]. For a detailed introduction to the concept of BP we refer the reader to [19,29]. The messages are updated according to the following rule:

$$\mu_{i,j}^{n+1}(x_j) = \sum_{x_i \in \mathbb{S}} \Phi_{X_i, X_j}(x_i, x_j) \Phi_{X_i}(x_i) \prod_{X_k \in (\Gamma(X_i) \backslash \{X_j\})} \mu_{k,i}^{n}(x_i), \qquad (2)$$

where $\mu_{i,j}^{n}(x_j)$ is the message from X_i to X_j of state x_j at iteration n.[1] Loosely speaking this means that X_i collects all messages from its neighbors $\Gamma(X_i)$ except for X_j. This product is then multiplied with the pairwise and local potentials $\Phi_{X_i, X_j}(x_i, x_j)$ and $\Phi_{X_i}(x_i)$. Finally the sum over all states of X_i is sent to X_j.

---

[1] Note that without loss of generality we will drop the superscript n where no ambiguities occur.

The marginals (or beliefs) $P(X_i = x_i)$ are obtained from all incoming messages according to

$$P(X_i = x_i) = \frac{1}{Z}\Phi_{X_i}(x_i) \prod_{X_k \in \Gamma(X_i)} \mu_{k,i}(x_i), \tag{3}$$

where $Z \in \mathbb{R}^+$ is the normalization constant ensuring that $\sum_{x_i \in \mathbb{S}} P(X_i = x_i) = 1$. When the specific realization is not relevant we use the shorthand notation $P(X_i)$ instead.

There is a rich history of statistical physicists studying the interaction in Ising models. The Edwards-Anderson model or Ising spin glass is an elegant abstraction that allows both, ferromagnetic and antiferromagnetic Ising models [14, p. 44]. Following the terminology of the Edwards-Anderson model we define the potentials of the model, such that we have a coupling $J_{i,j} \in \mathbb{R}$ and a local field $\theta_i \in \mathbb{R}$. Let the potentials be $\Phi_{X_i}(x_i) = \exp(\theta_i x_i)$ and $\Phi_{X_i,X_j}(x_i, x_j) = \exp(J_{i,j} x_i x_j)$. Plugging these potentials into (1), the Ising spin glass model defines the joint probability

$$P(\mathbf{X} = \mathbf{x}) = \frac{1}{Z}\exp\left(\sum_{(i,j)\,:\,e_{i,j} \in \mathbf{E}} J_{i,j} x_i x_j + \sum_{i=1}^N \theta_i x_i \right), \tag{4}$$

where the sum over $(i, j)\colon e_{i,j} \in \mathbf{E}$ runs over all edges of G and the second sum runs over all nodes. Spin glasses in this form offer a powerful generalization of the Ising model that allow for frustration.[2] Such models have been used to relate the convergence problem to the occurrence of phase transitions [4]. One can consequently derive a sharp bound for the parameter set $(J_{i,j}, \theta_i)$ and relate it to the convergence of loopy BP [18,25,26].

When analyzing the graph convergence over time, it is remarkable that certain subgraphs are almost converged after few iterations, while other regions are less stable. More formally we can introduce two subgraphs such that $G = G_c \cup G_{\bar{c}}$. We define the almost converged subgraph as $G_c = (\mathbf{X_c}, \mathbf{E_c})$, i.e. for all $(X_i, X_j) : e_{i,j} \in \mathbf{E_c}$ we have $\mu_{i,j}^{n+1}(x_j) \approx \mu_{i,j}^n(x_j)$. The second subgraph $G_{\bar{c}} = (\mathbf{X_{\bar{c}}}, \mathbf{E_{\bar{c}}})$ is less stable, i.e. $\mu_{i,j}^{n+1}(x_j) \not\approx \mu_{i,j}^n(x_j)$. Note that $G_{\bar{c}}$ may even include frustrated cycles such that convergence can never be reached.

3   Scheduling

For a given graph $G = (\mathbf{X}, \mathbf{E})$ we can define any message passing algorithm by basic operations on the alphabet of messages (cf. [14, p. 316]). The algorithm is converged if two successive messages show approximately the same value, i.e. $\mu_{i,j}^{n+1}(x_j) \approx \mu_{i,j}^n(x_j)$. At that point, updating the messages does not change their values, therefore we can also speak of a fixed point solution.

Note that in the original implementation of BP all messages are synchronously updated, i.e. to compute $\mu_{i,j}^{n+1}$ *all* messages at iteration n are used.

---

[2] Frustrated cycles have an overall parametrization, such that it is impossible to simultaneously satisfy all local constraints, i.e. convergence can never be achieved.

Substituting the synchronous update rule by a sequential update rule, we obtain a flexibility in developing variants of BP. Exploiting this flexibility and changing the update schedule significantly influences the performance in practice, as reported in [2,14]. We are essentially interested in the advantages of different update schedules, therefore we solely consider sequential (or asynchronous) scheduling for the remainder of the work.

All variants of BP are compared to the performance of asynchronous belief propagation (ABP). ABP is based on a rudimentary sequential update rule, where all messages are considered equally important. Messages are selected according to round robin scheduling, i.e. according to a fixed order. Although no assumptions are made on a smart choice of the order, it can be observed that this simple message scheduling concept improves the convergence behavior [10,22].

We propose two modifications to BP to improve convergence properties. Either we change the calculation of the *message values* directly (NIBP), or we utilize alternative *message scheduling* (WDBP). In the following we describe these modifications in detail. Experimental results demonstrating the effectiveness of the proposed modifications can be found in Section 4.

3.1 Residual Belief Propagation

Residual belief propagation (RBP) utilizes a priority measure for each message and introduces dynamic scheduling [2]. The underlying assumption is that any message passed along an edge $e_{i,j} \in \mathbf{E_c}$ in the already converged subgraph does not contribute to the overall convergence. Thus focusing on the subgraph that has not converged $G_{\bar{c}} = (\mathbf{X_{\bar{c}}}, \mathbf{E_{\bar{c}}})$ is beneficial for convergence of the overall graph. As $G_{\bar{c}}$ is not converged, messages along edges $\bar{e}_{i,j} \in \mathbf{E_{\bar{c}}}$ vary considerably in every step.

This leads to the update rule of RBP, where the residual $r_{i,j}^n = |\mu_{i,j}^{n+1}(x_j) - \mu_{i,j}^n(x_j)|$ measures the distance between two messages.[3] The indices that maximize the residual

$$(k,l) = \underset{(i,j)}{\operatorname{argmax}} \, r_{i,j}^n \tag{5}$$

identify the message to update next, i.e.

$$\tilde{\mu}_{k,l}^{n+1}(x_l) = \mu_{k,l}^{n+1}(x_l). \tag{6}$$

Compared to ABP the number of graphs where RBP converges increases significantly [2]. Still, RBP computes all residuals although only the message with the most significant residual is sent. To further increase the convergence rate, the authors of [24] bound and approximate the message values for the estimation of the residual.

---

[3] Ultimately one would be interested in the distance to the fixed point, if it exists, $\lim_{n \to \infty} \mu_{i,j}^n(x_j)$. However, since $\lim_{n \to \infty} \mu_{i,j}^n(x_j)$ is not known, the time variation of the messages offers a valid surrogate (cf. [2]).

3.2   Noise Injection Belief Propagation

Investigating graphs with random Ising factors, where RBP fails to converge, we observe that a large part of the PGM is almost converged. We observe local frustrated cycles in $G_{\bar{c}}$, where the same message values are passed around repeatedly along the edges $\bar{e}_{i,j} \in \mathbf{E}_{\bar{c}}$. Noise injection belief propagation (NIBP) compares the current message $\mu_{i,j}^n$ to the last $L \in \mathbb{Z}^+$ messages for duplicate values. If older messages are in an δ-neighborhood, i.e. $|\mu_{i,j}^n - \mu_{i,j}^{n-l}| < \delta$ for any $l \in \{1, 2, \ldots, L\}$, although these messages are not converged, i.e. $\mu_{i,j}^{n+1} \not\approx \mu_{i,j}^n$, we conclude that the message values oscillate. If no oscillations are detected NIBP does not change the scheduling of RBP. Therefore, NIBP always converges if RBP does. If, however, message values oscillate Gaussian noise $\mathcal{N}(0, \sigma^2)$ is added to the message $\mu_{k,l}^{n+1}$ that is selected according to (5). The new update rule is then given as

$$\tilde{\mu}_{k,l}^{n+1}(x_l) = \mu_{k,l}^{n+1}(x_l) + \mathcal{N}(0, \sigma^2), \tag{7}$$

where X_k and X_l are the nodes that maximize the residual in (5) and $\mathcal{N}(0, \sigma^2)$ is the normal distribution with zero mean and standard deviation σ.

Loosely speaking we aim to introduce a relevant change to the system by injecting noise to the message selected for update $\mu_{k,l}^{n+1}$. Adding noise to the most influential part of the PGM, we assume that this minor change of one message propagates through the whole graph and leads to a stable fixed point. Pseudocode of the implementation can be found in Appendix A.

3.3   Weight Decay Belief Propagation

As mentioned above RBP fails to converge if message values oscillate. Obviously, repeatedly sending around the same messages along the same path does not contribute to achieving convergence. Weight decay belief propagation (WDBP) penalizes this behavior by damping the residual of messages along $\bar{e}_{i,j}$. Consequently, WDBP increases the relevance of G_c and further refines the parametrization of this subgraph. In doing so, messages $\mu_{i,j}$ between both subgraphs, where $X_i \in \mathbf{X_c}$ and $X_j \in \mathbf{X_{\bar{c}}}$ are re-evaluated, such that convergence can be achieved on the overall graph G.

In particular, we damp the residual of all messages of a node X_i based on the number of times a message has already been scheduled. More formally we rewrite (5), such that the indices of the selected message $\tilde{\mu}_{k,l}^{n+1}$ are given to

$$(k, l) = \operatorname*{argmax}_{(i,j)} \frac{r_{i,j}^n}{\sum_{m=1}^{n} \mathbf{1}_{\mu_{i,j}^m}}, \tag{8}$$

where the indicator function $\mathbf{1}_{\mu_{i,j}^m} = 1$ if and only if $\mu_{i,j}^m = \tilde{\mu}_{i,j}^m$. Hence, the residual is divided by a factor corresponding to how often a certain message was selected. A detailed implementation is presented in Appendix A.

4 Experiments

In this section we evaluate the proposed methods and compare them to ABP and RBP. We evaluate all different types of scheduling with respect to the following measures: first the number of configurations where the algorithm converges will be considered, secondly we consider the rate of convergence, and finally we evaluate the quality of the marginals. To evaluate the marginals we obtain the approximate marginal distributions $\tilde{P}(X_i)$ and compare them to the exact marginal distributions $P(X_i)$, obtained by applying the junction tree algorithm [12,16]. Although the junction tree algorithm is intractable in general, the considered PGMs are simple enough to make exact inference computationally feasible. We quantify the quality of the marginals by computing the mean squared error (MSE) over all marginals. Note that the potential functions are identical for all compared methods.

Statistical physic provides exact statements regarding the performance of BP on Ising spin glasses, therefore such models are commonly used for evaluation of BP variants. In this work we perform message passing on Ising spin glasses of varying size with uniform and random coefficients.

For NIBP, the parameters of the additive Gaussian noise were optimized for different initialization and are zero mean and $\sigma = 0.25$. Simulations were either stopped after k_{max} iterations or if all messages converged, i.e. $\max_{i,j} |\mu_{i,j}^{n+1}(x_j) - \mu_{i,j}^n(x_j)| < \epsilon$ for all $i, j : i \neq j$, where $\epsilon = 10^{-3}$. Experiments on Ising grids with uniform parameters were stopped after $k_{max} = 4 \cdot 10^5$ iterations, whereas the experiments on Ising grids with random factors were stopped after $k_{max} = 2.5 \cdot 10^5$ iterations.

4.1 Fully Connected Graph with Uniform Parameters

We consider a fully connected Ising spin glass with $|\mathbf{X}| = 4$ binary spins, and uniform coupling $J_{i,j}$ and field θ_i among the four vertices. In the case of uniform parameters we introduce the shorthand notation (J, θ). Using such a model allows to compare our results to similar numerical experiments performed on this type of graphs in [18,25,26]. Figure 1 shows the complete graph for $|\mathbf{X}| = 4$.

Applying BP to this graph one can benefit from the rich history of statistical physics literature to discuss the effect of different messages schedules. For a fully connected Ising spin glass with uniform parameters the Gibbs measure is unique and the solution of BP is exactly equal to the one obtained by optimizing the Bethe approximation [26]. That is, there are certain regions in the 2-dimensional parameter-space (J, θ) where BP is guaranteed to converge. Nonetheless, there is a phase transition in the parameter space where BP does not converge. If $J \geq 0$ the model is known to be ferromagnetic and in fact reduces to the standard Ising model. The antiferromagnetic behavior is observed for $J < 0$, respectively [14].

In Figure 2a we show convergence of ABP and the transition to configurations (J, θ) where messages oscillate. The color encodes the logarithm of the number of iterations until convergence. We observe that reducing J increases the difficulty of finding an equilibrium state. This, however, is intuitive since, the more negative J is, the more one node X_i tries to push its neighbors $\Gamma(X_i)$ into the opposite state.

Looking at Figure 2b we observe how RBP pushes the transition boundary and increases the set of coefficients where convergence is achieved. Finally Figure 2c and 2d show the performance of NIBP and WDBP respectively. Notably, both methods further increase the region of convergence. It can be seen that these boundaries are heavily blurred. For specific parameter configurations our proposed methods result in equilibrium state after many runs, where established methods fail to converge.

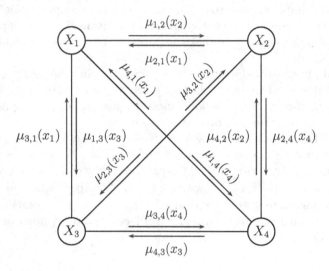

Fig. 1. 2x2 Ising Spin Glass.

4.2   Ising Grids with Random Factors

From the experiments in Figure 2 we can hardly make any concrete statements regarding the convergence behavior. Hence, to further investigate the influence of WDBP and NIBP we consider Ising grids with many loops and random parameters. These graphs are often used for evaluation of the performance of BP, since BP is prone to diverge on those graphs. We consider grid graphs of size $N = |\mathbf{X}| = K \times K$ with binary spins and randomly initialized parameters $(J_{i,j}, \theta_i)$. Depending on the grid size K these parameters are uniformly sampled such that both $(J_{i,j}, \theta_i) \in [-\frac{K}{2}, \frac{K}{2}]$. Thus, besides increasing the size of the graph, the difficulty is implicitly increased as well.

The larger the values of the coupling and the local field are, the harder the resulting constraints for convergence are. Thus, although there is no structural change of the grid, inference becomes easier by reducing the range of the parameters. According to [24] the parameters have to provide sufficient difficulty to be of interest for analyzing convergence properties.

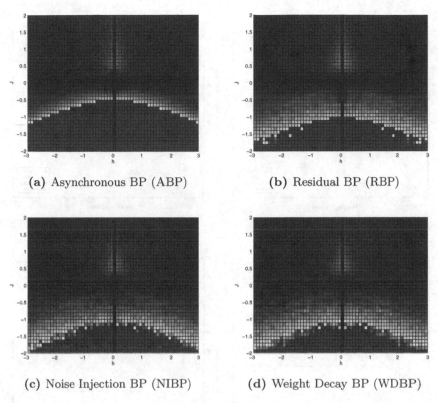

(a) Asynchronous BP (ABP)   (b) Residual BP (RBP)

(c) Noise Injection BP (NIBP)   (d) Weight Decay BP (WDBP)

Fig. 2. Convergence of various BP variants for a fully connected binary Ising spin glass with uniform parameters (J, θ). The color encodes the logarithm of the number of iterations until convergence; blue corresponds to convergence and red means that the method did not converge after $4 \cdot 10^5$ iterations. 2a shows the phase transition of ABP. Note how the RBP variants in 2b-2d increase the region of convergence.

The proportion of converged runs for different schedules is shown in Figure 3. We can see that ABP finds a fixed point in less scenarios than any other of the proposed variants, demonstrating the advantage of dynamic message scheduling. Looking at the overall performance we observe that NIBP converges most often throughout all experiments. The more complex the network, the better NIBP performs compared to other variants. WDBP outperforms RBP on all experiments and shows the best performance on the 7×7 graph, although the harder the problem, the slower it converges. Specifically, WDBP has a lower convergence rate than RBP. This observation is expected as damping the residual reduces the propagation of relevant messages even for relatively easy configurations.

It shall be noted that applying WDBP requires changing the residual to (8), where damping the residual implies some computational overhead. This overhead can be reduced partially with approximation of the messages according to [24].

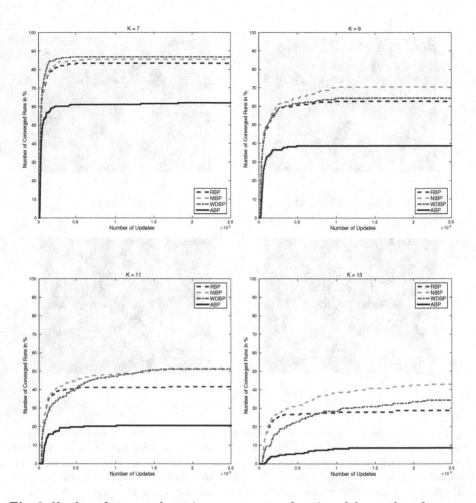

Fig. 3. Number of converged runs in percentage as a function of the number of message updates. All results were obtained by averaging over 233 random grid graphs. On graphs of the size $|\mathbf{X}| = K \times K$ we compare WDBP and NIBP to RBP and ABP.

All results were averaged over 233 runs with different random initialization of the pairs $(J_{i,j}, \theta_i)$ for $K \in \{7, 9, 11, 13\}$.

4.3   Quality of Marginals

Currently the influence of message scheduling was only evaluated in terms of the convergence rate and the number of graphs where BP converges. Here, we also evaluate the correctness of the approximated marginals $\tilde{P}(X_i)$, averaging the mean squared error (MSE) of all $N = K \times K = |\mathbf{X}|$ nodes, such that

$$MSE = \frac{1}{N} \sum_{i=1}^{N} \sum_{a \in \mathbb{S}} |\tilde{P}(X_i = a) - P(X_i = a)|^2, \tag{9}$$

where $P(X_i)$ are the exact marginals. Note that all RVs are binary and both, $\tilde{P}(X_i)$ and $P(X_i)$ are valid probability mass function, i.e. $\sum_{a \in \mathbb{S}} \tilde{P}(X_i) = 1$. Further applying symmetry properties it then follows that

$$MSE = \frac{2}{N} \sum_{i=1}^{N} |\tilde{P}(X_i = 1) - P(X_i = 1)|^2. \tag{10}$$

In Table 1 we present quantitative performance measures for all experiments. Solely considering the number of converged runs we can recapitulate the observation from Figure 3 that RBP converges in at least 20% of all experiments, where ABP did not. Both our proposed methods are able to further increase the convergence; throughout all experiments NIBP converges most often.

Still, in practice we are not only interested in the number of converged runs but also in the quality of the marginals. First we estimate the overall MSE based on (10) and average over all 233 randomly initialized graphs (MSE overall). Secondly, we average the MSE over all runs where the individual methods converged(MSE converged) – for ABP we estimate the MSE only for easy configurations, whereas the MSE for other variants includes harder configurations. Therefore, we finally estimate the MSE of all methods for those configurations where ABP converges to a fixed point(MSE ABP conv.).

It can be seen that ABP consistently shows the lowest MSE in terms of converged runs, i.e. averaging over all runs that converged with this method. This comes as no surprise as ABP converges only on graphs with relatively easy configurations. For these configurations we expect $\tilde{P}(X_i)$ to give a good approximation to the exact marginals $P(X_i)$. However, estimating the MSE of different methods for graphs where ABP is known to converge (MSE ABP conv.), we are surprised by the observations that the approximate marginals obtained by RBP or NIBP are consistently worse than the ones found by ABP. Still solely considering these easy graphs it is remarkable how well WDBP performs in terms of the MSE.

Note that by using an update rule based on RBP a lot of effort is put into locally complying with the constraints of $G_{\bar{c}}$ whereas ABP still puts a significant amount of resources into refining G_c. This clearly reduces the convergence rate but potentially boosts the correctness of the approximation.

We would expect the overall MSE, i.e. averaging over all 233 runs, is reduced using dynamic message scheduling. Despite ABP converges in less runs it still results in surprisingly good overall approximations of the marginals. In fact the obtained quality of the marginals is similar for ABP, RBP, and NIBP, supporting the empirical observations that ABP performs reasonable well on many graphs. Notably, it can also be seen that WDBP consistently reduces the overall MSE resulting in the best approximation of the marginals.

Looking at Table 1 we want to emphasize the superior overall performance of WDBP. The number of converged runs is significantly increased in comparison to ABP while a proper approximation accuracy is maintained.

Table 1. Performance of different BP schedules on Ising spin glasses of size $|\mathbf{X}| = K \times K$. The MSE is estimated between approximated $\tilde{P}(X_i)$ and exact $P(X_i)$ marginals. We average over all 233 runs (MSE overall), over runs where the individual methods converged (MSE converged), and over runs where ABP converged (MSE ABP conv.) We compare asynchronous BP (ABP), residual BP (RBP), noise injection BP (NIBP), and weight decay BP (WDBP).

Grid Size	Error Metric		**ABP**	**RBP**	**NIBP**	**WDBP**
$K = 7$	MSE	overall	0.0514	0.041	0.0382	**0.0330**
		converged	**0.0164**	0.0208	0.0218	0.0202
		ABP conv.	0.0164	0.0182	0.0150	**0.0130**
	Converged Runs		61.8%	83.26%	85.41%	**86.7%**
$K = 9$	MSE	overall	0.0706	0.0622	0.0538	**0.0486**
		converged	**0.0078**	0.0256	0.026	0.0230
		ABP conv.	**0.0078**	0.0190	0.0144	0.0112
	Converged Runs		38.63%	62.66%	**70.39%**	64.38%
$K = 11$	MSE	overall	0.0830	0.0914	0.0750	**0.0618**
		converged	**0.0106**	0.0340	0.0386	0.0258
		ABP conv.	**0.0106**	0.0262	0.0268	0.0152
	Converged Runs		20.6%	41.63%	**51.5%**	51.07%
$K = 13$	MSE	overall	0.1126	0.1274	0.1102	**0.0840**
		converged	**0.0286**	0.0642	0.0632	0.0314
		ABP conv.	0.0286	0.0746	0.0590	**0.0282**
	Converged Runs		8.58%	28.76%	**42.92%**	34.33%

5 Related Work

On trees and chains BP is guaranteed to converge, moreover BP obtains the optimal maximum a posterior assignment for PGMs with a single loop [27]. However, many graphs that represent a domain of the real world have an arbitrary structure, including loops. There is no general guarantee for BP to converge on such complex graphs [3,27]. Yet, it was shown empirically that BP can still give good results when applied to graphs with a complicated structure.

There are various approaches that aim to correct for the presence of loops such as loop correction [17] or the truncated loop series introduced in [6]. There are also many publications relating the fixed points of BP to extrema of approximate free energy functions from statistical physics [7,28]. It was shown in [28] how extrema of the Bethe free energy approximations correspond to the fixed points of BP. Using the generalization, the Kikuchi free energy function, generalized BP (GBP) was introduced in [28], which significantly improves the number of converged runs and the convergence rate compared to standard BP. Applying a concave- convex procedure to the Bethe and Kikuchi free energies the CCCP algorithm is introduced in [30] and results in slightly better results than those found by GBP. Convexified free energies [13] come with good convergence properties but still lack the empirical success. Linear programming relaxation can be used to deal with frustrated cycles as well [23]. Long range correlations often lead to failure of BP [14] but can be handled through the cavity method [15].

6   Conclusion

In this paper, we introduced two novel methods for dynamic message scheduling. Refining the ideas of residual belief propagation (RBP), we further improve the number of converged runs on various difficult graphs.

The first method, noise injection belief propagation (NIBP) detects if RBP fails to find a fixed point, i.e. message values oscillate. Gaussian noise is then added to the selected message such that the overall configuration is modified to achieve convergence. Our assumption is that this noise injection propagates to other parts of the network and counteracts the oscillations. Still if RBP converges, NIBP is guaranteed to converge as well.

The second method, weight decay belief propagation (WDBP) obviates the need for oscillation detection. Each time a message is selected for an update, the importance of the message for potential future updates is reduced. Thus, WDBP implicitly reduces the priority of subgraphs that oscillate and forces the overall graph to a fixed point.

Both proposed methods are applied to various Ising grids and are evaluated in comparison to other sequential message passing algorithms. Our main evaluation is based on convergence properties and the correctness of the marginals. In all experiments both methods, NIBP and WDBP converge more often than RBP and asynchronous belief propagation (ABP).

NIBP achieves the highest convergence rate and number of converged runs. Still, considering the mean squared error of the marginals we notice that ABP leads to surprisingly good marginals. Applying RBP and NIBP to increase the number of converged runs comes with a sacrifice of the approximation accuracy of the marginals.

We further compare the MSE between the exact and the approximated marginals in different scenarios. This quality aspect has not been mentioned in previous work. Only considering easy graphs, where ABP converges, we are surprised by the observation that ABP consistently outperforms RBP or NIBP in terms of the quality of the approximated marginals. The quality of the marginals obtained by WDBP on these graphs is remarkable and superior to all compared methods.

By all means the above results highlight how the message passing scheduling influences the performance of belief propagation. Still, the convergence rate of both, NIBP and WDBP can potentially be further improved by using an estimation of the residual [24] instead of computing the messages for every step.

Acknowledgments. This work was supported by the Austrian Science Fund (FWF) under the project number P25244-N15.

Appendix A: Pseudocode

We present the pseudocode for NIBP and WDBP. Removing the if then else clause in line 8 to 11 of NIBP and substituting it with $\mu_u^{old} \leftarrow \mu_u^{new}$ reduces

Algorithm 1 to RBP. The maximum number of iterations is denoted by $k_{max} = 2.5 \cdot 10^5$ and $\epsilon = 10^{-3}$. NrOfMessages denotes the overall number of messages in the graph.

Algorithm 1. Noise Injection Belief Propagation (NIBP)

---

input : Graph $G = (\mathbf{X}, \mathbf{E})$
output: Converged messages μ^{old}

1 initialization
2 **for** $i \leftarrow 1$ **to** NrOfMessages **do**
3     $\mu_i^{new} \leftarrow$ ComputeUpdate(μ_i^{old})
4     $r_i \leftarrow |\mu_i^{old} - \mu_i^{new}|$
5 $k \leftarrow 1$
6 **while** $k < k_{max}$ **and** $\max |\mu^{old} - \mu^{new}| > \epsilon$ **do**
7     $u \leftarrow \text{argmax}_i\, r$
8   **if** OscillationDetection(μ_u^{old}, L) **then**
9        $\mu_u^{old} \leftarrow \mu_u^{new} + \mathcal{N}(0, \sigma)$
10   **else**
11       $\mu_u^{old} \leftarrow \mu_u^{new}$
12   **for** $j \leftarrow 1$ **to** NrOfMessages **do**
13       $\mu_j^{new} \leftarrow$ ComputeUpdate(μ_j^{old})
14       $r_j \leftarrow |\mu_j^{new} - \mu_j^{old}|$
15    $k = k + 1$

Algorithm 2. Weight Decay Belief Propagation (WDBP)

---

input : Graph $G = (\mathbf{X}, \mathbf{E})$
output: Converged messages μ^{old}

1 initialization
2 **for** $i \leftarrow 1$ **to** NrOfMessages **do**
3     $\mu_i^{new} \leftarrow$ ComputeUpdate(μ_i^{old})
4     $r_i \leftarrow |\mu_i^{old} - \mu_i^{new}|$
5   NrUpdates (i) $\leftarrow 1$
6 $k \leftarrow 1$
7 **while** $k < k_{max}$ **and** $\max |\mu^{old} - \mu^{new}| > \epsilon$ **do**
8     $u \leftarrow \text{argmax}_i\, r$
9     $\mu_u^{old} \leftarrow \mu_u^{new}$
10   NrUpdates (u) $\leftarrow$ NrUpdates (u) $+ 1$
11   **for** $j \leftarrow 1$ **to** NrOfMessages **do**
12       $\mu_j^{new} \leftarrow$ ComputeUpdate(μ_j^{old})
13       $r_j \leftarrow \dfrac{|\mu_j^{new} - \mu_j^{old}|}{\text{NrUpdates(j)}}$
14    $k = k + 1$

References

1. Cooper, G.F.: The computational complexity of probabilistic inference using Bayesian belief networks. Artificial Intelligence **42**(2), 393–405 (1990)
2. Elidan, G., McGraw, I., Koller, D.: Residual belief propagation: Informed scheduling for asynchronous message passing. In: Conference on Uncertainty in Artificial Intelligence (UAI) (2006)
3. Frey, B.J., MacKay, D.J.: A revolution: Belief propagation in graphs with cycles. In: Neural Information Processing Systems (NIPS), pp. 479–485 (1998)
4. Georgii, H.O.: Gibbs Measures and Phase Transitions, vol. 9 (2011)
5. Goldberger, J., Kfir, H.: Serial schedules for belief-propagation: analysis of convergence time. IEEE Transactions on Information Theory **54**(3), 1316–1319 (2008)
6. Gómez, V., Mooij, J.M., Kappen, H.J.: Truncating the loop series expansion for belief propagation. The Journal of Machine Learning Research (2007)
7. Heskes, T.: On the uniqueness of loopy belief propagation fixed points. Neural Computation **16**(11), 2379–2413 (2004)
8. Ising, E.: Beitrag zur Theorie des Ferromagnetismus. Zeitschrift für Physik A Hadrons and Nuclei **31**(1), 253–258 (1925)
9. Jordan, M.I.: Graphical models. Statistical Science, pp. 140–155 (2004)
10. Kfir, H., Kanter, I.: Parallel versus sequential updating for belief propagation decoding. Physica A: Statistical Mechanics and its Applications **330**(1)
11. Koller, D., Friedman, N.: Probabilistic Graphical Models: Principles and Techniques. MIT press (2009)
12. Lauritzen, S.L., Spiegelhalter, D.J.: Local computations with probabilities on graphical structures and their application to expert systems. Journal of the Royal Statistical Society. Series B (Methodological), 157–224 (1988)
13. Meshi, O., Jaimovich, A., Globerson, A., Friedman, N.: Convexifying the bethe free energy. In: Conference on Uncertainty in Artificial Intelligence (UAI), pp. 402–410. AUAI Press (2009)
14. Mezard, M., Montanari, A.: Information, Physics, and Computation. Oxford University Press (2009)
15. Mézard, M., Parisi, G.: The Bethe lattice spin glass revisited. The European Physical Journal B-Condensed Matter and Complex Systems **20**(2), 217–233 (2001)
16. Mooij, J.M.: libdai: A free and open source c++ library for discrete approximate inference in graphical models. The Journal of Machine Learning Research **11** (2010)
17. Mooij, J.M., Kappen, H.J.: Loop corrections for approximate inference on factor graphs. Journal of Machine Learning Research **8**, 1113–1143 (2007)
18. Mooij, J.M., Kappen, H.J.: Sufficient conditions for convergence of the sum-product algorithm. IEEE Transactions on Information Theory **53**(12), 4422–4437 (2007)
19. Pearl, J.: Probabilistic Reasoning in Intelligent Systems: Networks of Plausible Inference. Representation and Reasoning Series. Morgan Kaufmann Publishers (1988)
20. Pernkopf, F., Peharz, R., Tschiatschek, S.: Introduction to Probabilistic Graphical Models (2014)
21. Pretti, M.: A message-passing algorithm with damping. Journal of Statistical Mechanics: Theory and Experiment **2005**(11), P11008 (2005)
22. Sharon, E., Litsyn, S., Goldberger, J.: Efficient serial message-passing schedules for LDPC decoding. IEEE Transactions on Information Theory **53**(11), 4076–4091 (2007)

23. Sontag, D., Choe, D.K., Li, Y.: Efficiently searching for frustrated cycles in MAP inference. In: Conference on Uncertainty in Artificial Intelligence (UAI) (2012)
24. Sutton, C.A., McCallum, A.: Improved dynamic schedules for belief propagation. In: Conference on Uncertainty in Artificial Intelligence (UAI) (2007)
25. Taga, N., Mase, S.: On the convergence of belief propagation algorithm for stochastic networks with loops. Citeseer (2004)
26. Tatikonda, S.C., Jordan, M.I.: Loopy belief propagation and Gibbs measures. In: Conference on Uncertainty in Artificial Intelligence (UAI) (2002)
27. Weiss, Y.: Correctness of local probability propagation in graphical models with loops. Neural Computation **12**(1), 1–41 (2000)
28. Yedidia, J.S., Freeman, W.T., Weiss, Y.: Bethe free energy, Kikuchi approximations, and belief propagation algorithms. Neural Information Processing Systems (NIPS) **13** (2001)
29. Yedidia, J.S., Freeman, W.T., Weiss, Y.: Understanding belief propagation and its generalizations. Exploring Artificial Intelligence in the New Millennium **8**, 239–269 (2003)
30. Yuille, A.L.: CCCP algorithms to minimize the Bethe and Kikuchi free energies: Convergent alternatives to belief propagation. Neural Computation **14**(7) (2002)

Output-Sensitive Adaptive Metropolis-Hastings for Probabilistic Programs

David Tolpin[✉], Jan-Willem van de Meent, Brooks Paige, and Frank Wood

Department of Engineering Science, University of Oxford, Oxford, England
{dtolpin,jwvdm,brooks,fwood}@robots.ox.ac.uk

Abstract. We introduce an adaptive output-sensitive Metropolis-Hastings algorithm for probabilistic models expressed as programs, Adaptive Lightweight Metropolis-Hastings (AdLMH). This algorithm extends Lightweight Metropolis-Hastings (LMH) by adjusting the probabilities of proposing random variables for modification to improve convergence of the program output. We show that AdLMH converges to the correct equilibrium distribution and compare convergence of AdLMH to that of LMH on several test problems to highlight different aspects of the adaptation scheme. We observe consistent improvement in convergence on the test problems.

Keywords: Probabilistic programming · Adaptive MCMC

1 Introduction

One strategy for improving convergence of Markov Chain Monte Carlo (MCMC) samplers is through online adaptation of the proposal distribution [1,2,15]. An adaptation scheme must ensure that the sample sequence converges to the correct equilibrium distribution. In a componentwise updating Metropolis-Hastings MCMC sampler, i.e. Metropolis-within-Gibbs [5,8,10], the proposal distribution can be decomposed into two components:

1. A stochastic schedule (probability distribution) for selecting the next random variable for modification.
2. The kernels from which new values for each of the variables are proposed.

In this paper we concentrate on the first component—adapting the schedule for selecting a variable for modification. Our primary interest in this work is to improve MCMC methods for probabilistic programming [6,7,11,13,17]. Probabilistic programming languages facilitate development of probabilistic models using the expressive power of general programming languages. The goal of inference in such programs is to reason about the posterior distribution over random variates that are sampled during execution, conditioned on observed values that constrain a subset of program expressions.

Lightweight Metropolis-Hastings (LMH) samplers [16] propose a change to a single random variable at each iteration. The program is then rerun, reusing

© Springer International Publishing Switzerland 2015
A. Appice et al. (Eds.): ECML PKDD 2015, Part II, LNAI 9285, pp. 311–326, 2015.
DOI: 10.1007/978-3-319-23525-7_19

previous values and computation where possible, after which the new set of sample values is accepted or rejected. While re-running the program each time may waste some computation, the simplicity of LMH makes developing probabilistic variants of arbitrary languages relatively straightforward.

Designing robust adaptive MCMC methods for probabilistic programming is complicated because of diversity of models that can be expressed as probabilistic programs. Ideally, a single adaptation scheme should perform well in different programs without requiring manual tuning of parameters. Here we present an adaptive variant of LMH that dynamically adjusts the schedule for selecting variables for modification. First, we review the general structure of a probabilistic program. We discuss convergence criteria with respect to the program output and propose a scheme for tracking the "influence" of each random variable on the output. We then adapt the selection probability for each variable, borrowing techniques from the upper confidence bound (UCB) family of algorithms for multi-armed bandits [3]. We show that the proposed adaptation scheme preserves convergence to the target distribution under reasonable assumptions. Finally, we compare original and Adaptive LMH on several test problems to show how convergence is improved by adaptation.

2   Preliminaries

2.1   Probabilistic Program

A probabilistic program is a stateful deterministic computation $\mathcal{P}$ with the following properties:

- Initially, $\mathcal{P}$ expects no arguments.
- On every call, $\mathcal{P}$ returns either a distribution and an address (F, α), a distribution and a value (G, y), a value z, or $\perp$.
- Upon returning F, $\mathcal{P}$ expects a value x drawn from F as the argument to the next call.
- Upon returning (G, y) or z, $\mathcal{P}$ is invoked again without arguments.
- Upon returning $\perp$, $\mathcal{P}$ terminates.

A program is run by calling $\mathcal{P}$ repeatedly until termination.

A program need not generate the same sequence of random variables in every execution. For this reason we assume that each random variable x is assigned a unique label α, which we call an address, that induces a correspondence between variables in different executions. Every execution implicitly produces a sequence of triples of distributions, values of *latent* random variables, and addresses, (F_i, x_i, α_i). We call this sequence a *trace* and denote it by $\boldsymbol{x}$. A trace induces a sequence of pairs (G_j, y_j) of distributions and values of *observed* random variables. We call this sequence an *image* and denote it by $\boldsymbol{y}$. For notational simplicity we assume that the program always generates the same ordered set $\boldsymbol{z}$ of *output* values z_k.

The target density $\pi(\boldsymbol{x}) := \gamma(\boldsymbol{x})/Z$ of a program is defined in terms of the product of the probabilities of all random choices $\boldsymbol{x}$ and the likelihood of all observations $\boldsymbol{y}$

$$\gamma(\boldsymbol{x}) := \prod_{i=1}^{|\boldsymbol{x}|} p_{F_i}(x_i) \prod_{j=1}^{|\boldsymbol{y}|} p_{G_j}(y_j). \tag{1}$$

The objective of inference in probabilistic program $\mathcal{P}$ is to discover the distribution of $\boldsymbol{z}$.

2.2 Adaptive Markov Chain Monte Carlo

MCMC methods generate a sequence of samples $\{\boldsymbol{x}^t\}_{t=1}^{\infty}$ by simulating a Markov chain using a transition operator that leaves a target density $\pi(\boldsymbol{x})$ invariant. In MH the transition operator is implemented by drawing a new sample $\boldsymbol{x}'$ from a parameterized proposal distribution $q_\theta(\boldsymbol{x}'|\boldsymbol{x}^t)$ that is conditioned on the current sample $\boldsymbol{x}^t$. The proposed sample is then accepted with probability

$$\rho = \min\left(\frac{\pi(\boldsymbol{x}')q_\theta(\boldsymbol{x}^t|\boldsymbol{x}')}{\pi(\boldsymbol{x}^t)q_\theta(\boldsymbol{x}'|\boldsymbol{x}^t)}, 1\right). \tag{2}$$

If $\boldsymbol{x}'$ is rejected, $\boldsymbol{x}^t$ is re-used as the next sample.

The convergence rate of MH depends on parameters θ of the proposal distribution q_θ. The parameters can be set either offline or online. Variants of MCMC in which the parameters are continuously adjusted based on the features of the sample sequence are called adaptive. Challenges in design and analysis of Adaptive MCMC methods include optimization criteria and algorithms for the parameter adaptation, as well as conditions of convergence of adaptive MCMC to the correct equilibrium distribution [14]. Continuous adaptation of parameters of the proposal distribution is a well-known research subject [1,2,15].

In a componentwise MH algorithm [10] that targets a density $\pi(\boldsymbol{x})$ defined on an N-dimensional space $\mathcal{X}$, the components of a sample $\boldsymbol{x} = \{x_1, \ldots, x_N\}$ are updated individually, in either random or systematic order. Assuming the component i is selected at the step t for modification, the proposal $\boldsymbol{x}'$ sampled from $q_\theta^i(\boldsymbol{x}|\boldsymbol{x}^t)$ may differ from $\boldsymbol{x}^t$ only in that component, and $x_j' = x_j^t$ for all $j \neq i$. Adaptive componentwise Metropolis-Hastings (Algorithm 1) chooses different probabilities for selecting a component for modification at each iteration. Parameters of this scheduling distribution may be viewed as a subset of parameters θ of the proposal distribution q_θ, and adjusted according to optimization criteria of the sampling algorithm.

Varying selection probabilities based on past samples violates the Markov property of $\{\boldsymbol{x}^t\}_1^{\infty}$. However, provided the change in selection probabilities decreases to zero as t approaches ∞, then under suitable regularity conditions for the target density (see Section 4) an adaptive componentwise MH algorithm will be ergodic [8], and the distribution on $\boldsymbol{x}$ induced by Algorithm 1 converges to π.

Algorithm 1. Adaptive componentwise MH

1: Select initial point $\boldsymbol{x}^0$.
2: Set initial selection probabilities $\boldsymbol{w}^0$.
3: **for** $t = 1 \ldots \infty$ **do**
4:     $\boldsymbol{w}^t \leftarrow f^t(\boldsymbol{w}^{t-1}, \boldsymbol{x}^0, \boldsymbol{x}^1, \ldots, \boldsymbol{x}^t)$.
5:     Choose $k \in \{1, \ldots, N\}$ with probability w_k^t.
6:     Generate $\boldsymbol{x}' \sim q_\theta^k(\boldsymbol{x}|\boldsymbol{x}^t)$.
7:     $\rho \leftarrow \min\left(\frac{\pi(\boldsymbol{x}')q_\theta^k(\boldsymbol{x}^t|\boldsymbol{x}')}{\pi(\boldsymbol{x}^t)q_\theta^k(\boldsymbol{x}'|\boldsymbol{x}^t)}, 1\right)$
8:     $\boldsymbol{x}^{t+1} \leftarrow \boldsymbol{x}'$ with probability ρ, $\boldsymbol{x}^t$ otherwise.
9: **end for**

2.3 Lightweight Metropolis-Hastings

LMH [16] is a sampling scheme for probabilistic programs where a single random variable drawn in the course of a particular execution of a probabilistic program is modified via a standard MH proposal. LMH differs from componentwise MH algorithms in that other random variables may also have to be modified, depending on the structural dependencies in the probabilistic program.

LMH initializes a proposal by selecting a single variable x_k at address α_k from an execution trace $\boldsymbol{x}$ and resampling its value x_k' either using a reversible kernel $\kappa(x_k'|x_k)$ or from the conditional prior F_k. The remainder of the program is then rerun to generate a new trace $\boldsymbol{x}'$. When generating a variable x_j' at address α_j' in the new trace, the value x_i from the previous trace such that $\alpha_j' = \alpha_i$ is reused, provided it exists and still lies in the support of F_j'. When no value can be rescored, a new value x_j' is sampled from F_j'. The acceptance probability ρ_{LMH} is obtained by substituting (1) into (2):

$$\rho_{\text{LMH}} = \min\left(1, \frac{p(\boldsymbol{y}'|\boldsymbol{x}')p(\boldsymbol{x}')q(\boldsymbol{x}|\boldsymbol{x}')}{p(\boldsymbol{y}|\boldsymbol{x})p(\boldsymbol{x})q(\boldsymbol{x}'|\boldsymbol{x})}\right). \tag{3}$$

We here further simplify LMH by assuming x_i' is sampled from the conditional prior F_i and that all variables are selected for modification with equal probability. Under these assumptions, ρ_{LMH} takes the form [17]

$$\rho_{\text{LMH}} = \min\left(1, \frac{p(\boldsymbol{y}'|\boldsymbol{x}')p(\boldsymbol{x}')|\boldsymbol{x}|p(\boldsymbol{x} \setminus \boldsymbol{x}'|\boldsymbol{x} \cap \boldsymbol{x}')}{p(\boldsymbol{y}|\boldsymbol{x})p(\boldsymbol{x})|\boldsymbol{x}'|p(\boldsymbol{x}' \setminus \boldsymbol{x}|\boldsymbol{x}' \cap \boldsymbol{x})}\right), \tag{4}$$

where $\boldsymbol{x}' \setminus \boldsymbol{x}$ denotes the resampled variables, and $\boldsymbol{x}' \cap \boldsymbol{x}$ denotes the variables which have the same values in both traces.

3 Adaptive Lightweight Metropolis-Hastings

We develop an adaptive variant of LMH that dynamically adjusts the probabilities of selecting variables for modification (Algorithm 2). Let $\boldsymbol{x}^t$ be the trace at iteration t of Adaptive LMH. We define the probability distribution of selecting variables for modification in terms of an indexed set of weights $\boldsymbol{W}^t$ that we

---

Algorithm 2. Adaptive LMH

1: Initialize W_α^0 to a constant for all addresses α.
2: Run the program.
3: **for** $t = 1 \ldots \infty$ **do**
4:     Randomly select a variable x_k^t according to $\boldsymbol{W}^t$.
5:     Propose a value for x_k^t.
6:     Run the program, accept or reject the trace.
7:     **if** accepted **then**
8:         Compute $\boldsymbol{W}^{t+1}$ based on the program output.
9:     **else**
10:         $\boldsymbol{W}^{t+1} \leftarrow \boldsymbol{W}^t$
11:     **end if**
12: **end for**

---

adapt, such that the probability w_k^t of selecting the variable at address α_k for modification is

$$w_k^t := W_{\alpha_k}^t \Big/ \sum_{i=1}^{|\boldsymbol{x}^t|} W_{\alpha_i}^t. \tag{5}$$

Just like LMH, Adaptive LMH runs the probabilistic program once and then selects variables for modification randomly. However, the acceptance ratio ρ_{AdLMH} must now include selection probabilities w_k and w_k' of the resampled variable in the current and the proposed sample

$$\rho_{\text{AdLMH}} = \min\left(1, \frac{p(\boldsymbol{y}'|\boldsymbol{x}')p(\boldsymbol{x}')w_k'p(\boldsymbol{x}\setminus\boldsymbol{x}'|\boldsymbol{x}\cap\boldsymbol{x}')}{p(\boldsymbol{y}|\boldsymbol{x})p(\boldsymbol{x})w_kp(\boldsymbol{x}'\setminus\boldsymbol{x}|\boldsymbol{x}'\cap\boldsymbol{x})}\right). \tag{6}$$

This high-level description does not detail how $\boldsymbol{W}^t$ is computed for each iteration. Indeed, this is the most essential part of the algorithm. There are two different aspects here — on one hand, the influence of a given choice on the output sequence must be quantified in terms of convergence of the sequence to the target distribution. On the other hand, the influence of the choice must be translated into re-computation of weights of random variables in the trace. Both parts of re-computation of $\boldsymbol{W}^t$ are explained below.

3.1   Quantifying Influence

Extensive research has been devoted to criteria for tuning parameters of adaptive MCMC [1,2,15]. The case of inference in probabilistic programs is different: the user of a probabilistic program is often interested in fast convergence of the program output $\{\boldsymbol{z}^t\}$ rather than the trace $\{\boldsymbol{x}^t\}$.

In adaptive MCMC variants the acceptance rate can be efficiently used as the optimization objective [15]. However, for convergence of the output sequence an accepted trace that produces the same output is indistinguishable from a rejected trace. Additionally, while optimal values of the acceptance rate [1,15] can be used to tune parameters in adaptive MCMC, in Adaptive LMH we do not change the parameters of proposal distributions of individual variables, and assume that they are fixed. However, proposing a new value for a random variable may or may not change the output even if the new trace is accepted. By changing variable selection probabilities we attempt to maximize the change in the output sequence so that it converges faster. In the pedagogical example

$$x_1 \sim \text{Bernoulli}(0.5), \quad x_2 \sim \mathcal{N}(x_1, 1),$$
$$z_1 \leftarrow (x_1, x_2),$$

selecting the Bernoulli random choice for modification changes the output only when a different value is sampled, while selecting the normal random choice will change the output almost always.

Based on these considerations, we quantify the influence of sampling on the output sequence by measuring the change in the output z of the probabilistic program. Since programs may produce output of any type, we chose to discern between identical and different outputs only, rather than to quantify the distance by introducing a type-dependent norm. In addition, when $|z| > 1$, we quantify the difference by the fraction of components of z with changed values.

Formally, let $\{z^t\}_1^\infty = \{z^1, \ldots, z^{t-1}, z^t, \ldots\}$ be the output sequence of a probabilistic program. Then the *influence* of a choice that produced z^t is defined by the total reward R^t, computed as normalized Hamming distance

$$R^t = \frac{1}{|z^t|} \sum_{l=1}^{|z^t|} \mathbf{1}(z_l^t \neq z_l^{t-1}). \tag{7}$$

The reward is used to adjust the variable selection probabilities for the subsequent steps of Adaptive LMH by computing W^{t+1} (line 8 of Algorithm 2). It may seem sufficient to assign the reward to the last choice and use average choice rewards as their weights, but this approach will not work for Adaptive LMH. Consider the generative model

$$x_1 \sim \mathcal{N}(1, 10), \quad x_2 \sim \mathcal{N}(x_1, 1),$$
$$y_1 \sim \mathcal{N}(x_2, 1),$$
$$z_1 \leftarrow x_1,$$

where we observe the value $y_1 = 2$. Modifying x_2 may result in an accepted trace, but the value of $z_1 = x_1$, predicted by the program, will remain the same as in the previous trace. Only when x_1 is also modified, and a new trace with the updated values for both x_1 and x_2 is accepted, the earlier change in x_2 is indirectly reflected in the output of the program. In the next section, we discuss propagation of rewards to variable selection probabilities in detail.

3.2   Propagating Rewards to Variables

Both LMH and Adaptive LMH modify a single variable per trace, and either re-use or recompute the probabilities of values of all other variables

Algorithm 3. Propagating Rewards to Variables

1: **for** l in $1, \ldots, |z^t|$ **do**
2:     Append α to history h_l
3:     **if** $z_l^{t+1} \neq z_l^t$ **then**
4:         $\delta \leftarrow 1/|h_l|$
5:         **for** α' in h_l **do**
6:             $r_{\alpha'} \leftarrow r_{\alpha'} + \delta,\ c_{\alpha'} \leftarrow c_{\alpha'} + \delta$
7:         **end for**
8:         Flush h_l.
9:     **else**
10:         $c_\alpha \leftarrow c_\alpha + 1$
11:     **end if**
12: **end for**

(except those absent from the previous trace or having an incompatible distribution, for which new values are also sampled). Due to this updating scheme, the influence of modifying a variable on the output can be delayed by several iterations. We propose the following propagation scheme: for each unique random variable x at address α, the reward r_α and count c_α are kept in a data structure used to compute W. A list of addresses selected for modification since the last change in output, which we call the history h_l, is maintained for each component z_l of the output z. When the value of z_l changes, the reward is distributed between all of the addresses in the history h_l, which is then emptied. When z_l does not change, the selected variable is penalized by zero reward. This scheme is shown in Algorithm 3 which expands line 8 of Algorithm 2.

Rewarding all of the variables in the history ensures that while variables which cause changes in the output more often get a greater reward, variables with lower influence are still selected for modification sufficiently often. This, in turn, ensures ergodicity of sampling sequence, and helps establish conditions for convergence to the target distribution, as we discuss in Section 4.

Let us show that under certain assumptions the proposed reward propagation scheme has a non-degenerate equilibrium for variable selection probabilities. Indeed, assume that for a program with two variables, x_1, and x_2, *probability matching*, or selecting a choice with the probability proportional to the unit reward $\rho_i = \frac{r_i}{c_i}$, is used to compute the weights, that is, $W_i = \rho_i$. Then, the following lemma holds:

Lemma 1. *Assume that for variables x_i, where $i \in \{1, 2\}$:*

- w_i *is the selection probability;*
- β_i *is the probability that the new trace is accepted given that the variable was selected for modification;*
- γ_i *is the probability that the output changed given that the trace was accepted.*

Assume further that w_i, β_i, and γ_i are constant. Then $\gamma_1 = 1$, $\gamma_2 = 0$ implies:

$$0 < \frac{w_2}{w_1} \leq \frac{1}{3} \tag{8}$$

Proof. We shall prove the lemma in three steps. First, we will analyze a sequence of samples between two subsequent arrivals of x_1. Then, we derive a formula for the expected unit reward of x_2. Finally, we shall bound the ratio $\frac{w_2}{w_1}$.

Consider a sequence of k samples, for some k, between two subsequent arrivals of x_1, including the sample corresponding to the second arrival of x_1. Since a new value of x_1 always ($\gamma_1 = 1$) and x_2 never ($\gamma_2 = 0$) causes a change in the output, at the end of the sequence the history will contain k occurrences of x_2. Let us denote by Δr_i, Δc_i the increase of reward r_i and count c_i between the beginning and the end of the sequence. Noting that x_2 is penalized each time it is added to the history (line 10 of Algorithm 3), and k occurrences of x_2 are rewarded when x_1 is added to the history (line 6 of Algorithm 3), we obtain

$$\Delta r_1 = \frac{1}{k+1}, \; \Delta c_1 = \frac{1}{k+1} \quad \Delta r_2 = \frac{k}{k+1}, \; \Delta c_2 = k + \frac{k}{k+1} \quad (9)$$

Consider now a sequence of M such sequences. When $M \to \infty$, $\frac{r_{iM}}{c_{iM}}$ approaches the expected unit reward $\bar{\rho}_i$, where r_{iM} and c_{iM} are the reward and the count of x_i at the end of the sequence.

$$\bar{\rho}_i = \lim_{M \to \infty} \frac{r_{iM}}{c_{iM}} = \lim_{M \to \infty} \frac{\frac{r_{iM}}{M}}{\frac{c_{iM}}{M}} = \lim_{M \to \infty} \frac{\frac{\sum_{m=1}^{M} \Delta r_{im}}{M}}{\frac{\sum_{m=1}^{M} \Delta c_{im}}{M}} = \frac{\overline{\Delta r_i}}{\overline{\Delta c_i}} \quad (10)$$

Each variable x_i is selected randomly and independently and produces an accepted trace with probability

$$p_i = \frac{w_i \beta_i}{w_1 \beta_1 + w_2 \beta_2}. \quad (11)$$

Acceptances of x_1 form a Poisson process with rate $\frac{1}{p_1} = \frac{w_1 \beta_1 + w_2 \beta_2}{w_1 \beta_1}$, so k is geometrically distributed $\Pr[k] = (1 - p_1)^k p_1$. Since $\Delta r_1 = \Delta c_1$ for any k, the expected unit reward $\bar{\rho}_1$ of x_1 is 1. We substitute $\overline{\Delta r_i}$ and $\overline{\Delta c_i}$ into (10) to obtain the expected unit reward $\bar{\rho}_2$ of x_2:

$$\overline{\Delta r_2} = \sum_{k=0}^{\infty} \frac{k}{k+1}(1 - p_1)^k p_1$$

$$\overline{\Delta c_2} = \sum_{k=0}^{\infty} \left(k + \frac{k}{k+1}\right)(1 - p_1)^k p_1 = \underbrace{\frac{1 - p_1}{p_1}}_{\bar{k}} + \sum_{k=0}^{\infty} \frac{k}{k+1}(1 - p_1)^k p_1 \quad (12)$$

$$\bar{\rho}_2 = \frac{\overline{\Delta r_2}}{\overline{\Delta c_2}} = \frac{\sum_{k=0}^{\infty} \frac{k}{k+1}(1 - p_1)^k p_1}{\frac{1-p_1}{p_1} + \sum_{k=0}^{\infty} \frac{k}{k+1}(1 - p_1)^k p_1} = \frac{1 - \overbrace{\sum_{k=0}^{\infty} \frac{1}{k+1}(1 - p_1)^k p_1}^{A}}{\frac{1}{p_1} - \underbrace{\sum_{k=0}^{\infty} \frac{1}{k+1}(1 - p_1)^k p_1}_{A}} \quad (13)$$

For probability matching, selection probabilities are proportional to expected unit rewards:

$$\frac{w_2}{w_1} = \frac{\bar{\rho}_2}{\bar{\rho}_1} \tag{14}$$

To prove the inequality, we shall derive a closed-form representation for $\bar{\rho}_2$, and analyse solutions of (14) for $\frac{w_2}{w_1}$. We shall eliminate the summation A in (13):

$$A = \sum_{k=0}^{\infty} \frac{1}{k+1} (1 - p_1)^k p_1 = \frac{p_1}{1 - p_1} \sum_{k=0}^{\infty} \frac{1}{k+1} (1 - p_1)^{k+1}$$

$$= \frac{p_1}{1 - p_1} \sum_{k=0}^{\infty} \int_{p_1}^1 (1 - \xi)^k d\xi = \frac{p_1}{1 - p_1} \int_{p_1}^1 \sum_{k=0}^{\infty} (1 - \xi)^k d\xi = -\frac{p_1}{1 - p_1} \log p_1$$

$$\tag{15}$$

By substituting A into (13), and then $\bar{\rho}_1$ and $\bar{\rho}_2$ into (14), we obtain

$$\frac{w_2}{w_1} = \frac{\bar{\rho}_2}{\bar{\rho}_1} = \bar{\rho}_2 = \left. \begin{array}{c} 1 + \frac{p_1 \log p_1}{1 - p_1} \\ \frac{1}{p_1} + \frac{p_1 \log p_1}{1 - p_1} \end{array} \right\} B(p_1) \tag{16}$$

The right-hand side $B(p_1)$ of (16) is a monotonic function for $p_1 \in [0, 1]$, and $B(0) = 0$, $B(1) = \frac{1}{3}$. According to (11), $\frac{w_2}{w_1} = 0$ implies $p_1 = 1$, hence $\frac{w_2}{w_1} \neq 0$, and $0 < \frac{w_2}{w_1} \leq \frac{1}{3}$. $\square$

By noting that any subset of variables in a probabilistic program can be considered a single random variable drawn from a multi-dimensional distribution, Lemma 1 is generalized to any number of variables by Corollary 1:

Corollary 1. *For any partitioning of the set $\boldsymbol{x}$ of random variables of a probabilistic program, AdLMH with weights proportional to expected unit rewards selects variables from each of the partitions with non-zero probability.*

To ensure convergence of W^t to expected unit rewards in the stationary distribution, we use upper confidence bounds on unit rewards to compute the selection probabilities, an idea that we borrow from the UCB family of algorithms for multi-armed bandits [3]. Following UCB1 [3], we compute the upper confidence bound $\hat{\rho}_i$ as the sum of the unit reward and the exploration term

$$\hat{\rho}_\alpha = \rho_\alpha + C \sqrt{\frac{\log \sum_\alpha c_\alpha}{c_\alpha}}, \tag{17}$$

where C is an exploration factor. The default value for C is $\sqrt{2}$ in UCB1; in practice, a lower value of C is preferable. Note that variable selection in Adaptive LMH is different from arm selection in multi-armed bandits: unlike in bandits, where we want to sample the best arm at an increasing rate, in Adaptive LMH we expect W^t to converge to an equilibrium in which selection probabilities are proportional to expected unit rewards.

4 Convergence of Adaptive LMH

As adaptive MCMC algorithms may depend arbitrarily on the history at each step, showing that a given sampler correctly draws from the target distribution can be non-trivial. General conditions under which adaptive MCMC schemes are still ergodic, in the sense that the distribution of samples converges to the target π in total variation, are established in [14]. The fundamental criteria for validity of an adaptive algorithm are *diminishing adaptation*, which (informally) requires that the amount which the transition operator changes each iteration must asymptotically decrease to zero; and *containment*, a technical condition which requires that the time until convergence to the target distribution must be bounded in probability [4].

The class of models representable by probabilistic programs is very broad, allowing specification of completely arbitrary target densities; however, for many models the adaptive LMH algorithm reduces to an adaptive random scan Metropolis-within-Gibbs in Algorithm 1. To discuss when this is the case, we invoke the concept of *structural* versus *structure-preserving* random choices [18]. Crucially, a *structure-preserving* random choice x_k does not affect the existence of other x_m in the trace.

Suppose we were to restrict the expressiveness of our language to admit only programs with no structural random choices: in such a language, the LMH algorithm in Algorithm 2 reduces to the adaptive componentwise MH algorithm. Conditions under which such an adaptive algorithm is ergodic have been established explicitly in [8, Theorems4.10and5.5]. Given suitable assumptions on the target density defined by the program, it is necessary for the probability vector $||w^t - w^{t-1}|| \to 0$, and that for any particular component k we have probability $w_k^t > \epsilon > 0$. Both of these are satisfied by our approach: from Corollary 1, we ensure that the unit reward across each x_i converges to a positive fixed point.

While any theoretical result will require language restrictions such that programs only induce distributions satisfying regularity conditions, we conjecture that this scheme is broadly applicable across most non-pathological programs. We leave a precise theoretical analysis of the space of probabilistic programs in which adaptive MCMC schemes (with infinite adaptation) may be ergodic to future work. Empirical evaluation presented in the next section demonstrates practical convergence of Adaptive LMH on a range of inference examples, including programs containing structural random choices.

5 Empirical Evaluation

We evaluated Adaptive LMH on many probabilistic programs and observed consistent improvement of convergence rate compared to LMH. We also verified on a number of tests that the algorithm converges to the correct distribution obtained by independent exact methods. In this section, we compare Adaptive LMH to LMH on several representative examples of probabilistic programs. The rates in the comparisons are presented with respect to the number of samples, or

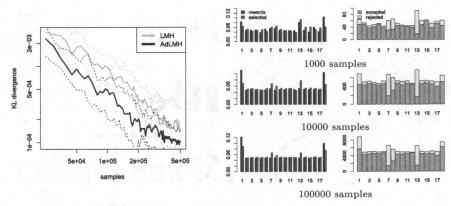

Fig. 1. HMM, predicting the 0th and 17th state

simulations, of the probabilistic programs. The additional computation required for adaptation takes negligible time, and the computational effort per sample is approximately the same for all algorithms. Our implementation of the inference engine is available at https://bitbucket.org/dtolpin/anglican.

In the following case studies differences between program outputs and target distributions are presented using Kullback-Leibler (KL) divergence, Kolmogorov-Smirnov (KS) distance, or L2 distance, as appropriate. In cases where target distributions cannot be updated exactly, they were approximated by running a non-adaptive inference algorithm for a long enough time and with a sufficient number of restarts. In each of the evaluations, all of the algorithms were run with 25 random restarts and 500 000 simulations of the probabilistic program per restart. The difference plots use the logarithmic scale for both axes. In the plots, the solid lines correspond to the median, and the dashed lines to 25% and 75% percentiles, taken over all runs of the corresponding inference algorithm. The exploration factor for computing upper confidence bounds on unit rewards (Equation 17) was fixed at $C = 0.5$ for all tests and evaluations.

The first example is a latent state inference problem in an HMM with three states, one-dimensional normal observations (0.9, 0.8, 0.7, 0, -0.025, 5, 2, 0.1, 0, 0.13, 0.45, 6, 0.2, 0.3, -1, -1) with variance 1.0, a known transition matrix, and known initial state distribution. There are 18 distinct random choices in all traces of the program, and the 0th and the 17th state are predicted. The results of evaluation are shown in Figure 1 as KL divergences between the inference output and the ground truth obtained using the forward-backward algorithm. In addition, bar plots of unit reward and sample count distributions among random choices in Adaptive LMH are shown for 1000, 10 000, and 100 000 samples.

As can be seen in the plots, Adaptive LMH (black) exhibits faster convergence over the whole range of evaluation, requiring half as many samples as LMH (cyan) to achieve the same approximation, with the median of LMH above the 75% quantile of Adaptive LMH.

In addition, the bar plots show unit rewards and sample counts for different random choices, providing an insight on the adaptive behavior of AdLMH.

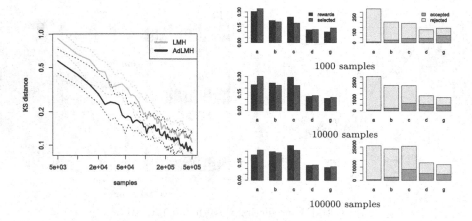

Fig. 2. Gaussian process hyperparameter estimation

On the left-hand bar plots, red bars are normalized unit rewards, and blue bars are normalized sample counts. On the right-hand bar plots, the total height of a bar is the total sample count, with green section corresponding to the accepted, and yellow to the rejected samples. At 1 000 samples, the unit rewards have not yet converged, and exploration supersedes exploitation: random choices with lower acceptance rate are selected more often (choices 7, 8 and 13 corresponding to states 6, 7 and 12). At 10 000 samples, the unit rewards become close to their final values, and choices 1 and 18, immediately affecting the predicted states, are selected more often. At 100 000 samples, the unit rewards converge, and the sample counts correspond closely to the equilibrium state outlined in Lemma 1.

The second case study is estimation of hyperparameters of a Gaussian Process. We define a Gaussian Process of the form

$$f \sim \mathcal{GP}(m, k),$$

$$\text{where } m(x) = ax^2 + bx + c, \quad k(x, x') = de^{-\frac{(x-x')^2}{2g}}.$$

The process has five hyperparameters, a, b, c, d, g. The program infers the posterior values of the hyperparameters by maximizing marginal likelihood of 6 observations $(0.0, 0.5)$, $(1.0, 0.4)$, $(2.0, 0.2)$, $(3.0, -0.05)$, $(4.0, -0.2)$, and $(5.0, 0.1)$. Parameters a, b, c of the mean function are predicted. Maximum of KS distances between inferred distributions of each of the predicted parameters and an approximation of the target distributions is shown in Figure 2. The approximation was obtained by running LMH with 2 000 000 samples per restart and 50 restarts, and then taking each 100th sample from the last 10 000 samples of each restart, 5000 samples total. Just as for the previous case study, bar plots of unit rewards and sample counts are shown for 1000, 10 000, and 100 000 samples.

Here as well, Adaptive LMH (black) converges faster over the whole range of evaluation, outperforming LMH by a factor 2 over the first 50 000 samples. Bar plots of unit rewards and sample counts for different number of choices,

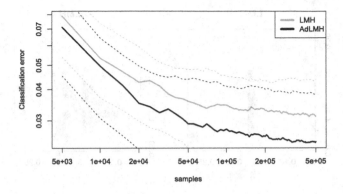

Fig. 3. Logistic regression on Iris dataset.

again, show the dynamics of sample allocation among random choices. Choices a, b, and c are predicted, while choices d and g are required for inference but not predicted. Choice a has the lowest acceptance rate (ratio between the total height of the bar and the green part on the right-hand bar plot), but the unit reward is close the unit reward of choices b and c. At 1 000 samples, choice a is selected with the highest probability. However, close to the converged state, at 100 000 samples, choices a, b, and c are selected with similar probabilities. At the same time, choices 4 and 5 are selected with a lower probability. Both the exploration-exploitation dynamics for choices a–c and probability matching of selection probabilities among all choices secure improved convergence.

The third case study involves a larger amount of data observed during each simulation of a probabilistic program. We use the well-known Iris dataset [9] to fit a model of classifying a given flower as of the species Iris setosa, as opposite to either Iris virginica or Iris versicolor. Each record in the dataset corresponds to an observation. For each observation, we define a feature vector x and an indicator variable z_i, which is 1 if and only if the observation is of an Iris setosa. We fit the model with five regression coefficients $\beta_1, \ldots, \beta_5$, defined as

$$\sigma^2 \sim \mathrm{InvGamma}(1,1),$$
$$\beta_j \sim \mathrm{Normal}(0, \sigma),$$
$$p(z_i = 1) = \frac{1}{1 + e^{-\beta^T x}}.$$

To assess the convergence, we perform shuffle split leave-2-out cross validation, selecting one instance belonging to the species Iris setosa and one belonging to a different species for each run of the inference algorithm. The classification error is shown in Figure 3 over 100 runs of LMH and Adaptive LMH. The results are consistent with other case studies: Adaptive LMH exhibits a faster convergence rate, requiring half as many samples to achieve the same classification accuracy as LMH.

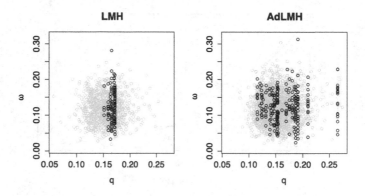

Fig. 4. Kalman filter, 500 samples after 10 000 samples of burn-in.

As a final case study we consider a linear dynamical system (i.e. a Kalman smoothing problem) that was previously described in [12]

$$x_t \sim \text{Norm}(A \cdot x_{t-1}, Q), \qquad y_t \sim \text{Norm}(C \cdot x_t, R).$$

In this problem we assume that 16-dimensional observations y_t are conditioned on 2-dimensional latent states x_t. We impose additional structure by assuming that the transition matrix A is a simple rotation with angular velocity ω, whereas the transition covariance Q is a diagonal matrix with coefficient q,

$$A = \left[\begin{array}{cc} \cos\omega & -\sin\omega \\ \sin\omega & \cos\omega \end{array} \right], \qquad Q = \left[\begin{array}{cc} q & 0 \\ 0 & q \end{array} \right].$$

We predict posterior values for ω, and q in a setting where C and R are assumed known, under mildly informative priors $\omega \sim \text{Gamma}(10, 2.5)$ and $q \sim \text{Gamma}(10, 100)$. Posterior inference is performed conditioned on a simulated sequence $y_{1:T}$ of $T = 100$ observations, with $\omega^* = 4\pi/T$, and $q^* = 0.1$. The observation matrix C and covariance R are sampled row-wise from symmetric Dirichlet distributions with parameters $c = 0.1$, and $r = 0.01$ respectively.

Figure 4 shows a qualitative evaluation of the mixing rate in the form of 500 consecutive samples (ω, q) from an LMH and AdLMH chain after 10 000 samples of burn-in. The LMH sequence exhibits good mixing over ω but is strongly correlated in q, whereas the AdLMH sequence obtains a much better coverage of the space.

To summarize, Adaptive LMH consistently attained faster convergence than LMH, measured by differences between the ongoing output distribution of the random program and the target independently obtained distribution, assessed using various metrics. Variable selection probabilities computed by Adaptive LMH are dynamically adapted during the inference, combining exploration of the model represented by the probabilistic program and exploitation of influence of random variables on program output.

6   Contribution and Future Work

In this paper we introduced a new algorithm, Adaptive LMH, for approximate inference in probabilistic programs. This algorithm adjusts sampling parameters based on the output of the probabilistic program in which the inference is performed. Contributions of the paper include

- A scheme of rewarding random choice based on program output.
- An approach to propagation of choice rewards to MH proposal scheduling parameters.
- An application of this approach to LMH, where the probabilities of selecting each variable for modification are adjusted.

Adaptive LMH was compared to LMH, its non-adaptive counterpart, and was found to consistently outperform LMH on several probabilistic programs, while still being almost as easy to implement. The time cost of additional computation due to adaptation was negligible.

Although presented in the context of a particular sampling algorithm, the adaptation approach can be extended to other sampling methods. We believe that various sampling algorithms for probabilistic programming can benefit from output-sensitive adaptation. Additional potential for improvement lies in acquisition of dependencies between predicted expressions and random variables. Exploring alternative approaches for guiding exploration-exploitation compromise, in particular, based on Bayesian inference, is another promising research direction.

Overall, output-sensitive approximate inference appears to bring clear advantages and should be further explored in the context of probabilistic programming models and algorithms.

Acknowledgments. This work is supported under DARPA PPAML through the U.S. AFRL under Cooperative Agreement number FA8750-14-2-0004. The U.S. Government is authorized to reproduce and distribute reprints for Governmental purposes notwithstanding any copyright notation heron. The views and conclusions contained herein are those of the authors and should be not interpreted as necessarily representing the official policies or endorsements, either expressed or implied, of DARPA, the U.S. Air Force Research Laboratory or the U.S. Government.

References

1. Andrieu, C., Thoms, J.: A tutorial on adaptive MCMC. Statistics and Computing **18**(4), 343–373 (2008)
2. Atchadé, Y., Fort, G., Moulines, E., Priouret, P.: Adaptive markov chain monte carlo: theory and methods. In: Barber, D., Cemgil, A.T., Chiappa, S. (eds.) Bayesian Time Series Models, pp. 32–51. Cambridge University Press (2011)
3. Auer, P., Cesa-Bianchi, N., Fischer, P.: Finite-time analysis of the Multiarmed Bandit problem. Machine Learning **47**(2–3), 235–256 (2002)

4. Bai, Y., Roberts, G.O., Rosenthal, J.S.: On the containment condition for adaptive Markov chain Monte Carlo algorithms. Advances and Applications in Statistics **21**(1), 1–54 (2011)
5. Gamerman, D., Lopes, H.F.: Markov Chain Monte Carlo: Stochastic Simulation for Bayesian Inference. Chapman and Hall/CRC (2006)
6. Goodman, N.D., Mansinghka, V.K., Roy, D.M., Bonawitz, K., Tenenbaum, J.B.: Church: a language for generative models. In: UAI, pp. 220–229 (2008)
7. Gordon, A.D., Henzinger, T.A., Nori, A.V., Rajamani, S.K.: Probabilistic programming. In: ICSE (FOSE track) (2014)
8. Łatuszyński, K., Roberts, G.O., Rosenthal, J.S.: Adaptive Gibbs samplers and related MCMC methods. Annals of Applied Probability **23**(1), 66–98 (2013)
9. Lauritzen, S.: Graphical Models. Clarendon Press (1996)
10. Levine, R.A., Yu, Z., Hanley, W.G., Nitao, J.J.: Implementing componentwise hastings algorithms. Computational Stastistics & Data Analysis **48**(2), 363–389 (2005)
11. Mansinghka, V.K., Selsam, D., Perov, Y.N.: Venture: a higher-order probabilistic programming platform with programmable inference. CoRR abs/1404.0099 (2014)
12. van de Meent, J.W., Yang, H., Mansinghka, V., Wood, F.: Particle Gibbs with Ancestor Sampling for Probabilistic Programs. In: AISTATS, pp. 986–994 (2015)
13. Nori, A.V., Hur, C.K., Rajamani, S.K., Samuel, S.: R2: An efficient mcmc sampler for probabilistic programs. In: AAAI, pp. 2476–2482 (2014)
14. Roberts, G.O., Rosenthal, J.S.: Coupling and ergodicity of adaptive MCMC. Journal of Applied Probability **44**, 458–475 (2007)
15. Roberts, G.O., Rosenthal, J.S.: Examples of adaptive MCMC. Journal of Computational and Graphical Statistics **18**(2), 349–367 (2009)
16. Wingate, D., Stuhlmüller, A., Goodman, N.D.: Lightweight implementations of probabilistic programming languages via transformational compilation. In: AISTATS, pp. 770–778 (2011)
17. Wood, F., van de Meent, J.W., Mansinghka, V.: A new approach to probabilistic programming inference. In: AISTATS, pp. 1024–1032 (2014)
18. Yang, L., Hanrahan, P., Goodman, N.D.: Generating efficient MCMC kernels from probabilistic programs. In: AISTATS, pp. 1068–1076 (2014)

Planning in Discrete and Continuous Markov Decision Processes by Probabilistic Programming

Davide Nitti[✉], Vaishak Belle, and Luc De Raedt

Department of Computer Science, KU, Leuven, Belgium
{davide.nitti,vaishak.belle,luc.deraedt}@cs.kuleuven.be

Abstract. Real-world planning problems frequently involve mixtures of continuous and discrete state variables and actions, and are formulated in environments with an unknown number of objects. In recent years, probabilistic programming has emerged as a natural approach to capture and characterize such complex probability distributions with general-purpose inference methods. While it is known that a probabilistic programming language can be easily extended to represent Markov Decision Processes (MDPs) for planning tasks, solving such tasks is challenging. Building on related efforts in reinforcement learning, we introduce a conceptually simple but powerful planning algorithm for MDPs realized as a probabilistic program. This planner constructs approximations to the optimal policy by importance sampling, while exploiting the knowledge of the MDP model. In our empirical evaluations, we show that this approach has wide applicability on domains ranging from strictly discrete to strictly continuous to hybrid ones, handles intricacies such as unknown objects, and is argued to be competitive given its generality.

1 Introduction

Real-world planning problems frequently involve mixtures of continuous and discrete state variables and actions. Markov Decision Processes (MDPs) [28] are a natural and general framework for modeling such problems. However, while significant progress has been made in developing robust planning algorithms for discrete and continuous MDPs, the more intricate hybrid (i.e., mixtures) domains and settings with an unknown number of objects have received less attention.

The recent advances of probabilistic programming languages (e.g., BLOG [15], Church [6], ProbLog [10], distributional clauses [7]) has significantly improved the expressive power of formal representations for probabilistic models. While it is known that these languages can be extended for decision problems [27,29], including MDPs, it is less clear if the inbuilt general-purpose inference system can cope with the challenges (e.g., scale, time constraints) posed by actual planning problems, and compete with existing state-of-the-art planners.

In this paper, we consider the problem of effectively planning in domains where reasoning and handling unknowns may be needed in addition to coping with mixtures of discrete and continuous variables. In particular, we adopt

© Springer International Publishing Switzerland 2015
A. Appice et al. (Eds.): ECML PKDD 2015, Part II, LNAI 9285, pp. 327–342, 2015.
DOI: 10.1007/978-3-319-23525-7_20

dynamic distributional clauses (DDC) [17,18] (an extension of distributional clauses for temporal models) to describe the MDP and perform inference. In such general settings, exact solutions may be intractable, and so approximate solutions are the best we can hope for. Popular approximate solutions include Monte-Carlo methods to estimate the expected reward of a policy (i.e., policy evaluation). Monte-Carlo methods provide state-of-the-art results in probabilistic planners [9,11]. Monte-Carlo planners have been mainly applied in discrete domains (with some notable exceptions, such as [1,13], for continuous domains). Typically, for continuous states, function approximation (e.g., linear regression) is applied. In that sense, one of the few Monte-Carlo planners that works in arbitrary MDPs with no particular assumptions is Sparse Sampling (SST) [8]; but as we demonstrate later, it is often slow in practice. We remark that most, if not all, Monte-Carlo methods require only a way to sample from the model of interest. While this property seems desirable, it prevents us from exploiting the actual probabilities of the model, as discussed (but unaddressed) in [9].

In this work, we introduce HYPE: a conceptually simple but powerful planning algorithm for a given MDP in DDC. However, HYPE can be adapted for other languages, such as RDDL [22]. The proposed planner exploits the knowledge of the model via importance sampling to perform policy evaluation, and thus, policy improvement. Importance sampling has been used in off-policy Monte-Carlo methods [20,24,25], where policy evaluation is performed using trajectories sampled from another policy. We remark that standard off-policy Monte-Carlo methods have been used in reinforcement learning, which are essentially model-free settings. In our setting, given a planning domain, the proposed planner introduces a new off-policy method that exploits the model and works, under weak assumptions, in discrete, continuous, hybrid domains as well as those with an unknown number of objects.

We provide a detailed derivation on how the approximation is obtained using importance sampling. Most significantly, we test the robustness of the approach on a wide variety of probabilistic domains. Given the generality of our framework, we do not challenge the plan times of state-of-the-art planners, but we do successfully generate meaningful plans in all these domains. We believe this performance is competitive given the algorithm's applicability. Indeed, the results show that our system at best outperforms SST [8] and at worst produces similar results, where SST is an equally general planner; in addition, it obtains reasonable results with respect to state-of-the-art discrete (probabilistic) planners.

2 Preliminaries

In a MDP, a putative agent is assumed to interact with its environment, described using a set S of *states*, a set A of *actions* that the agent can perform, a *transition function* $p : S \times A \times S \to [0,1]$, and a *reward function* $R : S \times A \to \mathbb{R}$. That is, when in state s and on doing a, the probability of reaching s' is given by $p(s' \mid s, a)$, for which the agent receives the reward $R(s, a)$. The agent is taken to operate over a finite number of time steps $t = 0, 1, \ldots, T$, with the goal of maximizing the expected reward: $\mathbb{E}[\sum_{t=0}^{T} \gamma^t R(s_t, a_t)]$, where s_0 is the start state, a_0 the first action, and $\gamma \in [0, 1]$ is a discount factor.

This paper focuses on maximizing the reward in a finite horizon MDP; however the same ideas are extendable for infinite horizons. This is achieved by computing a (deterministic) policy $\pi : S \times D \to A$ that determines the agent's action at state s and remaining steps d (horizon). The expected reward starting from state s_t and following a policy π is called the *value function* (V-function):

$$V_d^\pi(s_t) = \mathbb{E}\left[\sum_{k=t}^{t+d} \gamma^{k-t} R(s_k, a_k) \mid s_t, \pi\right]. \tag{1}$$

Furthermore, the expected reward starting from state s_t while executing action a_t and following a policy π is called the *action-value function* (Q-function):

$$Q_d^\pi(s_t, a_t) = \mathbb{E}\left[\sum_{k=t}^{t+d} \gamma^{k-t} R(s_k, a_k) \mid s_t, a_t, \pi\right]. \tag{2}$$

Since $T = t + d$, in the following formulas we will use T for compactness. An *optimal policy* π^* is a policy that maximizes the V-function for all states. A sample-based planner uses Monte-Carlo methods to solve an MDP and find a (near) optimal policy. The planner simulates (by sampling) interaction with the environment in episodes $E^m = <s_0^m, a_0^m, s_1^m, a_1^m, ..., s_T^m, a_T^m>$, following some policy π. Each episode is a trajectory of T time steps, and we let s_t^m denote the state visited at time t during episode m. (So, after M episodes, $M \times T$ states would be explored). After or during an episode generation, the sample-based planner updates $Q_d(s_t^m, a_t^m)$ for each t according to a backup rule, for example, averaging the total rewards obtained starting from (s_t^m, a_t^m) till the end. The policy is improved using a strategy that trades-off exploitation and exploration, e.g., the ϵ-greedy strategy. In this case the policy used to sample the episodes is not deterministic; we indicate with $\pi(a_t|s_t)$ the probability to select action a_t in state s_t under the policy π. Under certain conditions, after a sufficiently large number of episodes, the policy converges to a (near) optimal policy, and the planner can execute the greedy policy $argmax_a Q_d(s, a)$.

3   Dynamic Distributional Clauses

We assume some familiarity with standard terminology of statistical relational learning and logic programming [2]. We represent the MDP using *dynamic distributional clauses* [7,17], an extension of logic programming to represent continuous and discrete random variables. A *distributional clause* (DC) is of the form $h \sim \mathcal{D} \leftarrow b_1, \ldots, b_n$, where the b_i are literals and $\sim$ is a binary predicate written in infix notation. The intended meaning of a distributional clause is that each ground instance of the clause $(h \sim \mathcal{D} \leftarrow b_1, \ldots, b_n)\theta$ defines the random variable $h\theta$ as being distributed according to $\mathcal{D}\theta$ whenever all the $b_i\theta$ hold, where θ is a substitution. Furthermore, a term $\simeq(d)$ constructed from the reserved functor $\simeq/1$ represents the value of the random variable d.

Example 1. Consider the following clauses:

$$\text{n} \sim \text{poisson}(6). \tag{3}$$

$$\text{pos}(\text{P}) \sim \text{uniform}(1, 10) \leftarrow \text{between}(1, \simeq(\text{n}), \text{P}). \tag{4}$$

$$\text{left}(\text{A}, \text{B}) \leftarrow \simeq(\text{pos}(\text{A})) >\simeq(\text{pos}(\text{B})). \tag{5}$$

Capitalized terms such as P, A and B are logical variables, which can be substituted with any constant. Clause (3) states that the number of people n is governed by a Poisson distribution with mean 6; clause (4) models the position pos(P) as a random variable uniformly distributed from 1 to 10, for each person P such that P is between 1 and $\simeq$(n). Thus, if the outcome of n is two (i.e., $\simeq(n) = 2$) there are 2 independent random variables pos(1) and pos(2). Finally, clause (5) shows how to define the predicate left(A, B) from the positions of any A and B. Ground atoms such as left(1, 2) are binary random variables that can be true or false, while terms such as pos(1) represent random variables that can take concrete values from the domain of their distribution.

A *distributional program* is a set of distributional clauses (some of which may be deterministic) that defines a distribution over possible worlds, which in turn defines the underlying semantics. A possible world is generated starting from the empty set $S = \emptyset$; for each distributional clause $\text{h} \sim \mathcal{D} \leftarrow \text{b}_1, ..., \text{b}_n$, whenever the body $\{\text{b}_1\theta, ..., \text{b}_n\theta\}$ is true in the set S for the substitution θ, a value v for the random variable $\text{h}\theta$ is sampled from the distribution $\mathcal{D}\theta$ and $\simeq(h\theta) = v$ is added to S. This is repeated until a fixpoint is reached, i.e., no further variables can be sampled. *Dynamic distributional clauses* (DDC) extend distributional clauses in admitting temporally-extended domains by associating a time index to each random variable.

Example 2. Let us consider an object search scenario (*objsearch*) used in the experiments, in which a robot looks for a specific object in a shelf. Some of the objects are visible, others are occluded. The robot needs to decide which object to remove to find the object of interest. Every time the robot removes an object, the objects behind it become visible. This happens recursively, i.e., each new uncovered object might occlude other objects. The number and the types of occluded objects depend on the object covering them. For example, a box might cover several objects because it is big. This scenario involves an unknown number of objects and can be written as a partially observable MDP. However, it can be also described as a MDP in DDC where the state is the type of visible objects; in this case the state grows over time when new objects are observed or shrink when objects are removed without uncovering new objects. The probability of observing new objects is encoded in the state transition model, for example:

$$\text{type}(\text{X})_{t+1} \sim \text{val}(\text{T}) \leftarrow \simeq(\text{type}(\text{X})_t) = \text{T}, \text{not}(\text{removeObj}(\text{X})). \tag{6}$$

$$\text{numObjBehind}(\text{X})_{t+1} \sim \text{poisson}(1) \leftarrow \simeq(\text{type}(\text{X})_t) = \text{box}, \text{removeObj}(\text{X}). \tag{7}$$

$$\text{type}(\text{ID})_{t+1} \sim \text{finite}([0.2 : \text{glass}, 0.3 : \text{cup}, 0.4 : \text{box}, 0.1 : \text{can}]) \leftarrow$$

$$\simeq(\text{type}(\text{X})_t) = \text{box}, \text{removeobj}(\text{X}), \simeq(\text{numObjBehind}(\text{X})_{t+1}) = \text{N}, \text{getLastID}(\text{Last})_t,$$

$$\text{NewID is Last} + 1, \text{EndNewID is NewID} + \text{N}, \text{between}(\text{NewID}, \text{EndNewID}, \text{ID}). \tag{8}$$

Clause (6) states that the type of each object remains unchanged when we do not perform a remove action. Otherwise, if we remove the object, its type is removed from the state at time $t + 1$ because it is not needed anymore. Clauses (7) and (8) define the number and the type of objects behind a box X, added to the state when we perform a remove action on X. Similar clauses are defined for other types. The predicate getLastID(Last)$_t$ returns the highest object ID in the state and is needed to make sure that any new object has a different ID.

To complete the MDP specification we need to define a reward function $R(s_t, a_t)$, the terminal states that indicate when the episode terminates, and the applicability of an action a_t is a state s_t as in PDDL. For *objsearch* we have:

$$\text{stop}_t \leftarrow \simeq(\text{type}(\text{X})_t) = \text{can}.$$
$$\text{reward}(20)_t \leftarrow \text{stop}_t.$$
$$\text{reward}(-1)_t \leftarrow \text{not}(\text{stop}_t).$$

That is, a state is terminal when we observe the object of interest (e.g., a can), for which a reward of 20 is obtained. The remaining states are nonterminal with reward -1. To define action applicability we use a set of clauses of the form

$$\text{applicable}(\text{action})_t \leftarrow \text{preconditions}_t.$$

For example, action removeobj is applicable for each object in the state, that is when its type is defined with an arbitrary value Type:

$$\text{applicable}(\text{removeobj}(\text{X}))_t \leftarrow \simeq(\text{type}(\text{X})_t) = \text{Type}.$$

4   Planning by Importance Sampling

Our approach to plan in MDPs described in DDC is an *off-policy strategy* [28] based on importance sampling and *derived* from the transition model. Related work is discussed more comprehensively in Section 5, but as we note later, sample-based planners typically only require a generative model (a way to generate samples) and do not exploit the declarative model of the MDP (i.e., the actual probabilities) [9]. In our case, this knowledge leads to an effective planning algorithm that works in discrete, continuous, hybrid domains, and domains with an unknown number of objects under weak assumptions.

In a nutshell, the proposed approach samples episodes E^m and stores for each visited state s_t^m an estimation of the V-function (e.g., the total reward obtained from that state). The action selection follows an ϵ-greedy strategy, where the Q-function is estimated as the immediate reward plus the weighted average of the previously stored V-function points at time $t + 1$. This is justified by the means of importance sampling as explained later. The essential steps of our planning system HYPE (= *hybrid episodic planner*) are given in Algorithm 1.

The algorithm realizes the following key ideas:

– $\tilde{Q}$ and $\tilde{V}$ denote approximations of the Q and V-function respectively.

Algorithm 1. HYPE

1: **function** SAMPLEEPISODE(d, s_t^m, m)   ▷ Horizon d, state s_t^m in episode m
2:    **if** $d = 0$ **then**
3:        **return** 0
4:    **end if**
5:    **for** each applicable action a in s_t^m **do**   ▷ Q-function estimation
6:        $\tilde{Q}_d^m(s_t^m, a) \leftarrow R(s_t^m, a) + \gamma \frac{\sum_{i=0}^{m-1} w^i \tilde{V}_{d-1}^i(s_{t+1}^i)}{\sum_{i=0}^{m-1} w^i}$
7:    **end for**
8:    sample $u \sim uniform(0, 1)$ ▷ ϵ-greedy strategy
9:    **if** $u < 1 - \epsilon$ **then**
10:        $a_t^m \leftarrow argmax_a \tilde{Q}_d^m(s_t^m, a)$
11:    **else**
12:        $a_t^m \sim uniform(actions\ applicable\ in\ s_t^m)$
13:    **end if**
14:    sample $s_{t+1}^m \sim p(s_{t+1} \mid s_t^m, a_t^m)$ ▷ sample next state
15:    $G_d^m \leftarrow R(s_t^m, a_t^m) + \gamma \cdot$ SAMPLEEPISODE $(d - 1, s_{t+1}^m, m)$ ▷ recursive call
16:    $\tilde{V}_d^m(s_t^m) \leftarrow G_d^m$
17:    store $(s_t^m, \tilde{V}_d^m(s_t^m), d)$
18:    **return** $\tilde{V}_d^m(s_t^m)$ ▷ V-function estimation for s_t^m at horizon d
19: **end function**

- Lines 14-17 sample the next step and recursively the remaining episode of total length T, then stores the total discounted reward G_d^m starting from the current state s_t^m. This quantity can be interpreted as a sample of the expectation in formula (1), thus an estimator of the V-function. For this and other reasons explained later, G_d^m is stored as $\tilde{V}_d^m(s_t^m)$.
- Lines 8-13 implement an ϵ-greedy exploratory strategy for choosing actions.
- Most significantly, line 6 approximates the Q-function using the *weighted average* of the stored $\tilde{V}_{d-1}^i(s_{t+1}^i)$ points:

$$\tilde{Q}_d^m(s_t^m, a) \leftarrow R(s_t^m, a) + \gamma \frac{\sum_{i=0}^{m-1} w^i \tilde{V}_{d-1}^i(s_{t+1}^i)}{\sum_{i=0}^{m-1} w^i}, \tag{9}$$

where w^i is a *weight function* for episode i at state s_{t+1}^i. The weight exploits the transition model and is defined as:

$$w^i = \frac{p(s_{t+1}^i \mid s_t^m, a)}{q(s_{t+1}^i)} \alpha^{(m-i)}. \tag{10}$$

Here, for evaluating an action a at the current state s_t, we let w^i be the ratio of the transition probability of reaching a stored state s_{t+1}^i and the probability used to sample s_{t+1}^i, denoted using q. Recent episodes are considered more significant than previous ones, and so α is a parameter for realizing this. We provide a detailed justification for line 6 below.

We note that line 6 requires us to go over a finite set of actions, and so in the presence of continuous action spaces (e.g., real-valued parameter for a move action), we can discretize the action space or sample from it. More sophisticate approaches are possible [5, 26].

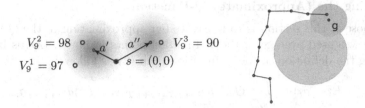

$$V_9^2 = 98 \quad \circ \quad a' \qquad a'' \qquad \circ \quad V_9^3 = 90$$
$$V_9^1 = 97 \quad \circ \qquad \qquad s = (0,0)$$

g

Fig. 1. Left: weight computation for the objpush domain. Right: a sampled episode that reaches the goal (blue), and avoids the undesired region (red).

Example 3. As a simple illustration, consider the following example called *obj-push*. We have an object on a table and an arm that can push the object in a set of directions; the goal is to move the object close to a point g, avoiding an undesired region (Fig. 1). The state consists of the object position (x, y), with push actions parameterized by the displacement (DX, DY). The state transition model is a Gaussian around the previous position plus the displacement:

$$\texttt{pos(ID)}_{t+1} \sim \texttt{gaussian}(\simeq(\texttt{pos(ID)}_t) + (\text{DX}, \text{DY}), \text{cov}) \leftarrow \texttt{push(ID}, (\text{DX}, \text{DY})). \quad (11)$$

The clause is valid for any object ID; nonetheless, for simplicity, we will consider a scenario with a single object. The terminal states and rewards in DDC are:

$$\texttt{stop}_t \leftarrow \texttt{dist}(\simeq(\texttt{pos(A)}_t), (0.6, 1.0)) < 0.1.$$
$$\texttt{reward}(100)_t \leftarrow \texttt{stop}_t.$$
$$\texttt{reward}(-1)_t \leftarrow \texttt{not(stop}_t), \texttt{dist}(\simeq(\texttt{pos(A)}_t), (0.5, 0.8)) >= 0.2.$$
$$\texttt{reward}(-10)_t \leftarrow \texttt{not(stop}_t), \texttt{dist}(\simeq(\texttt{pos(A)}_t), (0.5, 0.8)) < 0.2. \quad (12)$$

That is, a state is terminal when there is an object close to the goal point $(0.6, 1.0)$ (i.e., distance lower than 0.1), and so, a reward of 100 is obtained. The nonterminal states have reward -10 whether inside an undesired region centered in $(0.5, 0.8)$ with radius 0.2, and $R(s_t, a_t) = -1$ otherwise.

Let us assume we previously sampled some episodes of length $T = 10$, and we want to sample the $m = 4$-th episode starting from $s_0 = (0, 0)$. We compute $\tilde{Q}_{10}^m((0,0), a)$ for each action a (line 6). Thus we compute the weights w^i using (10) for each stored sample $\tilde{V}_9^i(s_1^i)$. For example, Figure 1 shows the computation of $\tilde{Q}_{10}^m((0,0), a)$ for action $a' = (-0.4, 0.3)$ and $a'' = (0.9, 0.5)$, where we have three previous samples $i = \{1, 2, 3\}$ at depth 9. A shadow represents the likelihood $p(s_1^i | s_0 = (0, 0), a)$ (left for a' and right for a''). The weight w^i (10) for each sample s_1^i is obtained by dividing this likelihood by $q(s_1^i)$ (with $\alpha = 1$). If $q(s_1^i)$ is uniform over the three samples, sample $i = 2$ with total reward $\tilde{V}_9^2(s_1^2) = 98$

will have higher weight than samples $i = 1$ and $i = 3$. The situation is reversed for a''. Note that we can estimate $\tilde{Q}_d^m(s_t^m, a)$ using episodes i that may never encounter s_t^m, a_t provided that $p(s_{t+1}^i | s_t^m, a_t) > 0$.

Computing the (Approximate) Q-Function

The purpose of this section is to motivate our approximation to the Q-function using the weighted average of the V-function points in line 6. Let us begin by expanding the definition of the Q-function from (2) as follows:

$$Q_d^\pi(s_t, a_t) = R(s_t, a_t) + \gamma \int_{s_{t+1:T}, a_{t+1:T}} G_{d-1} p(s_{t+1:T}, a_{t+1:T} | s_t, a_t, \pi) ds_{t+1:T}, a_{t+1:T}, \quad (13)$$

where G_{d-1} is the total (discounted) reward from time $t + 1$ for $d - 1$ steps: $G_{d-1} = \sum_{k=1}^{d-1} \gamma^{k-1} R(s_{t+k}, a_{t+k})$. Given that we sample trajectories from the target distribution $p(s_{t+1:T}, a_{t+1:T} | s_t, a_t, \pi)$, we obtain the following approximation to the Q-function equaling the true value in the sampling limit:

$$Q_d^\pi(s_t, a_t) \approx R(s_t, a_t) + \frac{1}{N} \gamma \sum_i G_{d-1}^i. \quad (14)$$

Policy evaluation can be performed sampling trajectories using another policy, this is called *off-policy* Monte-Carlo [28]. For example, we can evaluate the greedy policy while the data is generated from a randomized one to enable exploration. This is generally performed using (normalized) *importance sampling* [25]. We let w^i be the ratio of the target and proposal distributions to restate the sampling limit as follows:

$$Q_d^\pi(s_t, a_t) \approx R(s_t, a_t) + \frac{1}{\sum w^i} \gamma \sum_i w^i G_{d-1}^i. \quad (15)$$

In standard off-policy Monte-Carlo the proposal distribution is of the form:

$$p(s_{t+1:T}, a_{t+1:T} | s_t, a_t, \pi') = \prod_{k=t}^{T-1} \pi'(a_{k+1} | s_{k+1}) p(s_{k+1} | s_k, a_k)$$

The target distribution has the same form, the only difference is that the policy is π instead of π'. In this case the weight becomes equal to the policy ratio because the transition model cancels out. This is desirable when the model is not available, for example in model-free Reinforcement Learning. The question is whether the availability of the transition model can be used to improve off-policy methods. This paper shows that the answer to that question is positive.

We will now describe the proposed solution. Instead of considering only trajectories that start from s_t, a_t as samples, we consider all sampled trajectories from time $t+1$ to T. Since we are ignoring steps before $t+1$, the proposal distribution for sample i is the marginal

$$p(s_{t+1:T}, a_{t+1:T} | s_0, \pi^i) = q(s_{t+1}) \pi^i(a_{t+1} | s_{t+1}) \prod_{k=t+1}^{T-1} \pi^i(a_{k+1} | s_{k+1}) p(s_{k+1} | s_k, a_k),$$

where q is the marginal probability $p(s_{t+1}|s_0, \pi^i)$. To compute $\tilde{Q}_d^m(s_t^m, a)$ we use (15), where the weight w^i (for $0 \le i \le m-1$) becomes the following:

$$\frac{p(s_{t+1}^i|s_t^m, a)\pi^m(a_{t+1}^i|s_{t+1}^i)\prod_{k=t+1}^{T-1}\pi^m(a_{k+1}^i|s_{k+1}^i)p(s_{k+1}^i|s_k^i, a_k^i)}{q(s_{t+1}^i)\pi^i(a_{t+1}^i|s_{t+1}^i)\prod_{k=t+1}^{T-1}\pi^i(a_{k+1}^i|s_{k+1}^i)p(s_{k+1}^i|s_k^i, a_k^i)}$$

$$= \frac{p(s_{t+1}^i|s_t^m, a)}{q(s_{t+1}^i)}\frac{\prod_{k=t}^{T-1}\pi^m(a_{k+1}^i|s_{k+1}^i)}{\prod_{k=t}^{T-1}\pi^i(a_{k+1}^i|s_{k+1}^i)} \tag{16}$$

$$\approx \frac{p(s_{t+1}^i|s_t^m, a)}{q(s_{t+1}^i)}\alpha^{(m-i)}. \tag{17}$$

Thus, we obtain line 6 in the algorithm given that $\tilde{V}_{d-1}^i(s_t^i) = G_{d-1}^i$. In our algorithm the target (greedy) policy π^m is not explicitly defined, therefore the policy ratio is hard to compute. We replace the unknown policy ratio with a quantity proportional to $\alpha^{(m-i)}$ where $0 < \alpha \le 1$; thus, formula (16) is replaced with (17). The quantity $\alpha^{(m-i)}$ becomes smaller for an increasing difference between the current episode index m and the i-th episode. Therefore, the recent episodes are weighted (on average) more than the previous ones, as in *recently-weighted average* applied in on-policy Monte-Carlo [28]. This is justified because the policy is improved over time, thus recent episodes should have higher weight.

Since we are performing policy improvement, each episode is sampled from a different policy. It has been shown [20,25] that samples from different distributions can be considered as sampled from a single distribution that is the mixture of the true distributions. Therefore, for a given episode

$$q(s_{t+1}^i) = \frac{1}{m-1}\sum_j p(s_{t+1}^i|s_0, \pi_j) = \frac{1}{m-1}\sum_j \int_{s_t}\int_{a_t} p(s_{t+1}^i|s_t, a_t)p(s_t, a_t|s_0, \pi_j)ds_t da_t$$

$$\approx \frac{1}{m-1}\sum_j p(s_{t+1}^i|s_t^j, a_t^j),$$

where for each j the integral is approximated with a single sample (s_t^j, a_t^j) from the available episodes. Since each episode is sampled from $p(s_{0:T}, a_{0:T}|s_0, \pi_j)$, samples (s_t^j, a_t^j) are distributed as $p(s_t, a_t|s_0, \pi_j)$ and are used in the estimation of the integral.

The likelihood $p(s_{t+1}^i|s_t^m, a)$ is required to compute the weight. This probability can be decomposed using the chain rule, e.g., for a state with 3 variables we have: $p(s_{t+1}^i|s_t^m, a) = p(v_3|v_2, v_1, s_t^m, a)p(v_2|v_1, s_t^m, a)p(v_1|s_t^m, a)$, where $s_{t+1}^i = \{v_1, v_2, v_3\}$. In DDC this is performed evaluating the likelihood of each variable in v_i following the topological order defined in the DDC program. The target and the proposal distributions might be mixed distributions of discrete and continuous random variables; importance sampling can be applied in such distributions as discussed in [19, Chapter 9.8].

To summarize, for each state s_t^m, $Q(s_t^m, a_t)$ is evaluated as the immediate reward plus the weighted average of stored G_{d-1}^i points. In addition, for each state s_t^m the total discounted reward G_d^m is stored. We would like to remark that we can estimate the Q-function also for states and actions that have never been visited, as shown in example 1. This is possible without using function approximations (beyond importance sampling).

Extensions

Our derivation follows a Monte-Carlo perspective, where each stored point is the total discounted reward of a given trajectory: $\tilde{V}_d^m(s_t^m) \leftarrow G_d^m$. However, following the Bellman equation, $\tilde{V}_d^m(s_t^m) \leftarrow max_a \tilde{Q}_d^m(s_t^m, a)$ can be stored instead. The Q estimation formula in line 6 is not affected; indeed we can repeat the same derivation using the Bellman equation and approximate it with importance sampling:

$$Q_d^\pi(s_t, a_t) = R(s_t, a_t) + \gamma \int_{s_{t+1}} V_{d-1}^\pi(s_{t+1}) p(s_{t+1}|s_t, a_t) ds_{t+1}$$

$$\approx R(s_t, a) + \gamma \sum \frac{w^i}{\sum w^i} \tilde{V}_{d-1}^i(s_{t+1}^i) = \tilde{Q}_d^m(s_t, a_t), \qquad (18)$$

with $w^i = \frac{p(s_{t+1}^i|s_t, a_t)}{q(s_{t+1}^i)}$ and s_{t+1}^i the state sampled in episode i for which we have an estimation of $\tilde{V}_{d-1}^i(s_{t+1}^i)$, while $q(s_{t+1}^i)$ is the probability with which s_{t+1}^i has been sampled. This derivation is valid for a fixed policy π; for a changing policy we can make similar considerations to the previous approach and add the term $\alpha^{(m-i)}$. If we choose $\tilde{V}_{d-1}^i(s_{t+1}^i) \leftarrow G_{d-1}^i$, we obtain the same result as in (9) and (17) for the Monte-Carlo approach. Instead of choosing between the two approaches we can use a linear combination, i.e., we replace line 16 with $\tilde{V}_d^m(s_t^m) \leftarrow \lambda G_d^m + (1-\lambda) max_a \tilde{Q}_d^m(s_t^m, a)$. The analysis from earlier applies by letting $\lambda = 1$. However, for $\lambda = 0$, we obtain a local value iteration step, where the stored $\tilde{V}$ is obtained maximizing the estimated $\tilde{Q}$ values. Any intermediate value balances the two approaches (this is similar to, and inspired by, TD(λ) [28]). Another strategy consists in storing the maximum of the two: $\tilde{V}_d^m(s_t^m) \leftarrow max(G_d^m, max_a \tilde{Q}_d^m(s_t^m, a))$. In other words, we alternate Monte-Carlo and Bellman backup according to which one has the highest value. This strategy works often well in practice; indeed it avoids a typical issue in Monte Carlo methods: bad policies or exploration lead to low rewards, averaged in the estimated Q/V-function. For this reason it may occur that optimal actions are rarely chosen. The mentioned strategy avoids this, and a high ϵ value (line 9) is possible without affecting the performance.

5   Related Work

There is an extensive literature on MDP planners, we will focus mainly on Monte-Carlo approaches. The most notable sample-based planners include Sparse Sampling (SST) [8], UCT [11] and their variations. SST creates a lookahead tree of depth D, starting from state s_0. For each action in a given state, the algorithm samples C times the next state. This produces a near-optimal solution with theoretical guarantees. In addition, this algorithm works with continuous and discrete domains with no particular assumptions. Unfortunately, the number of samples grows exponentially with the depth D, therefore the algorithm is extremely slow in practice. Some improvements have been proposed [31], although the worst-case performance remains exponential. UCT [11] uses *upper confidence bound* for multi-armed bandits to trade off between exploration and exploitation in the tree search, and inspired successful Monte-Carlo tree search methods. Instead of building the full tree, UCT chooses the action a that maximizes an upper confidence bound of $Q(s, a)$, following the principle of optimism in the face of uncertainty. Several improvements and extensions for UCT have been proposed, including handling continuous actions [13] (see [16] for a review), and continuous states [1] with a simple Gaussian distance metric; however the knowledge of the probabilistic model is not directly exploited. For continuous states, parametric function approximation is often used (e.g., linear regression), nonetheless the model needs to be carefully tailored for the domain to solve [32].

There exist algorithms that exploit instance-based methods (e.g. [3,5,26]) for model-free reinforcement learning. They basically store Q-point estimates, and then use e.g., neighborhood regression to evaluate $Q(s, a)$ given a new point (s, a). While these approaches are effective in some domains, they require the user to design distance metric that takes into account the domain. This is straightforward in some cases (e.g., in Euclidean spaces), but it might be harder in others. We argue that the knowledge of the model can avoid (or simplify) the design of a distance metric in several cases, where the importance sampling weights and the transition model, can be considered as a kernel.

The closest related works include [20,21,24,25], they use importance sampling to evaluate a policy from samples generated with another policy. Nonetheless, they adopt importance sampling differently without the knowledge of the MDP model. Although this property seems desirable, the availability of the actual probabilities cannot be exploited, apart from sampling, in their approaches. The same conclusion is valid for practically any sample-based planner, which only needs a sample generator of the model. The work of [9] made a similar statement regarding PROST, a state-of-the-art discrete planner based on UCT, without providing a way to use the state transition probabilities directly. Our algorithm tries to alleviate this, exploiting the probabilistic model in a sample-based planner via importance sampling.

For more general domains that contain discrete and continuous (hybrid) variables several approaches have been proposed under strict assumptions. For example, [23] provide exact solutions, but assume that continuous aspects of the transition model are deterministic. In a related effort [4], hybrid MDPs are solved using dynamic programming, but assuming that transition model and reward is piecewise constant or linear. Another planner HAO* [14] uses heuristic search to find an optimal plan in hybrid domains with theoretical guarantees. However, they assume that the same state cannot be visited again (i.e., they assume plans do not have loops, as discussed in [14, sec.5]), and they rely on the availability of methods to solve the integral in the Bellman equation related to the continuous part of the state. Visiting the same state in our approach is a benefit and not a limit; indeed a previous visited state s' is useful to evaluate $Q_d(s, a)$, when the weight is positive (i.e., when s' is reachable from s with action a).

There exists several languages specific for planning, the most recent is RDDL [22]. A RDDL domain can be mapped in DDC and solved with HYPE. Nonetheless, RDDL does not support a state space with an unknown number of variables as in Example 2. Some planners are based on probabilistic logic programming, for example DTProbLog [29] and PRADA [12], though they only support discrete action-state spaces. For domains with an unknown number of objects, some probabilistic programming languages such as BLOG [15], Church [6], and DC [7] can cope with such uncertainty. To the best of our knowledge DTBLOG [27] and [30] are the only proposals that are able to perform decision making in such domains using a POMDP framework. Furthermore, BLOG is one of the few languages that explicitly handles data association and identity uncertainty. The proposed paper does not focus on POMDP, nor on identity uncertainty; however, interesting domains with unknown number of objects can be easily described as an MDP that HYPE can solve.

Among the mentioned sample-based planners, one of the most general is SST, which does not make any assumption on the state and action space, and only relies on Monte-Carlo approximation. In addition, it is one of the few planners that can be easily applied to any DDC program, including MDPs with an unknown number of objects. For this reason SST was implemented for DDC and used as baseline for our experiments.

6 Experiments

This section answers the following questions: (Q1) Does the algorithm obtain the correct results? (Q2) How is the performance of the algorithm in different domains? (Q3) How does it compare with state-of-the-art planners?

The algorithm was implemented in YAP Prolog and C++, and run on a Intel Core i7 Desktop.

To answer (Q1) we tested the algorithm on a nonlinear version of the hybrid mars rover domain (called *simplerover1*) described in [23] for which the exact V-function is available (depth $d = 3$ and 2 variables: a two-dimensional continuous position and one discrete variable to indicate if the picture was taken). We choose 31 initial points and ran the algorithm for 100 episodes each. Each point took on average $1.4s$. Fig. 2 shows the results where the line is the exact V, and dots are estimated V points. The results show that the algorithm converges to the optimal V-function with a negligible error. This domain is deterministic, and so, to make it more realistic we converted it to a probabilistic MDP adding Gaussian noise to the state transition model. The resulting MDP (*simplerover2*) is hard (if not impossible) to solve exactly. Then we performed experiments for different horizons, number of pictures points (1 to 4, each one is a discrete variable) and summed the rewards. For each instance the planner searches for an optimal policy and executes it, and after each executed action it samples additional episodes to refine the policy (replanning). The proposed planner is compared with SST

Table 1. Experiments: d is the horizon used by the planner, T the total number of steps, M is the maximum number of episodes sampled for HYPE, while C is the SST parameter (number of samples for each state and action). Time limit of $1800s$ per instance. PROST results refer to IPPC2011.

Planner		game1 $T=40$	game2 $T=40$	sysadmin1 $T=40$	sysadmin2 $T=40$
HYPE	reward	0.87 ± 0.11	0.77 ± 0.22	0.94 ± 0.07	0.87 ± 0.11
	time (s)	622	608	422	475
	param	$M=1200$ $d=5$	$M=1200$ $d=5$	$M=1200$ $d=5$	$M=1200$ $d=5$
SST	reward	0.34 ± 0.15	0.14 ± 0.20	0.47 ± 0.13	0.31 ± 0.12
	time (s)	986	1000	1068	1062
	param	$C=1$ $d=5$	$C=1$ $d=5$	$C=1$ $d=5$	$C=1$ $d=5$
HYPE	reward	$\mathbf{0.89\pm0.07}$	0.76 ± 0.19	$\mathbf{0.98\pm0.06}$	0.86 ± 0.11
	time (s)	312	582	346	392
	param	$M=1200$ $d=4$	$M=1200$ $d=4$	$M=1200$ $d=4$	$M=1200$ $d=4$
SST	reward	0.79 ± 0.08	0.27 ± 0.22	0.66 ± 0.08	0.46 ± 0.12
	time (s)	1538	1528	1527	1532
	param	$C=2$ $d=4$	$C=2$ $d=4$	$C=2$ $d=4$	$C=2$ $d=4$
PROST	reward	0.99 ± 0.02	1.00 ± 0.19	1.00 ± 0.05	0.98 ± 0.09

Planner		objpush $T=30$	simplerover2 $d=T$	marsrover $T=40$	objsearch $d=T$
HYPE	reward	83.7 ± 7.6	11.8 ± 0.2	249.8 ± 33.5	2.53 ± 1.03
	time (s)	472	38	985	13
	param	$M=4500$ $d=9$	$M=200$ $d=T=8$	$M=6000$ $d=6$	$M=500$ $d=T=5$
SST	reward	82.7 ± 2.7	11.4 ± 0.3	227.7 ± 27.3	1.46 ± 1.0
	time (s)	330	48	787	45
	param	$C=1$ $d=9$	$C=1$ $d=T=8$	$C=1$ $d=6$	$C=5$ $d=T=5$
HYPE	reward	86.4 ± 1.0	11.7 ± 0.2	269.0 ± 29.4	$\mathbf{3.64\pm1.09}$
	time (s)	1238	195	983	17
	param	$M=4500$ $d=10$	$M=500$ $d=T=9$	$M=6000$ $d=7$	$M=600$ $d=T=5$
SST	reward	82.4 ± 1.9	11.3 ± 0.3	N/A	2.48 ± 1.0
	time (s)	1574	238	timeout	138
	param	$C=1$ $d=10$	$C=1$ $d=T=9$	$C=1$ $d=7$	$C=6$ $d=T=5$
HYPE	reward	$\mathbf{87.5\pm0.5}$	11.9 ± 0.3	$\mathbf{296.3\pm19.5}$	3.3 ± 1.6
	time (s)	373	218	1499	20
	param	$M=2000$ $d=12$	$M=500$ $d=T=10$	$M=4000$ $d=10$	$M=600$ $d=T=6$
SST	reward	N/A	11.2 ± 0.3	N/A	0.58 ± 1.4
	time (s)	timeout	1043	timeout	899
	param	$C=1$ $d\geq11$	$C=1$ $d=T=10$	$C=1$ $d\geq8$	$C=5$ $d=T=6$

that requires replanning every step. The results for both planners are always comparable, which confirms the empirical correctness of HYPE (Table 1).

To answer (Q2) and (Q3) we studied the planner in a variety of settings, from discrete, to continuous, to hybrid domains, to those with an unknown number of objects. We performed experiments in a more realistic mars rover domain that is publicly available[1], called *marsrover* (Fig. 2). In this domain we consider one robot and 5 picture points that need to be taken, the movement of the robot causes a negative reward proportional to the displacement and the pictures can be taken only close to the interest point. Each taken picture provides a different reward. Other experiments were performed in the continuous *objpush* MDP described in Section 4 (Fig. 1), and in discrete benchmark domains of the IPPC 2011 competition. In particular, we tested a pair of instances of game of life and sysadmin domains. The results are compared with PROST [9], the IPPC 2011 winner, and shown in table 1 in terms of scores, i.e., the average reward normalizated with respect to IPPC 2011 results; score 1 is the highest result obtained, score 0 is the maximum between the random and the no operation policy.

As suggested by [9], limiting the horizon of the planner increases the performance in several cases. We exploited this idea for HYPE as well as SST (*simplerover2* excluded). For SST we were forced to use small horizons to keep plan time under 30 minutes. In all experiments we followed the IPPC 2011

---

[1] http://users.cecs.anu.edu.au/~ssanner/IPPC_2014/index.html

schema, that is each instance is repeated 30 times (*objectsearch* excluded), the results are averaged and the 95% confidence interval is computed. However, for every instance we replan from scratch for a fair comparison with SST. In addition, time and number of samples refers to the plan execution of one instance. The results (Table 1) highlight that our planner obtains generally better results than SST, especially at higher horizons. HYPE obtains good results in discrete domains but does not reach state-of-art results (score 1) for two main reasons. The first is the lack of a heuristic, that can dramatically improve the performance, indeed, heuristics are an important component of PROST [9], the IPPC winning planner. The second reason is the time performance that allows us to sample a limited number of episodes and will not allow to finish all the IPPC 2011 domains in 24 hours. This is caused by the expensive Q-function evaluation; however, we are confident that heuristics and other improvements will significantly improve performance and results.

Finally, we performed experiments in the *objectsearch* scenario (Section 3), where the number of objects is unknown. The results are averaged over 400 runs, and confirm better performance for HYPE with respect to SST.

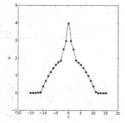
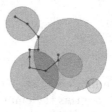

Fig. 2. *V*-function for different rover positions (with fixed $X = 0.16$) in *simplerover1* domain (left). A possible episode in *marsrover* (right): each picture can be taken inside the respective circle (red if already taken, green otherwise).

7   Practical Improvements

In this section we briefly discuss issues and improvements of HYPE. To evaluate the Q-function the algorithm needs to query all the stored examples, making the algorithm potentially slow. This issue can be mitigated with solutions used in instance-based learning, such as hashing and indexing. For example, in discrete domains we avoid multiple computations of the likelihood and the proposal distribution for samples of the same state. In addition, assuming policy improvement over time, only the N_{store} most recent episodes are kept, since older episodes are generally sampled with a worse policy.

The algorithm HYPE relies on importance sampling to estimate the Q-function, thus we should guarantee that $p > 0 \Rightarrow q > 0$, where p is the target and q is the proposal distribution. This is not always the case, like when we sample the first episode. Nonetheless we can have an indication of the estimation reliability. In our algorithm we use $\sum w^i$ with expectation equals to the number of samples: $\mathbb{E}[\sum w^i] = m$. If $\sum w^i < thres$ the samples available are considered insufficient to compute $Q_d^m(s_t^m, a)$, thus action a can be selected to perform exploration.

A more problematic situation is when, for some action a_t in some state s_t, we always obtain null weights, that is $p(s_{t+1}^i | s_t, a_t) = 0$ for each of the previous episodes i, no matter how many episodes are generated. This issue is solved by adding noise to the state transition model, e.g., Gaussian noise for continuous random variables. This is equivalent to adding a smoothness assumption to the V-function. Indeed the Q-function is a weighted sum of V-function points, where the weights are proportional to a noisy version of the state transition likelihood.

8 Conclusions

We proposed a sample-based planner for MDPs described in DDC under weak assumptions, and showed how the state transition model can be exploited in off-policy Monte-Carlo. The experimental results show that the algorithm produces good results in discrete, continuous, hybrid domains as well as those with an unknown number of objects. Most significantly, it challenges and outperforms SST. For future work, we will consider heuristics and hashing to improve the implementation.

References

1. Couetoux, A.: Monte Carlo Tree Search for Continuous and Stochastic Sequential Decision Making Problems. Université Paris Sud - Paris XI, Thesis (2013)
2. De Raedt, L., Kersting, K.: Probabilistic inductive logic programming. In: De Raedt, L., Frasconi, P., Kersting, K., Muggleton, S.H. (eds.) Probabilistic Inductive Logic Programming. LNCS (LNAI), vol. 4911, pp. 1–27. Springer, Heidelberg (2008)
3. Driessens, K., Ramon, J.: Relational instance based regression for relational reinforcement learning. In: Proc. ICML (2003)
4. Feng, Z., Dearden, R., Meuleau, N., Washington, R.: Dynamic programming for structured continuous Markov decision problems. In: Proc. UAI (2004)
5. Forbes, J., André, D.: Representations for learning control policies. In: Proc. of the ICML Workshop on Development of Representations (2002)
6. Goodman, N., Mansinghka, V.K., Roy, D.M., Bonawitz, K., Tenenbaum, J.B.: Church: A language for generative models. In: Proc. UAI, pp. 220–229 (2008)
7. Gutmann, B., Thon, I., Kimmig, A., Bruynooghe, M., De Raedt, L.: The magic of logical inference in probabilistic programming. Theory and Practice of Logic Programming (2011)
8. Kearns, M., Mansour, Y., Ng, A.Y.: A Sparse Sampling Algorithm for Near-Optimal Planning in Large Markov Decision Processes. Machine Learning (2002)
9. Keller, T., Eyerich, P.: PROST: probabilistic planning based on UCT. In: Proc. ICAPS (2012)

10. Kimmig, A., Santos Costa, V., Rocha, R., Demoen, B., De Raedt, L.: On the efficient execution of problog programs. In: Garcia de la Banda, M., Pontelli, E. (eds.) ICLP 2008. LNCS, vol. 5366, pp. 175–189. Springer, Heidelberg (2008)
11. Kocsis, L., Szepesvári, C.: Bandit based monte-carlo planning. In: Fürnkranz, J., Scheffer, T., Spiliopoulou, M. (eds.) ECML 2006. LNCS (LNAI), vol. 4212, pp. 282–293. Springer, Heidelberg (2006)
12. Lang, T., Toussaint, M.: Planning with Noisy Probabilistic Relational Rules. Journal of Artificial Intelligence Research **39**, 1–49 (2010)
13. Mansley, C.R., Weinstein, A., Littman, M.L.: Sample-Based planning for continuous action markov decision processes. In: Proc. ICAPS (2011)
14. Meuleau, N., Benazera, E., Brafman, R.I., Hansen, E.A., Mausam, M.: A heuristic search approach to planning with continuous resources in stochastic domains. Journal of Artificial Intelligence Research **34**(1), 27 (2009)
15. Milch, B., Marthi, B., Russell, S., Sontag, D., Ong, D., Kolobov, A.: BLOG: probabilistic models with unknown objects. In: Proc. IJCAI (2005)
16. Munos, R.: From Bandits to Monte-Carlo Tree Search: The Optimistic Principle Applied to Optimization and Planning. Foundations and Trends in Machine Learning, Now Publishers (2014)
17. Nitti, D., De Laet, T., De Raedt, L.: A particle filter for hybrid relational domains. In: Proc. IROS (2013)
18. Nitti, D., De Laet, T., De Raedt, L.: Relational object tracking and learning. In: Proc. ICRA (2014)
19. Owen, A.B.: Monte Carlo theory, methods and examples (2013)
20. Peshkin, L., Shelton, C.R.: Learning from scarce experience. In: Proc. ICML, pp. 498–505 (2002)
21. Precup, D., Sutton, R.S., Singh, S.P.: Eligibility traces for off-policy policy evaluation. In: Proc. ICML (2000)
22. Sanner, S.: Relational Dynamic Influence Diagram Language (RDDL): Language Description (unpublished paper)
23. Sanner, S., Delgado, K.V., de Barros, L.N.: Symbolic dynamic programming for discrete and continuous state MDPs. In: Proc. UAI (2011)
24. Shelton, C.R.: Policy improvement for POMDPs using normalized importance sampling. In: Proc. UAI, pp. 496–503 (2001)
25. Shelton, C.R.: Importance Sampling for Reinforcement Learning with Multiple Objectives. Ph.D. thesis, MIT (2001)
26. Smart, W.D., Kaelbling, L.P.: Practical reinforcement learning in continuous spaces. In: Proc. ICML (2000)
27. Srivastava, S., Russell, S., Ruan, P., Cheng, X.: First-order open-universe POMDPs. In: Proc. UAI (2014)
28. Sutton, R.S., Barto, A.G.: Reinforcement Learning: An Introduction. MIT Press (1998)
29. Van den Broeck, G., Thon, I., van Otterlo, M., De Raedt, L.: DTProbLog: a decision-theoretic probabilistic prolog. In: Proc. AAAI (2010)
30. Vien, N.A., Toussaint, M.: Model-Based relational RL when object existence is partially observable. In: Proc. ICML (2014)
31. Walsh, T.J., Goschin, S., Littman, M.L.: Integrating sample-based planning and model-based reinforcement learning. In: Proc. AAAI (2010)
32. Wiering, M., van Otterlo, M.: Reinforcement learning: state-of-the-art. In: Adaptation, Learning, and Optimization. Springer (2012)

Simplifying, Regularizing and Strengthening Sum-Product Network Structure Learning

Antonio Vergari[✉], Nicola Di Mauro, and Floriana Esposito

University of Bari "Aldo Moro", Bari, Italy
{antonio.vergari,nicola.dimauro,floriana.esposito}@uniba.it

Abstract. The need for feasible inference in Probabilistic Graphical Models (PGMs) has lead to tractable models like Sum-Product Networks (SPNs). Their highly expressive power and their ability to provide exact and tractable inference make them very attractive for several real world applications, from computer vision to NLP. Recently, great attention around SPNs has focused on structure learning, leading to different algorithms being able to learn both the network and its parameters from data. Here, we enhance one of the best structure learner, Learn-SPN, aiming to improve both the structural quality of the learned networks and their achieved likelihoods. Our algorithmic variations are able to learn simpler, deeper and more robust networks. These results have been obtained by exploiting some insights in the building process done by LearnSPN, by hybridizing the network adopting tree-structured models as leaves, and by blending bagging estimations into mixture creation. We prove our claims by empirically evaluating the learned SPNs on several benchmark datasets against other competitive SPN and PGM structure learners.

1 Introduction

Probabilistic Graphical Models (PGMs) [13] use a graph-based representation eliciting the conditional independence assumptions among a set of random variables, thus providing a compact encoding of complex joint probability distributions. The most common task one want to solve using PGMs is *inference*, a task that becomes intractable for complex networks, a difficulty often circumvented by adopting approximate inference. For instance, computing the exact marginal or conditional probability of a query is a #P-complete problem [27].

However, there are many recently proposed PGMs where inference becomes tractable. They include graphs with low *treewidth*, such as *tree-structured graphical models* where each variable has at most one parent in the network structure [7], and their extensions with mixtures [18] or latent variables [6], or *Thin Junction Trees* [4], allowing controlled treewidths. Being more general than all of these models and yet preserving tractable and exact inference, *Sum-Product Networks* (SPNs) [23] provide an interesting model, successfully employed in image reconstruction and recognition [1,9,23], speech recognition [21] and NLP [5] tasks. Similarly to *Arithmetic Circuits* (ACs) [16], to which they are equivalent for finite domains [26], they compile a high treewidth network into a deep

© Springer International Publishing Switzerland 2015
A. Appice et al. (Eds.): ECML PKDD 2015, Part II, LNAI 9285, pp. 343–358, 2015.
DOI: 10.1007/978-3-319-23525-7_21

probabilistic architecture. By layering inner nodes, sum and product nodes, they encode the probability density function over the observed variables, represented as leaf nodes. SPNs guarantee inference in time linear to their network size [23], and they possibly becomes more expressively efficient as their depth increases [17].

Recently the attention around SPNs has focused on structure learning algorithms as ways to automate latent interaction discovery among observed variables and to avoid the cost of parameter learning [8,10,19,26]. While many of these efforts concentrated on optimizing the likelihoods of the models, little attention has been devoted to the structural quality of such models, or to understand how data quality effects the learning process.

In this paper we extend and simplify one of the state-of-the-art SPN structure learning algorithm, LearnSPN [10], providing several improvements and insights. We show how to a) learn simpler SPNs, i.e. ones with less edges, parameters and more layers, b) stop the building process earlier while preserving goodness of fit, and c) be more robust and resilient in estimating the dependencies from data. In order to accomplish this we limit the number of node children when building the network, we introduce tractable multivariate distributions, in the form of Chow-Liu trees [7], as leaves of an hybrid architecture without adding complexity to the network, and we enhance the mixture models of an SPN via bootstrap samples, i.e. by applying bagging for the likelihood function estimation.

We produced different algorithmic variants incorporating one or more of these enhancements, and thoroughly evaluated them on standard benchmark datasets, both under the structure quality perspective and the more usual data likelihood gain. We compared them against the original algorithm, the best SPN structure learner up to now, ID-SPN [26], and MT [18], learning mixture of trees, reported to be the second best algorithm in [26] on the same datasets.

2   Sum-Product Networks

Sum-Product Networks have been introduced in [23] as a general architecture efficiently encoding an unnormalized probability distribution over a set of random variables $\mathbf{X} = \{X_1, \ldots, X_n\}$. The graphical representation of an SPN consists of a rooted DAG, S, whose leaves correspond to univariate distributions of observable variables in $\mathbf{X}$, while internal nodes are *sum* or *product* nodes. The *scope* of each internal node i, denoted as $\mathbf{X}_{\psi_i}$, is defined as the set of variables appearing as its descendants. The sub-network S_i rooted at node i encodes the unnormalized distribution over its scope. The *parameters* of the network are the positive weights w_{ij} associated to each edge $i \rightarrow j$ in S, where i is a sum node. As in [10,26] we will refer to the whole network as S, and, for a given state $\mathbf{x}$ of the variables $\mathbf{X}$, we will indicate as $S(\mathbf{x})$ the unnormalized probability of $\mathbf{x}$ according to the SPN S , i.e., the value of S's root when the network is evaluated after $\mathbf{X} = \mathbf{x}$ is observed. Intuitively, sum nodes encode mixtures over probability distributions whose coefficients are the children weights, while product nodes identify factorizations over independent distributions. Examples of different SPNs are shown in Figure 1. For the sake of simplicity, we are considering $\mathbf{X}$ to be discrete valued random variables (the extension to the continuous case is straightforward [23]).

An SPN is said to be *decomposable* if the scopes of the children of product nodes are disjoint, and *complete* when the scopes of sum nodes children are the same. Decomposability and completeness imply *validity* [23], i.e. the property of correctly and exactly computing each evidence probability by evaluating the network, that is, for a network S and a state $\mathbf{x}$, $P(\mathbf{X} = \mathbf{x}) = S(\mathbf{x})/Z$, where Z is the *partition function*, defined as $Z = \sum_{\mathbf{x}} S(\mathbf{x})$. From now on, we will assume the SPNs we are considering to be valid.

To compute $S(\mathbf{x})$, the whole network is evaluated bottom-up. For a leaf node i, representing the variable X_k, $S_i(x)$ corresponds to univariate distribution values for x, i.e. $S_i(x) = P(X_k = x)$. While for a generic internal node i, a) if it is a product node, then $S_i(\mathbf{x}_{\psi_i}) = \prod_{i \rightarrow j \in S} S_j(\mathbf{x}_{\psi_j})$; b) if it is a sum node, then $S_i(\mathbf{x}_{\psi_i}) = \sum_{i \rightarrow j \in S} w_{ij} S_j(\mathbf{x}_{\psi_j})$. If the weights of each sum node i sum to one, $\sum_j w_{ij} = 1$, and the leaf distributions are normalized, then the network will compute the exact, normalized, probability, i.e. $\forall \mathbf{x}, P(\mathbf{X} = \mathbf{x}) = S(\mathbf{X})$. For the rest of the paper we will assume SPNs being normalized in this way. Following these considerations, it can be demonstrated that all the marginal probabilities, the partition function and all MPE queries and states can be computed in time linear in the *size* of the network, i.e. its number of edges [10].

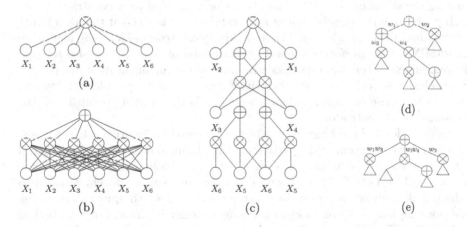

Fig. 1. Examples of SPNs: a naive factorization over 6 random variables (1a), a shallow mixture standing for a pointwise kernel density estimation (1b) and a deeper architecture (1c) over the same scope (weights are omitted for simplicity). An SPN with alternated layers of nodes of the same kind (1e) obtained from pruning the one in (1d) and yet encoding the same distribution.

Note that this does not automatically imply the tractability of inference, for it to be feasible the number of edges should be polynomial in the number of random variables. One way to control the number of edges is to layer the nodes into a deep architecture, where parameters are reused across the levels. We define the *depth* of a network as the longest path from the root to a leaf node in networks with strictly interleaving layers of nodes of the same kind. Note that it is always possible to convert an SPN in such a layered architecture. A node c having parents $\{p_i\}_{i=1}^{I}$ sharing its same type can be pruned, and c's children, $\{g_j\}_{j=1}^{J}$, can be directly attached to each p_i. If c was a sum node, the new weights for each g_j can be computed as $w_{p_i c} w_{c g_j}$, see Figures 1d and 1e. For a visual comparison of how the depth impacts the network size, see Figure 1c, where local dependencies are exploited compared to the shallow SPN in Figure 1b where they are not and each leaf is fully connected to the upper layer. Moreover, it has been shown that increasing the depth of a network makes it more expressively efficient [17]. Even the number of weights in a network impacts weight learning feasibility [9], when hand-crafted SPNs are employed, usually some sparsity constraint is applied during learning to prune as many edges as possible [23]. The opportunity to directly learn the structure of an SPN offers a way to govern the time future inferences and learning stages will take. However, up to now, the focus of structure learning algorithms has not been the quality of the learned architectures in terms of depth, number of parameters and edges, but the ability to better capture the data probability distribution in terms of the achieved likelihood scores.

2.1 Structure Learning

Learning the structure of SPNs has always been tackled as a constraint-based search problem exploiting heuristics and statistical tests to elicit the local latent relationships in the data [8,19]. The first principled structure learning algorithm is LearnSPN [10], it performs a greedy construction of treed SPNs, also referred to as *formula* SPN [17], i.e. networks with inner nodes having at most one parent. Nevertheless it is still the most attractive for its simplicity, parallelizability and ability to learn the networks weights as well. It is the starting point for all the extensions we will introduce.

The core idea behind LearnSPN, which is sketched in Algorithm 1, is to grow a tree top down by recursively partitioning the input data matrix consisting of a set T of rows as i.i.d instances, over V, the set of columns, i.e. the features. For each call of LearnSPN on a submatrix, column splits add child nodes to product nodes, while those on rows extend sum nodes. To split columns, the corresponding features are checked for independency by means of a statistical test in the splitFeatures procedure, while clusterInstances is employed to aggregate rows together by a similarity criterion. The weights of sum nodes children represent the proportions of instances falling into the computed clusters (line 13). Termination is achieved in two cases, when the current submatrix contains only one column (line 3) or when the number of its rows falls under a certain threshold m (line 5). In the former, a leaf node, standing for a univariate distribution, is introduced by estimating it from the submatrix data entries, i.e., for categorical random variables by counting their values occurrences while applying Laplace smoothing with parameter α. In the latter, the random variables for the submatrix columns are modeled with a naive factorization, i.e. they are

---

Algorithm 1. LearnSPN(T, V, α, m)

1: **Input:** a set of row instances T over a set of column features V; m: minimum number of instances to split; α: Laplace smoothing parameter
2: **Output:** an SPN S encoding a pdf over V learned from T
3: **if** $|V| == 1$ **then**
4:     $S \leftarrow$ univariateDistribution(T, V, α)
5: **else if** $|T| < m$ **then**
6:     $S \leftarrow$ naiveFactorization(T, V, α)
7: **else**
8:     $\{V_j\}_{j=1}^{C} \leftarrow$ splitFeatures(V, T)
9:     **if** $C > 1$ **then**
10:         $S \leftarrow \prod_{j=1}^{C}$ LearnSPN(T, V_j, α, m)
11:     **else**
12:         $\{T_i\}_{i=1}^{R} \leftarrow$ clusterInstances(T, V)
13:         $S \leftarrow \sum_{i=1}^{R} \frac{|T_i|}{|T|}$ LearnSPN(T_i, V, α, m)
    return S

Algorithm 2. naiveFactorization(T, V, α)

1: **Input:** a set of row instances T over a set of column features V, α Laplace smoothing parameter
2: **Output:** an SPN S encoding a product of factors V estimated from T
3: **return** $S \leftarrow \prod_{j=1}^{|V|}$ univariateDistribution(T, V_j, α)

---

considered to be independent and a product node is put over a set of univariate leaf nodes, as in Algorithm 2.

It is worth noting that the two splitting procedures depend on each other: the quality of row clusterings is likely to be enhanced by column splits correctly identifying dependent features. At the same time, well done instance splits would allow for finer independence tests in the next call of the algorithm. The likelihood on the data is never computed explicitly, the search for local hidden relationships leads to small submatrices whose likelihood can be easily estimated via naive factorizations. In [10] it is said that if the two splitting procedures are able to exactly separate the columns into independent groups and the rows by similarity, they would lead to locally optimal structures in the terms of data likelihood.

While the presented version of LearnSPN allows for different kinds of splitting procedures, the way they are implemented is crucial. In [10] a G-Test is used to check for the independence of pairs of random variables in splitFeatures: if the test p-value is less than a threshold ρ, then the two features are considered to be independent. Two subsets of approximately dependent features are produced by exploring features starting from one randomly chosen. Rows are aggregated into an adaptive number of clusters l, by employing the hard online EM algorithm. Columns V_j are assumed to be independent given the row clusters C_i, i.e., $P(V) = \sum_i P(C_i) \prod_j P(V_j|C_i)$. To control the cluster numbers, an exponential prior on clusterings in the form of $e^{-\lambda l|V|}$ is used, with λ as a tuning parameter. In this concrete formulation, the algorithm searches for treed SPNs in the hyperparameter space formed by m, α, ρ and λ. If the algorithm finds all the features in the complete data matrix to be independent, it would build an SPN representing a naive factorization consisting in a single product node over $|V|$ leaves (like the one in Figure 1a over $X_1, \ldots, X_6$). However, this degeneration is prevented by forcing a split on the rows during the first call of the algorithm. On the other hand, in the case of each cluster containing a single instance, the network would be similar to a pointwise kernel density estimation (see Figure 1b for a graphical example, where six instances are considered).

3   Contributions

Being greedy by design, LearnSPN, is highly prone to learn far from optimal structures, both in terms of likelihood scores and network quality. This is particularly true when the training data is noisy and/or scarce. The statistical tests implemented by the splitting procedures can easily be mislead into wrong choices, and, worst of all, overfitting could lead to overcomplex networks for which inference can be an issue. Given these shortcomings, our contributions will affect them in several ways: here we show how limiting the number of node children while splitting leads to deeper and simpler networks, how more complex and yet tractable factorizations as leaves are able to reduce network complexity favoring early termination, and how model averaging by bagging can be blended into the definition of SPNs in order to get more robust models.

3.1   Deepening by Limiting Node Splits

Our first contribution is to limit the number of node children while learning, resulting in networks that will be deeper, and potentially with less edges and parameters. Basically, we fix to two the number of submatrices the current matrix can split into, for each call of LearnSPN, i.e., $C \leq 2$ and $R \leq 2$ on lines 8, 12 of Algorithm 1. This is already achieved in LearnSPN when splitFeatures is implemented to decompose the features into two subsets, thus our variation will effectively limit only row splits.

   This simplifying idea is based on a number of observations. The first one is that checking earlier and more often for independency among features enhances the quality of row splits, reduces the average number of sum node children making the network less wide but deeper. Successive splits along the same axis can be seen as a hierarchical divisive clustering process whose termination is achieved when splits along the other axis happen. The aim is to slow down this greedy decision process to make it take the most out of data, while trying to exactly determine all the splits along one axis at once would lead to local optima faster. Secondly, we can observe that other splits on the same axis could always be done in the following iterations, but only if necessary. As noted in the previous section, in this way we are not limiting the expressive power of the learned SPNs at all, since after pruning nodes whose parent has their same type, the number of children per node can be more than two. This variation can be seen as an application of the simplicity bias criterion. By not committing to complex choices too early, the algorithm remains able to explore structures where the splits on the features could lead to better networks. Moreover, it is also more robust to overclustering in noisy situations since in those cases it can receive the valuable help from the anticipated independence tests. As our experiments suggested, this is particularly true when a row split is forced on the first call of LearnSPN.

Under these observations, to implement clusterInstances in our simplified version one could still use any clustering algorithm, but limiting the number of clusters to two, thus resulting in one hyperparameter less to tune.

3.2 Regularization by Tractable Multivariate Distribution Hybridization

As our second contribution, we tackle the problem of fitting tractable multivariate distributions as leaves of an SPN. By substituting them to the naive factorizations we aim at a twofold objective: improving the network likelihood by capturing finer local dependencies when estimating leaf distributions, and being able to stop the building process earlier.

To balance complexity and expressive power, we chose tree-structured distributions as the simplest tractable distributions that are able to model more dependencies than a naive factorization while not adding complexity to the structure by adding additional parameters or layers. *Directed Tree distributions* [18] are Bayesian Networks whose nodes, standing for the random variables X_j, have at most one parent each, Pa_j, which leads to the following factorization for the joint distribution $P(\mathbf{X}) = \prod_j P(X_j | Pa_j)$. From this formulation, it is easy to see how we preserve the same complexity for computing inference on complete evidences: it reduces again to a product of the same number of factors. Differently from a naive factorization, each factor here provides the valuable information about the conditional dependency between parent and child variables. Moreover, tree distributions guarantee that marginalizations, and MPE inference as well, can be computed in time linear to the number of factors [18]. The validity of the network is also preserved, row and column splits guarantee that the scope of the multivariate leaves will not compromise decomposability nor completeness.

The classic Chow-Liu algorithm [7] can be used to learn the tree distribution that best approximates the underlying joint distribution in terms of the Kullback-Liebler divergence. The algorithm builds a maximum spanning tree over the graph formed by the pairwise Mutual Information (MI) values among each pair of columns in the current submatrix. It then turns the undirected tree into a directed one by randomly selecting a root and traversing it. In practice, we substitute the procedure naiveFactorization from line 6 of Algorithm 1 with the procedure LearnCLT as shown in Algorithm 3. In our hybrid architecture now leaves can be simple univariate distributions like before *or* subnetworks S_t encoding multivariate tree distributions over the set $\mathbf{X}_{\psi_t}$ of random variables. Computing $S_t(\mathbf{x}_{\psi_t})$ equals to evaluate the Chow-Liu tree on $\mathbf{x}_{\psi_t}$.

Algorithm 3. LearnCLT(T, V, α)

1: **Input:** a set of instances T over a set of features V, α Laplace smoothing parameter
2: **Output:** a Chow-Liu tree S encoding a pdf over V learned from T
3: $M \leftarrow \mathbf{0}_{|V| \times |V|}$
4: **for each** $X_u, X_v \in V$ **do**
5:     $M_{u,v} \leftarrow$ estimateMutualInformation(X_u, X_v, α)
6: $\mathcal{T} \leftarrow$ maximumSpanningTree(M)
7: **return** $S \leftarrow$ traverseTree($\mathcal{T}$)

The complexity of learning a Chow-Liu tree is quadratic in the number of features taken into account, however efficient implementations can lower it to sub-quadratic times [18]. Note that we limit this stage to the last steps of the algorithm, where submatrices have usually only few features.

The hyperparameter α is still needed to smooth the marginals in Algorithm 3, line 5. If we consider the original formulation of LearnSPN, m and α offer the only simplistic forms of regularization, however, when using naive factorizations, m is not as valuable in terminating the search earlier as it is when tree distributions are employed. In fact, to have the naive independency assumption hold, one has to let the search continue up to small submatrices, where even larger ones can be equally or better approximated by a Chow-Liu tree. As we will show in the Section 5, by doing a grid search over the hyperparameter m, the best likelihood wise structures on a validation set, employing naive factorizations, would prefer smaller values for m, while the best ones introducing tree distributions would likely be the ones learned with larger values for m. In this way one is able to prefer even simpler models, possibly avoiding overfitting.

3.3 Strengthening by Model Averaging

While on the previous sections we concentrated on improving the structure quality, our next contribution will focus on directly increasing the likelihood estimation capability of LearnSPN. In order to do so, we leverage a very well know statistical tool for robust parameter estimation: *bagging* [12].

Before performing a split on rows, we draw k bootstrapped samples T_{B_i} from the current submatrix (sampling rows with replacements) and on each of those we call clusterInstances, thus leading to k learned SPNs, S_{B_i}. We then build the resulting network as a sum node as the parent of all the other sum nodes representing the roots of the networks S_{B_i}. We introduce $1/k$ as the weight for these nodes, as in usual parameter estimation by bagging[1]. The bagged SPN would result in this more robust estimation: $\hat{S} = \sum_{i=1}^{k} \frac{1}{k} S_{B_i}$. Note that while this approach is theoretically applicable at each stage of the algorithm before learning the mixture components, it will eventually build a network with an exponential number of edges, making inference unfeasible in practice. Hence we

---

[1] We have experimented with weights proportional to the likelihood score obtained by each bootstrapped component, however the gain over uniform ones is negligible.

Algorithm 4. baggingSPN(T, V, α, m, k)

1: **Input:** a set of row instances T over a set of column features V; m: minimum number of instances to split; α: Laplace smoothing parameter
2: **Output:** an SPN S encoding a pdf over V learned from T
3: $\{T_{B_i}\}_{i=1}^{k} \leftarrow$ bootstrapSamples(T, k)
4: **return** $S \leftarrow \sum_{i=1}^{k} \frac{1}{k}$LearnSPN($T_{B_i}, V, \alpha, m$)

---

limit this step to the first recursive call, where the split on rows is mandatory and it is more likely to improve the model estimation globally. The procedure, which we call baggingSPN is shown in Algorithm 4.

Our approach is similar to the one used in the discriminative framework for tasks like regression, when model averaging is applied over a set of bootstrapped weak learners, e.g. forests of regression trees [12]. In our case it is worth noting that the resulting architecture is still an SPN: by pruning the roots of the SPNs S_{B_i} as shown in the previous section, we end up with a single sum node averaging local and possibly perturbed distributions over the same scopes; as a matter of fact the validity of the network is preserved as well as the ability to use it as a generative model. Inference is tractable as long as the number of edges remains polynomial. Indeed, the newly introduced complexity grows linearly in the number of the bootstrapped components, k. To limit the growth of the number of edges and parameters in the network, it would be possible to merge identical subtrees to compact the model; or, by separating the bootstrapped SPNs aggregation from their learning phases, one could use a more informative procedure, i.e. a L1-regularized gradient descent, to select only the components consistently contributing to the likelihood increase.

4   Related Works

The first SPN structure learner has been proposed in [8]. It splits the data matrix top-down, however splits and their meaning are different from LearnSPN: instances are clustered only once, at the start, and feature clustering is achieved by K-Means, which is not able to locate independencies correctly. Arbitrary sum nodes are then introduced as product nodes parents. The EM algorithm is needed to learn the network weights. On the other hand, [19] proceeds bottom-up by selecting the features to merge iteratively with a Bayesian-Dirichlet test, then sum nodes and their parameters are learned by maximizing the MI through the Information Bottleneck method, however considering only the best likelihood scoring features to reduce the high complexity of the approach. Like in [8] the learned SPNs are not tree structured, while the overall approach is still greedy.

The recent ID-SPN algorithm [26], by exploiting both indirect and direct interactions among variables, unifies works on learning SPNs through top-down clustering with works on learning tractable Markov networks [16]. ID-SPN learns Sum-Products of Arithmetic Circuits (SPACs) models which consist of sum and product nodes as inner nodes, and ACs as leaf nodes. ID-SPN consistently outperforms the previous SPN and several other PGM structure learners [26]. AC leaves can potentially better approximate more complex distributions than our Chow-Liu tree leaves, however, at the cost of increasing structural complexity. Note that our approach differs from ID-SPN not only on the choice of tractable distributions to model leaves, but also on governing the greedy process. Starting from a single AC, ID-SPN splits each leaf into two new ACs only if this improves directly the likelihood on data, while we let the search be guided indirectly by the splitting procedures, estimating leaf distributions only at the end. This, combined with the high complexity of the base algorithm to learn ACs [16], makes ID-SPN very slow in practice.

Another search approach based on directly maximizing the likelihood is found in [20] where less expressive SPNs, named *Selective SPNs*, allow the efficient optimization of a closed form of the likelihood function by stochastic local search.

On the side of mixtures of generative models, a very competitive structure learner algorithm is MT [18]. MT learns a distribution of the form: $Q(\mathbf{x}) = \sum_{i=1}^{k} \lambda_i T_i(\mathbf{x})$, where the distributions T_i, learned with the Chow-Liu algorithm [7], are the mixture components and $\lambda_i \geq 0, \sum_{i=1}^{k} \lambda_i = 1$ are their coefficients. [18] finds the best components and weights as (local) likelihood maxima by using EM, with k fixed in advance. MT is reported as the second most accurate model after ID-SPN in [26]. The hybrid SPNs we propose can express more latent interactions than a single mixture of Chow-Liu trees, moreover, they allow leaf scopes to consist of single random variables or subsets of the whole scope. Hybrid architectures like ours are referred to as *Generalized SPNs* in [22].

While leading to potentially long learning times, the EM algorithm is still the preferred choice to learn mixtures of generative models, like in the recent case of mixtures of *Cutset Networks* (CNets) [24]. CNets are weighted probabilistic model trees with Chow-Liu trees as leaves. Their inner nodes are OR nodes conditioning on a variable, thus they do not represent latent features and despite the depth of the tree they are shallow architectures. Similar works, applying bagging to a generative scenario like ours, are those from [25] and [2]. In the former, bagging is used to regularize the variant of EM proposed to determine the number of components in boosting a mixture of density estimators. In the latter, again applied to density estimation, a perturbing strategy derived from bootstrapped or totally random samples lead to more robust mixtures of tree distributions. A further work by the same authors relaxes the Chow-Liu algorithm on random subspaces to further differentiate mixture components [3].

5   Experiments

To empirically evaluate our enhancements, namely Binary row clustering, Tree distributions as leaf nodes, and model averaging through Bagging, we consider these algorithmic variations of LearnSPN: SPN-B, implementing the first one,

SPN-BT adding to that the second one, SPN-BB including the clustering fix and bagging, and SPN-BTB, which incorporates all three of them.

For the original LearnSPN we used the publicly available Java implementation[2] from [10]. For the aforementioned ID-SPN, we used the implementation from the Libra toolkit [14]. However, since we were not able to reproduce the results shown in [26], we used the best learned models scores, kindly provided by the authors (which we thank). As a third competitor, we used MT, whose implementation can also be found in the Libra package. We implemented all our variations in Python[3], taking advantage of the scikit-learn version of EM used for Gaussian Mixture Models[4] for our variant of the clusterInstances procedure.

Table 1. Datasets used and their statistics.

	$|V|$	$|T_{train}|$	$|T_{val}|$	$|T_{test}|$		$|V|$	$|T_{train}|$	$|T_{val}|$	$|T_{test}|$
NLTCS	16	16181	2157	3236	DNA	180	1600	400	1186
MSNBC	17	291326	38843	58265	Kosarek	190	33375	4450	6675
KDDCup2k	65	180092	19907	34955	MSWeb	294	29441	3270	5000
Plants	69	17412	2321	3482	Book	500	8700	1159	1739
Audio	100	15000	2000	3000	EachMovie	500	4525	1002	591
Jester	100	9000	1000	4116	WebKB	839	2803	558	838
Netflix	100	15000	2000	3000	Reuters-52	889	6532	1028	1540
Accidents	111	12758	1700	2551	BBC	1058	1670	225	330
Retail	135	22041	2938	4408	Ad	1556	2461	327	491
Pumsb-star	163	12262	1635	2452					

We evaluated the inferred networks comparing both the learned structures quality and their likelihood scores on an array of 19 datasets, firstly introduced in [15] and [11], now a standard to compare graphical model structure learning algorithms [10,15,16,26]. They are binarized versions of datasets from tasks like classification, frequent itemset mining, recommendation. The training, validation and test splits statistics are reported in Table 1. Their features range from 16 to 1556, while training instances from 1670 to 291326, making them very suitable to evaluate how we improve LearnSPN under our dimensions on different scenarios.

Our first experimental objective is to verify whether SPN-B and SPN-BT do learn deeper and more compact structures compared to LearnSPN, and to do this we measure the number of edges, layers and parameters of each learned model. Secondly, we compare all algorithms in terms of average test data likelihoods to verify if structural improvements damage likelihood scores and how much bagging, on the other hand, improves them. ID-SPN does not appear in the first confrontation since we were provided only the model scores.

---

[2] http://spn.cs.washington.edu/learnspn/
[3] Code is available at http://www.di.uniba.it/~vergari/code/spyn.html
[4] http://goo.gl/HNYjfZ

5.1   Experimental Design

For each algorithm, we selected the best parameter configurations based on the average validation log-likelihood scores, then evaluated such models on the test sets. For LearnSPN we performed an exhaustive grid search for $\rho \in \{5, 10, 15, 20\}$, $\lambda \in \{0.2, 0.4, 0.6, 0.8\}$, $m \in \{10, 50, 100, 500\}$ and $\alpha \in \{0.1, 0.2, 0.5, 1, 2\}$, leaving EM restarts to the default 4. For SPN-B and SPN-BT we use the same parameter space for ρ, m and α, to make the comparison as fair as possible. We leave all the default parameters for scikit-learn's EM unchanged, with the exception of the number of restarts which we set to 3. Note that in [10], α and m were not considered hyperparameters, we are introducing them to show effective regularization in the form of early stopping achieved when not naive factorizations are employed. For ID-SPN, please refer to the original article for its complete experimental settings; here we point out that its overparametrization, which is likely a key factor in its performance, required a uniform random search in the parameter space, since an exhaustive one would have been unfeasible. Concerning MT, we learned a number of components k from 2 up to 30, with increments of 2, rerunning each experiment five times to mitigate EM random initializations.

To reduce the complexity of the experiments for both SPN-BB and SPN-BTB, we did not employed a grid search, but we used the best validation values for ρ, m and α as previously found by SPN-B and SPN-BT respectively. We learned $k = 50$ bootstrapped components for each of the two, then, we composed the models by adding one component at a time, selecting as the resulting composite model the one with the best validation score.

5.2   Results and Discussion

In Table 2 are reported the edge, the layers and the parameters statistics for the best models learned by LearnSPN, SPN-B and SPN-BT. As it can be seen, the introduction of the limited Binary row clustering always makes the networks deeper and significantly reduces the number of edges for both variants, except for SPN-B on Netflix. It is worth noting that on datasets like BBC, Reuters-52 and MSWeb, while the number of parameter increases, the networks grow deeper and not wider, preventing edge explosions. When SPN-BT yields smaller networks than SPN-B, e.g. on Plants, Audio, Netflix, Kosarek and Book, the gain is huge in terms of edges and parameters saved, while considerable depths are preserved. On the other hand, on cases like NLTCS, MSNBC, KDDCup2K, Jester and BBC no structural improvement is observed if we add to the count the number of edges in the Chow-Liu trees. Table 3 reports the average test log likelihoods; the scores in bold are significantly better than all others under a Wilcoxon signed rank test with p-value of 0.05. Figure 2a shows the total number of times one algorithm wins under this same test. SPN-B proves no worst than the original algorithm on all but two datasets, scoring even six victories, the same value achieved by ID-SPN. The addition of the Chow-Liu trees variant in SPN-BT improves SPN-B scores on 13 datasets, confirming the ability of trees to better capture local dependencies at low levels.

Table 2. Structural quality results for the best validation models for LearnSPN, SPN-B and SPN-BT as the number of edges, layers and parameters. For SPN-BT are reported the number of edges considering those in the Chow-Liu leaves and without considering them (in parenthesis).

	# edges			# layers			# params		
	LearnSPN	SPN-B	SPN-BT	LearnSPN	SPN-B	SPN-BT	LearnSPN	SPN-B	SPN-BT
NLTCS	7509	1133	1133 (1125)	4	15	15	476	275	275
MSNBC	22350	4258	4258 (3996)	4	21	21	1680	1071	1071
KDDCup2k	44544	4272	4272 (4166)	4	25	25	753	760	760
Plants	55668	13720	5948 (1840)	6	23	20	3819	2397	490
Audio	70036	16421	4059 (478)	8	23	15	3389	2631	105
Jester	36528	10793	10793 (8587)	4	19	19	563	1932	1932
Netflix	17742	25009	4132 (203)	4	25	14	1499	4070	82
Accidents	48654	12367	10547 (6687)	6	25	26	5390	2708	1977
Retail	7487	1188	1188 (1153)	4	23	23	171	224	224
Pumsb-star	15247	12800	9984 (6175)	8	25	23	1911	2662	1680
DNA	17602	3178	4225 (2746)	6	13	12	947	884	1113
Kosarek	7993	8174	2216 (1311)	6	27	21	781	1462	242
MSWeb	17339	9116	7568 (6797)	6	27	34	620	1672	1446
Book	42491	9917	3503 (3485)	4	15	13	1176	1351	430
EachMovie	52693	20756	20756 (17861)	8	23	23	1010	2637	2637
WebKB	52498	45620	8796 (6874)	8	23	16	1712	6087	1128
Reuters-52	307113	77336	77336 (59197)	12	31	31	3641	8968	8968
BBC	318313	63723	63723 (41247)	16	27	27	1134	6147	6147
Ad	70056	23606	23606 (20079)	16	59	59	1060	1222	1222

Table 3. Average test log likelihoods for all algorithms.

	LearnSPN	SPN-B	SPN-BT	ID SPN	SPN-BB	SPN-BTB	MT
NLTCS	-6.110	-6.048	-6.048	**-5.998**	-6.014	-6.014	-6.008
MSNBC	-6.099	-6.040	-6.039	-6.040	**-6.032**	-6.033	-6.076
KDDCup2k	-2.185	-2.141	-2.141	-2.134	-2.122	**-2.121**	-2.135
Plants	-12.878	-12.813	-12.683	-12.537	-12.167	**-12.089**	-12.926
Audio	-40.360	-40.571	-40.484	-39.794	-39.685	**-39.616**	-40.142
Jester	-53.300	-53.537	-53.546	**-52.858**	-52.873	-53.600	-53.057
Netflix	-57.191	-57.730	-57.450	**-56.355**	-56.610	**-56.371**	-56.706
Accidents	-30.490	-29.342	-29.265	**-26.982**	-28.510	-28.351	-29.692
Retail	-11.029	-10.944	10.942	**-10.846**	-10.858	-10.858	**-10.836**
Pumsb-star	-24.743	-23.315	-23.077	**-22.405**	-22.866	-22.664	-23.702
DNA	-80.982	-81.913	-81.840	-81.211	-80.730	**-80.068**	-85.568
Kosarek	-10.894	-10.719	-10.685	-10.599	-10.690	**-10.578**	-10.615
MSWeb	-10.108	-9.833	-9.838	-9.726	-9.630	**-9.614**	-9.819
Book	-34.969	-34.306	-34.280	-34.136	-34.366	**-33.818**	-34.694
EachMovie	-52.615	-51.368	-51.388	-51.512	**-50.263**	-50.414	-54.513
WebKB	-158.164	-154.283	-153.911	-151.838	-151.341	**-149.851**	-157.001
Reuters-52	-85.414	-83.349	-83.361	-83.346	**-81.544**	-81.587	-86.531
BBC	-249.466	-247.301	-247.254	-248.929	**-226.359**	-226.560	-259.962
Ad	-19.760	-16.234	-15.885	-19.053	-13.785	**-13.595**	-16.012

To consistently beat ID-SPN one has to give up simpler structures and make more robust ones with SPN-BB and SPN-BTB, which score 11 and 13 wins respectively. The likelihoods obtained on the datasets with fewer instances and many more features, are much higher than the ones from ID-SPN and MT. This proves how bagging can be effectively embedded as a very cheap way to strengthen sum node mixtures. In our experiments for SPN-BB and SPN-BTB, we found a monotonic behavior, resulting in $k = 50$ as the best validation parameter, while

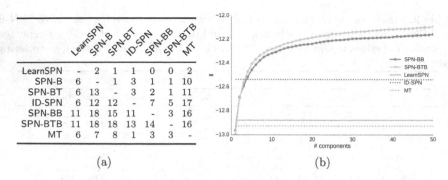

	LearnSPN	SPN-B	SPN-BT	ID-SPN	SPN-BB	SPN-BTB	MT
LearnSPN	-	2	1	1	0	0	2
SPN-B	6	-	1	3	1	1	10
SPN-BT	6	13	-	3	2	1	11
ID-SPN	6	12	12	-	7	5	17
SPN-BB	11	18	15	11	-	3	16
SPN-BTB	11	18	18	13	14	-	16
MT	6	7	8	1	3	3	-

(a)   (b)

Fig. 2. Numbers of statistically significant victories (Wilcoxon signed rank test, p-value= 0.05) for the algorithms on the rows compared to those on columns in 2a. Average test likelihood values (y-axis) on Plants for SPN-BB and SPN-BTB at the increase of k (x-axis), and the values from the best models from LearnSPN, ID-SPN and MT are reported in 2b.eps

for MT this value highly varies, implying that tuning is unavoidable. Trying to balance inference complexity and likelihood accuracy one can limit k for SPN-BB and SPN-BTB to smaller values. In Figure 2b we show an example of the test likelihood gain by increasing k for both algorithms on Plants, plotting the best value achieved by LearnSPN, ID-SPN and MT as a comparison. SPN-BB and SPN-BTB, with $k = 50$, score faster learning times than ID-SPN (run with default parameters) on all but four datasets, and sometimes even than MT which has less components, e.g. for Accidents, the times in seconds are 15472, 9198, 8280, 14073 for ID-SPN, SPN-BB, SPN-BTB and MT, respectively. Additional results including p-values, times, best parameter configurations on validation and the plots for the remaining datasets are availabe in the supplementary material[5].

6   Conclusions

We focused on enhancing LearnSPN, a state-of-the-art SPN structure learner, by proposing three algorithmic variations to improve the network quality in terms of the numbers of edges, layers and parameters on the one hand and the likelihood score on the other. We showed how limiting the number of node children while splitting yields simpler and deeper networks; how the introduction of Chow-Liu trees as multivariate leaf nodes leads to even more compact SPNs allowing an early interruption of the building process; and how embedding bagging into sum node mixtures can result in more robust models. An extensive empirical evaluation on standard datasets proved our enhancements to be effective, suggesting a number of future investigations: studying other ways of hybridizing SPNs with tractable multivariate distributions as building blocks; how to apply other

---

[5] http://www.di.uniba.it/~vergari/code/spyn.html

ensembling methods like arcing and boosting to sum node splits and the possible application of input corruption techniques to make this generative model even more robust. To balance structure compactness with likelihood gains we could try to prune or graft subtrees from different bootstrapped SPNs encoding exactly or approximately equal distributions. Furthermore, applying these ideas to other tractable models structure learning algorithms could also be an opportunity.

Acknowledgments. Work supported by the project PUGLIA@SERVICE (PON02 00563 3489339) financed by the Italian Ministry of University and Research (MIUR) and by the European Commission through the project MAESTRA, grant no. ICT-2013-612944. Experiments executed on the resources made available by two projects financed by the MIUR: ReCaS (PONa3_00052) and PRISMA (PON04a2_A).

References

1. Amer, M.R., Todorovic, S.: Sum-product networks for modeling activities with stochastic structure. In: 2012 IEEE Conference on (CVPR), pp. 1314–1321. IEEE (2012)
2. Ammar, S., Leray, P., Defourny, B., Wehenkel, L.: Probability density estimation by perturbing and combining tree structured markov networks. In: Sossai, C., Chemello, G. (eds.) ECSQARU 2009. LNCS, vol. 5590, pp. 156–167. Springer, Heidelberg (2009)
3. Ammar, S., Leray, P., Schnitzler, F., Wehenkel, L.: Sub-quadratic markov tree mixture learning based on randomizations of the Chow-Liu algorithm. In: Proceedings of the 5th European Workshop on Probabilistic Graphical Models, pp. 17–24 (2010)
4. Bach, F.R., Jordan, M.I.: Thin junction trees. In: Advances in Neural Information Processing Systems 14, pp. 569–576. MIT Press (2001)
5. Cheng, W., Kok, S., Pham, H.V., Chieu, H.L., Chai, K.M.A.: Language modeling with sum-product networks. In: 15th Annual Conference of the International Speech Communication Association, pp. 2098–2102 (2014)
6. Choi, M.J., Tan, V.Y.F., Anandkumar, A., Willsky, A.S.: Learning latent tree graphical models. Journal of Machine Learning Research **12**, 1771–1812 (2011)
7. Chow, C., Liu, C.: Approximating discrete probability distributions with dependence trees. IEEE Transactions on Information Theory **14**(3), 462–467 (1968)
8. Dennis, A., Ventura, D.: Learning the architecture of sum-product networks using clustering on varibles. In: Advances in Neural Information Processing Systems 25, pp. 2033–2041. Curran Associates, Inc. (2012)
9. Gens, R., Domingos, P.: Discriminative learning of sum-product networks. In: Advances in Neural Information Processing Systems 25, pp. 3239–3247. Curran Associates, Inc. (2012)
10. Gens, R., Domingos, P.: Learning the structure of sum-product networks. In: Proceedings of the 30th International Conference on Machine Learning, pp. 873–880. JMLR Workshop and Conference Proceedings (2013)
11. Haaren, J.V., Davis, J.: Markov network structure learning: A randomized feature generation approach. In: Proceedings of the 26th Conference on Artificial Intelligence. AAAI Press (2012)
12. Hastie, T., Tibshirani, R., Friedman, J.: The Elements of Statistical Learning. Springer (2009)

13. Koller, D., Friedman, N.: Probabilistic Graphical Models: Principles and Techniques. MIT Press (2009)
14. Lowd, D., Rooshenas, A.: The Libra Toolkit for Probabilistic Models. CoRR abs/1504.00110 (2015)
15. Lowd, D., Davis, J.: Learning markov network structure with decision trees. In: Proceedings of the 10th IEEE International Conference on Data Mining, pp. 334–343. IEEE Computer Society Press (2010)
16. Lowd, D., Rooshenas, A.: Learning markov networks with arithmetic circuits. In: Proceedings of the 16th International Conference on Artificial Intelligence and Statistics. JMLR Workshop Proceedings, vol. 31, pp. 406–414 (2013)
17. Martens, J., Medabalimi, V.: On the expressive efficiency of sum product networks. CoRR abs/1411.7717 (2014)
18. Meilă, M., Jordan, M.I.: Learning with mixtures of trees. Journal of Machine Learning Research 1, 1–48 (2000)
19. Peharz, R., Geiger, B.C., Pernkopf, F.: Greedy part-wise learning of sum-product networks. In: Blockeel, H., Kersting, K., Nijssen, S., Železný, F. (eds.) ECML PKDD 2013, Part II. LNCS, vol. 8189, pp. 612–627. Springer, Heidelberg (2013)
20. Peharz, R., Gens, R., Domingos, P.: Learning selective sum-product networks. In: Workshop on Learning Tractable Probabilistic Models. LTPM (2014)
21. Peharz, R., Kapeller, G., Mowlaee, P., Pernkopf, F.: Modeling speech with sum-product networks: Application to bandwidth extension. In: International Conference on Acoustics, Speech and Signal Processing, pp. 3699–3703. IEEE (2014)
22. Peharz, R., Tschiatschek, S., Pernkopf, F., Domingos, P.: On theoretical properties of sum-product networks. The Journal of Machine Learning Research (2015)
23. Poon, H., Domingos, P.: Sum-product network: a new deep architecture. In: NIPS 2010 Workshop on Deep Learning and Unsupervised Feature Learning (2011)
24. Rahman, T., Kothalkar, P., Gogate, V.: Cutset networks: a simple, tractable, and scalable approach for improving the accuracy of chow-liu trees. In: Calders, T., Esposito, F., Hüllermeier, E., Meo, R. (eds.) ECML PKDD 2014, Part II. LNCS, vol. 8725, pp. 630–645. Springer, Heidelberg (2014)
25. Ridgeway, G.: Looking for lumps: Boosting and bagging for density estimation. Computational Statistics & Data Analysis 38(4), 379–392 (2002)
26. Rooshenas, A., Lowd, D.: Learning sum-product networks with direct and indirect variable interactions. In: Proceedings of the 31st International Conference on Machine Learning, pp. 710–718. JMLR Workshop and Conference Proceedings (2014)
27. Roth, D.: On the hardness of approximate reasoning. Artificial Intelligence 82(1–2), 273–302 (1996)

Sparse Bayesian Recurrent Neural Networks

Sotirios P. Chatzis$^{(\boxtimes)}$

Department of Electrical Engineering, Computer Engineering and Informatics,
Cyprus University of Technology, 33 Saripolou Str., 3036 Limassol, Cyprus
sotirios.chatzis@cut.ac.cy

Abstract. Recurrent neural networks (RNNs) have recently gained renewed attention from the machine learning community as effective methods for modeling variable-length sequences. Language modeling, handwriting recognition, and speech recognition are only few of the application domains where RNN-based models have achieved the state-of-the-art performance currently reported in the literature. Typically, RNN architectures utilize simple linear, logistic, or softmax output layers to perform data modeling and prediction generation. In this work, for the first time in the literature, we consider using a *sparse Bayesian* regression or classification model as the output layer of RNNs, inspired from the automatic relevance determination (ARD) technique. The notion of ARD is to continually create new components while detecting when a component starts to overfit, where overfit manifests itself as a *precision hyperparameter posterior* tending to infinity. This way, our method manages to train sparse RNN models, where the number of effective ("active") recurrently connected hidden units is selected in a data-driven fashion, as part of the model inference procedure. We develop efficient and scalable training algorithms for our model under the stochastic variational inference paradigm, and derive elegant predictive density expressions with computational costs comparable to conventional RNN formulations. We evaluate our approach considering its application to challenging tasks dealing with both regression and classification problems, and exhibit its favorable performance over the state-of-the-art.

1 Introduction

Many naturally occurring phenomena such as music, speech, or human motion are inherently sequential. As a consequence, the problem of sequential data modeling is an important area of machine learning research. Recurrent neural networks (RNNs) [22] are among the most powerful models for sequential data modeling. As shown in [12], RNNs possess the desirable property of being universal approximations, as they are capable of representing any measurable sequence to sequence mapping to arbitrary accuracy. RNNs incorporate an internal memory module designed with the goal to summarize the entire sequence history in the form of high dimensional *state vector* representations. This architectural selection allows for RNNs to model and represent long-term dependencies in the observed data, which is a crucial merit in the context of sequential data modeling applications.

© Springer International Publishing Switzerland 2015
A. Appice et al. (Eds.): ECML PKDD 2015, Part II, LNAI 9285, pp. 359–372, 2015.
DOI: 10.1007/978-3-319-23525-7_22

A major challenge RNN-based architectures are confronted with concerns the fact that it is often the case that gradient-based optimization results in error signals either blowing up or decaying exponentially for events many time steps apart, rendering RNN training largely impractical [6,19]. A great deal of research work has been devoted to the amelioration of these difficulties, usually referred to as the problem of *vanishing and exploding gradients*. One first attempt towards this end consisted in coming up with special architectures, robust to the vanishing and exploding gradients problem. Long short-term memory (LSTM) [13] networks constitute the most successful architecture developed for this purpose, having been shown to yield the state-of-the-art performance in speech and handwriting recognition tasks [11]. In a different vein, [15] proposed the echo-state network (ESN) architecture, which consists in completely abandoning gradient-based training of the recurrent connection weights (which gives rise to the vanishing and exploding gradients problem). As such, ESNs solely rely on sensible initializations of the recurrent connection weights, and limit training to the connection weights of the output (*readout*) layer of the network. Finally, a recent breakthrough in the literature of RNNs has been accomplished in the landmark publication of [17], where it was shown that even standard RNN architectures can be successfully trained with the right optimization method. While a sophisticated Hessian-free optimizer for RNNs was developed therein, further research has shown that carefully designed conventional first-order methods can find optima of similar or slightly worse quality in the context of RNN training [24].

Despite these advances in the field of RNN research, a problem that has not been tackled by the machine learning community concerns data-driven model selection. The problem of model selection consists in coming up with model treatments allowing for an RNN model to automatically infer the number of necessary hidden units, as part of its training procedure. Our work constitutes the first attempt towards addressing these shortcomings. To achieve our goals, we introduce the concept of training RNN models under a Bayesian inferential procedure. Specifically, we consider imposing appropriate *sparsity-promoting prior distributions on the output connection weights* of RNN models. Under this construction, we essentially give rise to a Bayesian inference procedure for RNNs that yields sparse posterior distributions over the output connection weights, and associated predictive posteriors that characterize the output variables of the model. Under a different regard, our approach can be viewed as introducing a *sparse Bayesian* regression or classification model as the *output layer* of RNNs, resulting in a sparse *Bayesian treatment* of the whole RNN architecture.

Sparsity in the context of our model is induced by adopting a prior model configuration inspired from an inference technique widely known as automatic relevance determination (ARD) [10]. The notion of ARD is to continually create new model components while detecting when a component starts to overfit. Overfit manifests itself as a *precision hyperparameter posterior* tending to infinity, indicating that a single data value is being modeled by the component. In the case of the proposed model, the ARD mechanism is implemented by imposing an appropriate hierarchical prior over the weights of the output layer connections, which results in an efficient mechanism for *automatically inferring the effective number* of (*"active"*) recurrently connected hidden units, in a data-driven fashion. We derive an efficient and scalable training algorithm for our model under the *stochastic variational* inference (SVI) paradigm [14], exploiting the most recent advances on gradient-based RNN training algorithms. We dub our approach sparse Bayesian RNN (SB-RNN).

The layout of the paper is as follows. In Section 2, we briefly introduce the concept of RNNs, and discuss modern RNN training algorithms that obviate the vanishing and exploding gradients problem. In Section 3, we present our method and derive efficient model training and inference algorithms. In Section 4, we perform an extensive experimental evaluation of our approach considering several challenging benchmark tasks; we compare the obtained performance of our approach to related state-of-the-art approaches. Finally, in Section 5 we summarize our results, discussing the shortcomings and advantages of the proposed model, and conclude this work.

2   Recurrent Neural Networks

RNNs are neural network architectures designed for modeling sequential data with long temporal dynamics. RNNs operate by simulating a discrete-time dynamical system presented with M-dimensional inputs $\{x_t\}_{t=1}^T$, with corresponding N-dimensional outputs $\{y_t\}_{t=1}^T$, where T is the length (time-duration) of the observed sequences. The postulated dynamical system is defined by

$$y_t = \phi_0(\boldsymbol{V}\boldsymbol{h}_t) \tag{1}$$

where ϕ_0 is the activation function of the output units, $\boldsymbol{V} \in \mathbb{R}^{N \times H}$ is the parameters (weights) matrix of the connections of the output layer of the model, and $\boldsymbol{h}_t \in \mathbb{R}^H$ is the hidden state vector of the model. In essence, $\boldsymbol{h}_t$ is the vector of the activations of the hidden units of the network, which encodes the history of observations presented to the system up to time t in the form of a high-dimensional data point representation. Typically, the expressions of the hidden state vectors of the postulated dynamical system are considered to be given by the following expression:

$$\boldsymbol{h}_t = \begin{cases} \phi_h(\boldsymbol{W}\boldsymbol{h}_{t-1} + \boldsymbol{\Omega}\boldsymbol{x}_t), t > 0 \\ \boldsymbol{0}, t = 0 \end{cases} \tag{2}$$

In Eq. (2), ϕ_h is the activation function of the hidden (recurrently connected) units that capture the temporal dynamics in the modeled data, $\boldsymbol{W} \in \mathbb{R}^{H \times H}$ is the recurrent connections weights matrix of the model, and $\boldsymbol{U} \in \mathbb{R}^{H \times M}$ is the input connections weights matrix. Regarding selection of the hidden and output unit activation functions, typically a saturating nonlinear function is used, such as a logistic sigmoid function or a hyperbolic tangent function.

RNN model training, i.e. optimal estimation of the parameter values of the model given a set comprising S training sequences, $D = \{(\boldsymbol{x}_1^s, \boldsymbol{y}_1^s), \ldots, (\boldsymbol{x}_{T_s}^s, \boldsymbol{y}_{T_s}^s)\}_{s=1}^S$, can be performed by minimization of the cost function

$$J(\Theta) = \frac{1}{S} \sum_{s=1}^S \sum_{t=1}^{T_s} d(\boldsymbol{y}_t^s, \phi_0(\boldsymbol{V}\boldsymbol{h}_t^s)) \tag{3}$$

where Θ is the parameters set of the RNN, $\Theta = \{\boldsymbol{V}, \boldsymbol{W}, \boldsymbol{U}\}$, $\boldsymbol{h}_t^s = \phi_h(\boldsymbol{W}\boldsymbol{h}_{t-1}^s + \boldsymbol{U}\boldsymbol{x}_t^s)$, and $d(\boldsymbol{a}, \boldsymbol{b})$ is a suitable divergence measure, appropriate for the learning problem at hand (e.g., a Euclidean distance function when dealing with regression problems, and a cross-entropy function in cases of classification problems).

To effect the minimization task of the RNN objective function $J(\Theta)$ in a scalable manner, stochastic gradient descent (SGD) algorithms are typically used, with the gradient of the cost function in Eq. (3) computed by means of backpropagation through time (BPTT) [22]. However, as we already discussed in the introduction of this paper, conventional BPTT-based RNN training is often confronted with the problem of the obtained error signals either blowing up or decaying exponentially for events many time steps apart, with detrimental effects to the outcome of the model training algorithm. The effort to mitigate these hurdles has recently triggered a significant corpus of new research in the field of RNN methods. Among this large corpus of works, we here focus on a very recent research result showing that first-order optimizers can, indeed, avoid the problem of exploding and vanishing gradients by: (i) performing an appropriate initialization of the model weights based on the principles of ESNs; and (ii) using Nesterov's Accelerated Gradient (NAG) [18] to perform model training instead of conventional SGD [24].

As discussed in Section 1, ESNs are RNN-based architectures where the recurrent connections weights matrix $\boldsymbol{W}$ is not trained but merely properly initialized. Specifically, ESN theory stipulates that the initialization of the matrix $\boldsymbol{W}$ should be performed in a way ensuring that the *largest absolute eigenvalue of the determinant* $|\boldsymbol{W}|$ *(spectral radius)* be close to one. This way, the dynamics of the network can be shown to become oscillatory and chaotic, allowing it to generate responses that are varied for different input histories. At the same time though, the gradients are no longer exploding (and if they do explode, then only "slightly so"), so learning may be possible for even the hardest sequential data modeling problems that conventional RNNs fail to address [15].

On the other hand, NAG has been the subject of much recent attention by the convex optimization community [9,16]. Like SGD, NAG is a first-order optimization method with better convergence rate guarantee than conventional SGD. NAG has been shown to be closely related to classical momentum-based SGD variants (e.g., [20]), with the only difference lying in the precise update expression of the velocity vector of the algorithm. These differences, although slight, can be of major significance to the asymptotic rates of local convergence of the optimization algorithm, as discussed in [24].

Inspired from these merits, in developing the training algorithm of our proposed model we shall rely on both performing an ESN-inspired initialization, and using NAG instead of mainstream SGD optimization algorithms. We shall introduce our model and elaborate on our selected training strategies in the following section.

3 Proposed Approach

We differentiate SB-RNN model formulation between regression and (multiclass) classification tasks, to allow for better handling the intricacies of each type of problems. In the following, we elaborate on our modeling strategies in both cases, and derive efficient training algorithms under the SVI paradigm.

3.1 Regression SB-RNN

Let us consider that the modeled output variables y_t are N-dimensional real vectors, i.e. $y_t \in \mathbb{R}^N$. In this case, we define a likelihood function for our model of the form

$$p(y_t|x_t; V, \beta) = \mathcal{N}(y_t|Vh_t, \beta^{-1}I) \tag{4}$$

where β is the precision (inverse variance) of a simple white noise model adopted in the context of our method, and h_t is the *state vector* of our model that encodes the history of past observations $\{x_\tau\}_{\tau=1}^{t-1}$. In essence, h_t consists the vector of the activations of the (recurrently connected) hidden units of our model; we consider that it is expressed by a discrete-time dynamical system of the form (2). In the same vein, V can essentially be perceived as the weight matrix of the output connections of our model.

Further, we introduce an appropriate prior distribution over the weights matrix V; we consider a zero-mean Gaussian prior of the form

$$p(V|\alpha) = \prod_{n=1}^{N} \prod_{u=1}^{H} \mathcal{N}(v_{nu}|0, \alpha_u^{-1}) \tag{5}$$

where α_u is the precision of the weights pertaining to the uth hidden unit, $\{v_{nu}\}_{n=1}^{N}$, and $\alpha = [\alpha_u]_{u=1}^{H}$. Finally, we impose a Gamma hyper-prior over the precision hyperparameters α_u, yielding

$$p(\alpha_u) = \mathcal{G}(\alpha_u|\eta_1, \eta_2) \tag{6}$$

This hierarchical prior configuration of our model essentially gives rise to an ARD mechanism that allows for data-driven inference over the number of hidden units, H. As we have already discussed, the notion of ARD is to continually create new components while detecting when a component starts to overfit. Overfit manifests itself as a *precision hyperparameter posterior* tending to infinity, indicating that a single data value is being modeled by the component. Hence, in the case of our SB-RNN model, the ARD mechanism is implemented by imposing a hierarchical prior over the output weights matrix V, to discourage large weight values, with the width of each prior being controlled by a *Gamma distributed precision hyperparameter*, α_u, as illustrated in Eqs. (5)-(6). If one of these precisions tends to infinity, $\alpha_u \to \infty$, then the outgoing weights will have to be very close to zero in order to maintain a high likelihood under this prior. This in turn leads the model to ignore the likelihood contribution of the corresponding hidden unit, which is effectively 'switched off' of the model.

We underline that this approach is substantially different from dropout [5,23], where, on each iteration of the training algorithm, *different* hidden units are *randomly* ignored, while *all* hidden units are used to perform prediction.

Model Training. To perform training of our model in a way scalable to massive datasets, we resort to the SVI paradigm [14]. SVI is an iterative stochastic optimization algorithm for mean-field variational inference that approximates the posterior distribution of a probabilistic model, and can handle massive datasets of observations. Indeed, SVI renders Bayesian inference scalable to massive datasets by splitting the observed data into small batches, and letting the inference algorithm operate only on one batch on each iteration.

SVI yields a lower bound to the log-evidence of the treated model (evidence lower bound, ELBO) expressed as a function of an approximate (variational) posterior distribution it seeks to optimally determine. Then, inference for a treated model consists in maximizing the corresponding ELBO over the variational posterior and the estimates of the associated hyper-parameters. For this purpose, SVI computes on each iteration a set of noisy estimates of the *natural* gradient [2] of the model's ELBO, and uses them in the context of a stochastic optimization scheme.

In the following, we denote as $\langle \cdot \rangle$ the posterior expectation of a quantity; the analytical expressions of these posteriors can be found in the Appendix. Let us consider that the training set $D = \{(\boldsymbol{x}_1^s, \boldsymbol{y}_1^s), \ldots, (\boldsymbol{x}_{T_s}^s, \boldsymbol{y}_{T_s}^s)\}_{s=1}^S$ comprises S sequences, and that we split this dataset into batches comprising S_b sequences each. Then, the SVI algorithm for the proposed SB-RNN regression model yields the following posterior over the output weights matrix $\boldsymbol{V}$:

$$q(\boldsymbol{V}) = \prod_{n=1}^N \mathcal{N}(\boldsymbol{v}_n | \bar{\boldsymbol{v}}_n, \bar{\boldsymbol{\Sigma}}) \tag{7}$$

where the posterior hyperparameters $\bar{\boldsymbol{v}}_n$ and $\bar{\boldsymbol{\Sigma}}$ are updated according to

$$\bar{\boldsymbol{\Sigma}}^{-1} \leftarrow (1 - \rho_k)\bar{\boldsymbol{\Sigma}}^{-1} + \rho_k \left[\langle \boldsymbol{A} \rangle + (\beta \tilde{\boldsymbol{H}}^T \tilde{\boldsymbol{H}}) \frac{S}{S_b} \right] \tag{8}$$

$$\bar{\boldsymbol{v}}_n \leftarrow (1 - \rho_k)\bar{\boldsymbol{v}}_n + \rho_k \frac{S}{S_b} \beta \bar{\boldsymbol{\Sigma}} \tilde{\boldsymbol{H}}^T \tilde{\boldsymbol{y}}_n \tag{9}$$

where $\langle \boldsymbol{A} \rangle = \mathrm{diag}(\langle \boldsymbol{\alpha} \rangle)$. In Eqs. (8)-(9), $\tilde{\boldsymbol{H}}$ denotes the matrix of the network state vectors pertaining to the sequences included in the current batch, while $\tilde{\boldsymbol{y}}_n$ denotes the (target) values of the nth model output pertaining to the sequences included in the current batch. On the other hand, ρ_k is the *learning rate* of the developed stochastic updating algorithm on the current (say, kth) iteration. Following common practice (e.g., [14]), in this work the learning rates ρ_k are updated on each algorithm iteration according to the rule

$$\rho_k = (k + \kappa)^{-f} \tag{10}$$

where the delay κ satisfies $\kappa \geq 0$, and the forgetting rate f satisfies $f \in (0.5, 1]$.

Further, the precision hyperparameters $\boldsymbol{\alpha}$ yield the hyperposteriors:

$$q(\boldsymbol{\alpha}) = \prod \mathcal{G}(\alpha_u | \bar{\eta}_{1u}, \bar{\eta}_{2u}) \tag{11}$$

where

$$\bar{\eta}_{1u} = \eta_1 + N \frac{S}{2} \tag{12}$$

$$\bar{\eta}_{2u} \leftarrow (1 - \rho_k)\bar{\eta}_{2u} + \rho_k \left[\eta_2 + \frac{S}{2S_b} \sum_{n=1}^{N} \langle v_{nu}^2 \rangle \right] \tag{13}$$

Interestingly, note that the updates of $\bar{\eta}_{1u}$ do not depend on the training data points of each batch; as such, the value of $\bar{\eta}_{1u}$ need not be updated on each algorithm iteration.

Finally, the input weights matrix $\boldsymbol{\Omega}$ and the recurrent connection weights matrix $\boldsymbol{W}$ of our model are updated as model hyperparameters, yielding point-estimates. This is effected by optimization of the ELBO of our model by application of a NAG-type optimization procedure, yielding:

$$\boldsymbol{W} \leftarrow (1 - \rho_k)\boldsymbol{W} + \rho_k \beta \delta_{\boldsymbol{W}} + \boldsymbol{\mu}_{\boldsymbol{W}} \tag{14}$$

$$\boldsymbol{\Omega} \leftarrow (1 - \rho_k)\boldsymbol{\Omega} + \rho_k \beta \delta_{\boldsymbol{\Omega}} + \boldsymbol{\mu}_{\boldsymbol{\Omega}} \tag{15}$$

In these equations, $\delta_{\boldsymbol{W}}$ and $\delta_{\boldsymbol{\Omega}}$ are the updates of the weights matrices $\boldsymbol{W}$ and $\boldsymbol{\Omega}$ obtained by application of *conventional BPTT [22]*, by setting the value of $\boldsymbol{V}$ equal to its posterior expectation $\bar{\boldsymbol{V}} = [\bar{v}_n]_{n=1}^{N}$. In addition, $\boldsymbol{\mu}_{\boldsymbol{W}}$ and $\boldsymbol{\mu}_{\boldsymbol{\Omega}}$ are the momentum-type terms introduced by adoption of the NAG optimization scheme (c.f. [24]), as discussed in Section 2. Initialization of the recurrent connection weights matrix $\boldsymbol{W}$ is performed by adopting the principles of ESN architectures, as also discussed in Section 2.

Predictive Density. Having found estimates of the model hyperparameters and parameter posteriors, we can now proceed to derive the expression of its predictive distribution over the output variables $\boldsymbol{y}_t$ for a new input $\boldsymbol{x}_t$, with corresponding observation history $\{\boldsymbol{x}_\tau\}_{\tau=1}^{t-1}$ and state vector $\boldsymbol{h}_t$. We have

$$q(\boldsymbol{y}_t | \{\boldsymbol{x}_\tau\}_{\tau=1}^{t}) = \mathcal{N}(\boldsymbol{y}_t | \bar{\boldsymbol{V}}\boldsymbol{h}_t, \sigma^2(\boldsymbol{x}_t)\boldsymbol{I}) \tag{16}$$

where

$$\sigma^2(\boldsymbol{x}_t) = \beta^{-1} + \boldsymbol{h}_t^T \bar{\boldsymbol{\Sigma}} \boldsymbol{h}_t \tag{17}$$

It is worthwhile to underline here a significant difference between our approach and conventional RNN formulations when it comes to prediction generation: Conventional RNNs only provide an estimate of the target (output variables); instead, our SB-RNN approach, apart from this estimate (taken as the mode $\hat{y} = \bar{V}h_t$ of the predictive distribution) does also yield a *predictive variance* estimate, given by $\sigma^2(x_t)$. The obtained predictive variance is in essence a measure of the confidence of the model in the obtained predictions $\hat{y}$, and can be utilized to provide error bars (or a reject option in safety-critical applications).

3.2   Classification SB-RNN

We now turn to the case of modeling a multiclass classification problem using our SB-RNN approach. Let us denote as $y_t \in \{0,1\}^N$ the output variables of the addressed problem. In this case, the nth component of vector y_t indicates whether class $n \in \{1, \ldots, N\}$ is on or off at time t. On this basis, to obtain a suitable construction for our model, we postulate a standard Multinomial likelihood assumption of the form:

$$p(y_t|x_t; V) = \prod_{n=1}^{N} \phi_0(v_n^T h_t)^{y_{tn}} \tag{18}$$

where v_n is the nth row of matrix V, and ϕ_0 is a *sigmoid* activation function. We impose the same hierarchical prior over the output weights matrix V as in the previously examined regression setting, given by Eqs. (5)-(6), to introduce the ARD mechanism into our model.

Model Training. To perform SB-RNN model training in the classification setting, we can again resort to the SVI inference paradigm. However, a major obstacle to the application of SVI in this setting concerns the fact that the imposed likelihood (18) does not yield a conjugate model formulation. This in turn prohibits obtaining closed-form analytical expressions for the (variational) posterior distribution over the weights matrix V. Specifically, we have

$$\log q(v_n) \propto \sum_t y_{tn} \log \phi_0(v_n^T h_t) + \langle \log p(v_n|\alpha) \rangle \ \forall n \tag{19}$$

To resolve these issues, in this work we resort to a Laplace approximation of the intractable posteriors $q(v_n)$. Laplace approximation consists in taking the second order Taylor expansion of $\log q(v_n)$ around its mode, resulting in the considered posterior distribution being conveniently approximated by a Gaussian. Specifically, our model yields

$$q(v_n) \approx \mathcal{N}(v_n|\bar{v}_n, \bar{\Sigma}_n) \tag{20}$$

where

$$\bar{\Sigma}_n^{-1} \leftarrow (1 - \rho_k)\bar{\Sigma}_n^{-1} + \rho_k \left[\langle A \rangle + (\tilde{H}^T \tilde{B}_n \tilde{H}) \frac{S}{S_b} \right] \tag{21}$$

$$\bar{v}_n \leftarrow (1 - \rho_k)\bar{v}_n + \rho_k \frac{S}{S_b} \bar{\Sigma}_n \tilde{H}^T \tilde{B}_n \tilde{y}_n \qquad (22)$$

and $\tilde{B}_n$ is the diagonal matrix of the set of quantities $\phi_0(v_n^T h_t)$ corresponding to the sequences in the current batch.

On the basis of the derivations (20)-(22), the updates of the hyperposteriors $q(\alpha)$, as well as the updates of the model weight matrices W and Ω, yield exactly the same expressions as in (11)-(15).

Predictive Density. Having obtained the training algorithm expressions of the SB-RNN model for the case of dealing with classification tasks, we now turn to deriving the corresponding predictive density expressions. Based on the preceding discussions, the predictive density of our model yields:

$$q(y_{tn} = 1 | \{x_\tau\}_{\tau=1}^t) \propto \langle \phi_0(v_n^T h_t) \rangle \qquad (23)$$

where the state vectors h_t are given by (2). Note that the posterior expectations in (23) cannot be computed analytically due to the nonlinear nature of the activation function $\phi_0(\cdot)$. For this reason, we resort to a Monte Carlo sampling-based approximation, yielding:

$$\langle \phi_0(v_n^T h_t) \rangle \approx \frac{1}{Z} \sum_{\zeta=1}^{Z} \phi_0((v_n^\zeta)^T h_t) \qquad (24)$$

where Z is the number of samples v_n^ζ drawn i.i.d. from the posterior $q(v_n)$, approximated by (20).

4 Experiments

We experimentally evaluate our approach in both regression and classification tasks. In all cases, we manually tune the hyperparameters of the learning rate schedule (10) for each dataset, as well as the hyperparameters of the momentum terms, similar to [24]. We developed our source codes in Python, using the Theano library [4][1]. We run our experiments on an Intel Xeon 2.5GHz Quad-Core server with 64GB RAM and an NVIDIA Tesla K40 GPU.

4.1 Human Motion Modeling

We begin by evaluating our method in a *regression* task. For this purpose, we use a publicly available benchmark, namely walking sequences from the CMU motion capture (MoCap) dataset [1]. The considered training sequences correspond to several different subjects included in the CMU MoCap database, following the experimental setup of [8]. After training, we use the obtained models to generate

---

[1] The source codes will be made available through our website, to allow for easier reproducibility of our results.

the human pose information in a different set of walking sequence videos, namely videos 35-03, 12-02, 16-21, 12-03, 07-01, 07-02, 08-01, and 08-02 of the same database[2]. The inputs presented to the evaluated algorithms are the positions of the tracked human joints, and their output is the predicted joint positions at a time point of interest. The dimensionality of the input space is equal to 62, similar to the output space.

To obtain some comparative results, apart from our method we also evaluate conventional RNNs trained as described in Section 2; we also cite the performance of ESNs, an ESN-driven formulation of Gaussian processes dubbed the ESGP method [8], and the Dynamic GP method [25]. In Table 1, we provide the RMSEs obtained by each one of the considered methods. These results correspond to optimal numbers of hidden units for the evaluated recurrent network-based methods; interestingly, this optimal model size turned out to be equal to 100 hidden units in all cases, as also observed in [8]. As we illustrate in Table 1, the SB-RNN method outperforms all the rest of the evaluated methods, both in average and in each single individual experimental case considered here.

Table 1. Human Motion Modeling: Obtained missing frames RMSEs.

Video ID	Dynamic GP	ESN	ESGP	RNN	SB-RNN
35-03	49.68	62.55	32.59	35.11	32.28
12-02	54.96	63.14	45.32	42.88	39.58
16-21	78.05	98.74	59.03	51.17	48.02
12-03	63.63	72.12	46.25	47.09	44.14
07-01	84.12	121.47	77.34	76.18	75.69
07-02	80.77	100.94	73.88	75.37	72.87
08-02	95.52	120.45	101.54	95.66	94.66
08-01	82.66	152.44	118.0	97.54	93.05
Average	73.67	98.98	69.24	65.13	62.54

4.2 Acoustic Novelty Detection

Further, in this experiment we perform evaluation of our approach in the context of a *classification* task, and under a setup that also allows for evaluating the *quality* of the obtained *predictive distributions*. Specifically, we consider the problem of novelty detection in acoustic signals. For the purposes of this experiment, we use a dataset composed of around three hours of recordings of a home environment, taken from the PASCAL CHiME speech separation and recognition challenge dataset [3]. Our dataset corresponds to a typical in-home scenario (a living room), recorded during different days and times; the inhabitants are two adults and two children that perform common actions, namely *talking, watching*

---

[2] All videos have been downsampled by a factor of 4, following the experimental setup of [8].

television, playing, and *eating.* On this basis, we use randomly chosen sequences to compose 100 minutes of background for training set, around 40 minutes for validation set, and another 30 minutes for test set. The validation and test sets were generated by randomly adding in the available sequences different kinds of sounds, namely *screams, alarms, falls,* and *fractures.* The total duration of *each novel* type is equal to 200 s.

Our experimental setup is the following: Initially, we train our model considering as input variables the auditory spectral features (ASF) computed by means of the short-time Fourier transformation (STFT); we use a frame size of 30 ms and a frame step of 10 ms. Each STFT yields the power spectrogram of the signal, which is eventually converted to the Mel-Frequency scale using a filter-bank with 26 triangular filters; we use a logarithmic representation of these features, to match the human perception of loudness. Finally, we also include the frame energy in our feature vectors, following standard practices in the literature.

Subsequently, we use the trained model to predict the class corresponding to each frame in the validation set. Since some frames correspond to *novel* classes which the model has not been trained to recognize, we are interested in examining how certain the model is for its predictions when these novel classes are actually the ones that appear in the data. Indeed, one would expect that the model should yield low predictive probability values for the *winner* class in cases where the actual class belongs to the set of *novel* ones. To examine whether this assumption does actually hold, we use the results obtained from our validation set to determine a *novelty threshold* for our model: if the predictive probability pertaining to the winner class is lower than this threshold, we consider that the current data frame actually belongs to a *novel class.* Determination of this threshold is performed on the basis of two different criteria: (i) maximization of the *precision* of the model in the task of novelty detection; (ii) maximization of the *recall* of the model in the task of novelty detection. Eventually, we utilize the so-obtained thresholds to measure the novelty detection precision and recall of our model using the available test set.

To obtain some comparative results, apart from our method we also evaluate conventional RNNs (trained as described in Section 2), and the state-of-the-art I/O-RNN-RBM and I/O-RNN-NADE methods presented in [7], under the same experimental setup. Our results are depicted in Table 2; these results our obtained for the best-performing model size in all cases. We observe that our method yields a very competitive result both in terms of the obtained precision and the yielded recall on the test set.

Table 2. Acoustic novelty detection: Precision and recall (%) of the evaluated models.

Model	Size	Precision	Recall
RNN	600	90.12	86.21
I/O-RNN-RBM	400	91.87	87.55
I/O-RNN-NADE	400	92.15	88.03
SB-RNN	600	92.30	88.56

4.3   Computational Complexity

Let us now turn to an analysis of the computational complexity of our method, and how it compares to conventional RNNs (trained as discussed in Section 2). We begin with the case of regression tasks: From the computational complexity perspective, the main difference between our approach and conventional RNNs concerns the fact that our approach also computes the quantities $\bar{\Sigma}$ and $\bar{\eta}_{2u}$, $\forall u$. However, the expressions of these approaches can be computed in time linear to the number of hidden units, as they do not entail any tedious calculations. Similar is the case when it comes to classification tasks. As such, one can expect that our method and conventional RNNs should share same computational complexity. To conclude, to provide some empirical evidence towards this direction, we here report the total training time of our method and conventional RNNs in the case of the acoustic novelty detection task (similar results can be obtained for the rest of our experimental scenarios). In our implementation, conventional RNNs took 6,855 sec to train, while our approach took 7,169 sec, that is a mere 4.58% extra computational time. Prediction generation took identical time in both cases. As such, we deduce that our approach offers a favorable performance/complexity trade-off over existing RNN formulations.

5   Conclusions and Future Work

In this paper, we proposed a sparse Bayesian formulation of RNNs, based on the introduction of a sparsity-inducing hierarchical prior over the output connection weights of the model. As we discussed, this model formulation introduces the ARD mechanism into the inferential procedures of RNNs, which allows for data-driven determination of the *effective* number of hidden (recurrently connected) units. We provided two alternative formulations of our model: one with likelihood function properly selected for handling regression tasks, and one designed for handling classification tasks. We devised simple and efficient inference algorithms for our model, scalable to massive datasets, for both the regression and (multi-class) classification settings. For this purpose, we resorted to the SVI paradigm.

To empirically evaluate the efficacy of our approach and how it compares to the competition, we conducted a number of experimental investigations dealing with human motion modeling using MoCap data and novelty detection in acoustic signals. In all cases, we used benchmark datasets in our experiments, and compared the performance of our method to state-of-the-art methods in the corresponding domains. As we observed, our approach yields a clear modeling performance advantage over the competition, without inducing notable overheads in terms of computational complexity.

In this work, posterior inference was conducted only for the output connection weights of the postulated RNNs, and the associated precision hyperparameters. In contrast, for the input and recurrent connection weight matrices of the model we obtained point-estimates, by maximization of the ELBO of the model over them. As such, one direction for future research concerns obtaining a fully Bayesian treatment of RNNs, with appropriate priors imposed over all the weight matrices of the model, and associated posterior distributions obtained during model inference. A challenge we expect to encounter working towards this direction concerns the nonlinear nature of the activation functions of the hidden units $\phi_h(\cdot)$, which may prevent us from obtaining closed-form expressions of the associated posteriors. Employing the black-box variational inference framework proposed in [21] to train our model might be a suitable possible candidate solution towards the amelioration of these issues.

Appendix

We have

$$\langle \boldsymbol{\alpha} \rangle = \left[\frac{\bar{\eta}_{1u}}{\bar{\eta}_{2u}} \right]_{u=1}^{H} \tag{25}$$

and

$$\langle v_{nu}^2 \rangle = \left[\langle \boldsymbol{v}_n \boldsymbol{v}_n^T \rangle \right]_u \tag{26}$$

where $[\cdot]_u$ stands for the uth element of a vector, and it holds

$$\langle \boldsymbol{v}_n \boldsymbol{v}_n^T \rangle = \begin{cases} \bar{\boldsymbol{v}}_n \bar{\boldsymbol{v}}_n^T + \bar{\boldsymbol{\Sigma}}, & \text{for regression tasks} \\ \bar{\boldsymbol{v}}_n \bar{\boldsymbol{v}}_n^T + \bar{\boldsymbol{\Sigma}}_n, & \text{for classification tasks} \end{cases} \tag{27}$$

Finally, the expression of $\langle \log p(\boldsymbol{v}_n|\boldsymbol{\alpha}) \rangle$ yields (ignoring constant terms)

$$\langle \log p(\boldsymbol{v}_n|\boldsymbol{\alpha}) \rangle = -\frac{1}{2} \sum_{u=1}^{H} \langle v_{nu}^2 \rangle \langle \alpha_u \rangle + \frac{1}{2} \sum_{u=1}^{H} \langle \log \alpha_u \rangle \tag{28}$$

where

$$\langle \log \alpha_u \rangle = \psi(\bar{\eta}_{1u}) - \log \bar{\eta}_{2u} \tag{29}$$

and $\psi(\cdot)$ is the Digamma function.

Acknowledgments. We gratefully acknowledge the support of NVIDIA Corporation with the donation of one Tesla K40 GPU used for this research.

References

1. The CMU MoCap database. http://mocap.cs.cmu.edu/
2. Amari, S.: Natural gradient works efficiently in learning. Neural Computation **10**(2), 251–276 (1998)
3. Barker, J., Vincent, E., Ma, N., Christensen, H., Green, P.: The Pascal Chime speech separation and recognition challenge. Computer Speech & Language **27**(3), 621–633 (2013)
4. Bastien, F., Lamblin, P., Pascanu, R., Bergstra, J., Goodfellow, I.J., Bergeron, A., Bouchard, N., Bengio, Y.: Theano: new features and speed improvements. In: Deep Learning and Unsupervised Feature Learning NIPS 2012 Workshop (2012)

5. Bayer, J., Osendorfer, C., Korhammer, D., Chen, N., Urban, S., van der Smagt, P.: On fast dropout and its applicability to recurrent networks. In: Proc. ICLR (2014)
6. Bengio, Y., Simard, P., Frasconi, P.: Learning long-term dependencies with gradient descent is difficult. IEEE Transactions on Neural Networks **52**(2), 157–166 (1994)
7. Boulanger-Lewandowski, N., Bengio, Y., Vincent, P.: High-dimensional sequence transduction. In: Proc. ICASSP, pp. 3178–3182 (2013)
8. Chatzis, S., Demiris, Y.: Echo state Gaussian process. IEEE Transactions on Neural Networks **22**(9), 1435–1445 (2011)
9. Cotter, A., Shamir, O., Srebro, N., Sridharan, K.: Better mini-batch algorithms via accelerated gradient methods. In: Proc. NIPS (2011)
10. Fokoue, E.: Stochastic determination of the intrinsic structure in Bayesian factor analysis. Tech. Rep. TR-2004-17, Statistical and Applied Mathematical Sciences Institute (2004)
11. Graves, A., Mohamed, A., Hinton, G.: Speech recognition with deep recurrent neural networks. In: Proc. ICASSP (2013)
12. Hammer, B.: On the approximation capability of recurrent neural networks. Neurocomputing **31**(1), 107–123 (2000)
13. Hochreiter, S., Schmidhuber, J.: Long short-term memory. Neural Computation **9**(8), 1735–1780 (1997)
14. Hoffman, M., Blei, D.M., Wang, C., Paisley, J.: Stochastic variational inference. Journal of Machine Learning Research **14**(5), 1303–1347 (2013)
15. Jaeger, H.: The "echo state" approach to analysing and training recurrent neural networks. Tech. Rep. 148, German National Research Center for Information Technology, Bremen (2001)
16. Lan, G.: An optimal method for stochastic composite optimization. Mathematical Programming, 1–33 (2010)
17. Martens, J., Sutskever, I.: Learning recurrent neural networks with hessian-free optimization. In: Proc. ICML (2011)
18. Nesterov, Y.: A method of solving a convex programming problem with convergence rate $o(1/sqr(k))$. Soviet Mathematics Doklady **27**, 372–376 (1983)
19. Pascanu, R., Mikolov, T., Bengio, Y.: On the difficulty of training recurrent neural networks. In: Proc. ICML (2013)
20. Polyak, B.: Some methods of speeding up the convergence of iteration methods. USSR Computational Mathematics and Mathematical Physics **4**(5), 1–17 (1964)
21. Ranganath, R., Gerrish, S., Blei, D.M.: Black box variational inference. In: Proc. AISTATS (2014)
22. Rumelhart, D., Hinton, G., Williams, R.: Learning internal representations by error propagation. In: Parallel Dist. Proc., pp. 318–362. MIT Press (1986)
23. Srivastava, N., Hinton, G., Krizhevsky, A., Sutskever, I., Salakhutdinov, R.: Dropout: A simple way to prevent neural networks from overfitting. J. Machine Learning Research **15**(6), 1929–1958 (2014)
24. Sutskever, I., Martens, J., Dahl, G., Hinton, G.: On the importance of initialization and momentum in deep learning. In: Proc. ICML (2013)
25. Wang, J.M., Fleet, D.J., Hertzmann, A.: Gaussian process dynamical models for human motion. IEEE Transactions on Pattern Analysis and Machine Intelligence **30**(2), 283–298 (2008)

Structured Prediction of Sequences and Trees Using Infinite Contexts

Ehsan Shareghi[1]([✉]), Gholamreza Haffari[1], Trevor Cohn[2], and Ann Nicholson[1]

[1] Monash University, Melbourne, Australia
{ehsan.shareghi,gholamreza.haffari,ann.nicholson}@monash.edu
[2] University of Melbourne, Melbourne, Australia
t.cohn@unimelb.edu.au

Abstract. Linguistic structures exhibit a rich array of global phenomena, however commonly used Markov models are unable to adequately describe these phenomena due to their strong locality assumptions. We propose a novel hierarchical model for structured prediction over sequences and trees which exploits global context by conditioning each generation decision on an *unbounded* context of prior decisions. This builds on the success of Markov models but without imposing a fixed bound in order to better represent global phenomena. To facilitate learning of this large and unbounded model, we use a hierarchical Pitman-Yor process prior which provides a recursive form of smoothing. We propose prediction algorithms based on A* and Markov Chain Monte Carlo sampling. Empirical results demonstrate the potential of our model compared to baseline finite-context Markov models on three tasks: morphological parsing, syntactic parsing and part-of-speech tagging.

Keywords: Structured prediction · Infinite markov model · Chinese restaurant process

1 Introduction

Markov models are widespread popular techniques for modelling the underlying structure of natural language, e.g., as sequences and trees. However local Markov assumptions often fail to capture phenomena outside the local Markov context, i.e., when the data generation process exhibits long range dependencies. A prime example is language modelling where only short range dependencies are captured by finite-order (i.e. n-gram) Markov models. However, it has been shown that going beyond finite order in a Markov model improves language modelling because natural language embodies a large array of long range depepndencies [Wood et al., 2009]. While *infinite* order Markov models have been extensively explored for language modelling [Gasthaus and Teh, 2010; Wood et al., 2011], this has not yet been done for structure prediction.

In this paper, we propose an infinite-order Markov model for predicting latent structures, namely tag sequences and trees. We show that this expressive model can be applied to various structure prediction tasks in NLP, such as syntactic and morphological parsing and part-of-speech tagging. We propose effective

© Springer International Publishing Switzerland 2015
A. Appice et al. (Eds.): ECML PKDD 2015, Part II, LNAI 9285, pp. 373–389, 2015.
DOI: 10.1007/978-3-319-23525-7_23

algorithms to tackle significant learning and inference challenges posed by the infinite Markov model.

More specifically, we propose an unbounded-depth, hierarchical, Bayesian non-parametric model for the generation of linguistic utterances and their corresponding structure (e.g., the sequence of POS tags or syntax trees). Our model conditions each decision in a tree generating process on an *unbounded* context consisting of the vertical chain of their ancestors, in the same way that infinite sequence models (e.g., ∞-gram language models) condition on an unbounded window of linear context [Mochihashi and Sumita, 2007; Wood et al., 2009].

Learning in this model is particularly challenging due to the large space of contexts and corresponding data sparsity. For this reason predictive distributions associated with contexts are smoothed using distribtions for successively smaller contexts via a hierarchical Pitman-Yor process, organised as a suffix trie. The infinite context makes it impossible to directly apply dynamic programing for structure prediction. We present two inference algorithms based on A* and Markov Chain Monte Carlo (MCMC) for predicting the best structure for a given input utterance.

The experiments on part-of-speech (POS) tagging show that our generative model obtains similar performance to the state-of-the-art Stanford POS tagger [Toutanova and Manning, 2000] for English and Swedish. For Danish, our model outperforms the Stanford tagger, which is impressive given the Stanford parser uses many more complex features and a discriminative training objective. Our experiments on morphological parsing and syntactic parsing show that our unbounded-context tree model adapts itself to the data to effectively capture sufficient context to outperform the PCFG baseline.

2   Background and Related Work

The parse tree of an utterance can be generated by combining a set of rules from a grammar, such as a context free grammar (CFG). A CFG is a 4-tuple $\mathcal{G} = (\mathcal{T}, \mathcal{N}, S, \mathcal{R})$, where $\mathcal{T}$ is a set of terminal symbols, $\mathcal{N}$ is a set of nonterminal symbols, $S \in \mathcal{N}$ is the distinguished root non-terminal and $\mathcal{R}$ is a set of productions (aka rewriting rules). A PCFG assigns a probability to each grammar rule, where $\sum_{B,C} P(A \rightarrow B\ C | A) = 1$. The grammar rules are often in Chomsky Normal Form (CNF), taking either the form $A \rightarrow B\ C$ or $A \rightarrow a$ where A, B, C are nonterminals, and a is a terminal.

Syntactic parsing is the task of predicting the parse tree of a given sentence. In syntactic parsing, the nonterminals of the underlying grammar are syntactic catergories, e.g. the input sentence (S), noun phrase (NP) and verb phrase (VP); the terminals are words. Morphological parsing is the task of breaking down an unsegmented input into words and their morphological structure. In this task, the grammar terminals are morphemes (smallest meaningful units of a language), and nonterminals represent the input Sequence, Word, prefix (P), etc. Tag sequences can also be represented as a tree structure, without loss of generality, in which rules take the form $A \rightarrow B\ a$ or $A \rightarrow a$ where A, B are POS

tags, and a is a word. This unified view to syntactic parsing, morphological parsing, and POS tagging will allow us to apply our model and inference algorithms
to these problems with only minor refinements (see Figure 1).

In PCFG, a tree is generated by starting with the root symbol and rewriting (substituting) it with a grammar rule, then continuing to rewrite frontier
non-terminals with grammar rules until there are no remaining frontier non-
terminals. When making the decision about the next rule to expand a frontier
non-terminal, the only conditioning context used from the partially generated
tree is the frontier non-terminal itself, i.e., the rewrite rule is assumed independent from the remainder of the tree given the frontier non-terminal. Our model
relaxes this strong independence assumptions by considering unbounded vertical
history when making the next inference decision. This takes into account a wider
context when making the next parsing decision.

Perhaps the most relevant work is on unbounded history language models
[Mochihashi and Sumita, 2007; Wood et al., 2009]. A prime work is Sequence
Memoizer [Wood et al., 2011] which conditions the generation of the next word
on an unbounded history of previously generated words. We build on these techniques to develop rich infinite-context models for structured prediction, leading
to additional complexity and challenges.

For syntactic parsing, several infinite extensions of probabilistic context free
grammars (PCFGs) have been proposed [Finkel et al., 2007; Liang et al., 2007].
These approaches achieve infinite grammars by allowing an unbounded set of
non-terminals (hence grammar rules), but still make use of a bounded history
when expanding each non-terminal. An alternative method allows for infinite
grammars by considering segmentation of trees into arbitrarily large tree fragments, although only a limited history is used to conjoin fragments [Cohn et al.,
2010; Johnson et al., 2006]. Our work achieves infinite grammars by growing the
vertical history needed to make the next parsing decision, as opposed to growing
the number of rules, non-terminals or states *horizontally*, as done in prior work.

Earlier work in syntactic parsing has also looked into growing both the history vertically and the rules horizontally, in a *bounded* setting. [Johnson, 1998]
has increased the history for the parsing task by parent-annotation, i.e., anno-
tating each non-terminal in the training parse trees by its parent, and then
reading off the grammar rules from the resulting trees. [Klein and Manning,
2003] have considered vertical and horizontal markovization while using the head
words' part-of-speech tag, and showed that increasing the size of the vertical
contexts consistently improves the parsing performance. [Petrov et al., 2006],
[Petrov and Klein, 2007] and [Matsuzaki et al., 2005] have treated non-terminal
annotations as latent variables and estimated them from the data.

Likewise, finite-state hidden Markov models (HMMs) have been extended
horizontally to have countably infinite number of states [Beal et al., 2001].
Previous works on applying Markov models to part-of-speech tagging
either considered finite-order Markov models [Chen, 2000], or finite-order
HMM [Thede and Harper, 1999]. We differ from these works by conditioning
both the emissions and transitions on their *full* contexts.

3   The Model

Our model relaxes strong local Markov assumptions in PCFG to enable capturing phenomena outside of the local Markov context. The model conditions the
generation of a rule in a tree on its unbounded vertical history, i.e., its ancestors
on the path towards the root of the tree (see Figure 1). Thus the probability of
a tree T is

$$P(T) = \prod_{(\mathbf{u},r)\in T} G_{[\mathbf{u}]}(r) \tag{1}$$

where r denotes the rule and $\mathbf{u}$ its history, and $G_{[\mathbf{u}]}(.)$ is the probability of the
next inference decision (i.e., grammar rule) conditioned on the context $\mathbf{u}$. In
other words, a tree T can be represented as a sequence of context-rule events
$\{(\mathbf{u},r) \in T\}$.

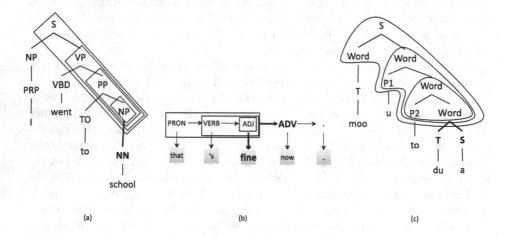

Fig. 1. Examples of infinite-order conditioning and smoothing mechanism. The
bold symbols (**NN, ADV, fine, T, S**) are the part of the structure being
generated, and the boxes correspond to the conditioning context. (a) Syntactic
Parsing, (b) Infinite-order HMM for POS tagging, (c) Morphological Parsing.

When learning such a model from data, a vector of predictive probabilities for
the next rule $G_{[\mathbf{u}]}(.)$ given each possible vertical context $\mathbf{u} \in \mathcal{U}$ must be learned,
where depending on the problem $\mathcal{U}$ can denote the set of spines of non-terminals
$\mathcal{N}^*$ (as in Fig. 1(a),(b)) or chains of rules $\mathcal{R}^*$ (as in Fig. 1(c)). As the context size
increases, the number of events observed for such long contexts in the training
data drastically decreases which makes parameter estimation challenging, particularly when generalising to unseen contexts. Assuming our unbounded-depth
model, we need suitable *smoothing* techniques to estimate conditional rule probabilities for large (and possibly infinite depth) contexts. We achieve smoothing

by placing a hierarchical Bayesian prior over the set of probability distributions $\{G_{[\mathbf{u}]}\}_{u \in \mathcal{U}}$. We smooth $G_{[\mathbf{u}]}$ with a distribution conditioned on a shorter context $G_{[\pi(\mathbf{u})]}$, where $\pi(\mathbf{u})$ is the suffix of $\mathbf{u}$ containing all but the earliest event. This ties parameters of longer histories to their shorter suffixes in a hierarchical manner, and leads to sharing statistical strengths to overcome sparsity issues. Figure 1 shows our infinite-order Markov model and the smoothing mechanism described here.

More specifically, we assume that a distribution with the full history $G_{[u]}$ is related to a distribution with the most recent history $G_{[\pi(u)]}$ through the Pitman-Yor process PYP [Wood et al., 2011]:

$$G_{[\varepsilon]} \mid d_{[\varepsilon]}, c_{[\varepsilon]}, H \ \sim \ PYP(d_0, c_0, H) \tag{2}$$

$$G_{[\mathbf{u}]} \mid d_{|\mathbf{u}|}, c_{|\mathbf{u}|}, G_{[\pi(\mathbf{u})]} \ \sim \ PYP(d_{|\mathbf{u}|}, c_{|\mathbf{u}|}, G_{[\pi(\mathbf{u})]}) \tag{3}$$

where H denotes the base (e.g. uniform) distribution, and ε denotes the empty context. The Pitman-Yor process $PYP(d, c, H)$ is a distribution over distributions, where d is the discount parameter, c is the concentration parameter, and H is the base distribution. Note that $G_{[u]}$ depends on $G_{[\pi(u)]}$ which itself depends on $G_{[\pi(\pi(u))]}$, etc. This leads to a hierarchical Pitman-Yor process prior where context-dependent distributions are *hidden*. The formulation of the hierarchical PYP over different length contexts is illustrated in Figure 2.

Figure 3 demonstrates the property of PYP and how its behavior depends on discount d, and concentration c parameters. Note that the PYP allows a good fit to data distribution compared to the Dirichlet Process ($d = 0$; as used in prior work) which cannot adequately represent the long tail of events.

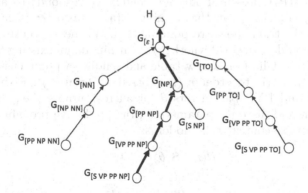

Fig. 2. Part of the smoothing mechanism corresponding to Figure 1(a). Each node represents a distribution G labeled with a context, and the directed edges demonstrate the direction of smoothing. The path in bold corresponds to the smoothing for the *rule* $NP \to NN$.

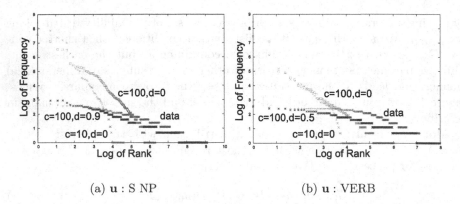

(a) **u** : S NP   (b) **u** : VERB

Fig. 3. log-log plot of rule frequency vs rank, illustrated for (a) syntactic parsing and (b) POS tagging. Besides the data distribution, we also show samples from three PYP distributions with different hyperparameter values, c, d.

4   Learning

Given a training tree-bank, i.e., a collection of utterances and their trees, we are interested in the posterior distribution over $\{G_{[\mathbf{u}]}\}_{\mathbf{u}\in\mathcal{U}}$. We make use of the approach developed in [Wood et al., 2011] for learning such suffix-based graphical models when learning infinite-depth language models. It makes use of Chinese Restaurant Process (CRP) representation of the Pitman-Yor process in order to marginalize out distributions $G_{[\mathbf{u}]}$ [Teh, 2006] and learn the predictive probabilities $P(r|\mathbf{u})$.

Under the CRP representation each context corresponds to a restaurant. As a new $(\mathbf{u}, r)$ is observed in the training data, a *customer* is entered to the restaurant, i.e., the trie node corresponding to **u**. Whenever a customer enters a restaurant, it should be decided whether to seat him on an existing table serving the *dish* r, or to seat him on a new table and sending a proxy customer to the parent node in the trie to order r (i.e., based on $(\pi(\mathbf{u}), r)$). Fixing a seating arrangement **S** and PYP parameters $\boldsymbol{\theta}$ for all restaurants (i.e., the collection of concentration and discount parameters), the predictive probability of a rule based on our infinite-context rule model is:

$$P(r|\epsilon, \mathbf{S}, \boldsymbol{\theta}) = H(r) \tag{4}$$

$$P(r|\mathbf{u}, \mathbf{S}, \boldsymbol{\theta}) = \frac{n_{r.}^{\mathbf{u}} - d_{|\mathbf{u}|} t_r^{\mathbf{u}}}{n_{..}^{|\mathbf{u}|} + c_{|\mathbf{u}|}} + \frac{c_{|\mathbf{u}|} + d_{|\mathbf{u}|} t_{..}^{\mathbf{u}}}{n_{..}^{\mathbf{u}} + c_{|\mathbf{u}|}} P(r|\pi(\mathbf{u}), \mathbf{S}, \boldsymbol{\theta}) \tag{5}$$

where $d_{|\mathbf{u}|}$ and $c_{|\mathbf{u}|}$ are the discount and concentration parameters, $n_{rk}^{\mathbf{u}}$ is the number of customers at table k served the dish r in the restaurant **u** (accordingly $n_{r.}^{\mathbf{u}}$ is the number of customers served the dish r and $n_{..}^{\mathbf{u}}$ is the number of customers), and $t_r^{\mathbf{u}}$ is the number of tables serving dish r in the restaurant **u** (accordingly $t_{..}^{\mathbf{u}}$ is the number of tables).

The seating arrangements (the state of all restaurants including their tables and customers sitting on each table) are hidden, so they need to be marginalized out:

$$P(r|\mathbf{u}, \mathcal{D}) = \int P(r|\mathbf{u}, \mathbf{S}, \boldsymbol{\theta}) P(\mathbf{S}, \boldsymbol{\theta}|\mathcal{D}) d(\mathbf{S}, \boldsymbol{\theta}) \tag{6}$$

where $\mathcal{D}$ is the training tree-bank. We approximate this integral by the so called "minimal assumption seating arrangement" and the MAP parameter setting $\boldsymbol{\theta}$ which maximizes the corresponding data posterior. Based on the minimal assumption, a new table is created only when there is no table serving the desired dish in a restaurant $\mathbf{u}$. That is, a proxy customer is created and sent to the parent node in the trie $\pi(\mathbf{u})$ for each unique dish type (sequence of events). This approximation is related to the well-known interpolated Kneser-Ney smoothing [Chen and Goodman, 1996], when applied to hierarchical Pitman-Yor process language models [Teh, 2006].

The parameter $\boldsymbol{\theta}$ is learned by maximising the posterior, given the seating arrangement corresponding to the minimal assumption. We put the following prior distributions over the parameters: $d_m \sim \text{Beta}(a_m, b_m)$ and $c_m \sim \text{Gamma}(\alpha_m, \beta_m)$. The posterior is the prior multiplied by the following likelihood term:

$$\prod_r H(r)^{n_r^0} \prod_{\mathbf{u}} \frac{[c_{|\mathbf{u}|}]_{d_{|\mathbf{u}|}}^{t_{\cdot\cdot}^{\mathbf{u}}}}{[c_{|\mathbf{u}|}]_1^{n_{\cdot\cdot}^{\mathbf{u}}}} \prod_r \prod_{k=1}^{t_{\cdot\cdot}^{\mathbf{u}}} [1 - d_{|\mathbf{u}|}]_1^{(n_{rk}^{\mathbf{u}} - 1)} \tag{7}$$

where $[a]_b^c$ denotes the generalised factorial function.[1] We maximize the posterior with the constraints $c_m \geq 0$ and $d_m \in [0, 1)$ using the L-BFGS-B optimisation method [Zhu et al., 1997], leading to the optimised discount and concentration for each context size.

5 Prediction

In this section, we propose algorithms for the challenging problem of predicting the highest scoring tree. The key ideas are to compactly represent the *space* of all possible trees for a given utterance, and then *search* for the best tree in this space in a *top-down* manner. By traversing the hyper-graph top-down, the search algorithms have access to the full history of grammar rules.

In the test time, we need to predict the tree structure of a given utterance $\mathbf{w}$ by maximizing the tree score:

$$\arg \max_T P(T|\mathcal{D}, \mathbf{w}) = \arg \max_T \prod_{(\mathbf{u}, r) \in T} P(r|\mathbf{u}, \mathcal{D}) \tag{8}$$

The unbounded context allowed by our model makes it infeasible to apply dynamic programming, e.g. CYK [Cocke and Schwartz, 1970], for finding the

---

[1] $[a]_b^0 = [a]_b^{-1} = 1$ and $[a]_c^b = \prod_{i=0}^{c-1}(a + ib)$.

highest scoring tree. CYK is a *bottom-up* algorithm which requires storing in a dynamic programming table the score of each utterance's sub-span conditioned on all possible contexts. Even truncating the context size to bound this term may be insufficient to allow CYK for prediction, due to the unreasonable computational complexity.

The space of all possible trees for a given utterance can be compactly represented as a *hyper-graph* [Klein and Manning, 2001]. Each hyper-graph node is labelled with a non-terminal and a sub-span of the utterance. There exists a hyper-edge from the nodes $B[i, j]$ and $C[j + 1, k]$ to the node $A[i, k]$ if the rule $A \to B\ C$ belongs to the grammar (Figure 4). Starting from the top node $S[0, N]$, our prediction algorithms search for the highest scoring tree sub-graph that covers all of the utterance terminals in the hyper-graph. Our top-down prediction algorithms have access to the full history needed by our model when deciding about the next hyper-edge to be added to the partial tree.

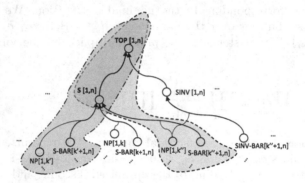

Fig. 4. Hyper-graph representation of the search space for a syntactic parsing example. The gray areas are examples of two partial hypotheses in A* priority queue.

5.1 A* Search

This algorithm incrementally expands frontier nodes of the best partial tree until a complete tree is constructed. In the expansion step, all possible rules for expanding all frontier non-terminals are considered and the resulting partial trees are inserted into a priority queue (see Figure 4), sorted based on the following score:

$$Score(T^+) = \log P(T) + \log G_{\mathbf{u}}(A \to B\ C) + h(T^+, A \to B\ C, i, k, j | G') \quad (9)$$

where T^+ is a partial tree *after* expanding a frontier non-terminal, $P(T)$ is the probability of the current partial tree, $G_{\mathbf{u}}(A \to B\ C)$ is the probability of expanding a non-terminal via a rule $A \to B\ C$ in the full context $\mathbf{u}$, and h is the heuristic function (i.e., the estimate of the score for the best tree completing T^+). We use various heuristic functions when expanding a node $A[i, j]$ in the hypergraph via a hyperedge with tails $B[i, k]$ and $C[k + 1, j]$:

- **Full Frontier:** which estimates the completion cost by

$$h(T^+, A \to B\ C, i, k, j|G') = \sum_{(A', i', j') \in \mathrm{Fr}(T^+)} \log P(A', i', j'|G') \qquad (10)$$

where $\mathrm{Fr}(T^+)$ is the set of frontier nodes of the partial tree, and G' is a *simplified grammar* admitting dynamic programming. Here we choose the PCFG used the base measure H in the root of the PYP hierarchy. Accordingly the $\log P$ terms can be computed cheaply using the PCFG inside probabilities.

- **Local Frontier:** which only considers the completion of the following frontier nodes, and uses the completion cost of the sub-span using the selected rule:

$$h(T^+, A \to B\ C, i, k, j|G') = \log P(B, i, k|G') + \log P(C, k+1, j|G') \quad (11)$$

The above heuristics functions are not admissible, hence the A* algorithm is not guaranteed to find the optimal tree. However the PCFG provides reasonable estimates of the completion costs, and accordingly with a sufficiently wide beam, search error is likely to be low.

5.2   MCMC Sampling

We make use of Metropolis-Hastings (MH) algorithm, which is a Markov chain Monte Carlo (MCMC) method, for obtaining a sequence of random trees. We then combine these trees to construct the predicted tree.

We use a *PCFG* as our *proposal* distribution Q and draw samples from it. Each sampled tree is then accepted/rejected using the following acceptance rate:

$$\alpha(T, T') = \min\left\{1, \frac{P(T')Q(T)}{P(T)Q(T')}\right\} \qquad (12)$$

where T' is the sampled tree, T is the current tree, $P(T')$ is the probability of the proposed tree under our model, and $Q(T')$ is its probability under the proposal PCFG. Under some conditions, i.e., detailed balance and ergodicity, it is guaranteed that the stationary distribution of the underlying Markov chain (defined by the MH sampling) is the distribution that our model induces over the space of trees P. For each utterence, we sample a fresh tree for the whole utterance from a PCFG using the approach of [Johnson et al., 2007], which works by first computing the inside lattice under the proposal model (computed once and reused), followed by top-down sampling to recover a tree. Finally the proposed tree is scored using the MH test, according to which the tree is randomly accepted as the next sample or else rejected in which case the previous sample is retained.

Once the sampling is finished, we need to choose a tree based on statistics of the sampled collection of trees. One approach is to select the most frequently sampled tree, however this does not work effectively in such large search spaces because of high sampling variance. Note that local Gibbs samplers might be able to address this problem, at least partly, through resampling subtrees instead of full tree sampling (as done here). Local changes would allow for more rapid mixing from trees with some high and low scoring subtrees to trees with uniformly high scoring sub-structures. We leave local sampling for future work, noting that the obvious local operation of resampling complete sub-trees or local tree fragments would compromise detailed balance, and thus not constitute a valid MCMC sampler [Levenberg et al., 2012].

To address this problem, we use a Minimum Bayes Risk (MBR) decoding method to predict the best tree [Goodman, 1996] as follows: For each pair of a nonterminal-span, we record the count in the collection of sampled trees. Then using the Viterbi algorithm, we select the tree from the hypergraph for which the sum of the induced pairs of nonterminal-span is maximized. Roughly speaking, this allows to make local corrections that result in higher accuracy compared to the best sampled trees.

6   Experiments

In order to evaluate the proposed model and prediction algorithms, we performed two sets of experiments on tasks with different structural complexity. The statistics of the tasks and datasets are provided in Table 1.

6.1   Morphological Parsing

We consider the problem of morphological parsing of unsegmented inputs, i.e. seeking to model words and their morphological structure in the input stream. A morphological structure of a word breaks it into its building blocks: Prefixes, Stem, and Suffixes. For example, for the word *"antidisestablishmentarianism"*, the terms "anti", "dis" are the prefixes, "establish" is the stem, and "ment", "arian", and "ism" are the suffixes.

For this experiment, we model the Sesotho language, a Bantu language which combines rich productive agglutinative morphology with relatively simple phonology. We use the dataset from [Johnson, 2008], which comprises utterances marked with word boundaries. In contrast to Johnson's approach, our method is supervised, and consequently we require treebanked input for training and

Table 1. Statistics for PTB syntactic Parsing and part-of-speech tagging, showing the number of training and test sentences, average sentence length in words and number of grammar rules. For morph the numbers are averaged over the 10 folds.

Task	Train	Test	Len	Rules
morph	36479	4000	5	3080
parse	33180	2416	24	31920
pos EN	38219	5462	24	29499
pos DN	3638	1000	20	5269
pos SW	10653	389	18	9739

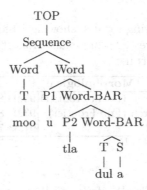

Fig. 5. Binarized morphological tree for the Sethoso sequence *"moo utladula"*.

evaluation. To form a proxy 'gold-standard', we augmented the input to include morphological trees with prefix (P), suffix (S) and stem (T) structure inferred automatically from segmented utterances using Johnson's Adaptor Grammar with his *word-smorph* grammar. In this grammar a word consists of a stem with an optional suffix, and zero to three prefixes: $Word \rightarrow (P1(P2(P3)))\ T\ (S)$, where $P1, P2, P3$ are prefixes, and T, and S are stem and suffix. An example input is shown in Figure 5. We right-binarized the trees and replaced segments with $count \leq 2$ with two categories of $OUT\text{-}V$ and $OUT\text{-}C$ depending on their initial character being a vowel or consonant. We applied 10-fold cross validation, and the predicted trees were evaluated using EVALB evaluation package.

As reported in Table 2, the best result is achieved by A* search with local frontier heuristic. It might seem surprising that considering the full frontier heuristic results in lower performance. We speculate that this is because the PCFG over estimates the completion cost, due to its reduced conditioning context which leads to higher entropy distributions and lower probability estimates. The reduced effect of the heuristic in the local method moderates this issue. The MCMC sampler obtains similar results to the baseline PCFG.

In the morphological parsing task, the grammar has 3080 rules, and the average sentence length is 5 words. This leads to a reasonably-small search space, with the net effect that A* search (with beam size 200) is an effective parsing strategy. The small grammar size of this task has allowed us to use grammar rules as fine-grained conditioning contexts. In the remaining tasks of syntactic parsing and POS tagging, we will condition only on the spine. This is due to the intractable magnitude of the spaces generated by infinite order rule conditioning, which are problematic for MCMC sampling and A* search based on our preliminary experiments.

6.2 Syntactic Parsing

For syntactic parsing, we use the Penn. treebank (PTB) dataset [Marcus et al., 1993]. We used the standard data splits for training and testing (train sec 2-21;

Table 2. Morphological parsing results, showing 10-fold cross validation evaluation for unlabelled F-Measure (F1) and exact bracketing match (ACC). MCMC results are averaged over 10 runs.

Parser (Morphological)	F1	ACC
A* Search (Local Frontier)	95.99	89.77
A* Search (Full Frontier)	93.08	85.04
MCMC	91.33	78.86
PCFG CYK	91.27	79.39

validation sec 22; test sec 23). We followed [Petrov et al., 2006] preprocessing steps by right-binarizing the trees and replacing words with $count \leq 1$ in the training sample with generic *unknown* word markers representing the tokens' lexical features and position. The results reported in Table 3 are produced by EVALB.

The results in Table 3 demonstrate the superiority of our model compared to the baseline PCFG. We note that the A* parser becomes less effective (even with a large beam size) for this task, which we attribute to the large search arising for the large grammar and long sentences. Our best results are achieved by MCMC, demonstrating the effectiveness of MCMC in large search spaces.

An interesting observation is how our results compare with those achieved by bounded vertical and horizontal Markovization reported in [Klein and Manning, 2003]. Our binarization corresponds to one of their simpler settings for horizontal markovization, namely $h = 0$ in their terminology, and note also that we ignore the head information which is used in their models. Despite this we still manage to equal their results obtained using vertical context of size 3 ($v = 3$), with 76.7 F1 score. Their best result, $F_1 = 79.74$, was achieved with $h \leq 2$, $v = 3$ (and tags for head words). We believe that our model would outperform theirs if we consider greater horizontal markovization and incorporate head word information. To facilitate a fair comparison with vertical markovization, we experimented with limiting the size of the vertical contexts to 2, 3 or 4 within our model. Using MCMC parsing we found that performance consistently improved as the size of the context was increased, scoring 68.1, 71.1, 75.0 F-measure respectively. This is below 76.7 F-measure of our unbounded-context model which adapts itself to data to effectively capture the right context.

Table 3. Syntactic parsing results for the Penn. treebank, showing labelled F-Measure (F1) and exact bracketing match (ACC).

Syntactic Parser	all		≤ 40	
	F1	ACC	F1	ACC
A* (Local Frontier)	75.33	16.12	76.21	16.85
A* (Full Frontier)	72.27	13.14	72.34	13.57
MCMC	76.74	18.23	78.21	18.99
PCFG CYK	58.91	4.11	60.25	4.42

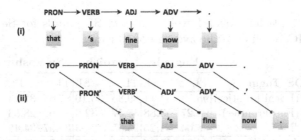

Fig. 6. The analogy between HMM (i) and our representation (ii) for the part-of-speech tags of the sentence *"that's fine now."*

The run-time of our parser under MCMC (with $30k$ samples) is $0.29 \times |S|$ secs, and under A*(Local) is $0.13 \times |S|$ secs, where $|S|$ is the length of the sentence. With a smaller number of samples the parsing time reduces linearly and the predictive accuracy only suffers slightly; for instance with 5k samples the F1 measure (all) falls by 0.8.

Overall our approach significantly outperforms the baseline PCFG, although note these results are well below the current state-of-the-art in parsing, which typically makes use of discriminative training with much richer features. We speculate that future enhancements could close the gap between our results and that of modern parsers, while offering the potential benefits of our generative model which allows further incorporation of different types of contexts (e.g., head words and n-gram lexical context).

6.3 Part-of-Speech Tagging

The part of speech (POS) corpora have been extracted from PTB (sections 0-18 for training and 22-24 for test) for English, and NAACL-HLT 2012 Shared task on Grammar Induction[2] for Danish and Swedish [Gelling et al., 2012]. We convert the sequence of part-of-speech tags for each sentence into a tree structure analogous to a Hidden Markov Model (HMM). For each POS tag we introduce a twin (e.g., ADJ' for ADJ) in order to encode HMM-like transition and emission probabilities in the grammar. As shown in Figure 6, this representation guarantees that all the rules in the structures are either in the form of $t_i \rightarrow t_j\ t'_j$ (transition) or $t' \rightarrow \texttt{word}$ (emission).

The tagging results are reported in Table 4, including comparison with the baseline PCFG ($\equiv$ HMM) and the state-of-the-art Stanford POS Tagger [Toutanova and Manning, 2000], which we trained and tested on these datasets. As illustrated in Table 4, our model consistently improves the PCFG baseline. While for Danish we outperform the state-of-the-art tagger, the results for English and Swedish we are a little behind the Stanford Tagger. This is a promising result since our model is only based on the rules and their contexts,

---

[2] http://wiki.cs.ox.ac.uk/InducingLinguisticStructure/SharedTask

Table 4. TL stands for Token-Level Accuracy, SL stands for Sentence-Level Accuracy. MCMC results are the average of 10 runs.

POS Tagger	English		Danish		Swedish	
	TL	SL	TL	SL	TL	SL
A*(Local Frontier)	95.50	54.11	89.85	35.10	87.04	32.13
A*(Full Frontier)	95.27	53.88	88.57	32.6	85.62	28.53
MCMC	96.04	54.25	95.55	72.93	89.97	34.45
PCFG CYK	94.69	47.22	89.04	31.7	89.76	33.93
Stanford Tagger	97.24	56.34	93.66	51.30	91.28	37.02

Table 5. (a) Percentage of the matched spines over the top-1000 frequent spines for each spine length in the trees predicted by our unbounded-context model $(v = \infty)$ and the baseline limited-context model $(v = 2)$. (b) The top-5 frequent contexts for NP, VP, DT, and JJ in the trees predicted by our model; the ones marked with (*) exist in the top-5 contexts in the gold standard trees as well.

	$v = \infty$				$v = 2$			
	parse	POS			parse	POS		
size	WSJ	EN	DN	SW	WSJ	EN	DN	SW
2	100	96	100	97	100	96	100	97
3	75	100	100	100	75	100	100	100
4	72	69	72	68	70	68	72	68
5	68	57	58	57	63	57	58	57
6	62	56	51	53	60	56	51	53
7	59	52	55	37	59	50	55	38
8	58	42	45	29	51	41	44	29
9	60	68	61	37	49	60	31	37
10	68	75	67	35	51	69	25	34

(a)

NP	VP
S*	S*
SINV*	S VP SBAR S*
S PP*	SINV*
S VP PP	S SBAR S
S VP*	S PRN
DT	**JJ**
S NP *	S NP*
S PP NP*	S PP NP*
S VP SBAR S NP*	S VP ADJP*
S VP PP NP*	SINV NP
S S-BAR NP	S VP SBAR S VP PP NP*

(b)

as opposed to the Stanford Tagger which uses complex hand-designed features and a complex form of discriminative training. Note the strong performance of MCMC sampling, which consistently outperforms A* search.

6.4 Analysis

For the analysis we focus on the syntactic parsing and POS tagging tasks. For each different spine size from 2 to 10, we extract the top-1000 frequent spines in the trees predicted based on our model, and compare them with those extracted from the gold standard trees. The numbers reported in Table 5(a), are the percentage of the intersection of these two sets. As reported in the table, in all cases (except one) the infinite order model $(v = \infty)$ outperforms the model with limited size context $(v = 2)$. Particularly in Danish POS tagging, our model predicts correctly 65% of top-1000 high-frequency spines of length 10 vs. 25% of the model with limited context. For syntactic parsing, the short range dependencies captured by limited context model $(v = 2)$ over the spines of size 2 and

3 matches the results of our unbounded context model ($v = \infty$); however, the gap becomes wider for longer spines.

Our next analysis looks into the contexts of 4 linguistic categories in syntactic parsing: NP (noun phrase), VP (verb phrase), DT (determiner), and JJ (adjective). data set. We chose NP and VP mainly because they tend to appear in higher levels of the tree and most probably often in shorter contexts, and DT and JJ for the opposite reason. A list of the most frequent contexts for these syntactic categories in the trees predicted by our model is provided in Table 5(b); the ones marked with (*) exist in the gold standard trees as well. Our model successfully retrieves most of the long and short high-frequency contexts for the aforementioned syntactic categories.

7 Conclusion and Future Work

We have proposed a novel hierarchical model over linguistic trees which exploits global context by conditioning the generation of a rule in a tree on an unbounded tree context consisting of the vertical chain of its ancestors.

To facilitate learning of such a large and unbounded model, the predictive distributions associated with tree contexts are smoothed in a recursive manner using a hierarchical Pitman-Yor process. We have shown how to perform prediction based on our model to predict the parse tree of a given utterance using various search algorithms, e.g. A* and Markov Chain Monte Carlo. This consistently improved over baseline methods in several tasks, and produced state-of-the-art results for Danish part-of-speech tagging.

In future, we would like to consider sampling the seating arrangements and model hyperparameters, and seek to incorporate several different notions of context besides the chain of ancestors.

Acknowledgments. The authors are grateful to National ICT Australia (NICTA) for generous funding, as part of collaborative machine learning research projects. This work was funded in part by the Australian Research Council. We would like to thank the anonymous reviewers for their constructive comments.

References

Beal, M.J., Ghahramani, Z., Rasmussen, C.E.: The infinite hidden markov model. In: Advances in Neural Information Processing Systems, Vancouver, British Columbia, Canada, pp. 577–584 (2001)

Brants, T.: Tnt - A statistical part-of-speech tagger. In: Proceedings of the Sixth Conference on Applied Natural Language Processing, pp. 224–231 (2000)

Chen, S.F., Goodman, J.: An empirical study of smoothing techniques for language modeling. In: Proceedings of the 34th Annual meeting on Association for Computational Linguistics, pp. 310–318. Association for Computational Linguistics (1996)

Cocke, J., Schwartz, J.T.: Programming languages and their compilers : preliminary notes. Technical report (1970)

Cohn, T., Blunsom, P., Goldwater, S.: Inducing tree-substitution grammars. The Journal of Machine Learning Research **11**, 3053–3096 (2010)

Finkel, J., Grenager, T., Manning, C.: The infinite tree. In: Proceedings of the 45th Annual Meeting of Association for Computational Linguistics, pp. 272–279 (2007)

Gasthaus, J., Teh, Y.W.: Improvements to the sequence memoizer. In: Advances in Neural Information Processing Systems, pp. 685–693 (2010)

Gelling, D., Cohn, T., Blunsom, P., Graca, J.: The PASCAL challenge on grammar induction. In: Proceedings of the NAACL-HLT Workshop on the Induction of Linguistic Structure, pp. 64–80. Association for Computational Linguistics (2012)

Goodman, J.: Parsing algorithms and metrics. In: Proceedings of the 34th Annual Meeting on Association for Computational Linguistics, ACL 1996, Stroudsburg, PA, USA, pp. 177–183. Association for Computational Linguistics (1996)

Johnson, M.: Pcfg models of linguistic tree representations. Computational Linguistics **24**(4), 613–632 (1998). ISSN 0891–2017

Johnson, M.: Unsupervised word segmentation for Sesotho using adaptor grammars. In: Proceedings of the 10th Meeting of ACL Special Interest Group on Computational Morphology and Phonology, Columbus, Ohio, pp. 20–27. Association for Computational Linguistics, June 2008

Johnson, M., Griffiths, T.L., Goldwater, S.: Adaptor grammars: A framework for specifying compositional nonparametric bayesian models. In: Advances in Neural Information Processing Systems 19, Proceedings of the Twentieth Annual Conference on Neural Information Processing Systems, Vancouver, British Columbia, Canada, December 4–7, pp. 641–648 (2006)

Johnson, M., Griffiths, T.L., Goldwater, S.: Bayesian inference for pcfgs via markov chain monte carlo. In: HLT-NAACL, pp. 139–146 (2007)

Klein, D., Manning, C.D.: Parsing and hypergraphs. In: Proceedings of the Seventh International Workshop on Parsing Technologies (IWPT-2001), October 17–19, Beijing, China (2001)

Klein, D., Manning, C.D.: Accurate unlexicalized parsing. In: Proceedings of the 41st Annual Meeting on Association for Computational Linguistics, vol. 1, pp. 423–430. Association for Computational Linguistics (2003)

Levenberg, A., Dyer, C., Blunsom, P.: A bayesian model for learning scfgs with discontiguous rules. In: Proceedings of the 2012 Joint Conference on Empirical Methods in Natural Language Processing and Computational Natural Language Learning, pp. 223–232. Association for Computational Linguistics (2012)

Liang, P., Petrov, S., Jordan, M., Klein., D.: The infinite PCFG using hierarchical dirichlet processes. In: Proceedings of the 2007 Joint Conference on Empirical Methods in Natural Language Processing and Computational Natural Language Learning (EMNLP-CoNLL), pp. 688–697 (2007)

Marcus, M.P., Marcinkiewicz, M.A., Santorini, B.: Building a large annotated corpus of english: The penn treebank. Computational Linguistics **19**(2), 313–330 (1993)

Matsuzaki, T., Miyao, Y., Tsujii, J.: Probabilistic cfg with latent annotations. In: Proceedings of the 43rd Annual Meeting on Association for Computational Linguistics, ACL 2005, Stroudsburg, PA, USA, pp. 75–82. Association for Computational Linguistics (2005). doi:10.3115/1219840.1219850

Mochihashi, D., Sumita, E.: The infinite markov model. In: Advances in Neural Information Processing Systems 20, Proceedings of the Twenty-First Annual Conference on Neural Information Systems, Vancouver, British Columbia, Canada (2007)

Petrov, S., Klein, D.: Learning and inference for hierarchically split PCFGs. In: Proceedings of the Twenty-Second AAAI Conference on Artificial Intelligence, Vancouver, British Columbia, Canada (2007)

Petrov, S., Barrett, L., Thibaux, R., Klein, D.: Learning accurate, compact, and interpretable tree annotation. In: Proceedings of the 21st International Conference on Computational Linguistics and the 44th Annual Meeting of the Association for Computational Linguistics, pp. 433–440. Association for Computational Linguistics (2006)

Teh, Y.W.: A hierarchical bayesian language model based on pitman-yor processes. In: Proceedings of the 21st International Conference on Computational Linguistics and the 44th Annual Meeting of the Association for Computational Linguistics, pp. 985–992. Association for Computational Linguistics (2006)

Thede, S.M., Harper, M.P.:. A second-order hidden markov model for part-of-speech tagging. In: Proceedings of the 37th Annual Meeting of the Association for Computational Linguistics on Computational Linguistics, ACL 1999, Stroudsburg, PA, USA, pp. 175–182. Association for Computational Linguistics (1999). ISBN 1-55860-609-3

Toutanova, K., Manning, M.D.: Enriching the knowledge sources used in a maximum entropy part-of-speech tagger. In: Proceedings of the 2000 Joint SIGDAT Conference on Empirical Methods in Natural Language Processing and Very Large Corpora, pp. 63–70. Association for Computational Linguistics (2000)

Wood, F., Archambeau, C., Gasthaus, J., James, L., Teh, Y.W.: A stochastic memoizer for sequence data. In Proceedings of the 26th Annual International Conference on Machine Learning, ICML 2009, Montreal, Quebec, Canada, June 14–18, p. 142 (2009a)

Wood, F., Archambeau, C., Gasthaus, J., James, L., Teh, Y.W.: A stochastic memoizer for sequence data. In: Proceedings of the 26th Annual International Conference on Machine Learning, pp. 1129–1136. ACM (2009b)

Wood, F., Gasthaus, J., Archambeau, C., James, L., Teh, Y.W.: The sequence memoizer. Commun. ACM 54(2), 91–98 (2011)

Zhu, C., Byrd, R.H., Lu, P., Nocedal, J.: Algorithm 778: L-bfgs-b: Fortran subroutines for large-scale bound-constrained optimization. ACM Transactions on Mathematical Software (TOMS) 23(4), 550–560 (1997)

Temporally Coherent Role-Topic Models (TCRTM): Deinterlacing Overlapping Activity Patterns

Evgeniy Bart[✉], Bob Price, and John Hanley

Palo Alto Research Center, Palo Alto, USA
{bart,bprice,jhanley}@parc.com

Abstract. The Temporally Coherent Role-Topic Model (TCRTM) is a probabilistic graphical model for analyzing overlapping, loosely temporally structured activities in heterogeneous populations. Such structure appears in many domains where activities have temporal coherence, but no strong ordering. For instance, editing a PowerPoint presentation may involve opening files, typing text, and downloading images. These events occur together in time, but without fixed ordering or duration. Further, several different activities may overlap – the user might check email while editing the presentation. Finally, the user population has subgroups; for example, managers, salespeople and engineers have different activity distributions. TCRTM automatically infers an appropriate set of roles and activity types, and segments users' event streams into high-level activity instance descriptions. On two real-world datasets involving computer user monitoring and debit card transactions we show that TCRTM extracts semantically meaningful structure and improves hold-out perplexity score by a factor of five compared to standard models.

1 Introduction

Models of user activities can be used to improve productivity and enable new services across a wide variety of domains such as finance, personal assistants, health care, and many others. However, such modeling is very challenging due to the complexity and variations in activities. We present a new generative model whose structure uniquely exploits properties of user activity streams in order to build better models of behavior in realistic contexts. Because these behavior models are generative, they can be used for a variety of classification and prediction tasks ranging from predicting future user needs, to detecting organizational saboteurs, to connecting users with common interests.

Real-world event streams (such as financial transaction streams or computer event logs) exhibit several forms of complexity. First, the latent structure is non-obvious, because semantically meaningful activities often manifest via groups of observed events that may be large, heterogeneous, and include significant variations in composition, order, and size. For example, editing a PowerPoint presentation may involve opening files, typing text, downloading images, and saving

© Springer International Publishing Switzerland 2015
A. Appice et al. (Eds.): ECML PKDD 2015, Part II, LNAI 9285, pp. 390–405, 2015.
DOI: 10.1007/978-3-319-23525-7_24

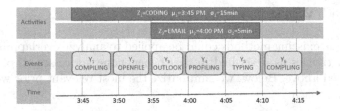

Fig. 1. A given user on a given day typically engages in multiple, possibly overlapping, activities. Each activity has a defined temporal extent. These activities in turn generate a sequence of events at specific times. Events from multiple activities may be interleaved.

files, but the exact order and frequency of these events varies significantly. Some users prefer saving their presentation more often than others; some presentations may involve a lot of text and not many images, while for others the opposite is true; some users download content using Firefox, while others prefer Chrome or Safari; and so on.

Second, the activities generating events can overlap in complex ways: a user email activity, a user powerpoint development activity and a background operating system update activity could all be active simultaneously. In real-world data streams, the events generated by these activities are intermixed in an extended history and are not segmented out or distinguished in any way.

Third, users typically comprise multiple distinct subgroups with very different behaviors. For instance, an office may include administrators, salespeople, and engineers; these will have very different computer activity distributions. Identifying these groups may be interesting in itself; in addition, attempting to create one model for all groups may result in poor performance.

As an example, consider the events associated with a workstation user who is coding and writing emails during the same interval (see Figure 1). These two activities result in interleaved operating system events such as compiling, opening of files and text entry. We note that while activities have temporal extents, the events generated by the activities do not have strong sequential orderings. Compiling, editing and file events are all part of coding, but the specific ordering of these is not highly determined.

A desirable goal is abstracting these complex raw event streams into a concise, high-level description of the user's typical activities. In this paper, we explain how our proposed model addresses the issues outlined above, and provide examples of its modeling capability on two challenging real-world datasets.

The remainder of this paper is organized as follows. In section 2, we summarize relevant prior work and highlight the differences from our proposed model. In section 3, we describe the proposed TCRTM model. The experiments in section 4 demonstrate that TCRTM successfully overcomes the challenges outlined above and significantly outperforms previously available methods. We conclude with some final remarks in section 5.

2   Related Work

Although some existing methods could be applied to analyze overlapping, loosely structured activities as described in this paper, none addresses all of the challenges outlined above. Below, we summarize the most relevant prior work, categorized by approach.

Handcrafted Models of Behavior. Behavior analysis has a long history in psychology and organizational theory [11], but these theories do not provide a formal computational model that can be used for prediction. The multi-agent systems community has applied computational agents to modeling of organizations [3]. These approaches are based on simulation, which allows us to see the implications of predefined behaviors. We find that our data changes too rapidly to admit handcrafted rules. Existing models from psychology and organizational behavior do not include an inferential component capable of directly extracting new behavioral insights from observational data.

Hidden Markov Models. Automatically learning human activity from observational data is a more recent endeavor but widely studied by many researchers. Many of these models employ variations on the Hidden Markov Model (HMM) [8]. Interestingly, in many transactional domains we find that the user activities have *temporal coherence* but do not have a strong sequential regularity. As illustrated in section 1, editing a PowerPoint presentation may involve events such as opening files, typing text, downloading images, and saving files, but the exact order and frequency of these events varies significantly from one case to another. Transitions between different activities are often not very structured as well; for example, some users may check email while working on a presentation, while others prefer to complete the current task before doing so. If such loosely structured activities were to be learned by an HMM, it would have to learn multiple different orderings of events and activities separately to account for all possible variations. This would combinatorially increase model complexity and training data requirements, and result in poor generalization.

Topic Models. Topic modeling is a key method for explaining discrete events in terms of a mixture of shared latent distributions. LDA [2] is perhaps the best-known example, although numerous generalizations exist (see below for some examples). Originally designed to simultaneously extract a set of topic vectors from a corpus of documents, and model the content of the documents as a mixture of extracted topics, LDA has been used in a wide variety of applications in which one wants to represent the content of the objects as a mixture of a small number of shared profiles.

A significant drawback of LDA is that it doesn't model time or temporal coherence. As a result, the topics it discovers may correspond not to a coherent user activity, but rather to a set of related events across multiple activities.

This is illustrated in section 4 and in Figures (3) and (4) giving examples of inferred activity types and LDA topics respectively. A detailed comparison of the proposed model to LDA is given in section 4.

Dynamic Topic Models. such as DTM [1] ignore local ordering of events (such as the order of words in a single document), and model how both the topics and mixture priors change over time (across multiple documents, typically spanning years). DTM could capture topic content changes such as the fact that articles about football have recently begun to include material on head injuries as well as capturing changes in preferences over topics such as a shift from football discussion topic to a cell-phone apps discussion topic over time. Variations of this model have been applied to to capture shopping behavior over time [5] and industrial chlorine sensor network streams [14]. The problems addressed by dynamic topic models are orthogonal to the problem discussed here; our focus is on exploiting local temporal coherence of events (within a single document) to improve the semantic meaning of extracted topics and achieve better fit to the data. The drift of these topics over time is not addressed in this paper, although that is certainly an interesting future direction. Returning to the example in Figure 1, we'd like our model to understand, for example, that the 'check email' activity started later than the 'coding' activity, rather than determine how the 'check email' activity evolved over 5 years.

Dynamic Processes Treated as Stationary Distribution. Topic models have been applied to activity recognition in video sequences [9]. Spatio-temporal interest points (small patches in time that capture visual and motion texture) are extracted from a short video segment. A topic model is used to find a small set of topics that explains the features extracted from a set of short video clips of various actions. The interpretability of the model can be increased by semi-supervised training in which the classes (identified with latent topics) are known and a subset of instances are labeled [15]. In cases where there are common behaviors and rare behaviors, the model can be augmented to share features between common and rare behaviors so that the model only has to model how rare behaviors are different from common behaviors [13] – reminiscent of hierarchical population model style inference when data is sparse. These models treat an entire clip as features drawn from a stationary distribution, so they will not work unless the data is already segmented into regions of coherent activity (which is the goal of this paper).

Encoding Time in Vocabulary Words. Topic models have been extended to include time by augmenting vocabulary words with time information [6]. In this model, the user's location behavior is categorized as being at home (H), work (W) or other (O). The sequence of locations sampled at 1/2 hours resolution is grouped into trigrams (e.g., HHH or WWO) which are then augmented with a "coarse day segment" number (early morning 0-7am=1, morning 7-11am =2, etc.)

to get segment-augmented trigrams such as HHH1 or HHH2. The augmented trigrams are fed into a generic author-topic model which learns that specific users have certain patterns such as being home early in the afternoon or going out in the evenings. This model does not have any notion of an activity independent of time – the observation distributions are directly coupled to coarse time segments. So it is not possible to learn about shopping behavior in the morning and use this to make inferences about a shopping trip planned in the afternoon.

Preprocessing via Topic Models for Dynamical Models. In some work, topic models are used as a preprocessing step for later stages of activity recognition [4]. Topic models can be used as part of a preprocessing step to compress or project high-dimensional signals down to a vector over a small set of topics. The topic indexes can then be fed to an HMM or classifier. These models combine the drawbacks of LDA and HMM: for the preprocessing step to succeed, LDA must extract relevant topics, which is difficult with loosely structured event streams (see section 4); while HMM at the subsequent stage will only succeed if the sequence of transitions between topics is highly structured.

Topic Models over Multiple Corpora. Practitioners have recognized that there may be distinct subpopulations which need to be modeled in different ways. Topic models have been extended to explicitly model the interests of authors [12]. Topic models have also be extended to cover multiple corpora over time [16] in order to expose commonalities and differences of different media over time. While these models capture subpopulations, they, like LDA, do not reflect temporal coherence.

Non-negative Matrix Factorization. Mixture models can be applied to data to pull out possibly overlapping subcomponents. Non-negative matrix factorization [7] has been a popular approach for factorizing data. The technique has been explicitly applied to signal separation [10]. While NNMF does utilize the idea of events being generated by mixtures, it doesn't exploit the temporal coherence of activities. As a result (just like with LDA, see above), the clusters extracted are not necessarily coherent activities, but just collections of events with similar properties. In addition, NNMF typically requires very structured input that can be organized in a matrix or a higher-dimensional tensor. This makes it difficult to apply to our data, where different users have a different number of events at different time points.

In summary, existing models do not handle the challenges of distinct subpopulations with loosely structured, temporally coherent activities found in many real-world datasets. In the next section, we develop a model that has elements of a mixture model but incorporates temporal coherence, subpopulations and a notion of activity instances to handle these challenges.

3 Model

The observations that we are interested in modeling generally consist of streams of discrete events. Each event description includes the user who performed it, the timestamp, and an 'event type'. This event type is a discrete category label such as 'file open', 'image download', and so on.

Two key components of our model are *activity type* and *activity instance*. *Activity type* is a general category of activity performed by users, such as 'checking email' or 'road trip'. *Activity instance* is a specific exemplar of that activity being performed by a given user at a given time, such as 'user 124 checking email at 10 am' or 'customer 71 taking a road trip to Las Vegas on June 14th'. In our model, activity types are modeled as multinomial parameter vectors ψ, with ψ_t specifying the distribution over event types for the given activity type t. For instance, ψ_{email} might be a distribution favoring events such as 'selecting message', 'sending email', 'typing up a response', etc. (cf. Figure 3(a)). Activity instances are modeled by selecting an activity type, as well as the mean and variance of event timestamps in the activity instance. The mean and variance parameters define the temporal extent of the activity. So an 'email' activity instance might be for '10:00 am ± 10 minutes', and the email-related events around this time (as determined by a Gaussian density) will be preferentially associated with this activity instance. The smoothness of the Gaussian likelihood will facilitate the sampler's exploration of assignments of events to instances.

One additional component of the proposed model addresses the fact that users can often be grouped into distinct subgroups based on their observed activity. For example, company employees have different job roles. These roles dictate the types of activities users typically engage in. In our model, roles are modeled as multinomial parameter vectors ϕ. Each parameter ϕ_r specifies the distribution over activity types that users in role r are likely to engage in. A software engineer role might have a high probability for coding-related activities such as code compilation, whereas a marketing role might have a high probability for email and presentation activities.

The plate diagram for the TCRTM model incorporating user roles, activity types and activity instances is shown in Figure 2, where the corresponding generative process is also summarized.

The proposed TCRTM model could be compared to standard LDA [2] as follows. Consider treating users as documents, and individual events as words. Then both LDA and TCRTM explain the observed event types in a document via a set of topics ψ_t. The difference is that LDA has no concept of activity instance; all events assigned to a topic t are treated equally, and their timestamps are ignored. In contrast, in TCRTM an event is assigned to a 'topic' via an intermediate I_{uj} variable that corresponds to a specific activity instance; therefore, an activity type (such as 'checking email') may be repeated multiple times by each user, and there is an explicit separation of these multiple instances. The timestamps in TCRTM are not ignored, but rather are used to encourage temporal coherence of individual instances.

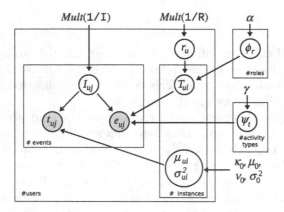

$Mult(1/I)$ $Mult(1/R)$ α

(a) Generative model

- For each activity type t, pick a multinomial distribution ψ_t over event types
- For each role r, pick a multinomial distribution ϕ_r over activity types
- For each user u:
  • Pick the role r_u for that user
  • For each of I activity instances:
    * Pick the activity type for the current activity instance from the role-specific distribution ϕ_{r_u}
    * Pick the mean and variance of timestamps for this activity instance
  • For each event j generated by the current user:
    * Pick an activity instance I_{uj}
    * Pick the event timestamp from instance-specific Gaussian distribution
    * Pick the event type from the activity for current activity instance

(b) Generative process

Fig. 2. (a): TCRTM plate diagram. Shaded nodes represent observable variables; variables not enclosed in circles represent hyperparameters. In the diagram, r_u is the role assigned to user u; ϕ_r is the distribution over activity types for role r; T_{ui} is the activity type for the i'th activity instance for user u; I_{uj} is the activity instance assigned to the j'th event of user u; e_{uj} is the event type and t_{uj} is the timestamp for the j'th event; ψ_t is the distribution of event types for activity type t; and μ_{ui} and σ_{ui}^2 represent the time and duration of activity instance i. The conditional distributions are as follows: $\phi_r \sim \mathrm{Dir}(\alpha)$; $r_u \sim \mathrm{Mult}(1/R)$; $T_{ui} \sim \mathrm{Mult}(\phi_{r_u})$; $I_{uj} \sim \mathrm{Mult}(1/I)$; $e_{uj} \sim \mathrm{Mult}(\psi_{T_{uI_{uj}}})$; $t_{uj} \sim N(\mu_{uI_{uj}}, \sigma_{uI_{uj}}^2)$; $\mu_{ui}, \sigma_{ui}^2 \sim NI\chi^2(\mu_0, \kappa_0, \nu_0, \sigma_0^2)$; $\psi_t \sim \mathrm{Dir}(\gamma)$. R: number of roles; I: number of instances per user. (b): the corresponding generative process.

Compared to hidden Markov models (HMMs), the TCRTM allows multiple simultaneous activities to take place. Such multitasking is common in many datasets; for example, workstation data exhibits considerable overlapping activity due to both the user's attempts at multi-tasking, as well as due to the system executing background processes during the course of the user's normal work. These simultaneous activities are separated from each other in

TCRTM (a process called 'deinterlacing'), and are grouped into coherent activity instances. In contrast, in standard HMM implementations multiple simultaneous activities are usually modeled by augmenting the latent space to include a cross-product of multiple activities – a process that increases the modeling complexity significantly. The TCRTM also loosely models temporal coherence without imposing ordering. Unlike an HMM, the TCRTM's activity instances prefer representations in which the events of an activity occur close together in time without requiring any specific ordering of these events. To get the same generalization power as a TCRTM, an HMM must be trained on enough data to learn each possible ordering.

Finally, in addition to the differences discussed above, TCRTM also incorporates the concept of roles. These determine the activity types users can engage in, but are otherwise not constrained by official job titles. As a result, informal, but significant subgroups of people will be allocated distinct roles; these could correspond to different job types when modeling an organization's computer logs, or to customer groups when modeling debit card transaction data. The advantages of the TCRTM are summarized below:

- Compared to HMM, TCRTM can deinterlace overlapping activities. This is important for many practical datasets.

- Compared to LDA, TCRTM can deal with observable events that are temporally coherent (as opposed to activity that occurs throughout an interval of time)

- Compared to HMM, TCRTM can deal with observable events that are not strictly ordered.

- Compared to both LDA and HMM, TCRTM models user roles. This allows finding coherent groups of people with similar behavior.

3.1 Inference

The goal of inference is to estimate the parameters of the model given a collection of observed events and the hyperparameters (α, γ, κ_0, etc.). The parameters of interest describe the inferred domain structure (for example, ψ_t describes activity types in terms of event types that are likely under that activity), as well as specific assignments of objects to clusters (for example, I_{uj} represents the activity instance to which the j'th event of user u is assigned).

Our overall approach is to use Gibbs sampling, which allows drawing samples from the posterior distribution of the model's parameters given the data. The parameters of interest can then be estimated from these samples. For efficiency, we use a collapsed Gibbs sampler, where the variables ϕ, ψ, μ, and σ^2 are integrated out, and the remaining discrete variables r_u, T_{ui}, and I_{uj} are sampled until convergence. Estimates for the integrated-out variables can then be obtained in terms of the discrete variables.

The expressions below are derived using standard methodology for Gibbs sampling; therefore, the derivations are omitted. The resulting conditional distributions are shown for completeness, as well as for intuition and for comparison to standard models.

As usual, the sampling distributions are expressed using count data. In our notation, N represents a count variable. Its superscript indicates what entities are being counted (for example, N^I is a count of activity instances and N^e is a count of individual events). The subscripts are indices of the relevant entities. A dot in place of an index indicates summation over that index. The current entity being sampled is omitted from the counts.

In the conditional sampling distributions below, N_{rt}^I is the number of activity instances that belong to users with role r that have activity type t, excluding the current instance. Similarly, N_{ut}^I is the number of instances of user u that have activity type t, and T is the total number of activity types.

The conditional sampling distribution for the role of user u is:

$$p(r_u = r_0 | \text{rest}) \propto \frac{\prod_t \Gamma(\alpha + N_{r_0 t}^I + N_{ut}^I)}{\Gamma(\alpha T + N_{r_0 \cdot}^I + N_{u \cdot}^I)} \cdot \frac{\Gamma(\alpha T + N_{r_0 \cdot}^I)}{\prod_t \Gamma(\alpha + N_{r_0 t}^I)}. \qquad (1)$$

The conditional sampling distribution for the activity type of the ith activity instance of user u is given next. Here, N_{te}^e is the number of events of type e assigned to activity type t, N_{uie}^e is the number of events of type e for user u assigned to activity instance i, and E is the total number of event types:

$$p(T_{ui} = t_0 | \text{rest}) \propto \frac{\alpha + N_{r_u t_0}^I}{\alpha T + N_{r_u \cdot}^I} \cdot \frac{\prod_e \Gamma(\gamma + N_{t_0 e}^e + N_{uie}^e)}{\Gamma(\gamma E + N_{t_0 \cdot}^e + N_{ui \cdot}^e)} \cdot \frac{\Gamma(\gamma E + N_{t_0 \cdot}^e)}{\prod_e \Gamma(\gamma + N_{t_0 e}^e)} \qquad (2)$$

The conditional sampler for the user u's j^{th} activity is given below. Here, t_ν is the Student's t distribution, and its parameters are $\nu_{ui_0} = \nu_0 + N_{ui_0 \cdot}^e$, $\kappa_{ui_0} = \kappa_0 + N_{ui_0 \cdot}^e$, $\mu_{ui_0} = \frac{\kappa_0}{\kappa_{ui_0}} \mu_0 + \frac{N_{ui_0 \cdot}^e}{\kappa_{ui_0}} \bar{t}_{ui_0}$, and

$$\sigma_{ui_0}^2 = \frac{1}{\nu_{ui_0}} \left[\nu_0 \sigma_0^2 + SS_{ui_0} - N_{ui_0 \cdot}^e \bar{t}_{ui_0}^2 + \frac{\kappa_0 N_{ui_0 \cdot}^e}{\kappa_0 + N_{ui_0 \cdot}^e} (\bar{t}_{ui_0} - \mu_0)^2 \right], \qquad (3)$$

where $\bar{t}_{ui_0}$ is the empirical mean and SS_{ui_0} is the empirical sum of squares of timestamps for user u, activity instance i_0.

$$p(I_{uj} = i_0 | \text{rest}) \propto \frac{\gamma + N_{T_{ui_0} e_{uj}}^e}{\gamma E + N_{T_{ui_0} \cdot}^e} \, t_{\nu_{ui_0}} \left(t_{uj} \mid \mu_{ui_0}, \frac{1 + \kappa_{ui_0}}{\kappa_{ui_0}} \sigma_{ui_0}^2 \right). \qquad (4)$$

It is interesting to compare these expressions with the corresponding sampling equation for regular LDA. In LDA, the probability of assigning an event e_{uj} to a topic z_0 is

$$p(z_{uj} = z_0 | \text{rest}) \propto \frac{\gamma + N_{z_0 e_{uj}}^e}{\gamma E + N_{z_0 \cdot}^e} \cdot \frac{\alpha + N_{u z_0}^e}{\alpha T + N_{u \cdot}^e}. \qquad (5)$$

In TCRTM, the equivalent of topics is activity types. Events, however, are not assigned to activity types directly; rather, events are assigned to activity instances via eq. (4), and the activity instance i is associated with an activity type given by T_{ui}. Comparing eq. (5) to eq. (4), we note that the first term for LDA is similar to the first term for TCRTM, except z_0 is replaced with T_{ui_0} (since the activity instance i_0 has activity type T_{ui_0}, which is equivalent to the topic z_0 in LDA). The second term in LDA is absent from eq. (4), but appears instead as the first term in eq. (2), except that individual users u are replaced with user roles r that combine multiple users, and the fact that in TCRTM, activity instances are counted instead of individual events. Finally, the last term in eq. (4) is absent from the LDA sampling because LDA doesn't model event time stamps. This term simply encourages individual events from a particular activity instance to be clustered in time.

4 Experiments

We have experimented with two datasets. The first dataset includes debit card transactions from over 300,000 users over a period of approximately 7 month. The users are the beneficiaries of various state government programs; once a month each card is loaded with an allotment of money which the users can subsequently spend. The total number of transactions is about 50 million. Each transaction includes a timestamp and a merchant code. This merchant code is a description of the general type of products sold or services provided, such as "Veterinary services" or "Hardware stores". This merchant code was used as the 'event type' in our model.

The second dataset includes data from monitoring user workstations at a large defense contractor. In this domain, the observables correspond to operating system primitives such as opening a file, executing a utility, or initiating a network connection. Each such primitive consists of two parts: the application that was used to perform the action (e. g., 'firefox.exe') and the action itself (e.g., 'ImageDownloadEvent'). About 5000 employees were monitored over one month, resulting in over 100 million individual events.

TCRTM is not very sensitive to the choice of hyperparameters. For our experiments, the following settings were used: $\alpha = 1$, $\gamma = 1$. These were selected using simple logarithmic grid search. In addition, μ_0 was set to the empirical mean of all the timestamps in each dataset, and $\kappa_0 = 0.0001$ was used to reduce influence of the prior mean on timestamp variance (eq. (3)). Further, σ_0^2 was set to $(D/I)^2$, where D is the duration of the modeled time period and I is the number of activity instances within that time period. This prior simply splits the entire time period into I intervals of roughly equal length (note that the posterior distribution will adjust this prior based on the actual observed data). Finally, ν_0 was set to 5.0, again, chosen using simple logarithmic grid search.

TCRTM was initialized at random, and then Gibbs sampling was run for 200 iterations. Examining the marginal likelihood revealed that the sampler converged typically after about 30 iterations (not shown).

nlnotes.exe:EmailViewed	68%
notes2.exe:GenericTextEvent	8%
nlnotes.exe:EmailSent	5%

(a) 'Checking email'

sshd.exe:FileReadEvent	39%
dropbox.exe:FileReadEvent	7%
httpd.exe:FileReadEvent	6%
searchprotocolhost.exe:FileReadEvent	3%

(b) 'Using dropbox'

onenote.exe:FileReadEvent	26%
wmplayer.exe:FileReadEvent	26%
firefox.exe:OtherDownloadEvent	3%

(c) 'Personal notes'

VRU BALANCE INQUIRY	46%
LOAD REGULAR DEPOSIT	10%
FINANCIAL INSTITUTIONS-AUTOMATED CASH DISBURSEMENT	9%
GROCERY STORES SUPERMARKETS	5%

(d) 'Cash operations'

FAST FOOD RESTAURANTS (QUICK PAY SERVICE PILOT)	18%
GROCERY STORES SUPERMARKETS	15%
VRU BALANCE INQUIRY	10%
DISCOUNT STORES	6%
SERVICE STATIONS WITH OR WITHOUT ANCILLARY SERVICE	6%

(e) 'Food'

RECORD SHOPS	9%
BOOK STORES	8%
DIRECT MARKETING-INBOUND TELEMARKETING MERCHANTS	8%
FAST FOOD RESTAURANTS (QUICK PAY SERVICE PILOT)	5%

(f) 'Culture'

Fig. 3. Example activity types learned automatically by TCRTM. For each activity type, top event types and corresponding probabilities are shown; the numbers are rounded to nearest integer. The captions are not part of the model and were given by the authors for illustration. (a)-(c): workstation dataset. (d)-(f): debit card dataset. As can be seen, TCRTM successfully identifies semantically related groups of events.

The remaining parameters of interest in TCRTM are the number of roles R, the number of activity types T, and the number of activity instances per user I. For the debit card dataset, we've used $R = 25$, $T = 25$, and $I = 14$. R was chosen by trial and error, T was chosen by observing that for settings of $T > 25$ duplicate activity types started appearing, and I was chosen so that there would be roughly two activity instances per month (so that beginning-of-the-month and end-of-the-month spending patterns could be separated).

For the workstation dataset, we've used $R = 10$ and $T = 100$. Since activity types in this dataset (such as 'checking email' or 'creating a presentation') are

powerpnt.exe:ProcessStopped	29%
powerpnt.exe:ProcessStarted	29%
excel.exe:ProcessStarted	2%
excel.exe:ProcessStopped	2%
winword.exe:ProcessStarted	2%
winword.exe:ProcessStopped	2%

(a) 'MS Office suite'

Fig. 4. Example topic learned by LDA on the workstation dataset. As can be seen, LDA grouped together a variety of events related to the MS Office suite. Although this grouping is understandable (as users who have Office installed typically use multiple applications within the suite), it is unlikely that they use all three applications simultaneously. Thus, the grouping reflects an artefact of software bundling rather than a semantically meaningful, coherent user activity.

of inherently much shorter average duration, we've used a setting of 10 activity instances per day, or 300 activity instances per month (which is our modeling period). While TCRTM could run with these settings as is, an additional observation is that activities are short and rarely span across day boundaries because of the way most people's work days are scheduled. Therefore, we modified the model in Figure 2 to treat each day separately, with the corresponding obvious modifications to the sampling equations. The effect of this change is that instead of selecting one of 300 global activity instances for each event (most of the 300 with very low probability), the model only needs to select one of 10 activity instances for a particular day. This makes the sampler computationally more efficient.

TCRTM was compared to standard LDA [2]. LDA was chosen as a basis for comparison because it is naturally suited to modeling observations that are generated by several distinct latent processes, as the observations in our datasets are. Alternative methods, such as HMM, are unlikely to perform well on our datasets because there are no natural fixed transitions between events and between activities.

4.1 Activity Types

Several activity types discovered automatically by TCRTM are shown in Figure 3. As can be seen, the model organizes event types into semantically meaningful groups. Note that these groups encompass events that occur together when a user performs a natural task; they are not limited to grouping together all events from a single executable file or all events that co-occur in temporal proximity. For example, in Figure 3(a), all event types that occur when working with email are identified, even though they are performed by separate executable files. Note that the executable names themselves were not available to TCRTM; the events in question were encoded as '1733077456:98', '2341822329:303', and '1733077456:96', providing no text similarity clues to the appropriate event

Table 1. Perplexity of LDA and TCRTM on the two datasets used. Lower values are better (note that perplexity measures the degree of surprise or confusion). Note that the values are on logarithmic scale. As can be seen, TCRTM significantly outperforms LDA on both datasets.

Method Dataset	LDA	TCRTM
Debit cards	11.44	**2.56**
Workstation data	10.12	**5.91**

groupings. Similarly, multiple executable files pertaining to using the Dropbox service were grouped together (Figure 3(b)), and preference of several users to listen to music while editing notes with OneNote was identified (Figure 3(c)).

In contrast, topics learned by LDA often reflected not a coherent activity, but rather related events across multiple activities. This is illustrated in Figure 4. As can be seen, LDA grouped together a variety of events related to the MS Office suite. Although this grouping is understandable (as users who have Office installed typically use multiple applications within the suite), it is unlikely that they use all three applications simultaneously. Thus, the grouping reflects an artefact of software bundling rather than a semantically meaningful, coherent user activity.

For debit card data, several interesting patterns were identified as well. For example, an activity type in Figure 3(d) groups together event types related to cash aspects of the card (checking the balance, receiving the monthly allotment, and withdrawing it as cash). Figure 3(e) shows an activity type related to buying food and other everyday items (such as gas).

The conclusion is that TCRTM can successfully identify semantically meaningful, coherent activity types.

4.2   Perplexity

Next, we compared TCRTM to LDA in terms of their ability to extract structure from data and anticipate future events. To perform the comparison, we split each dataset into a training and hold-out set (there was no overlap between the two sets). TCRTM and LDA were both fitted to the training set, and the perplexity of the corresponding models was then evaluated on the hold-out set. For this, the log-probability of each hold-out event was evaluated for each model under the parameters of an immediately preceding observed event. The results are reported in Table 1. As can be seen, TCRTM significantly outperforms LDA on both datasets.

(a) User histories in which each row of colored pixels denotes sequence of event types

(b) User histories ordered by inferred TCRTM roles

<table>
<tr><td>■ AUTOMATED CASH DISBURSEMENT</td><td>■ GROCERY STORES SUPERMARKETS</td></tr>
<tr><td>■ VRU BALANCE INQUIRY</td><td>■ FAST FOOD RESTAURANTS</td></tr>
<tr><td>■ LOAD REGULAR DEPOSIT</td><td>■ DISCOUNT STORES</td></tr>
<tr><td>■ EATING PLACES RESTAURANTS</td><td>■ DRUG STORES PHARMACIES</td></tr>
<tr><td>■ SERVICE STATIONS</td><td></td></tr>
</table>

Fig. 5. Effect of TCRTM roles. (a): unordered histories; (b): histories ordered by TCRTM roles. In the top half of subfigure (b) we see *blue* online grocery purchases over the whole monthly cycle whereas in the bottom half of (b) we see a concentration of *bright green* ATM cash withdrawls early in the monthly cycle (cf. Figure 3(d)) and few online transactions of any type mid month. Thus TCRTM automatically infers cash-based vs. online client types. (Best viewed on-screen, enlarged and in color.)

4.3   Effect of Roles

To visualize the effect of modeling user roles, we performed an experiment on a small subset of the debit card dataset that contained a randomly selected set of 287 users and 50,000 transactions. The same settings as for the main dataset were used, except we reduced the number of roles to 2 (this was done for easier visualization, as well as due to the small size of the subset). The results are shown in Figure 5 (the figure is best viewed on-screen, enlarged and in color). In both images, each row corresponds to a different user. Time flows from left to right. Each transaction type is color-coded according to the legend at the bottom. Thus, each row shows a snapshot of user's behavior over 3 months (note that the full 7 months of data were used for modeling, but the display was truncated to 3 months due to space considerations). In Figure 5(a), the users are shown in the original, random order. In Figure 5(b), the same set of users was rearranged by their role, as inferred automatically by TCRTM. Thus, the only difference between sub-figures 5(a) and 5(b) is that in Figure 5(b), rows are shuffled such that users with the same role are clustered together. As can be seen, TCRTM groups users with similar behavior into the same role. There are noticeable similarities in behavior between users with the same role, and significant differences in behavior across roles. For example, the bottom part in Figure 5(b) contains more bright green transactions, corresponding to cash-based activities (cf. Figure 3(d)), and is overall brighter, while the top part contains more blue transactions and is overall darker. Examining the conditional distributions inferred for the roles indeed confirms that for the 'AUTOMATED CASH DISBURSEMENT' transactions (color-coded bright green), the probability under role 2 (corresponding to the bottom part in Figure 5(b)) is 0.19, while for role 1 (corresponding to the top part) it is only 0.08. For the 'GROCERY STORES SUPERMARKETS' transaction (color-coded dark blue), the probability under role 1 (top part) is 0.12, while under role 2 (bottom part) it is only 0.06. The conclusion is that the role modeling aspect of TCRTM identifies semantically meaningful groups of users.

5   Conclusions

The experiments comparing temporally coherent role-topic model (TCRTM) to conventional models such as LDA suggest that TCRTM can exploit the local temporal coherence of events and population subgroup modeling to increase the predictive power of the model. The significant increase in modeling power makes us optimistic about the potential for TCRTM to improve a broad range of applications related to activity analysis such as prediction, recommendation and classification. As such, we argue that the TCRTM is an important milestone that will stimulate more accurate algorithms for real-world transactional activity analysis applications.

Acknowledgments. The authors gratefully acknowledge support for this work from DARPA through the ADAMS (Anomaly Detection At Multiple Scales) program funded project GLAD-PC (Graph Learning for Anomaly Detection using Psychological Context). Any opinions, findings, and conclusions or recommendations in this material are those of the authors and do not necessarily reflect the views of the government funding agencies.

References

1. Blei, D.M., Lafferty, J.D.: Dynamic topic models. In: Proceedings of the 23rd International Conference on Machine Learning, pp. 113–120. ACM (2006)
2. Blei, D.M., Ng, A.Y., Jordan, M.I.: Latent dirichlet allocation. In: NIPS, pp. 601–608 (2001)
3. Dignum, V. (ed.): Handbook of Research on Multi-Agent Systems: Semantics and Dynamics of Organizational Models (2009)
4. Huynh, T., Fritz, M., Schiele, B.: Discovery of activity patterns using topic models. In: Proceedings of the 10th International Conference on Ubiquitous Computing, UbiComp 2008 (2008)
5. Iwata, T., Watanabe, S., Yamada, T., Ueda, N.: Topic tracking model for analyzing consumer purchase behavior. In: IJCAI (2009)
6. Gatica-Perez, D., Farrahi, K.: Discovering routines from large-scale human locations using probabilistic topic models. In: ACM Transactions on Intelligent Systems and Technology (2011)
7. Lee, D.D., Seung, H.S.: Algorithms for non-negative matrix factorization. In: NIPS, pp. 556–562. MIT Press (2000)
8. Natarajan, P., Nevatia, R.: Coupled hidden semi markov models for activity recognition. In: Proceedings of the IEEE Workshop on Motion and Video Computing, WMVC 2007, Washington, DC, USA, p. 10. IEEE Computer Society (2007)
9. Niebles, J.C., Wang, H., Li, F.-F.: Unsupervised learning of human action categories using spatial-temporal words. International Journal of Computer Vision **79**(3), 299–318 (2008)
10. Plumbley, M.: Conditions for non-negative independent component analysis. IEEE Signal Processing Letters **9**(6), 177–180 (2002)
11. Robbins, S.P.: Organizational Behavior: Concepts, Controversies and Applications, 5th edn. Prentice-Hall (1991)
12. Rosen-Zvi, M., Griffiths, T., Steyvers, M., Smyth, P.: The author-topic model for authors and documents. In: Proceedings of the 20th Conference on Uncertainty in Artificial Intelligence, UAI 2004, Arlington, Virginia, USA, pp. 487–494. AUAI Press (2004)
13. Gong, S., Hospedales, T.M., Li, J.: Identifying rare and subtle behaviors: A weakly supervised joint topic model. In: Pattern Analysis and Machine Vision (2011)
14. Wei, X., Wang, X., Sun, J.: Dynamic mixture models for multiple time-series. In: IJCAI (2007)
15. Mori, G., Wang, Y.: Human action recognition by semilatent topic models. In: Pattern Analysis and Machine Intelligence (2009)
16. Zhang, J., Song, Y., Zhang, C., Liu, S.: Evolutionary hierarchical dirichlet processes for multiple correlated time-varying corpora. In: Proceedings of the 16th ACM (2010)

The Blind Leading the Blind: Network-Based Location Estimation Under Uncertainty

Eric Malmi[1,2]([✉]), Arno Solin[1,2], and Aristides Gionis[1,2]

[1] Helsinki Institute for Information Technology, Helsinki, Finland
{eric.malmi,arno.solin,aristides.gionis}@aalto.fi
[2] Department of Computer Science, Aalto University, Espoo, Finland

Abstract. We propose a probabilistic method for inferring the geographical locations of linked objects, such as users in a social network.
Unlike existing methods, our model does not assume that the exact locations of any subset of the linked objects, like neighbors in a social network, are known. The method efficiently leverages prior knowledge on the
locations, resulting in high geolocation accuracies even if none of the locations are initially known. Experiments are conducted for three scenarios:
geolocating users of a location-based social network, geotagging historical church records, and geotagging Flickr photos. In each experiment, the
proposed method outperforms two state-of-the-art network-based methods. Furthermore, the last experiment shows that the method can be
employed not only to network-based but also to content-based location
estimation.

1 Introduction

Observations recorded as data are typically associated with a location. Similarly, data attributes are often spatially correlated. For example, consider friendship relations in a social network that correlate with the geographical distances
between friends, or business types that cluster in different parts of a city, such
as restaurants and cafés being more concentrated in touristic areas.

On the other hand, there is a plethora of available datasets that lack explicit
location information, even though the data objects they contain are inherently
associated with a location (or a distribution of locations). For instance, consider
online social networks where only a small fraction of the users provide their
location explicitly. As a second example, motivated by the domain of historical
research and social sciences, consider historical documents, such as letters or public registry records. Such documents contain many pieces of valuable information,
but are often not accurately geolocated, either because their authors assumed
that the location is implicit, or because there is a reference to an uncertain
location, as the location of an old village can be uncertain.

As can be easily motivated from the previous examples, identifying the location of data objects is an important problem, and has compelling applications.
For example, as pointed out by Backstrom et al. [2], locating the users of an
online social network can be used to improve the network security by detecting

© Springer International Publishing Switzerland 2015
A. Appice et al. (Eds.): ECML PKDD 2015, Part II, LNAI 9285, pp. 406–421, 2015.
DOI: 10.1007/978-3-319-23525-7_25

"phishing" attempts, or to improve the user experience by offering personalized functionalities. Similarly, locating place names or whereabouts of people mentioned in historical documents is an extremely valuable tool for research in history or other social sciences. Yet another domain where geolocation can provide vital insights is the field of forensics where it has been used to pinpoint serial offenders [9].

In this paper, we propose a new method for inferring the geographical location of linked objects. In a nutshell, our method can be described as follows. We consider a set of objects for which certain attributes are known but the location information is missing. We assume that the known attributes can be used to provide two types of additional information:

1. *A prior distribution of each data object over a set of candidate locations.* For instance, in the social network scenario, known friend locations can be taken as candidate locations [2]. Or if we know the city where the object is located, we can simply define a grid over the city and impose a uniform prior over the grid cells, as done in the Flickr experiment in this paper.
2. *Links between data objects.* In the case of social networks, friendship relations between users are available, indicating that friends are more likely to be located in nearby locations. Similarly, in the photo-location application, two photos taken by the same user within a short time interval are more likely to be located in nearby locations.

Our method follows a probabilistic inference approach and gives predictions for the locations of the objects in the dataset by taking into account the prior distribution over locations and the links between objects.

The most closely-related work to our paper is the study of Backstrom et al. [2], who propose a probabilistic model for inferring the locations of Facebook users, given the location of their friends. However, our method extends and improves this prior work in the following ways:

– Our method does not assume that the exact locations for any of the linked data objects, like neighbors in a social network, are known. Instead we impose a prior distribution over these locations. This generalization makes the method very well suited to cope with the uncertainty that is present in most datasets. The case that the location of some objects is known, can also be naturally incorporated in our model.
– The proposed model offers a general abstraction that can be used to infer the locations of any kind of linked data with spatial dependencies. We demonstrate the generality of the model by applying it to three application scenarios: (*i*) geolocating users in a social network; (*ii*) geotagging historical church records from the 1600s to 1800s; and (*iii*) geotagging Flickr photos.
– The last experiment regarding Flickr photos shows that the proposed method can be adapted to content-based analysis, even though it is primarily designed for network-based geolocation.

Even though many relevant problems naturally fall under this problem setup as is demonstrated in the experiments, to our knowledge there is a lack of methods that would attempt to perform location estimation based on linked items whose locations are not known exactly.

The rest of the paper is organized as follows. In Section 2 we give an overview of previous related work. Our method is presented in Section 3, where we formalize the abstract geolocation problem and present the probabilistic algorithm for solving it. The three experiments discussed above are presented in detail in Section 4. Section 5 contains a final discussion and suggestions for future research directions.

2   Related Work

With the abundance of data gathered from all kinds of human activity, and the wide spread of social media applications that support collection of large amounts of user-generated content, problems related to geolocating various types of data have gained importance. As a result, a large number of related papers have appeared in machine learning, data mining, and web science venues. These pieces of work can be roughly categorized under network-based and content-based methods [10].

Network-Based Geolocation. In network-based methods, only the network structure and the location information about the other nodes are used for geolocation. Examples of this type of approach are the methods proposed by Backstrom et al. [2] and Jurgens [5]. These two methods will be further described in Section 3.4.

Rout et al. [10] approach the user geolocation problem as a classification problem and apply an SVM classifier. The classification is done on a city-level, and for each city a number of features, including the number of friends in the city and its total population, are extracted. The performance on the city prediction problem is better than the performance by the Backstrom et al. [2] method, but it is not clear how the classification approach scales down if we need to predict more fine-grained locations since the number of classes and sparsity of the data would both increase.

McGee et al. [8] build on the work of Backstrom et al., and they study the effect of incorporating information about tie strengths in the geolocation model. This line of thinking is complementary to our method. As we discuss later, our approach supports having different edge types, which can be learned separately. Nevertheless, our focus is in the general network-based geolocation problem, whereas many of the features used for inferring the tie strength in McGee et al. are Twitter specific, like the number followers and mentions in Twitter.

Sadilek et al. [11] propose a probabilistic method for location prediction based on dynamic Bayesian networks. This method is shown to provide high accuracy estimates, but the problem setting is more specific than ours. They assume that a time series of friend locations is provided as an input for the system.

Content-Based Geolocation. A different approach to the problem of geolocating users in social networks is to perform a more detailed analysis on the content generated by users. Most of the content-based methods from the recent years have focused on geolocating Twitter users [1,6,7,13].

The focus of our work is in network-based geolocation, but in our last experiment, we show that the proposed method can be used for content-based estimation as well. In that experiment, we aim to geotag Flickr photos based on textual annotations (tags). We show that the solution obtained using our framework corresponds to the method proposed by Serdyukov et al. [12]. Additionally, we present the idea of linking consecutive photos of a user in order to estimate their locations jointly and show that it slightly improves the geotagging accuracy. Another approach for geotagging Flickr photos has been proposed by Crandall et al. [4]. Their method also uses tags, but additionally they extract visual features (SIFT descriptors) from the photos. The photos are then geolocated using the resulting distribution over the joint feature space. However, the idea of estimating the locations of consecutive photos jointly is not explored in either of these works.

3   Methods

In this section we describe the general setting for the geolocation problem. As mentioned in the introduction, our model offers a unified framework that can fit various, seemingly very different types of geolocation problems. We present our solution in two steps. First we derive an exact solution for the general problem in the case of geolocating a single object. Then we extend to the multiple object case and show how to obtain an approximate solution. Later we show how the specific application scenarios fit under the general model.

3.1   Problem Setting

We consider a set V of items whose locations we want to find out. We assume that relations between items have been observed and represented by a set of edges E. Thus, the data items form a graph $G = (V, E)$. The neighbors of an item u in the graph G are denoted by $N(u) = \{v \mid \{u, v\} \in E\}$.

We also consider a discrete set of locations $\mathcal{L}$ which are the candidate locations to place the items in V. A distance function $d : \mathcal{L} \times \mathcal{L} \to \mathbb{R}$ is defined between locations, such that $d(\ell_1, \ell_2)$ denotes the distance between locations $\ell_1, \ell_2 \in \mathcal{L}$. In this paper, $d(\cdot, \cdot)$ is considered to be a geodetic distance, except for the photo geotagging case study where employ use the Manhattan (city block) distance since we consider the center of New York City. For each item $u \in V$ we write $\ell(u) \in \mathcal{L}$ to denote the location where to u is mapped. The mapping of all the items in V to locations in $\mathcal{L}$ is denoted by (boldface) vector $\boldsymbol{\ell}$, in other words, $\boldsymbol{\ell} = \langle \ell(u) \mid u \in V \rangle$.

We model uncertainty by considering a probability distribution $\Pr[\ell(u)]$ for item u over the space of possible locations. In our problem formulation we assume that initially, as part of the input, a prior distribution $\Pr[\ell(u)]$ for each item u is given. This is a fairly natural assumption for many applications as illustrated by

Observed edges   Prior distributions

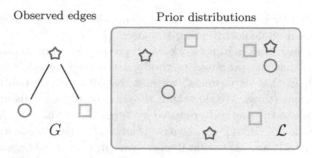

G $\mathcal{L}$

Fig. 1. A simple example of the GeoLocation problem with three items to be located, and a set of discrete candidate locations.

our experiments. For item u, we denote by $\mathcal{L}(u)$ the subset of locations in $\mathcal{L}$ for which the prior distribution $\Pr[\ell(u)]$ is non-zero. In other words, $\mathcal{L}(u)$ is the set of candidate locations where to place item u. Depending on the application, $\mathcal{L}(u)$ may be a significantly smaller set than $\mathcal{L}$. In practice, we can further prune the set $\mathcal{L}(u)$ by removing locations that have very small prior probability for u. If the exact location of an item u is known then $\Pr[\ell(u)]$ is a delta distribution. If no information about an item u is known then $\Pr[\ell(u)]$ is the uniform distribution. Note that we also assume that the set of candidate locations $\mathcal{L}$ is discrete. This is the case with the first two of our case studies. In cases where the set of locations is continuous, we can discretize it on a set of grid cells, as done in our Flickr photo geolocation case study.

We consider data for which spatial dependencies are present. The existence of an edge between two items u and v is thus assumed to depend on the location of the items. In our probabilistic model setting, the locations of the items are viewed as model parameters and edges as the observed data. Each edge $\{u, v\} \in E$ is assumed to be produced by a generative process that depends on the location of the two items u and v. Given two candidate locations $\ell(u)$ and $\ell(v)$ for items u and v we write $\Pr[\{u, v\} \in E \mid \ell(u), \ell(v)]$ for the likelihood of an edge between u and v given their candidate locations. To simplify the problem we assume that an edge depends only on the distance between the two candidate locations, and we write $\Pr[\{u, v\} \in E \mid d(\ell(u), \ell(v))]$ to denote this likelihood.

In the context of social networks, the generative process would correspond to the process of people forming social ties. Even for online social networks, formed in a virtual world, distance have been shown to play an important role in the process of relationship formation [2]. We also note that we may have different kinds of edges and the likelihood of an edge may depend also on the edge type. For instance, in our third case study, where we estimate the locations of Flickr photos based on their tags and consecutive photos, we have two types of edges: "photo-to-tag" and "photo-to-photo."

Our problem can now be defined as follows.

Problem 1. (*GeoLocation*) Consider a graph $G = (V, E)$ over items V, and a set of candidate locations $\mathcal{L}$. For each item $u \in V$ we are given a prior distribution

$\Pr[\ell(u)]$. The goal is to infer a mapping ℓ of items to locations in order to maximize the likelihood $\Pr[E \mid \ell]$ of observing the data given the inferred locations.

In some cases, we are interested in estimating the locations for a subset of the items $U \subseteq V$. In this case, we consider that for the rest of the items $V \setminus U$ the prior distributions are kept fixed. This special case can be easily incorporated in our framework. Unless stated otherwise, we assume that $U = V$.

An illustration of a very simple instance of the geolocation problem is shown in Figure 1. In this case there are three items (depicted by three different shapes: $\bigcirc$, $\heartsuit$, and $\square$) and two edges, as shown in the left side of the figure. The prior distribution of each item to 2 or 3 candidate locations is assumed to be uniform and it is shown in the right. If the edge probability increases as the distance between items decreases, then the maximum-likelihood estimate will give a solution according to which the items lie in the upper-right corner of the figure.

3.2   Estimating a Single Location

We start by deriving the *maximum a posteriori* (MAP) estimate for the location $\ell(u)$ of a single item $u \in V$ assuming that the location distributions of the other items are kept fixed. The likelihood function of the observed edges is given by

$$\Pr[E \mid \ell(u)] = \sum_{\boldsymbol{\ell}_{N(u)}} \Pr[E, \boldsymbol{\ell}_{N(u)} \mid \ell(u)]$$

$$= \sum_{\boldsymbol{\ell}_{N(u)}} \Pr[\boldsymbol{\ell}_{N(u)} \mid \ell(u)] \Pr[E \mid \ell(u), \boldsymbol{\ell}_{N(u)}]$$

$$= \sum_{\boldsymbol{\ell}_{N(u)}} \prod_{v:\{u,v\}\in E} \Pr[\ell(v)] \Pr[\{u,v\} \in E \mid \ell(u), \ell(v)],$$

where $\boldsymbol{\ell}_{N(u)} = \{\ell(v) : \{u,v\} \in E\}$ are the locations of the neighbors of u and the summation goes over all different candidate locations of each neighbor. In the above derivation we have assumed independence for the prior probabilities of different locations and for the different edges. As already discussed, we further assume that the likelihood of an edge being present, given the locations of the adjacent vertices, $\Pr[\{u,v\} \in E \mid \ell(u), \ell(v)]$, only depends on the distance $d(\ell(u), \ell(v))$ of the locations. By reordering the terms, we then get

$$\Pr[E \mid \ell(u)] = \prod_{v:\{u,v\}\in E} \sum_{\ell(v)} \Pr[\ell(v)] \Pr[\{u,v\} \in E \mid d(\ell(u), \ell(v))]. \qquad (1)$$

The maximum a posteriori estimate for $\ell(u)$ is then given by

$$\hat{\ell}(u) = \arg \max_{\ell(u)\in\mathcal{L}(u)} \Pr[\ell(u)] \Pr[E \mid \ell(u)].$$

If no information is provided for the items and the prior is set to the uniform distribution the MAP estimate corresponds to the *maximum-likelihood estimate* (MLE).

Alg. 1. Approximate maximum likelihood estimation for multiple locations.

Input: Graph $G = (V, E)$, prior distributions $\Pr[\ell(v)]$ for each $v \in V$, and items
     whose locations we want to estimate $U \subseteq V$.
Output: Locations for each $u \in U$.
Initialize a list of lists T ;   // Stores likelihood of each candidate location of each user
for $i \leftarrow 1$ **to** max_iter **do**
    foreach $u \in U$ **do**
        foreach $j \in \mathcal{L}(u)$ **do**
            $T_{u,j} \leftarrow \Pr[E \mid \ell(u) = j]$;   // Use Eq. (1)

    foreach $u \in U$ **do**
        $T_{u,:} \leftarrow \frac{T_{u,:}}{\sum_j T_{u,j}}$;   // Normalize distribution
        $\Pr[\ell(u)] \leftarrow T_{u,:}$;   // Update priors in a batch

return $\arg\max_{\ell(u)} \Pr[\ell(u)]$ for each $u \in V$;

In the above estimation we assumed that the term $\Pr[\{u, v\} \in E \mid d(\ell(u), \ell(v))]$ is known. This probability function can be learned from training data containing items with known locations and some edges between them. Learning the edge probability function includes also the case that there are edges of different types. In this case the edge probability function may depend on the edge type. It should be expected that the edge probability function is a monotonically decreasing function of the distance, and indeed, this is the case in all three of our case studies. However, the model does not require monotonicity.

3.3 Estimating Multiple Dependent Locations

We now show how to extend the method to find MAP or MLE location estimates for all items jointly. The likelihood of the edges in the graph, given all locations is given by

$$\Pr[E \mid \ell] = \prod_{\{u,v\} \in E} \Pr[\{u, v\} \in E \mid \ell] = \prod_{\{u,v\} \in E} \Pr[\{u, v\} \in E \mid d(\ell(u), \ell(v))].$$

This function is not maximized by simply computing the MLE for each location individually using Eq. (1). The reason is that if the distribution of $\ell(u)$ is updated, it will potentially change the MLEs of u's neighbors.

In order to get an approximate MLE, we use a simple iterative method. In each iteration, for each item u the prior $\Pr[\ell(u)]$ is recomputed using the current estimate for the locations of the neighbors of u. The recomputation of $\Pr[\ell(u)]$ is done by computing $\Pr[E \mid \ell(u) = j]$, using Eq. (1), for each candidate location $j \in \mathcal{L}(u)$ and normalizing. The method terminates when the estimates converge or when a maximum number of iterations is reached. The method is illustrated in Algorithm 1.

Note that an alternative way of getting an approximate solution for the locations of all items could be to define an inference problem for an *undirected graphical model*, where the items would correspond to the vertices of the graph, and use methods such as loopy belief propagation for estimating the locations. However, in our early experimentation we noticed that the estimation of conditional density functions $\Pr[\ell(u) \mid \ell(v)]$ used in graphical models is more challenging than the estimation of edge probabilities, since the former seems to depend more heavily on geographical characteristics, like oceans or metropolitan areas.

3.4 Baseline Methods

The method closest to our approach is the algorithm proposed by Backstrom et al. [2] for the problem of determining the locations of Facebook users. They also compute a MLE for each user and then iterate the computation step with the user locations updated in a batch. However, there are two key differences compared to our method: First, they consider only friends whose location is known exactly and after each iteration the users are assigned to their most likely locations. If there are multiple almost as probable locations for a user, assigning her to a single location seems harsh, which is why we have designed our method to work with location distributions. Second, they include an additional term which is a product over all edges not being present. However, they state that this term typically plays a small role and is expensive to compute, so in our experiments we run a slightly simplified version of their algorithm by omitting the additional term. This baseline method is referred to as BACKSTROM*.

As another baseline, we use the method proposed by Jurgens [5]. This method is designed for cases when only a small fraction of locations is known initially, and the idea is to propagate location "labels" in the network until all users have been geolocated. The author experimented with different ways for selecting the user location $\ell(u)$ based on the known neighbor locations $\ell(v)$, and the best performance was obtained by selecting the geometric median

$$\arg\min_{\ell(u)} \sum_{\ell(v)} d(\ell(u), \ell(v)).$$

We note that this term can be rewritten as

$$\arg\max_{\ell(u)} \prod_{\ell(v)} e^{-d(\ell(u), \ell(v))},$$

which, quite interestingly, shows that the method is equivalent to BACKSTROM* given that

$$\Pr[\{u, v\} \in E \mid d(\ell(u), \ell(v))] \propto e^{-d(\ell(u), \ell(v))}.$$

We refer to this baseline method as JURGENS.

4 Experiments

The problem of determining the geographical location of an entity—such as a person in a social network, a tweet, a photo, or virtually any piece of information—is an integral part of many services. We present three very different types of geolocation problems which can be all represented and solved under the same general framework described in Section 3.1. Our experiments are performed on publicly available data, and on data collected via public APIs. To facilitate reproducibility of our results, we have made the software used for the experiments available at: https://github.com/ekQ/geolocation.

4.1   Predicting Social Network User Home Locations

In this experiment, we use a similar setup that was used in Backstrom et al. [2] and vary the fraction of users, whose locations are initially known exactly. For the remaining users, we try to find the best location from a candidate set consisting of the known or the most probable locations of the neighboring users. Then we start iterating and update the candidate sets in the beginning of each iteration. This method is compared with BACKSTROM* and JURGENS presented in Section 3.4.

Data. We use a location-based social network called Brightkite [3]. This dataset contains $58\,228$ users, $214\,078$ edges between the users, and $4\,491\,143$ check-ins by the users. In order to estimate the ground truth locations of the users, we simply compute the median latitude and longitude of their check-ins. The users are randomly split into a training set (50%), used for fitting the models, and a testing set (50%), used for evaluating the geolocation performance. Any edges between the training and testing users are ignored.

Social Network Experimental Results. The proposed method, which keeps track of the uncertainty in the estimates, is compared to two other recently proposed methods which assign a single location to a user instead of a distribution. If a user already initially has some friends whose exact location is known, then it is not clear that keeping track of the location distributions should improve the estimation. And even if none of the user's friends have a known location, they will eventually be assigned one if the graph is connected, as the estimates will propagate throughout the graph, enabling location estimation for each user. Nevertheless, the results in Figure 2 show that the proposed method outperforms BACKSTROM* and JURGENS. Accuracy is defined as the fraction of users geolocated within 40 km from the ground truth location and the average accuracy improvement over BACKSTROM* is 0.5 percentage points. McNemar's test confirms that the improvements are statistically significant ($p < 0.001$) up till fraction 0.6. After this point, quantifying the uncertainty does not help anymore since most of the neighbor locations are known exactly. Compared to JURGENS, the difference is more clear, the average improvement being 4.7 percentage points. Figure 3 shows a power-law fit for the term $\Pr[\{u, v\} \in E \mid d(\ell(u), \ell(v))]$ employed in MLE and BACKSTROM*.

4.2   Geotagging Historical Church Records

We consider a big historical dataset containing digitized church records from Finland. The digitalization from the original hand-written documents has been

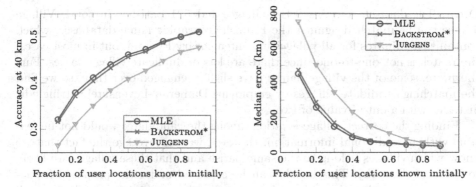

Fig. 2. Brightkite user geolocation performance with the proposed method (MLE) and two other recently proposed methods.

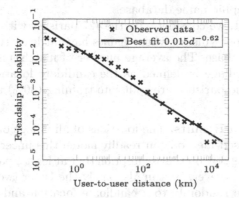

Fig. 3. Probability of two Brightkite users being friends given their distance and the power law fit to the data.

obtained by volunteers in the "HisKi" project.[1] This data can be considered an early population register, which was kept by the Evangelical Lutheran Church, the national church of Finland. The data contains millions of records of births, deaths, marriages and migration, spanning approximately three hundreds years from the 1600s to the late 1800s. The coverage of the church records was originally close to full, but the digitized material covers only parts of the complete dataset (some material is not digitized yet, and some is lost).

As part of the exploratory data analysis of the HisKi dataset, our interest in this paper is to attach geographical coordinates to the records in the data by geolocating the village associated with each record. This is not a trivial problem since most village names are not unique. In addition to the village name, the records contain the associated parish name, and the name of the estate/farm.

---

[1] The Genealogical Society of Finland has an online interface to the HisKi data: http://hiski.genealogia.fi/hiski?en

Data. The dataset contains 9 410 villages and 521 parishes in total. Village names can be matched against the Finnish geographic name database,[2] which contains coordinates for all villages in contemporary Finland, but in most cases, the match is not one-to-one since there are lots of duplicate village names. Furthermore, some of the village names have slightly changed over time, so we find the matching candidate villages by employing Damerau–Levenshtein string edit distance with a cutoff value of two.

Finding the correct village location among the candidates would not be feasible without additional information. However, we can leverage the fact that we know which villages belong to the same parish and that these villages are likely to be located nearby. This insight can be captured under the proposed framework by representing the village data as a graph, containing a node for each village and an edge between each pair of villages belonging to the same parish. The prior distribution for a village node is given by the locations of the matching villages in the geographic name database.

In this experiment, we consider only the 427 parishes with a known location and their 4 574 member villages whose names have at least two matches in the geographic name database. The average number of matches is 7.2. The ground truth location of a village is defined as the candidate location nearest to the associated parish. The parishes are split into training (30%) and testing (70%) parishes.

Village Geolocation Results. The locations of all village objects are initially unknown. The proposed method can readily model this uncertainty but BACK-STROM* and JURGENS rely on the assumption that at least a part of the locations are initially known. However, we can also apply the latter two methods by first assigning each village randomly to a candidate location and then running the methods. Since some of these initial guesses will be correct, the algorithms might be able to gradually find more and more correct locations. This is indeed what happens as is shown in Figure 4(a). The term $\Pr[\{u, v\} \in E \mid d(\ell(u), \ell(v))]$ is learned from the data for the proposed method and BACKSTROM*, whereas JURGENS inherently assumes that it follows an exponential distribution. Hence BACKSTROM* obtains a higher accuracy in the first iterations than JURGENS but they both stagnate to the same accuracy of 79%. The proposed method clearly outperforms these by achieving an 87% accuracy, which is remarkable given that none of the locations are initially known and randomly assigning the villages to candidate locations would yield an accuracy of only 25%.

Geolocating Parishes. In addition to villages, we can use the proposed method for geolocating the parishes whose coordinates are not recorded in the HisKi database. One way of achieving this, is to build a graph where each village is linked only to its parish. The HisKi database contains also information about neighboring parishes, so we can additionally draw edges between parishes.

---

[2] Open data provided by the National Land Survey of Finland: http://www. maanmittauslaitos.fi/en/digituotteet/geographic-names

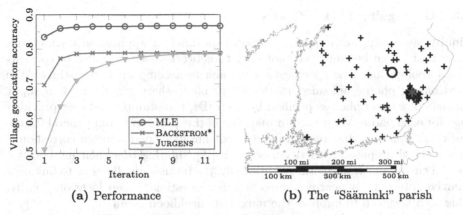

(a) Performance   **(b)** The "Sääminki" parish

Fig. 4. (a) Fraction of correctly geolocated villages considering the 4 574 villages with at least two village name matches. (b) Location of a historical parish called "Sääminki" as identified by our method (✳) and the location of an island called Sääminki in contemporary Finland (O). Candidate village locations are shown by black crosses.

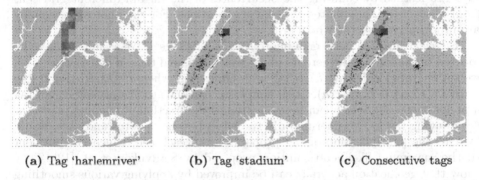

(a) Tag 'harlemriver'   **(b)** Tag 'stadium'   **(c)** Consecutive tags

Fig. 5. (ab) Prior distributions for Flickr tags 'harlemriver' (blue) and 'stadium' (black) around New York City with a grid size of 0.5 km. (c) A photo with the tag 'stadium' can be narrowed down to the neighborhood of the Yankee Stadium, using MLE, if the next photo has been tagged with 'harlemriver'.

Due to lack of space, we do not present the full results of this experiment here, but instead we provide some anecdotal evidence that the locations found by the method are useful. Let us study a parish called *Sääminki* for which there is only one match in the geographic name database. However, it turns out that this match is 200 km away from the location found by the algorithm. By searching for some background information on this parish, we find out that there used to be a municipality called Sääminki but nowadays it belongs to the city of Savonlinna. However, 200 km away from Savonlinna, there is still an island called Sääminki, where we might erroneously locate the parish if we do not consider the information about the member villages. The member village candidate locations and the two different Sääminki locations are visualized in Figure 4(b).

4.3   Geotagging Flickr Photos

Finally, we apply the proposed method to a different application domain and show that it can be employed, not only to network-based, but also to content-based geolocation. The specific task we aim to accomplish is to estimate the locations of photos uploaded in the Flickr photo-sharing system.[3] Our input consists of a set of photos specified by their IDs, timestamps, and user-provided tags for each photo. We present a mapping of this problem to our general framework, and find out that the obtained maximum likelihood solution corresponds to a method proposed by Serdyukov et al. [12], showing that their method is a special case of our framework. Additionally, the framework allows us to link consecutive, presumably nearby photos of a user to estimate their locations jointly. This idea is shown to improve the maximum likelihood estimate.

The set of items V consists of all photos and all tags. We only focus on estimating the photo locations, so the set U contains only the photo items. As the candidate locations for each photo and tag, we define an $N \times N$ grid over the city, giving a total of N^2 candidate locations. The prior distributions of the tags are learned by counting the occurrences of the tags in different grid cells from a set of training photos with known locations. For the photos, we employ a uniform prior over the whole grid.

Edges are created between a photo h and all its tags a. We assume a tag to be located in the same cell where its photo was taken, and thus, if $d(\ell(h), \ell(a)) > 0$, we set $\Pr[\{h, a\} \in E \mid d(\ell(h), \ell(a))] = 0$. Otherwise, when $d(\ell(h), \ell(a)) = 0$, the term $\Pr[\{h, a\} \in E \mid d(\ell(h), \ell(a))]$ simply corresponds to the probability of tag a at location $\ell(h)$, which can be estimated as the fraction of training photos at $\ell(h)$ having tag a. Considering only the edges between photos and tags, would lead to the multiplication of tag probabilities in each cell, which is equivalent to the maximum likelihood solution proposed by Serdyukov et al. [12]. They show that geolocation accuracy can be improved by applying various smoothing techniques, but for simplicity, we have only applied Laplace smoothing, adding a dummy count of 0.1 to each grid cell.

Furthermore, we create an edge between two consecutive photos u and v taken by the same user, within a 5-minute interval. The underlying assumption is that such photos have been taken in nearby locations. Term $\Pr[\{u, v\} \in E \mid d(\ell(u), \ell(v))]$ corresponds to the edge probability between two photos by a single user given their distance. This term peaks at distance 0 and then decreases monotonically. In this experiment, distances are measured by the Manhattan distance. Figure 5 illustrates the advantage of using edges between consecutive photos. Dots correspond to individual photos with a given tag and heatmaps show the estimated probabilities.

Computation. Next we show how to evaluate Eq. (1) used in Algorithm 1 conveniently using matrix operations due to shared candidate locations (recall that we assume that each photo can be located in any grid cell). First, we define a linear indexing from 1 to N^2 for the grid cells. Second, we have two different

---

[3] http://www.flickr.com/

types of edges: "photo–tag" and "photo–photo." For a photo h we denote by $N_t(h)$ the tags of the photo, and by $N_p(h)$ the neighboring photos of h (which can be 0, 1, or 2, as we assume a linear order induced by time with a 5-min cutoff). Then, the evaluation of Eq. (1) for a location $j \in \{1, \ldots, N^2\}$ takes the following form:

$$\Pr[E \mid \ell(h) = j] = \prod_{v \in N(h)} \sum_{\ell(v)} \Pr[\ell(v)] \Pr[\{h, v\} \in E \mid d(j, \ell(v))]$$

$$= \prod_{v_t \in N_t(h)} \sum_{i=1}^{N^2} \Pr[\ell(v_t) = i] \Pr[\{h, v_t\} \in E \mid d(j, i)]$$

$$\times \prod_{v_p \in N_p(h)} \sum_{i=1}^{N^2} \Pr[\ell(v_p) = i] \Pr[\{h, v_p\} \in E \mid d(j, i)]$$

$$=: A(h, j) \prod_{v_p \in N_p(h)} B(v_p, j).$$

Let us first look at term $A(h, j)$. As pointed out earlier in this section, the edge probability is nonzero only when $d(i, j) = 0$ and thus we get rid of the summation

$$A(h, j) = \prod_{v_t \in N_t(h)} \Pr[\ell(v_t) = j] \Pr[\{h, v_t\} \in E \mid d(j, j)] = C \prod_{v_t \in N_t(h)} \Pr[\ell(v_t) = j],$$

where C is a constant.

Let P be the number of photos and $\boldsymbol{A} \in \mathbb{R}^{P \times N^2}$ a matrix defined by $A_{v,j} = A(v, j)$. The matrix $\boldsymbol{A}$ can be precomputed before starting Algorithm 1, since the location distributions of the tags are not updated.

Then let matrix $\boldsymbol{T} \in \mathbb{R}^{P \times N^2}$ denote the uniform prior probabilities of the photo locations, given by $T_{v,j} = \Pr[\ell(v) = j] = \frac{1}{N^2}$, and $\boldsymbol{P} \in \mathbb{R}^{N^2 \times N^2}$ the edge probabilities given the locations of the adjacent vertices $P_{\ell(u),\ell(v)} = \Pr[\{u, v\} \in E \mid d(\ell(u), \ell(v))]$. Now we notice that we can compute $B(v_p, j)$ using the following vector multiplication

$$B(v_p, j) = \sum_{i=1}^{N^2} \Pr[\ell(v_p) = i] \Pr[\{h, v_p\} \in E \mid d(j, i)] = \boldsymbol{T}_{v_p,\cdot} \boldsymbol{P}_{\cdot,j}$$

and thus

$$\Pr[E \mid \ell(h) = j] = A(h, j) \prod_{v_p \in N_p(h)} B(v_p, j) = A_{h,j} \prod_{v_p \in N_p(h)} \boldsymbol{T}_{v_p,\cdot} \boldsymbol{P}_{\cdot,j}.$$

Using this formula, we can execute Algorithm 1 efficiently and conveniently, updating the matrix $\boldsymbol{T}$ at every iteration.

Data Preprocessing. The Flickr service allows users to geolocate their photos at different accuracy levels. The dataset we use contains only the photos with the highest accuracy level. Furthermore, in Flickr, it is possible to upload multiple photos with the same set of tags in a bulk. To reduce noise in the data due to bulk uploads, we only keep the first photo if there are multiple photos from the same user, on the same date, and with exactly the same tags. Additionally, we filter out the tags that have been used by fewer than three users. The photos originate from a 10 km × 10 km area in the center of New York City, and in total we have 727 457 photos, 42 519 users, and 59 190 unique tags, after the aforementioned preprocessing steps.

Geotagging Results. In our experimental setup, the grid size is set to 10×10 so that each cell is 1 km by 1 km. From each user, we take only the photos that have an edge to another photo to understand the effect of linking photos in more detail. A 10-fold cross-validation over users is employed to evaluate the performance of the methods.

A majority-vote baseline, computed by predicting the grid cell with the largest number of training photos, yields an accuracy of 23.5%. The method by Serdyukov et al. [12] obtains a clearly higher accuracy of 46.6%. This is further improved by the MLE method, which converges in two iterations, yielding an accuracy of 47.1%. The 0.5% improvement is statistically significant ($p < 0.001$) according to McNemar's test, suggesting that the information regarding consecutive photos could prove useful when designing methods tailored for the photo geotagging problem.

5   Conclusions and Discussion

We have presented a probabilistic framework for inferring the geographical locations of objects. We assume that the objects are linked in a graph structure, and a prior distribution of the object locations is available. We showed that these assumptions are mild, and many application scenarios fit the proposed setting. To demonstrate the generality of the proposed method and to evaluate its performance, we presented detailed experiments for three different types of geolocation problems. Our evaluation indicated that the proposed method outperforms two other recently proposed network-based geolocation methods, BACKSTROM* and JURGENS.

An important novelty of the method is that it can manage large degrees of uncertainty in the data. Unlike the existing approaches we are aware of, our method does not need to assume the exact locations for any of the objects in the data. This is convincingly demonstrated in all three of our case studies.

Several future directions are worth exploring. In some cases it is more natural to treat location as a continuous variable. Thus we could try to find the maximum likelihood estimate in the continuous case, employing gradient-based methods. Also, it would be interesting to compare the proposed approach with undirected graphical models (Markov random fields), in which approximate inference can be achieved, for instance, by adopting loopy belief propagation.

Acknowledgments. We would like to thank the team in the Genealogical Society of Finland and Jouni Malinen for making the HisKi data available to us. We also thank Géraud Le Falher for sharing the Flickr dataset.

References

1. Ahmed, A., Hong, L., Smola, A.: Hierarchical geographical modeling of user locations from social media posts. In: Proceedings of the 22nd International Conference on World Wide Web, pp. 25–36. ACM (2013)
2. Backstrom, L., Sun, E., Marlow, C.: Find me if you can: improving geographical prediction with social and spatial proximity. In: Proceedings of the 19th International Conference on World Wide Web, pp. 61–70. ACM (2010)
3. Cho, E., Myers, S.A., Leskovec, J.: Friendship and mobility: user movement in location-based social networks. In: Proceedings of the 17th ACM SIGKDD International Conference on Knowledge Discovery and Data Mining, pp. 1082–1090. ACM (2011)
4. Crandall, D.J., Backstrom, L., Huttenlocher, D., Kleinberg, J.: Mapping the world's photos. In: Proceedings of the 18th International Conference on World Wide Web, pp. 761–770. ACM (2009)
5. Jurgens, D.: That's what friends are for: Inferring location in online social media platforms based on social relationships. In: Proceedings of the 7th International AAAI Conference on Weblogs and Social Media, pp. 273–282 (2013)
6. Li, R., Wang, S., Deng, H., Wang, R., Chang, K.C.C.: Towards social user profiling: unified and discriminative influence model for inferring home locations. In: Proceedings of the 18th ACM SIGKDD International Conference on Knowledge Discovery and Data Mining, pp. 1023–1031. ACM (2012)
7. Mahmud, J., Nichols, J., Drews, C.: Where is this tweet from? inferring home locations of twitter users. In: Proceedings of the 6th International AAAI Conference on Weblogs and Social Media, pp. 511–514. ACM (2012)
8. McGee, J., Caverlee, J., Cheng, Z.: Location prediction in social media based on tie strength. In: Proceedings of the 22nd ACM International Conference on Information & Knowledge Management, pp. 459–468. ACM (2013)
9. O'Leary, M.: The mathematics of geographic profiling. Journal of Investigative Psychology and Offender Profiling 6(3), 253–265 (2009)
10. Rout, D., Bontcheva, K., Preoţiuc-Pietro, D., Cohn, T.: Where's @wally? a classification approach to geolocating users based on their social ties. In: Proceedings of the 24th ACM Conference on Hypertext and Social Media, pp. 11–20. ACM (2013)
11. Sadilek, A., Kautz, H., Bigham, J.P.: Finding your friends and following them to where you are. In: Proceedings of the 5th ACM International Conference on Web Search and Data Mining, pp. 723–732. ACM (2012)
12. Serdyukov, P., Murdock, V., Van Zwol, R.: Placing flickr photos on a map. In: Proceedings of the 32nd International ACM SIGIR Conference on Research and Development in Information Retrieval, pp. 484–491. ACM (2009)
13. Yamaguchi, Y., Amagasa, T., Kitagawa, H., Ikawa, Y.: Online user location inference exploiting spatiotemporal correlations in social streams. In: Proceedings of the 23rd ACM International Conference on Conference on Information and Knowledge Management, pp. 1139–1148. ACM (2014)

Weighted Rank Correlation: A Flexible Approach Based on Fuzzy Order Relations

Sascha Henzgen and Eyke Hüllermeier[✉]

Department of Computer Science, University of Paderborn, Paderborn, Germany
{shenzgen,eyke}@upb.de

Abstract. Measures of rank correlation are commonly used in statistics to capture the degree of concordance between two orderings of the same set of items. Standard measures like Kendall's tau and Spearman's rho coefficient put equal emphasis on each position of a ranking. Yet, motivated by applications in which some of the positions (typically those on the top) are more important than others, a few weighted variants of these measures have been proposed. Most of these generalizations fail to meet desirable formal properties, however. Besides, they are often quite inflexible in the sense of committing to a fixed weighing scheme. In this paper, we propose a weighted rank correlation measure on the basis of fuzzy order relations. Our measure, called *scaled gamma*, is related to Goodman and Kruskal's gamma rank correlation. It is parametrized by a fuzzy equivalence relation on the rank positions, which in turn is specified conveniently by a so-called scaling function. This approach combines soundness with flexibility: it has a sound formal foundation and allows for weighing rank positions in a flexible way. The usefulness of our class of weighted rank correlation measures is shown by means of experimental studies using both synthetic and real-world ranking data.

1 Introduction

Rank correlation measures such as Kendall's tau [11] and Spearman's rho [20], which have originally been developed in non-parametric statistics, are used extensively in various fields of application, ranging from bioinformatics [1] to information retrieval [21]. In contrast to numerical correlation measures such as Pearson correlation, rank correlation measures are only based on the ordering of the observed values of a variable. Thus, measures of this kind are not limited to numerical variables but can also be applied to non-numerical variables with an ordered domain (i.e., measured on an ordinal scale) and, of course, to rankings (permutations) directly.

In many applications, such as Internet search engines, one is not equally interested in all parts of a ranking. Instead, the top positions of a ranking (e.g., the first 10 or 50 web sites listed) are typically considered more important than the middle part and the bottom. Standard rank correlation measures, however, put equal emphasis on all positions. Therefore, they cannot distinguish disagreements in different parts of a ranking. This is why *weighted* variants have been

© Springer International Publishing Switzerland 2015
A. Appice et al. (Eds.): ECML PKDD 2015, Part II, LNAI 9285, pp. 422–437, 2015.
DOI: 10.1007/978-3-319-23525-7_26

proposed for some correlation measures, as well as alternative measures specifically focusing on the top of a ranking [6,7,10,16,21]. Most of these generalizations fail to meet desirable formal properties, however. Besides, they are often quite inflexible in the sense of committing to a fixed weighing scheme.

In this paper, we develop a general framework for designing weighted rank correlation measures based on the notion of *fuzzy order relation*, and use this framework to generalize Goodman and Kruskal's gamma coefficient [8].[1] Our approach has a sound formal foundation and allows for weighing rank positions in a flexible way. In particular, it is not limited to monotone weighing schemes that emphasize the top in comparison to the rest of a ranking. The key ingredients of our approach, to be detailed further below, are as follows:

- *Fuzzy order relations* [4] are generalizations of the conventional order relations on the reals or the integer numbers: SMALLER, EQUAL and GREATER. They enable a smooth transition between these predicates and allow for expressing, for instance, that a number x is smaller than y *to a certain degree*, while to some degree these numbers are also considered as being equal. Here, the EQUAL relation is understood as a kind of similarity relation that seeks to model the "perceived equality" (instead of the strict mathematical equality).
- *Scaling functions* for modeling fuzzy equivalence relations [12]. For each element x of a linearly ordered domain X, a scaling function $s(\cdot)$ essentially expresses the degree $s(x)$ to which x can be (or should be) distinguished from its neighboring values. A measure of distance (or, equivalently, of similarity) on X can then be derived via accumulation of local degrees of distinguishability.
- *Fuzzy rank correlation* [5,17] generalizes conventional rank correlation on the basis of fuzzy order relations, thereby combining properties of standard rank correlation (such as Kendall's tau) and numerical correlation measures (such as Pearson correlation). Roughly, the idea is to penalize the inversion of two items (later on called a *discordance*) depending on how dissimilar the corresponding rank positions are: the more similar (less distinguishable) the positions are according to the EQUAL relation, the smaller the influence of the inversion on the rank correlation.

The rest of the paper is organized as follows. In the next two sections, we briefly recall the basics of fuzzy order relations and fuzzy rank correlation, respectively. Our weighted rank correlation measure, called *scaled gamma*, is then introduced in Section 5, and related work is reviewed in Section 6. A small experimental study is presented in Section 7, prior to concluding the paper in Section 8.

2   Rank Correlation

Consider $N \geq 2$ paired observations $\{(x_i, y_i)\}_{i=1}^{N} \subset \mathbb{X} \times \mathbb{Y}$ of two variables X and Y, where $\mathbb{X}$ and $\mathbb{Y}$ are two linearly ordered domains; we denote

$$\boldsymbol{x} = (x_1, x_2, \ldots, x_N), \quad \boldsymbol{y} = (y_1, y_2, \ldots, y_N) \ .$$

---

[1] A preliminary version of this paper has been presented in [9], on the occasion of the German Workshop on Computational Intelligence, Dortmund, Germany, 2013.

In particular, the values x_i (and y_i) can be real numbers ($\mathbb{X} = \mathbb{R}$) or rank positions ($\mathbb{X} = [N] = \{1, 2, \ldots, N\}$). For example, $\boldsymbol{x} = (3, 1, 4, 2)$ denotes a ranking of four items, in which the first item is on position 3, the second on position 1, the third on position 4 and the fourth on position 2.

The goal of a (rank) correlation measure is to capture the dependence between the two variables in terms of their tendency to increase and decrease (their position) in the same or the opposite direction. If an increase in X tends to come along with an increase in Y, then the (rank) correlation is positive. The other way around, the correlation is negative if an increase in X tends to come along with a decrease in Y. If there is no dependency of either kind, the correlation is (close to) 0.

2.1   Concordance and Discordance

Many rank correlation measures are defined in terms of the number C of *concordant*, the number D of *discordant*, and the number T of *tied* data points. Let $\mathcal{P} = \{(i, j) \mid 1 \leq i < j \leq N\}$ denote the set of ordered index pairs. We call a pair $(i, j) \in \mathcal{P}$ concordant, discordant or tied depending on whether $(x_i - x_j)(y_i - y_j)$ is positive, negative or 0, respectively. Thus, let us define three $N \times N$ relations $\mathcal{C}$, $\mathcal{D}$ and $\mathcal{T}$ as follows:

$$\mathcal{C}(i, j) = \begin{cases} 1 & (x_i - x_j)(y_i - y_j) > 0 \\ 0 & \text{otherwise} \end{cases} \tag{1}$$

$$\mathcal{D}(i, j) = \begin{cases} 1 & (x_i - x_j)(y_i - y_j) < 0 \\ 0 & \text{otherwise} \end{cases} \tag{2}$$

$$\mathcal{T}(i, j) = \begin{cases} 1 & (x_i - x_j)(y_i - y_j) = 0 \\ 0 & \text{otherwise} \end{cases} \tag{3}$$

The number of concordant, discordant and tied pairs $(i, j) \in \mathcal{P}$ are then obtained by summing the entries in the corresponding relations:

$$C = \sum_{(i,j) \in \mathcal{P}} \mathcal{C}(i, j) = \frac{1}{2} \sum_{i \in [N]} \sum_{j \in [N]} \mathcal{C}(i, j)$$

$$D = \sum_{(i,j) \in \mathcal{P}} \mathcal{D}(i, j) = \frac{1}{2} \sum_{i \in [N]} \sum_{j \in [N]} \mathcal{D}(i, j)$$

$$T = \sum_{(i,j) \in \mathcal{P}} \mathcal{T}(i, j) = \frac{1}{2} \sum_{i \in [N]} \sum_{j \in [N]} \mathcal{T}(i, j) - \frac{N}{2}$$

Note that

$$\mathcal{C}(i, j) + \mathcal{D}(i, j) + \mathcal{T}(i, j) = 1 \tag{4}$$

for all $(i, j) \in \mathcal{P}$, and

$$C + D + T = |\mathcal{P}| = \frac{N(N - 1)}{2} . \tag{5}$$

2.2 Rank Correlation Measures

Well-known examples of rank correlation measures that can be expressed in terms of the above quantities include Kendall's tau [11]

$$\tau = \frac{C - D}{N(N-1)/2} \tag{6}$$

and Goodman and Kruskal's gamma coefficient [8]

$$\gamma = \frac{C - D}{C + D} . \tag{7}$$

As will be detailed in the following sections, our basic strategy for generalizing rank correlation measures such as γ is to "fuzzify" the concepts of concordance and discordance. Thanks to the use of fuzzy order relations, we will be able to express that a pair (i, j) is concordant or discordant to a certain degree (between 0 and 1). Measures like (7) can then be generalized in a straightforward way, namely by accumulating the degrees of concordance and discordance, respectively, and putting them in relation to each other.

3 Fuzzy Relations

3.1 Fuzzy Equivalence

The notion of a fuzzy relation generalizes the standard notion of a mathematical relation by allowing to express "degrees of relatedness". Formally, a (binary) fuzzy relation on a set $\mathbb{X}$ is characterized by a membership function $\mathcal{E} : \mathbb{X} \times \mathbb{X} \longrightarrow [0, 1]$. For each pair of elements $x, y \in \mathbb{X}$, $\mathcal{E}(x, y)$ is the degree to which x is related to y.

Recall that a conventional equivalence relation on a set $\mathbb{X}$ is a binary relation that is reflexive, symmetric and transitive. For the case of a fuzzy relation $\mathcal{E}$, these properties are generalized as follows:

- reflexivity: $\mathcal{E}(x, x) = 1$ for all $x \in \mathbb{X}$
- symmetry: $\mathcal{E}(x, y) = \mathcal{E}(y, x)$ for all $x, y \in \mathbb{X}$
- $\top$-transitivity: $\top(\mathcal{E}(x, y), \mathcal{E}(y, z)) \le \mathcal{E}(x, z)$ for all $x, y, z \in \mathbb{X}$

A fuzzy relation $\mathcal{E}$ having these properties is called a fuzzy equivalence relation [5]. While the generalizations of reflexivity and symmetry are rather straightforward, the generalization of transitivity involves a triangular norm (t-norm) $\top$, which plays the role of a generalized logical conjunction [13]. Formally, a function $\top : [0, 1]^2 \longrightarrow [0, 1]$ is a t-norm if it is associative, commutative, monotone increasing in both arguments, and satisfies the boundary conditions $\top(a, 0) = 0$ and $\top(a, 1) = a$ for all $a \in [0, 1]$. Examples of commonly used t-norms include the minimum $\top(a, b) = \min(a, b)$ and the product $\top(a, b) = ab$. To emphasize the role of the t-norm, a relation $\mathcal{E}$ satisfying the above properties is also called a $\top$-equivalence.

3.2   Fuzzy Ordering

The notion of an order relation $\leq$ is similar to that of an equivalence relation, with the important difference that the former is antisymmetric while the latter is symmetric. A common way to formalize antisymmetry is as follows: $a \leq b$ and $b \leq a$ implies $a = b$. Note that this definition already involves an equivalence relation, namely the equality $=$ of two elements. Thus, as suggested by Bodenhofer [2], a fuzzy order relation can be defined on the basis of a fuzzy equivalence relation. Formally, a fuzzy relation $\mathcal{L} : \mathbb{X} \times \mathbb{X} \longrightarrow [0,1]$ is called a *fuzzy ordering* with respect to a t-norm $\top$ and a $\top$-equivalence $\mathcal{E}$, for brevity $\top$-$\mathcal{E}$-*ordering*, if it satisfies the following properties for all $x, y, z \in \mathbb{X}$:

- $\mathcal{E}$-reflexivity: $\mathcal{E}(x,y) \leq \mathcal{L}(x,y)$
- $\top$-$\mathcal{E}$-antisymmetry: $\top(\mathcal{L}(x,y), \mathcal{L}(y,x)) \leq \mathcal{E}(x,y)$
- $\top$-transitivity: $\top(\mathcal{L}(x,y), \mathcal{L}(y,z)) \leq \mathcal{L}(x,z)$

Furthermore a $\top$-$\mathcal{E}$-ordering $\mathcal{L}$ is called *strongly complete* if

$$\max\big(\mathcal{L}(x,y), \mathcal{L}(y,x)\big) = 1$$

for all $x, y \in \mathbb{X}$. This is expressing that, for each pair of elements x and y, either $x \leq y$ or $y \leq x$ should be fully true.

A fuzzy relation $\mathcal{L}$ as defined above can be seen as a generalization of the conventional "smaller or equal" on the real or the integer numbers. What is often needed, too, is a "stricly smaller" relation $<$. In agreement with the previous formalizations, a relation of that kind can be defined as follows: A binary fuzzy relation $\mathcal{R}$ is called a *strict fuzzy ordering* with respect to a $\top$-norm and a $\top$-equivalence $\mathcal{E}$, or strict $\top$-$\mathcal{E}$-*ordering* for short, if it has the following properties for all $x, x', y, y', z \in \mathbb{X}$ [5]:

- irreflexivity: $\mathcal{R}(x,x) = 0$
- $\top$-transitivity: $\top(\mathcal{R}(x,y), \mathcal{R}(y,z)) \leq \mathcal{R}(x,z)$
- $\mathcal{E}$-extensionality: $\top(\mathcal{E}(x,x'), \mathcal{E}(y,y'), \mathcal{R}(x,y)) \leq \mathcal{R}(x',y')$

3.3   Practical Construction

The above definitions provide generalizations $\mathcal{E}$, $\mathcal{L}$ and $\mathcal{R}$ of the standard relations $=$, $\leq$ and $<$, respectively, that exhibit reasonable properties and, moreover, are coherent with each other. Practically, one may start by choosing an equivalence relation $\mathcal{E}$ and a compatible t-norm $\top$, and then derive $\mathcal{L}$ and $\mathcal{R}$ from the corresponding $\top$-equivalence.

More specifically, suppose the set $\mathbb{X}$ to be a linearly ordered domain, that is, to be equipped with a standard (non-fuzzy) order relation $\leq$. Then, given a $\top$-equivalence $\mathcal{E}$ on $\mathbb{X}$, the following relation is a coherent fuzzy order relation, namely a strongly complete $\top$-$\mathcal{E}$-ordering:

$$\mathcal{L}(x,y) = \begin{cases} 1 & \text{if } x \leq y \\ \mathcal{E}(x,y) & \text{otherwise} \end{cases}$$

Moreover, a strict fuzzy ordering $\mathcal{R}$ can be obtained from $\mathcal{L}$ by

$$\mathcal{R}(x,y) = 1 - \mathcal{L}(y,x) \tag{8}$$

The relations thus defined have a number of convenient properties. In particular, $\min(\mathcal{R}(x,y), \mathcal{R}(y,x)) = 0$ and

$$\mathcal{R}(x,y) + \mathcal{E}(x,y) + \mathcal{R}(y,x) = 1 \tag{9}$$

for all $x, y \in \mathbb{X}$. These properties can be interpreted as follows. For each pair of elements x and y, the unit mass splits into two parts: a degree $a = \mathcal{E}(x,y)$ to which x and y are equal, and a degree $1 - a$ to which either x is smaller than y or y is smaller than x.

4   Fuzzy Relations on Rank Data

Since we are interested in generalizing rank correlation measures, the underlying domain $\mathbb{X}$ is given by a set of rank positions $[N] = \{1, 2, \ldots, N\}$ (equipped with the standard $<$ relation) in our case. As mentioned before, this domain could be equipped with fuzzy relations $\mathcal{E}$, $\mathcal{L}$ and $\mathcal{R}$ by defining $\mathcal{E}$ first and deriving $\mathcal{L}$ and $\mathcal{R}$ afterward. Note, however, that the number of degrees of freedom in the specification of $\mathcal{E}$ is of the order $O(N^2)$, despite the constraints this relation has to meet.

4.1   Scaling Functions on Rank Positions

In order to define fuzzy relations even more conveniently, while emphasizing the idea of weighing the importance of rank positions at the same time, we leverage the concept of a *scaling function* as proposed by Klawonn [12]. Roughly speaking, a scaling function $w : \mathbb{X} \longrightarrow \mathbb{R}_+$ specifies the dissimilarity of an element x from its direct neighbor elements, and the dissimilarity between any two elements x and y is then obtained via integration of the local dissimilarities along the chain from x to y. In our case, a scaling function can be defined as a mapping $w : [N-1] \longrightarrow [0,1]$ or, equivalently, as a vector

$$\boldsymbol{w} = \Big(w(1), w(2), \ldots, w(N-1)\Big) \in [0,1]^{N-1} . \tag{10}$$

Here, $w(n)$ can be interpreted as the degree to which the rank positions n and $n-1$ are distinguished from each other; correspondingly, $1 - w(n)$ can be seen as the degree to which these two positions are considered to be equal. From the local degrees of distinguishability, a global distance function is derived on $\mathbb{X}$ by defining

$$d(x,y) = \min \left(1, \sum_{i=\min(x,y)}^{\max(x,y)-1} w(i) \right) . \tag{11}$$

Put in words, the distance between x and y is the sum of the degrees of distinguishability between them, thresholded at the maximal distance of 1. In principle, accumulations of the degrees of distinguishability other than the sum are of course conceivable. For example, the maximum could be used as well:

$$d(x,y) = \max\left\{ w(i) \mid i \in \{\min(x,y),\ldots,\max(x,y) - 1\}\right\} . \tag{12}$$

In general, $d(x, y)$ is supposed to define a pseudo-metric on $\mathbb{X}$. Under this condition, it can be shown that the fuzzy relation $\mathcal{E}$ defined as

$$\mathcal{E}(x,y) = 1 - d(x,y)$$

for all $x, y \in \mathbb{X}$ is a $\top_L$-equivalence, where $\top_L$ is the Łukasiewicz t-norm $\top_L(a, b) = \max(0, a + b - 1)$ [3]. Relations $\mathcal{L}$ and $\mathcal{R}$ can then be derived from $\mathcal{E}$ as described in Section 3.3. In particular, we obtain

$$\mathcal{R}(x,y) = \begin{cases} d(x,y) & \text{if } x < y \\ 0 & \text{otherwise} \end{cases}$$

According to our discussion so far, the only remaining degree of freedom is the scaling function s. Obviously, this function can also be interpreted as a *weighing function*: the more distinguishable a position n from its neighbor positions, i.e., the larger $w(n - 1)$ and $w(n)$, the higher the importance of that position.

An example of a scaling function for $N = 12$ is shown in Figure 1. This function puts more emphasis on the top and the bottom ranks and less on the middle part. According to (11), the distinguishability between the positions 4 and 7 is $d(4, 7) = 0.4 + 0.2 + 0.2 = 0.8$ (sum of the weights $w(i)$ in the shaded region). Thus, 4 is strictly smaller than 7 to the degree of $\mathcal{R}(4, 7) = 0.8$, while both positions are considered equal to the degree $\mathcal{E}(4, 7) = 0.2$.

Note that, with $w(i) = \llbracket i < k \rrbracket$, we also cover the top-$k$ scenario as a special case. Here, the standard $<$ relation is recovered for all elements on the first k positions, whereas the remaining positions are considered as fully equivalent, i.e., these elements form an equivalence class in the standard sense.

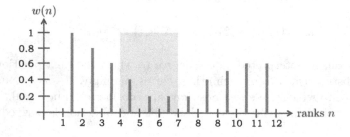

Fig. 1. Example of a scaling function.

5 Weighted Rank Correlation

Our approach to generalizing rank correlation measures is based on the "fuzzification" of the relations (1–3) and, correspondingly, the number of concordant, discordant and tied item pairs. The tools that are needed to do so have already been introduced in the previous sections. In particular, suppose a fuzzy equivalence relation $\mathcal{E}$ and a "strictly smaller" relation $\mathcal{R}$ to be derived from a scaling function w on $\mathbb{X}$, based on the procedure outlined above. For notational convenience, we assume the same scaling function (and hence the same relations) to be used on both domains $\mathbb{X}$ and $\mathbb{Y}$. In principle, however, different functions w_X and w_Y (and hence relations $\mathcal{E}_X$, $\mathcal{R}_X$ and $\mathcal{E}_Y$, $\mathcal{R}_Y$) could be used.

Now, according to (1), a pair $(i, j) \in \mathcal{P}$ is concordant if both x_i is (strictly) smaller than x_j and y_i is smaller than y_j, or if x_j is smaller than x_i and y_j is smaller than y_i. Using our fuzzy relation $\mathcal{R}$ and a t-norm $\top$ as a generalized conjunction, this can be expressed as follows:

$$\tilde{\mathcal{C}}(i, j) = \top\big(\mathcal{R}(x_i, x_j), \mathcal{R}(y_i, y_j)\big) + \top\big(\mathcal{R}(x_j, x_i), \mathcal{R}(y_j, y_i)\big) \tag{13}$$

The discordance relation can be expressed analogously:

$$\tilde{\mathcal{D}}(i, j) = \top\big(\mathcal{R}(x_i, x_j), \mathcal{R}(y_j, y_i)\big) + \top\big(\mathcal{R}(x_j, x_i), \mathcal{R}(y_i, y_j)\big) \tag{14}$$

Finally, the degree to which (i, j) is tied is given by

$$\tilde{\mathcal{T}}(i, j) = \bot\big(\mathcal{E}(x_i, x_j), \mathcal{E}(y_i, y_j)\big) \ ,$$

where $\bot$ is the t-conorm associated with $\top$ (i.e., $\bot(u, v) = 1 - \top(1 - u, 1 - v)$), serving as a generalized logical disjunction. Generalizing (4), the three degrees sum up to 1, i.e.,

$$\tilde{\mathcal{C}}(i, j) + \tilde{\mathcal{D}}(i, j) + \tilde{\mathcal{T}}(i, j) \equiv 1 \ , \tag{15}$$

and either $\tilde{\mathcal{C}}(i, j) = 0$ or $\tilde{\mathcal{D}}(i, j) = 0$. In other words, a pair (i, j) that has originally been concordant (discordant) will remain concordant (discordant), at least to some extent. However, since $\mathcal{E}$ may introduce a certain indistinguishability between the positions x_i and x_j or the positions y_i and y_j, the pair could also be considered as a partial tie.

Given the above fuzzy relations, the number of concordant, discordant and tied data points can be obtained as before, namely by summing over all ordered pairs $(i, j) \in \mathcal{P}$:

$$\tilde{C} = \sum_{(i,j) \in \mathcal{P}} \tilde{\mathcal{C}}(i, j) \ , \quad \tilde{D} = \sum_{(i,j) \in \mathcal{P}} \tilde{\mathcal{D}}(i, j) \ , \quad \tilde{T} = \sum_{(i,j) \in \mathcal{P}} \tilde{\mathcal{T}}(i, j) \ .$$

According to (15),

$$\tilde{C} + \tilde{D} + \tilde{T} = |\mathcal{P}| = \frac{N(N-1)}{2} \ ,$$

which generalizes (5). Using these quantities, rank correlation measures expressed in terms of the number of concordant and discordant pairs can be generalized in a straightforward way. In particular, a generalization of the gamma coefficient (7) is obtained as

$$\tilde{\gamma} = \frac{\tilde{C} - \tilde{D}}{\tilde{C} + \tilde{D}} \, . \tag{16}$$

It is worth mentioning that the weighted rank correlation measure thus defined exhibits a number of desirable formal properties, which it essentially inherits from the general fuzzy extension of the gamma coefficient; we refer to [17], in which these properties are analyzed in detail.

6   Related Work

Weighted versions of rank correlation measures have not only been studied in statistics but also in other fields, notably in information retrieval [6,10,16,21]. Most of them are motivated by the idea of giving a higher weight to the top-ranks: in information retrieval, important documents are supposed to appear in the top, and a swap of important documents should incur a higher penalty than a swap of unimportant ones.

Kaye [10] introduced a weighted, non-symmetric version of Spearman's rho coefficient. Costa and Soares [6] proposed a symmetric weighted version of Spearman's coefficient resembling the one of Kaye. Another approach, based on average precision and called *AP correlation*, was introduced by Yilmaz *et al.* [21]. Maturi and Abdelfattah [16] define weighted scores $W_i = w^i$ with $w \in (0,1)$ and compute the Pearson correlation coefficient on these scores. All four measures give higher weight to the top ranks.

Two more flexible measures, not restricted to monotone decreasing weights, have been proposed by Shieh [19] and Kumar and Vassilivitskii [14]. In the approach of Shieh [19], a weight is manually given to every occurring concordance or discordance through a symmetric weight function $w : [N] \times [N] \longrightarrow \mathbb{R}_+$:

$$\tau_w = \frac{\sum_{i<j} w_{ij} C_{ij} - \sum_{i<j} w_{ij} D_{ij}}{\sum_{i<j} w_{ij}} = \frac{\sum_{i<j} w_{ij}(C_{ij} - D_{ij})}{\sum_{i<j} w_{ij}}. \tag{17}$$

The input parameter for w are the ranks of a reference ranking π_{ref}, which is assumed to be the natural order $(1, 2, 3, \ldots, N)$. Therefore, this approach is not symmetric. To handle the quadratic number of weights, Shieh proposed to define them as $w_{ij} = v_i v_j$ with v_i the weight of rank i.

Kumar and Vassilivitskii [14] introduce a generalized version of Kendall's distance. Originally, they proposed three different weights: element weights, position weights, and element similarities. The three weights are defined independently of each other, and each of them can be used by its own for weighting discordant pairs. Here, we focus on the use of position weights. Like in our approach, Kumar and Vassilivitskii define $N - 1$ weights $\delta_i \geq 0$, which are considered as costs for swapping two elements on adjacent positions $i + 1$ and i. The accumulated cost of changing from position 1 to $i \in \{2, \ldots, N\}$ is $p_i = \sum_{j=1}^{i-1} \delta_j$, with $p_1 = 0$. Moreover,

$$\bar{p}_i(\pi_1, \pi_2) = \frac{p_{\pi_1(i)} - p_{\pi_2(i)}}{\pi_1(i) - \pi_2(i)} \tag{18}$$

is the average cost of moving element i from position $\pi_1(i)$ to position $\pi_2(i)$; if $\pi_1(i) = \pi_2(i)$ then $\bar{p}_i = 1$. The weighted discordance of a pair (i, j) is then defined in terms of the product of the average costs for index i and j:

$$\hat{D}_\delta(i, j) = \begin{cases} \bar{p}_i(\pi_1, \pi_2)\bar{p}_j(\pi_1, \pi_2) & \text{if } (i, j) \text{ is discordant} \\ 0 & \text{otherwise} \end{cases} . \tag{19}$$

Finally, the weighted Kendall distance K_δ is given by

$$K_\delta = \tilde{D}_\delta = \sum_{i=1}^{N-1} \sum_{i+1}^{N} \hat{D}_\delta(i, j) . \tag{20}$$

Note that (20) is indeed a distance and not a correlation measure. To enable a comparison with τ_ω and $\tilde{\gamma}$ in the next section, we define

$$\hat{C}_\delta(i, j) = \begin{cases} \bar{p}_i(\pi_1, \pi_2)\bar{p}_j(\pi_1, \pi_2) & \text{if } (i, j) \text{ is concordant} \\ 0 & \text{otherwise} \end{cases} \tag{21}$$

as the weighted concordance of a pair (i, j), and finally another weighted version of gamma:

$$\tilde{\gamma}_\delta = \frac{\tilde{C}_\delta - \tilde{D}_\delta}{\tilde{C}_\delta + \tilde{D}_\delta} .$$

7   Experiments

Needless to say, an objective comparison of weighted rank correlation measures is very difficult, if not impossible. Even in the case of standard measures, one cannot say, for example, that Kendall's tau is "better" than Spearman's rho. Instead, these are simply different measures trying to capture different types of correlation in the data.

Nevertheless, we conducted some controlled experiments with synthetic data, for which there is a natural expectation of how the measures are supposed to behave and what results they should ideally produce. We compare our approach with those of Shieh as well as Kumar and Vassilivitskii, since these are able to handle non-monotone weight functions, too. For the purpose of these experiments, our measure $\tilde{\gamma}$ was instantiated with the maximum in (12) and the product t-norm in (13) and (14).[2]

---

[2] Of course, other instantiations are conceivable; however, tuning our measure by optimizing the choice of operators was beyond the scope of the experiments.

7.1   First Study

In a first experiment, we generated rank data by sampling from the Plackett-Luce (PL) model, which is a parameterized probability distribution on the set of all rankings over N items. It is specified by a parameter vector $v = (v_1, v_2, \ldots v_N) \in \mathbb{R}_+^N$, in which v_i accounts for the "skill" of the ith item. The probability assigned by the PL model to a ranking represented by a permutation π is given by

$$\mathbf{P}(\pi \,|\, v) = \prod_{i=1}^{N} \frac{v_{\pi^{-1}(i)}}{v_{\pi^{-1}(i)} + v_{\pi^{-1}(i+1)} + \ldots + v_{\pi^{-1}(N)}} \,, \tag{22}$$

where $\pi(i)$ is the position of item i in the ranking, and $\pi^{-1}(j)$ the index of the item on position j. This model is a generalization of the well-known Bradley-Terry model [15], a model for the pairwise comparison of alternatives, which specifies the probability that "a wins against b" in terms of $v_a/(v_a + v_b)$. Obviously, the larger v_a in comparison to v_b, the higher the probability that a is chosen. Likewise, the larger the parameter v_i in (22) in comparison to the parameters v_j, $j \neq i$, the higher the probability that the ith item appears on a top rank. Moreover, the more similar the skill parameters, the more likely two items are reversed. Thus, a ranking drawn from a PL model is more stable, and hence more "reliable", in regions in which the difference between the skill values (sorted in decreasing order from highest to lowest) is large, and less stable in regions in which this difference is small.

Instead of defining the skills v directly, it is more convenient to define them via the representation of PL as a Thurstone model with scores following a Gumble distribution. The means μ_i of this distribution translate into PL-parameters via $v_i = \exp(\frac{\mu_i}{\beta})$, with β the scaling parameter of the Gumble distribution.

For our experimental study, we generated mixtures of $c = 4$ PL distributions, i.e., data sets consisting of four clusters. To this end, c reference rankings were first generated by sampling from the PL distribution with $\mu = (30, 29, \ldots, 1)$ and $\beta = 0.3$, i.e., these references are perturbations of the identity $\pi_{id} = (1, 2, \ldots, n)$. Then, a score vector

$$\mu^{(0)} = (18, \ldots 14, 13.1, 12.3, \ldots, 9.6, \ldots, 9.6, 9.5, 9.3, \ldots, 5.9, 5, \ldots, 1)$$

is defined, which reflects high stability in the top and bottom ranks, and low stability in the middle ranks, and new score vectors $\mu^{(i)}$, $i = 1, \ldots, c$, are generated by permuting $\mu^{(0)}$ according to the reference rankings; each of these rankings $\mu^{(i)}$ defines the center of a cluster. Finally, 200 rankings are sampled from each of the PL models with parameter $\mu^{(i)}$ and $\beta = 0.3$, and these rankings are assigned label i.

We produced 100 such data sets with rankings of length 30. As a weight vector, which corresponds to the scaling function (10) for $\tilde{\gamma}$ and defines the transitions costs δ_i for $\tilde{\gamma}_\delta$, we used

$$w = (1, 1, 1, 1, 0.9, \ldots, 0.1, 0, 0, 0, 0.1, \ldots, 0.9, 1, 1, 1, 1) \,.$$

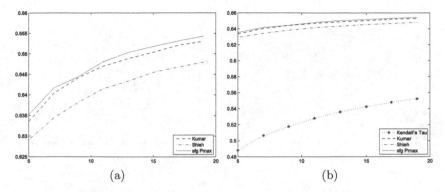

Fig. 2. Both plots show the average accuracy against the neighborhood size k. The right plot additionally shows the results for classical Kendall's tau.

This weight vector seeks to account for the fact that, according to our construction, the middle positions of the observed rankings are less reliable and, therefore, should have a lower weight in the computation of similarities between rankings. The weight vector for τ_ω is derived from w, so as to make it maximally comparable, and is given by $v_i = (w_{i-1} + w_i)/2$ with $w_{-1} = 1$ and $w_n = 1$.

For each correlation measure and each data set, we applied a k-nearest neighbor classifier with the correlation as a similarity measure. Here, the idea is the following: the better the similarity between rankings is reflected by a correlation measure, the stronger the performance of the classifier is supposed to be. The classifiers were validated by averaging one hundred repetitions of a 10-fold cross validation. In the end, we also averaged over all data sets. The results are shown in Figure 2. As can be seen, $\tilde{\gamma}_\delta$ and $\tilde{\gamma}$ are performing more or less on par, with a slight advantage for $\tilde{\gamma}$. Moreover, they both outperform τ_ω, which is nevertheless much better than the classical Kendall's tau (Figure 2(b)).

In Figure 3, two exemplary data sets are visualized using a kernel-PCA [18] for dimensionality reduction, using the different correlation measures to produce the similarity matrices. Every data point is colored according to its original class membership. As can be seen, the classical Kendall's tau is hardly able to separate the classes, whereas $\tilde{\gamma}_\delta$, $\tilde{\gamma}$, and τ_ω are at least able to separate three of the four classes. Despite following quite different approaches, the results of these three measures appear to be surprisingly similar.

7.2 Second Study

The second experiment is meant to explore the behavior of the rank correlation coefficients when comparing two rankings of a specific type. We compared a ranking $\pi_{id} = (1, 2, \ldots, 11, 12)$ with rankings $\pi_{i\rightarrow 1} = (2, 3, \ldots i, 1, i+1, \ldots, 12)$ in which the ith item is moved from rank i to rank 1 and all items with index smaller i are shifted one position to the right. Each time the index i is incremented, another discordant pair is created, hence the similarity between π_{id} and

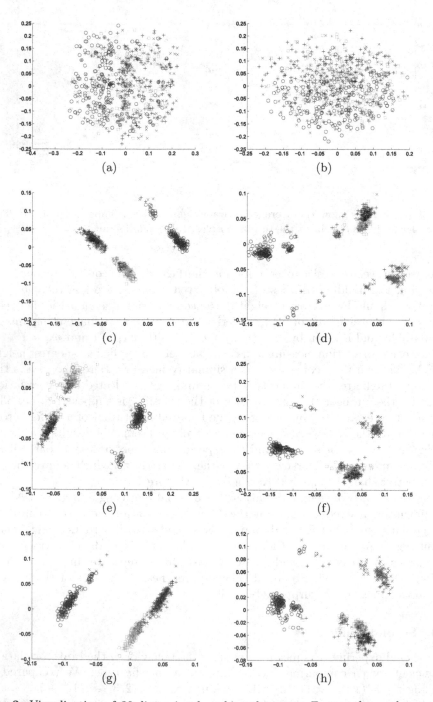

Fig. 3. Visualization of 30-dimensional ranking data sets. Every column shows one data set, every row one rank correlation coefficient. (a) – (b) Kendall's tau, (c) – (d) $\tilde{\gamma}_\delta$, (e) – (f) τ_ω, (g) – (h) $\tilde{\gamma}$.eps

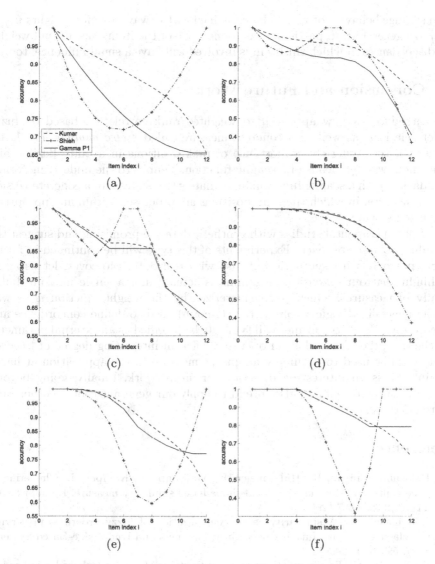

Fig. 4. Behavior of $\tilde{\gamma}_\delta$, τ_ω, and $\tilde{\gamma}$ in the "item i on rank 1" setting. The following weight vectors are used: (a) $(1\,0.9,\ldots,0.1,0)$, (b) $(1\,0.8,\ldots,0,0.2,\ldots,1)$, (c) $(1,1,1,1,0,0,0,1,1,1,1)$, (d) $(0,0.1,\ldots,0.9,1)$, (e) $(0,0.2,\ldots,1,0.8,\ldots,0)$, (f) $(0,0,0,1,1,1,1,1,0,0,0)$

$\pi_{i\rightarrow 1}$ should be monotone decreasing in i. Moreover, the higher the weight of the position i, the more pronounced the decrease should be.

Only $\tilde{\gamma}$ meets this expectation for all 6 weight vectors that have been considered (Figure 4). For instance, in Figure 4(a), $\tilde{\gamma}_\delta$ shows an increasing weighting of discordance with an increasing item index, although the weights are decreasing.

The strange behavior of τ_ω can be explained by the way in which weights $w_{ij} = v_i v_j$ are generated. In particular, as soon as one of the items has a small weight, all discordances in which this item is involved will have a small influence, too.

8   Conclusion and Future Work

We introduced a new approach to weighted rank correlation based on fuzzy order relations, as well as a concrete measure called *scaled gamma*. The latter allows for specifying the importance of rank positions in a quite flexible and convenient way by means of a scaling function. Thanks to the underlying formal foundation, such a scaling function immediately translates into a concrete version of our measure, in which the rank positions are processed within an appropriate weighting scheme.

First experimental studies with synthetic data are promising and suggest the usefulness of our approach. Experiments of this type will be continued in future work, not only with synthetic but also with real data. Moreover, let us again highlight that our extension of gamma is actually not a single measure but a family of measures, which is parameterized by the weight function w as well as the generalized logical conjunction (t-norm) used to define concordance and discordance. While the former will typically be specified as an external parameter by the user, the (fuzzy) logical operators offer an interesting degree of freedom that could be used to optimally adapt the measure to the application at hand. Again, this is an interesting direction for future work. Finally, going beyond the gamma coefficient, we also intend to apply our generalization to other rank correlation measures.

References

1. Balasubramaniyan, R., Hüllermeier, E., Weskamp, N., Kämper, J.: Clustering of gene expression data using a local shape-based similarity measure. Bioinformatics **21**(7), 1069–1077 (2005)
2. Bodenhofer, U.: A similarity-based generalization of fuzzy orderings preserving the classical axioms. Int. J. Uncertainty, Fuzziness and Knowledge-Based Systems **8**(5), 593–610 (2000)
3. Bodenhofer, U.: Representations and constructions of similarity-based fuzzy orderings. Fuzzy Sets and Systems **137**, 113–136 (2003)
4. Bodenhofer, U., Demirci, M.: Strict fuzzy orderings with a given context of similarity. Int. J. of Uncertainty, Fuzziness and Knowledge-Based Systems **16**(2), 147–178 (2008)
5. Bodenhofer, U., Klawonn, F.: Robust rank correlation coefficients on the basis of fuzzy orderings: Initial steps. Mathware & Soft Computing **15**, 5–20 (2008)
6. Pinto da Costa, J., Soares, C.: A weighted rank measure of correlation. Australian & New Zealand Journal of Statistics **47**(4), 515–529 (2005)
7. Fagin, R., Kumar, R., Sivakumar, D.: Comparing top k lists. SIAM Journal on Discrete Mathematics **17**(1), 134–160 (2003)
8. Goodman, L.A., Kruskal, W.H.: Measures of Association for Cross Classifications. Springer-Verlag, New York (1979)

9. Henzgen, S., Hüllermeier, E.: Weighted rank correlation measures based on fuzzy order relations. In: Hoffmann, F., Hüllermeier, E. (eds.) Proceedings 23. Workshop Computational Intelligence, pp. 227–236. KIT Scientific Publishing, Dortmund, Germany (2013)
10. Kaye, D.: A weighted rank correlation coefficient for the comparison of relevance judgements. Journal of Documentation **29**(4), 380–389 (1973)
11. Kendall, M.G.: Rank correlation methods. Charles Griffin, London (1955)
12. Klawonn, F.: Fuzzy sets and vague environment. Fuzzy Sets and Systems **66**, 207–221 (1994)
13. Klement, E.P., Mesiar, R., Pap, E.: Triangular Norms. Kluwer Academic Publishers (2002)
14. Kumar, R., Vassilvitskii, S.: Generalized distances between rankings. In: Proc. WWW, 19. International Conference on World Wide Web, pp. 571–580 (2010)
15. Marden, J.I.: Analyzing and Modeling Rank Data. CRC Press (1996)
16. Maturi, T.A., Abdelfattah, E.H.: A new weighted rank correlation. Journal of Mathematics and Statistics **4**(4), 226 (2008)
17. Dolorez Ruiz, M., Hüllermeier, E.: A formal and empirical analysis of the fuzzy gamma rank correlation coefficient. Information Sciences **206**, 1–17 (2012)
18. Schölkopf, B., Smola, A., Müller, K.R.: Kernel principal component analysis. In: Advances in Kernel Methods: Support Vector Learning, pp. 327–352. MIT Press (1999)
19. Shieh, G.S.: A weighted Kendall's tau statistic. Statistics & Probability Letters **39**(1), 17–24 (1998)
20. Spearman, C.: The proof and measurement for association between two things. Amer. Journal of Psychology **15**, 72–101 (1904)
21. Yilmaz, E., Aslam, J.A., Robertson, S.: A new rank correlation coefficient for information retrieval. In: Proc. 31st Annual International ACM SIGIR Conference on Research and Development in Information Retrieval, pp. 587–594. ACM (2008)

Rich Data

Concurrent Inference of Topic Models and Distributed Vector Representations

Debakar Shamanta[1]([✉]), Sheikh Motahar Naim[1], Parang Saraf[2],
Naren Ramakrishnan[2], and M. Shahriar Hossain[1]

[1] Department of Computer Science, University of Texas at El Paso,
El Paso, TX 79968, USA
{dshamanta,snaim}@miners.utep.edu, mhossain@utep.edu
[2] Department of Computer Science, Virginia Tech, Arlington, VA 22203, USA
{parang,naren}@cs.vt.edu

Abstract. Topic modeling techniques have been widely used to uncover dominant themes hidden inside an unstructured document collection. Though these techniques first originated in the probabilistic analysis of word distributions, many deep learning approaches have been adopted recently. In this paper, we propose a novel neural network based architecture that produces distributed representation of topics to capture topical themes in a dataset. Unlike many state-of-the-art techniques for generating distributed representation of words and documents that directly use neighboring words for training, we leverage the outcome of a sophisticated deep neural network to estimate the topic labels of each document. The networks, for topic modeling and generation of distributed representations, are trained concurrently in a cascaded style with better runtime without sacrificing the quality of the topics. Empirical studies reported in the paper show that the distributed representations of topics represent intuitive themes using smaller dimensions than conventional topic modeling approaches.

Keywords: Topic modeling · Distributed representation

1 Introduction

The representation of textual datasets in vector space has been a long-standing central issue in data mining with a veritable cottage industry devoted to representing domain-specific information. Most representations consider features as localized chunks as a result of which the interpretation of the features might lack generalizability. Researchers have recently become interested in distributed representations [8,12,14,19] because distributed representations generalize features based on the facts captured from the entire dataset rather than one single object or a small group of objects. Moreover, modern large and unstructured datasets involve too many heterogeneous entries for which local subspaces cannot capture relationships between the features. For example, publication datasets nowadays come with a substantial number of features like author information, scientific

© Springer International Publishing Switzerland 2015
A. Appice et al. (Eds.): ECML PKDD 2015, Part II, LNAI 9285, pp. 441–457, 2015.
DOI: 10.1007/978-3-319-23525-7_27

area, and keywords along with the actual text for each document. News article datasets have author information, time stamp data, category, and sometimes tweets and comments posted against the articles. Movie clips are accompanied by synopsis, production information, rating, and text reviews. The focus of this paper is on the design of a flexible mechanism that can generate multiple types of features in the same space. We show that the proposed method is not only able to generate feature vectors for labeled information available with the datasets but also for discovered information that are not readily available with the dataset as labels, for example, topics. Current state-of-the-art of distributed representations for unstructured text datasets can model two different types of elements in the same hyperspace, as described by Le and Mikolov [16]. Le and Mikolov's framework generates distributed vectors of documents (or paragraphs) and words in the same space using a deep neural network. Further generalization, that we have described in this paper, can provide distributed representations for heterogeneous elements of a dataset in the same hyperspace. However, the problem of creating distributed representations becomes more challenging when the label information is not contained within the dataset. The focus of this paper is on the generation of topical structures and their representations in the same space as documents and words. The capability of representing topics, documents, words, and other labeled information in the same space opens up the opportunity to compute syntactic and semantic relationships between not only words but also between topics and documents by directly by using simple vector algebra.

Estimating the topic labels for documents is another challenge while using distributed representations. Earlier topic modeling techniques [9,13] used to define a document as a mixture of topics and estimate the probability $p(t|d)$ of a topic (t) of a document (d) through probabilistic reasoning. More recently, topic models are seen from a neural network point of view [6,15,26] where these probabilities are generated from the hidden nodes of a network. Such neural networks require compact numeric representations of words and documents for effective training, which are not easy to estimate with traditional vector space based document modeling techniques that represent the documents using a very high dimensional space. There have been attempts to use the compact distributed representations of words and documents learned from a general purpose large dataset [6] but the precomputed vectors may not be always appropriate for many new domain specific datasets. Furthermore, the vocabulary shifts in a new direction over time resulting in changes in the distributed representations.

Specific contributions of this paper are as follows.

1. We formulate the problem of computing distributed representation of topics in the same space as documents and words using a novel fusion of a neural network based topic modeling and a distributed representation generation technique.
2. The tasks of computing topics for documents and generating distributed representations are simultaneous in the proposed method unlike closely related state-of-the-art techniques where precomputed distributed vectors of words are leveraged to compute topics. Additionally, none of the state-of-the-art

methods generates distributed representation of topics to the best of our
knowledge.

3. Our proposed method generates the distributed vectors using a smaller number of dimensions than the actual text feature space. Even if the space is
of lower number of dimensions, the vectors capture syntactic and semantic
relationships between language components.

4. We demonstrate that the generated topic vectors explain domain specific
properties of datasets, help identify topical similarities, and exhibit topic-
specific relationships with document vectors.

2 Related Work

Distributed representations have been used in diverse fields of scientific research
with notable success due to their superiority in capturing generalized view of
information over local representations. Rumelhart et al. [22] designed a neural
network based approach for distributed representation of words which has been
followed by many efforts in language modeling. One such model is the neural
probabilistic model [2] proposed by Bengio et al. This framework uses a sliding
window based context of a word to generate compact representations. Mikolov
et al. [17] brings in continuous bag-of-words (CBOW) and skip-gram models to
compute continuous vector representations of words efficiently from very large
data sets. The skip-gram model was significantly improved in [18], which includes
phrase vectors along with words. Le and Mikolov [16] extended the CBOW
model to learn distributed representation of higher level texts like paragraphs
and documents. Our proposed model further enriches the literature by including
the capability to generate (1) vectors for arbitrary labels in the dataset and
(2) vectors for topics for which a text dataset does not contain any labeled
information.

Finding hidden themes in a document collection has been of great interest
to data mining and information retrieval researchers for more than two decades.
An earlier work in the literature is latent semantic indexing (LSI) [9] that maps
document and terms in a special "latent semantic" space by applying dimensionality reduction on traditional bag-of-words vector space representations of
documents. A probabilistic version of LSI, pLSI [13], introduces a mixture model
where each document is represented by a mixing proportion of hidden "topics". Latent Dirichlet Allocation (LDA) [5], a somewhat generalized but more
sophisticated version of pLSI, is one of the most notable ones in the literature.
It provides a generative probabilistic approach for document modeling assuming
a random process by which the documents are created. LDA spawned a deluge
of work exploring different aspects of topic modeling. For example, the Dynamic
Topic Model (DTM) [4] captures the evolution of topics in a time-labeled corpus.
Online LDA (OLDA) [1] handles streams of documents with dynamic vocabulary,
Wallach [25] and Griffiths et al. [11] exploit the sentence structures of documents
and Correlated Topic Model (CTM) [3] captures the correlation between topics.

More recently, neural network based models have received great attention
from the data mining community. Wan [26] et al. introduce a hybrid model

in computer vision settings; DocNADE [15] provides an autoregressive neural network for topic modeling; Cao et al. [6] propose a neural topic model (NTM) with supervised extension. The latter work has close resemblance to a part of our proposed model that focuses on generating topics for each document.

3 Problem Formulation

Let $D = \{d_1, d_2, \ldots, d_N\}$ be a text dataset containing N documents taking terms from the set of M words $W = \{w_1, w_2, \ldots, w_M\}$. Each document can contain an arbitrary number of words in any sequence. The objective is to generate a universal distributed representation for the labeled items (e.g., words and documents) and latent topics of each document of dataset D. Let $T = \{t_1, t_2, \ldots, t_K\}$ be the set of topics. Consider that the expected number of dimensions in the distributed representation of words, documents, and topics is L. L should be much smaller than the number of words M. Word vectors $W \in \mathbb{R}^{M \times L}$, document vectors, $D \in \mathbb{R}^{N \times L}$ and topic vectors, $T \in \mathbb{R}^{K \times L}$ generated in the same L-dimensional space should maintain two specific properties: (1) distributed representation of each type should be capable of capturing the semantic, syntactic, and topical aspect of conventional language models, and (2) all types of vectors (topics, documents, and words) organized in the L-dimensional hyperspace must be comparable to each other.

The first property aligns the framework with the objectives of any language model where features are generated for most common data mining tasks likes clustering and classification. The second property, however, is unique and specific to relating vectors of different types of entities like topics, documents, and vectors. In word2vec [17], the authors show that distributed representations of word can retrieve linguistic similarities between pairs of words. For example, $W_{\text{King}} - W_{\text{Man}}$ is close to $W_{\text{Queen}} - W_{\text{Woman}}$. The ability to model topics in the same hyperspace extends this property by capturing similarity between relationships among topics and documents. For example, if two documents d_i and d_j are drawn from the same topic t_p then $T_p - D_i$ should be closer to $T_p - D_j$. Similarly, if two documents d_i and d_j are drawn from two different topics t_p and t_q, then $T_p - D_i$ should tend to be different than $T_p - D_j$.

4 Methodology

The main objective of the proposed framework is to generate a compact distributed representation for topics, documents, and words of a document collection in the same hyperspace in such a way that all these heterogeneous objects are comparable to each other and capture the semantic, syntactic and thematic properties. The proposed framework has three major components. First, we adopt a generic neural network that can generate distributed vectors for documents, words, and any given labels. Second, we propose a deep neural network based topic modeling that can take distributed representations of words and documents, and estimate topic distribution for each document. Finally, we convolute

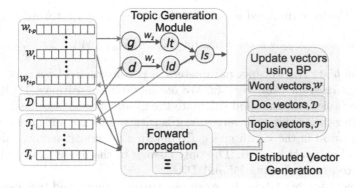

Fig. 1. The proposed framework.

both these networks so that they can share information and train simultaneously. Fig. 1 shows the proposed framework. The following subsections describe the model in a sequence.

4.1   Distributed Representation of Heterogeneous Entities

Inferring a distributed representation $\mathcal{W}$ for the words of a document collection D having vocabulary W is based on predicting a word given other words in the same context. The objective of such a word representation model is to maximize the average log probability

$$\frac{1}{M} \sum_{m=p}^{M-p} \log p(w_m | w_{m-p}, \ldots, w_{m+p}) \tag{1}$$

The individual probabilities in Equation 1 are estimated by training a multi-class deep neural network, such as softmax. They can be computed as:

$$p(w_m | w_{m-p}, \ldots, w_{m+p}) = \frac{e^{y_m}}{\sum_i e^{y_i}} \tag{2}$$

Algorithm 1. *LearnDistRep* – algorithm for learning topic vectors

input   : Document id, d
          Set of topics in d, T_d
          Word to predict, w
          Context of w, C_w
parameter: Distributed representations $\mathcal{D}$, $\mathcal{W}$ and $\mathcal{T}$

1 Calculate y using Equation 4 ;
2 Calculate gradient gr using stochastic gradient descent ;
3 Update document vector $\mathcal{D}_d$, topic vectors $\mathcal{T}_{T_d}$ and word vectors $\mathcal{W}_{C_w}$ using gr;

---

where y_i is the unnormalized log-probability for every output word w_i.

$$y_i = b + Uh(w_m|w_{m-p}, \ldots, w_{m+p}; \mathcal{W}) \tag{3}$$

Here, U and b are the softmax parameters. h is constructed by a concatenation or average of relevant word vectors. We use hierarchical softmax [17] instead of softmax for faster training, and calculate the gradient using stochastic gradient descent. After the training converges, words with similar meaning are mapped to a similar position in the vector space. To obtain a document vector, a document is thought of as another word. The only change in the model is in Equation 3, where h is constructed using $\mathcal{W}$ and $\mathcal{D}$.

Inclusion of further labels, for example, authors, topic, and tags can be done the same way document vectors are added. Our focus in this paper is to incorporate topics instead of additional labels. Incorporation of topic vectors is challenging because the topics are not given and rather should be generated using the documents and words. For the time being, let us assume that topic is just a given label that comes with the data. In contrast to the word vector matrix $\mathcal{W}$ that is shared across all the documents, a topic vector can be shared only across the documents which contain that particular topic. Considering topic vectors along with the vectors for words and documents, Equation 3 is modified to:

$$y = b + Uh(w_{t-k}, \ldots, w_{t+k}, d_q, t_{r_1}, t_{r_2}, \ldots, t_{r_s}; \mathcal{W}, \mathcal{D}, \mathcal{T}) \tag{4}$$

For the training purpose, we use sampling of variable-length contexts using a sliding window over each document. Such a sliding window is commonly referred to as n-gram. We use n-grams instead of single words (unigrams) since n-grams produce representative contexts around each word [18]. A procedure for training this generic network for topic, documents, and words is explained in Algorithm 1.

4.2   Estimating Topic Labels of Documents

As stated earlier, the generic model described in Section 4.1 requires topic as labels of each document. This section focuses on a topic modeling technique that can generate topic labels taking document vectors and word vectors into account. For effective and efficient generation of topic vectors, the topic modeling technique must synchronize with the iterations of the distributed vector generation part. Several topic modeling techniques have been proposed in the literature to find topic distribution of documents of such unlabeled datasets. In a general topic model, each document is seen as a mixture of topics, and each topic is represented as a probability distribution over the vocabulary of the entire corpus. The conditional probability $p(w|d)$ of a word and a document is computed from word-topic distribution and topic-document distribution as $p(w|d) = \sum_{i=1}^{K} p(w|t_i)p(t_i|d)$, where K is the number of topics and t_i is a latent topic. This equation can be re-written as

$$p(w|d) = \phi(w) \times \theta^T(d) \tag{5}$$

Algorithm 2. *LearnTopic* – algorithm for learning topic distribution.

input   : Document id, d
            N-gram or context id, g
parameter: Distributed representations $\mathcal{D}$, $\mathcal{W}$ and $\mathcal{T}$
            Weight matrices W_1 and W_2
output   : Updated weight matrices

1 Calculate $ls(g, d)$ using equations for lt and ld ;
2 Determine error in output node with respect to the ideal value:
  $\delta^{(3)} = ls(g, d) - 1$;
3 Compute the error in n-gram-topic hidden node:
  $\delta_1^{(2)} = (\delta^{(3)} \times ld(d)) \cdot (lt(g) \cdot (1 - lt(g)))$;
4 Update W_2: $W_2 = W_2 + \alpha[\delta_1^{(2)} \times \mathcal{W}_g + \lambda \times W_2]$;
5 Compute error in the document-topic hidden node:
  $new\_ld(d) = ld(d) + \alpha[\delta^{(3)} \times lt(g) + \lambda \times ld(d)]$;
6 $\delta_2^{(2)} = new\_ld(d) - ld(d)$;
7 Update W_1: $W_1 = W_1 + \alpha[\delta_2^{(2)} \times \mathcal{D}_d + \lambda \times W_1]$;

---

where $\phi(w) = [p(w|t_1), p(w|t_2), \ldots, p(w|t_K)]$ is the conditional probabilities of w with all the topics and $\theta(d) = [p(t_1|d), p(t_2|d), \ldots, p(t_K|d)]$ is the topic distribution of d.

We can view topic models from a neural network perspective considering the formation of Equation 5. Let us consider a neural network with two input nodes for sliding window with n-gram g and document d, two hidden nodes lt (representing $\phi(g)$) and ld (representing $\theta(d)$), and one output node ls producing the conditional probability $p(g|d)$. The topic-document node $ld \in \mathbb{R}^{1 \times K}$ computes the topic distribution of a document (similar to θ in topic models) using the weight matrix $W_1 \in \mathbb{R}^{L \times K}$. It is computed by the equation $ld(d) = softmax(\mathcal{D}_d \times W_1)$ which uses a softmax function to maintain the probabilistic constraint on topic distribution that all the topic probabilities of a document must sum up to 1.

The n-gram-topic node $lt \in \mathbb{R}^{1 \times K}$ stands for the topic representation of the input n-grams, and calculated as $lt(g) = sigmoid(\mathcal{W}_g \times W_2)$ where $W_2 \in \mathbb{R}^{L \times K}$ denotes the weight matrix between the n-gram input node and the n-gram-topic node. This vector follows a probabilistic form similar to ϕ in topic models.

The output node $ls \in \mathbb{R}$ gives the matching score of an n-gram g and a document d by computing the dot product of $lt(g)$ and $ld(d)$. The outputted score $ls(g, d) = lt(g) \times ld(d)^T$ is a value between 0 and 1, similar to the conditional probability of $p(g|d)$.

The n-gram-document probability $p(g|d)$, which initially is expected to be very different from the ideal value, is estimated by performing a forward propagation in the network. Algorithm 2 describes the training procedure for the neural topic model part of our proposed model. For each n-gram-document pair (g, d) the expected output value is 1 due to the fact that g is taken from document d. The weights are updated using backpropagation to mitigate that error (Steps 3 to 7 in Algorithm 2).

4.3 Concurrent Training

The training process runs concurrently for both topic modeling and distributed vector generation. Fig. 1 shows the proposed combination of two networks. Notice the training is simultaneous unlike NTM [6] where already trained word vectors are used for topic modeling. All the weights (W_1 and W_2 matrices) and vectors ($\mathcal{W}, \mathcal{D}$ and $\mathcal{T}$ matrices) in both the networks are initialized with random values (Step 1 and 2 of Algorithm 3). As shown in the loop at Step 3 of Algorithm 3, the combined framework reads each document in sequence of n words (context) using a continuous window. For a particular document, the topic modeling network gives its topic distribution as the output of the hidden node ld. We select k most probable topics from this distribution – with an assumption that a document is made up of k number of topics – and provide them as input to the distributed vector generation network. The call to the method `LearnTopics` in Step 7 of Algorithm 3 accomplishes this task. The corresponding word, document and topic vectors are updated using method `LearnDistRep` in Step 8 of Algorithm Algorithm 3. Method `LearnTopics` and `LearnDistRep` are explained in Algorithms 2 and 1, respectively.

Notice that the document and word vectors of context (n-gram) generated by Algorithm 1 are provided as input to the topic modeling network of Algorithm 2. Also the top k topics generated for each document using Algorithm 2 are provided to the distributed vector generation part (Algorithm 1). Algorithm 3 combines all these steps.

Algorithm 3. *ConcurrentTrain* – algorithm for simultaneous training of both networks

---

    input   : Document collection D
    parameter: Distributed representations $\mathcal{D}, \mathcal{W}$ and $\mathcal{T}$
                 Weight matrices W_1 and W_2 of topic modeling network
    output   : $\mathcal{D}, \mathcal{W}$ and $\mathcal{T}$

1 Randomly initialize $\mathcal{D}, \mathcal{W}$ and $\mathcal{T}$;
2 Randomly initialize W_1 and W_2 ;
3 **for** *each document $d \in D$* **do**
4   Topics in d, $T_d \leftarrow$ top k topics from $ld(d)$;
5   **for** *each word w of d* **do**
6          $C_w \leftarrow$ context of w ;
7   LearnTopics(d, C_w) ;
8   LearnDistRep(d, T_d, w, C_w) ;
9   **end**
10 **end**

---

5   Complexity Analysis

Although both the neural networks in our proposed framework are concurrently trained, we analyze their complexities separately for simplicity. For every example during the training of the distributed vector generation network, there are P words (context length), k topics and one document as input resulting in $I = P+k+1$ input nodes. These inputs are projected into a L dimensional space. Although there are $V = N+M+K$ output nodes, this part of the network needs to update only $O(\log V)$ nodes using the gradient vector since the model uses hierarchical softmax. I input nodes get updated during backpropagation making the complexity for training a single example, $C_{dr} = I \times L + O(\log V) \times L$.

The topic modeling network takes the same document and input words. Calculating $\mathcal{W}_g$ from the words in n-gram g takes $O(P \times L)$ time. Calculating each of ld and lt takes $O(L \times K)$ operations and ls requires $O(K)$ operations. Backpropagation (step 3 to 7 of Algorithm 2) runs in $O(L \times K)$ time incurring a total cost of $C_{tm} = O(P \times L) + O(L \times K) + O(K) + O(L \times K)$, or $C_{tm} = O(L \times K)$ given $K > P$, for every example. Therefore, the cost of training the combined network for each example is $C = C_{tm} + C_{dr}$.

6   Evaluation

We use a number of metrics to evaluate the quality of our results. Some of these metrics are generally used to evaluate clustering results when ground truth labels are not available. Two such evaluations are the Dunn Index (DI) [10] and the Average Silhouette Coefficient (ASC) [21]. DI measures the separation between groups of vectors and larger values are better. ASC is a measure that takes both cohesion and separation of groups into account (higher values are better). In our experiments, we utilize ASC and DI together to evaluate the final topic assignments of the documents. Topics are analogous to clusters in those evaluations. ASC and DI give us an idea about how crisply the topics are distributed across the documents.

In the presence of ground truth labels, we evaluated the assigned topics using Normalized Mutual Information (NMI) [7], Adjusted Rand Index (ARI) [24], and the hypergeometric distribution-based enrichment. Both NMI and ARI estimates the agreement between two topic assignments, irrespective of permutations. Higher values are better for NMI and ARI. Hypergeometric enrichment [23] maps topics to available ground truth labels. This allows us to measure a significance based on hypergeometric distribution of the topic assignments over the already known labels. Higher number of enriched topics is better.

Our proposed model is able to generate topic and document vectors in the same hyperspace. In an ideal case, all angles between a topic vector and each document vector assigned to this topic should be similar and the standard deviation of those angles should be small. We use this concept to compute alignment between a topic vector and a given set of document vectors. Given a topic vector $\mathcal{T}_i$ of topic t_i, and a set of document vectors $\mathcal{D}^{t_j}$ that are assigned a topic t_j, we compute alignment using the following formula:

$$A(\mathcal{T}_i, \mathcal{D}^{t_j}) = \sqrt{\frac{1}{|\mathcal{D}^{t_j}|} \sum_{m=1}^{|\mathcal{D}^{t_j}|} \left(\frac{\mathcal{T}_i.\mathcal{D}_m^{t_j}}{\|\mathcal{T}_i\|\|\mathcal{D}_m^{t_j}\|} - \mu \right)^2} \tag{6}$$

where $\mathcal{D}_m^{t_j}$ refers to the document vector of mth document in topic t_j, and

$$\mu = \frac{1}{|\mathcal{D}^{t_j}|} \sum_{m=1}^{|\mathcal{D}^{t_j}|} \frac{\mathcal{T}_i.\mathcal{D}_m^{t_j}}{\|\mathcal{T}_i\|\|\mathcal{D}_m^{t_j}\|} \tag{7}$$

Notice that Equation 6 is the standard deviation between the cosine angles between the topic vectors and the document vectors. Lower values are expected when $t_i = t_j$ and higher values are expected when $t_i \neq t_j$.

7   Experiments

In this section, we seek to answer the following questions to justify the capabilities and correctness of the proposed model.

1. Can our framework establish relationships between distributed representations of topics and documents? (Section 7.1)
2. Are the generated topic vectors expressive enough to capture similarity between topics and to distinguish difference between them? (Section 7.2)
3. How do our topic modeling results compare with the results produced by other topic modeling algorithms? (Section 7.3)
4. Do the generated topics bring documents with similar domain-specific themes together? (Section 7.4)
5. How does the runtime of the proposed framework scale with the size of the distributed representations, increasing number of documents, and increasing number of topics? (Section 7.5)

We used seven different text datasets[1] with different number of documents and words. The datasets are listed in Table 1. Some of these datasets are widely used in the text processing literature (e.g., Reuters , WebKB, and 20Newsgroups datasets), while we have collected most of the other corpora from the public domain. The PubMed dataset is collected from publicly available citation databases for biomedical literature provided by the US National Library of Medicine. The PubMed dataset contains abstracts of *cancer*-related publications. The *Spanish news* dataset was collected as a part of the *EMBERS* [20] project. The articles covered news stories from 207 countries around the world.

7.1   Analysis of Distributed Representations of Topics and Documents

The topic and document vectors generated by the proposed framework maintain consistent relationships that can be leveraged in many applications to study the

---

[1]Data and software source codes are provided here: http://dal.cs.utep.edu/projects/tvec/.

Table 1. Summary of the datasets.

Dataset	#Docs	#Words	Additional information
Synthetic	400	40,000	Four lower and two upper level groups.
20 Newsgroups	18,821	2,654,769	20 categories in seven groups.
Reuters R8	7,674	495,226	Eight category labels.
Reuters R52	9,100	624,456	52 groups.
WebKB	4,199	559,984	Four overlapping categories
PubMed	1.3 million	220 million	Publication abstracts related to cancer.
Spanish news	3.7 million	3 billion	News articles from 2013 and 2014.

topics of a stream of unseen documents. To be able to develop such applications, a relationship between a topic vector $\mathcal{T}_i$ and any of its document vectors $\mathcal{D}_p^{t_i}$ should be different than the relationship between another topic $\mathcal{T}_j$ and a document vector $\mathcal{D}_q^{t_j}$.

In contrast, such topic-document relationships should be similar for two documents of the same topic. Each plot of Fig. 2 shows a heat map of alignment between a topic vector $\mathcal{T}_i$ of topic t_i and all document vectors $\mathcal{D}^{t_j}$ of topic t_j using Equation 6. Fig. 2 shows the heat map with four topics of the synthetic dataset. In this heat map, lower alignment values result in darker cells depicting stronger topic-document alignment for topic and document vectors of the same topic, whereas weaker alignments are exhibited when document vectors are chosen from a different topic. This indicates that our proposed framework captures topical structures as well as it models relationships between topics and documents in the same hyperspace.

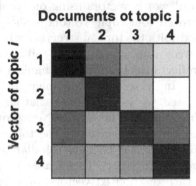

Fig. 2. Heat map of variance of cosine similarity between topic vector i and all documents of topic j.

7.2 Expressiveness of Topic Vectors

As described in Section 4.3, k-best topics generated by the topic modeling part of the proposed model are selected as input to the distributed representation generation part. We set $k = 1$ for all our experiments including the ones described in this subsection . To examine how expressive our distributed topic vectors are, we prepared a synthetic corpus containing documents with term from seven sets as illustrated by Fig. 3(a). Four groups of documents contains terms specific to each group. The same dataset can be divided into two groups of documents because each group contains terms from a specific group set of words. Additionally, all sets of documents share a common set of terms. We generated topic, document,

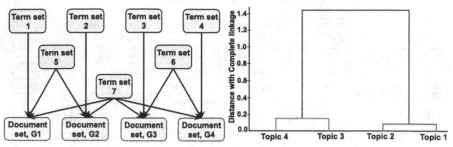

(a) The synthetic dataset has four groups of documents.

(b) Dendrogram generated using topic vectors.

Fig. 3. Experiment with a synthetic dataset. (a) Sets of terms used to prepare the synthetic text corpus, (b) Dendrogram generated from the topic vectors.

and words vectors using our proposed framework. A dendrogram for the generated four topic vectors is shown in Fig. 3(b). As expected, the dendrogram exhibits the topical structure where two topic vectors separately and then those two groups merge at the top of the hierarchy. The dendrogram of topic vectors reflects the grouping mechanism we used to create the dataset.

In a second experiment in this space, we used a dataset that already has category labels (20 Newsgroups) to verify how intuitive the topic vectors are in bringing similar categories together. To be able to generate distributed vectors for existing categories along with document and word vectors, we directly provided the known labels to the distributed representation generation part of the model as an inputs as opposed to providing topics generated by the topic modeling network. The official site for the 20 News Groups dataset reports that some of the newsgroups are very closely related to each other (e.g. *comp.sys.ibm.pc.hardware* and *comp.sys.mac.hardware*), while others

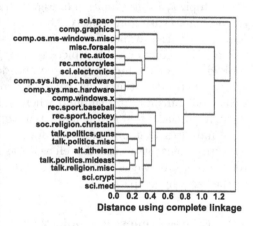

Fig. 4. Dendrogram prepared with the 20 category vectors of 20 Newsgroups dataset.

may be highly unrelated (e.g *misc.forsale* and *soc.religion.christian*). Our target is to verify if the generated category vectors can provide insights about how the topics should be merged. Fig. 4 shows the dendrogram prepared for the 20 category vectors of 20 Newsgroups dataset. There are some differences between the official grouping and the grouping we have discovered using the category

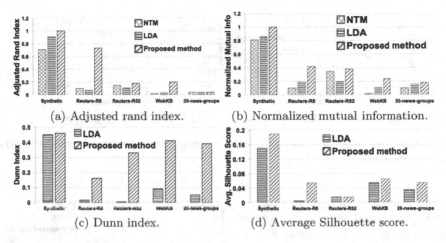

(a) Adjusted rand index.   (b) Normalized mutual information.

(c) Dunn index.   (d) Average Silhouette score.

Fig. 5. Evaluation using benchmark labels (a & b) and locality of the topics (c & d).

vectors, for example, sci.electronics is grouped with *comp.sys.mac.hardware* and *comp.sys.ibm.pc.hardware*. The label *sci.electronics* is far away from *sci.space* even though they have a common prefix "sci". Our observation is that *sci.electronics* has many documents containing hardware related discussions. As a result, *sci.electronics* has greater similarity with hardware than *sci.space*. Similar evidences are found for the rec.* groups. For example, rec.sport.* groups are different from rec.motorcycles and rec.autos but the latter two groups are closely related, as evident in the dendrogram.

7.3 Comparison of Quality of Generated Topics

Fig. 5 shows a comparison of results generated by our framework and two other topic modeling methods, LDA and NTM, when applied on four classification datasets — synthetic, Reuters-R8, Reuters-R52, WebKB, and 20 Newsgroups. Fig. 5 (a) and (b) use adjusted Rand index (ARI) and normalized mutual information (NMI) to compare the topic assignments of the documents with the expected classes. ARI and NMI are larger for the proposed methods for all the datasets. This implies that our framework realizes the expected themes of the collections better than LDA and NTM. Not only the expected categories better match with the topic assignments, but also the generated topics are local in the corresponding space of our framework. Higher Dunn index and higher average silhouette coefficient for all the datasets, as depicted in Fig. 5(c) and (d), imply that our model provides high quality local topics. Notice that Fig. 5(c) and (d) do not have NTM. This is because Dunn index and average silhouette coefficient require document vectors, but NTM [6] does not directly use any document vector; rather, it uses precomputed word vectors only.

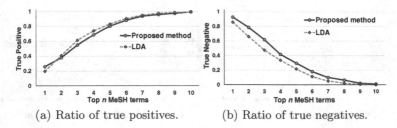

|(a) Ratio of true positives. | (b) Ratio of true negatives.|

Fig. 7. Comparison of our method and LDA using MeSH terms associated with the PubMed abstracts.

We also used a hypergeometric distribution based procedure to map each topic to a class label. Fig. 6 shows that the topic assignments suing our framework have higher number of enriched topics than any other method. This indicates that the topics generated by our methods has higher thematic resemblance with the benchmark labels.

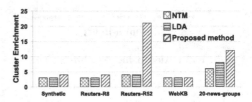

Fig. 6. Comparison of numbers of topics enriched by hypergeometric distribution.

All these datasets described so far, in this subsection are labeled and are widely used a ground truths in many data mining and machine learning evaluations. In addition to these datasets, we used our EMBERS data containing around 3.7 million news articles to compare locality of the topics with other methods. Table 2 shows that our method produces topics with greater Dunn index and average silhouette score than other methods. This indicates that our method performs even better when the datasets are very large.

7.4 Evaluation using Domain Specific Information

In this experiment, we used the PubMed dataset to compute overlap of domain specific information for documents in the same topic (i.e., true positive) and lack of such overlap for a pair of documents from two different topics (i.e., true negative). In the PubMed dataset, each abstract is provided with some major

Table 2. Evaluation using the EMBERS news article dataset.

Method	Evaluation metric	
	Dunn index	Silhouette score
NTM	0.04	0.01
LDA	0.01	-0.015
Proposed method	**0.1**	**0.05**

Medical Subject Header (MeSH) terms which come from a predefined ontology. We used these MeSH terms as domain specific information to evaluate the topics. It is expected that the sets of MeSH terms of two documents of the same topic will have some common entries, where as the sets of MeSH terms of two documents from two different topics will have lesser or no overlapping records. For each abstract, we ordered the MeSH terms based on Jaccard similarity between a MeSH terms and the abstract. Notice that if we pick up n best

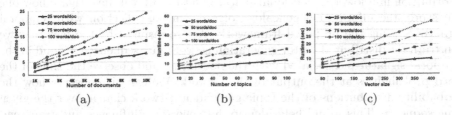

Fig. 8. Execution time with varying (a) number of documents, (b) number of topics, and (c) vector size.

MeSH terms for two documents from the same topic the chance that these two sets of n best MeSH terms have common entries increases with larger n. This trend is observed in Fig. 7(a) for both our framework and LDA. The true positive ratio quickly becomes around 80% with only five best MeSH terms for each pair of documents. Now, the top n MeSH terms of two documents from two different topics should have higher absence of overlapping terms with smaller n since the topical similarity of these two documents is minimal. As n increases the true negative ratio will decrease due to inclusion of more general entries in the lists of n best MeSH terms. Fig. 7(b) shows the expected trend for both LDA and our framework. We selected random 5,000 pairs of documents from same topics and another 5,000 pairs from different topics for the two plots, Fig. 7(a) and (b) respectively. Fig. 7(a) and (b) demonstrate that our method follows an expected trend of sharing domain specific information. Although the true positive values are slightly lower than LDA in our method in some cases, the true negative values are always greater than LDA. This indicates that our model generates topics containing similar biological themes while documents of different topics, as expected, have lesser similarity in domain specific information.

7.5 Runtime Characteristics

Fig. 8 depicts the runtime behavior of our proposed framework with varying number of documents, topics, and vector size. The runtime increases almost linearly with each of these variables. This indicates our proposed framework is scalable with large amount of data. The experiments in this space were done using synthetic data with different number of words in each document as depicted by multiple lines in each of the plots of Fig. 8.

8 Conclusion

We have presented a framework to generate distributed vectors for elements in a corpus as well as the underlying latent topics. All types of vectors — topics, documents, and words — share the same space allowing the framework to compute relationships between all types of elements. Our results show that the framework can efficiently discover latent topics and generate distributed vectors simultaneously. The proposed framework is expressive and able to capture domain specific information in a lower-dimensional space. In future, we will investigate how one can study the information genealogy of a document collection with temporal signatures using the proposed framework. We are inspired by the fact that we can train the distributed vector generation network in a sequence as found in the temporal signatures associated with the documents and observe the shift of the word probabilities at the output of the network. We can also observe how the probability distributions of the topic generation network change over the given time sequence. This would help identify how one topic influence and transcend another and how the topical vocabulary shifts over time.

Acknowledgments. This work is supported in part by M. S. Hossain's startup grant at UTEP, University Research Institute (URI, Office of Research and and Sponsored Projects, UTEP), and the Intelligence Advanced Research Projects Activity (IARPA) via DoI/NBC contract number D12PC000337. The funders had no role in study design, data collection and analysis, decision to publish, or preparation of the manuscript. The US Government is authorized to reproduce and distribute reprints of this work for Governmental purposes notwithstanding any copyright annotation thereon.

References

1. AlSumait, L., Barbará, D., Domeniconi, C.: On-line lda: adaptive topic models for mining text streams with applications to topic detection and tracking. In: ICDM 2008, pp. 3–12 (2008)
2. Bengio, Y., Ducharme, R., Vincent, P., Janvin, C.: A neural probabilistic language model. Machine Learning Research **3**, 1137–1155 (2003)
3. Blei, D., Lafferty, J.: Correlated topic models. Advances in Neural Information Processing Systems **18**, 147 (2006)
4. Blei, D.M., Lafferty, J.D.: Dynamic topic models. In: ICML 2006, pp. 113–120 (2006)
5. Blei, D.M., Ng, A.Y., Jordan, M.I.: Latent dirichlet allocation. Machine Learning Research **3**, 993–1022 (2003)
6. Cao, Z., Li, S., Liu, Y., Li, W., Ji, H.: A novel neural topic model and its supervised extension. In: AAAI 2015 (2015)
7. Chaitin, G.J.: Algorithmic information theory. Wiley Online Library (1982)
8. Chalmers, D.J.: Syntactic transformations on distributed representations. In: Connectionist Natural Language Processing, pp. 46–55. Springer (1992)
9. Deerwester, S.C., Dumais, S.T., Landauer, T.K., Furnas, G.W., Harshman, R.A.: Indexing by latent semantic analysis. American Society for Information Science **41**(6), 391–407 (1990)
10. Dunn, J.C.: A fuzzy relative of the isodata process and its use in detecting compact well-separated clusters (1973)
11. Griffiths, T.L., Steyvers, M., Blei, D.M., Tenenbaum, J.B.: Integrating topics and syntax. In: NIPS 2004, pp. 537–544 (2004)
12. G. E. Hinton. Learning distributed representations of concepts. In: CogSci 1986, vol. 1, p. 12 (1986)
13. Hofmann, T.: Probabilistic latent semantic indexing. In: SIGIR 1999, pp. 50–57. ACM (1999)
14. Hummel, J.E., Holyoak, K.J.: Distributed representations of structure: A theory of analogical access and mapping. Psychological Review **104**(3), 427 (1997)

15. Larochelle, H., Lauly, S.: A neural autoregressive topic model. In: NIPS 2012, pp. 2708–2716 (2012)
16. Le, Q.V., Mikolov, T.: Distributed representations of sentences and documents. In: ICML 2014, pp. 1188–1196 (2014)
17. Mikolov, T., Chen, K., Corrado, G., Dean, J.: Efficient estimation of word representations in vector space (2013). arXiv preprint arXiv:1301.3781
18. Mikolov, T., Sutskever, I., Chen, K., Corrado, G.S., Dean, J.: Distributed representations of words and phrases and their compositionality. In: NIPS 2013, pp. 3111–3119 (2013)
19. Pollack, J.B.: Recursive distributed representations. Artificial Intelligence **46**(1), 77–105 (1990)
20. Ramakrishnan, N., et al.: 'Beating the news' with EMBERS: Forecasting civil unrest using open source indicators. In: SIGKDD 2014, pp. 1799–1808 (2014)
21. Rousseeuw, P.J.: Silhouettes: A graphical aid to the interpretation and validation of cluster analysis. Computational and Applied Mathematics **20**, 53–65 (1987)
22. Rumelhart, D.E., Hinton, G.E., Williams, R.J.: Learning representations by back-propagating errors. Cognitive Modeling **5**, (1988)
23. Rumelhart, D.E., Hinton, G.E., Williams, R.J.: Enrichment or depletion of a go category within a class of genes: which test? Bioinformatics **23**(4), 401–407 (2007)
24. Steinley, D.: Properties of the hubert-arable adjusted rand index. Psychological Methods **9**(3), 386 (2004)
25. Wallach, H.M.: Topic modeling: beyond bag-of-words. In: ICML 2006, pp. 977–984 (2006)
26. Wan, L., Zhu, L., Fergus, R.: A hybrid neural network-latent topic model. In: AISTATS 2012, pp. 1287–1294 (2012)

Differentially Private Analysis of Outliers

Rina Okada$^{(\boxtimes)}$, Kazuto Fukuchi, and Jun Sakuma

University of Tsukuba, 1-1-1 Tennodai, Tsukuba, Ibaraki 305-8577, Japan
{rina,kazuto}@mdl.cs.tsukuba.ac.jp, jun@cs.tsukuba.ac.jp

Abstract. This paper presents an investigation of differentially private analysis of distance-based outliers. Outlier detection aims to identify instances that are apparently distant from other instances. Meanwhile, the objective of differential privacy is to conceal the presence (or absence) of any particular instance. Outlier detection and privacy protection are therefore intrinsically conflicting tasks. In this paper, we present differentially private queries for counting outliers that appear in a given subspace, instead of reporting the outliers detected. Our analysis of the global sensitivity of outlier counts reveals that regular global sensitivity-based methods can make the outputs too noisy, particularly when the dimensionality of the given subspace is high. Noting that the counts of outliers are typically expected to be small compared to the number of data, we introduce a mechanism based on the smooth upper bound of the local sensitivity. This study is the first trial to ensure differential privacy for distance-based outlier analysis. The experimentally obtained results show that our method achieves better utility than global sensitivity-based methods do.

Keywords: Differential privacy · Outlier detection · Smooth sensitivity

1 Introduction

Data mining technologies are now becoming increasingly influential in our daily life. When data mining is processed over personal data collected from individuals, the acquired knowledge might be used to infer private information. In this paper, we investigate differentially private outlier analysis.

Outlier detection is a task to identify instances that are apparently distant from the remaining instances. The objective of differential privacy [3] is to prevent adversaries from learning of the presence (or absence) of any particular instance from released information. Outlier detection and privacy protection are therefore intrinsically conflicting tasks. It presents a challenging difficulty. To overcome this difficulty, instead of identifying outliers, we consider reporting information which helps to recognize the occurrence of anomalous situations. More specifically, we examine the problem of counting outliers that appear in a given subspace with a guarantee of differential privacy.

© Springer International Publishing Switzerland 2015
A. Appice et al. (Eds.): ECML PKDD 2015, Part II, LNAI 9285, pp. 458–473, 2015.
DOI: 10.1007/978-3-319-23525-7_28

Related Works. We introduce existing studies of privacy aspects of outlier analysis. Secure multiparty computation (SMC) is a cryptographic tool that facilitates the evaluation of a specified function over their private inputs jointly, while maintaining these inputs as private. Vaidya et al. [20] introduced a SMC for distance-based outlier detection from horizontally and vertically partitioned private databases using random shares. Xue et al. [21] investigated a SMC for spatial outlier detection. Dung et al. [1] presented a SMC for distance-based outlier detection with the Mahalanobis distance. Li et al. [12] presented a SMC for density-based outlier detection. The objective of these works is to detect outliers securely without mutually sharing privately distributed data; privacy invasion caused by observing detected outliers is not considered.

Studies of differential privacy for outlier analysis are few, presumably because of its intrinsic difficulty, as described. Only one report in the literature [5] describes a study that considers the differential privacy of outlier analysis. This study was conducted to detect anomalous changes from a time series under a guarantee of differential privacy. The objective of this study is closely related to ours, whereas this method releases a one-dimensional time series with differential privacy; outlier detection is applied to the released data as a post process. Consequently, the approach differs from ours.

Lui et al. [14] introduced a novel privacy notion, outlier privacy, as a generalization of differential privacy. Outlier privacy measures an individual's privacy parameter by how much of an "outlier" the individual is. The objective of this study is to define privacy using the notion of outliers, but not for differentially private outlier analysis.

Our Contribution. We examine the problem of counting outliers that appear in a given subspace with a guarantee of differential privacy (Section 2). Randomization of query responses based on the global sensitivity analysis is the most straightforward approach for realization of differential privacy [4]. We derive the lower and upper bound of the global sensitivity of outlier counts (Section 4.1). From the derived bounds, we reveal that the global sensitivity-based randomization can make the outputs too noisy, particularly when the dimensionality of the given subspace is high. We specifically examine the observation that the counts of outliers are expected to be small compared to the number of data in typical datasets. Taking advantage of this, we develop a randomization mechanism for the counts of outliers based on the smooth upper bound of local sensitivity [18] (Section 4.2). A randomization mechanism based on the smooth upper bound typically has better utility because of its data-dependency. However, its evaluation is often costly. To alleviate this, we provide an efficient algorithm for evaluation of the smooth upper bound for counting outliers (Section 4.2). We demonstrated our methods with synthesized datasets and real datasets (Section 5). The experimentally obtained results demonstrate that our methods achieve better utility than that achieved using global sensitivity-based methods.

2 Differential Privacy

Let $X = \{x_1, x_2, \ldots, x_N\} \in \mathbb{R}^{d \times N}$ be a database. An *analyst* issues a query $q : \mathbb{R}^{d \times N} \to \mathcal{T}$; then the database returns an output, where $\mathcal{T}$ denotes the range of the outputs. *Differential privacy* measures the privacy breach of database X caused by releasing output $T \in \mathcal{T}$ with no assumptions of the background knowledge of adversaries. The outputs are typically modified using a randomization *mechanism* $\mathcal{A} : \mathbb{R}^{d \times N} \to \mathcal{T}$ before release to preserve differential privacy.

Let $H(X, X') = |\{i : x_i \neq x'_i\}|$ denote the Hamming distance, the number of different records in X and X'. If $H(X, X') = 1$, then it can be said that X and X' are neighbor databases. In the following, we presume $|X| = |X'| = N$. Then, mechanism $\mathcal{A}$ guarantees (ϵ, δ)-differential privacy if, $\forall X' : H(X, X') = 1$ and $\forall T \subseteq \mathcal{T}$,

$$Pr[\mathcal{A}(X) \in T] \leq e^\epsilon Pr[\mathcal{A}(X') \in T] + \delta.$$

The parameter ϵ and δ are designated as privacy parameters. Randomization based on the global sensitivity is the most straightforward realization of differential privacy for continuous outputs [3].

Global Sensitivity. Presuming that the output domain of query q is in $\mathbb{R}^p$, then randomization based on the global sensitivity [3] provides a mechanism that guarantees differential privacy for queries of any type, as long as its global sensitivity is evaluable. The ℓ_2 global sensitivity of query $q : \mathbb{R}^{d \times N} \to \mathbb{R}^p$ is defined by $GS_q = \max_{X, X' : H(X, X') = 1} \|q(X) - q(X')\|_2$ where $\| \cdot \|$ denotes ℓ_2 norm of vectors. Given the global sensitivity GS_q for query q, the following mechanism $\mathcal{A}$ that randomizes the output of the query by eq. (1) provides (ϵ, δ)-differential privacy [2]:

$$\mathcal{A}_q(X) = q(X) + Y, \tag{1}$$

where Y is an sample drawn from the Gaussian distribution with mean 0 and variance $\frac{GS_q^2 \cdot 2 \log (2/\delta)}{\epsilon^2}$.

Smooth Sensitivity. For some functions, the global sensitivity can be impractically large even when the sensitivities are small with almost all neighboring pairs. This large sensitivity occurs because it is evaluated as the greatest difference of outputs among possible neighboring pair of databases. For example, the global sensitivity of median is N, the whole sample size, but this arises only in a pathological situation. Randomization based on the smooth sensitivity [18] enables the use of moderate sensitivity for such overly sensitive queries. For a given database X, the ℓ_2 *local sensitivity* for query q is defined as the greatest difference of outputs for $\forall X'$ s.t. $X' : H(X, X') = 1$:

$$LS_q(X) = \max_{X' : H(X, X') = 1} \|q(X) - q(X')\|_2.$$

It is noteworthy that $GS_q = \max_{X \in \mathbb{R}^{d \times N}} LS_q(X)$ holds.

Nissim et al. presented the *smooth sensitivity* [18], which is a class of smooth upper bounds to the local sensitivity. Given $\beta > 0$, the smooth sensitivity of query $q : \mathbb{R}^{d \times N} \to \mathbb{R}^p$ is defined by

$$S_{q,\beta}^*(X) = \max_{X' \in \mathbb{R}^{d \times N}} (LS_q(X') \cdot e^{-\beta H(X,X')}).$$

[18] also showed that adding noise proportional to the smooth sensitivity yields a differentially private mechanism if the noise distribution satisfies some properties. Let Y be a noise generated from the Gaussian distribution with mean 0 and variance 1. Let $S_{q,\beta}$ be a β-smooth upper bound of query q. Then, if $\alpha = \frac{\epsilon}{5\sqrt{2\ln 2/\delta}}$ and $\beta = \frac{\epsilon}{4(p+\ln 2/\delta)}$, mechanism $\mathcal{A}_q$ guarantees (ϵ, δ)-differential privacy [18]:

$$\mathcal{A}_q(X) = q(X) + \frac{S_{q,\beta}(X)}{\alpha} \cdot Y.$$

3   Problem Statement

Our objective is to analyze outliers that are included in a private database in a differentially private manner. Outlier detection is a problem to identify an instance that is significantly distant from other instances. Therefore, the result of outlier detection is fundamentally privacy-invasive in terms of differential privacy. In order to understand the behavior of the outliers in the target dataset without identifying outliers, we investigate counting outliers in a given subspace under the constraint of differential privacy.

3.1   Counting Outliers

In this study, we use distance-based outliers [9]. Presuming that records are real-valued vectors, $x_i \in \mathbb{R}^d$, and letting $X = \{x_i\}_{i=1}^N$ denote the database, we let $S \in \{1, 2, \ldots, d\}$ denote a subspace. The Euclidean distance between $x, y \in \mathbb{R}^d$ in subspace S is denoted by $dist_S(x, y) = \sqrt{\frac{\sum_{i \in S}(x_i - y_i)^2}{|S|}}$ [7]. Let $r > 0$ and $k \in \{1, \ldots, N\}$. Then, the set of neighborhood vectors of x in subspace S is defined by

$$N_S(X, r, x) = \{y \in X : dist_S(x, y) \le r, x \ne y\}.$$

With this definition of the neighboring vectors, the outliers in subspace S are defined by

$$O_S(X, k, r) = \{x \in X : |N_S(X, r, x)| < k\}.$$

Then, the task of the outlier count is to find the number of outliers in S:

$$q_{count}(X, k, r, S) = |O_S(X, k, r)|.$$

If the subspace is not specified, then $O(X, k, r)$ denotes the set of outliers in the full space. Distance-based outliers are definable with any type of object and distance defined for the corresponding objects, but we presume that the objects are represented as real vectors and that the Euclidean distance is used as the distance definition.

3.2   Differential Privacy of Outlier Analysis

We introduce several typical scenarios of differentially private outlier analysis using query q_{count}.

Scenario 1. Given threshold k and radius r, presume that the objective is to inspect that the outliers exists in the given dataset. The analyst issues a query $z = q_{count}(X, k, r)$; then checking $z > \theta$ yields the final result where θ denotes a prescribed threshold parameter for outlier counts. Let $z' = q_{count}(X', k, r)$. For guarantee of (ϵ, δ)-differential privacy, we require, for $\forall X' : H(X, X') = 1$ and $\forall T \in \mathcal{T}$,

$$Pr[T = \mathcal{A}(z)] \leq e^{\epsilon} Pr[T = \mathcal{A}(z')] + \delta.$$

Scenario 2. Let the data dimension be $d = 3$. Given threshold k and radius r, presume that the objective is to identify the subspaces that cause the largest numbers of outliers. Then, the target subspace set is $\mathcal{S} = \{\{1\}, \{2\}, \{3\}, \{1, 2\}, \{1, 3\}, \{2, 3\}, \{1, 2, 3\}\}$. The analyst issues query $q_{count}(X, k, r, S_i)$ for each $S_i \in \mathcal{S}$. Let $z_i = q_{count}(X, k, r, S_i)$. For the guarantee of (ϵ, δ)-differential privacy, we require, $\forall X' : H(X, X') = 1$ and $\forall T \in \mathcal{S}$,

$$Pr[T = \mathcal{A}(z_1, \ldots, z_7)] \leq e^{\epsilon} Pr[T = \mathcal{A}(z_1, \ldots, z_7)] + \delta.$$

4   Differentially Private Count of Outliers

As explained in this section, we investigate the problem of differentially private count of outliers in a given subspace. The discussion herein holds for any subspace including the full space. Therefore, for this discussion, we presume that the outlier is counted in the full space.

4.1   Difficulties in Global Sensitivity Method

Analytical evaluation of the global sensitivity of determination of q_{count} is not trivial, partly because it needs the kissing number. The kissing number K_d is the largest number of hyperspheres with same radius in $\mathbb{R}^d$ that can touch equivalent hyperspheres with no intersections [15–17]. The kissing numbers in $d = 1$ and $d = 2$ are readily derived respectively as $K_1 = 2$ and $K_2 = 6$ (see Fig. 1 for $K_2 = 6$). However, finding the kissing number in $d \geq 3$ is not trivial. In addition, the kissing number in general dimensions remains as an open problem [15–17]. We derive the upper and lower bound of the global sensitivity of q_{count} presuming that the kissing number in general dimensions is given.

Theorem 1 (Upper and lower bound on the global sensitivity of q_{count}). *Let K_d be the kissing number in $\mathbb{R}^d$. Then, the upper and lower bound on the global sensitivity of q_{count} is*

$$\min(N, 2dk + 1) \leq GS_{q_{count}, d}(k) \leq \min(N, kK_d + 1). \tag{2}$$

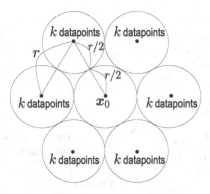

Fig. 1. This figure shows an example of the upper bound of the global sensitivity in two dimension. Six surrounding hyperspheres can be packed around the center hypersphere because the kissing number is $K_2 = 6$. We here suppose k datapoints exist at the center of each surrounding hypersphere and no datapoint exists at x_0, the center of the center hypersphere. Then, kK_2 outliers become inliers by adding a point to x_0. Suppose the added point is an outlier, Then, the added point can be changed from an outlier to an inlier, too. The upper bound of the global sensitivity for two dimension is thus $kK_2 + 1 = 6k + 1$.

Sketch of Proof. The lower bound is trivial so we omit the proof. We show the sketch of the proof for the upper bound. Suppose the radius of the center hypersphere and the hyperspheres touching the center hyperspheres (referred to as the surrounding hyperspheres) are $r/2$. Let x_0 be the center of the center hypersphre. Note that intersection between the surrounding hyperspheres does not exist. We further suppose k datapoints exist at the center of each surrounding hypersphere. These datapoints are outliers by definition, and become inliers by adding a point to the center x_0 of the center hypersphere. By definition of the kissing number, the number of the surrounding hyperspheres that do not touch or intersect mutually is at most K_d. No more surrounding hyperspheres can be packed around x_0, so $kK_d + 1$ is the upper bound of the outlier count. Since the global sensitivity is at most N, we can conclude that $GS_{q_{count},d}(k) \leq \min(N, kK_d + 1)$.

We empirically investigate the tightness of the bound in low dimensions. In $d = 1$ and $d = 2$, the global sensitivity is given respectively as $GS_{q_{count},1}(k) = 2k + 1$ and $GS_{q_{count},2}(k) = 5k + 1$. Noting that $K_1 = 2$ and $K_2 = 6$, the bound is tight in $d = 1$ but not in $d = 2$. Fig. 2 shows the upper and lower bounds of the global sensitivity of q_{count} evaluated using known upper bounds on the kissing number [15–17]. As the figure shows, the upper bound of the global sensitivity grows exponentially with respect to the dimensionality, which indicates that the guarantee of differential privacy by perturbation based on the global sensitivity can be impractical, especially when the dimensionality of the target subspace is large.

The global sensitivity can be prohibitively large simply because the global sensitivity is evaluated considering the worst case. However, one can typically

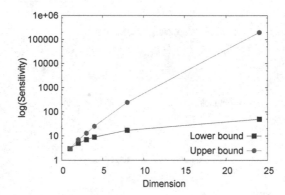

Fig. 2. The bounds of the global sensitivity for counting outliers

expect that the number of outliers in the database is much smaller than the number of instances. To improve the utility of the count query, we introduce the smooth sensitivity, which is a sensitivity definition depending on the database.

4.2   Local Sensitivity and Smooth Sensitivity

For convenience of discussion later, several notations are introduced here. Given radius r, $deg(\boldsymbol{x})$ denotes the size of neighborhoods of $\boldsymbol{x}$:

$$deg(X, r, \boldsymbol{x}) = |N(X, r, \boldsymbol{x})|.$$

We say that the degree of $\boldsymbol{x}$ is k if $deg(X, r, \boldsymbol{x}) = k$. A set of vectors in X whose degree is exactly k is denoted as

$$V(X, k, r) = \{\boldsymbol{x} \in X : deg(\boldsymbol{x}) = k\}.$$

Unless specifically stated otherwise, the radius r and target database X is fixed. Therefore, they are omitted as $deg(\boldsymbol{x})$ and $V(k)$. Finally, a set of degree-k neighborhoods of $\boldsymbol{x}$ in X is denoted as

$$CV(X, \boldsymbol{x}, k, r) = B(\boldsymbol{x}, r) \cap V(k),$$

where $B(\boldsymbol{x}, r)$ denotes the sphere with radius r and centered at $\boldsymbol{x}$.

Local Sensitivity. Given database X, let X_1 be a database s.t. $H(X, X_1) = 1$. Then, following the definition of the local sensitivity in Section 2, the local sensitivity of q_{count} is defined as

$$LS^{(0)}_{q_{count}}(X, k, r) = \max_{X_1 : H(X, X_1) = 1} \|q_{count}(X_0, k, r) - q_{count}(X_1, k, r)\|.$$

Exact evaluation of the exact local sensitivity is intractable. Instead, the following theorem gives the upper bound of the local sensitivity.

Theorem 2. *Given X, the local sensitivity of q_{count} for X is bounded above as*

$$LS^{(0)}_{q_{count}}(X, k, r) \leq$$

$$\max \left\{ \max_{\boldsymbol{x} \in X} \{|CV(X, \boldsymbol{x}, k, r)|\}, \max_{\boldsymbol{x} \in \mathbb{R}^d} \{|CV(X, \boldsymbol{x}, k-1, r)|\} \right\} + 1.$$

Proof. $CV(X, \boldsymbol{x}, k, r)$ is the set of non-outliers that become outliers if $\boldsymbol{x}$ is removed; $CV(X, \boldsymbol{x}, k-1, r)$ is the set of outliers that become inliers if a vector is placed at $\boldsymbol{x}$. Thus, if vector $\boldsymbol{x}_0 \in X$ is moved to $\boldsymbol{x}'_0$, the number of outliers increases by $|CV(X, \boldsymbol{x}_0, k, r)|$ by removing $\boldsymbol{x}_0$ and the number of inliers decreases by $|CV(X, \boldsymbol{x}'_0, k-1, r)|$ by adding $\boldsymbol{x}'_0$. With this understanding, the local sensitivity is given as:

$$LS^{(0)}_{q_{count}}(X, k, r)$$

$$= \max_{X_1 : H(X, X_1) = 1} \|q_{count}(X, k, r) - q_{count}(X_1, k, r)\|$$

$$\leq \max_{\boldsymbol{x}_0 \in X, \boldsymbol{x}'_0 \in \mathbb{R}^d} |CV(X, \boldsymbol{x}_0, k, r) \setminus CV(X, \boldsymbol{x}'_0, k-1, r)| + 1$$

$$\leq \max_{\boldsymbol{x}_0 \in X, \boldsymbol{x}'_0 \in \mathbb{R}^d} \max \{|CV(X, \boldsymbol{x}_0, k, r)|, |CV(X, \boldsymbol{x}'_0, k-1, r)|\} + 1$$

$$= \max \left\{ \max_{\boldsymbol{x} \in X} \{CV(X, \boldsymbol{x}, k, r)\}, \max_{\boldsymbol{x}'_0 \in \mathbb{R}^d} \{CV(X, \boldsymbol{x}, k-1, r)\} \right\} + 1.$$

Naive evaluation of the local sensitivity is intractable. An algorithm to evaluate this upper bound is presented in Section 4.3.

Smooth Sensitivity. Given database X, let X_t be a database s.t. $H(X, X_t) = t$. By definition, the smooth sensitivity of q_{count} is given as

$$S^*_{q_{count}}(X) = \max_{t=0,1,\dots,N} e^{-t\beta} LS^{(t)}_{q_{count}}(X),$$

where

$$LS^{(t)}_{q_{count}}(X) = \max_{X_t : H(X, X_t) = t} LS^{(0)}_{q_{count}}(X_t).$$

The function $LS^{(t)}_q(X)$ returns the largest local sensitivity among the datasets of which t records differ from X. Similarly to $LS^{(0)}_{q_{count}}(X)$, exact evaluation of $LS^{(t)}_{q_{count}}(X)$ is intractable because the variation of X_t can increase exponentially with respect to t. Instead, we derive the upper bound on $LS^{(t)}_{q_{count}}(X)$ using $CV(X, \boldsymbol{x}, k, r)$.

Theorem 3. *Given X, for $t \geq 0$, $LS^{(t)}_{q_{count}}(X)$ is bounded above as*

$$LS^{(t)}_{q_{count}}(X) \leq \max_{\boldsymbol{x} \in \mathbb{R}^d} \left\{ \max\{C^{(t)}(X, \boldsymbol{x}, k, r), C^{(t)}(X, \boldsymbol{x}, k-1, r)\} + t + 1 \right\}, \quad (3)$$

where

$$C^{(t)}(X, \boldsymbol{x}, k, r) = \left| \bigcup_{i=-t}^{t} CV(X, \boldsymbol{x}, k+i, r) \right|.$$

For the proof of this theorem, we use the following helper lemma.

Lemma 1. *Let* $t \geq 0$ *be an integer, and let* X *and* X_t *be databases such that* $H(X, X_t) = t$. *Then, for any* $\boldsymbol{x} \in \mathbb{R}^d$, *threshold* k, *and radius* r,

$$|CV(X_t, \boldsymbol{x}, k, r)| \leq \left| \bigcup_{i=-t}^{t} CV(X, \boldsymbol{x}, k+i, r) \right| + t.$$

Proof. We first consider the case $t = 1$. Suppose $\boldsymbol{x} \in X$ is moved from $\boldsymbol{x}$ to $\boldsymbol{x}_1$, and X_1 is given as $X_1 = X \setminus \{\boldsymbol{x}\} \cup \{\boldsymbol{x}_1\}$. The degree of records in $X \setminus \{\boldsymbol{x}\}$ around $\boldsymbol{x}$ decreases by one by removing $\boldsymbol{x}$, and the degree of records in $X \setminus \{\boldsymbol{x}\}$ around $\boldsymbol{x}_1$ increases by one by adding $\boldsymbol{x}_1$. Since the degree of the records in $V(X, k+1, r)$ and $V(X, k-1, r)$ may become k in X_1, $V(X_1, k, r)$ is thus a subset of $V(X, k+1, r) \cup V(X, k, r) \cup V(X, k-1, r) \cup \{\boldsymbol{x}_1\}$. When $t > 1$, for the same reason, $V(X_t, k, r)$ is a subset of $\bigcup_{i=-t}^{t} V(X, k+i, r) \cup \{\boldsymbol{x}_1, \boldsymbol{x}_2, ..., \boldsymbol{x}_t\}$ where $\boldsymbol{x}_1, ..., \boldsymbol{x}_t$ are the records moved from X to X_t. Thus, the size of $CV(X_t, \boldsymbol{x}, r, k)$ is bounded above as

$$|CV(X_t, \boldsymbol{x}, r, k)| \leq \left| B(\boldsymbol{x}, r) \cap \left\{ \bigcup_{i=-t}^{t} V(X, k+i, r) \cup \{\boldsymbol{x}_1, \boldsymbol{x}_2, ..., \boldsymbol{x}_t\} \right\} \right|$$

$$\leq \left| \bigcup_{i=-t}^{t} B(\boldsymbol{x}, r) \cap V(X, k+i, r) \right| + |\{\boldsymbol{x}_1, \boldsymbol{x}_2, ..., \boldsymbol{x}_t\}|$$

$$\leq \left| \bigcup_{i=-t}^{t} CV(X, \boldsymbol{x}, k+i, r) \right| + t.$$

Sketch of Proof (of Theorem 3). From Theorem 2 and exchangeability of max, letting

$$C_{\text{out}}^{(t)}(X, k, r) = \max_{X_t : H(X, X_t) = t} \max_{\boldsymbol{x} \in X_t} |CV(X_t, \boldsymbol{x}, r, k)| \text{ and}$$

$$C_{\text{in}}^{(t)}(X, k-1, r) = \max_{X_t : H(X, X_t) = t} \max_{\boldsymbol{x} \in \mathbb{R}^d} |CV(X_t, \boldsymbol{x}, r, k-1)|$$

yields

$$LS_{q_{count}}^{(t)}(X) \leq \max\{C_{\text{out}}^{(t)}(X, k, r), C_{\text{in}}^{(t)}(X, k-1, r)\} + 1.$$

We derive the bound on $C_{\text{out}}^{(t)}(X, k, r)$ using $C_{\text{in}}^{(t)}(X, k-1, r)$, and the bound on $C_{\text{in}}^{(t)}(X, k-1, r)$ using Lemma 1.

4.3   Efficient Computation of Smooth Sensitivity Bound

For randomization by the mechanism of Theorem 3, it is necessary to evaluate the smooth upper bound. Naive evaluation of the smooth upper bound of eq. (3) is intractable because it requires an exhaustive search over continuous domain to evaluate $LS^{(t)}_{q_{count}}(X)$. To alleviate this, we first show an efficient algorithm that evaluates the upper bound of $LS^{(t)}_{q_{count}}(X)$ shown derived by Theorem 3. Then using the algorithm, we derive the algorithm that calculates the smooth sensitivity upper bound.

Algorithm for Local Sensitivity Bound. To evaluate the upper bound of $LS^{(t)}_{q_{count}}(X)$, we need to calculate

$$\max_{\boldsymbol{x}\in\mathbb{R}^d} C^{(t)}(X,\boldsymbol{x},k,r) = \max_{\boldsymbol{x}\in\mathbb{R}^d} \left| \bigcup_{i=-t}^{t} V(X,k+i,r) \cap B(\boldsymbol{x},r) \right|, \text{ and} \quad (4)$$

$$\max_{\boldsymbol{x}\in\mathbb{R}^d} C^{(t)}(X,\boldsymbol{x},k-1,r) = \max_{\boldsymbol{x}\in\mathbb{R}^d} \left| \bigcup_{i=-t}^{t} V(X,k+i-1,r) \cap B(\boldsymbol{x},r) \right|. \quad (5)$$

Letting $P = \bigcup_{i=-t}^{t} V(X,k+i,r)$ (resp. $P = \bigcup_{i=-t}^{t} V(X,k+i-1,r)$), we can obtain the value of eq. (4) (resp. eq. (5)) by finding the largest subset $C \subseteq P$ that is enclosed by a ball with radius r. To check whether or not a given subset $C \subseteq P$ is enclosed by the ball, we use the algorithm that solves the *smallest enclosing ball* (seb) problem [6]. The goal of the problem is to find the smallest ball that encloses the given points. The given subset $C \subseteq P$ is enclosed by a ball with radius r if $\mathrm{seb}(C) \leq r$ where $\mathrm{seb}(C)$ denotes the radius of the resultant ball of the smallest enclosing ball problem of C.

Algorithm 1 shows the recursive algorithm that calculates eq. (4) or eq. (5) for given $P = \bigcup_{i=-t}^{t} V(X,k+i,r)$ or $P = \bigcup_{i=-t}^{t} V(X,k+i-1,r)$. $P[i]$ denotes the i-th element of the set P. Algorithm 1 searches for the largest subsets $C \subseteq P$ that is enclosed by a ball with radius r with the breadth-first search. In the algorithm, the calls of seb can be skipped for efficiency by using the fact that the radius of the enclosing ball of C_2 is larger than one of C_1 if $C_1 \subseteq C_2 \subseteq P$. The computational cost of Algorithm 1 is $\mathcal{O}(2^{|P|})$ of the calls of seb.

Algorithm for Smooth Sensiticity Bound. Algorithm 1 costs exponential time with respect to $|P|$ and the size of P increases monotonically as t increases. However, because of exponential decrease of $e^{-t\beta}$, maximization of $e^{-t\beta} LS^{(t)}_{q_{count}}(X)$ is attained by small t in most cases. Taking account of this property, we provide Algorithm 2 that calculates the smooth sensitivity bound with avoiding evaluation of $LS^{(t)}_{q_{count}}(X)$ of large t.

Proposition 1. *For any t and $t' < t$, $LS^{(t)}_{q_{count}}$ is bounded above as*

$$LS^{(t)}_{q_{count}}(X) \leq \min\{N, \max\{U^{(t)}_{t'}(X,k,r), U^{(t)}_{t'}(X,k-1,r)\} + t + 1\},$$

Algorithm 1. Calculation of $\max_{\boldsymbol{x} \in \mathbb{R}^d} C^{(t)(X, \boldsymbol{x}, k, r)}$ (eq. (4) and eq. (5))

Input: Records P and radius r.
Output: The value of eq. (4) or eq. (5).
Initialization: $C = \emptyset$ and $i = 1$

```
1  Function E(r, P, C, i)
2  |   br ← 0
3  |   if C ≠ ∅ then
4  |   |   br ← seb(C)
5  |   end
6  |   if br ≤ r then
7  |   |   m ← |C|
8  |   |   if i ≤ |P| then
9  |   |   |   b₁ ← E(r, P, C ∪ {P[i]}, i + 1)
10 |   |   |   b₂ ← E(r, P, C, i + 1)
11 |   |   |   m ← max{m, b₁, b₂}
12 |   |   end
13 |   |   return m
14 |   end
15 |   else
16 |   |   return 0
17 |   end
18 end
```

---

where

$$U_{t'}^{(t)}(X, k, r) = \max_{\boldsymbol{x} \in \mathbb{R}^d} C^{(t')}(X, \boldsymbol{x}, k, r) + \left| \bigcup_{i \in \{-t, \ldots, -t'-1\} \cup \{t'+1, \ldots, t\}} V(X, k + i, r) \right|.$$

Sketch of Proof. For any database X, because the number of outliers does not exceed the number of the records in X, the local sensitivity is less than N. In addition, using the fact that $CV(X, \boldsymbol{x}, k, r) \subseteq V(X, k, r)$ for any $\boldsymbol{x} \in \mathbb{R}^d$, we can derive $\max_{\boldsymbol{x} \in \mathbb{R}^d} C^{(t)}(X, \boldsymbol{x}, k, r) \leq U_{t'}^{(t)}(X, k, r)$ for any t and $t' < t$.

Using the bound in Proposition 1, we have the upper bound of $e^{-t\beta} LS_{q_{count}}^{(t)}(X)$ as

$$e^{-t\beta} LS_{q_{count}}^{(t)}(X) \leq e^{-t\beta} \min\{N, \max\{U_{t'}^{(t)}(X, k, r), U_{t'}^{(t)}(X, k - 1, r)\} + t + 1\}$$
$$=: S_{UB}^{t', t}(X).$$

Letting $S_{UB}^t(X) = \max_{i=1, \ldots, N-t} S_{UB}^{t, t+i}(X)$, we can obtain the following proposition.

Proposition 2. *If there exists U_T such that $\max_{t=0, \ldots, T} e^{-t\beta} LS_{q_{count}}^{(t)}(X) \leq U_T$ and $S_{UB}^T(X) \leq U_T$, then $S_{q_{count}}^*(X) \leq U_T$.*

Algorithm 2. Calculation of the smooth sensitivity of q_{count}

Input: Database X, threshold k, radius r and smooth parameter ϵ.
Output: The smooth sensitivity upper bound of query q_{count} for database X.
Initialization: $S_{\max} = 0$ and
$$\max_{\boldsymbol{x} \in \mathbb{R}^d} C^{(-1)}(X, \boldsymbol{x}, k, r) = \max_{\boldsymbol{x} \in \mathbb{R}^d} C^{(-1)}(X, \boldsymbol{x}, k-1, r) = 0.$$

1 **for** $t = 0$ *to* N **do**
2   Calculate S_{UB}^{t-1} by Proposition 2
3   **if** $S_{\mathrm{UB}}^{t-1} \leq S_{\max}$ **then**
4 | **return** $S_{\max}$
5   **end**
6      $S_{\max} \leftarrow \max\{S_{\max}, e^{-t\beta} LS_{q_{count}}^{(t)}(X)\}$
7   Store $\max_{\boldsymbol{x} \in \mathbb{R}^d} C^{(t)}(X, \boldsymbol{x}, k, r)$ and $\max_{\boldsymbol{x} \in \mathbb{R}^d} C^{(t)}(X, \boldsymbol{x}, k-1, r)$ for
     calculating S_{UB}^t in next loop
8 **end**
9 **return** $S_{\max}$

Proof. If $S_{\mathrm{UB}}^T(X) = \max_{i=1,\ldots,N-T} S_{\mathrm{UB}}^{T,T+i}(X) \leq U_T$, since $e^{-t\beta} LS_{q_{count}}^{(t)}(X) \leq$ $S_{\mathrm{UB}}^{T,t}(X)$ for any $t > T$, we have $e^{-t\beta} LS_{q_{count}}^{(t)}(X) \leq U_T$, $\forall t > T$. Thus, we have $\max_{t=0,\ldots,T} e^{-t\beta} LS_{q_{count}}^{(t)}(X) \leq U_T$ and $\max_{t>T} e^{-t\beta} LS_{q_{count}}^{(t)}(X) \leq U_T$.

Proposition 2 shows that if the largest upper bound in Theorem 3 for $t = 0, \ldots, T$ can be bounded above by $S_{\mathrm{UB}}^T(X)$, then the calculation of the upper bound in Theorem 3 for $t > T$ can be skipped. Algorithm 2 shows the calculation of the smooth sensitivity of q_{count} with this skip by following Proposition 2.

5 Experiments

In this section, we show the empirical evaluation of the utility of the mechanism for counting outliers query.

5.1 Settings

We used a synthetic dataset and a real dataset (adult). The synthetic dataset consists with 50 samples of 2 dimensional real vectors. The dataset contains 45 inliers which are sampled from $\mathcal{N}(\mathbf{0}, \mathbb{I})$ where $\mathbb{I}$ represents an identity matrix. The 5 outliers are sampled from $\mathcal{N}(\boldsymbol{\mu}, \Sigma)$, where $\mu_1 = \mu_2 = 20$ and Σ is a diagonal matrix such that $\Sigma_{11} = \Sigma_{22} = 100$.

    A real dataset (adult) was chosen from UCI Machine Learning Repository [13]. We removed two categorical attributes, "category" and "fnlwgt". The dataset was scaled so that the average and variance of each attribute is 0 and 1, respectively. The dataset is originally prepared for classification tasks. For our outlier analysis, following [19,22], 45 samples with the positive label are treated as inliers and 5 samples with negative labels were treated as outliers (See Table 1 for the detail). We changed the privacy parameter from $\epsilon = 0.1$ to 0.9; δ was

Table 1. Sumarry of datasets

	synthetic	adult
The number of outliers	5	5
The number of inliers	45	45
The number of samples N	50	50
Dimension d	2	7
Treshold k	3	3
Radious r	1.1	0.35

fixed as $\delta = 0.01$. See Table 1 for the parameters of the outliers. We partitioned the instances into two classes: one is "true", indicating the instance detected as an outlier; the other is "false". For each dataset, we tuned the radius r so that the *Accuracy* given by eq. (6) is maximized:

$$Accuracy = \frac{TP + TN}{TP + FP + FN + TN},\tag{6}$$

where TP, TN, FP and FN respectively denote true positive, true negative, false positive, and false negative. For implementation, we used [11] to solve the smallest enclosing ball problem.

5.2   Count Outliers

Following the Scenario 1 described in Section 3.2, we evaluated the utility of the mechanisms of q_{count} on the synthetic dataset. As the criterion of the utility of the mechanisms, we show the standard deviation of the noise added to the query. We compared the standard deviation of the noise of the mechanism based on the smooth sensitivity upper bound in eq. (3) with the mechanism based on the global sensitivity lower bound in eq. (2). Fig. 3 shows the output values and the standard deviations for each mechanism in various ϵ. In Fig. 3, "Global" and "Smooth" respectively present the global sensitivity-based mechanism and the smooth sensitivity-based mechanism.

It is apparent that the standard deviation of the noise of the smooth sensitivity-based mechanism is significantly lower than that of the global sensitivity-based mechanism. Indeed, the standard deviation of the noise of global sensitivity-based mechanism is approximately 10-30 times larger than that of the smooth sensitivity-based mechanism even though the global sensitivity-based mechanism uses the lower bound. In addition, the smooth sensitivity-based mechanism achieves the noise of which standard deviation is lower than 7 for $\epsilon \geq 0.7$ for each datasets. The reason why we got these results is our approach depends only on the number of outliers, not on the number of dimensions. From these results, we can conclude that our framework is sufficiently practical in this setting.

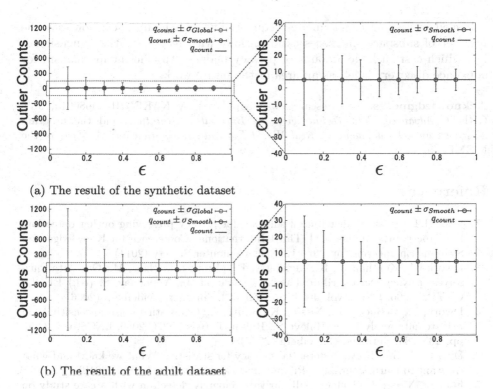

(a) The result of the synthetic dataset

(b) The result of the adult dataset

Fig. 3. Experimental results for the global sensitivity-based mechanism and the smooth sensitivity-based mechanism on each dataset. The right panel is obtained by scaling the left panel so that the error bars of the smooth sensitivity-based mechanism are visible. The horizontal axis denotes the privacy parameter ϵ. The vertical axis denotes the output value of the query without randomization. The error bars denote the standard deviation of the noise added by the mechanisms.

6   Conclusion and Future Works

We present the differentially private distance-based outlier analysis for the query that counts outliers in a given subspace. Taking advantage of the smooth sensitivity [18], the resulting output of the mechanism can be less noisy than that of the global sensitivity-based mechanism. Although the evaluation of the smooth upper bound is often costly, we provide an efficient algorithm for the evaluation of the smooth upper bound for the problem for outlier counting. This paper describes an initial step towards differentially private outlier analysis, and the experimental evaluation is performed with relatively small-size datasets. In our algorithm, we invoke the smallest enclosing ball algorithm that takes as input the power set of instances. Because of this construction, we need a more efficient algorithm for application to larger size datasets.

Subspace discovery for outlier analysis has been investigated as a major topic of outlier detection [7,8,10]. Differentially private subspace discovery can be achieved by issuing count queries sequentially to each subspace; however, the number of subspaces increases exponentially with respect to the dimensionality, which costs a large amount of privacy budget. An efficient mechanism for subspace discovery is left as an area of the future work.

Acknowledgments. This research was supported by KAKENHI 24680015, JST CREST *Advanced Core Technologies for Big Data Integration*, and the program *Research and Development on Real World Big Data Integration and Analysis* of the MEXT, Japan.

References

1. Bao, H.T., et al.: A distributed solution for privacy preserving outlier detection. In: Proceedings of the 2011 Third International Conference on Knowledge and Systems Engineering, pp. 26–31. IEEE Computer Society (2011)
2. Dwork, C., Kenthapadi, K., McSherry, F., Mironov, I., Naor, M.: Our data, ourselves: privacy via distributed noise generation. In: Vaudenay, S. (ed.) EUROCRYPT 2006. LNCS, vol. 4004, pp. 486–503. Springer, Heidelberg (2006)
3. Dwork, C., McSherry, F., Nissim, K., Smith, A.: Calibrating noise to sensitivity in private data analysis. In: Halevi, S., Rabin, T. (eds.) TCC 2006. LNCS, vol. 3876, pp. 265–284. Springer, Heidelberg (2006)
4. Dwork, C., Smith, A.: Differential privacy for statistics: What we know and what we want to learn. Journal of Privacy and Confidentiality $1(2)$, 2 (2010)
5. Fan, L., Xiong, L.: Differentially private anomaly detection with a case study on epidemic outbreak detection. In: Proceedings of the 2013 IEEE 13th International Conference on Data Mining Workshops, pp. 833–840. IEEE Computer Society (2013)
6. Fischer, K., Gärtner, B., Kutz, M.: Fast smallest-enclosing-ball computation in high dimensions. In: Di Battista, G., Zwick, U. (eds.) ESA 2003. LNCS, vol. 2832, pp. 630–641. Springer, Heidelberg (2003)
7. Keller, F., Müller, E., Böhm, K.: Hics: high contrast subspaces for density-based outlier ranking. In: IEEE 28th International Conference on Data Engineering (ICDE 2012), Washington, DC, USA (Arlington, Virginia), 1–5 April, 2012, pp. 1037–1048. IEEE Computer Society (2012)
8. Keller, F., Müller, E., Wixler, A., Böhm, K.: Flexible and adaptive subspace search for outlier analysis. In: 22nd ACM International Conference on Information and Knowledge Management, CIKM 2013, San Francisco, CA, USA, October 27 - November 1, 2013, pp. 1381–1390. ACM (2013)
9. Knorr, E.M., Ng, R.T.: Algorithms for mining distance-based outliers in large datasets. In: Proceedings of the 24rd International Conference on Very Large Data Bases. pp. 392–403. VLDB 1998, Morgan Kaufmann Publishers Inc., San Francisco, CA (1998)
10. Knorr, E.M., Ng, R.T.: Finding intensional knowledge of distance-based outliers. In: Proceedings of the 25th International Conference on Very Large Data Bases, pp. 211–222. VLDB 1999, Morgan Kaufmann Publishers Inc., San Francisco, CA (1999)
11. Kutz, M., Kaspar, F., Bernd, G.: A java library to compute the miniball of a point set. https://github.com/hbf/miniball, last Accessed Time: February 2, 2015

12. Li, L., Huang, L., Yang, W., Yao, X., Liu, A.: Privacy-preserving lof outlier detection. Knowledge and Information Systems **42**(3), 579–597 (2015)
13. Lichman, M.: UCI machine learning repository (2013). http://archive.ics.uci.edu/ml
14. Lui, E., Pass, R.: Outlier privacy. In: Dodis, Y., Nielsen, J.B. (eds.) TCC 2015, Part II. LNCS, vol. 9015, pp. 277–305. Springer, Heidelberg (2015)
15. Mittelmann, H.D., Vallentin, F.: High-accuracy semidefinite programming bounds for kissing numbers. Experimental Mathematics **19**(2), 175–179 (2010)
16. Musin, O.R.: The kissing problem in three dimensions. Discrete & Computational Geometry **35**(3), 375–384 (2006)
17. Musin, O.R.: The kissing number in four dimensions. Annals of Mathematics **168**(1), 1–32 (2008)
18. Nissim, K., Raskhodnikova, S., Smith, A.: Smooth sensitivity and sampling in private data analysis. In: Proceedings of the Thirty-ninth Annual ACM Symposium on Theory of Computing, pp. 75–84. STOC 2007. ACM, New York (2007)
19. Pham, N., Pagh, R.: A near-linear time approximation algorithm for angle-based outlier detection in high-dimensional data. In: Proceedings of the 18th ACM SIGKDD International Conference on Knowledge Discovery and Data Mining, pp. 877–885. KDD 2012. ACM, New York (2012)
20. Vaidya, J., Clifton, C.: Privacy-preserving outlier detection. In: The Fourth IEEE International Conference on Data Mining, pp. 233–240. IEEE Computer Society, Brighton (2004)
21. Xue, A., Duan, X., Ma, H., Chen, W., Ju, S.: Privacy preserving spatial outlier detection. In: Proceedings of the 9th International Conference for Young Computer Scientists, pp. 714–719. IEEE Computer Society (2008)
22. Zhang, K., Hutter, M., Jin, H.: A new local distance-based outlier detection approach for scattered real-world data. In: Theeramunkong, T., Kijsirikul, B., Cercone, N., Ho, T.-B. (eds.) PAKDD 2009. LNCS, vol. 5476, pp. 813–822. Springer, Heidelberg (2009)

Inferring Unusual Crowd Events from Mobile Phone Call Detail Records

Yuxiao Dong[1], Fabio Pinelli[2], Yiannis Gkoufas[2], Zubair Nabi[2],
Francesco Calabrese[2(✉)], and Nitesh V. Chawla[1]

[1] Department of Computer Science and Engineering,
University of Notre Dame, Notre Dame, USA
{ydong1,nchawla}@nd.edu
[2] IBM Research, Mulhuddart, Ireland
{fabiopin,yiannisg,zubairn,fcalabre}@ie.ibm.com

Abstract. The pervasiveness and availability of mobile phone data offer
the opportunity of discovering usable knowledge about crowd behavior
in urban environments. Cities can leverage such knowledge to provide
better services (e.g., public transport planning, optimized resource allocation) and safer environment. Call Detail Record (CDR) data represents
a practical data source to detect and monitor unusual events considering
the high level of mobile phone penetration, compared with GPS equipped
and open devices. In this paper, we propose a methodology that is able to
detect unusual events from CDR data, which typically has low accuracy
in terms of space and time resolution. Moreover, we introduce a concept
of unusual event that involves a large amount of people who expose an
unusual mobility behavior. Our careful consideration of the issues that
come from coarse-grained CDR data ultimately leads to a completely
general framework that can detect unusual crowd events from CDR data
effectively and efficiently. Through extensive experiments on real-world
CDR data for a large city in Africa, we demonstrate that our method
can detect unusual events with 16% higher recall and over 10× higher
precision, compared to state-of-the-art methods. We implement a visual
analytics prototype system to help end users analyze detected unusual
crowd events to best suit different application scenarios. To the best of
our knowledge, this is the first work on the detection of unusual events
from CDR data with considerations of its temporal and spatial sparseness and distinction between user unusual activities and daily routines.

1 Introduction

The ubiquity of mobile devices offers an unprecedented opportunity to analyze the trajectories of movement objects in an urban environment, which can

We wish to thank the Orange D4D Challenge (http://www.d4d.orange.com) organizers for releasing the data we used for testing our algorithms. Research was sponsored
in part by the Army Research Laboratory under Cooperative Agreement Number
W911NF-09-2-0053 and the U.S. Air Force Office of Scientific Research (AFOSR)
and the Defense Advanced Research Projects Agency (DARPA) grant #FA9550-12-
1-0405.

A. Appice et al. (Eds.): ECML PKDD 2015, Part II, LNAI 9285, pp. 474–492, 2015.
DOI: 10.1007/978-3-319-23525-7_29

have a significant effect on city planning, crowd management, and emergency response [4]. The big data generated from mobile devices, thus, provides a new powerful social microscope, which may help us to understand human mobility and discover the hidden principles that characterize the trajectories defining human movement patterns. Cities can leverage the results of the analytics to better provide and plan services for citizens as well as to improve their safety. For example, during the occurrence of expected or chaotic events such as riots, parades, big sport events, concerts, the city should be able to provide a proactive response in allocating the correct amount of resources, adapt public transport services, and more generally adopt all possible actions to safely handle such events. Many methods have been proposed in the literature to detect groups of people moving together from a trajectory database [14,16,18,32], specifically the GPS data. However, only a very little percentage of people currently carry GPS devices, and share their movement trajectories with a central entity that can use them to identify crowd events.

In this paper, we study the problem of unusual event detection from mobile phone data that is opportunistically collected by telecommunication operators, in particular the Call Detail Records (CDR). In 2013, the number of mobile-phone subscriptions reached 6.8 billion, corresponding to a global penetration of 96%. The pervasiveness of mobile phones is spreading fast, with the number of subscriptions reaching 7.3 billion by 2014, from a recent report by International Telecommunications Union (ITU) at 2013 Mobile World Congress [7]. Therefore, CDR data represents a practical data source to detect and monitor unusual events considering the high level of mobile phone penetration. This is specifically useful in developing countries where other methodologies to gather crowd movement data (e.g., GPS or cameras) are very expensive to be installed.

The task of detecting unusual events from CDR data is very different from previous work on fine-grained trajectory data, such as GPS data, and presents several unique challenges. **Temporal sparseness:** CDR data only records the user location when a call or text message is made or received, thus is temporally sparse since call or message frequency of users is usually low and unpredictable. **Spatial sparseness:** The location information of users when they make a call or message is recorded as the location of the antenna, which brings the spatial sparseness of CDR data. **Non-routine events:** Our objective is to detect unusual crowd events from human daily movements, which mostly consist of usual routines. Thus, it is necessary to discriminate unusual crowd movements from routine trajectories.

To address these challenges, we aim to estimate the location of users in absence of spatio-temporal observations (i.e., the users don't make phone calls), detect groups of people moving together, and proactively discover unusual events. We propose a general framework to infer unusual crowd events from mobile phone data. Specifically, our contributions can be summarized as follows:

- We first define the cylindrical cluster to capture sparse spatio-temporal location data and provide practical methods to extract crowd events from CDR data, and further formalize the unusual crowd event detection problem by

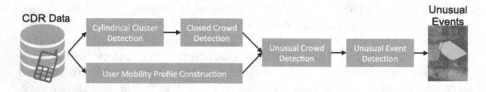

Fig. 1. Process flow of the system to detect unusual events.

considering the similarity between individuals' trajectories and their historical mobility profiles.

- We provide a Visual Analytics Prototype System to help the end user (e.g. a city manger or analyst) analyze the detected crowd events and set the values of the parameters to best suit an application scenario.
- Finally, we evaluate our proposed framework on a real-world CDR dataset and demonstrate its effectiveness and efficiency. Our method significantly outperforms ($10\times$ precision and $+16\%$ recall) previous event detection methods on GPS data with verification on real-world unusual crowd events.

The mobile phone CDR data used in this work is collected from Cote d'Ivoire over five months, from December 2011 to April 2012. During that period, this Africa country faced the Second Ivorian Civil War and political crisis[1]. From the election of new president and parliament, continued outbreaks of post-election conflicts happened, including boycott, violence and protest etc. The experimental results on this real-world dataset deliver the effectiveness of our proposed methodologies, which demonstrates the significant importance of our work in the supervision of unusual crowds and events for city and country management. Moreover, through the proposed method, city mangers and officials can gain insights into non-ticketed events taking place in public spaces, which could lead to estimating the number of attendees and to estimating the event's success. A particular instance of such method has been recently implemented to help the police and the event organizers monitor visitors to the Mons 2015 - European Capital of Culture Opening Ceremony [3].

2   Unusual Event Detection Problem

Given the nature of CDR data, we face three major challenges in extracting accurate individual trajectories. First, we can only record user locations when they make calls or receive calls (text messages). As most mobile users do not make phone calls frequently and periodically, positions are not regularly sampled, as opposed to GPS navigation systems. Moreover, mobile users do not follow a call pattern consistently with others in the group. Second, when a user makes a call, CDR data only records the base station she is using, providing very low quality location information. Finally, the scenario that we are considering—e.g.,

---

[1] http://en.wikipedia.org/wiki/2010-11_Ivorian_crisis

going to a protest—is not consistent with an individual's daily activity pattern such as going from home to office, thus we cannot leverage the previous history of the user to enrich his trajectory to make it more accurate.

We formally define the problem of unusual event detection and decompose the problem in different steps that enable us to solve the challenges brought by CDR data. Figure 1 shows the process flow to detect unusual events. The system receives CDR data as input, extracts clusters, and detects crowds from the sequences of clusters. Then, the system verifies some constraints for each crowd and labels them as unusual if necessary. Subsequently, one or more unusual crowds compose unusual events.

Let $DB_{CDR} = \{call_1, call_2, \cdots, call_n\}$ denote the set of all calls collected from a mobile phone network. We define a call as a tuple $call_i = < t_i, v_j, l_k >$, which means a user v_j makes or receives a call at location l_k at timestamp t_i, where $v_j \in V, t_i \in T, l_k \in L$. V is the set of all users and T denotes all possible timestamps. Specifically, l_k stands for the geographical location of the kth mobile network antenna and L means the set of locations of all antennas found in DB_{CDR}. We define the individual mobility trajectory [10,27] for each user as follows:

Definition 1. Individual Trajectory: A user v_j's mobility trajectory from start time t_p to end time t_q is defined as a sequence of spatio-temporal tuples $s_j^{pq} = \bigcup(t_i, l_k)$, where $t_p \leq t_i \leq t_q$ and $s_j^{pq} \in S$. S stands for the set of user trajectory sequences.

Cylindrical Cluster. The first step to identify crowd events from individual trajectories is to find, at any specific timestamp, clusters of individuals that are close in space. However, since CDR data is very sparse on the time scale (i.e., users do not make calls regularly), we propose the concept of *cylindrical cluster* in coarse-grained spatio-temporal data. Finer grain clustering methods, such as density-based clustering [8], cannot be applied as the antenna is the lowest level of spatial resolution available in the data. Indeed, users are already clustered by association to the antenna they use at each call (which defines a specific coverage area in the city, ranging from a few hundred squared meters to a few kilometers).

Definition 2. Cylindrical Cluster: Given a CDR database DB_{CDR} which contains individual calls with time and antenna information, and a scale threshold ϵ_n, the cylindrical cluster CC_t at timestamp t is a non-empty subset of users $V_t \subseteq V$ satisfying the following conditions:

- Connectivity. $\forall v_i \in V_t$, v_i makes at least one call by using antenna a_x, in the interval $[t - \epsilon_t, t + \epsilon_t]$.
- Scale. The number of users $|V_t|$ in CC_t is no less than ϵ_n.

Figure 2(a) shows an illustrative example for cylindrical clusters. Given a timestamp t_1, we can see that $user1, user2, user3$ and $user4$ make calls during time interval $[t_1 - \epsilon_t, t_1 + \epsilon_t]$. Also, $user3, user1$ and $user2$ use the same antenna which is different from user $user4$'s. Then they are clustered into two groups. One potential issue is that there may exist multiple locations for one single user if she/he makes multiple calls during time interval $[t_1 - \epsilon_t, t_1 + \epsilon_t]$. A number of methods can be considered to assign one single location from multiple locations, such as the central position or the most common position. We use the most common position due to its ease of calculation and understanding.

Crowd. In order to detect crowds lasting for a certain amount of time we need to consider shared characteristics between clusters detected in consecutive timestamps.

Definition 3. Crowd: Given a CDR database DB_{CDR} with individual trajectories, a lifetime threshold ϵ_{lt}, a consecutive intersection threshold ϵ_{ci} and a commitment probability threshold ϵ_p, a crowd C is a sequence of consecutive cylindrical clusters $\{CC_{t_m}, CC_{t_{m+1}}, \cdots, CC_{t_n}\}$ which satisfy the following constraints:

- Movement. The number of total locations in one crowd is more than one.
- Durability. The lifetime of C, $C.lt$, namely the number of consecutive clusters, is greater than ϵ_{lt}, i.e., $C.lt \geq \epsilon_{lt}$ where $C.lt = n - m + 1$.
- Commitment. At least ϵ_{ci} users appear in each cylindrical cluster with existence probability ϵ_p.

The movement and durability characterizations specify the types of crowd we are interested in. The commitment instead characterizes the fact that a certain subset of users needs to participate to all clusters. Again, due to the spatio-temporal sparsity of the CDR data, the computation of the commitment of an user requires some further considerations. Therefore, we propose the concept of existence probability, which is designed to overcome CDR sparsity. Indeed, as an individual is not constantly making calls, consecutive timestamps could not see all users in the cluster making calls.

We design the existence probability of one user locating in a cluster at timestamp t as the proportion of the number of users in CC_t to the number of users in CC_{t-1}. The intuition for the definition of existence probability is that the user has conformity to follow others in the group that she or he was assigned to [24]. For example, the existence probability of $user3$ in Figure 2(a) at time t_2 is $1/3$. In timestamp t_1, $user1$, $user2$, and $user3$ stay in cluster CC_{t_1}, and one of them, user $user2$, goes to cluster CC_{t_2} at timestamp t_2. $user1$ and $user3$ do not make calls in timestamp t_2, which results in the uncertainty of their locations. Thus, we assign them the probability to stay with $user2$, which is in cluster CC_{t_2}. Furthermore, we make the existence probability decay over time, i.e., if a user does not appear in consecutive timestamps, such as user $user4$ in timestamp t_3 and t_4. Her existence probabilities in Figure 2(a) are $[0, 1, \frac{1}{2}, \frac{1}{2} \times \frac{2}{3}]$ at each timestamp, respectively.

Considering that a crowd is a sequence of clusters, we use the standard terminology of sequential pattern mining and affirm that: a crowd C is called a closed crowd if it has no super crowds, which means there does not exist super sequences containing C.

Unusual Crowd. Usually, people have their own mobility trajectories in daily lives, such as going from home to work place everyday. When people go to attend a concert or a protest, their trajectories differ from their usual ones.

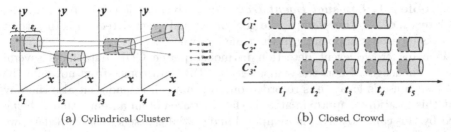

(a) Cylindrical Cluster   (b) Closed Crowd

Fig. 2. Illustrative Examples of Cylindrical Cluster and Closed Crowd.

The definition of crowd given above includes both usual daily trajectories (e.g., commuting) as well as unusual event trajectories (e.g., protests). This is, for instance, what the method in [32] aims to do. As we will show in the experiments section, such method generates an enormous amount of events, as opposed to what a city would need in order to identify specific unusual events. Here we define the concept of mobility profile to capture people's normal movement behaviors, by comparing with which we can detect abnormal mobility behaviors.

Definition 5. Mobility Profile: Given a CDR database DB_{CDR} with individual trajectories, one's mobility profile is the groups of locations she/he visited for each time unit (hour) in every day. Notice that a location here corresponds to an antenna.

Definition 6. Unusual Crowd: Given the mobility profiles of users, a similarity threshold ϵ_{si}, a closed crowd C is said to be an unusual crowd UC if the average similarity between the trajectory of each user in the crowd and her/his mobility profile in corresponding time intervals is less than ϵ_{si}.

Unusual Event Detection. Due to the inaccuracy of CDR data and to the introduction of the existence probability concept, it is possible that two or more crowds share users and thus they represent the same event. Moreover, it is possible that many crowds might correspond to the same large event (e.g., two parades converging to the same square). To group together these unusual crowds, we define the concept of unusual event:

Definition 7. Unusual Event: Given two unusual crowds UC_i and UC_j, UC_i and UC_j are connected into one unusual event if they satisfy the following principles:

- Overlapping: The ending time $C_i.t_{end}$ of crowd C_i is temporally close to the beginning time $C_j.t_{begin}$ of other crowd C_j, w.r.t. $C_j.t_{begin} < C_i.t_{end}$.
- Sharing: The number of common users, $|C_i \bigcap C_j|$, is larger than or equal to half of the total users $|C_i \bigcup C_j|$.

An unusual event is a set of unusual crowds $E = \{UC_1, UC_2, \cdots, UC_n\}$ in which any two unusual crowds are connected to each other by a path. Here one separate unusual crowd is also an unusual event, if it does not connect with others. Based on the discussed concepts above, we formalize the unusual event detection problem as follows.

Problem 1. *Unusual Event Detection:* Given all detected crowds during the interval of two timestamps, the goal of unusual event detection is to extract all unusual events happening in the time interval.

Unusual crowd event detection in mobile phone CDR data faces several unique challenges. First, the sparseness of CDR data comes from not only the fact that a user's location is recorded only when a call is made but also the way that this location is approximated as the cover area of an antenna that is being used by this call. To solve the temporal and spatial sparseness of CDR data, we propose to define user existence probability that can overcome the fact that a user's location is recorded only when a call is made, and also to leverage the idea of cylindrical cluster to address the coarseness of user locations as they are recorded as the cover area of involved antenna. Moreover, the problem is targeted at inferring unusual events rather than people daily routines. To achieve so, we propose the concept of mobility profile to distinguish unusual crowding behavior from daily movements.

3   Unusual Event Detection Framework

Given the formal definitions above, we describe now an innovative and efficient framework to detect unusual crowd events from CDR data. Our framework is composed of four parts: cylindrical cluster detection, closed crowd detection, unusual crowd detection, and unusual event detection.

Cylindrical Cluster Detection. Given the database of the individual calls with the respective time and antenna information, a duration threshold ϵ_t, and a scale threshold ϵ_n, the Cylindrical Cluster Detection algorithm maintains at each timestamp t the set of users observed from each antenna a, in the time interval $[t - \epsilon_t, t + \epsilon_t]$. Then, for each timestamp it returns all the set of users whose size is larger than ϵ_n. All the detected cylindrical clusters are stored in *ClusterDB*.

Closed Crowd Detection. The input for crowd detection is a set of cylindrical clusters *ClusterDB* extracted at each timestamp. There are three constraint thresholds considered in our crowd definition: *movement, durability, commitment*. Explicitly, if the subcrowd of one crowd meets the *durability* and *movement* constraints, it will satisfy the *commitment* constraint also. Thus the crowd definition satisfies the requirement of downward closure property, and then it is unnecessary to output all crowds, including the subcrowds of closed crowds. To avoid the redundancy resulted from outputting subcrowds, we can follow the Lemma 1 to decide if a crowd is closed or not.

Lemma 1. *A crowd C with clusters $\{CC_{t+m}, CC_{t+m+1}, \cdots, CC_{t+n}\}$ is a closed crowd, if there does not exist CC_{t+m-1} or CC_{t+n+1} that can be added to crowd C such that a new crowd is formed.*

The restriction of closed crowd contains two conditions, one is that no suffixed cluster can be appended into it and the other is that no prefixed cluster can be merged in its front. To discover closed crowds in cluster database at current timestamp t, the first condition is easy to check: if there exist clusters in next timestamp $t+1$ that can be appended to current crowd C, then the process will continue; if not, we only need to verify whether current crowd C is the subcrowd of crowds formed at current timestamp t. It is not necessary to check every crowd at previous timestamps because that current crowd at timestamp t can only be the subcrowd of crowds ending at timestamp t.

Figure 2(b) shows an illustrative example for this process. Suppose that crowds C_1 and C_2 are found as closed crowds, if there is no cluster at timestamp t_6 that can be appended to crowd C_3, then we need to further check whether it is the subcrowd of previous crowds. It is obvious that it is impossible for C_3 to be the subcrowd of crowds ending at t_4 or earlier timestamps, such as C_1, but it is possible to be the subcrowd of crowds ended at t_5, such as C_2.

To find all closed crowds in $ClusterDB$, we start with iterating each times-tamp in an increasing order. At each timestamp t, we check whether each can-didate crowd at timestamp $t-1$ can be appended by clusters at timestamp t. If the candidate crowd satisfies the *movement* and *durability* constraints, and at the same time it is not the subcrowd of crowds ending at timestamp $t-1$, then we can output the current candidate as a closed crowd. The current candidate crowd can then be appended by one more cluster to form a new candidate crowd at t. The candidate crowd set contains all crowds which can be appended by a new cluster at t. Then we put all clusters at timestamp t to it to form a new candidate crowd set at t. This order of adding candidate crowd to candidate set guarantees that we only need to check whether the potential crowd is the subcrowd of closed crowds ending at the same timestamp.

Complexity: The extraction of closed crowds is similar to the extraction of closed frequent sequential patterns whose complexity in the worst case can be approximated with $\mathcal{O}(|A|^2 * |T|)$ where $|A|$ is the number of antennas (i.e. clus-ters) and $|T|$ is the number of timestamps.

Unusual Crowd Detection. With the detected closed crowds, we further verify whether their users present unusual or regular behaviors. As introduced in Section 2, we use mobility profile to decide whether users' movement trajectories are unusual.

To generate the mobility profiles, we scan the historic CDR data once to record the specific locations a user visited at each timestamp during every time period. For example, $user4$'s existence probability vector in correspond-ing crowd is $\mathbf{w}_c = [0, 1, \frac{1}{2}, \frac{1}{2} \times \frac{2}{3}]$ in Figure 2(a). His profile vector is extracted from his mobility profile at corresponding timestamps (from t_1 to t_4), i.e. $\mathbf{w}_m = [\frac{0}{3}, \frac{2}{2+5}, \frac{1}{4+1}, \frac{2}{2+1}]$. There are several ways to define the similarity between user's mobility profile and his trajectory in the crowd. We use cosine similarity to calculate the similarity score, because of its ease of understanding and imple-mentation. The cosine similarity between two vectors $\mathbf{w}_c$ and $\mathbf{w}_m$ is defined as: $CosSim(\mathbf{w}_c, \mathbf{w}_m) = \frac{\mathbf{w}_c \cdot \mathbf{w}_m}{\|\mathbf{w}_c\|\|\mathbf{w}_m\|}$.

The Unusual Crowd Detection algorithm first calculates for each user in the crowd the similarity between her trajectory and her own mobility profile.

Then the similarities obtained are averaged, and the obtained value is greater than the *similarity* parameter ϵ_{si}, it is an unusual crowd.

Complexity: The mobility profile construction requires a scan of the dataset, therefore its complexity is $\mathcal{O}(DB_{CDR})$. The detection of Unusual Crowds requires for each crowd the computation of the cosine similarity for all the users being part of a crowd, thus its complexity is $\mathcal{O}(|C|*|V|)$ where $|C|$ is the number of crowds and $|V|$ the number of users.

Unusual Event Detection. With discovered unusual crowds, we finally detect their relationships and connect them into one event if they meet the requirements of Definition 7. In this step, we use graph theory to find and generate unusual events. First if two unusual crowds satisfy both *overlapping* and *sharing* principles, we create an edge to connect them. With this generated graph, where each node is one unusual crowd and an edge indicates that two crowds belong to the same event, the event detection is to generate all components in the graph. Note that this graph may not only be disjoint but also include single nodes. Each component or single node is an unusual event that is our final goal of this work. The first part of this algorithm checks if two unusual crowds can be connected to each other by parameters *overlapping* and *sharing*. The second part generates all the components in the unusual crowds graph, where any graph algorithm can be used. The detected event contains the users in each cluster and its corresponding timestamp and location.

Complexity: The detection of Unusual Events requires a pair-wise comparison between all the Unusual Crowds, therefore the complexity of this procedure is $\mathcal{O}(|UC|^2)$ where $|UC|$ is the number of Unusual Crowds.

4   Experiments

4.1   Experimental Setup

CDR Data. The D4D Orange challenge made available data collected in Cote d'Ivoire over a five-month period, from December 2011 to April 2012. The datasets describe call activities of 50,000 users chosen randomly from every two weeks. Specifically, the data contains the cell phone tower and a timestamp at which the user sent or received a text message or a call in the form of tuple <UserID, Day, Time, Antenna>. Each antenna is associated with location information. To avoid privacy issues, the data has been anonymized by D4D data provider.

From the CDR data, we find that about 63% users do not make calls in consecutive hours and 19% users make calls in only two consecutive hours. The pattern demonstrates the necessity of existence probability for user's location estimation, as most of users do not make regular and consecutive calls at each timestamp. We also observe that the probability that there is one hour between one user's two calls is more than 75% and that is 8% for a two-hour interval. In total, there are more than 80% two consecutive calls whose intervals are at most two hours. These observations demonstrate the challenges of CDR data's

Table 1. Comparison of the unusual event detection (UE) and gathering detection (GAT) [32].

Period	Date	Event Name [22]	UE	#UE	GAT	#GAT
Dec. 05 - Dec. 18	Dec. 07	Anniversary of Felix Death	√		√	
	Dec. 11	Parliament election	√	20	×	287
	Dec. 17	Violence	√		√	
Dec. 19 - Jan. 01	Dec. 25	Christmas day	×		×	
	Dec. 31	New year eve	√	36	√	56
	Jan. 1	New year day	√		×	
Jan. 02 - Jan. 16	Jan. 08	Baptism of Lord Jesus	√	31	√	176
	Jan. 14	Arbeen Iman Hussain	√		×	
Jan. 17 - Jan. 29	Jan. 17	Visit of Hilary Clinton	√	15	√	481
	Jan. 18	Visit of Kofi Annan	√		√	
Jan. 30 - Feb. 12	Jan. 30	ACNF 2012 vs Angola	√		√	
	Feb. 04	ACNF 2012 vs Equatorial Guinea	√		√	
	Feb. 04	Mawlid an Nabi Sunni	√		√	
	Feb. 05	Yam	√	58	×	310
	Feb. 08	ACNF 2012 Semi Final VS. Mali	×		√	
	Feb. 09	Mawlid an Nabi Shia	√		√	
	Feb. 12	ACNF 2012 Final VS. Zambia	√		×	
Feb. 13 - Feb. 26	Feb. 13	Post African Cup of Nations Recovery	√	52	√	152
	Feb. 22	Ash Wednesday	√		√	
Feb. 27 - Mar. 10	None			26		269
Mar. 11 - Mar 25	Mar. 12	Election of National Assembly President	√	17	√	342
	Mar. 13	Election of National Prime Minister	√		√	
Mar. 26 - Apr 08	Apr. 01-04	Education International Congress	√	75	√	1220
	Apr. 06	Good Friday	√		√	
Apr. 09 - Apr. 22	Apr. 09	Easter Monday	√	10	√	33
	Apr. 13-14	Assine fashion days	√		√	
Total			23/25	340	19/25	3326
Precision			**0.0676**		**0.0057**	
Recall			**0.9200**		**0.7600**	

spatio-temporal sparseness, which makes the design for degenerative existence probability reasonable for the coarse-grained CDR data.

Comparison Methods. To the best of our knowledge, this is the first work to detect unusual crowds and events in spatio-temporal data, and it is also the first time that we discover moving clusters in CDR data. We compare the results of our approach with GAT described in [32] and MOV in [17], as the methods employed in these work are also able to identify moving crowds. However, those methods are not designed to work on CDR and have to be adapted to perform the comparison. GAT defines a method to detect the gatherings in a trajectory dataset. A gathering is a sequence of spatial clusters with a certain number of committed users being member of an enough number of clusters. We use the same setting with GAT for parameters that indicate the same physical meanings in both methods. Clearly by following our intuition and goal of problem design, there should not exist any crowd or event at most days. Based on the results of parameter analysis in Section 4.3 and the developed Visual Analytics System, we selected the following parameters $\epsilon_n=20$, $\epsilon_{lt}=4$, $\epsilon_{ci}=10$, $\epsilon_p=0.2$, and $\epsilon_{si}=0.2$. Since they correspond to a probability to find unusual crowds to be around 10-15%, which helps us focus on rare events (as opposed to business as usual events).

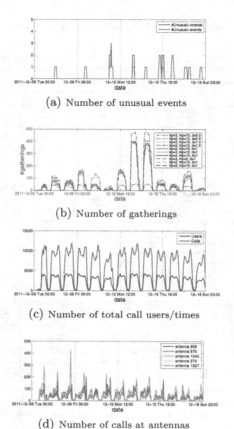

(a) Number of unusual events

(b) Number of gatherings

(c) Number of total call users/times

(d) Number of calls at antennas

Fig. 3. Time series of unusual events, gatherings, users, and calls in the first two-week.

4.2   Experimental Results

Detected Unusual events.   Table 1 reports a series of events occurred in Abidjan in the different periods covered by the datasets. In order to perform a fair study of the effectiveness of our method in comparison with GAT, we selected a third part set of events reported in [22]. To limit the explosion in the number of detected gatherings, we have set the most restrictive values for the remaining parameters: $d = 0.0$, $k_p = 2$, and $m_p = 5$. Moreover, we report the total number of generated events by both methods for each two-week period and subsequently generate *Precision* and *Recall* scores for both algorithms. It is possible to notice that our method detects a lower number of unusual events w.r.t. GAT. This is reflected in a higher value of Precision. Although, our method reports a lower number of events, it is able to detect a greater number of ground truth events, and this corresponds to a higher value of Recall. Notice that the two measures represent an estimation of precision and recall since the ground truth is not given. Indeed the list of events in [22] is not comprehensive of all events that happened in Ivory Coast in the monitored 5-month period, and this explains the low precision of both methods. This is the reason why we did not try to find the optimal values of the parameters to maximize Precision and Recall, but instead set such values based on the general criteria to find unusual crowds only in around 10-15% of the hours. However, the list in Table 1 gives a good basis for comparison and shows that our method is 10 times more precise than GAT.

We further perform comparisons with MOV, where the authors introduce the concept of moving clusters. Extracting moving clusters is equivalent to run our method with the parameter ϵ_p (the probability of a user to be committed) and ϵ_{si} (the similarity threshold between the mobility profile and the trajectories in the event) to 1. With these parameter settings the algorithm was not able to find any moving clusters. This is due to the fact that our method is able to handle the spatial and temporal sparsity of CDR data, while the MOV method is designed to work with GPS trajectories.

Time-Series. We report the time series of the numbers of unusual events and gatherings detected with different input parameter settings by using our method and the GAT algorithm in Figures 3(a) and 3(b). We try to match the same parameters we used in our method. For the minimum lifetime as well as the minimum number of objects that should belong to a cluster, we choose the same values adopted in the study of the effectiveness of our method ($\epsilon_{lt} = 4$ and $\epsilon_n = 20$). The rest of the input parameters of the algorithm to detect gatherings are the following: d is the minimum distance necessary to connect clusters detected in two consecutive time snapshots; k_p is the minimum number of time snapshots required to consider an user as a participant; m_p is the minimum number of participants to create a gathering. For these input parameters, we tried different enumerations to span the full admissible ranges. As it is possible to see, the number of detected gatherings is very high even if the parameters are chosen to be very restrictive. All the graphs show a daily trend, demonstrating that this method is not able to find unusual events our proposed method. Indeed, GAT can detect a large number of gatherings every day, which might not correspond to specific unusual events.

To further evaluate our discoveries, we check the total communication volumes and the specific antenna activities. We can clearly see that between Dec. 06 and Dec. 18, in Figure 3(c) there exist periodic patterns on each day without obvious peak values corresponding to the discovery of crowds—**anniversary of Felix death** on Dec. 07 and **Parliament election** on Dec. 11. Furthermore, the events on the day of parliament election involved five antennas. Their communication activities are plotted in Figure 3(d). Obviously there do not exist correlations between corresponding antenna activities and our unusual crowd/event output. These two regular and stable time series of communication activities further confirm the effectiveness of our problem design. These examples show that detecting unusual events is a complex task, which cannot be easily accomplished by looking at outliers in call time series. Thus, methods like [6, 21] are not directly applicable.

Spatial Distributions. Another comparison performed against GAT regards the spatial distribution of the detected events/gatherings. For both methods, we

(a) Detected unusual events   (b) Detected gatherings

Fig. 4. The unusual events (a) by our methods and the gatherings (b) by GAT detected on December 11th. Colors range from green to red as function of the number of detected participants

select the results obtained on December 11th. For the GAT method we select the results with lowest number of detected gatherings. In Figures 4(a) and 4(b) we report the detected events and detected gatherings respectively. Notice that a Voronoi tessellation has been applied in order to associate a covering area to each antenna. Our method detect 2 events, *Event 1* (left) covers 3 antennas and lasts for 4 hours. *Event 2* (right) covers 2 antennas and also lasts for 4 hours. For the same day, the GAT algorithm detects many gatherings (25) occurring in different regions of the city. This is probably due to the fact that the typical mobility profiles of the users are not taken into account in the process and thus recurring and unusual events are both detected. Moreover, if we consider the lifetime of gatherings occurring in the same locations of our events, we notice that it is generally longer. For example, a gathering, covering the same antennas of *Event 1*, lasts for 14 hours. Another characteristic of the gatherings is that they happen in the same location at different times. Instead, in our model, we define a method to consider those as one large event. In summary, with our method it is possible to identify events that occur occasionally in a precise zone of the city and happen in a precise period of time, while the other method detects several events without any distinctions between the periodical and the unusual ones.

4.3   Efficiency and Parameters

Our algorithms are implemented in Python 2.7.5, and all experiments were performed on a laptop running Windows 7 with Intel(R) Core(TM) i7-2720QM CPU@2.20GHz (2 cores) and 8GB memory. All related experiments are running on the first two-week dataset, which contains about two million CDR historic data. We simulate each experiment with specific parameter setting for 100 times to get both the average running time and standard deviation. In general, the algorithms for detecting unusual crowd events are efficient, in the fact that it only takes about 30 seconds to two minutes on two million CDR data. Furthermore, the execution of our methods is stable among different runs.

We evaluate the effect of parameter setting on the number of detected unusual crowds and discuss the guidelines for determining parameter settings. We find that the algorithm is particularly sensitive to ϵ_{lt}, ϵ_p, and ϵ_{si}. ϵ_{lt} is indicative of

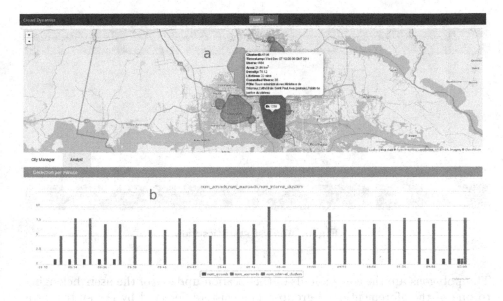

Fig. 5. Real-time view of city map and statistics.

the duration of moving crowds. Based on the definition of *commitment*, a larger ϵ_p can produce more compact crowds, which have much higher probabilities to be unusual events. Finally, the lower *similarity* ϵ_{si} threshold between regular mobility profiles and specific trajectory we set, the more crowds will consist of people whose mobility behavior differs from their typical profiles.

We would like to point out that there is not an unique way to optimally select the values of the parameters, as this strongly depends on the end-user application. For instance, if a city manager is interested in monitoring visitors to a museums, she might set different values of ϵ_{lt} and ϵ_p, compared to the monitoring on a protest. We develop a visual analytics tool described in Section 5 to help end users explore and test the detected results under different parameter settings for different applications.

5   Visual Analytics Prototype System

We have developed a visual analytics system to support the exploration of the unusual crowd events based on the proposed framework. The system allows end users—such as analysts and city managers—to analyze the formation and evolution of crowds, and study the impact of different parameters on the obtained results, heuristically suggesting possible changes to get more meaningful results depending on the desired application. The interface consists mainly of two components: the map overview of the observed city (Figure 5 (a)) and the statistics of users, crowds, and events (Figure 5 (b)).

Map View.   In the map, the system visualizes the latest clusters, crowds and unusual events detected in the form of polygons as shown in Figure 5 (a).

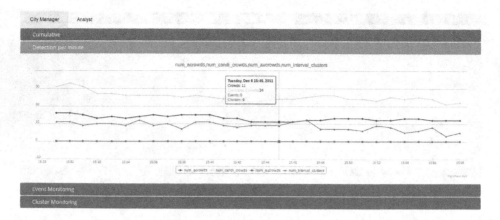

Fig. 6. Analyst statistics view

The polygons are the convex hulls of the location updates of the users belonging to one of the aforementioned groups. The clusters detected by the system at a given timestamp are visualized on the map as green polygons. On mouse-over, the UI shows a pop-up window with the cluster attributes, including 1) timestamp— the timestamp when the cluster was detected, 2) #users—the number of users that are a part of the cluster, 3) area—the area covered by the Cluster polygon in square kilometers, 4) density—the ratio between the number of users in the Cluster and its area, and 5) POIs—the list of Points Of Interest located within the cluster area. The detected Crowds and the Unusual events are visualized on the map as blue and red polygons, respectively. Similar to the cluster visualization, a set of properties can be shown in a pop-up window.

Statistics View. We envision the tool to be used by two different types of actors: the Analyst, which is in charge of setting up the analytics system, and the city manager, who has to take actions based on the identified unusual crowd events. The City manager tab contains a single time-series graph representing the number of Clusters, Crowds, Unusual Events detected at every timestamp as displayed in Figure 5 (b). This tab contains the most crucial outcome of the analytics performed and it provides an intuitive way to represent the most recent mobility patterns of the city. The Analyst tab contains a richer set of statistics in Figure 6. In order to make efficient use of available space we fit the graphs into collapsible panels into groups of semantically relevant statistics, including 1) Cumulative—the cumulative trends of the detected clusters, crowds and unusual events, 2) Detection per minute—the time-series of the number of clusters, candidate crowds, crowds, and unusual events, 3) Event monitoring— the time-series of maximum and minimum value of lifetime, number of committed users, total number of users, and similarity of the candidate crowds, 4) Cluster monitoring—the time-series of the maximum size of the detected clusters, and their minimum spatial radius. In addition, a red dashed-line corresponding to each parameter is shown to depict parameter efficacy (e.g. event monitoring, cluster monitoring). This allows the Analyst to understand the role of each parameter on the obtained results and to set the most appropriate parameters for the specific application in scope.

6 Related Work

The availability of mobility data has offered researchers the opportunity to analyze both individual's and group's moving behaviors. In [12,14], the authors defined a methodology to extract dense areas in spatio-temporal databases, thus identifying where and when dense areas of mobile objects appear. A similar definition was proposed in [17], where the authors introduced the concept of *moving clusters*. Following these ideas, several group and cluster mobility pattern models have been proposed [1,16,19,31]. For instance, in [2,11,28,29] the concept of *flock* is widely investigated. A flock is a group of objects that travel together within a disc of some user-specified size for at least k consecutive timestamps. The main limit of this model is that a simple circular shape does not reflect natural grouping in reality. A marginal different concept *convoy* was introduced in [15,16], where a density-based clustering is adopted instead of the radius of the disk. Li at al. [18] proposed a more general type of trajectory patterns: *swarm*. The swarm is a cluster of moving objects that lasts for at least k timestamps, possibly non-consecutive. Another group mobility model was introduced in [32], called *gathering*, whose novelty regards the introduction of the concept of commitment.

Other interesting works dealing with the detection of anomalies in city traffic flow are presented in [6,21,25]. In [21], the authors use likelihood ratio test statistic (LRT) on GPS trajectories of taxis to detect traffic flow anomalies. Transportation model detection problem is studied in mobile phone data and GIS data [23]. A passive route sensing framework is introduced to monitor users' significant driving routes with low-power sensors in mobile phones [20]. However, these works do not address the problem detecting unusual events considering people mobility but are more focused on traffic flow analysis through an aggregation of the information. On the contrary, in this paper we are interested in detecting such events that involve a large number of people whose current mobility differs from their typical one.

All the above works are designed and tested on high-resolution trajectory data, such as the one provided by GPS systems. Low-resolution location data collected from telecommunication operators, on the other hand, is much more pervasive resulting in a much larger sample of the population being monitored, see [4,5,9,13]. In this paper, we propose a new method to mine coarse grain mobile phone data (in the form of CDR) to detect unusual crowd events. Indeed, the aim of our work is to detect events that involve a large number of people performing unusual activities. To do so, we compute a similarity between the mobility profile of the users and their trajectories in group pattern. This extends, thus, the concept of commitment since users need to be committed and have trajectories that differ from the ones in their mobility profiles. Our method however is able to identify moving events that span several locations over time, and involve a subset of committed users, something that could not be detected by using the methods in [26,30].

7 Conclusion and Future Work

In this paper, we formally define the problem of inferring unusual crowd events from mobility data. Previous work on event detection is limited on inferring the usual event from the fine-grained GPS data. Our problem definition differs by characterizing the unusual crowd events and we present a new methodology to extract them from coarse-grained CDR data. The main contributions of this paper w.r.t. existing methods are the ability to analyze temporally and spatially sparse data as CDRs and the definition of a subclass of events which are unusual to its attendees. Our experimental results demonstrate the effectiveness of our proposed method in real-world mobile datasets.

Despite the promising results of the present work, there is still much room left for future work. First, while this proposed method relies on Visual Analytics to help end users set parameters, we are planning to design algorithms to determine parameters for specific applications of interest as well as an optimization procedure for evaluation metrics. Moreover, we are working toward combining mobile and social media data together to detect unusual events. In doing so, we have the potential to detect and monitor crowding activities in real time, and eventually yield a better and smarter planet.

References

1. Appice, A., Malerba, D.: Leveraging the power of local spatial autocorrelation in geophysical interpolative clustering. DMKD **28**(5–6), 1266–1313 (2014)
2. Benkert, M., Gudmundsson, J., Hbner, F., Wolle, T.: Reporting flock patterns. Computational Geometry **41**(3), 111–125 (2008)
3. Calabrese, F., Di Lorenzo, G., McArdle, G., Pinelli, F., Van Lierde, E.: Real-time social event analytics. In: Netmob 2015 (2015)
4. Calabrese, F., Ferrari, L., Blondel, V.: Urban sensing using mobile phones network data: a survey of research. ACM Comput. Surv. (2014)
5. Calabrese, F., Pereira, F.C., Di Lorenzo, G., Liu, L., Ratti, C.: The geography of taste: analyzing cell-phone mobility and social events. In: Floréen, P., Krüger, A., Spasojevic, M. (eds.) Pervasive 2010. LNCS, vol. 6030, pp. 22–37. Springer, Heidelberg (2010)
6. Chawla, S., Zheng, Y., Hu, J.: Inferring the root cause in road traffic anomalies. In: IEEE ICDM 2012, pp. 141–150 (2012)
7. Dong, Y., Yang, Y., Tang, J., Yang, Y., Chawla, N.V.: Inferring user demographics and social strategies in mobile social networks. In: KDD 2014, pp. 15–24. ACM (2014)
8. Ester, M., Peter Kriegel, H., Sander, J., Xu, X.: A density-based algorithm for discovering clusters in large spatial databases with noise. In: ACM SIGKDD 1996, pp. 226–231 (1996)

9. Georgiev, P., Noulas, A., Mascolo, C.: The call of the crowd: Event participation in location-based social services. In ICWSM 2014 (2014)
10. Giannotti, F., Nanni, M., Pinelli, F., Pedreschi, D.: Trajectory pattern mining. In: ACM SIGKDD 2007, pp. 330–339. ACM, New York (2007)
11. Gudmundsson, J., van Kreveld, M.: Computing longest duration flocks in trajectory data. In: ACM GIS 2006, pp. 35–42. ACM, New York (2006)
12. Hadjieleftheriou, M., Kollios, G., Gunopulos, D., Tsotras, V.J.: On-line discovery of dense areas in spatio-temporal databases. In: Proc. of the 7th International Conference on Advances in Spatial and Temporal Databases, SSTD 2003, pp. 306–324 (2003)
13. Isaacman, S., Becker, R., Cáceres, R., Martonosi, M., Rowland, J., Varshavsky, A., Willinger, W.: Human mobility modeling at metropolitan scales. In: Proceedings of the 10th International Conference on Mobile Systems, Applications, and Services, MobiSys 2012, pp. 239 252. ACM, New York (2012)
14. Jensen, C.S., Lin, D., Ooi, B.C., Zhang, R.: Effective density queries on continuously moving objects. In: IEEE ICDE 2006, pp. 71–81. Washington, DC, USA (2006)
15. Jeung, H., Shen, H.T., Zhou, X.: Convoy queries in spatio-temporal databases. In: IEEE ICDE 2008, pp. 1457–1459. IEEE Computer Society, Washington (2008)
16. Jeung, H., Yiu, M.L., Zhou, X., Jensen, C.S., Shen, H.T.: Discovery of convoys in trajectory databases. Proc. VLDB Endow. 1(1), 1068–1080 (2008)
17. Kalnis, P., Mamoulis, N., Bakiras, S.: On discovering moving clusters in spatio-temporal data. In: Medeiros, C.B., Egenhofer, M., Bertino, E. (eds.) SSTD 2005. LNCS, vol. 3633, pp. 364–381. Springer, Heidelberg (2005)
18. Li, Z., Ding, B., Han, J., Kays, R.: Swarm: mining relaxed temporal moving object clusters. Proc. VLDB Endow. 3(1-2), 723–734 (2010)
19. Lu, E.H.-C., Tseng, V.S., Yu, P.S.: Mining cluster-based temporal mobile sequential patterns in location-based service environments. IEEE Trans. on Knowl. and Data Eng. 23(6), 914–927 (2011)
20. Nawaz, S., Mascolo, C.: Mining users' significant driving routes with low-power sensors. In: ACM SenSys 2014, pp. 236–250. ACM (2014)
21. Pang, L.X., Chawla, S., Liu, W., Zheng, Y.: On detection of emerging anomalous traffic patterns using gps data. Data Knowl. Eng. 87, 357–373 (2013)
22. Paraskevopoulos, P., Dinh, T.-C., Dashdorj, Z., Palpanas, T., Serafini, L.: Identification and characterization of human behavior patterns from mobile phone data. In: International Conference on the Analysis of Mobile Phone Datasets (NetMob 2013) (2013)
23. Stenneth, L., Wolfson, O., Yu, P.S., Xu, B.: Transportation mode detection using mobile phones and gis information. In: GIS 2011, pp. 54–63. ACM, New York (2011)
24. Tang, J., Wu, S., Sun, J.: Confluence: conformity influence in large social networks. In: ACM SIGKDD 2013, pp. 347–355. ACM, New York (2013)
25. Telang, A., Deepak, P., Joshi, S., Deshpande, P., Rajendran, R.: Detecting localized homogeneous anomalies over spatio-temporal data. DMKD 28(5–6), 1480–1502 (2014)
26. Traag, V.A., Browet, A., Calabrese, F., Morlot, F.: Social event detection in massive mobile phone data using probabilistic location inference. In: SocialCom 2011, pp. 625–628. IEEE (2011)
27. Trasarti, R., Pinelli, F., Nanni, M., Giannotti, F.: Mining mobility user profiles for car pooling. In: ACM SIGKDD 2011, pp. 1190–1198. ACM, New York (2011)

28. Vieira, M.R., Bakalov, P., Tsotras, V.J.: On-line discovery of flock patterns in spatio-temporal data. In: ACM GIS 2009, pp. 286–295. ACM, New York (2009)
29. Wachowicz, M., Ong, R., Renso, C., Nanni, M.: Finding moving flock patterns among pedestrians through collective coherence. International Journal of Geographical Information Science **25**(11), 1849–1864 (2011)
30. Witayangkurn, A., Horanont, T., Sekimoto, Y., Shibasaki, R.: Anomalous event detection on large-scale gps data from mobile phones using hidden markov model and cloud platform. In: ACM UbiComp 2013 Adjunct, pp. 1219–1228. ACM, New York (2013)
31. Wu, M., Jermaine, C., Ranka, S., Song, X., Gums, J.: A model-agnostic framework for fast spatial anomaly detection. ACM Trans. Knowl. Discov. Data **4**(4), 1–30 (2010)
32. Zheng, K., Zheng, Y., Yuan, N.J., Shang, S.: On discovery of gathering patterns from trajectories. In: IEEE ICDE 2013, pp. 242–253, Washington, DC, USA (2013)

Learning Pretopological Spaces for Lexical Taxonomy Acquisition

Guillaume Cleuziou[1](✉) and Gaël Dias[2]

[1] Université D'Orléans, INSA Centre Val de Loire, LIFO EA 4022, Orléans, France
cleuziou@univ-orleans.fr
[2] Université de Caen Basse-Normandie, GREYC UMR 6072, Caen, France

Abstract. In this paper, we propose a new methodology for semi-supervised acquisition of lexical taxonomies. Our approach is based on the theory of pretopology that offers a powerful formalism to model semantic relations and transforms a list of terms into a structured term space by combining different discriminant criteria. In order to learn a parameterized pretopological space, we define the Learning Pretopological Spaces strategy based on genetic algorithms. In particular, rare but accurate pieces of knowledge are used to parameterize the different criteria defining the pretopological term space. Then, a structuring algorithm is used to transform the pretopological space into a lexical taxonomy. Results over three standard datasets evidence improved performances against state-of-the-art associative and pattern-based approaches.

1 Introduction and Related Work

By coding the semantic relations between terms, lexical taxonomies (LTs) such as WordNet [7] have enriched the reasoning capabilities of applications in information retrieval and natural language processing. However, the globalized development of semantic resources is largely limited by the efforts required for their construction [5]. As a consequence, many research studies have been appearing to automatically learn LTs. Instead of manually creating LTs, learning them from texts has undeniable advantages. First, they may fit the semantic component neatly and directly within a given domain. Second, the cost per entry is greatly reduced, giving rise to much larger resources.

The two main stages for the automatic construction of LTs are term extraction (TE) and term structuring (TS). A substantial amount of works exist on TE [8], but the present study exclusively focuses on the TS stage. Within this context, similarity-based, pattern-based, set-theoretical and associative approaches have traditionally been proposed.

Similarity-based or clustering-based approaches [9,10] hierarchically cluster terms based on similarities of their meanings usually represented by a vector of quantifiable features. They have the main advantage that they are able to discover relations which do not explicitly appear in text. They also avoid the problem of inconsistent chains by addressing the structure of a taxonomy globally from the outset. However, it is generally believed that these methods can

© Springer International Publishing Switzerland 2015
A. Appice et al. (Eds.): ECML PKDD 2015, Part II, LNAI 9285, pp. 493–508, 2015.
DOI: 10.1007/978-3-319-23525-7_30

not generate relations as accurate as pattern-based approaches [5]. *Pattern-based strategies* [5,15] define lexical- syntactic patterns for semantic relations and use these patterns to discover instances of relations. They are known for their high accuracy in recognizing instances of relations if the patterns are carefully chosen, either manually or via automatic boostrapping. However, this approach suffers from sparse coverage of patterns in specific corpora, especially technical domain ones. Moreover, it may evidence inconsistent concept chains as instances are extracted in pairs and gathered to form taxonomy hierarchies. *Set-theoretic approaches* [3] use formal concept analysis that naturally structures terms with intensional inclusion relations within a concept lattice. Such term organization differs from usual lexical taxonomies that provide semantic relations between terms rather than inclusion relations between formal concepts. This strategy usually highlights low performance as contextual vector seldom overlap in large open uncontrolled domains. Finally, *associative frameworks* [12] use asymmetric similarities between terms to model the subsumption relation. For that purpose, distributions of terms over document collections are used to discover general/specific noun relationships. The main drawback of this approach is that the subsumption model implicitly hypotheses that general terms are always more frequent than their specific terms, which is not always satisfied in practice.

Note that these methodologies rely on one exclusive criterion to model the subsumption (is-a) relation and build the respective taxonomy. In order to take advantage of multiple criteria, two important works have been proposed [14,16]. Both methodologies first learn an ontology metric, which models the is-a relation based on vectors of discriminant criteria (e.g. contextual, cooccurrence, syntactic dependency or patterns). This step is obtained by supervised learning over existing taxonomies. The logistic regression is used by [14] and [16] applies the ridge regression. Then, the ontology metric guides the incremental taxonomy acquisition process modeled as an optimization task: 1-objective for [14] and 2-objectives for [16]. The main advantage of these approaches is to model the is-a relation between terms based on multiple criteria, thus greatly avoiding data sparseness and low coverage. However, both proposals depend on a supervised learning stage that relies on large known ontologies such as WordNet or Open Directory Project. However, in real-world situations, this knowledge is not accessible and only partial (usually small) knowledge of the domain can be accessed. Moreover, note that these large resources are mainly available for the English general language. As such, language/domain/genre adaptability is not ensured.

In this paper, we propose a new semi-supervised multi-criteria strategy for taxonomy induction. The overall idea is (1) to learn a propagation metric[1] based on a set of relevant associative and pattern-based features constrained by small (yet accessible) pieces of knowledge of the domain and (2) to induce the taxonomy based on a pretopological framework which transforms the pretopological term space into a directed acyclic graph, the output taxonomy. To achieve these objectives, we consider pretopology on the multi-criteria analysis point of view, where criteria are statistical indices (associative approach) and linguistic

---

[1] As opposed to the ontology metric.

patterns (pattern-based approach) retrieved from a corpus. In particular, we define the concept of *parameterized pretopological space* (P-space), where parameters express the confidence that exists over each criterion. As such, LT induction can be viewed as learning the set of parameters (confidences), which best (1) approximate the expected LT structure and (2) verify a given number of linguistic patterns constraints. In order to learn the parameters, we define a new *Learning Pretopological Spaces* (LPS) strategy based on genetic algorithms, which leads to induce a LT from an "optimized" P-space. The main advantages of the LT acquisition methodology presented in this paper, when compared to state-of-the-art methodologies are enunciated as follows:

(1) We learn a propagation metric, which directly models the is-a relation into the taxonomy induction process in contrast to [16] and [14] who propose a two-steps process,
(2) Linguistic patterns, which embody small (yet accessible) pieces of knowledge of the domain constrain the semi-supervised learning process but are also used as relevant criteria,
(3) We deal with both general and specialized domains where linguistic patterns fail to retrieve any relation,
(4) Our framework is quasi-independent regarding to language as only few and simple linguistic patterns and raw texts are required.

In the remainder of this paper, we first define the required notions of our pretopological framework and its usage for multi-criteria analysis (Section 2). Then, in Section 3, we define the concept of parameterized pretopological space (P-space) and propose the learning pretopological spaces (LPS) strategy based on genetic algorithms in the context of taxonomy induction. In Section 4, we evaluate our framework on the LT reconstruction task, considering both general (i.e. WordNet) and specialized domains (i.e. UMLS). Finally, in Section 5, concluding statements are enunciated.

2   Pretopological Framework

Pretopology [1] is a theory that generalizes both topology and graph theories and is commonly used to model complex propagation phenomena thanks to a pseudo-closure function. Let's consider a non-empty set E and its powerset $\mathcal{P}(E)$. A pretopological space[2] is noted (E, a), where $a(.)$ is a pseudo-closure function described in Definition 1.

Definition 1 (Pseudo-closure). *A pseudo-closure is a function* $a(.) : \mathcal{P}(E) \rightarrow \mathcal{P}(E)$, *which respects the following three conditions:*

i) $a(\emptyset) = \emptyset$,
ii) $\forall A \in \mathcal{P}(E), A \subseteq a(A)$,

---

[2] Note that in this paper, we always consider V-type spaces, as they present good structuring properties.

iii) $\forall A, B \in \mathcal{P}(E), A \subseteq B \Rightarrow a(A) \subseteq a(B)$.

So, the pseudo-closure function behaves as an expansion operator that enlarges any non-empty subset $A \subset E$. As a consequence, successive applications of $a(.)$ on A lead to a fix-point called *closed subset* and noted F_A. Two other concepts are required to introduce our model: *elementary closed subset* and *maximal elementary closed subset* formalized in Definitions 2 and 3 respectively.

Definition 2 (Elementary Closed Subset). *An elementary closed subset* $F_{\{x\}}$, *is the closure of a singleton* $\{x\}$ *with* $x \in E$.

Definition 3 (Maximal Elementary Closed Subset). *A maximal elementary closed subset is an elementary closed subset, maximal in terms of inclusion with respect to all possible elementary closed subsets in E.*

These definitions give us two key concepts on a structuring point of view: (1) an elementary closed subset $F_{\{x\}}$ refers to the subset of items reachable from x and (2) when $F_{\{x\}}$ is maximal, it means that x is only reachable from items y with an identical elementary closed subset ($F_{\{x\}} = F_{\{y\}}$), thus capturing a kind of equivalence class.

2.1   Pretopology and Multi-criteria Analysis

Pretopology can be used in the context of multi-criteria analysis since it allows complex but efficient aggregation of several criteria at the pseudo-closure function level. So, considering (1) a set of K criteria providing different views on the manner a discrete set E is structured and (2) each criterion defining one neighborhood relation on E and $N_k(x)$ the kth neighborhood of x, the family of neigborhoods $\mathcal{N} = \{N_1, \ldots, N_K\}$ suggests a multi-criteria environment. Note that to be consistent with the formal definition of neighborhoods [1], we constrain any $N_k(x)$ to contain x itself:

$$\forall k = 1, \ldots, K, \quad \forall x \in E, \quad x \in N_k(x). \tag{1}$$

A usual pseudo-closure definition for neighborhood aggregation, which satisfies the V-type space conditions is given by

$$\forall A \in \mathcal{P}(E), a(A) = \{x \in E | \forall N_k \in \mathcal{N}, N_k(x) \cap A \neq \emptyset\}. \tag{2}$$

Such a pseudo-closure expands a subset A to an item x if and only if all neighborhoods (criteria) of x intersect A. It is important to note that when A is not reduced to a singleton, the agreement can be reached by intersections that concern different items of A. Thus, a complex propagation process is defined at the subset level rather than at the element level and there is no way to reproduce such a process on a single neighborhood structure that would result from the *a priori* aggregation of the different criteria[3].

---

[3] Proof of this statement is out of the scope of this paper.

2.2   Pretopology and LT Acquisition

It is well-established that known LTs such as WordNet or Cyc share some specific common structure. As a consequence, a learned LT should ideally satisfy the following two structural requirements:

(1) a DAG structure: each node must be characterized by two disjoint sets of predecessors and successors with no cycles,
(2) aggregating nodes: each node must contain one or several terms from the vocabulary E.

Such a structure can be obtained based on a pretopological term space with the structuring algorithm proposed by [6]. In our specific case, we propose a top-down version of this algorithm. So, instead of considering minimal closed subsets, we consider maximal ones. The basic idea of the algorithm for LT induction is defined in Figure 1 and illustrated in Figure 2[4].

---

1. Determine elementary closed subsets associated to each element x of E giving rise to the family of closures $\mathcal{F}e(E, a)$.
2. Find the family of maximal elementary closed subsets $\mathcal{F}M(E, a)$. This means enumerating all the maximal elementary closed subsets by inclusion in $\mathcal{F}e(E, a)$. Any element $F \in \mathcal{F}M(E, a)$ is then a core.
3. Within each core, recursively determine the largest elementary closed subsets of E in terms of inclusion, until no other can longer be found. The recursive process allows to generate, from each core, a set of homogeneous parts by successive reductions and outputs the final LT.

Fig. 1. LT induction algorithm.

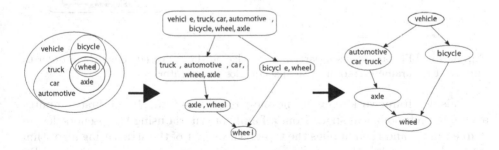

Fig. 2. Top-down structuring inducing a DAG from a pretolopological term space.

---

[4] More details can be found in [4].

2.3   Current Limitations

Despite its interesting properties for multi-criteria analysis as evidenced in [4] for LT induction, in its current form, the pretopological LT process evidences two main limitations that make it under-efficient:

(1) it is sensitive to unreliable criteria,
(2) it only allows a limited number of criteria to combine.

Both issues are due to the definition of the pseudo-closure operator itself that requires that all criteria must satisfy the intersection property in order to start the propagation process from elementary sets.

3   Learning Pretopological Spaces

To overcome previous limitations, we propose in this paper a new learning pre-toplogical spaces (LPS) framework based on a more flexible pseudo-closure definition. It is illustrated in Figure 3.

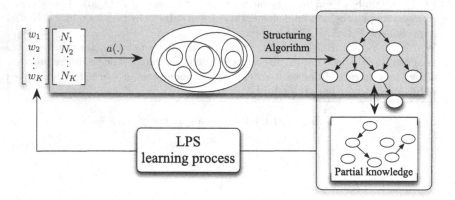

Fig. 3. The LPS process uses partial knowledge on the expected structure in order to improve the parameterization of the pseudo-closure operator.

This new framework relies on the one-pass process from [4] that first computes a unique pretopological space from a family of criteria using the pseudo-closure defined in (2) and then applies the top-down variant of the structuring algorithm from [6]. But, rather than providing the resulting structure as output, the LPS framework consists in comparing the built structure to some partial knowledge and modifying the pseudo-closure operator in order to improve the final structuring. This is achieved by an iterative semi-supervised learning process. Such a framework requires to introduce new concepts into the pretopology theory, especially the concept of *parameterized pretopological space* (P-Space).

3.1   Parameterized Pretopological Space

To relax the constraint that requires the agreement on all criteria to allow the propagation process, we propose to introduce a parameter p that indicates a requirement on the minimum number of neighborhoods that must intersect a subset A in order to expand it. Its formalization is given in Equation 3 with $p \in \{1, \ldots, K\}$ and $\mathbb{1}_{N_k(x) \cap A \neq \emptyset} = 1$ if the neighborhood $N_k(x)$ intersects A and 0 otherwise.

$$\forall A \in \mathcal{P}(E), a(A) = \{x \in E \mid \sum_{N_k \in \mathcal{N}} \mathbb{1}_{N_k(x) \cap A \neq \emptyset} \geq p\} \tag{3}$$

To express the combination model as a learning model, we define the notion of *parameterized pretopological space* (P-Space) that introduces supplementary parameters to manage the reliability of the criteria in Definition 4.

Definition 4 (P-Space). *A P-space* $(E, a, \mathbf{w})$ *is a V-type pretopological space with a parameterized pseudo-closure* $a(.)$ *defined by*

$$\forall A \in \mathcal{P}(E), a(A) = \{x \in E \mid \sum_{N_k \in \mathcal{N}} w_k \cdot \mathbb{1}_{N_k(x) \cap A \neq \emptyset} \geq w_0\} \tag{4}$$

such that (1) $w_0 > 0$, (2) $\sum_{k=1}^{K} w_k \geq w_0$ *and* (3) $\forall k, \ w_k \geq 0$.

Note that conditions (1), (2) and (3) over the set of parameters $\mathbf{w}$ are defined to respectively ensure the three conditions i), ii) and iii) expressed in Definition 1 over the V-type spaces. In particular, each parameter w_k in Equation (4) quantifies the reliability on the kth criteria and w_0 represents a global required confidence to expand the subset. Thus, a subset A will be expanded to an element x only if the sum of the confidences on the criteria in agreement with the expansion exceeds the global required confidence w_0.

The P-Space concept evidences two strong advantages: (1) it overcomes the limitations about reliability and multiplication of the criteria and (2) it extends significantly the possibilities of combination, passing from a single conjunctive decision rule to a set of logical decision rule (without negation). But the noticeable improvement on the model makes a new challenging question to appear: **How to parameterize a P-Space?**

3.2   Semi-supervised Learning of P-Spaces

We propose a semi-supervised strategy to learn the parameters of a P-Space. So, if S is a given source providing a true **partial** structuring on E^5 and considering that a V-type pretopological space induces a unique DAG, the Learning P-Space (LPS) process aims to find a P-Space inducing a DAG that satisfies:

---

[5] Note that in the context of LT acquisition S is usually a small number of "evident" subsumption relations between terms.

(1) the constraints implied by the partial knowledge S and
(2) a taxonomy-like structuring.

The following $Score(.,.)$ quantifies such a satisfaction:

$$Score(\mathbf{w}, S) = F_{Measure}(\mathbf{w}, S) \times I_{taxonomy}(\mathbf{w}). \tag{5}$$

The $F_{Measure}$ is the usual external validation index [11] that, in our context, combines *precision* and *recall* calculated over the pairs of elements linked in the partial knowledge S only. More precisely, given a DAG $D_\mathbf{w}$ induced by the P-Space with parameters $\mathbf{w}$ and the partial knowledge S also formalized as a (more sparse) DAG, we first operate a closure operation on both graphs (resulting in $\bar{D}_\mathbf{w}$ and $\bar{S}$) in order to make any implicit (indirect) edge to emerge before computing *precision*, *recall* and $F_{Measure}$. Metrics are defined in Equations 6 where $\bar{S}^t$ denotes the graph opposite to $\bar{S}$, which must be considered in order to count the false positive relations.

$$precision = \frac{|\{(x,y) \in \bar{D}_\mathbf{w} \cap \bar{S}\}|}{|\{(x,y) \in \bar{D}_\mathbf{w} \cap (\bar{S} \cup \bar{S}^t)\}|}$$

$$recall = \frac{|\{(x,y) \in \bar{D}_\mathbf{w} \cap \bar{S}\}|}{|\{(x,y) \in \bar{S}\}|} \tag{6}$$

$$F_{Measure} = \frac{2.precision.recall}{precision+recall}$$

The $I_{taxonomy}$ term is used to control the structural properties of the induced DAG $D_\mathbf{w}$ (independently to S). Although, the structure of a taxonomy is not formally defined, one can observe that a taxonomy usually looks like more to a tree (with one parent per node - except for the root) than to a lattice structure (for example). In order to favor tree-like structures, we compute on $D_\mathbf{w}$ its average ascendant degree (i.e. average number of parents per node) $Ad(D_\mathbf{w})$, and we use it to penalize a DAG moving away from a tree-like structure. This constraint is formalized in Equation 7.

$$I_{taxonomy}(\mathbf{w}) = e^{-(Ad(D_\mathbf{w})-1)^2} \in [0, 1] \tag{7}$$

The final satisfaction measure $Score(\mathbf{w}, S)$ reaches a maximum value of 1 for a DAG that (1) fits exactly to the knowledge source S and (2) structures the elements with an average ascendant degree of 1 (taxonomy).

This measure is used to guide the exploration of the space of solutions through a learning strategy based on a Genetic Algorithm (GA). GAs are stochastic exploration methods inspired from the natural selection principle [13]. Given a $fitness(.)$ function over the solution space, they simulate a natural evolution process by iteratively (1) generating populations of solutions (with mutation and crossover operators) and then (2) selecting the ones with highest fitness. The LPS approach uses such a stochastic exploration process based on the following fitness function:

$$fitness(\mathbf{w}) = \begin{cases} Score(\mathbf{w}, S) & \text{if } \mathbf{w} \text{ satisfies Def. (4)} \\ 0 & \text{otherwise.} \end{cases} \tag{8}$$

Figure 4 gives an overview of the general LPS process. Each iteration of the algorithm leads to an ensemble of P-Spaces, evaluated and selected as regard to their ability to induce a taxonomy-like structure satisfying partial knowledge requirements. Ultimately, only the best P-Spaces are returned. Let us notice that, in addition to the expected taxonomies the final P-Spaces allow to induce, the returned P-Spaces are themselves of high interest since they hold a learned propagation process that could be reused in an incremental context[6].

---

1. Given:
   a set of elements E,
   a family of criteria $\mathcal{N} = \{N_1, \ldots, N_K\}$,
   a partial knowledge S,
   a maximum number of iterations t_{max}.
2. Build an initial population $\mathbf{W}^0 = \{\mathbf{w} \in [0,1]^{K+1}\}$.
3. $t \leftarrow 0$
4. For each solution $\mathbf{w} \in \mathbf{W}^t$
      - Build the induced DAG $D_{\mathbf{w}}$,
      - Evaluate its $fitness(\mathbf{w})$
5. Select the best P-Spaces
6. if $(t < t_{max})$ then
      $t \leftarrow t + 1$,
      - Generate a new population $\mathbf{W}^t$ by mutation
        and crossover,
      - GoTo step 4
7. else return the selection.

Fig. 4. LPS general algorithm.

4   Experiments on LT Acquisition

The objective of the present proposal is to combine associative and pattern-based methods for LT acquisition by applying our multi-criteria LPS algorithm. Two situations have been considered in the following experiments:

When the linguistic patterns succeed in retrieving (maybe few) accurate relations. This is usually the case for generic domains. In this case, the set of term relations automatically extracted from a given set of patterns plays the role of the partial knowledge S. LT acquisition is thus performed in an *auto-supervised* context since no expert intervention is needed,

When the linguistic patterns fail to provide any reliable piece of knowledge that could guide the structuring process. This situation frequently occurs for specialized domains and makes most of the existing pattern-based approaches [5,15] totally inoperative. An expert is so required to give at least a couple of term relations (S) as in a usual *semi-supervised* learning context.

---

[6] This is out of the scope of this paper.

4.1   Experimental Setups

For each LT construction experiment, the list of terms to structure E comes from a reference R (in english) and the acquired taxonomies are compared to this reference using the $F_{Measure}$ as defined in (6) but with R that stands in for S. Indeed, S is only a set of term relations that helps the learning process and R is the complete gold standard reference taxonomy, to which the induced LT must be compared.

The linguistic patterns used are limited to the following list of four simple and usual ones [5]: "X such as Y", "X including Y", "X like Y" and "Y are X that". For any pair of terms (x, y) from the list, each pattern is tested on en.wikipedia.org and each time a pattern is observed between x and y, an edge $x \leftarrow y$ (x subsumes y) is added to S.

The english subpart of wikipedia.org (i.e. en.wikipedia.org) is also used as corpus for frequency counts extraction. For each pair of terms (x, y), we retrieve the number of wikipedia pages where both terms occur ($hits(x, y)$) in the corresponding sub-domain of wikipedia. Sub-domains are artificially identified by introducing the root term of the taxonomy into the wikipedia query. For example, $hits(cars, trucks)$ is retrieved with the following query ["cars" AND "trucks" AND "vehicle"], $vehicle$ being the root of the taxonomy to reconstruct.

From the frequency counts, three kinds of associative criteria are built in order to serve as basis neighborhoods for the P-Spaces:

N_{kSand} corresponds to the subsumption relation modeled by [12] : $y \in N_{kSand}(x)$ iff $P(y|x) \approx \frac{hits(x,y)}{hits(x)} \geq \sigma_k \wedge P(y|x) > P(x|y)$.

N_{kNP} associates to each term x its k Nearest Parents in the sense of $P(y|x)$: $y \in N_{kNP}(x)$ iff $P(y|x)$ is one of the k best $\{P(z|x)\}_{z \in E}$.

N_{kNC} associates to each term x its k Nearest Children: $y \in N_{kNC}(x)$ iff $P(y|x)$ is one of the k best $\{P(y|z)\}_{z \in E}$.

All criteria depend of the parameter k that controls the number of selected relations. In particular, we adjust the threshold σ_k in such a way that N_{kSand} selects as many relations as the two other criteria for a same value of k (i.e. $k.|E|$ relations). So, each type of criterion provides several effective criteria depending of the parameter k. In the following experiments, each criterion will be used with three different values ($k \in \{1\dots3\}$) leading to families containing respectively three, six and nine effective criteria.

Let us notice that, unlike the two first criteria, N_{kNC} has a strong weakness as it tends towards a non-taxonomic structure. In particular, it will be used to illustrate the behaviour of our approach in the context of an existing unreliable criterion when compared to previous studies.

To conclude on the preliminaries, let us mention the following operational details. The LPS algorithm has been implemented using the R package "GA" [13] with default configurations for crossover and mutation operators. We fixed the size of the population in the range $\{25\dots1000\}$ depending of the number of terms to structure and a maximum number of iterations to 25. As GAs are stochastic methods, we select in the coming results the best learned P-Space (in terms of $fitness(\mathbf{w})$) over a set of 5 runs.

4.2 LT Acquisition with Auto-supervision

The LT construction task is experimented on three domains extracted from WordNet, *Vehicles*, *Plants* and *Food*, with respectively 108, 554 and 1485 terms. The first two datasets are usually used as gold standards on LT induction [5,15] and the *Food* dataset is provided by the recent SEMEVAL 2015 contest [2].

Table 1 reports in the three top parts, the scores obtained (and the corresponding best parameters k) using purely associative approaches (without LPS), with or without aggregation of two or three statistical criteria.

Table 1. Quantitative evaluation of reconstructed lexical taxonomies on the domains *Vehicles*, *Plants* and *Food*.

Criteria	Vehicles				Plants				Food			
	Prec.	Rec.	FM.	k	Prec.	Rec.	FM.	k	Prec.	Rec.	FM.	k
[12] $N_{kSand.}$	0.75	0.35	0.48	2	0.55	0.32	0.40	2	0.28	0.20	0.23	4
N_{kNP}	0.44	0.45	0.44	2	0.57	0.29	0.38	1	0.50	0.23	0.31	1
N_{kNC}	0.06	0.26	0.10	2	0.04	0.02	0.03	1	0.01	0.03	0.01	10
2-Criteria Combinations (without LPS)												
$N_{kSand.} \wedge N_{kNP}$	0.77	0.34	0.47	2	0.70	0.31	0.43	2	**0.72**	0.19	0.30	8
$N_{kSand.} \vee N_{kNP}$	0.42	0.46	0.44	2	0.57	0.29	0.38	1	0.43	0.23	0.30	1
[4] $N_{kSand.} \diamond N_{kNP}$	0.77	0.34	0.47	2	0.70	0.31	0.43	2	**0.72**	0.19	0.30	8
3-Criteria Combinations (without LPS)												
$\wedge$ Combination	0.31	0.41	0.36	14	0.15	0.07	0.10	15	0.26	0.03	0.05	20
$\vee$ Combination	0.33	0.37	0.35	1	0.28	0.30	0.29	1	0.24	0.23	0.24	1
[4] $\diamond$ Comb.	0.45	0.36	0.40	6	0.16	0.34	0.22	14	0.26	0.03	0.05	20
LPS based on associative criteria only												
3 criteria	0.77	0.34	0.47	2	0.95	0.25	0.40	1	0.50	0.23	0.31	1
6 criteria	0.77	0.34	0.47	1..2	0.58	0.32	0.41	1..2	0.49	0.23	0.31	1..2
9 criteria	0.76	0.36	0.49	1..3	0.64	0.32	0.43	1..3	0.44	0.25	**0.32**	1..3
LPS based on associative criteria + the linguistic criteria S												
4 criteria	**0.84**	0.37	0.52	1	**0.96**	0.31	0.47	1	0.50	0.23	0.31	1
7 criteria	0.77	0.42	0.55	1..2	0.58	**0.40**	0.47	1..2	0.49	0.23	0.31	1..2
10 criteria	0.74	**0.48**	**0.58**	1..3	0.62	**0.40**	**0.49**	1..3	0.43	**0.27**	**0.32**	1..3

The N_{2Sand} criterion corresponding to the methodology of [12] clearly outperforms all other single criteria in terms of *precision* while N_{kNP} evidences increased *recall* compared to all other criteria for the *Vehicles* domain. Note that this situation is reversed for the *Plants* and *Food* domains, which indicates that the subsumption relation can be described differently depending on the studied domain. This is an important issue when compared to [14,16], who suppose that the is-a relation can universally be learned from WordNet. Expectedly, N_{kNC} shows poor performance due to its non-taxonomic nature.

When the best two criteria are joined into a non-guided (without LPS) aggregation strategy, results show similar performance (with slight improvements for the *Food* domain), especially for the conjunctive ($\wedge$) and pretopological ($\diamond$) [4] aggregations. However, the disjunctive ($\vee$) aggregation operator leads to worst results as the subsumption definition is not enough constrained. Note that the disjunctive and conjunctive aggregations consist in generating one new criterion from initial ones by considering respectively their union and their intersection i.e. the neighborhood family is thus reduced to a single neighborhood.

Finally, when the three criteria (including a non-performant one, i.e. N_{kNC}) are gathered in the multi-criteria framework without LPS, all aggregations fail and performance drastically drops. The difference is even higher for the *Plants* and *Food* domains, which are known to be well-structured. These experiments clearly show the incapacity of this previous model to handle unreliable criteria. The next experiments aim to evidence the superiority of the LPS strategy.

As pattern-based methods succeed in extracting reliable relations from the three domains, we performed our LPS approach in an *auto-supervised* way. In particular, for *Vehicles*, 93 relations were found corresponding to a recall of 17.6% and with a rate of 78.5% in precision as regards the reference. For *Plants*, 332 relations were foundwith a recall of 10.2% and a precision of 61.9%, and for *Food* only 244 relations were extracted, resulting in a low recall (3%) and with a small precision (36%). So, the fourth sub-table of Table 1 shows how the LPS methodology allows to learn new P-Spaces by selecting and combining more efficiently three, six or nine associative criteria and reaches slightly improved results to the ones presented by [12], which are the best up to now in terms of associative frameworks. Interestingly, higher precision is obtained to the detriment of recall, especially for the *Plants* domain.

Let us mention that if S can be used as a partial supervisor in the LPS method, it can also be used, without reserve, as a new criterion in the family of neighborhoods $\mathcal{N}$. In Table 1, the bottom part reports the scores obtained by introducing the pattern-based feature S as a supplementary criterion to consider in the construction of the combination rule. This experiment evidences the efficiency of LPS with such a mixture of pattern-based and associative criteria that makes reachable new P-Spaces inducing strongly improved taxonomies (e.g. up to 9% $F_{Measure}$ for *Vehicles* and 6% for *Plants*). To illustrate the P-Space learned by the four criteria (N_{1Sand}, N_{1NP}, N_{1NC} and S) on the *Vehicles* domain, we derive the DNF rule from the final parameters $\mathbf{w}$ and we obtain the following (simplified) expansion strategy:

$$\delta_S \vee (\delta_{N_{1Sand}} \wedge \delta_{N_{1NP}}) \vee (\delta_{N_{1Sand}} \wedge \delta_{N_{1NC}}), \tag{9}$$

which means that a subset A can be expanded with an element x if one of the following properties are satisfied:

(1) x is in relation with at least one element from A in the partial knowledge S,
(2) both neighborhoods $N_{1Sand}(x)$ and $N_{1NP}(x)$ intersect A,
(3) both neighborhoods $N_{1Sand}(x)$ and $N_{1NC}(x)$ intersect A.

(a) Obtained LT with $N_{kSand.}$ best configuration ($F_{Measure} = 0.48$ and $k = 2$)

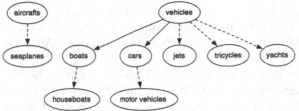

(b) Obtained LT with 10 criteria best configuration ($F_{Measure} = 0.58$)

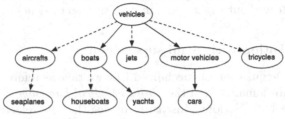

Fig. 5. Examples of induced LT with $N_{kSand.}$ and LPS for *Vehicles*.

It is also interesting to visualize well-chosen results in order to understand the different behaviours between approaches. In Figure 5, we present LT subparts respectively resulting from the best configurations of the methodology proposed by [12] and the 10 criteria LPS learning approach. In particular, continuous edges are present in S and dashed ones are learned relations for *Vehicles*.

Finally, in order to propose a meaningful evaluation, we summarize in Table 2 the comparative results with most reproducible state-of-the-art approaches[7]: [12] for the associative paradigm, [4] for the initial pretopological framework and [5] for the pattern-based approach.

Table 2. Comparison of LT acquisition methodologies on *Vehicles*, *Plants* and *Food*.

Method/Approach	Vehicles			Plants			Food		
	Prec.	Rec.	FM.	Prec.	Rec.	FM.	Prec.	Rec.	FM.
[12] associative	0.75	0.35	0.48	0.55	0.32	0.40	0.28	0.20	0.23
[4] pretopological	0.45	0.36	0.40	0.16	0.34	0.22	0.26	0.03	0.05
[5] pattern-based	**0.79**	0.18	0.29	**0.62**	0.10	0.18	0.36	0.03	0.05
LPS framework	0.74	**0.48**	**0.58**	**0.62**	**0.40**	**0.49**	**0.43**	**0.26**	**0.32**

Let us mention that in order to compare these methods within similar conditions, the final structuring approach from [5] has been performed on the partial

---

[7] Note that [16] evidence a $F_{Measure}$ of 0.82 on a non-available WordNet dataset, where training and testing are performed over the same data, thus invalidating any conclusion. Note also that [14] only present experiments for populating an existing ontology and direct comparison cannot be evidenced.

knowledge S extracted from en.wikipedia.org. Results obtained by [5] on the same datasets are higher but they are obtained using the (non-free) Yahoo!Boss search engine. Moreover, note that results of [5] are very similar to the ones of [15] who use different evaluation metrics in their paper. As a consequence, only results of [5] are reported here.

Results in Table 2 clearly reveal the benefice of mixing both statistical information and linguistic patterns within a unified learning process. The pattern-based approach obtains, as expected, better precision but significantly fails to retrieve most of the relations, whereas the LPS framework outperforms any other method on *recall* without drastic loss in *precision* so evidencing high $F_{Measure}$.

4.3   LT Acquisition with Semi-supervision

To deal with the acquisition of specialized LTs, we take as reference the concatenation of four sub-domains from the *Unified Medical Language System*[8] (UMLS) produced by the U.S. National Institutes of Health. The selected sub-domains are cardiovascular system, digestive system, nervous system and respiratory system. Their concatenation results in a list of 128 specialized terms like *upper gastrointestinal tract* or *blood-retinal barrier*.

Over this term list, none of the four considered lexical patterns retrieved any relation on en.wikipedia.org, whereas terms actually occur with an average of 2225 counts per term. A pattern-based extraction test performed on the more specialized corpus PubMed[9] has led to the same statement. Thus, we used our LPS methodology in a semi-supervised context. In particular, we simulated the input of expert knowledge S by randomly extracting relations from the reference.

Figure 6 (left) shows the performances in reconstructing the UMLS subpart with LPS as regards to the number of given external relations. The means and standard deviations on 5 trials (different sets of randomly extracted relations) are reported and the area in gray represents the benefice of the LPS process with respect to the structuring based on the external knowledge only.

First, we can notice that the obtained $F_{Measures}$ are much lower than the scores obtained on the previous general domains. This statement reveals the complexity of the task that is reinforced by the fact that UMLS is not structured with only is-a relations but also with part-of subsumptions. Despite that, we clearly observe that the acquired taxonomies take advantage of the proposed semi-supervised LPS methodology, especially in situations where the given external knowledge is lacking. This situation matches with a more realistic practical context of use.

Finally, we performed the same experiment on the Nervous system sub-domain that contains 28 terms mainly structured with is-a relations. We can observe in Figure 6 (right) the same tendency but with strongly increased benefits although on a small dataset.

---

[8] http://www.nlm.nih.gov/research/umls/
[9] http://www.nlm.nih.gov/research/pubmed/

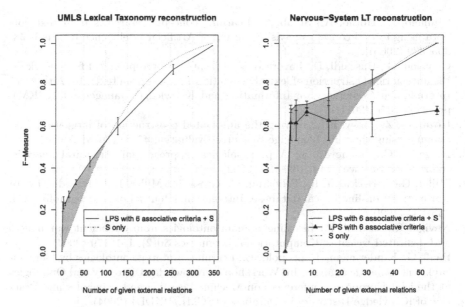

Fig. 6. Quantitative evaluation of reconstructed lexical taxonomies on the medical domain (UMLS).

5   Conclusions

In this paper, we proposed a new learning strategy to efficiently combine linguistic and statistical features for lexical taxonomy acquisition. This methodology uses the pretopological formalism into which we defined the new concept of P-Space that relies on a parameterized pseudo-closure operator formalized in a multi-criteria analysis context. Then, we developed a semi-supervised strategy called LPS to learn P-Spaces in the taxonomy induction perspective. Experiments confirmed our expectations on both general and specialized domains. In particular, significant $F_{Measure}$ improvements are obtained for the auto-supervised context when compared to recent works [5]. Moreover, where pattern-based methodologies [5,15] fail to learn LTs due to the absence of pattern evidences (usually for specialized domains), the introduction of external knowledge combined with statistical features allows the construction of LTs with reasonable accuracy.

References

1. Belmandt, Z.T.: Basics of Pretopology. Hermann (2011)
2. Bordea, G., Buitelaar, P., Faralli, S., Navigli, R.: Semeval-2015 task 17: Taxonomy extraction evaluation (texeval). In: Proceedings of the 9th International Workshop on Semantic Evaluation. Association for Computational Linguistics (2015)

3. Cimiano, P., Hotho, A., Staab, S.: Learning concept hierarchies from text corpora using formal concept anaylsis. Journal of Artificial Intelligence Research **24**, 305–339 (2005)
4. Cleuziou, G., Buscaldi, D., Levorato, V., Dias, G.: A pretopological framework for the automatic construction of lexical-semantic structures from texts. In: 20th ACM International Conference on Information and Knowledge Management (CIKM), pp. 2453–2456 (2011)
5. Kozareva, Z., Hovy. E.: Tailoring the automated construction of large-scale taxonomies using the web. Language Resource Evaluation **47**(3) (2013)
6. Largeron, C., Bonnevay, S.: A pretopological approach for structural analysis. Information Sciences **144**, 169–185 (2002)
7. Miller, G.A., Beckwith, R., Fellbaum, C., Gross, D., Miller, K.J.: Introduction to wordnet: An on-line lexical database. International Journal of Lexicography **3**(4), 235–244 (1990)
8. Navigli, R., Velardi, P.: Learning domain ontologies from document warehouses and dedicated websites. Computational Linguistics **30**(2), 151–179 (2004)
9. Paaß, G., Kindermann, J., Leopold, E.: Learning prototype ontologies by hierachical latent semantic analysis. In: Workshop on Knowledge Discovery and Ontologies at the joint European Conferences on Machine Learning and Principles and Practice of Knowledge Discovery in Databases (ECML/PKDD) (2004)
10. Pereira, F., Tishby, N., Lee, L.: Distributional clustering of english words. In: 31st Annual Meeting on Association for Computational Linguistics (ACL), pp. 183–190 (1993)
11. Van Rijsbergen, C.J.: Information Retrieval, 2nd edn. Butterworth-Heinemann, Newton (1979)
12. Sanderson, M., Croft, B.: Deriving concept hierarchies from text. In: 22nd Annual International ACM SIGIR Conference on Research and Development in Information Retrieval (SIGIR), pp. 206–213 (1999)
13. Scrucca, L.: Ga: A package for genetic algorithms in r. Journal of Statistical Software **53**(4), 1–37 (2013)
14. Snow, R., Jurafsky, D., Ng, Y.A.: Semantic taxonomy induction from heterogenous evidence. In: 21st International Conference on Computational Linguistics and 44th Annual Meeting of the Association for Computational Linguistics, pp. 801–808 (2006)
15. Velardi, P., Faralli, S., Navigli, R.: OntoLearn Reloaded: A Graph-Based Algorithm for Taxonomy Induction. Computational Linguistics **39**(3), 665–707 (2013)
16. Yang, H., Callan, J.: A metric-based framework for automatic taxonomy induction. In: Joint Conference of the 47th Annual Meeting of the ACL and the 4th International Joint Conference on Natural Language Processing of the AFNLP, pp. 271–279 (2009)

Multidimensional Prediction Models
When the Resolution Context Changes

Adolfo Martínez-Usó$^{(\boxtimes)}$ and José Hernández-Orallo

DSIC, Universitat Politècnica de València, Camí de Vera s/n, 46022 València, Spain
{admarus,jorallo}@dsic.upv.es

Abstract. Multidimensional data is systematically analysed at multiple
granularities by applying aggregate and disaggregate operators (e.g., by
the use of OLAP tools). For instance, in a supermarket we may want to
predict sales of tomatoes for next week, but we may also be interested
in predicting sales for all vegetables (higher up in the product hierarchy)
for next Friday (lower down in the time dimension). While the domain
and data are the same, the *operating context* is different. We explore
several approaches for multidimensional data when predictions have to
be made at different levels (or contexts) of aggregation. One method
relies on the same resolution, another approach aggregates predictions
bottom-up, a third approach disaggregates predictions top-down and a
final technique corrects predictions using the relation between levels. We
show how these strategies behave when the resolution context changes,
using several machine learning techniques in four application domains.

Keywords: Multidimensional data · Operating context aggregation ·
Disaggregation · OLAP cubes · Quantification

1   Introduction

Most existing algorithms in machine learning only manipulate data at an individual level (flat data tables), not considering the case of multiple abstract levels
for the given data set. However, in many applications, data contains structured
information that is multidimensional (or multilevel) in nature, such as retail-
ing, geographic, economic or scientific data. The multidimensional model is a
widely extended conceptual model originated in the database literature that
can be used to properly capture the multiresolutional character of many data
sets [1,5,13,26]. Multidimensional databases arrange data into fact tables and
dimensions. A fact table includes instances of facts at the lowest possible level.
Each row represents a fact, such as "The sales of product 'Tomato soup 500ml'
in store '123' on day '20/06/2014' totalled 25 units". The features (or fields) of
a fact table are either measures (indicators such as units, euros, volumes, etc.)
or references to dimensions. A dimension is here understood as a particular variable that has predefined (and hopefully meaningful) levels of aggregation, with
a hierarchical structure.

© Springer International Publishing Switzerland 2015
A. Appice et al. (Eds.): ECML PKDD 2015, Part II, LNAI 9285, pp. 509–524, 2015.
DOI: 10.1007/978-3-319-23525-7_31

Figure 1 shows several examples of dimensions and hierarchies. Using the hierarchies, the data can be aggregated or disaggregated at different granularities. Each of this set of aggregation choices for all dimensions is known as a *data cube* [6], which provides an easy understanding and offers flexibility for visualisation (aggregated tables and cubes). OLAP technology, for instance, has been developed to handle large volumes of multidimensional data in a highly efficient way, and moving through the space of cubes by the use of roll-up, drill-down, slice&dice and pivoting operators.

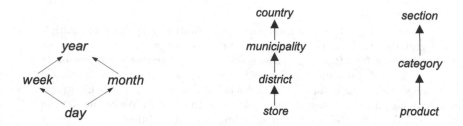

Fig. 1. Examples of dimension hierarchies. Left: Time dimension, Middle: Location dimension, Right: Product dimension.

Despite the success of multidimensional schemas and its widespread use for data warehouses for about two decades, a full integration of machine learning and multidimensional datasets has not taken place. Even in business intelligence tools, which aim at integrating data warehouses, OLAP technology and data mining tools, the usual procedure is to select a cube using an OLAP query or operator, and derive a view from it. Next, this 'minable view' is transferred to the data mining tool to apply machine learning or statistical techniques to this flat, traditional view of the data.

When we analyse the problem more carefully, we see that the main issue for a successful integration is that we would like to use off-the-shelf machine learning techniques but taking full potential of the hierarchical information. Machine learning models are not designed to take hierarchical attributes. Consequently, we need to do something different whenever the cube we want to predict for changes. In other words, the predictions for tomatoes and weeks will be different than the predictions for vegetables and Fridays. These two situations represent *operating contexts*. In principle, a model that has been obtained for one context cannot be *directly* applied to a different context.

This leads us to two major alternatives. On the one hand, we can learn one model for each operating context and apply it for that level of aggregation, which means *retraining* the model for each operating context. On the other hand, we can learn one, more versatile, model at the lowest operating context (highest resolution) and then aggregate their predictions, as in a quantification problem [2,3,11]. This second point of view results in *reframing* the model for each operating context. In addition to these major views, it is worth exploring other models such as *disaggregation*, where a reframing philosophy is addressed in the opposite way, that is, working at an upper level (lower resolution) and then disaggregating the predictions until the working level of granularity is reached. Finally, there also exists the possibility of *correcting the predictions* worked out for each operating context by means of using the coarse information from upper levels of granularity for improving (correcting somehow) finer predictions, as done by [10] in a multilevel (but not multidimensional) scenario.

In this paper we analyse all these approaches systematically with several machine learning techniques in four different application environments.

The rest of the paper is organised as follows. Section 2 formalises the notion of multidimensional context and properly defines the two main approaches that we will study: the same-level (retraining) approach and the low-level (reframing) approach. Furthermore, the disaggregation model and the same-level correction model are also defined in this section. Section 3 discusses how datamarts have to be understood when models are required to predict some of the measures of the fact table and also states some measurement considerations. Section 4 presents the techniques, datamarts and error measure that will be used in the experiments. After that, results for each approach are analysed. Section 5 discusses some related work and section 6 closes the paper with some take-away messages and some future work.

2   Multidimensional Contexts

We consider a multidimensional data set D (or datamart) of schema $\langle X, Y \rangle$ where $X = \{X_1, \ldots, X_d\}$ is the set of d dimensions (used as predictor attributes or features) and Y, which is the *target attribute* (one measure or indicator that can be numeric or nominal). We use D_A to denote the projection of dataset for attribute A. Note that datasets and projections are multisets (i.e., they can have repeated values). Each dimension X_i has an associated hierarchy $h(X_i)$ of m_i elements or *levels* $\{X_i^{(1)}, \ldots, X_i^{(m_i)}\}$ with a strict partial order $<$. In this paper we will assume that hierarchies are linear, so the partial order becomes a total order from the lowest level $X_i^{(1)}$ to the highest level $X_i^{(m_i)}$. This is not a strong restriction, as a non-linear dimension can be converted into several linear dimensions (one for each possible pathway in the lattice). For instance, if $X_2 = $ location, as in Figure 1 (middle), we have $X_2^{(1)} = $ store, $X_2^{(2)} = $ district, $X_2^{(3)} = $ municipality and $X_2^{(4)} = $ country with store $<$ district $<$ municipality $<$ country and their transitive closure. We will consider that the top level m_i for every hierarchy is all-i, such that for every $l \in h(X_i)$, $l < $ all-i. Non-hierarchical attributes are just special cases, by just considering that $m_i = 2$ (the bottom and the top all-i level). These dimensions then just become regular attributes but with the possibility of aggregating them to the top level all.

Each level $X_i^{(j)}$ of a hierarchy $h(X_i)$ has an associated domain $\mathcal{X}_i^{(j)}$, which can be nominal or numeric. We will assume that there are no levels with the same name in the same or different hierarchies. In this way, if the name of a level is name then we can just refer to the level by $X^{(\text{name})}$ and the associated domain by $\mathcal{X}^{(\text{name})}$. For instance, the domain of the level country for dimension location, i.e., $\mathcal{X}_2^{(4)}$, or $\mathcal{X}^{(\text{country})}$, might be the set with values {UK, Spain, France}. For every pair of consecutive levels $X_i^{(j)}$ and $X_i^{(j+1)}$ in a hierarchy we define a regrouping function ϕ_i^j between the values of $X_i^{(j)}$ to the values of $X_i^{(j+1)}$. For instance, $\phi_2^3(\text{Valencia}) = \text{Spain}$. We denote by $\phi_i^{j:k}$, with $j \leq k$ the successive application of ϕ from j to k, i.e., $\phi_i^{j:k}(v) = \phi_i^k(...\phi_i^{j+1}(\phi_i^j(v))...)$. Given a value v at a level $X_i^{(k)}$ of the dimension i, we denote by $\perp(v)$ the set of all the values at the lowest level of that hierarchy that belongs to, i.e., $\{w \in \mathcal{X}_i^{(1)} \mid \phi_i^{1:k}(w) = v\}$. For instance, $\perp(\text{Valencia})$ would be all the stores of all the districts of Valencia.

Definition 1. *A multidimensional operating context or resolution is a d-tuple of levels $\langle l_1, \ldots, l_d \rangle$, with each $l_i \in h(X_i)$. A multidimensional context determines the level for every dimension of the dataset.*

Definition 2. *Given a multidimensional context, a selection of D at a context $\langle l_1, \ldots, l_d \rangle$ with values $\langle v_1, \ldots, v_d \rangle$ is defined as follows:*

$$\sigma_{[l_1=v_1,\ldots,l_d=v_d]}(D) \triangleq \{\langle x_1, \ldots, x_d, y \rangle \mid x_1 \in \perp(v_1), \ldots, x_d \in \perp(v_d)\} \quad (1)$$

For instance, if we have three dimensions $X_1 = $ product, $X_2 = $ location and $X_3 = $ time, and Y is representing units, we could select all the facts for the context $\langle \text{item}, \text{municipality}, \text{year} \rangle$ with values tomato, Valencia and 2013 respectively with $\sigma_{[\text{item}=\text{Tomato},\text{munic.}=\text{Valencia},\text{year}=2013]}(D)$.

Finally, we can define an aggregation operator as follows:

Definition 3. *Given an aggregation function, agg, as a function from sets to real numbers, the aggregation of a datamart D for a context $\langle l_1, \ldots, l_d \rangle$ is defined as follows:*

$$\gamma_{[l_1,\ldots,l_d]}^{agg}(D) \triangleq \{\langle x_1, \ldots, x_d, z \rangle \mid x_1 \in \mathcal{X}^{(l_1)}, \ldots, x_d \in \mathcal{X}^{(l_d)},$$

$$z = agg(\{y \mid \langle v_1, \ldots, v_d, y \rangle \in \sigma_{[l_1=x_1,\ldots,l_d=x_d]}(D)\})\}$$

The above aggregation is extended for unlabelled datasets with no y attribute.

For instance, $\gamma_{[\text{item},\text{municipality},\text{year}]}^{sum}(D)$ returns all the tuples for each possible combination of values at the level item in the dimension product, at the level municipality in the dimension location and at the level year in the dimension time, where the output variable is constructed by summing all the y of the corresponding rows according to the hierarchies.

Given the above notation, now we consider a predictive problem from X to Y. For instance, how many tomatoes we expect to sell in Valencia next week? Assuming we have a training dataset, how would we train our model? As a first idea, it seems reasonable to aggregate the training data using the γ operator above for the context $c = \langle \text{item}, \text{municipality}, \text{week} \rangle$, producing a model M

that will be applied to the deployment data with the same context. However, if some time later we are interested in predicting sales for all vegetables for next Friday, what would we do? We could aggregate the training data for the context $c' = \langle$category, municipality, day$\rangle$, learn a new model M' and predict for the deployment data. This is what we see in Figure 2 (top). We refer to this approach as the *Same-Level* (SL) approach or the *retraining* approach.

Definition 4. *Given a training data T with measure Y and a deployment data D, a model learnt for measure Y at the* same *level, denoted by* SL*, in context $\langle l_1, \ldots, l_d \rangle$ is defined as follows. We first aggregate T for that context, i.e., $T^{\Delta} \triangleq \gamma^{agg}_{[l_1,\ldots,l_d]}(T)$. Then we train a model $M^{\Delta} : X^{\Delta} \to Y^{\Delta}$. For the deployment data D we also aggregate the original data as $D^{\Delta} \triangleq \gamma^{agg}_{[l_1,\ldots,l_d]}(D)$. We finally add an attribute $\hat{Y}$ to D^{Δ} by setting it equal to the predictions of the model M^{Δ} for each of these aggregated rows, so producing $\hat{D}^{\Delta}$. This yields pairs $\langle X, \hat{Y} \rangle$ at that context.*

An alternative approach goes as follows. Consider that we train a predictive model M for the lowest level in D. Once a new multidimensional appears, we apply the model to the deployment data and *aggregate the predictions*. With this approach, one model is used for every possible context. This is illustrated in Figure 2 (middle). We refer to this approach as the *Lowest-Level* (LL) approach or the *reframing* approach.

Definition 5. *Given a training data T with measure Y and a deployment data D, a model learnt for measure Y at the* lowest *level, denoted by* LL*, and deployed at a context $\langle l_1, \ldots, l_d \rangle$ is defined as follows. We first train a model $M : X \to Y$ for the whole training dataset T. Now, for each row at the lowest level in the deployment data D we apply M. We add a new attribute $\hat{Y}$, and set it to the result of the model for each row, giving a new dataset $\hat{D}$. Finally, given an aggregation function agg, we now calculate the predictions for a context $\langle l_1, \ldots, l_d \rangle$ as $\hat{D}^{\Delta} \triangleq \gamma^{agg}_{[l_1,\ldots,l_d]}(\hat{D})$, which produces pairs $\langle X, \hat{Y} \rangle$ at that context.*

Another alternative is to disaggregate predictions from a higher level of granularity. Figure 2 (bottom), trains a predictive model M for a higher level and keeps the frequencies or proportions that are shared on the lower levels for the training set. Then with a new prediction at a higher level, the frequencies are used for the disaggregation. We refer to this approach as the *Disaggregation* (dAg) approach. Figure 3 shows a disaggregation example taking into account the two first levels of the product dimension from Fig.1. This example takes into account one dimension and only one level of disaggregation. However, it could be extended to more than one level of disaggregation, for instance from section to product in the same dimension (2 levels), and for more than one dimension, for instance taking into account dimensions product and location, which would also imply working out all the combinations. Nonetheless, for simplicity, in the rest of the paper, we have limited the disaggregation approach to the following setting:

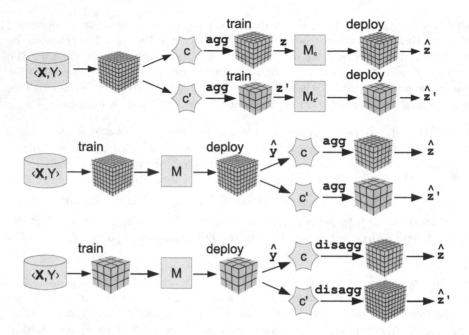

Fig. 2. Retraining (*Same-Level* approach), reframing (*Lowest-Level* approach) and disaggregation for two different multidimensional contexts c and c'. Retraining (top) needs to convert the training data for the two contexts c and c' and then aggregating the output into z or z' respectively. Two models M_c and $M_{c'}$ are learnt (one for each context) at the same level the predictions must be done. Reframing (middle) shows how the training data is used just once at the lowest level to create a single model M that is applied to different operating contexts c or c' by aggregating the outputs appropriately. Disaggregation (bottom) shows how the training data is used just once at a higher level to create a single model M that is applied to different operating contexts c or c' by disaggregating the outputs appropriately.

- Only one level per dimension has been disaggregated for each result, that is, we just disaggregate to the level immediately below.
- All the dimensions are taken into account, although we only disaggregate one dimension for each result.

Definition 6. *Given a training data T with measure Y and a deployment data D_{lo}, a model learnt for measure Y using* disaggregation, *denoted by* dAg, *and deployed at a context $\langle l_1, \ldots, l_d \rangle$, is defined as follows. Let us suppose the context $\langle l'_1, \ldots, l'_d \rangle$ as the context at the level immediately below to $\langle l_1, \ldots, l_d \rangle$. We first aggregate T for the upper context, i.e., $T_{hi}^{\Delta} \triangleq \gamma_{[l_1,\ldots,l_d]}^{agg}(T)$ and for its immediately context below, i.e., $T_{lo}^{\Delta} \triangleq \gamma_{[l'_1,\ldots,l'_d]}^{agg}(T)$. We define a function F that counts the number of observations that fall into each of the disjoint categories. This is done as an aggregation function as in Definition 3 but using count as the aggregation function. Let us also suppose that we have predictions for the deployment data D_{hi}^{Δ} at the high level using the* SL *approach. Now, for each row at the lower*

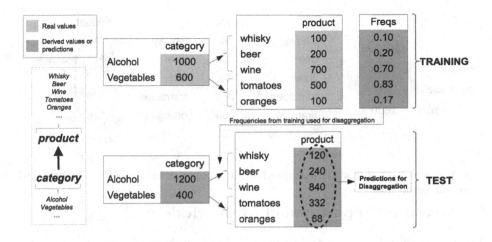

Fig. 3. Disaggregation example for the product dimension. The frequency of each product (whisky, beer, tomatoes, etc.) within each category (Alcohol, Vegetables) is learned during the training. These frequencies are then applied to the predictions made for a higher level of granularity resulting in the disaggregated predictions for the lower level. Light blue cells represent real values whereas light red represent derived values (such as frequencies) or predictions.

context in the deployment data D_{lo}^A we apply F on the predictions for D_{hi}^A to create the predictions for the lower level.

Finally, in [10], the authors presented a method for improving the current predictions using the coarse information from upper levels of granularity. Their methodology uses the approximation values of the aggregated targets (in our case the predictions) and the predictions of the individual targets for producing new modified predictions. Figure 4 shows an example of this procedure taking into account the two first levels of the product dimension from Fig.1. As it can be seen, we work out a correction ϵ as the relation between the sum of the predictions at the category level and the sum of the predictions at the product level, using then this value for uniformly distributing the differences among the predictions of the lower level. This approach is actually a correction of the same-level model and we thus refer to this approach as the *same-level correction* model (SLc).

Definition 7. *Let us consider the deployment data at two different contexts, D_{lo} and D_{hi}, where deployment data D_{lo} is defined at the context immediately below to D_{hi}. $\hat{D}_{lo}^A$ and $\hat{D}_{hi}^A$ have been produced applying the SL model to each deployment data respectively and, therefore, $\hat{Y}_{lo}$ and $\hat{Y}_{hi}$ attributes with predictions can be found for each context respectively. Thus, the same-level correction model, denoted by SLc, is defined as follows. We first work out $s\hat{Y}_{lo} = \sum_i \hat{Y}_{lo}[i]$ and $s\hat{Y}_{hi} = \sum_i \hat{Y}_{hi}[i]$ and let us define $\epsilon = \frac{s\hat{Y}_{hi}}{s\hat{Y}_{lo}}$. For the deployment data D_{lo} we correct the attribute $\hat{Y}_{lo}$ for each row by means of multiplying each $\hat{Y}_{lo}$ value by ϵ.*

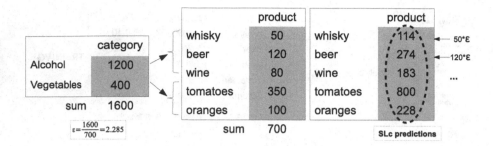

Fig. 4. Same-level correction example for the product dimension. Cells in light red represent predictions.

3   Measure Properties and Mean Models

The first thing we need to consider is the kind of machine learning tasks that are common with multidimensional data. The way the information is arranged in a multidimensional schema, with a fact table containing *measures* suggests that many machine learning tasks, especially predictive ones, are usually focussed on predicting the measures. For instance, if facts are sales, consumptions, failures, usages, etc., it is common to become interested in predicting some of the measures in these tables (e.g., units, dollars, hours, etc.) from past data. As measures are usually numerical, many problems will turn out to be regression problems. Nonetheless, some measures can be nominal, such as whether a purchase has been satisfactory or not. In that case, however, the measure becomes a percentage, i.e., a number, when we aggregate, so binary nominal measures can also be taken as numbers.

The time dimension is found in most datamarts. In a predictive scenario, the time dimension becomes slightly special: predictions are about future facts, so training is usually performed with available data up to a given time and the model is then used to extrapolate from that point on (next week, next month, next year, depending on the resolution). This occurs in three out of four datamarts used in this work. Some other databases are sparse and only collect positive cases (e.g., 0 sales are not included) and have to be preprocessed to contain this information for them to be meaningful.

Another important issue about multidimensional schemas is whether the measures we want to predict are *additive*, *semi-additive* or *non-additive*. A measure is *additive* when, for any dimension and any set of values S at level j that we want to aggregate up to level $n + k$, the summation of these values using any partition of S is equivalent, i.e., gives consistent results. For instance, units sold in a supermarket aggregate well for all dimensions, i.e., the result is independent of the way it has been aggregated. However, percentages do not aggregate well, as the denominator is not known when performing the aggregation. Therefore percentages are *non-additive*. Finally, the term *semi-additive* measure is used for those measures that aggregate well for some dimensions but not for others. For instance, measures that accumulate or depend on the state, such as stock levels are usually semi-additive.

The aggregated function that is used for aggregating datamarts, as in Definition 3, does not have to always be $sum(S) \triangleq \sum_{s \in S} s$. For instance, it could be an average, $avg(S) \triangleq \frac{sum(S)}{|S|}$. Some functions just work for some measures. For instance, consumption (e.g., in kWh) can be aggregated by averaging it. However, we have to be very careful about how this aggregation is performed. For instance, avg is not *composable*.

In regression tasks, we usually look at a baseline method that consists in averaging the values for the training data and apply these values systematically during deployment. This is known as the *mean* or *constant* model. In this work, we define our baseline method as follows:

Definition 8. *Given a training data T with measure Y and a deployment data D, the MEAN model for measure Y at the same level in context $\langle l_1, \ldots, l_d \rangle$, denoted by SL.MEAN, is defined as follows. We first aggregate T for that context, i.e., $T^\Delta \triangleq \gamma^{agg}_{[l_1, \ldots, l_d]}(T)$. Then we calculate $\overline{Y^\Delta} \triangleq avg(T^\Delta{}_Y)$, the average of the measure Y for this context. For the deployment data D we also aggregate the original data as $D^\Delta \triangleq \gamma^{agg}_{[l_1, \ldots, l_d]}(D)$. We finally add an attribute $\hat{Y}$ to D^Δ by setting it equal to $\overline{Y^\Delta}$ for every row in D^Δ, so producing $\hat{D}^\Delta$. This produces pairs $\langle X, \hat{Y} \rangle$ at that context.*

4 Experimental Setting and Results

The MEAN approach is useful as a baseline, but we of course are interested in the use of machine learning methods to get good predictions. We have also considered other four techniques: LRW (linear regression using RWeka in R [14,22]), M5P (regression tree using RWeka), SVM (package e1071 in R, linear kernel) and KNN (package kknn in R). The datamarts used are now presented:

- **GENOMICS**: Originally, this human genome dataset contains genomic data (HGDB) from several public and private research databases, including information about genes, chromosomes, mutations, diseases, etc. structured in 20 (numerical and nominal) attributes [20]. Data goes from years 1970 to 2012 and the output variable is the number of variations. We converted it into a multidimensional datamart, where each fact showed the number of variations according to five different dimensions (hierarchies in parenthesis): SPEC (Eff < All), GENOTYPE (ID < Chrom < All), PHENOTYPE (Name < ICD10 < ICD10.Cat < All), DBANK (Dbnk < All) and DATE (Year). Note that as we use the DATE dimension to split the data we only consider one level here. The number of possible multidimensional contexts is then $2 \times 3 \times 4 \times 2 \times 1 = 48$.
- **AROMA**: This is an artificial dataset constructed from IBM sales information. It contains sales data for coffee and tea products sold in stores across the United States [16]. The data is almost directly converted into a multidimensional datamart where each fact describes the sales of products using two measures (units and dollars, although we will only use dollars as the

output variable) according to five dimensions (hierarchies in parenthesis): PROMO (KeyPromo < PromoType < All), CLASS (KeyClass < All), PRODUCT (KeyProduct < All), STORE (KeyStore < KeyMKT < MKT-HQ-City < MKT-HQ-State < MKT-District < MKT-Region < All) and PERIOD (Year). Note that as we use the PERIOD dimension to split the data we only consider one level here. Data goes from years 2004 to 2006 and the number of possible multidimensional contexts is $3 \times 2 \times 2 \times 7 \times 1 = 84$.

– **CARS**: This is a dataset for car fuel consumption and emissions which is created as a reduced representation of [9] (some attributes are removed) in order to construct a datamart. It describes fuel consumption in cars from years 2000 to 2013, being published by the UK's Vehicle Certification Agency (VCA). The target variable is car fuel emissions (CO_2) and we have six dimensions (hierarchies in parenthesis): CAR (Man.Model.Description < Man.Model < Manufacturer < All), ENGINE (EngineCapacity < All), TRANS (Transmission < TransType < All), EURO (EuroSTD < All), FUEL (FuelType < All) and TIME (Year). Note that as we use the TIME dimension to split the data we only consider one level here. The number of possible multidimensional contexts is $4 \times 2 \times 3 \times 2 \times 2 \times 1 = 96$.

– **UJIndoorLoc**: This is a dataset for benchmarking indoor localisation algorithms [25]. UJIndoorLoc contains the Wi-Fi access points readings for all the spaces (offices, laboratories, etc.) of the School of Technology and Experimental Sciences of the University Jaume I. The average signal intensity has been used as the target variable ($INTENSITY$). We have three dimensions in this case (hierarchies in parenthesis):WHERE (Space < Floor < Building < All), TIME (Hour < Day < WeekDay < All), PHONE (Model < Manufacturer < All), being the number of possible multidimensional contexts $4 \times 4 \times 3 = 48$.

We split GENOMICS, AROMA and CARS datasets into training and test on the basis of a *split-year*. Parameter *split-year* has been set to 2006 for all the datasets, being the *split-year* included in the test set. On the other hand, for UJIndoorLoc dataset, the authors already provide the training and evaluation subsets (see [25]). No rows with zeros were added to CARS and UJIndoorLoc datasets, as missing cases are just absence of information. The target variable is a ratio, so the aggregation function that makes sense for these datasets is *avg*, which is neither additive nor associative. Finally, we clip the predictions of all methods to 0 if they are negative, as in the four datamarts the measures cannot be negative. This is important for methods that could potentially predict negative values such as M5P or LRW.

As the four datamarts have led to regression problems, we may use the Mean Squared Error (MSE) as the error measure. However, for the two datamarts that use *sum* as aggregating function, the magnitude of the error will be much higher for highly aggregated contexts, and the values will be difficult to compare. A good way of getting rid of this problem is to divide SE (or MSE) by the SE (or MSE) of the MEAN model. Interestingly, in a classical regression setting, the MSE of the MEAN model equals its error variance. So, actually, what we are doing is to show the *MSE* by some kind of error variance. We use the SL.MEAN model, as it ensures that it is constant for the deployment multidimensional context.

Definition 9. *The normalised squared error (NSE) of a technique* TECH *is defined as* $\frac{MSE(\text{TECH})}{MSE(\text{SL.MEAN})}$.

Table 1. Comparison *NSE* among LL, SL, dAg and SLc.

	GENOMICS				AROMA				CARS				UJIndoorLoc			
	SL	LL	dAg	SLc	SL	LL	dAg	SLc	SL	LL	dAg	SLc	SL	LL	dAg	SLc
MEAN	1.00	0.57	0.46	0.62	1.00	0.54	0.28	0.53	1.00	0.92	1.39	1.73	1.00	0.97	2.51	6.99
SVM	0.49	0.50	0.44	0.63	0.11	0.03	0.02	0.05	0.73	0.46	1.25	1.64	1.02	1.21	2.49	7.00
M5P	0.92	0.14	0.42	0.85	0.87	0.04	0.34	0.40	0.79	0.77	1.31	1.62	1.06	2.81	2.49	7.07
LRW	1.02	0.58	0.44	0.82	1.09	0.67	0.33	0.56	0.84	0.92	1.31	1.66	1.07	0.82	2.49	7.05
KNN	0.87	0.08	0.36	0.71	1.07	0.03	0.27	0.34	0.79	0.48	1.51	1.88	1.08	1.31	2.53	6.95

Table 2. Rank summary for the four datasets.

	GENOMICS				AROMA				CARS				UJIndoorLoc			
	SL	LL	dAg	SLc	SL	LL	dAg	SLc	SL	LL	dAg	SLc	SL	LL	dAg	SLc
MEAN	5.00	2.80	1.53	3.97	5.00	3.22	1.97	3.57	2.37	1.66	3.86	2.62	1.88	1.49	3.18	3.81
SVM	3.48	3.31	1.58	4.83	4.34	3.37	2.11	3.71	2.48	1.06	3.73	3.33	1.38	2.39	2.81	3.85
M5P	4.74	1.00	2.80	4.02	5.00	1.01	2.95	3.83	2.35	1.83	3.67	2.78	1.34	3.24	2.38	3.65
LRW	4.98	2.66	1.71	3.94	5.00	3.92	1.89	3.01	2.07	2.48	3.50	2.67	2.30	1.09	3.05	3.96
KNN	4.58	1.00	2.53	4.28	5.00	1.00	2.88	3.74	2.20	1.15	3.84	3.40	1.43	2.28	2.93	3.81
Overall	4.56	2.15	2.03	4.21	4.87	2.50	2.36	3.57	2.29	1.64	3.72	2.96	1.67	2.10	2.87	3.82

Table 1 compares all the techniques for the four datasets in terms of the normalised squared error (*NSE*). For the GENOMICS and AROMA datasets we see that the LL and dAg approaches are better. LL continues its good behaviour for CARS and UJIndoorLoc, being clearly the best one (or close to the best one when it is the second), however a very different picture happens for the disaggregation strategy, which shows a quite poor performance on these datamarts, especially on UJIndoorLoc.

In order to offer a more comprehensive perspective of the results, Table 2 rank all the strategies within each technique for each dataset individually. The lower the better on these ranks, being the ranks averaged when they tie.

Focusing on the ranking results, for the AROMA and CARS datasets we see that the LL approach is better, obtaining a very good ranking when the KNN technique is used. Moreover, LL obtains the second best rank in the other datasets. SL obtains the best rank in UJIndoorLoc dataset. However, it shows a quite poor performance for GENOMICS and AROMA. dAg obtains good results in GENOMICS (the best) and AROMA, however its rank is not so good for the other datasets.

Finally, as mentioned in Sect.2, there exist many operating contexts where several dimensions could be disaggregated at the same time. Instead of just comparing the results when only one dimension is disaggregated, we could have taken into account all the possible disaggregating operations for that context and averaged the results (as an ensemble) for comparing them to the SL and LL approaches. This option, called dAg^{avg}, was analysed experimentally and showed worse results in general. Table 3 shows the overall rankings for each strategy and for each dataset when all the possible dimensions in each cube have been taken into account by means of averaging their predictions. The overall rankings for the dAg^{avg} methodology are always worse than the ones shown in Table 2 except for the CARS dataset whereas in UJIndoorLoc dataset, dAg and dAg^{avg} practically obtain the same rank.

Table 3. Summary results for all the strategies and for each dataset when all the disaggregation operations for each cube have been performed and their results have been averaged.

	SL	LL	dAg^{avg}	SLc
GENOMICS	3.70	1.93	2.42	4.37
AROMA	3.60	1.60	3.10	4.39
CARS	2.12	1.40	3.63	3.54
UJIndoorLoc	1.49	1.84	2.89	4.12

5   Related Work

As we mentioned in the introduction, the efforts for a full integration of data mining and OLAP tools have not been as common as originally expected. There are, though, some significant contributions for descriptive models. For instance, multidimensional association rules were firstly introduced in [17] and, since then, some related approaches have appeared in areas such as hierarchical association rules, subgroup discovery, granular computing [18] and others [7].

'Prediction cubes' [7,8], despite the term 'predictive' in their name, actually perform subgroup discovery or exploratory mining [23], where we want to have a metric (e.g., predictive accuracy) for a model on a given subset of the data (a cell in a cube) and see whether some cells have different metrics than others (hence being special). It is important to note that "Prediction Cubes" are not meant to aggregate outputs. They are not actually used to make predictions at several resolution levels of values that are unknown. In fact, they always work with a labelled test set to which they compare to get the metrics.

When looking at predictive modelling, the usual approach in the literature has been the *same level* approach (i.e., generating a view for the resolutions at hand). There is no versatile model that can work for the whole hierarchy in every dimension. A significant exception is the area of multilevel modelling (MLM) [4,12], also known as hierarchical (linear) modelling (HLM) [24], among other names. This is an extension of linear, and non-linear, models such that the variables are measured at different levels of a global, usually linear, hierarchy. The first and key difference between a multilevel modelling problem and a multidimensional problem is hence that in the latter all the measurements take place at the lowest level (e.g., they come from facts in a multidimensional data warehouse). However, in multilevel modelling, measurements may take place at any level. As a result, in multilevel modelling, putting all the variables at the lowest level does not make sense, as it means that some of the input variables would have to be disaggregated (or repeated). The second difference is that in multilevel modelling hierarchies apply to *all* attributes. In other words, there is an orthogonal hierarchy, which can be applied to each attribute, depending on the level at which the value has been measured. So it is not actually applicable to a multidimensional database, where each attribute can be aggregated independently. The third difference is that in multilevel modelling the predictions are still made generally at the lowest level. In a multidimensional setting we want predictions at whatever level of aggregation. In addition, multilevel modelling has usually been addressed by linear (and occasionally non-linear) regression models with several assumptions about normality, homoscedasticity, independence, etc. Despite the differences, in [19], multilevel models are applied to a datamart. However, we still see a separate concept hierarchy that is applied to all dimensions, instead of having a particular hierarchy for each dimension, as usual in datamarts and OLAP tools, so it is not actually a multidimensional database.

Some connections have also been found with the work of Perlich and Provost in [21], where the authors introduce new aggregates that capture more information about the distributions. That is, instead of using simpler aggregates such as the MEAN, MODE or SUM, this work presents novel aggregates that empirically improve predictive modelling in high-dimensional (categorical) domains. The setting of hierarchies is different to our multidimensional setting, and our approach considers a given hierarchy where the natural aggregates in our case are MEAN and SUM. Nonetheless, it could be worth studying where more aggregates or statistics about the distributions (e.g., SKEW, VAR) could help to correct the MEAN and SUM aggregates.

About the disaggregation approaches, the dAg and SLc models were inspired by [10]. Actually, they do not disaggregate as it can be understood in a cube-space data mining [23], but this work made us realise that the disaggregation approach is not properly covered within the OLAP-style multidimensional data analysis. Our proposed SLc model is not equivalent to the approach presented in [10], since in their work authors needed the *exact*, or at least *accurate enough*, values for the aggregated target.

As a result, the problem of having several hierarchies, one for each dimension and seeing the problem (including predictions) at any possible resolution, is new. Also there is no general approach about how to apply any data mining technique to this kind of problem (and not only linear regression models or non-linear variants). So, the multidimensional approach presented in this paper is more general in at least these two aspects.

6   Conclusions and Future Work

Multidimensional data is a rich and complex scenario where the same task can change significantly depending on the level of aggregation over some of the dimensions. This is the 'multidimensional context'. The approaches we have analysed are very general, and applicable to any set of off-the-shelf machine learning techniques. Three of the approaches can be considered as retraining approaches, whereas the LL approach the only reframing approach. From this distinction, we see that resources are an important criterion, as retraining a model again and again may become infeasible for some applications, and reframing a single, versatile model may be a much better option in cost-effective terms. Also the results are generally better for the LL approach. It may be the case that there are some criteria to choose the best option at each granularity. From our analysis, however, we have not found any clear pattern to make a different take-away recommendation other than the LL approach. In order to facilitate repeatability of the experiments, the software associated to this work is available at http:// users.dsic.upv.es/~admarus/sw.html.

This work suggests many avenues for future work. One area we are undertaking is a modification of the LL approach where the aggregation function is substituted by a quantification procedure [2,11]. As quantification is able to correct some aggregation problems, we hope some quantification techniques (especially those for regression using crisp regression models [3] or soft regression models [15]) to be beneficial for the LL approach. The set of predictions from different approaches could also be used as an ensemble, hopefully leading to better results. Another further improvement could be strengthen the existing relationship among the different approaches by means of more experimentation, reinforcing in this way our knowledge about dependencies in predictions, in which particular circumstances any of the approaches is better or the different efficiencies for each methodology, which is critical in OLAP scenarios. Finally, even if the approaches analysed in this paper are general to work for any off-the-shelf machine learning techniques, there may be room for improvement if specific techniques are developed for the multidimensional setting: multidimensional KNN, multidimensional decision trees and multidimensional Naive Bayes.

Acknowledgments. This work was supported by the Spanish MINECO under grants TIN 2010-21062-C02-02 and TIN 2013-45732-C4-1-P, and the REFRAME project, granted by the European Coordinated Research on Long-term Challenges in Information and Communication Sciences Technologies ERA-Net (CHIST-ERA), and funded by MINECO in Spain (PCIN-2013-037) and by Generalitat Valenciana PROMETEOII2015/013.

References

1. Agrawal, R., Gupta, A., Sarawagi, S.: Modeling multidimensional databases. In: Proceedings of the Thirteenth International Conference on Data Engineering, ICDE 1997, pp. 232–243. IEEE Computer Society (1997)
2. Bella, A., Ferri, C., Hernández-Orallo, J., Ramírez-Quintana, M.: Quantification via probability estimators. In: IEEE ICDM, pp. 737–742 (2010)
3. Bella, A., Ferri, C., Hernández-Orallo, J., Ramírez-Quintana, M.J.: Aggregative quantification for regression. DMKD **28**(2), 475–518 (2014)
4. Bickel, R.: Multilevel analysis for applied research: It's just regression! Guilford Press (2012)
5. Cabibbo, L., Torlone, R.: A logical approach to multidimensional databases. In: Schek, H.-J., Saltor, F., Ramos, I., Alonso, G. (eds.) EDBT 1998. LNCS, vol. 1377, p. 183. Springer, Heidelberg (1998)
6. Chaudhuri, S., Dayal, U.: An overview of data warehousing and OLAP technology. ACM Sigmod Record **26**(1), 65–74 (1997)
7. Chen, B.C.: Cube-Space Data Mining. ProQuest (2008)
8. Chen, B.C., Chen, L., Lin, Y., Ramakrishnan, R.: Prediction cubes. In: Proc. of the 31st Intl. Conf. on Very Large Data Bases, pp. 982–993 (2005)
9. Datahub: Car fuel consumptions and emissions 2000–2013 (2013). http://datahub.io/dataset/car-fuel-consumptions-and-emissions
10. Dhurandhar, A.: Using coarse information for real valued prediction. Data Mining and Knowledge Discovery **27**(2), 167–192 (2013)
11. Forman, G.: Quantifying counts and costs via classification. Data Min. Knowl. Discov. **17**(2), 164–206 (2008)
12. Goldstein, H.: Multilevel Statistical Models, vol. 922. John Wiley & Sons (2011)
13. Golfarelli, M., Maio, D., Rizzi, S.: The dimensional fact model: a conceptual model for data warehouses. Intl. J. of Coop. Information Systems **7**, 215–247 (1998)
14. Hall, M., Frank, E., Holmes, G., Pfahringer, B., Reutemann, P., Witten, I.H.: The WEKA data mining software: An update. SIGKDD Explor. **11**(1), 10–18 (2009)
15. Hernández-Orallo, J.: Probabilistic reframing for cost-sensitive regression. ACM Transactions on Knowledge Discovery from Data **8**(3) (2014)
16. IBM Corporation: Introduction to Aroma and SQL (2006). http://www.ibm.com/developerworks/data/tutorials/dm0607cao/dm0607cao.html
17. Kamber, M., Jenny, J.H., Chiang, Y., Han, J., Chiang, J.Y.: Metarule-guided mining of multi-dimensional association rules using data cubes. In: KDD, pp. 207–210 (1997)
18. Lin, T., Yao, Y., Zadeh, L.: Data Mining, Rough Sets and Granular Computing. Studies in Fuzziness and Soft Computing. Physica-Verlag HD (2002)
19. Páircéir, R., McClean, S., Scotney, B.: Discovery of multi-level rules and exceptions from a distributed database. In: Proc. of the 6th ACM SIGKDD Intl. Conf. on Knowledge discovery and data mining, pp. 523–532. ACM (2000)
20. Pastor, O., Casamayor, J.C., Celma, M., Mota, L., Pastor, M.A., Levin, A.M.: Conceptual Modeling of Human Genome: Integration Challenges. In: Düsterhöft, A., Klettke, M., Schewe, K.-D. (eds.) Conceptual Modelling and Its Theoretical Foundations. LNCS, vol. 7260, pp. 231–250. Springer, Heidelberg (2012)
21. Perlich, C., Provost, F.: Distribution-based aggregation for relational learning with identifier attributes. Machine Learning **62**(1–2), 65–105 (2006)
22. Team, R., et al.: R: A language and environment for statistical computing. R Foundation for Statistical Computing, Vienna, Austria (2012)

23. Ramakrishnan, R., Chen, B.C.: Exploratory mining in cube space. Data Mining and Knowledge Discovery **15**(1), 29–54 (2007)
24. Raudenbush, S.W., Bryk, A.S.: Hierarchical linear models: applications and data analysis methods, vol. 1. Sage (2002)
25. UCI Repository: UJIIndoorLoc data set (2014). http://archive.ics.uci.edu/ml/datasets/UJIIndoorLoc
26. Vassiliadis, P.: Modeling multidimensional databases, cubes and cube operations. In: Proc. of the 10th SSDBM Conference, pp. 53–62 (1998)

Semi-supervised Subspace Co-Projection for Multi-class Heterogeneous Domain Adaptation

Min Xiao and Yuhong Guo[✉]

Department of Computer and Information Sciences,
Temple University, Philadelphia, PA 19122, USA
{minxiao,yuhong}@temple.edu

Abstract. Heterogeneous domain adaptation aims to exploit labeled training data from a source domain for learning prediction models in a target domain under the condition that the two domains have different input feature representation spaces. In this paper, we propose a novel semi-supervised subspace co-projection method to address multi-class heterogeneous domain adaptation. The proposed method projects the instances of the two domains into a co-located latent subspace to bridge the feature divergence gap across domains, while simultaneously training prediction models in the co-projected representation space with labeled training instances from both domains. It also exploits the unlabeled data to promote the consistency of co-projected subspaces from the two domains based on a maximum mean discrepancy criterion. Moreover, to increase the stability and discriminative informativeness of the subspace co-projection, we further exploit the error-correcting output code schemes to incorporate more binary prediction tasks shared across domains into the learning process. We formulate this semi-supervised learning process as a non-convex joint minimization problem and develop an alternating optimization algorithm to solve it. To investigate the empirical performance of the proposed approach, we conduct experiments on cross-lingual text classification and cross-domain digit image classification tasks with heterogeneous feature spaces. The experimental results demonstrate the efficacy of the proposed method on these heterogeneous domain adaptation problems.

1 Introduction

Domain adaptation is the task of exploiting labeled training data in a *label-rich source* domain to train prediction models in a *label-scarce target* domain, aiming to greatly reduce the manual annotation effort in the target domain. Recently, heterogeneous domain adaptation, which generalizes the standard domain adaptation into a more challenging scenario where the source domain and the target domain have different feature spaces, has attracted a lot attention in the research community [6,10,16]. Heterogeneous domain adaptation techniques have applications in many different areas, including image classification in computer vision

© Springer International Publishing Switzerland 2015
A. Appice et al. (Eds.): ECML PKDD 2015, Part II, LNAI 9285, pp. 525–540, 2015.
DOI: 10.1007/978-3-319-23525-7_32

[10,16], drug efficiency prediction in biotechnology [16], cross-language text classification [6] and cross-lingual text retrieval [17] in natural language processing.

A fundamental challenge in heterogeneous domain adaptation lies in the disjoint feature representation spaces of the two domains; with the disjoint feature spaces, a prediction model trained in the source domain cannot be applied in the target domain. A number of representation learning methods have been developed in the literature to address this challenge, including the instance projection methods [6,16] which project instances in the two domains into a common feature space, and the instance transformation methods [10,12] which transform instances from one domain into the other one. These methods however conduct representation learning either in a fully unsupervised manner [16] without exploiting the label information, or in a fully supervised manner [6,10,12] without exploiting the available unlabeled instances. Moreover, some works [16,18] perform representation learning and prediction model training separately, leading to non-optimal representations for the target classification task.

In this paper, we propose a novel semi-supervised subspace co-projection method to address heterogeneous domain adaptation problems, which overcomes the drawbacks of the previous methods mentioned above. The proposed method projects instances in the source and target domains from domain-specific feature spaces to a co-located low-dimensional representation space, while *simultaneously* training prediction models in the projected feature space with labeled instances from the two domains. Moreover, the unlabeled instances are exploited to promote cross-domain instance co-projection by enforcing the empirical mean distributions of the projected source instances and the projected target instances to be similar. Furthermore, we exploit Error-Correcting Output Code (ECOC) schemes [5] to cast a cross-domain multi-class classification task into a large number of cross-domain binary prediction tasks, aiming to increase the stability and discriminative informativeness of the subspace co-projection and enhance cross-domain multi-class classification. The overall semi-supervised learning process is formulated as a joint minimization problem, and solved using an alternating optimization procedure. To evaluate the proposed learning method, we conduct cross-lingual text classification experiments on multilingual Amazon product reviews and cross-domain digit image classification experiments on the UCI handwritten digits data. The experimental results demonstrate the efficacy of the proposed approach for multi-class heterogeneous domain adaptation.

2   Related Work

In this section, we provide a brief review over the related works on heterogeneous domain adaptation, including latent subspace learning methods, instance transformation methods, and auxiliary resources assisted learning methods.

A group of works address heterogeneous domain adaptation by developing latent subspace learning methods that project instances from the domain-specific feature spaces into a common latent subspace [6,13,16,17,20]. In particular, Shi et al. [16] proposed a heterogeneous spectral mapping (HeMap) method, which learns two projection matrices and projects instances via spectral transformation. Wang et al. [17] proposed a manifold alignment (DAMA) method, which learns projection matrices by using manifold alignment and similarity/dissimilarity constraints constructed on pairs of instances with same/different labels. Duan et al. [6] proposed a heterogeneous feature augmentation (HFA) method, which first projects instances into a common subspace and uses the projected latent features to augment the original features of the instances, and then trains a classification model with the feature-augmented instances. Later, Li et al. [13] extended the HFA method into a semi-supervised HFA (SHFA) method by incorporating unlabeled target training data. Wu et al. [20] proposed to address heterogeneous domain adaptation by performing heterogeneous transfer discriminant analysis of canonical correlations, which maximizes/minimizes the intra/inter-class canonical correlations of the projected instances while simultaneously reducing the data distribution mismatch between the original data and the projected data. Our proposed approach shares similarities with these subspace learning methods on projecting original instances into common representation subspaces. But different from these previous works, our approach exploits both labeled and unlabeled instances and simultaneously learns the projection matrices and the prediction models. Moreover, our approach can naturally exploit error-correcting output code schemes to promote label informative subspace co-projection.

Another group of works developed instance transformation methods to address heterogeneous domain adaptation, which learn asymmetric mapping matrices to transform instances from the source domain to the target domain or vice versa [10,12,18,21]. Kulis et al. [12] proposed an asymmetric regularized cross-domain transformation method that learns an asymmetric feature transformation matrix by performing nonlinear metric learning with similarity/dissimilarity constraints constructed on all pairs of labeled instances. Wang et al. [18] proposed a two-step feature mapping method based on Hilbert-Schmidt Independence Criterion (HSIC) [8] for heterogeneous domain adaptation. It first selects features in each domain based on the HSIC between the instance feature kernel matrix and the instance label kernel matrix, and then maps the selected features across domains based on HSIC. Hoffman et al. [10] proposed a Max-Margin Domain Transforms (MMDT) method to learn domain-invariant image representations. It transforms target instances into the source domain and trains a prediction model in the source domain with the original labeled instances and the transformed labeled instances. Xiao and Guo [21] proposed a semi-supervised kernel matching method for heterogeneous domain adaptation. It learns a prediction function on the labeled source data while mapping the target data points to similar source data points by matching the target kernel matrix to a sub-matrix of the source kernel matrix based on a Hilbert Schmidt Independence Criterion.

In addition to the two groups of methods mentioned above, some other works exploit different types of auxiliary resources to build connections between the source features and the target features, including the ones that use bilingual dictionaries [4,9,19], and the ones that use additional unlabeled image and documents [22]. However, these auxiliary resource based learning methods are typically designed for specific applications and may have difficulty to be applied on other application tasks.

3   Semi-supervised Multi-class Heterogeneous Domain Adaptation

In this paper, we focus on multi-class heterogeneous domain adaptation problems. We assume in the source domain we have plenty of labeled instances while in the target domain we only have a small number of labeled instances. The two domains have disjoint input feature spaces, $\mathcal{X}_s = \mathbb{R}^{d_s}$ and $\mathcal{X}_t = \mathbb{R}^{d_t}$, where d_s is the dimensionality of the source domain feature space and d_t is the dimensionality of the target domain feature space, but share the same multi-class output label space $\mathcal{Y} = \{-1, 1\}^L$, where L is the number of classes. In particular, let $X_s = [X_s^\ell; X_s^u] \in \mathbb{R}^{n_s \times d_s}$ denote the data matrix in the source domain, where each instance is represented as a row vector. $X_s^\ell \in R^{\ell_s \times d_s}$ is the labeled source data matrix with a corresponding label matrix $Y_s \in \{-1, 1\}^{\ell_s \times L}$, and $X_s^u \in \mathbb{R}^{u_s \times d_s}$ is the unlabeled source data matrix. Each row of the label matrix contains only one positive 1, which indicates the class membership of the corresponding instance. Similarly, let $X_t = [X_t^\ell; X_t^u] \in \mathbb{R}^{n_t \times d_t}$ denote the data matrix in the target domain, where $X_t^\ell \in R^{\ell_t \times d_t}$ is the labeled target data matrix with a corresponding label matrix $Y_t \in \{-1, 1\}^{\ell_t \times L}$ and $X_t^u \in \mathbb{R}^{u_t \times d_t}$ is the unlabeled target data matrix. The number of labeled target domain instances ℓ_t is small and the number of labeled source domain instances ℓ_s is much larger than ℓ_t.

In this section, we present a semi-supervised subspace co-projection method to address heterogeneous multi-class domain adaptation under the setting described above. We formulate a co-projection based discriminative subspace learning method to simultaneously project the instances from both domains into a co-located subspace and train a multi-class classification model in the projected subspace, while exploiting the available unlabeled data to enforce a maximum mean discrepancy criterion across domains in the projected subspace. We further exploit ECOC schemes to enhance the discriminative informativeness of the projected subspace while directly addressing multi-class classification problems.

3.1   Semi-supervised Learning Framework

With the disjoint feature spaces across domains, traditional machine learning methods and homogeneous domain adaptation methods cannot be directly applied in the heterogeneous domain adaptation setting. However, if we can transform the two disjoint feature spaces $\mathcal{X}_s$ and $\mathcal{X}_t$ into a common subspace $\mathcal{Z} = \mathbb{R}^m$ with two transformation functions $\psi_s : \mathcal{X}_s \longrightarrow \mathcal{Z}$ and $\psi_t : \mathcal{X}_t \longrightarrow \mathcal{Z}$, we can then build a unified prediction model in the common subspace to adapt information across domains. Since the same multi-class prediction task is shared across the source domain and the target domain, i.e., the two domains have the same output label space, we can identify a useful common subspace representation of the data by enforcing the discriminative informativeness of the subspace representation of the labeled data in both domains for the common multi-class prediction task. Based on this motivation, we propose to project the instances from the source domain and the target domain into a common subspace using two projection matrices U_s and U_t respectively such that $\psi_s(X_s) = X_s U_s$ and $\psi_t(X_t) = X_t U_t$, while simultaneously training shared cross-domain prediction models using the projected data. This process can be formulated as the following minimization problem over the projection matrices and the prediction model parameters

$$\min_{U_s, U_t, W} \frac{1}{\ell_s + \beta \ell_t} \mathcal{L}\left(f(X_s^\ell U_s, W), \phi(Y_s)\right) + \frac{\alpha_s}{2} R(U_s) +$$
$$\frac{\beta}{\ell_s + \beta \ell_t} \mathcal{L}\left(f(X_t^\ell U_t, W), \phi(Y_t)\right) + \frac{\alpha_t}{2} R(U_t) + \frac{\gamma}{2} R(W) \qquad (1)$$

where $U_s \in \mathbb{R}^{d_s \times m}$ and $U_t \in \mathbb{R}^{d_t \times m}$ are two projection matrices that transform the input data in the source domain and target domain respectively to a common and low dimensional feature space, such that $m < \min(d_s, d_t)$; $f(\cdot, \cdot)$ is a prediction function for both domains in the projected common feature space and $W \in \mathbb{R}^{m \times K}$ is the prediction model parameter matrix; $R(\cdot)$ denotes a regularization function; $\phi(\cdot)$ denotes a label transformation function, which transforms the multi class label vectors from the original space $\{-1, 1\}^L$ to a new space $\{-1, 1\}^K$; $\mathcal{L}(\cdot, \cdot)$ is a loss function; and $\{\beta, \alpha_s, \alpha_t, \gamma\}$ are trade-off parameters. We introduce the label transformation function $\phi(\cdot)$ to provide a mechanism for incorporating label encoding schemes later.

Since the same prediction model is shared across the two domains, we expect that the discriminative subspace learning framework above can successfully identify a common subspace representation if there are sufficient labeled instances in both domains to enforce the predictive consistency of the subspace projections. However, there are typically only a small number of labeled instances in the target domain, which might lead to poor subspace identification in the target domain. To overcome this potential problem, we further incorporate unlabeled instances to assist the subspace co-projection across domains. Specifically, we assume the empirical marginal instance distributions of the two domains in the projected subspace should be similar, i.e., $P(\psi(X_s))$ and $P(\psi(X_t))$ are similar, and hence the prediction model built in the projected subspace using the labeled source domain instances can work well for the target domain. We thus propose to minimize the distance between the means of the projected instances (both labeled and unlabeled) in the two domains, $\mathcal{D}(\overline{\psi}(X_s), \overline{\psi}(X_t))$. The empirical mean vector $\overline{\psi}(X_s)$ in the source domain can be expressed as $\overline{\psi}(X_s) = \frac{1}{n_s} \mathbf{1}_{n_s}^\top X_s U_s$, where $\mathbf{1}_{n_s}$ denotes a column vector of 1s with length n_s. Similarly, the empirical mean vector $\overline{\psi}(X_t)$ in the target domain can be expressed as $\overline{\psi}(X_t) = \frac{1}{n_t} \mathbf{1}_{n_t}^\top X_t U_t$, where $\mathbf{1}_{n_t}$ denotes a column vector of 1s with length n_t. By incorporating the empirical mean vector distance measure into our formulation above, we produce the following semi-supervised heterogeneous domain adaptation framework

$$\min_{U_s, U_t, W} \frac{1}{\ell_s + \beta \ell_t} \mathcal{L}\left(f(X_s^\ell U_s, W), \phi(Y_s)\right) + \frac{\alpha_s}{2} R(U_s) +$$
$$\frac{\beta}{\ell_s + \beta \ell_t} \mathcal{L}\left(f(X_t^\ell U_t, W), \phi(Y_t)\right) + \frac{\alpha_t}{2} R(U_t) +$$
$$\frac{\gamma}{2} R(W) + \eta \mathcal{D}\left(\frac{1}{n_s} \mathbf{1}_{n_s}^\top X_s U_s, \frac{1}{n_t} \mathbf{1}_{n_t}^\top X_t U_t\right) \tag{2}$$

This framework will ensure the common subspace identified across domains to be informative for the shared prediction model in the two domains, while enforcing the two domains have similar marginal instance distributions in the projected subspace to facilitate information adaptation across domains.

We expect the semi-supervised formulation above to provide a general framework for identifying discriminative common subspace representations for effective information adaptation across domains. Nevertheless, to produce a specific learning problem, we need to consider specific prediction functions, loss functions, regularization functions and distance functions. In this work, we use a linear prediction function $f(x, w) = xw$, a least squares loss function $\mathcal{L}(\hat{y}, y) = (\hat{y} - y)^2$, and a squared L2-norm regularization function $R(w) = \|w\|_2^2$. We consider an Euclidean distance function $\mathcal{D}(\cdot, \cdot)$, which leads to a *maximum mean discrepancy criterion* [2]. The maximum mean discrepancy criterion has been used in the literature to induce similar marginal instance distributions across domains in homogeneous domain adaptation setting, and it has been shown to be effective in bridging the domain divergence gaps [3,14]. We expect such an empirical distribution based criterion can be useful for learning the common subspace across heterogeneous domains in our setting. These specific components together lead to the following semi-supervised learning problem

$$\min_{U_s, U_t, W} \frac{1}{\ell_s + \beta \ell_t} \left\| X_s^\ell U_s W - \phi(Y_s) \right\|_F^2 + \frac{\alpha_s}{2} \|U_s\|_F^2 +$$
$$\frac{\beta}{\ell_s + \beta \ell_t} \left\| X_t^\ell U_t W - \phi(Y_t) \right\|_F^2 + \frac{\alpha_t}{2} \|U_t\|_F^2 +$$
$$\frac{\gamma}{2} \|W\|_F^2 + \eta \left\| \frac{1}{n_s} \mathbf{1}_{n_s}^\top X_s U_s - \frac{1}{n_t} \mathbf{1}_{n_t}^\top X_t U_t \right\|_2^2 \tag{3}$$

where $\|.\|_F$ denotes the Frobenius norm, $\|.\|_2$ denotes the L2 norm, and $\{\alpha_s, \alpha_t, \beta, \gamma, \eta\}$ are trade-off parameters.

The label transformation function $\phi(\cdot)$ allows one to use different multi-class classification schemes within the proposed framework above. For example, if we use the standard one-vs-all (OVA) scheme to address multi-class classification, i.e., training one binary predictor for each label class, we then will have an identical label transformation function $\phi(Y) = Y$, and set $K = L$ for the size of the prediction model parameter matrix W.

3.2 Multi-class Classification with ECOC Schemes

In addition to the one-vs-all (OVA) scheme for multi-class classification, we further exploit the general error-correcting output code (ECOC) [5] schemes for multi-class classification. There are two reasons to use ECOC schemes in our learning framework. First, ECOC schemes have the capacity of encoding a multi-class classification problem into many more binary classification problems than the OVA scheme. More cross-domain binary classification tasks can help to increase the stability and prediction informativeness of the subspace co-projection in the proposed approach above, and lead to more robust domain adaptation performance. Second, ECOC schemes have been used in the literature to robustly solve multi-class classification problems with good empirical results [5]. Incorporating an ECOC scheme in our learning framework will benefit our multi-class classification task.

An ECOC scheme has two components: encoding process and decoding process. Given a L-class classification problem, in the encoding process, an ECOC scheme assigns a codeword from $\{-1, +1\}^K$ to each of the L classes, where K is the length of the codeword. All the codewords for the L classes can then form a codeword matrix $M \in \{-1, +1\}^{L \times K}$, whose each row contains the codeword for one of the L classes. Based on such a codeword matrix, the label transformation function $\phi(\cdot)$ can transform any given label vector from the one-vs-all form into a new label vector with length K, while converting the L-class classification problem to K binary classification problems, each of which corresponds to one column of the codeword matrix M. In the decoding process, one can simply compare the predicted codeword with the codewords in the codeword matrix M to determine the predicted class (one of the L classes). In this work, we use the Euclidean distance based loss decoding [7].

There are different ECOC schemes proposed in the literature. One standard scheme is the exhaustive ECOC [5], which constructs codewords with length $K = 2^{L-1} - 1$. Dense random encoding [1] is another simple ECOC encoding scheme. For a given codeword length K, the random encoding constructs the codeword vectors for the L classes by randomly filling the vectors with 1s and -1s, and then selects the codeword matrix with the largest sum of column separation and row separation from the results of multiple random repeats.

4 Training Algorithm

The semi-supervised learning problem in Eq (3) is a non-convex joint minimization problem over the three parameter matrices, U_s, U_t, and W. But the problem is convex in each individual parameter matrix given the other two fixed, and has closed-form solutions.

First, given fixed U_t and W, the optimization problem over U_s in Eq (3) is simply a least squares minimization problem. By setting the derivative of the objective function regarding U_s to zeros, we obtain the following closed-form solution

$$\text{vec}(U_s) = \left((WW^\top) \otimes A_s + I \otimes B_s \right)^{-1} \text{vec}(Q_s) \tag{4}$$

where $\otimes$ denotes the Kronecker product operator, vec($\cdot$) is the matrix vectorization operator, I is an identity matrix with proper size in the given context, and

$$A_s = \frac{2}{\ell_s + \beta\ell_t} X_s^{\ell\top} X_s^\ell,$$

$$B_s = \alpha_s I + \frac{2\eta}{n_s^2} X_s^\top 1_{n_s} 1_{n_s}^\top X_s,$$

$$Q_s = \frac{2}{\ell_s + \beta\ell_t} X_s^{\ell\top} \phi(Y_s) W^\top + \frac{2\eta}{n_s n_t} X_s^\top 1_{n_s} 1_{n_t}^\top X_t U_t,$$

Similarly, given fixed U_s and W, the optimization problem over U_t in Eq (3) has the following closed-form solution

$$\text{vec}(U_t) = \left((WW^\top) \otimes A_t + I \otimes B_t\right)^{-1} \text{vec}(Q_t) \tag{5}$$

where

$$A_t = \frac{2\beta}{\ell_s + \beta\ell_t} X_t^{\ell\top} X_t^\ell,$$

$$B_t = \alpha_t I + \frac{2\eta}{n_t^2} X_t^\top 1_{n_t} 1_{n_t}^\top X_t,$$

$$Q_t = \frac{2\beta}{\ell_s + \beta\ell_t} X_t^{\ell\top} \phi(Y_t) W^\top + \frac{2\eta}{n_s n_t} X_t^\top 1_{n_t} 1_{n_s}^\top X_s U_s.$$

Finally, the optimization problem over W given fixed U_s and U_t has the following closed-form solution

$$W = \left(\frac{2N_x}{\ell_s + \beta\ell_t} + \gamma I\right)^{-1} \left(\frac{2N_y}{\ell_s + \beta\ell_t}\right) \tag{6}$$

where

$$N_x = U_s^\top X_s^{\ell\top} X_s^\ell U_s + \beta U_t^\top X_t^{\ell\top} X_t^\ell U_t,$$

$$N_y = U_s^\top X_s^{\ell\top} \phi(Y_s) + \beta U_t^\top X_t^{\ell\top} \phi(Y_t).$$

Given these closed-form solutions for each individual subproblem, we use an alternating procedure to solve the optimization problem in Eq (3) in an iterative manner. After a random initialization over $\{U_s, U_t, W\}$, in each iteration the alternating procedure sequentially updates U_s, U_t and W according to equations (4), (5) and (6) respectively to minimize the objective function. We stop the iteration until a local optimal objective has been reached. On high-dimensional data, where the closed-form solutions in (4) and (5) involve large matrix inversions, we use a conjugate gradient descent algorithm to solve the subproblems over U_s and U_t to achieve scalability.

5   Experiments

We conducted experiments on cross-lingual text classification tasks and digit image classification tasks with heterogeneous feature spaces. In this section we report the experimental settings and the empirical results.

5.1   Datasets and Methods

We conducted experiments on two types of data, text data and image data, using Amazon product reviews [15] and UCI handwritten digits [11] respectively. The Amazon product review dataset is a multilingual sentiment classification dataset. It contains reviews from three different categories (Books, DVD and Music), written in four different languages (*English (E)* , *French (F)* , *German (G)* and *Japanese (J)*), where each review is represented as a term-frequency feature vector. With this dataset, we constructed 12 cross-lingual multi-class classification tasks with the three categories {*Books, DV D, Music*} as classes, one for each source-target language pair. For example, the task *E2F* uses *English* as the source language and *French* as the target language. For each task, there are 4000 views for each class in each language domain.

The UCI handwritten digits dataset contains 2000 digit images, evenly distributed among ten digit classes (from zero to nine). We randomly split the dataset into two subsets with equal size as two domains. Images in one domain are represented using the feature set of the Zernike moments (Zer), while images in the other domain are represented using the feature set of the profile correlations (Fac). We then constructed two heterogeneous domain adaptation tasks, *Fac2Zer* and *Zer2Fac*, one for each ordered source-target domain pair.

Methods: For each constructed heterogeneous domain adaptation task, we compared the following methods: (1) *TB* - this is a target baseline method that trains a classifier using only the labeled instances in the target domain. (2) *HeMap* - this is an unsupervised representation learning method for heterogeneous domain adaptation [16], which first learns two projection matrices for the two domains and then trains a classifier using the projected labeled instances from the two domains. (3) *DAMA* - this is a semi-supervised heterogeneous domain adaptation method proposed in [17], which performs representation learning and model training in separate steps. (4) *MMDT* - this is a maximum margin domain transform method for heterogeneous domain adaptation [10]. (5) *SHFA* - this is a semi-supervised heterogeneous feature augmentation-based domain adaptation method [13]. (6) *SCP-OVA* - this is the proposed subspace co-projection method with the one-verse-all (OVA) scheme for multi-class classification. (7) *SCP-ECOC* - this is the proposed subspace co-projection method with the exhaustive ECOC scheme for multi-class classification. The DAMA method [17] cannot handle the original high-dimensional features of the review data, we thus applied PCA to reduce the dimensionality of the input features in each language domain to 1000, as suggested in the *SHFA* work [13]. The alternating training algorithm for our proposed approaches is very efficient, and it typically converges within 30 iterations in our experiments.

5.2   Cross-lingual Text Classification

For each of the 12 cross-lingual multi-class classification tasks on Amazon product reviews, there are 4000 instances for each of the three classes in each domain.

Table 1. Average test accuracy ($\pm$ standard deviations) (%) over 10 runs for cross-lingual text classification tasks.

TASK	TB	HeMap	DAMA	MMDT	SHFA	SCP-OVA	SCP-ECOC
E2F	73.8$\pm$0.5	73.8$\pm$0.4	74.2$\pm$0.5	78.2$\pm$0.5	78.4$\pm$0.4	79.2$\pm$0.5	**80.6$\pm$0.4**
E2G	72.4$\pm$0.5	76.5$\pm$0.5	77.0$\pm$0.4	79.2$\pm$0.4	79.4$\pm$0.4	81.0$\pm$0.4	**82.2$\pm$0.3**
E2J	66.8$\pm$0.5	67.3$\pm$0.5	67.6$\pm$0.5	72.7$\pm$0.5	70.6$\pm$0.8	73.4$\pm$0.6	**74.4$\pm$0.6**
F2E	72.8$\pm$0.6	79.3$\pm$0.6	80.3$\pm$0.5	82.2$\pm$0.4	82.4$\pm$0.4	84.3$\pm$0.3	**85.6$\pm$0.2**
F2G	72.4$\pm$0.5	76.3$\pm$0.4	77.7$\pm$0.6	79.4$\pm$0.4	79.5$\pm$0.4	80.9$\pm$0.4	**82.2$\pm$0.3**
F2J	66.8$\pm$0.5	67.9$\pm$0.8	68.4$\pm$0.4	72.6$\pm$0.5	70.5$\pm$0.8	73.4$\pm$0.7	**74.5$\pm$0.6**
G2E	72.8$\pm$0.6	79.8$\pm$0.4	80.6$\pm$0.6	82.2$\pm$0.4	82.4$\pm$0.4	84.5$\pm$0.3	**85.5$\pm$0.2**
G2F	73.8$\pm$0.5	73.9$\pm$0.4	75.0$\pm$0.5	78.2$\pm$0.5	78.4$\pm$0.4	79.4$\pm$0.5	**80.6$\pm$0.4**
G2J	66.8$\pm$0.5	65.8$\pm$1.0	67.5$\pm$0.6	72.6$\pm$0.5	70.5$\pm$0.8	73.3$\pm$0.7	**74.4$\pm$0.6**
J2E	72.8$\pm$0.6	81.0$\pm$0.4	81.2$\pm$0.4	82.2$\pm$0.4	82.5$\pm$0.5	84.2$\pm$0.2	**85.5$\pm$0.2**
J2F	73.8$\pm$0.5	74.8$\pm$0.3	75.1$\pm$0.7	78.3$\pm$0.5	78.3$\pm$0.4	79.3$\pm$0.5	**80.5$\pm$0.4**
J2G	72.4$\pm$0.5	76.4$\pm$0.4	77.1$\pm$0.6	79.2$\pm$0.4	79.3$\pm$0.4	81.0$\pm$0.4	**82.2$\pm$0.4**

We conducted experiments in the following way. In the source domain, we randomly selected 2000 instances from each class as labeled data and used the remaining 2000 instances as unlabeled data. In the target domain, we randomly selected 100 instances and 2900 instances from each class as labeled and unlabeled data respectively. We used all these selected instances for training, and used the remaining 3000 instances (1000 for each class) in the target domain as testing data. For the comparison approaches, *HeMap, DAMA, SCP-OVA, SCP-ECOC*, which involve low dimensional subspaces, we set the dimension of the latent subspaces, m, as 100. Then we performed empirical parameter selection using the first task *E2F* with three runs. For the proposed approaches, *SCP-OVA* and *SCP-ECOC*, we chose α_s and α_t from $\{0.01, 0.1, 1, 10, 100\}$, β from $\{1, 2, 5, 10, 100\}$, η from $\{0.01, 0.1, 1, 10, 100\}$, and chose γ from $\{0.01, 0.1, 1, 10, 100\}$. We picked the parameter setting with the best test classification accuracy for each approach, $\{\alpha_s = 0.1, \alpha_t = 0.1, \beta = 1, \eta = 10, \gamma = 0.1\}$ for *SCP-OVA* and $\{\alpha_s = 10, \alpha_t = 0.1, \beta = 1, \eta = 10, \gamma = 0.1\}$ for *SCP-ECOC*. We conducted parameter selection for the other comparison approaches, *HeMap, DAMA, MMDT, SHFA*, in the same way. Using the selected parameters, for each of the 12 tasks we then repeatedly ran all the comparison methods for 10 times with different random selections of the training instances. The comparison results in terms of average test accuracy in the target domain are reported in Table 1.

From Table 1, we can see that the *TB* baseline method performs poorly across all the twelve tasks, which shows that the 100 labeled target training instances from each class are far from enough to obtain a good classification model in the target language domain. By exploiting the labeled training data from the source language domain, the *HeMap* method improves the prediction performance on most tasks. However, its improvements over *TB* are very small on some tasks and it even performs worse than *TB* on the task *G2J*. The *DAMA* method on the other hand consistently outperforms both *TB* and *HeMap*. The explanation

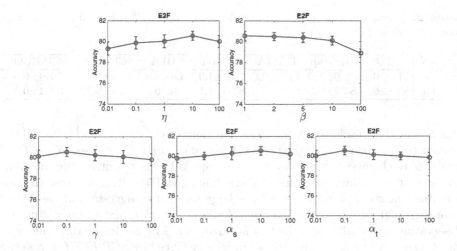

Fig. 1. Parameter sensitivity analysis over trade-off parameters $\{\eta, \beta, \gamma, \alpha_s, \alpha_t\}$.

is that *HeMap* conducts representation learning in a fully unsupervised manner while *DAMA* learns more informative representations in a semi-supervised manner with constraints constructed from the label information. By exploiting the label information *directly* for representation learning and prediction model training, the supervised method *MMDT* and semi-supervised method *SHFA*, further outperform *DAMA* on all the twelve tasks. Nevertheless, our proposed approaches, *SCP-OVA* and *SCP-ECOC*, outperform all the other comparison methods across all the tasks. This suggests that the proposed learning framework, which exploits both labeled and unlabeled training data to simultaneously perform subspace representation learning and prediction model training, is an effective model for heterogeneous domain adaptation. Between the two variants of the proposed model, *SCP-ECOC* consistently outperforms *SCP-OVA* across all the tasks, which suggests that the exhaustive error-correcting output coding is more effective than the one-vs-all coding scheme in our learning framework, while our proposed learning framework has the nice property of naturally incorporating different ECOC schemes.

5.3 Parameter Sensitivity Analysis

Next, we conducted parameter sensitivity analysis for the proposed *SCP-ECOC* approach over the trade-off parameters $\{\eta, \beta, \gamma, \alpha_s, \alpha_t\}$ using the first cross-lingual text classification task, *E2F*. We used the same experimental setting as above, and empirically investigated how the values of the trade-off parameters $\{\eta, \beta, \gamma, \alpha_s, \alpha_t\}$ affect the heterogeneous cross-domain prediction performance. We first conducted sensitivity analysis over η, which controls the relative weight for the mean discrepancy term in the proposed objective function. We conducted experiments with different η values from $\{0.01, 0.1, 1, 10, 100\}$, while fixing the other trade-off parameters as the selected values in the section above. For each η value, we repeated the

Table 2. Average test accuracy (± standard deviations) (%) over 10 runs for digit image classification tasks.

TASK	TB	HeMap	DAMA	MMDT	SHFA	SCP-OVA	SCP-ECOC
Fac2Zer	71.9±0.7	72.0±1.0	72.5±0.6	73.4±1.0	73.8±0.6	75.0±0.8	**76.6±0.5**
Zer2Fac	83.8±0.9	84.2±0.9	85.4±0.6	87.0±1.1	87.6±0.7	88.7±0.7	**90.4±0.5**

experiment 10 times based on random partitions of the dataset and reported the average test performance in the top left figure of Figure 1. We can see $SCP\text{-}ECOC$ produces the highest test accuracy when η equals 10. As η controls the contribution weight of the maximum mean discrepancy (MMD) criterion across the two domains, the good performance of the large value of η suggests that the MMD term is helpful for improving the cross-domain prediction performance. Another observation is that although the test accuracy varies as we change the value of η, the changes are small and the test accuracies produced by $SCP\text{-}ECOC$ across the whole range of different η values are all higher than the other comparison methods, TB, $HeMap$, $DAMA$, $MMDT$ and $SHFA$ (see both Figure 1 and Table 1). This suggests that the proposed $SCP\text{-}ECOC$ is not very sensitive to η within the studied range of values.

We next studied how β affects cross-lingual test classification accuracy. Note that β can be viewed as the relative weight ratio between a labeled target domain instance and a labeled source domain instance regarding their contribution to the training loss. As we have many more labeled training instances in the source domain than in the target domain and we aim to learn a classification model that works well in the target domain, it is reasonable to give a target domain instance larger (or equal) weight than a source domain instance and consider $\beta \geq 1$. In particular, we conducted experiments with different β values from $\{1, 2, 5, 10, 100\}$ while fixing all the other trade-off parameters as the selected values in the previous section. The average test classification results over 10 repeated runs are reported in the top right figure of Figure 1. We can see that the performance of $SCP\text{-}ECOC$ is quite stable with β values changing from 1 to 10. However, if placing too much weights ($e.g.$, $\beta = 100$) on the target instances, the test performance degrades. These results suggest that the performance of the proposed $SCP\text{-}ECOC$ is quite robust to β within a range of reasonable values.

We finally investigated the three trade-off parameters $\{\gamma, \alpha_s, \alpha_t\}$ used for the Frobenius norm regularization terms over W, U_s, and U_t respectively. We conducted experiments similarly as above. For each of the three parameters, we repeated the experiment 10 times for each of its values in $\{0.01, 0.1, 1, 10, 100\}$ while fixing all the other trade-off parameters as previously selected values. We reported the average test accuracy results in the bottom three figures of Figure 1 for the three parameters $\{\gamma, \alpha_s, \alpha_t\}$ respectively. We can see although the performance of the proposed $SCP\text{-}ECOC$ changes with the value change for each of the three parameters, the performance variations are very small. The performance of $SCP\text{-}ECOC$ is quite robust to the values of $\gamma, \alpha_s, \alpha_t$ within the range of values considered in the experiments.

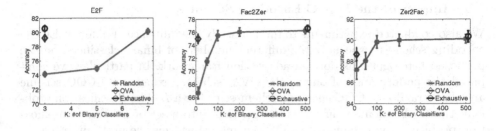

Fig. 2. Empirical comparison of different ECOC schemes.

5.4 Experimental Results on UCI Dataset

We have also conducted experiments using the UCI handwritten digits dataset. The two tasks we constructed on the UCI handwritten digits dataset have different feature spaces across domains, and have 100 instances from each class, i.e., 1000 instances in total, in each domain. For each task, in the source domain, we randomly chose 50 instances from each class (500 in total) as the labeled training data and used the remaining 500 instances as the unlabeled training data. In the target domain, we randomly chose 10 and 70 instances from each class as the labeled and unlabeled training data respectively, and used the remaining instances as the testing instances. For the approaches that involve subspaces, we set the dimension of the subspace as 20. We then used the same parameter selection procedure as before to select values for the trade-off parameters of all the comparison methods using the task *Fac2Zer*. For our proposed approaches, we got $\{\alpha_s = 0.1, \alpha_t = 0.1, \beta = 10, \eta = 0.1, \gamma = 10\}$ for *SCP-OVA* and $\{\alpha_s = 1, \alpha_t = 1, \beta = 1, \eta = 0.1, \gamma = 10\}$ for *SCP-ECOC*. With the selected parameters, for each task, we ran the comparison methods for 10 times with different random selections of the training and testing data. The average test accuracy results are reported in Table 2.

We can see that by exploiting the existing labeled data from the auxiliary source domain, all the heterogeneous domain adaptation methods outperform the baseline method on learning prediction models in the target domain. This again shows the importance of performing heterogeneous domain adaptation. Nevertheless, these few methods used in our experiments also demonstrated different efficacies on heterogeneous domain adaptation. *HeMap* displays similar performance as in the cross-lingual text classification experiments, with limited improvements over the baseline *TB*. The methods *DAMA, SHFA* and *MMDT* outperform *HeMap*, while our proposed two approaches outperform all the other comparison methods. Between the two proposed approaches, again *SCP-ECOC* outperforms *SCP-OVA*. All these results again verified the efficacy of the proposed learning framework.

5.5   Impact of the ECOC Encoding Schemes

We also conducted experiments to further study the influence of different ECOC encoding schemes, especially the different numbers of binary classifiers, on the proposed heterogeneous domain adaptation framework. In particular, we compared the performance of one-vs-all (OVA) scheme, exhaustive ECOC scheme and dense random ECOC encoding schemes [1]. For a L-class classification problem, the OVA scheme transforms the problem into a set of L binary classification problems, the exhaustive ECOC scheme transforms the problem into a set of $(2^{L-1} - 1)$ binary classification problems, while the random ECOC encoding scheme transforms the problem into a given number of K binary classification problems.

We conducted experiments on the first cross-lingual text classification task, *E2F* and the two tasks on UCI digits dataset, *Fac2Zer* and *Zer2Fac*. The *E2F* is a 3-class classification task, and we tested the random encoding ECOC scheme with different K values from $\{3, 5, 7\}$. The *Fac2Zer* and *Zer2Fac* are 10-class classification tasks, and we tested the random encoding ECOC scheme with different K values from $\{10, 50, 100, 200, 500\}$. The experimental results are reported in Figure 2. We can see that the exhaustive ECOC encoding scheme demonstrates the best performance on all the three tasks, even though its codeword length is smaller than the random schemes in some cases on the *E2F* task where the class number is small. This is reasonable since the codeword matrix generated by the exhaustive ECOC scheme typically has much better row and column separations than randomly generated codeword matrix. With the same codeword length, even the OVA scheme produces better performance than the random scheme. But with the increasing of the number of binary classifiers, i.e., the codeword length K, the performance of the proposed approach based on random encoding ECOC improves quickly. In particular, on *Fac2Zer* and *Zer2Fac*, when K increases from 10 to 100, the performance of the proposed approach increases dramatically. Similar performance is observed on *E2F* as well. This observation verifies our hypothesis that incorporating more binary classification tasks can help to increase the stability and usefulness of the subspace co-projection in the proposed learning framework and induce better domain adaptation performance.

6   Conclusion

In this paper, we developed a novel semi-supervised subspace co-projection approach to address multi-class heterogeneous domain adaptation problems, where the source domain and the target domain have disjoint input feature spaces. The proposed method projects instances in the two domains into a co-located latent subspace, while *simultaneously* training prediction models in the projected feature space. It also exploits the unlabeled data to promote the consistency of subspace co-projection from the two domains. Moreover, the proposed learning framework can naturally exploit error-correcting output codes for multi-class classification to enforce the informativeness of the subspace co-projection. We formulated the overall semi-supervised learning process as a joint minimization problem, and solved it using an alternating optimization procedure. To investigate the empirical performance of the proposed approach, we conducted cross-lingual text classification experiments on the Amazon product reviews and cross-domain image classification experiments on the UCI digits dataset. The empirical results demonstrated the effectiveness of the proposed approach comparing to a number of state-of-the-art heterogeneous domain adaptation methods.

Acknowledgments. This research was supported in part by NSF grant IIS-1065397

References

1. Allwein, E., Schapire, R., Singer, Y.: Reducing multiclass to binary: A unifying approach for margin classifiers. Journal of Machine Learning Research (JMLR) **1**, 113–141 (2001)
2. Borgwardt, K., Gretton, A., Rasch, M., Kriegel, H., Schölkopf, B., Smola, A.: Integrating structured biological data by kernel maximum mean discrepancy. In: Proceedings of the International Conference on Intelligent Systems for Molecular Biology (2006)
3. Chattopadhyay, R., Fan, W., Davidson, I., Panchanathan, S., Ye, J.: Joint transfer and batch-mode active learning. In: Proceedings of the International Conference on Machine Learning (ICML) (2013)
4. Dai, W., Chen, Y., Xue, G., Yang, Q., Yu, Y.: Translated learning: transfer learning across different feature spaces. In: Advances in Neural Information Processing Systems (NIPS) (2008)
5. Dietterich, T., Bakiri, G.: Solving multiclass learning problems via error-correcting output codes. Journal of Artificial Interlligence Research (JAIR) **2**(1), 263–286 (1995)
6. Duan, L., Xu, D., Tsang, I.: Learning with augmented features for heterogeneous domain adaptation. In: Proceedings of the International Conference on Machine Learning (ICML) (2012)
7. Escalera, S., Pujol, O., Radeva, P.: On the decoding process in ternary error-correcting output codes. IEEE Transactions on Pattern Analysis and Machine Intelligence (TPAMI) **32**(1), 120–134 (2010)
8. Gretton, A., Bousquet, O., Smola, A.J., Schölkopf, B.: Measuring statistical dependence with Hilbert-Schmidt norms. In: Jain, S., Simon, H.U., Tomita, E. (eds.) ALT 2005. LNCS (LNAI), vol. 3734, pp. 63–77. Springer, Heidelberg (2005)
9. He, J., Liu, Y., Yang, Q.: Linking heterogeneous input spaces with pivots for multi-task learning. In: Proceedings of SIAM International Conference on Data Mining (SDM) (2014)
10. Hoffman, J., Rodner, E., Donahue, J., Darrell, T., Saenko, K.: Efficient learning of domain-invariant image representations. In: Proceedings of the International Conference on Learning Representations (ICLR) (2013)
11. Jain, A., Duin, R., Mao, J.: Statistical pattern recognition: A review. IEEE Transactions on Pattern Analysis and Machine Intelligence (TPAMI) **22**(1), 4–37 (2000)
12. Kulis, B., Saenko, K., Darrell, T.: What you saw is not what you get: domain adaptation using asymmetric kernel transforms. In: Proceedings of the IEEE Conference on Computer Vision and Pattern Recognition (CVPR) (2011)

13. Li, W., Duan, L., Xu, D., Tsang, I.: Learning with augmented features for supervised and semi-supervised heterogeneous domain adaptation. IEEE Transactions on Pattern Analysis and Machine Intelligence (TPAMI) **36**(6), 1134–1148 (2014)
14. Pan, S., Tsang, I., Kwok, J., Yang, Q.: Domain adaptation via transfer component analysis. In: Proceedings of the International Joint Conference on Artificial Intelligence (IJCAI) (2009)
15. Prettenhofer, P., Stein, B.: Cross-language text classification using structural correspondence learning. In: Proceedings of the Annual Meeting of the Association for Computational Linguistics (ACL) (2010)
16. Shi, X., Liu, Q., Fan, W., Yu, P., Zhu, R.: Transfer learning on heterogenous feature spaces via spectral transformation. In: Proceedings of the IEEE International Conference on Data Mining (ICDM) (2010)
17. Wang, C., Mahadevan, S.: Heterogeneous domain adaptation using manifold alignment. In: Proceedings of the International Joint Conference on Artificial Intelligence (IJCAI) (2011)
18. Wang, H., Yang, Q.: Transfer learning by structural analogy. In: Proceedings of the AAAI Conference on Artificial Intelligence (AAAI) (2011)
19. Wei, B., Pal, C.: Heterogeneous transfer learning with rbms. In: Proceedings of the AAAI Conference on Artificial Intelligence (AAAI) (2011)
20. Wu, X., Wang, H., Liu, C., Jia, Y.: Cross-view action recognition over heterogeneous feature spaces. In: Proceedings of the IEEE International Conference on Computer Vision (ICCV) (2013)
21. Xiao, M., Guo, Y.: Feature space independent semi-supervised domain adaptation via kernel matching. IEEE Transactions on Pattern Analysis and Machine Intelligence (TPAMI) **37**(1), 54–66 (2014)
22. Zhu, Y., Chen, Y., Lu, Z., Pan, S., Xue, G., Yu, Y., Yang, Q.: Heterogeneous transfer learning for image classification. In: Proceedings of the AAAI Conference on Artificial Intelligence (AAAI) (2012)

Towards Computation of Novel Ideas
from Corpora of Scientific Text

Haixia Liu[1]([⊠]), James Goulding[2], and Tim Brailsford[1]

[1] School Of Computer Science, University of Nottingham Malaysia Campus,
Jalan Broga, 43500 Semenyih, Selangor Darul Ehsan, Malaysia
{khyx3lhi,tim.brailsford}@nottingham.edu.my
[2] Horizon Digital Economy Research, School of Computer Science,
University of Nottingham, Nottingham NG7 2TU, UK
james.goulding@nottingham.ac.uk

Abstract. In this work we present a method for the computation of novel 'ideas' from corpora of scientific text. The system functions by first detecting concept noun-phrases within the titles and abstracts of publications using Part-Of-Speech tagging, before classifying these into sets of *problem* and *solution* phrases via a target-word matching approach. By defining an idea as a co-occurring <*problem,solution*> pair, *known-idea* triples can be constructed through the additional assignment of a relevance value (computed via either phrase co-occurrence or an 'idea frequency-inverse document frequency' score). The resulting triples are then fed into a collaborative filtering algorithm, where problem-phrases are considered as *users* and solution-phrases as the *items* to be recommended. The final output is a ranked list of novel idea candidates, which hold potential for researchers to integrate into their hypothesis generation processes. This approach is evaluated using a subset of publications from the journal *Science*, with precision, recall and F-Measure results for a variety of model parametrizations indicating that the system is capable of generating useful novel ideas in an automated fashion.

Keywords: Idea mining · Text mining · Natural language processing · Recommender systems · Collaborative filtering

1 Introduction

The process of attacking problems by first canvassing participants for spontaneous ideas, collating their responses and distilling the results, is often referred to as *brainstorming*. The term, as popularized by Osborn [26] and expanded upon by Kling [22] and Jessop [19], now corresponds to a well-known set of guidelines for generating creative solutions that entail: discussion of the problem; unconstrained consideration as to how best to solve the problem; screening of the contributions; and, finally, commitment to action. While this approach to problem solving has traditionally required active human participation, in this paper we explore the following challenge: *given the inordinate amount of scientific literature now accessible via the web, is it possible to automate the brainstorming process via machine learning?*

© Springer International Publishing Switzerland 2015
A. Appice et al. (Eds.): ECML PKDD 2015, Part II, LNAI 9285, pp. 541–556, 2015.
DOI: 10.1007/978-3-319-23525-7_33

While the idea of supporting the ideation process via technology is not new (the term *Computer-Assisted Brainstorming* was coined three decades ago [17]), prior research has focussed on visualization tools, organizational applications and associated Human-Computer Interaction challenges [5,6,14]. However, text mining and computational linguistic techniques have now progressed to the point that notions of automatically extracting information from text and recognizing the links between underlying topics and concepts has become commonplace [2,12,31]. This brings with it opportunity to not only provide support tools for ideation process, but to actually generate *ideas* themselves.

Generating novel ideas from the automated processing of mass corpora of scientific text requires us to address several conceptual problems. First, we face the issue that the term *'idea'* itself is not at all well-defined from a comprehension perspective [15,20]. Second, new ideas are built upon domain knowledge that is extremely hard, if not impossible, to formalize [37]. Third, ideas from different domains exhibit widely varying characteristics; and finally, commonly used methods for ideation, such as the Gordon technique and expertness [32] are very difficult to computerize. These issues imply that obtaining perfect solutions to problems without human input is unrealistic. However, there is much potential in addressing the sub-task of generating *idea candidates*. Using a functional definition of an "idea" as a *<problem,solution>* pair (in the vein of [37]), we present an algorithmic approach to idea formulation. Our method breaks the task at hand into the following components: 1. a stage of text mining and linguistic processing of mass scientific corpora; 2. a supervised classification stage to isolate problem and solution concepts; and 3. a stage of re-combination via collaborative filtering, which outputs novel idea pairs for researchers to consider. This approach is evaluated using a subset of publications from the journal *Science*, and both statistical and qualitative evaluations indicate encouraging results. With a corpus of papers that cut across multiple disciplines, it is hoped that some of the idea candidates produced by the system will assist with the sort of cross-disciplinary ideation that is difficult to generate by conventional means.

2   Related Work

The concept of Computer-Assisted Brainstorming (CAB) was established by Hollander [17] in the 1980s, and envisioned interactive computer programs designed to enhance creative thinking. It was several decades later, however, before researchers successfully developed software tools to support brainstorming. Hardenberg *et al.* [14] introduced a *Bare-hand HCI* system, which integrated optical finger tracking into a two-phase brainstorming scenario. Phase 1 involved the collection a large number of ideas from participants and display on a video wall, with phase 2 seeing participants freely and simultaneously rearrange these items via touch manipulation. More recently, Biemann *et al* [5,6] developed *SemanticTalk*, software for visualizing brainstorming sessions and thematic concept trails that acted as a visual memory with both spoken dialogs and text documents being captured on a two-dimensional plane.

One of the key features of *Semantic Talk* was its ability to automatically generate associations between terms within the text identified as being important. This process of identifying key concept terms is being extended by the nascent field of *Idea Mining* [33], which focuses on the task of extracting reified idea structures that are embedded within text - whether that be in websites [42,43], patents [34], databases [27], blogs [35,38] or scientific literature [36]. A general approach for Idea Mining was introduced by Thorleuchter *et al.* in [39], which defined a technological idea as being represented by a combination of a *purpose* and a corresponding *means*, before going on to semi-automatically discover novel idea patterns in unstructured technological texts.

While the systems described above offer valuable digital support for the iteration component in real-world brainstorming, they are all limited in one important respect: they still rely heavily on human input to generate the novel ideas themselves. A possible way to attack this issue is to generate new links between problems and solutions - and a plausible approach to doing this is to harness the success of collaborative filtering (CF) techniques. CF algorithms [7,28] have proven to be extremely effective in generating novel recommendations, both in scientific research and real-world applications. CF uses the known preferences of a group of users to make recommendations (i.e. predictions) of the unknown preferences for other users [30] and CF techniques generally fall into one of three main categories: memory-based, model-based, and hybrid. In this work we focus on the memory-based CF, a method whose most critical component is the mechanism of finding similarities between items and/or users [30]. Many different methods exist to compute similarity [1], and in this work we focus on three that have proven effective in our experiments - log-likelihood, City Block and Tanimoto, all of which are detailed in [13].

Motivated by previous findings in CAB, the idea mining methodology currently being developed in the literature and the established effectiveness of recommender system techniques, we present a new algorithm to generate novel idea candidates. This approach automatically extracts <*problem,solution*> pairs from the titles and abstracts of scientific publications and uses these to computes novel ideas via a CF algorithm. While aimed at helping researchers to conduct scientific research via novel hypothesis generation, our main contribution is to demonstrate the possibility of automating ideation processes via CF techniques.

3   Defining an "Idea"

Young *et al.* [44] describe two principles for producing novel ideas:
- An idea is nothing more or less than a new combination of old elements.
- The capacity to bring old elements into new combinations depends largely on the ability to see relationships.

Based on these principles, we argue that novel idea candidates can be established by uncovering the relationships between *problems* and *solutions* within scientific texts. These components can then be intelligently recombined into previously unforeseen <*problem,solution*> pairs ready for consideration by researchers.

This constructive definition of an "idea" echoes Thorleuchter *et al*'s use of the term, who themselves reference the definitions in [41] in their attempts to identify concepts within various text corpora. In [39] they define an idea as a combination of a *means* and an appertaining *ends*, using unconstrained term vectors to represent each of these entities. In contrast, we represent *problems* and *solutions* using noun-phrases. This assumption is based on previous studies [16, 21] which indicate that while a sentence's main conceptual information is usually expressed by both noun- and verb-phrases, its primary concepts are predominantly carried by the noun-phrases.

Considering a document T, represented as an ordered set of N words, where $T = \langle w_1, w_2, \ldots, w_N \rangle$, then the functional definitions used to construct our representation of an idea are as follows:

Noun-phrase: A *noun-phrase*, ϕ, is an ordered subset of the text, extracted from T (in our case extracted from the titles or abstracts of publications using part-of-speech tagging technique):

$$\phi = \langle w_1, w_2, \ldots, w_n \rangle. \tag{1}$$

P-phrase: A *p-phrase* is defined as a noun-phrase determined to be a *scientific problem*. We define $\mathcal{P}_T$ as the set of m p-phrases extracted from a document, T (where $m \leq N$):

$$\mathcal{P}_T = \{\phi_a, \phi_b, \phi_c, \ldots\} \tag{2}$$

S-phrase: An *s-phrase* is defined as a noun-phrase that has been categorized as a *technical solution* or a *methodological approach*. We then define $\mathcal{S}_T$ to be the set of q p-phrases extracted from the document, T (where $q \leq N$):

$$\mathcal{S}_T = \{\phi_d, \phi_e, \phi_f, \ldots\}, \, where \, \mathcal{S}_T \cap \mathcal{P}_T = \emptyset \tag{3}$$

Idea: A specific *idea*[1] can then be defined as a combination of some p-phrase and s-phrase extracted from dataset, D:

$$idea = \langle p\text{–}phrase, s\text{–}phrase \rangle \tag{4}$$

Known Idea: A *known-idea* is defined as a combination of any p-phrase and s-phrase that are found in the same document, T:

$$\exists T \, known\text{–}idea \in \mathcal{P}_T \times \mathcal{S}_T \tag{5}$$

Known-ideas may additionally be attributed a relevance value, representing some measure of the idea's significance within the literature. In this work we evaluate four statistics to measure this significance, described in more detail in §4.5.

Novel Idea: A *novel-idea* is the combination of some p-phrase and s-phrase from the dataset, but which do not co-occur in the same document:

$$\exists T \exists U \, (novel\text{–}idea \in \mathcal{P}_T \times \mathcal{S}_U) \wedge (T \neq U) \tag{6}$$

Novel-ideas may also be assigned a value that reflects the strength of the relationship between its *p-phrase* and *s-phrase* components (as discussed in §4.6).

---

[1] We of course do not claim that a <problem, solution> pairs represents a universal definition of an idea, but a related pragmatic construct amenable to computation.

4   Methodology

Our method focuses on discovering problem-solution relationships between noun-phrases as detected in the abstracts (together with their titles) of scientific papers. An abstract is a fully self-contained, capsule description of a paper [23]. The noun-phrases it contains should reflect the issue(s) that the author(s) wish to address, so a list of noun-phrases extracted from it provides an ideal foundation for our seed pool of *p-phrases*. If an s-phrase is detected in the same piece of text, semantic relationships between it and neighbouring p-phrases are established[2]. Based on this premise, our approach to subsequently computing novel idea candidates can be broken down into six stages:

1. Noun-phrase extraction from a training-set corpora using Part-Of-Speech tagging.
2. Phrase filtering to remove stop words and text with low information content.
3. Classification of noun-phrases into *p-phrases* (problems) and *s-phrases* (solutions).
4. Aggregation of highly co-occurring <*p-phrase,s-phrase,relevance*> known-idea triples.
5. Processing of this set of known-idea triples via a collaborative filtering mechanism.
6. Assessment of the resulting ranked list of novel idea candidates that is output.

Several of the steps in this automated process analogise to specific stages in traditional brainstorming sessions. This is demonstrated in Fig. 1, which shows the process of the novel idea computation system on the right, and the corresponding steps in real-world ideation sessions on the left. We examine each of the stages in our method in more detail below.

4.1   Noun-Phrase Extraction

The first step in our method involves the detection of noun-phrases within the titles and abstracts of the publications that make up our training set. This is undertaken using a standard Part-of-speech (POS) tagging algorithm[3]. While there exist more complex linguistic indicators of an "idea", there are numerous advantages in assuming that noun-phrases are sufficient to represent the informational content of concepts: they are computationally parsimonious; their detection is well understood algorithmically; and studies show that such n-grams preserve far more semantic content than individual term extraction [12]. Recalling our definition of a noun-phrase, $\phi = \langle w_1, w_2, \ldots, w_n \rangle$, for each document, T, we are able to produce a list of noun-phrases: $\Phi_T = \langle \phi_1, \phi_2, \ldots \rangle$.

---

[2] For example, consider the sentence *"a dynamic panel data estimation technique is used to examine effects of internal demand on domestic credit"*. The n-gram "dynamic panel data estimation technique" will be recognised as an *s-phrase*, and associated with co-occurring *p-phrases* such as "effects of internal demand on domestic credit".

[3] In this study we have used the *CiteSpace* application for POS tagging [10], which we found performed better than other options such as the *TextBlob* Python library.

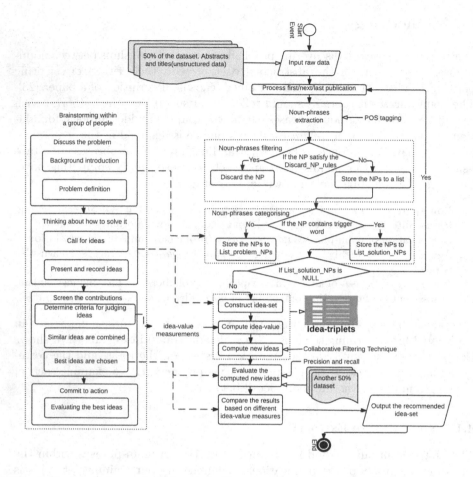

Fig. 1. System flow - each of the method's processing steps are listed on the right hand side, with corresponding stages in real-life brainstorming sessions mapped on the left.

4.2   Noun-Phrase Filtering

Some noun-phrases generated by the POS tagger are not suitable for inclusion within idea construction. In particular, some concepts will be semantically redundant (i.e. they will have minimal information content in the same vein as *stop words* in traditional informational retrieval tasks). Examples in scientific abstracts are n-grams such as "we present a ***novel model***..." or "our ***general approach*** is tested through an evaluation procedure that...". This stage aims to eliminate such phrases, thus streamlining the method's subsequent processing steps. To this end we employ two filtering steps. First, given a set of hand-crafted "danger" terms W, we remove bi-grams that feature any of its elements:

$$\Phi_{filtered} = \{\phi \in \Phi_T : |\phi| = 2 \rightarrow \phi \cap W = \varnothing\} \tag{7}$$

The version of W used in our experiments is listed in Table 7 in the appendix. In addition, we enforce a threshold on the frequency of retained noun-phrases. For this we used Jenks natural breaks classification method [18], assigning p-phrases into five categories according to their frequency across the corpus, eliminating all noun-phrases in the most frequent category. This step is based on the assumption that phrases that exhibit extremely high frequency either have low information content, or reflect noun-phrases that offer no untapped research value.

4.3 Noun-Phrases Categorization

Now we have a filtered set of noun-phrases we must categorize them into two groups: *p-phrases* and *s-phrases* (representing problems and solutions respectively). There are numerous possible approaches to achieve this task, ranging from *named entity recognition techniques* [25] to the application of *linguistic structure matching* [9]. Unfortunately, in order to utilise these techniques a vast amount of annotated data is required, data that is as-yet-unavailable. Therefore, and in lieu of a fully supervised machine learning approach, we fall back on a rule based pattern-matching approach to identify *s-phrases*. Filtered noun-phrases, $\Phi_{filtered}$, are then compared with a compact bag of trigger words, $\mathcal{G}$, in order to explicitly identify *s-phrases*.

Examples of these cue terms contained in $\mathcal{G}$ might be "method", "approach" and "theory" (the set of trigger words used in our experiments and method of derivation is detailed in Table 6 of the Appendix). Those noun-phrases which remain unmatched are subsequently designated as *p-phrases*[4]. The result is that for a document, T, we produce a set of *s-phrases*, S, and *p-phrases*, P, where:

$$\mathcal{S}_T = \{\phi \in \Phi_{filtered} : \exists w \in \mathcal{G} \, [w \in \phi]\} \tag{8}$$

$$\mathcal{P}_T = \Phi_{filtered} - \mathcal{S}_T \tag{9}$$

This stage of classification has analogies to the real-world brainstorming processes discussed in §2 - in the "discuss the problem" stage of the process [26], if no specific problem angle is specified, participants are instructed to conjure up any noun-phrases that are parts of the problem (i.e. people, places, entities, etc.) in a free-form fashion.

4.4 Known-Idea Construction

The algorithm must now enumerate *known-ideas* before it will be able to generate novel idea candidates through their recombination. It does this by pairing *p-phrases* and *s-phrases* deemed to be associated with each other. In linguistic processing the specific relation types that are extant between noun-phrases can be uncovered using a range of extraction techniques such as kernel methods; dependency trees [11]; text pattern or structure creation [33]; semantic graphs,

---

[4] In some ways this is an algorithmic rendition of the arguable expression: "if you are not part of the solution, you are part of the problem".

topic templates and ontologies (e.g. WordNet) [4,40]. However, due to the general qualities of a good abstract [3] - i.e. it should be a condensed and concentrated version of the full text of the research manuscript - we are able to assume that the concepts introduced in a single abstract are all related with each other regarding a specific topic domain. This assumption means we can postulate valid idea-pairs simply by observing the co-occurrence of a problem and solution within the same abstract. We note that this approach may generate some unexpected pairings - this, however, is still in line with the general rules of brainstorming, where the pairing non-obvious components can expand the creativity of a real-world ideation session. The corresponding expansion of the idea pool can increase the chances of producing a radical and effective solution, and as such, we currently neglect some traditional linguistic processing constraints:

1. we do not integrate details of relationship types between noun-phrases.
2. nor distances between the *root* and other nodes in the *Parse Tree*.

Once co-occurring $<p\text{-}phrase,s\text{-}phrase>$ pairs have been identified they are assigned a score reflecting their "interestingness" or relevance to the corpus. This value, v, is necessarily subjective, and as such we examine several approaches to determining it, as described in more detail in §4.5. Whatever value is selected, the result of the idea construction process is the set of known-ideas, $\mathcal{K} = \{idea_1, idea_2, ...\}$, as summarised by algorithm 1 below.

Algorithm 1

1: **procedure** EXTRACT_KNOWN_IDEAS($\mathcal{D}$)   ▷ $\mathcal{D}$ is the full document set
2:     $\mathcal{K} \leftarrow \emptyset$ ▷ Container for the results
3:     **for each** T in $\mathcal{D}$ **do**
4:         $\Phi \leftarrow$ extract_noun_phrases(T)
5:         $\Phi \leftarrow$ filter(Φ)
6:         $\mathcal{S}_T \leftarrow$ categorize_s-phrases(Φ)
7:         $\mathcal{P}_T \leftarrow$ categorize_p-phrases(Φ)
8:         **for each** s in $\mathcal{S}_T$ **do**
9:             **for each** p in $\mathcal{P}_T$ **do**
10:                 $v \leftarrow$ compute_idea_value(p, s)
11:                 $idea \leftarrow \langle p, s, v \rangle$
12:                 $\mathcal{K} = \mathcal{K} \cup \{idea\}$
13:     **return** $\mathcal{K}$ ▷ The output known-idea set

4.5   Relevance Values for *Known-Ideas*

In this study we have implemented four statistics which attempt to measure the relevance of a *known-ideas* to future recommendations (and which analogise to the *rating* a user has assigned to an item in traditional collaborative filtering). Each statistic is described below, with examples illustrated in Table 1:

OCC: the simplest way of estimating the relationship intensity between a *p-phrase* and *s-phrase* is to count the number of distinct documents in which they both appear (based on the assumption that the more times they co-occur, the stronger the relationship there is between them):

$$OCC(p,s) = \left| \{ T \in \mathcal{D} : p \in P_T, s \in \mathcal{S}_T \} \right| \tag{10}$$

FREQ: idea frequency is similar to document occurrences, but also takes into account the frequency of idea-pair occurrences *within* documents:

$$FREQ(p,s) = \sum_{T \in \mathcal{D}} \left| \{ \langle p,s \rangle \in P_T \times S_T \} \right| \tag{11}$$

CON: In order to address the fact that term counts alone cannot reflect the fact that some problems have numerous lines of attacks, while others have only a limited solution pool, we have defined the statistic *contribution*. This is a normalization that divides the number of times a certain idea pair co-occurs by the total number of s-phrases used to address the same problem:

$$CON(p,s) = \frac{OCC(p,s)}{\left| \{ T \in \mathcal{D} : s \in S_T \} \right|} \tag{12}$$

IF-IDF: *idea frequency / inverse document frequency* is an adaptation of the traditional *tf-idf* statistic that we have designed to addresses two observations: 1. the more times an *s-phrase* occurs in any given document, the more likely it is to be a 'key' solution to the document's *p-phrases*, so we wish to favour it; and 2. if an idea-pair crops up across the whole corpus the less likely it is to be "interesting" - either its research value has been saturated, or it is semantically redundant pairing in the same vein as a stop word. IF-IDF balances these two conflicting issues via the following formula[5]:

$$IF\text{-}IDF(p,s) = FREQ(p,s) \times \log \left(\frac{|\mathcal{D}|}{OCC(p,s)} \right) \tag{13}$$

Table 1. Examples of known-idea triplets from Journal Science:

p-phrase	s-phrase	OCC	FREQ	CON	IF-IDF
global warming	climate model	3	24	0.21	19.77
neuropsychiatric disorders	mouse model	3	16	0.50	13.18
impurity atoms	three-dimensional atom probe technique	2	6	0.5	3.00
nickel catalysis	photoredox-metal catalysis approach	1	4	1.00	5.20
impulsive optical excitation	first-principle theoretical simulation	1	3	1.00	3.90

---

[5] N.b. Idea Frequency (IF) differs from traditional Term Frequency (TF) in that it counts the idea's support over the *whole* corpus, and not just for a single document, resulting in a global statistic.

4.6   Computation of *Novel-Idea* Pairs

We now address the prediction of new links between problems and solutions through comparison of *s-phrase* and *p-phrase* patterns that span across different domains. There are numerous ways to measure similarity between such patterns, but the strategy at the heart of our techniques is based upon a collaborative filtering [7]. Collaborative filtering and the recommendation systems they underpin are based on imputation - a target user's past behaviours are first modelled and then compared to the habits of other users. Items favoured by similar users, but which do not yet exist in the target user's history, are then used as the basis for new recommendations. Our technique considers *p-phrases* as analogs to users, and *s-phrases* as items. Traditional recommender systems can be formulated as *user-based* or *item-based* algorithms [28] and we assess both approaches. In the generation of novel-ideas, our collaborative filtering task consists of the following steps:

1. **Construct a Preference Matrix:** in our case, each row represents a *p-phrase* and each column represents an *s-phrase* with the numerical value at the intersection of a row and a column represents the idea's relevance value, v (as selected from one of the statistics in §4.4).

2. **Compute Similarity Scores**: for a specific problem vector (i.e. a row in the preference matrix), u, iterate through every other problem vector, w, and compute a similarity s between u and w and retain the k nearest neighbours, $\mathcal{N}$. In our experiments we optimize N for each of the following distance metrics: *Tanimoto, Loglikelihood* and *CityBlock*[6].

3. **Generate Novel-Idea Pairs:** this is achieved by recommending novel solutions to existing problems. For each potential solution, i, that the current problem has no entry for, we consider every vector, w, in the neighbourhood, N, and add its relevance score for solution i to a running average, weighted by the vector's similarity score s. Finally, results are sorted, producing is a ranked list of novel *s-phrases* to the *p-phrase* under consideration. The top n *s-phrases* are combined with the p-phrase under consideration as our novel-idea prediction (in our experiments n is drawn from 2, 5, 10).

5   Experimental Evaluation

A collection of the titles and abstracts was studied, extracted randomly from 3,665 English language articles published in the journal *Science*. The dataset, covering the years 1998-2015, was partitioned so that half of the articles formed our training set, $\mathcal{D}$, and the other half our test set. After noun-phrase filtering, Φ contained 57,621 noun-phrases and noun-phrase categorization resulted in 54,073 p-phrases and 3,548 s-phrases. From this the algorithm constructed 90,212 unique <*p-phrase,s-phrase*> known-idea pairs[7].

---

[6] please refer to http://mahout.apache.org for implementation details.

[7] The restricted number of known-ideas is because no cogent s-phrases could be extracted for many abstracts, even though numerous noun-phrases were identified.

For each model parametrization a set of novel idea candidates were generated from the training set. These were then evaluated to determine if recommended idea-candidates *actually occurred* in the test set, with results being summarized for each *p-phrase* using traditional precision, recall and F-measure scores. This process was repeated for both user- and item-based collaborative filtering, using relevance value statistics drawn from {OCC, CON, IF-IDF} (n.b. FREQ is not reported due to its similarity to OCC) and varying the size of the recommended solution list for each p-phrase ($n \in \{2, 5, 10\}$). Overall mean precision, recall and F-measure scores were produced for each of the 54 model parameterizations (we report the results for each model using an optimized neighbourhood).

5.1 Results

Every stage of our approach has the potential for future refinement. Despite this, the novel-idea candidates output by the system's first iteration were highly encouraging. Examples of the system's output in Table 2, taken from a range of categories in the Science corpus, illustrate the cogent recommendations the system can produce. Precision and recall results are similarly positive - full results in Tables 3-5 indicate how the number of items in the recommendation set influences results for each of the relevance values we tested (with the size of the neighbourhood being optimized for each recommendation set size).

Table 2. Examples of novel idea-pairs generated from journal Science.

problem	old-solution	proposed-solution
first stars	cosmological simulation	nucleosynthesis model
creep damage	diffraction analysis	thermodynamic analysis
ancestral state reconstruction	likelihood-based approach	fluorescence technique
primary dendritic cells	unbiased approach	genome-wide location analysis
large void volumes coincides	diffraction analysis	thermodynamic analysis

Because we are assessing the efficacy of our idea recommendation approach as a whole rather than contrasting results for different collaborative filtering parameterizations, let us first consider the system's top *2-recommendations*. The results tables illustrate that across the board the system's top two novel idea recommendations match our test set over 90% of the time (with a maximum recall of 0.941 when using the CityBlock similarity measure and a relevance value based on OCC - see Table 2 for example idea pairs). While these statistical results are highly encouraging we note that extensive human evaluation of output ideas is required before we can be confident that these results could be translated into hypothesis generation processes. Additionally - and as one might expect - as the size of our recommendation list increases, results drop off starkly (by the time we have reached 10-recommendations the F-measure of our recommendations has fallen by almost half). This indicates that the system currently works optimally only for its highest ranked recommendations.

In a comparison of the distance metric used to determine CF neighbourhoods, the *CityBlock* measure is the clear winner. This represents absolute distance between solution vectors, and for all parameterizations of the model it

Table 3. OCC performance (precision/recall/F-measure)

metric	2-recommendations	5-recommendations	10-recommendations
user Loglikelihood	0.900/0.928/0.913	0.557/0.734/0.633	0.374/0.527/0.437
user CityBlock	0.951/0.941/0.946	0.730/0.765/0.747	0.474/0.685/0.560
user Tanimoto	0.901/0.929/0.915	0.561/0.735/0.636	0.366/0.536/0.435
item Loglikelihood	0.606/0.635/0.620	0.480/0.716/0.575	0.374/0.824/0.515
item CityBlock	0.209/0.208/0.208	0.027/0.027/0.027	0.013/0.126/0.023
item Tanimoto	0.423/0.437/0.430	0.259/0.403/0.315	0.267/0.611/0.372

Table 4. CON performance (precision/recall/F-measure)

metric	2-recommendations	5-recommendations	10-recommendations
user Loglikelihood	0.890/0.926/0.908	0.557/0.743/0.637	0.401/0.591/0.478
user CityBlock	0.943/0.939/0.941	0.730/0.761/0.745	0.493/0.674/0.570
user Tanimoto	0.892/0.928/0.910	0.562/0.747/0.641	0.393/0.600/0.475
item Loglikelihood	0.601/0.637/0.618	0.486/0.713/0.578	0.414/0.840/0.554
item CityBlock	0.208/0.210/0.209	0.034/0.050/0.040	0.027/0.135/0.045
item Tanimoto	0.422/0.443/0.432	0.267/0.407/0.323	0.305/0.655/0.416

Table 5. IF-IDF performance (precision/recall/F-measure)

metric	2-recommendations	5-recommendations	10-recommendations
user Loglikelihood	0.890/0.926/0.908	0.600/0.743/0.639	0.401/0.591/0.478
user CityBlock	0.943/0.940/0.941	0.730/0.761/0.745	0.493/0.674/0.570
user Tanimoto	0.892/0.928/0.910	0.565/0.747/0.643	0.393/0.600/0.475
item Loglikelihood	0.602/0.637/0.619	0.489/0.713/0.580	0.414/0.840/0.554
item CityBlock	0.207/0.210/0.210	0.034/0.050/0.041	0.027/0.135/0.045
item Tanimoto	0.422/0.443/0.432	0.267/0.407/0.323	0.305/0.655/0.416

consistently returns the highest F-measure results (this is down mostly to its superior precision results, with recall being relatively consistent across all distance measures).

A clear contrast also exists between results for user- and item-based collaborative filtering approaches, with the former performing far better than the latter in all cases. We conclude from these results that it is far better to recommend new solutions to old problems, than to try and bring new problems to old solutions. In many ways this is an intuitive result, as it is far more likely that extant solutions will be immediately attempted when new research problems arise.

Finally we consider the effectiveness of the three idea relevance scores tested. Despite being the least complex statistic implemented, OCC (focusing on City-Block measurement with 2 recommendations) provides the strongest results. Results for CON and IF-IDF are almost indistinguishable, and examination of idea recommendations for each problem indicate an extremely high crossover (in fact 44% of problems received identical recommendation sets for all sizes of recommendation list). These results appear to indicate that simply counting idea-pair occurrences in the dataset is a sufficient basis to assess the significance of a solution to any given problem.

6 Discussion

This study demonstrated the plausibility of generating novel idea-candidates in an automated fashion. User-based CF offered the best performance and, while different distance measures produced comparable results, OCC provided a simple method to achieve the most effective performance. Nonetheless, there is scope for further research at each of the stages of the idea-generation process.

First, there is potential to improve the filtering of noun-phrases identified by POS tagging (based perhaps upon a more formalized information-theoretic approach to detecting 'semantically redundant' terms). Second, our approach to classifying s-phrases and p-phrases remains relatively coarse, using a pattern matching approach based on trigger words. A further investigation of this processing stage would be of particular interest and numerous options seem viable.

It is our aim, for example, to implement a supervised classification model to improve detection of *s-phrases* and *p-phrases*. The input features for this model could be generated from language models [29], lexical cohesion [24] and linguistic grammar-based techniques [8], in addition to the statistical features already used. Training would need to be performed on ground truth annotations of scientific abstracts, but these could be collated in a crowd-sourced fashion by presenting abstracts to domain experts and allowing them to manually identify problem and solution term patterns within the text. The goal here would be to directly address some of the limitations with our current approach, such as the fact that *p-phrases* and *s-phrases* are overly dependent upon their context (for example, a p-phrase in one document might be an s-phrase in another).

Additional areas of interest lie not only in investigating other similarity measures from the collaborative filtering literature, but also in exploring other external indicators of a known-idea's relevance value. These might include the number of citations generated by the paper the idea appears in, or the impact factor of its parent publication, or indeed any of the host of methods that are used to assess the relevance of a paper within the scientific literature.

Finally, and perhaps of greatest importance, there is a good deal of room to extend our evaluation of the efficacy of the ideas generated by the system. Currently, we assess a novel-idea candidate's merit based upon whether it occurs in (or is absent from) future literature. This neglects two factors: 1. the comprehensibility and interestingness of generated idea-pairs (a situation which can only be addressed by a programme of human evaluation of the system's outputs), and 2. any assessment of an idea's *inventiveness*. Currently, if a recommended idea does not appear in our test set, it is deemed as a false positive out of hand, whereas it may be the case that the idea is simply yet to be researched.

7   Conclusion

In this study, we have presented a first approach for generating novel idea candidates from corpora of scientific text, that is decomposable into six distinct stages. Noun-phrases are extracted from the abstracts of scientific papers via POS tagging; a filtering process occurs to remove redundant concepts; the results set of phrases are subsequently categorized into problem and solution; co-occurring pairs are assigned a relevance score (based on number of co-occurrences, contribution to a problem's overall support or an idea frequency/inverse document frequency score); and finally a collaborative filtering algorithm generates new idea recommendations. This process illustrates the ability to transform unstructured textual data into structured idea pairs, and the potential to manipulate that structure computationally to generate new idea candidates. The approach was evaluated using a subset of publications from the journal *Science*, and both statistical and qualitative evaluations indicate strongly encouraging results, with an OCC relevance value combined with a (user-based) CityBlock similarity measure offering the best performance. Our hope is that in establishing this modular approach to automated idea generation, each stage may be honed by the broader research community to ultimately produce a system that has real utility to hypothesis generation.

Acknowledgments. This work was jointly supported by CFFRC-PLUS PhD scholarship scheme, the RCUK Horizon Digital Economy Research Hub grant, EP/G065802/1 and the EPSRC Neodemographics grant, EP/L021080/1.

References

1. Ahn, H.J.: A new similarity measure for collaborative filtering to alleviate the new user cold-starting problem. Information Sciences **178**(1), 37–51 (2008)
2. Allan, J., Carbonell, J.G., Doddington, G., Yamron, J., Yang, Y.: Topic detection and tracking pilot study final report (1998)
3. Andrade, C.: How to write a good abstract for a scientific paper or conference presentation. Indian Journal of Psychiatry **53**(2), 172 (2011)
4. Banko, M., Etzioni, O., Center, T.: The tradeoffs between open and traditional relation extraction. In: Proceedings of 46th Annual Meeting of the Association for Computational Linguistics: Human Language Technologies, vol. 8, pp. 28–36 (2008)
5. Biemann, C., Böhm, K., Heyer, G., Melz, R.: Semantictalk: software for visualizing brainstorming sessions and thematic concept trails on document collections. In: Boulicaut, J.-F., Esposito, F., Giannotti, F., Pedreschi, D. (eds.) PKDD 2004. LNCS (LNAI), vol. 3202, pp. 534–536. Springer, Heidelberg (2004)
6. Biemann, C., Böhm, K., Heyer, G., Melz, R.: Automatically building concept structures and displaying concept trails for the use in brainstorming sessions and content management systems. In: Böhme, T., Larios Rosillo, V.M., Unger, H., Unger, H. (eds.) IICS 2004. LNCS, vol. 3473, pp. 157–167. Springer, Heidelberg (2006)
7. Breese, J.S., Heckerman, D., Kadie, C.: Empirical analysis of predictive algorithms for collaborative filtering. In: Proceedings of the Fourteenth conference on Uncertainty in artificial intelligence, pp. 43–52. Morgan Kaufmann (1998)
8. Brown, P.F., deSouza, P.V., Mercer, R.L., Pietra, V.J.D., Lai, J.C.: Class-based n-gram models of natural language. Comput. Linguist. **18**(4), 467–479 (1992)
9. Bybee, J.L., Hopper, P.J.: Frequency and the emergence of linguistic structure, vol. 45. John Benjamins Publishing (2001)
10. Chen, C.: Citespace ii: Detecting and visualizing emerging trends and transient patterns in scientific literature. Journal of the American Society for information Science and Technology **57**(3), 359–377 (2006)
11. Culotta, A., Sorensen, J.: Dependency tree kernels for relation extraction. In: Proceedings of the 42nd Annual Meeting on Association for Computational Linguistics, p. 423. Association for Computational Linguistics (2004)

12. Ding, W., Chen, C.: Dynamic topic detection and tracking: A comparison of hdp, c-word, and cocitation methods. Journal of the Association for Information Science and Technology (2014)
13. Guo, S., et al.: Analysis and evaluation of similarity metrics in collaborative filtering recommender system (2014)
14. von Hardenberg, C., Bérard, F.: Bare-hand human-computer interaction. In: Proceedings of the 2001 Workshop on Perceptive User Interfaces, PUI 2001, pp. 1–8. ACM, New York (2001)
15. Hare, V.C., Milligan, B.: Main idea identification: Instructional explanations in four basal reader series. Journal of Literacy Research 16(3), 189–204 (1984)
16. Hildreth, P.M., Kimble, C.: Knowledge networks: Innovation through communities of practice. IGI Global (2004)
17. Hollander, S.: Computer-assisted Creativity and the Policy Process. Thayer School of Engineering (1984)
18. Jenks, G.F.: The data model concept in statistical mapping. International Yearbook of Cartography 7(1), 186–190 (1967)
19. Jessop, J.L.: Expanding our students' brainpower: Idea generation and critical thinking skills. IEEE Antennas and Propagation Magazine 44(6), 140–144 (2002)
20. Jitendra, A.K., Cole, C.L., Hoppes, M.K., Wilson, B.: Effects of a direct instruction main idea summarization program and self-monitoring on reading comprehension of middle school students with learning disabilities. Reading & Writing Quarterly: Overcoming Learning Difficulties 14(4), 379–396 (1998)
21. Kamp, H.: A theory of truth and semantic representation. Formal semantics-the essential readings, 189–222 (1981)
22. Kling, H.: Get more out of group projects by using structured brainstorming. Quality Progress 23(3), 136–136 (1990)
23. Koopman, P.: How to write an abstract. Carnegie Mellon University. Retrieved May 31, 2013 (1997)
24. Morris, J., Hirst, G.: Lexical cohesion computed by thesaural relations as an indicator of the structure of text. Computational Linguistics 17(1), 21–48 (1991)
25. Nadeau, D., Sekine, S.: A survey of named entity recognition and classification. Lingvisticae Investigationes 30(1), 3–26 (2007)
26. Osborn, A.: Applied Imagination - Principles and Procedures of Creative Problem-Solving. Charles Scribner's Sons (1953)
27. Park, Y., Lee, S.: How to design and utilize online customer center to support new product concept generation. Expert Systems with Applications 38(8) (2011)
28. Sarwar, B., Karypis, G., Konstan, J., Riedl, J.: Item-based collaborative filtering recommendation algorithms, pp. 285–295 (2001)
29. Song, F., Croft, W.B.: A general language model for information retrieval. In: Proceedings of the Eighth International Conference on Information and Knowledge Management, pp. 316–321. ACM (1999)
30. Su, X., Khoshgoftaar, T.M.: A survey of collaborative filtering techniques. Advances in Artificial Intelligence 2009, 4 (2009)
31. Tan, A.H., et al.: Text mining: the state of the art and the challenges. In: Proceedings of the PAKDD 1999 Workshop on Knowledge Disocovery from Advanced Databases, pp. 65–70 (1999)
32. Taylor, J.W.: How to create new ideas. Prentice-Hall (1961)
33. Thorleuchter, D.: Finding new technological ideas and inventions with text mining and technique philosophy. In: Data Analysis, Machine Learning and Applications, pp. 413–420 (2008)

34. Thorleuchter, D., den Poel, D.V., Prinzie, A.: A compared r&d-based and patent-based cross impact analysis for identifying relationships between technologies. Technological Forecasting and Social Change **77**(7), 1037–1050 (2010)
35. Thorleuchter, D., Van den Poel, D.: Companies website optimising concerning consumer's searching for new products. In: 2011 International Conference on Uncertainty Reasoning and Knowledge Engineering (URKE), vol. 1. IEEE (2011)
36. Thorleuchter, D., Van den Poel, D.: Semantic technology classificationa defence and security case study. In: 2011 International Conference on Uncertainty Reasoning and Knowledge Engineering (URKE), vol. 1, pp. 36–39. IEEE (2011)
37. Thorleuchter, D., Van den Poel, D.: Extraction of ideas from microsystems technology. In: Jin, D., Lin, S. (eds.) Advances in CSIE, Vol. 1. AISC, vol. 168, pp. 563–568. Springer, Heidelberg (2012)
38. Thorleuchter, D., Van den Poel, D., Prinzie, A.: Extracting consumers needs for new products-a web mining approach. In: Third International Conference on Knowledge Discovery and Data Mining, WKDD 2010, pp. 440–443. IEEE (2010)
39. Thorleuchter, D., den Poel, D.V., Prinzie, A.: Mining ideas from textual information. Expert Systems with Applications **37**(10), 7182–7188 (2010)
40. Trampuš, M., Mladenic, D.: Constructing domain templates with concept hierarchy as background knowledge. Information Technology And Control **43**(4) (2014)
41. Wallas, G.: The art of thought (1926)
42. Wang, C., Lu, J., Zhang, G.: Mining key information of web pages: A method and its application. Expert Systems with Applications **33**(2), 425–433 (2007)
43. Yoon, J.: Detecting weak signals for long-term business opportunities using text mining of web news. Expert Systems with Applications **39**(16), 12543–12550 (2012)
44. Young, J.W.: A technique for producing ideas. NTC Business Books (1975)

A   Appendix

Table 6. S-phrase cue terms - "method" was used as a seed term, with trigger words being expanded through synonym extraction via *www.thesaurus.com* and isolating nearest neighbours using *Word2vec* (see https://code.google.com/p/word2vec/).

approach, technique, scheme, algorithm, analysis, model, modelling, methodology, strategy, framework, tool, procedure, structure, processing, heuristic, mechanism, architecture, theory, paradigm, formalism, platform, simulation

Table 7. Noun-Phrase filtering terms

overall, primary, key, valuable, excellent, potential, essential, unique, numerous, important, prior, practical, basic, different, simple, successful, current, possible, previous, existing, well-established, independent, particular, usual, new, old, powerful, main, common, detailed, efficient, good, acceptable, effective, novel, state-of-the-art, useful, modern, unreliable, additional, methodological, available, recent, general, specific, creative, brief, critical, major, second, reasonable, various, personal, latest , interesting

Social and Graphs

Discovering Audience Groups
and Group-Specific Influencers

Shuyang Lin[1]([✉]), Qingbo Hu[1], Jingyuan Zhang[1], and Philip S. Yu[1,2]

[1] University of Illinois at Chicago, Chicago, IL, USA
{slin38,qhu5,jzhan8,psyu}@uic.edu
[2] Institute for Data Science, Tsinghua University, Beijing, China

Abstract. Recently, user influence in social networks has been studied extensively. Many applications related to social influence depend on quantifying influence and finding the most influential users of a social network. Most existing work studies the global influence of users, i.e. the aggregated influence that a user has on the entire network. It is often overlooked that users may be significantly more influential to some audience groups than others. In this paper, we propose *AudClus*, a method to detect *audience groups* and identify *group-specific influencers* simultaneously. With extensive experiments on real data, we show that AudClus is effective in both the task of detecting audience groups and the task of identifying influencers of audience groups. We further show that AudClus makes possible for insightful observations on the relation between audience groups and influencers. The proposed method leads to various applications in areas such as viral marketing, expert finding, and data visualization.

Keywords: Social influence · Influencer detection · Audience group

1 Introduction

Quantifying influence to find the most influential users in social networks is a fundamental problem of social network studies. Many important applications such as influencer detection and viral marketing rely on this problem. Most existing studies quantify the influence of a user as a globally aggregated influence value. An observation that is often overlooked by these studies is that a social network contains various groups of users, and the strength of influence of a user varies drastically over different groups. On one hand, most users have their influence limited to a small part of the social network. Even the globally most influential users of a social network have their influence concentrated to some specific groups of audience. On the other hand, different groups of users in a social network have their own specific sets of influencers.

Based on this observation, in this paper, we explore group-specific influence. We attempt to (1) detect audience groups and (2) identify influencers of audience groups. The tasks have two major challenges. **First**, to make the results meaningful, audience groups should reflect natural boundaries of influence, i.e. users

© Springer International Publishing Switzerland 2015
A. Appice et al. (Eds.): ECML PKDD 2015, Part II, LNAI 9285, pp. 559–575, 2015.
DOI: 10.1007/978-3-319-23525-7_34

in the same audience group share similar influencers, while users in different audience groups have different influencers. However, most existing community detection algorithm is not specifically optimized for detecting such audience groups. **Second**, current methods for detecting global influencer do not work well for the task of detecting group-specific influencers. One can certainly consider each audience group as a network and apply existing influencer detection algorithms to each network. This naive approach, however, yields to suboptimal results, because it can only detect influencers within groups, but an influencer of an audience group may actually be outside that group.

To solve these problems, in this paper, we propose **AudClus**, a probabilistic mixture model based method to detect audience groups and group-specific influencers simultaneously. By using information diffusion data, it groups users into different audience groups according to the users who are influential to them, and simultaneously quantifies the influence of users with respect to each audience group. Both the tasks of detecting audience groups and identifying group-specific influencers benefit from the simultaneous inference.

The main contributions of this paper are summarized as follows.

– We propose AudClus, a probabilistic mixture model based method, to capture audience groups and group-specific influence. We design an EM algorithm to infer audience groups and users' influence over audience groups simultaneously.
– AudClus is very flexible in capturing group-specific influence in social networks. It does not rely on the structure of social networks or any specific information diffusion model. It can capture both direct and indirect influence.
– AudClus provides a new tool for analysis and visualization of social influence. It leads to interesting observations and insightful understandings on the structure of influence in social networks. It facilitates applications such as finding experts in specific areas, and targeting specific groups of audience for viral marketing.

2   Preliminary

A social network is often considered as a graph, with users as nodes, and links between users as edges. By considering a social network as a graph, graph clustering or community detection algorithms can be used to detect groups from the social network. For a given social network, different community detection algorithms may lead to substantially different clustering results. In this paper, we are interested in detecting **audience groups**, which reflect users' behaviors of being influenced in information diffusion processes. Specifically, users in the same audience group are influenced by a similar set of influencers, while users in different groups are influenced by different influencers. To serve this purpose well, we propose **AudClus**, a clustering method, which detect groups of users from **information diffusion data**, instead of from social network structure.

When information diffuses in a social network, it is carried by actions of users in the network. An **action** of a user is, for example, posting a tweet in Twitter, or publishing a paper in the citation network. Each action comes with some information. For example, a tweet can talk about some news and events, and a paper can propose or adopt some techniques. Information carried by an action may be introduced into the social network by the current action itself, or it may be introduced by some previous actions and adopted by the current action. If a current action adopts information from some previous actions, we say that information propagates from the previous actions to the current action, and we define it by an **information propagation link** from each previous action to the current action. In this paper, we study the case that the information propagation links are observable in the data. For example, in Twitter, a retweeting or a replying can be considered as a information propagation link from the original tweet to the current tweet, while in the citation network, a citing can be considered as a link from the reference paper to the current paper.

A **diffusion pathway graph** contains a set of actions and the information propagation links between them. Formally, we define a diffusion pathway graph as follows.

Definition 1. *A diffusion pathway graph $D = (\mathcal{A}_D, \mathcal{L}_D)$ is a DAG (directed acyclic graph) of actions. Each action $a_i \in \mathcal{A}_D$ is taken by a user denoted by v_{a_i}. Directed links in $\mathcal{L}_D \subset \mathcal{A}_D \times \mathcal{A}_D$ define the information diffusion links between actions. A directed link $(a_i, a_j) \in \mathcal{L}_D$ means that action a_j is directly influenced by action a_i. Links in $\mathcal{L}_D$ should be acyclic. If $(a_i, a_j) \in \mathcal{L}_D$, we say a_i is a parent of a_j.*

The above definition of diffusion pathway graph is very general and flexible in the sense that it does not make any assumption on the underlying diffusion process. It can therefore be applied to various information diffusion models, and different information diffusion models may add different constraints to this general definition. For example, for an IC model [4,10], the diffusion pathway graph is actually a forest, since any action can be triggered by only one previous action, while the diffusion pathway graph for a LT model [5,9] can be any DAG. Besides, a diffusion pathway graph is not limited to describe the diffusion of one single piece of information. When the pieces of information are not explicitly available, it means less effort in preprocessing data. For example, we can directly construct a diffusion pathway graph from a citation network, with papers as actions, and citation relations as information diffusion links. We do not need to extract the pieces of information that is actually spread between papers.

The first goal of AudClus is to detect audience groups based on diffusion pathway graphs. More formally, given a set of users $\mathcal{V}$ and a set of diffusion pathway graphs $\mathcal{D} = \{D_1, \cdots, D_m\}$ with these users, it detects a set of audience groups $\mathcal{C}$, such that each user $u \in \mathcal{V}$ is assigned to a group $c \in \mathcal{C}$. Notice that the setting is different from that of traditional community detection problem: we detect groups of users from diffusion pathway graphs, instead of from the social network. Unlike social networks, nodes in diffusion pathway graphs are actions, not users. The difference in problem setting means traditional community detection algorithms cannot be directly applied to the task of detecting audience groups.

Table 1. Notations

	Description		Description
$\mathcal{V}$	Set of all users.	k	Number of audience groups.
$\mathcal{C}$	Set of all audience groups.	n	Number of users.
$\mathcal{D}$	Set of all diffusion pathway graphs.	θ_{vc}	The influence user v has on group c.
$\mathcal{A}_D$	Set of actions in diffusion pathway graph D.	η_{vc}	The conditional probability that v belongs to c.
$\mathcal{L}_D$	Set of links in diffusion pathway graph D.	ϕ_c	The prior probability of group c.
$z(v)$	The audience group user v is assigned to.	q	Transfer rate parameter in influence backtracing method.

The second goal of AudClus is to identify influencers who are specific to each audience group. To achieve that, AudClus quantifies the influence from each user $v \in \mathcal{V}$ to each group $c \in \mathcal{C}$, then it can identify users who are the most influential to a specific group. As we will show in the next section, the two goals of AudClus can be achieved simultaneously under a mixture model framework.

We summarize notations in Table 1.

3   The AudClus Method

In this section, we introduce a probabilistic mixture model based method to detect audience groups of a social network and to quantify group-based influence for users simultaneously. We will first study a simple case, which we call single-direct case. In the single-direct case, each action is either spontaneous or influenced by exactly one previous action, i.e. each action either has no parent in the diffusion pathway graph or has one single parent, and only the influence from the parent is considered. We will first show a probabilistic mixture model for audience clustering for the single-direct case, and then extend the model for more general cases.

3.1   Audience Clustering for the Single-direct Case

The intuition for audience clustering is that users who are influenced by similar sets of influencers should be assigned to the same group. The proposed clustering method originates from the probabilistic mixture model proposed in [15], which decides the group of a user according to the neighbors whom he is linked to and assigns users who are linked to similar sets of neighbors into the same group. The original model in [15] was designed for undirected graph, but we extend it to make it work with directed graphs such as diffusion pathway graphs. Besides, in the proposed model, the groups of users are decided by their influencers, not by their neighbors.

The basic concepts with regard to the proposed clustering model are as follows. From a set of users $\mathcal{V}$, each user u is assigned to a group $c \in \mathcal{C}$, denoted by $z(u)$. For each group $c \in \mathcal{C}$ and each user $v \in \mathcal{V}$, θ_{vc} defines the influence that user v has on group c. More specifically, for an action taken by users in group c, θ_{vc} is the probability that the action is influenced by some previous actions taken by user v. Notice that we are considering the single-direct case that each action is either spontaneous or influenced by exactly one parent action. For each action a_i that is influenced by a parent action, we regard the user who takes the parent action as the influencer of a_i, denoted by $r(a_i)$. $r(a_i)$ is generated from a categorical distribution as follows.

$$r(a_i) \sim Categorical_{|\mathcal{V}|}(\boldsymbol{\theta}_{\cdot z(v_{a_i})})$$

where v_{a_i} is the user who takes action a_i, and $z(v_{a_i})$ is the group that v_{a_i} belongs to. $\boldsymbol{\theta}_{\cdot c} = \{\theta_{vc}\}_{v \in \mathcal{V}}$ denotes the influence from users in $\mathcal{V}$ to group c.

For any group c, $\boldsymbol{\theta}_{\cdot c}$ are the parameters for a categorical distribution, which should satisfy the following normalization condition.

$$\sum_{v \in \mathcal{V}} \theta_{vc} = 1, \ \forall c \in \mathcal{C}$$

We consider the clustering of users as a probabilistic mixture model. The prior probability for group c is denoted by ϕ_c, satisfying the following normalization condition.

$$\sum_{c \in \mathcal{C}} \phi_c = 1$$

We denote with $\mathcal{Z}$ the multivariate random variable that consists of $z(v)$ for all $v \in \mathcal{V}$, i.e. $\mathcal{Z} = \{z(v)\}_{v \in \mathcal{V}}$. Similarly, we have $\Theta = \{\theta_{vc}\}_{v \in \mathcal{V}, c \in \mathcal{C}}$ and $\Phi = \{\phi_v\}_{v \in \mathcal{V}}$.

Given the parameters Θ and Φ, the joint probability of $\mathcal{D}$ and $\mathcal{Z}$ is the product of two probabilities: the probability that each user v is assigned to the group $z(v)$, and the probability that each action a_i influenced by the influencer $r(a_i)$. Formally, the likelihood function of parameters Θ and Φ, are given as follows.

$$\mathcal{L}(\Theta, \Phi; \mathcal{D}, \mathcal{Z}) = \left(\prod_{D \in \mathcal{D}} \prod_{a_i \in A_D} \theta_{r(a_i)z(v_{a_i})} \right) \left(\prod_{v \in \mathcal{V}} \phi_{z(v)} \right)$$
$$= \left(\prod_{v \in \mathcal{V}} \prod_{u \in \mathcal{V}} \theta_{vz(u)}^{A_{vu}} \right) \left(\prod_{v \in \mathcal{V}} \phi_{z(v)} \right) \tag{1}$$

where

$$A_{vu} = \sum_{D \in \mathcal{D}} \sum_{\substack{a_i \in A_D, \\ v(a_i) = u}} I_{r(a_i) = v} \tag{2}$$

denotes the number of actions of user u that are influenced by user v in all diffusion pathway graphs.

Parameter Estimation. We estimate the parameters Θ and Φ by their maximum likelihood estimation. Notice that the group of each user $z(u)$ is the missing data that also needs to be inferred. Therefore, the problem of finding maximum likelihood estimation is formalized as follows:

$$\max_{\Theta, \Phi} \sum_{\mathcal{Z}} p(\mathcal{D}, \mathcal{Z} | \Theta, \Phi)$$

We solve this problem by EM algorithm. In the E-step, we calculate expected value of log-likelihood function, with respect to the conditional distribution $\mathcal{Z}$. The expected value is defined as follows.

$$E_{\mathcal{Z}|\Theta^{(t)}, \Phi^{(t)}} \log \mathcal{L}(\Theta, \Phi; \mathcal{D}, \mathcal{Z}) = \sum_{v \in \mathcal{V}} \sum_{u \in \mathcal{V}} \sum_{c \in \mathcal{C}} A_{vu} \eta_{uc}^{(t)} \log \theta_{vc} + \sum_{v \in V} \sum_{c \in \mathcal{C}} \eta_{vc}^{(t)} \log(\phi_c)$$
$$\tag{3}$$

where

$$\eta_{uc}^{(t)} = \left(\phi_c^{(t)} \prod_{v \in V} (\theta_{vc}^{(t)})^{A_{vu}} \right) / \left(\sum_{c' \in C} \phi_{c'}^{(t)} \prod_{v \in V} (\theta_{vc'}^{(t)})^{A_{vu}} \right) \tag{4}$$

is the conditional probability that $z(u) = c$ given $\mathcal{D}$ under the current estimations of parameters $\Theta^{(t)}$ and $\Phi^{(t)}$.

In the M-step, we update estimation of Θ and Φ to maximize the expected log-likelihood. By taking partial derivatives of Equation 3, we can find out that the expected log-likelihood is maximized by the following values of parameters.

$$\theta_{vc}^{(t+1)} = \left(\sum_{u \in V} A_{vu} \eta_{uc}^{(t)} \right) / \left(\sum_{w \in V} \sum_{u \in V} A_{wu} \eta_{uc}^{(t)} \right) \tag{5}$$

and

$$\phi_c^{(t+1)} = \left(\sum_{u \in V} \eta_{uc}^{(t)} \right) / \left(\sum_{c' \in C} \sum_{u \in V} \eta_{uc'}^{(t)} \right) \tag{6}$$

We repeat the E-step and the M-step until it converges. When it converges, the **influence** of user v on group c is defined by θ_{vc}, the value that $\theta_{vc}^{(t)}$ converges to, while the **belongingness** of user v to group c is defined by η_{vc}, the value that $\eta_{vc}^{(t)}$ converges to. When a non-probabilistic clustering is needed, we assign user v to the group c with the largest η_{vc}, i.e. $z(v) = \arg\max_{c \in C} \eta_{vc}$.

3.2 Generalized Model

In the previous section, we have introduced the audience clustering algorithm for the single-direct case that each action either has no parent or has one single parent in the cascade, and only the influence from the parent is considered. Many real applications, however, do not satisfy this condition for two reasons. First, in many diffusions, each action may have multiple parent actions. For example, in the citation network, each action (paper) can cite multiple previous papers, thus has multiple parent actions. Second, in many applications, both direct and indirect influence is important and should be considered. For example, in the citation network, an influential paper should not only have a large citation number itself, but also inspires some innovative papers which also have plenty of citations.

We first propose a partial credit method to generalize the clustering method, so that actions can have arbitrary number of parents in the diffusion pathway graph, and then further propose a influence backtracking method to incorporate both direct and indirect influence under the same model.

Partial Credit Method. In the single-direct case, each action a_i has one single influencer $r(a_i)$. If each action can have multiple parents, the assumption will be violated. However, similar to the partial credit method in [9], we can generalize the model by letting all parents of an action share the "credit" of influencing that action. Notice that the likelihood function in Equation 1 actually depends on A_{vu}, the number of times that user v influences user u. Thus, we can replace A_{vu} in Equation 2 with following value.

$$A_{vu} = \sum_{\substack{D \in \mathcal{D} \\ }} \sum_{\substack{a_j \in \mathcal{A}_D, \\ v(a_j)=v}} \sum_{\substack{a_i \in \mathcal{A}_D, \\ v(a_i)=u}} \frac{1}{|F(a_i)|} I_{a_j \in F(a_i)} \tag{7}$$

where $F(a_i)$ is set of parents of action a_i. In this equation, each parent of action a_i gets $\frac{1}{|F(a_i)|}$ credit for influencing action a_i.

Influence Backtracking Method. We now introduce an influence backtracking method to measure the influence between actions in a diffusion pathway graph. The benefits of influence backtracking method are as follows: (1) the same as the partial credit method, each action can have arbitrary number of parents; (2) both direct and indirect influence is captured by the same measurement. By incorporating the influence backtracking method, the AudClus model can be generalized to all diffusion pathway graphs under the flexible definition as in Definition 1.

The intuition of influence backtracking is measuring influence by the amount of information that is brought into the social network by an action and is adopted by following actions. Consider the scenario of an author writing a blog post. The author gets some information from some other blog posts, and brings in some new ideas at the same time. Therefore, some information in this new blog post originates from previous blog posts, which are listed as references of this post, and some may be traced back even further to references of the references. For each piece of information carried by action a_i, we can trace back the diffusion pathway graph to find out which action brings that piece of information to the network.

To simulate this process, for each piece of information in an action a_i, we use a reverse random walk to trace back which action brings this piece of information to the network. The reverse random walk is defined as follows:

1. It starts at the node a_i.
2. When it arrives at a node a_j with no parents, the random walk terminates at a_j.
3. When it arrives at a node a_j with some parents, with probability $1 - q$ the random walk terminates at the node a_j, and with probability q the random walk continues. If it continues, it has equal probability to walk to each parent of a_j. If the random walk terminates at the action a_j, it represents that the piece of information originates from action a_j. We call q the transfer rate parameter of influence backtracking.

Since the diffusion pathway graph is an acyclic graph of actions, it is easy to calculate the probability that the random walk terminates at action a_j. The probability can be calculated recursively by the equation as follows.

$$Q(a_j, a_i) = \begin{cases} 1 & \text{if } F(a_i) = \emptyset, i = j \\ 0 & \text{if } F(a_i) = \emptyset, i \neq j \\ 1 - q & \text{if } F(a_i) \neq \emptyset, i = j \\ \frac{q}{|F(a_i)|} \sum_{a_k \in F(a_i)} Q(a_j, a_k) & \text{if } F(a_i) \neq \emptyset, i \neq j. \end{cases}$$

where $F(a_i)$ is set of parents of a_i.

Suppose that there are M pieces of information in an action a_i. We use the random walk to trace back where each piece of information is from. When M is large enough, the fraction of information pieces that are carried by action a_i and originate from action a_j is approximately $Q(a_j, a_i)$. Thus, for an action a_i and a previous action a_j, we regard $Q(a_j, a_i)$ as the part of a_i that is influenced by a_j, and replace A_{vu}, the number of actions of user u that are influenced by actions of v in Equation 2, by the value as follows.

$$A_{vu} = \sum_{D \in \mathcal{D}} \sum_{\substack{a_j \in \mathcal{A}_D, \\ v(a_j)=v}} \sum_{\substack{a_i \in \mathcal{A}_D, \\ v(a_i)=u}} Q(a_j, a_i) \tag{8}$$

4   Experiment

4.1   Experiment Setup

Datasets We experiment with two real-world datasets.

- **Citation Dataset.** This is the citation network dataset released by Arnet-Miner [17]. The publication data are extracted from DBLP, ACM and other sources. Since the influence of authors changes over time, to reflect current landscape of influence, we have removed publications before year 2000. We have also filtered out authors who have less than 5 publications. After pre-processing, the dataset contains 368,101 publications of 113,006 authors, and 592,889 citation links. It also contains conference information of publications, which we use for clustering evaluation. The original dataset considers venues such as CoRR as "conferences". It also contains some less competitive conferences which have far more larger number of accepted papers than normal conferences. We use the conference list provided by Microsoft academic search (http://academic.research.microsoft.com/) to clean the data. We only consider the top 100 computer science conferences listed by Microsoft academic search in the experiment.
- **Meme Dataset.** This is the meme dataset from Memetracker [11]. It contains posts from news websites and blogs, and links between then. We use the meme dataset crawled at August 2008. We consider websites as nodes ("users") of the network, and posts in websites as actions. Diffusion pathway graphs are generated from links between posts. We have removed websites with less than 5 posts. After preprocessing, the dataset contains 40,072 websites, 394,636 pages, and 1,394,710 links.

Methods. For quantitative evaluation, we compare following proposed methods and baselines.

- **AC-i.** AudClus with information backtracking.
- **AC-p.** AudClus with partial credit.
- **MD.** Mixture model proposed in [15]. We construct a directed network of users by adding a directed link from user v to user u, if there is some actions of user u linked to some actions of user v in some diffusion pathway graphs.

- **FG.** Fast greedy algorithm in [6]. It is a modularity based community detection algorithm. The algorithm is implemented in the igraph network analysis package [7]. Similar to **MD**, we construct a directed network of users from diffusion pathway graphs, but each edges in the user network weighted by the number of links between actions of the two users.

4.2 Qualitative Analysis and Case Studies

Citation Dataset. We begin with a qualitative analysis on the citation dataset. As we will show later, **AC-i** achieves best quantitative evaluation result among all algorithms, and it achieve best result with $q = 0.3$ and $k = 35$ for the citation dataset. We conform to this setting for this part of experiment.

In Table 2, we show an overview of the audience groups. For each group, we show its top 3 most common conferences, and its top one influencer. We assign each user v to the group that he has the largest belongingness η_{vc}. We identify frequent conferences of a group by counting the number of users in that group who have published papers in each conference, and finding the top 3 conferences with the largest counts. We identify the top influencer of a group by finding the user with the largest influence θ_{vc}.

The list of top frequent conferences for groups provides intuitive observations on the clustering quality. First, the frequent conferences for a certain group are usually conferences that are related to the same research area. For example, the top 3 frequent conferences for group 5 (KDD, CIKM, ICDE) are conferences related to the data mining area, while those for group 8 (ICIP, CVPR, ICCV) are related to the computer vision area. Second, different groups tend to have different list of frequent conferences. No two groups share exactly the same frequent conference list. Most overlaps of frequent conferences between groups can be explained by the phenomenon that a conference is often related to more than one research areas. For example, the top 3 frequent conferences of group 3 are ICDE, VLDB and CIKM, and CIKM and ICDE are also the 2nd and 3rd most frequent conferences for group 5, respectively. The explanation for it is as follows: (1) group 3 reflects the database area, while group 5 reflects the data mining area. (2) CIKM and ICDE accept papers in both the database and data mining areas.

To get a detailed observation on the extracted audience groups, we show longer lists for top influencers and top frequent conferences for group 5 in Table 5. Values in parentheses after names of influencers are θ_{vc} for that influencer. Values in parentheses after conference are the percentage of users in the group who have publications in the conference. It is very obvious that the top frequent conferences are related to the data mining area and the top influencers are indeed influential researchers in that area.

In Figure 1, we display another case study. In Figures 1(a) and 1(b), we show the influence spread (θ_{vc}) and belongingness distribution (η_{vc}) of author *Jiawei Han* over all groups. As shown by the figures, the belongingness of Jiawei Han almost completely concentrates to group 5. However, the influence of Jiawei Han spreads over several groups, although his influence on group 5 is much larger

Table 2. Audience groups

	Most common conf.	Top Influencer
0	SC, ICS, ICPP	Jos E. Moreira
1	ICC, WCNC, ICASSP	Anil K. Jain
2	ICSE, HICSS, ECOOP	Barry W. Boehm
3	ICDE, VLDB, CIKM	David J. DeWitt
4	PODC, ICDCS, ICC	Sanjay Jain
5	KDD, CIKM, ICDE	Jiawei Han
6	CDC, EUROCRYPT,CRYPTO	Moni Naor
7	SODA, ICRA, ICIP	Joseph S. B. Mitchell
8	ICIP, CVPR, ICCV	David J. Hawkes
9	CAV, CONCUR, LICS	Moshe Y. Vardi
10	IJCAI, AAAI, KR	Endre Boros
11	NIPS, ICML, IROS	Kalyanmoy Deb
12	ICRA, IROS, CHI	David H. Laidlaw
13	DAC, ICCAD, DATE	David Blaauw
14	ACL, SIGIR, COLING	Andrew McCallum
15	IJCAI, HICSS, AAAI	Gheorghe Paun
16	CHI, CSCW, UIST	Benjamin B. Bederson
17	ICIP, ICPR, ICASSP	Anil K. Jain
18	ITC, DATE, DAC	Krishnendu Chakrabarty
19	ICRA, IROS, ICPR	Sebastian Thrun
20	DATE, DAC, ISCAS	Margaret Martonosi
21	HPDC, SC, ICPP	Ian T. Foster
22	IJCAI, AAAI, ICALP	Jack H. Lutz
23	AAAI, IJCAI, ICRA	Milind Tambe
24	SC, POPL, LICS	William Gropp
25	ICC, IROS, INFOCOM	Edward R. Dougherty
26	ICASSP, ISCAS, ICC	Aapo Hyvrinen
27	WWW, CIKM, ICDE	Ian Horrocks
28	HICSS, ICC, ICRA	Viswanath Venkatesh
29	CDC, ICC, HICSS	Wil M. P. van der Aalst
30	ICC, INFOCOM, WCNC	Donald F. Towsley
31	ICIP, ICPR, ICC	Etienne E. Kerre
32	ICASSP, ACL, IROS	Andreas Stolcke
33	OR, SODA, ICC	David E. Goldberg
34	SODA, STOC, FOCS	Christos H. Papadimitriou

Table 3. Running time of algorithms (in seconds)

	AC-i	AC-p	MD	FG
Preprocessing time	263.5	123.2	69.9	77.4
Inference time	104.94	39.5	34.3	7182.4

Table 4. Case study: top influential websites

First group(political news)		
	website	description
1	telegraph.co.uk	newspaper
2	foxnews.com	news channel
3	nydailynews.com	newspaper
4	msnbc.msn.com	news channel
5	timesonline.co.uk	newspaper
6	nypost.com	newspaper
7	news.bbc.co.uk	news channel
8	cnn.com	news channel
9	politicalticker.blogs.cnn.com	political news blog
10	troktiko.blogspot.com	political news blog
Second group (technology)		
	website	description
1	blog.wired.com	technology magazine
2	gizmodo.com	technology blog
3	digg.com	news aggregator (technology)
4	telegraph.co.uk	newspaper
5	arstechnica.com	technology news
6	universetoday.com	technology news
7	sciencedaily.com	technology news
8	engadget.com	technology blog
9	msnbc.msn.com	news channel
10	scienceblogs.com	technology blog

Table 5. Case study: data mining group (group 5)

	Top influencers	Frequent conference
1	Jiawei Han (0.0251)	KDD (38.5%)
2	Jian Pei (0.0194)	CIKM (33.0%)
3	Charu C. Aggarwal (0.0096)	ICDE (29.2%)
4	Mohammed Javeed Zaki (0.0075)	VLDB (15.2%)
5	Philip S. Yu (0.0060)	WWW (11.8%)
6	Ramesh C. Agarwal (0.0060)	AAAI (9.4%)
7	Johannes Gehrke (0.0035)	ICML (8.8%)
8	Ke Wang (0.0043)	IJCAI (8.7%)
9	George Karypis (0.0041)	SIGIR (8.0%)
10	Rakesh Agrawal (0.0035)	ICC (6.0%))

than his influence on other groups. Similar observations hold for other users in the network as well: a user who is influential to a group does not have to be a member of that class; a user can have influence over more than one groups; however, the strength of influence can be very different in different groups.

Meme Dataset. For the meme dataset, we use **AC-i** with $q = 0.3$ and $k = 10$. Since there is no information like conferences in the citation network, it is hard to summarize the category or area of each group. Instead, we show a case study with that dataset. In Table 4, we show the top influencer of two groups. For the first group, the top 8 influencers are either newspapers or news channels, and political news is a major topic covered by these newspapers and channels. The last 2 are political news blogs. For the second group, most of the top influencers are technology news websites, except for msnbc.msn.com. Notice that msnbc.msn.com appears in the top influencer lists of both groups, but it is more influential on the first group than on the second group. It conforms to the intuition that MSNBC is a more influential source for political news than it is for technology news.

4.3   Quantitative Analysis

Clustering Evaluation. In this section, we evaluate the quality of clustering quantitatively. For both datasets, the ground truth of groups is not available. However, for the citation dataset, we can use the conferences of an author's papers to roughly identify the group of authors. Thus, we can conduct quantitative evaluation of clustering on the *citation dataset*.

We use three measurements for the evaluation of clustering: purity, mutual information (MI), and normalized mutual information (NMI). The three measurements are often used for evaluation of clustering quality. They are usually defined for datasets with a set of explicit classes and each node is assigned to exactly one class. In the citation datasets, conferences of authors' identify the areas of study and can be used for the evaluation. However, unlike datasets with explicit classes, each author in the citation dataset can publish in multiple conferences. To make the evaluations work for the citation dataset, we use modified definitions of purity, MI, and NMI as follows:

- **Purity.** We first find most frequent conferences for each group as we did in the last section, and then defined $purity(m)$ as the fraction of users who have published in at least one of the top m most frequent conferences of their groups.
- **MI.** For each conference e, we divide users into positive class (users who published in this conference), and negative class (users who did not publish in this conference). We then calculate the mutual information between positive/negative classes and groups of users. Formally, the mutual information is defined as $I(f,\mathcal{C}) = \sum_{l_e=\{0,1\}} \sum_{c\in\mathcal{C}} \frac{N_{l_e c}}{N} \log \frac{N_{l_e c}N}{N_{l_e}.N._c}$, where $l_e = 1$ and $l_e = 0$ represent the positive and negative classes, respectively. $N_{l_e c}$ is the number of users in group c that belong to the positive or negative class. $N_{l_e}. = \sum_{c\in\mathcal{C}} N_{l_e c}$ is the total number of users belong to the positive or negative class. $N._c = \sum_{l_e\in\{0,1\}} N_{l_e c}$ is the total number of users in group c. N is the total number of users. For each conference e, we calculate the mutual information according to the equation above, and then we calculate the average value over all conferences.
- **NMI.** When the number of groups k is large enough, it is easy for a clustering method to achieve high values of purity and MI. Normalized mutual information (NMI) can be used to tradeoff the clustering quality against the increasing number of groups. For a conference e, the normalized mutual information between it and clustering $\mathcal{C}$ is defined as $NMI(e,\mathcal{C}) = I(e,\mathcal{C})/\sqrt{H(e)H(\mathcal{C})}$, where $H(e) = -\sum_{l_e=\{0,1\}} \frac{N_{l_e}.}{N} \log(\frac{N_{l_e}.}{N})$ is the entropy for conference e, and $H(\mathcal{C}) = -\sum_{c\in\mathcal{C}} \frac{N._c}{N} \log(\frac{N._c}{N})$ is the entropy for clustering. Similar to MI, we use the average value over all conferences to evaluate the clustering.

Since the result clusterings of **AC-i**, **AC-p** and **MD** are influenced by the random values we use to initialize parameters, we run each of them 10 times for each setting and show both mean value and standard deviation.

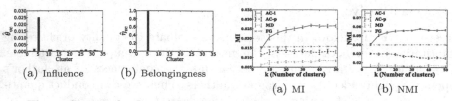

(a) Influence   (b) Belongingness

Fig. 1. Case study: Jiawei Han

(a) MI   (b) NMI

Fig. 2. MI and NMI

Figure 3 illustrates purity evaluation for all algorithms (for **AC-i**, q is set to 0.3). In Figures 3(a), 3(b) and 3(c), we show $purity(m)$ with $m = 1, 3, 5$, respectively, for varying number of groups. In each case, when k increases, $purity(m)$ for **AC-i**, **AC-p** and **MD** first increases with increasing k, and then stays around a certain value or even drops slightly when k is large. For **FG**, $purity(m)$ increases slightly when k increases. For each case, **AC-i** and **AC-p** consistently outperform **MD**. It suggests that by considering diffusion pathway graphs of actions, the proposed mixture model based algorithms **AC-i** and **AC-p** improves over **MD**, which is also a mixture model based algorithms but considers the links between users only. Further more, **AC-i** achieves better clustering quality than **AC-p** and **MD**. That is because **AC-i** quantifies both the direct and indirect influence simultaneously, while **AC-p** and **MD** consider direct influence only. Clustering quality of **AC-p** and **FG** are comparable, with **AC-p** slighter better for $purity(1)$, and **FG** better for $purity(3)$ and $purity(5)$. This is is more clearer in Figure 3(d), in which we show $purity(m)$ with varying m for $k = 35$. As illustrated Figure 3(d), when m, the number of top conferences of each group, increases, $purity(m)$ for all algorithms goes up steadily. For each m, **AC-i** always achieves best $purity(m)$ among four algorithms, while **MD** always has lowest $purity(m)$. **AC-p** and **FG** has similar $purity(m)$ for $m = 1$. When n increases, **FG** achieves better clustering quality than **AC-p**. However, as we will show later, **FG** is much slower than mixture model based algorithms (**AC-p**, **AC-i**, and **MD**).

Figure 2(a) illustrates the MI measurement for algorithms. In the figure, k is illustrated on the X-axis, while MI is illustrated on the Y-axis. As shown by the figure, when k increases, the MI for **AC-i**, **AC-p** and **MD** first increases and then stays stable. while the MI for **FG** only slightly increases as k increases from 5 to 50. **AC-i** achieves best clustering quality among the three algorithms, and **AC-p** and **FG** outperform **MD**.

Figure 2(b) illustrates the NMI measurement. Comparing with the MI measurement in Figure 2(a), the normalization makes NMI favors clustering with smaller k. For **AC-i**, the maximum of NMI is achieved at $k = 35$, while for **AC-p**, it is achieved at $k = 15$. Comparing with the curve of **FG** in Figure 2(a), we can find out that the clustering of **FG** actually has an almost fixed entropy $H(\mathcal{C})$ when k increases. That suggests that **FG** does not make fully use of increasing number of groups k, and assign most of users to a handful of groups.

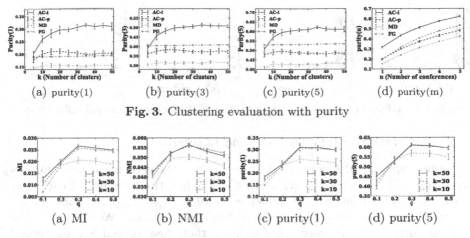

(a) purity(1)   (b) purity(3)   (c) purity(5)   (d) purity(m)

Fig. 3. Clustering evaluation with purity

(a) MI   (b) NMI   (c) purity(1)   (d) purity(5)

Fig. 4. Selection of parameter q

Running Time. Table 3 lists the running time for four algorithms. The running time for each algorithm has two parts: the preprocessing time and the inference time. For **AC-i** and **AC-p**, the preprocessing time is the time spent on calculating A_{vu} for all pair of users. For **MD** and **FG**, the preprocessing time is the time spent on constructing user networks. For each algorithm, the inference time is the time spent on generating the clustering. In Table 3, we show the preprocessing time and the inference time for the citation network for the case $k = 50$. As shown in the table, comparing with **MD**, both the preprocessing time and the inference time of **AC-i** and **AC-p** are increased as they consider actions for clustering inference. **AC-i** takes more time than **AC-p** because both direct and indirect influences are considered by **AC-i**. Although the preprocessing time of **FG** is similar to that of **MD**, the inference time is significantly longer than those of other algorithms. That is because, as a modularity-based community detection algorithm, **FG** invokes time-consuming calculation on graphs. Nevertheless, as we showed in the previous section, the clustering quality of **AC-p** is similar to **FG**, and the clustering quality of **AC-i** is significantly better than that of **FG**.

Selection of Parameter q. The clustering quality of **AC-i** depends on the transfer rate parameter q. In Figure 4, we illustrate the clustering quality of **AC-i** with varying q. Figures 4(a), 4(b),4(c) and 4(d) show the clustering evaluation with MI, NMI, $purity(1)$, and $purity(5)$, respectively. For each measurement, we show the curves for $k = 10, 30$ and 50. In each case, when q varies from 0.1 to 0.5, the clustering quality first increases then decreases. The explanation is as follows: when q is too small, the indirect influence is underestimated; when q is too large, the indirect influence is overestimated by the algorithm. For each case, **AC-i** achieves the best clustering quality when $q = 0.3$.

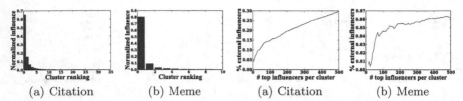

(a) Citation   (b) Meme   (a) Citation   (b) Meme

Fig. 5. User influence spread over groups   **Fig. 6.** Fraction of external influencers

4.4   Observations on Group-Specific Influence

Influence Spread of Users. First, we study how the influence of users spreads over groups. To illustrate that, we first normalize the influence of each user v over groups with his overall influence, i.e. $\theta_{vc}/\sum_{c'\in\mathcal{C}}\theta_{vc'}$, and rank groups according to the normalized influence v has on the groups. We then show the average normalized influence users have on their first ranked groups, second ranked groups, etc. As illustrated by the figure, the influence spread of users tends to concentrate to the first ranked groups. For the citation dataset, on average, more than 65% of a user's influence concentrates to the first ranked group. Nevertheless, users can still have significant influence on a few other groups. For example, in the citation dataset, on average, the second ranked group for a user has 15% influence of that user. These results confirm the intuition that the strength of influence of a user varies drastically over audience groups, but the influence is not limited to the group that the user belongs to.

The second question we are interested in is whether the overall influence of users is correlated with the extent that their influence spreads on different groups. To answer this question, we first quantify the extent of influence spread of a user by the entropy of user influence distribution. With larger entropy, the influence of that user tends to spread over different groups more evenly. In Figure 7, we illustrate the entropy of influence spread for users with increasing overall influence. Each point in the figure represents a small range of user overall influence, illustrated on the X-axes. The average entropy for users whose overall influence is within that range is illustrated on the Y-axes. As shown by the figure, comparing with users who has larger overall influence, users with smaller overall influence are more likely to have their influence concentrated to fewer groups.

Fraction of External Influencers. The other question about the influencers of groups that we study is: given an audience group, among the top m influencers of the group, how many of them belong to this group, and how may of them belong to other groups. Figure 6 illustrates the fraction of external influencers. In the figure, X-axes illustrate m, the number of top influencers of a group, while Y-axes illustrate the fraction of external influencers, i.e. the fraction of top influencers who does not belong to the group. (We show average value of that fraction over all groups). For both the citation and meme datasets, the fraction of external influencers increases as m increases, which suggests that influencers with larger influence on a group are more likely to be a member of that group. Moreover, the fraction of external influencers is larger in the meme dataset

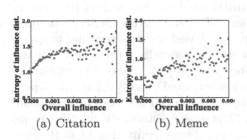

(a) Citation   (b) Meme

Fig. 7. Entropy of influence spread distribution vs. overall influence

Fig. 8. Visualization for influence between some groups in the citation dataset

than in the citation dataset. The difference can be explained by the inherent difference between citation networks and website networks: academic authors usually specialize in one or a few areas and seldom have large influence outside the area they specialize in, while many influential websites are comprehensive websites that cover various topics.

Visualization of Influence between Groups. At last, as an example for possible applications of AudClus in influence visualization, in Figure 8, we show the influence between 7 groups in the citation dataset. To quantify the influence from group c_i to group c_j, we calculate the average influence users in c_i have on group c_j, i.e. $Inf(c_i, c_j) = \sum_{z(v)=c_i} \theta_{vc_j}/|N_{c_i}|$ where N_{c_i} is the total number of users in c_i. We selected the groups that are related to database, data mining, and machine learning areas. For each group, we show its index, as well as the top 3 most frequent conferences. The complicated relation between those areas is clearly illustrated by the figure. For example, the research area of data mining (group 5) is strongly influenced by the areas of database (group 3), natural language processing (group 11), and machine learning theory (group 14). On the other hand, the area of data mining (group 5) also has large influence back to the database area (group 3), while it has less influence to the natural language processing and machine learning theory areas.

5   Related Work

Quantifying Influence. There has been extensive work on the problem of quantifying influence and detecting the most influential users. Some work regarded influence as the outcome of information diffusion processes, like the independent cascade (IC) model [4,10] and the linear threshold (LT) model [5,9]. This line of work proposed methods for finding a set of users such that the expected influence is maximized under a given information diffusion model. Another line of work conducted empirical studies to quantify the influence of users [1,3]. Topic-dependent influence was also studied frequently in recent years [13,16].

Information Diffusion and Community. Recently, researchers have taken notice of the relation between information diffusion and community structures in social networks. [12,18] proposed community-based greedy algorithms to speed up influence maximization on social networks. [8] generalized the influence maximization problem to the group level. Latest work in [2,14] analyzed social influence on the community level, which were closely related to our work in this paper. [14] proposed a hierarchical method to summarize social influence by reciprocal influence strength between communities. [2] proposed a stochastic mixture membership generative model to detect cascade-based community. The tasks that we work on are different from [2,14] in following aspects. **First**, our method focuses on detecting audience groups, while the previous models grouped users based on how they influence others and how they are influenced by others mixedly. **Second**, our method captures the influence of each user to each audience group. Therefore, it works for the task of identifying the most influential users to groups. The previous models only captured the influence between groups, and could not be used to identify the most influential users.

Since influence between users is inherently asymmetric, most traditional community detection algorithms designed for undirected network do not work well for the study of social influence. [15] proposed a community detection method based on probabilistic mixture model. It is a very flexible model that can be naturally extended to detect role-based groups, such as audience groups.

6   Conclusion

In this paper, we study audience groups in networks and group-specific influence of users. We propose AudClus, a mixture model based algorithm to detect audience groups and quantify group specific influence simultaneously. We also invent an influence backtracking method to capture both direct and indirect influence. We show qualitative and quantitative evaluations on real-world datasets. The proposed AudClus algorithm provides a new approach to understand the structure of influence in social networks, which leads to many insightful observations.

Acknowledgments. This work is supported in part by NSF through grants CNS-1115234, Google Research Award, the Pinnacle Lab at Singapore Management University, and Huawei grants.

References

1. Bakshy, E., Hofman, J.M., Watts, D.J., Mason, W.A.: Everyone's an influencer: quantifying influence on twitter. In: WSDM (2011)
2. Barbieri, N., Bonchi, F., Manco, G.: Cascade-based community detection. In: WSDM (2013)
3. Cha, M., Haddadi, H., Benevenuto, F., Gummadi, K.: Measuring user influence in twitter: the million follower fallacy. In: ICWSM (2010)
4. Chen, W., Wang, Y.: Efficient influence maximization in social networks. In: KDD (2009)

5. Chen, W., Yuan, Y., Zhang, L.: Scalable influence maximization in social networks under the linear threshold model. In: ICDM (2010)
6. Clauset, A., Newman, M.E.J., Moore, C.: Finding community structure in very large networks. Phys. Rev. E **70**(6 Pt 2), 066111 (2004)
7. Csardi, G., Nepusz, T.: The igraph software package for complex network research. InterJournal, Complex Systems **1695** (2006)
8. Eftekhar, M., Ganjali, Y., Koudas, N.: Information cascade at group scale. In: KDD (2013)
9. Goyal, A., Bonchi, F., Lakshmanan, L.V.: Learning influence probabilities in social networks. In: WSDM (2010)
10. Kempe, D., Kleinberg, J., Tardos, E.: Maximizing the spread of influence through a social network. In: KDD (2003)
11. Leskovec, J., Backstrom, L., Kleinberg, J.: Meme-tracking and the dynamics of the news cycle. In: KDD (2009)
12. Lin, S., Hu, Q., Wang, G., Yu, P.S.: Understanding community effects on information diffusion. In: Cao, T., Lim, E.-P., Zhou, Z.-H., Ho, T.-B., Cheung, D., Motoda, H. (eds.) PAKDD 2015, Part I. LNCS (LNAI), vol. 9077, pp. 82–95. Springer, Heidelberg (2015)
13. Liu, L., Tang, J., Han, J., Jiang, M., Yang, S.: Mining topic-level influence in heterogeneous networks. In: CIKM (2010)
14. Mehmood, Y., Barbieri, N., Bonchi, F., Ukkonen, A.: CSI: community-level social influence analysis. In: Blockeel, H., Kersting, K., Nijssen, S., Železný, F. (eds.) ECML PKDD 2013, Part II. LNCS, vol. 8189, pp. 48–63. Springer, Heidelberg (2013)
15. Newman, M.E.J., Leicht, E.A.: Mixture models and exploratory analysis in networks. PNAS **104**(23), 9564–9569 (2007)
16. Tang, J., Sun, J., Wang, C., Yang, Z.: Social influence analysis in large-scale networks. In: KDD (2009)
17. Tang, J., Zhang, J., Yao, L., Li, J., Zhong, L., Su, Z.: Arnetminer: extraction and mining of academic social networks. In: KDD (2008)
18. Wang, Y., Cong, G., Song, G., Xie, K.: Community-based greedy algorithm for mining top-k influential nodes in mobile social networks. In: KDD (2010)

Estimating Potential Customers Anywhere and Anytime Based on Location-Based Social Networks

Hsun-Ping Hsieh[1,2,3]([✉]), Cheng-Te Li[1,2,3], and Shou-De Lin[1,2,3]

[1] Graduate Institute of Networking and Multimedia, National Taiwan University,
Taipei, Taiwan
{d98944006,sdlin}@csie.ntu.edu.tw, ctli@citi.sinica.edu.tw
[2] Research Center for Information Technology Innovation, Academia Sinica,
Taipei, Taiwan
[3] Department of Computer Science and Information Engineering, National Taiwan
University, Taipei, Taiwan

Abstract. Acquiring the knowledge about the volume of customers for places and time of interest has several benefits such as determining the locations of new retail stores and planning advertising strategies. This paper aims to estimate the *number of potential customers* of arbitrary query locations and any time of interest in modern urban areas. Our idea is to consider existing established stores as a kind of *sensors* because the near-by human activities of the retail stores characterize the geographical properties, mobility patterns, and social behaviors of the target customers. To tackle the task based on store sensors, we develop a method called *Potential Customer Estimator* (PCE), which models the spatial and temporal correlation between existing stores and query locations using geographical, mobility, and features on location-based social networks. Experiments conducted on NYC Foursquare and Gowalla data, with three popular retail stores, Starbucks, McDonald's, and Dunkin' Donuts exhibit superior results over state-of-the-art approaches.

Keywords: Customer prediction · Retail stores · Location-based social network · Check-in data · Store sensor

1 Introduction

Modern big cities, such as New York City, London, Paris, and Taipei, are densely and crowded areas, where not only million of people live but also a great number of business established. As time proceeds, people move around in such urban areas in either a periodic or unpredicted manner. Various kinds of retail stores (e.g. Starbucks, McDonald's and Dunkin' Donuts) usually choose the locations possessing higher potential to attract more customers to construct new venues expecting more people can bring more revenue. In other words, the number of potential customers becomes one of the most important factor for business to

© Springer International Publishing Switzerland 2015
A. Appice et al. (Eds.): ECML PKDD 2015, Part II, LNAI 9285, pp. 576–592, 2015.
DOI: 10.1007/978-3-319-23525-7_35

determine their geographical placement or launch campaign events. It would be very practical and useful to acquire the knowledge about where and when a particular business can attract more consumers or audience.

In this paper, we aim to estimate the potential customers of an arbitrary location and at a given time in an urban area. By referring the number of potential customers as the number of people visited there, we propose to exploit the check-in records, place information, and social connections in location-based social networks (e.g. Foursquare and Gowalla) for estimating potential customers. The central idea is to consider existing stores of the target retail business (e.g. Starbucks) as a kind of *sensors* to estimate its potential customers at other locations without stores at any time. We use Figure 1 to elaborate our idea of estimating potential customers anywhere and anytime. Take the stores of Starbucks in New York City as examples, as marked by red circles. From location-based services, .we might know the historical customers (i.e., the number of check-ins) of each store month by month, illustrated by the histogram of check-in numbers in terms of months. Now Starbucks would like to construct a new store or hold a marketing campaign, with two arbitrary locations in mind, as labeled as A and B. The problem is to estimate the number of potential customers of such two locations month by month so that it is possible to acquire the knowledge which one can bring more profit when a new business or event is launched. Given the potential customers over time, Starbucks can further understand which months are more profitable.

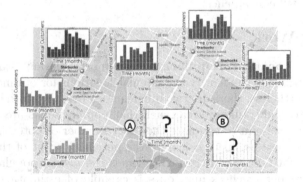

Fig. 1. An example for estimating potential customers anywhere and anytime.

Estimating the number of potential customers for arbitrary locations over time is a challenging problem. The characteristics of a location's geo-spatial neighborhood is usually one of the major factors that determine the potential customers. Such geo-spatial characteristics include population, spatial density, traffic flows, competitiveness (i.e., number of the same category of retail chain), how people interact and transit between different categories of venues, and the structure of underlying social connections in the near-by area. One major challenge of this task lies in how to model the complex composition of venues and various moving behaviors of people in its geo-neighborhood. On the other hand, the number of potential customers of a location might change and evolve over time. Both the temporal factors of periodic growth and decline (e.g. high vs. slack seasons and weekdays vs. weekend) and special activities (e.g. anniversary and seasonal discount campaign) can affect the customer numbers. Consequently, there might not exist regular patterns to be used to predict the potential customers of a location over time. This work tries to bring such temporal impact into our estimation model.

Given some locations in a city, certain time periods of interest, and a set of stores of a target retail chain (e.g. Starbucks) that already has venues established with historical check-in data, our goal is to estimate the number of potential customers for the given locations at the designated time periods (e.g. weeks or months). To deal with this problem, we devise a model called *Potential Customer Estimator* (PCE), whose idea is three-fold. First, we construct a *Correlation Graph* (CG), which is a *multi-layer* graph, to represent the spatial and temporal correlation between existing stores and the query locations. We investigate three categories of features, *geographical*, *mobility*, and *social*, to model the correlation between locations in *CG*. Second, since different features have different effects on the estimation target, we estimate the *location correlation* separately by investigating the predictability of each feature. The correlation values derived are represented as edge weights in *CG*. Third, based on the *CG* with location correlation values on edges, we develop a *Customer Inference Algorithm* (CIA), which iteratively adjusts the estimated number of potential customers of the query locations till convergence.

2   Related Work

Investigating Location Popularity. The most relevant study is GEO-SPOTTING [8], which is to identify the popular locations for optimal retail store placement. Nevertheless, there are two differences. First, they formulate the task as a *ranking* problem: ranking areas such that popular areas are at the top of the list. However, we aim to estimate the exact number of potential customers, which might be more useful to calculate the potential profit of the placement. Second, while what GEO-SPOTTING considers is the *overall* popularity (i.e., the accumulated number of check-ins), ours is capable of estimating the number of location check-ins for a particular week, month, and season, which can be regarded as a kind of reflection of the weekly, monthly, and seasonal revenues, to facilitate the advertising strategy for the retail chain. Though the studies of Li et al. [11] investigate the common characteristics of popular locations, Kisilevich et al. [9] analyze the geo-spatial properties of attractive areas, Tiwari and Kaushik [20] design a new popularity measure based on user category, visiting frequency, and stay time, they do not make the prediction of future popularity. For other relevant work, Fu et al. [5] propose to rank the residential real estate based on investment values by mining the opinions from online user reviews and offline human mobility. Chen et al. [2] use the road network data to find locations to set up new servers such that the cost of clients being served by nearest

servers is minimized. Liu et al. [13] leverage the technique of matrix factorization to recommend locations by modeling the geographical characteristics of
their neighborhoods. In addition, Hsieh et al. [6] develop a graph-based model to
infer miss sensor values through learning the correlation between heterogeneous
features and air quality values.

Human Mobility Prediction. Human mobility prediction is to predict the
next locations that the user might visit before. Monreale et al. [16] predict
the next location of a moving object with an assumption: people tend to follow common paths. With mined frequent trajectory patterns capturing common
paths, they construct a decision tree-like structure, T-pattern Tree, as a predictor of the next location. Ying et al. [21] leverage the semantic information,
which describes the activities (in the form of tags and types) of locations. Given
the recent moves of a user, they compute the matching score geographically and
semantically between mined frequent sub-trajectories and the given moves to
find the the next location. Sadilek et al. [18] predict the most likely location of
a user at any time, given the historical trajectories of his/her friends. They use
the discrete dynamic Bayesian network to model the motion patterns of users
from their friends.

3   Problem Statement

We define the number of potential customers, followed by the problem definition.

Definition 1: Number of Potential Customers. The *number of potential
customers* $pc(v)$ of a location $v \in L$ is the number of check-ins performed at v,
where $pc(v)$ is a positive integer, and L is the set of locations in the check-in data.
In addition, the *number of potential customers* of location v at time t_i, denoted
by $pc(v(t_i))$, is the number of potential customers derived from a certain time
period t_i (e.g. a week or a month).

   We use Foursquare check-in data in the work. We denote the maximum number of potential customers in a check-in data to be pc_{max}. Note that throughout
this paper we use terms "number of potential customers", "number of check-ins",
and "popularity" interchangeably.

**Problem Definition: Estimating Potential Customer Number Any-
where & Anytime.** Given a target retail chain and a set of its stores geographically established in the city, with the historical check-in data of the stores
in time periods $T = \langle t_1, t_2, ..., t_n \rangle$, the set L of all venues in the city, the underlying social network $\mathbb{G} = (\mathbb{V}, \mathbb{E})$ among people ($\mathbb{V}$ is the set of users and $\mathbb{E}$ is the
set of social relationships between $\mathbb{V}$), an arbitrary query locations v in the city
($v \notin L$), the goal is to estimate the number of potential customers $pc(v(t_i))$ of
location v in each time period $t_i \in T$.

4   Dataset

We aim to estimate the number of potential customers of query location by utilizing the check-in and venue data from the most well-known location-based service Foursquare[1] and the commonly-used location-based social network data Gowalla[2]. Since Foursquare had been launched in 2009, the volume of users, check-in records, and venue information are accumulated rapidly. Up to the end of 2013, there are 45 million users, 5 billion check-ins, and 60 million venues. Although Foursquare does not allow developers to directly access the check-in data, they allow users to share their check-ins publicly on Twitter[3]. Therefore, with the help of GEO-SPOTTING [8], we have the check-in data from Twitter and the venue data from Foursquare. To have adequate data for the experiments, we focus on New York City where Foursquare was launched and thus has significantly more users than any city in the world. The collected data in New York City contains 47,581 geo-tagged venues and 4,337,663 check-ins in a period of ten months (December 2010 to September 2011), i.e., forty weeks in total. Note that this data subset of NYC accounts for approximately 55% of all venues collected. As for the Gowalla location-based social network, which is collected by Cho et al. [3], there are totally 196,591 users, 950,327 social connections between users, and 6,442,890 check-in records collected from February 2009 to October 2010.

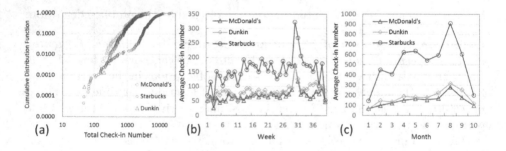

Fig. 2. Data Statistics: (a) Cumulative distribution function (CDF) of total check-ins per store for three retail chains. (b) The average check-ins of stores over weeks. (c) The average check-in of stores over months.

We target at the stores of three popular retail chains, Starbucks (SB), McDonald's (MC), and Dunkin' Donuts (DD). The statistics of each retail chain is reported as – the number of stores: 245, 89, and 149 for SB, MC, and DD; the total number of check-ins: 1,051,398, 100,520, and 187,704 for SB, MC, and DD respectively. The cumulative distribution of check-ins are also shown

---

[1] https://foursquare.com/

[2] https://snap.stanford.edu/data/loc-gowalla.html

[3] https://twitter.com/

in Figure 2(a). We can find that the check-in patterns of coffee shops (e.g. Starbucks) are different from those of fast food restaurants (e.g. McDonald's). Starbucks has the most number of stores as well as the most number of check-ins, in which its average number of check-in per store is almost four times than the other two chains. In addition, about 60% of Starbucks stores have check-in numbers higher than 3,000, which is significantly more than the other two as well. We believe it is because the time people take to stay in coffee shops is usually longer than that in fast food restaurants, and longer staying time would lead to higher possibility of performing check-ins. Since our goal is to estimate the number of potential customers, i.e., the evolution of check-in numbers over time, in Figure 2(b) and 2(c) we report the average number of check-ins per store over time in terms of weeks and months respectively. In general the check-in behaviors of three chains are different, except for a burst in the thirty week and in the eighth month. The average potential customer numbers of weeks fluctuate more significantly, comparing to those of months. These statistics show the difficulty of estimating potential customers.

5 Potential Customer Estimator (PCE)

5.1 Geographical, Mobility, and Social Features

We consider the following features to estimate potential customers. We calculate the features from the set of venues $N(v) = \{u : dist(u, v) < r\}$ in the near-by area with a disk of radius r centered at location v to be estimated, where $dist(u, v)$ is the geographical distance between venue u and location v. Note that we refer a venue to be a place that some kind of business has been established. We choose the radius r to be 200 meters in default according to the optimal neighborhood size suggested by the urban planning community [14]. There are three main categories of features. The first is the geographical features (GF), which describe the category distribution and the geographical interactions between venues. The specific feature items include:

- *Density* is the number of venues in the geographical neighborhood $N(v)$ of location v. The formation definition of density of location v is given by: $Density(v) = |\{u \in L : dist(u, v) < r\}|$, where L is the set of all venues in the data.
- *Neighbor Entropy* measures the heterogeneity of venue categories in $N(v)$. By denoting the set of venues with category c_i in the neighborhood of location v as $N_{c_i}(v)$ and the entire set of venue categories as C, the neighbor entropy can be defined as: $NbrEntropy(v) = -\sum_{c_i \in C} \frac{|N_{c_i}(v)|}{|N(v)|} \cdot \log \frac{|N_{c_i}(v)|}{|N(v)|}$.
- *Competitiveness* is the proportion of venues whose categories are the same as the category of the target store (e.g. "fast food restaurant" for McDonald's). Given the category of location v, denoted by c_v, its competitiveness is given by: $Compete(v) = -\frac{|N_{c_v}(v)|}{|N(v)|}$. Locations with lower competitiveness scores tend to be promising ones.

- *Attractiveness* is to capture the deployment and interactions between venue categories. If a location of a certain venue category can attract more locations with other venue categories in its neigborhood, such location is said to be more attractive. The attractiveness of location v is defined as: $Attract(v) = \sum_{c_i \in C} \log(\kappa_{c_i \to c_v}) \cdot (|N_{c_i}(v)| - |N'_{c_i}(v)|)$, where $|N'_{c_i}(v)|$ is the average number of neighboring locations of category c_i in the neighborhood of all the locations of category c_v. In addition, $\kappa_{c_i \to c_v}$ denotes the *inter-category coefficient* from category c_i to c_v. Such inter-category coefficient can be defined as: $\kappa_{c_i \to c_v} = \frac{|N| - |N_{c_i}|}{|N_{c_i}| \cdot |N_{c_v}|} \sum_{u \in L} \frac{|N_{c_v}(u)|}{|N(u)| - |N_{c_u}(u)|}$, where N is the entire set of locations in the dataset, N_{c_u} is the set of near-by locations of c_u.

The second category is the mobility features, which aim to model how users move and transit between venues. The specific feature items include:

- *Area Popularity* is the total number of check-ins for venues in $N(v)$. The formal definition of area popularity for location v is given by: $AreaPop(v) = |\{(CI(u) \in M : dist(u, v) < r\}|$, where $CI(u)$ denotes the set of check-in records at location u and M is the set of all check-ins in the data.
- *Transition Density* is the density of transitions between venues within $N(v)$. By denoting the set of consecutive check-in transitions between each pair of locations x and y as $TS((x, y) \in TS)$, the measure of transition density is defined as: $TransDensity(v) = |\{(x, y) \in TS : dist(x, v) < r \land dist(y, v) < r\}|$.
- *Incoming Flow* estimates the transitions from venues outside $N(v)$ to those in $N(v)$. The incoming flow score of location v is given by: $InFlow(v) = |\{(x, y) \in TS : dist(x, v) > r \land dist(y, v) < r\}|$.
- *Transition Attractiveness* is designed to estimate the probability of transitions between all other types of venues and venues of the same type as the target store. That says, assume people prefer to travel from locations of category c_u to locations of category c_v, if the near-by locations $u \in N(v)$ of location v can gather higher check-in numbers, then the transition attractiveness of location v tends to be high. The transition attractiveness is given by: $TransAttract(v) = \sum_{u \in N(v)} \rho_{c_u \to c_v} \cdot M_u$, where M_u is the set of check-ins at location u, and $\rho_{c_u \to c_v}$ is the probability of transitions from category c_u to category c_v. Such inter-category transition probability can be defined by the average percentage of all the check-ins from c_u to c_v: $\rho_{c_u \to c_v} = |\{(x, y) \in TS : x = u \land c_y = c_v\}| \cdot \frac{1}{|M_u|}$.

For the mobility features, we further consider two feature sets according to the time periods used to compute the feature values: based on the current time period to be estimated and based on the cumulative time periods from past to now. Therefore, we have two mobility feature sets: temporal mobility features (TMF), and cumulative mobility features (CMF). The third category is the social features, which characterize the social interactions for users who had ever visited the near-by area of location v. The specific feature items include:

- *Cohesiveness* is to model the structure connectivity of the graph $\mathbb{G}[N(v)]$ induced by users who had ever visit $N(v)$. This is designed to characterize the extent of cohesion or separation for people who live or visit locations within

$N(v)$. We employ the density and the clustering coefficient of $\mathbb{G}[N(v)]$ as the feature values. In addition, we also compute the number of components in $\mathbb{G}[N(v)]$ to be another indicator of cohesiveness.

- *Social Groups* estimates the number of groups of potential customers on the location-based social network $\mathbb{G}$. We consider a *community* as a group of potential customers, and calculate the number of communities for people who had ever visited $N(v)$ as the feature value. The *Louvian method* [1] is used for community detection.
- *Network Centrality* measures the importance of users on the location-based social network $\mathbb{G}$. We compute the values of degree, closeness, betweenness, PageRank, and SimRank, and consider the maximum, minimum, and average scores over users who had ever visit $N(v)$ as feature values.
- *Geo-Social Metrics* aim to quantify the geo-social influence of a user within the near-by area of location v. We employ four well-known geo-social metrics: *spatial degree centrality* [12], *spatial closeness centrality* [12], *node novelty* [19], and *geographic clustering coefficient* [19]. We compute the maximum, minimum, and average values over users who had ever visited $N(v)$.

5.2   Correlation Graph

We construct the *Correlation graph* to model the spatial and temporal correlations between existing and query locations. What follows first defines and elaborates the correlation graph, and then describes how to exploit the features extracted above to derive the *correlation* between locations as edge weights.

Definition: Correlation Graph (CG). A CG is a multi-layer weighted connected graph $G = \langle G^{t_1}, G^{t_2}, ..., G^{t_n} \rangle$, in which t is the total number of layers for time periods $t_1, t_2, ..., t_n$, and $G^{t_i} = (V, E, W)$ is the layer graph in at the t_i-th time period, where V is the set of locations, E is the set of edges between locations, and $W = W^{t_i} + W^{t_i, t_j}$ is the matrix representing edge weights, where W^{t_i} and W^{t_i, t_j} are edge weights learned from nodes within time period t_i and across time periods t_i and $t_j (i \neq j)$ respectively. The node set V consists of (a) existing locations of retail stores whose potential customer numbers have known, denoted by labeled nodes $V_\bullet$, and (b) the query locations, denoted by unlabeled nodes $V_\circ$, where $V = V_\bullet \cup V_\circ$. Each labeled node $v_\bullet \in V_\bullet$ is associated with its potential customer numbers $pc(v_\bullet(t_i))$. The edge set E also consists of two parts: the set of edges $E_\circlearrowleft$ connecting nodes within each layer graph G^{t_i}, and the set of edges $E_\frown$ connecting the same nodes across different layer graphs G^{t_i} and $G^{t_j} (i \neq j)$, where $E = E_\circlearrowleft \cup E_\frown$.

The construction of correlation graph consists of three parts. First, because we aim to use existing stores to estimate the numbers of potential customers for query locations, we connect each unlabeled node $v_\circ \in V_\circ$ to all the labeled nodes $u_\bullet \in V_\bullet$ within each time period. Second, owing to the fact that the potential customer number of a store is highly correlated to its historical values, we connect each unlabeled node within time period t_j to the corresponding unlabeled node within each of its previous time period $t_i (i < j)$. Third, since the potential

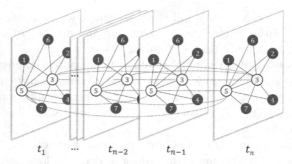

Fig. 3. An illustration to the correlation graph.

customer numbers of near-by locations have higher possibility to be close to one another (due to sharing similar volume of crowds), each unlabeled node $v_o \in V_o$ is connected to the near-by unlabeled ones $u_o \in V_o$ within a geographical radius r ($r = 200$ meters), where $dist(u_o, v_o) < r$.

We illustrate the correlation graph using Figure 3. There are seven locations as nodes $V = \{v_1, v_2, ..., v_7\}$, in which five are labeled nodes $V_\bullet = \{v_1, v_2, v_4, v_6, v_7\}$ and two are unlabeled nodes $V_o = \{v_3, v_5\}$. Since there are n time periods $t_1, t_2, ..., t_n$, we construct n layer graphs sharing the same node set V. In each layer graph G^{t_i}, a set of internal edges $E_\circlearrowleft$ are constructed, as illustrated using bold lines. For any two layer graphs G^{t_i} and $G^{t_j} (i \neq j)$, we construct a set of external edges $e_{ij} \in E_\frown$ connecting the same unlabeled nodes $v_o \in V_o$ between G^{t_i} and G^{t_j}, as shown using dash lines. We will describe the way to determine edge weights W^{t_i} through *feature-aware location correlation* in the following.

5.3   Location Correlation

Learning edge weights in CG from location features plays a key role in the estimation of potential customer numbers for unlabeled nodes. We aim to model the *correlation* between a labeled and an unlabeled node as their edge weight. The idea is that for a certain time period, if two locations with higher *correlation*, they tend to have *closer* potential customer numbers. In other words, for an unlabeled node $v \in V_o$, its number of potential customer will be close to that of the location with higher correlation to each other. The geographical, mobility, and social features are exploited to characterize the *correlation* between locations. In general, two locations whose features have lower difference should share closer potential customer numbers, while higher difference should make their potential customer numbers far away. Nevertheless, various feature might have different degree of effect on the correlation of potential customer numbers between the feature difference. For example, though lower feature differences of *Area Popularity* and *Density* between two locations make their potential customer numbers close, one should have more significant effect on the other. Therefore, the importance of separate feature should be considered.

For a certain store, we estimate the *feature-aware location correlation* based on their differences with respect to each feature. Then we combine the values of feature-aware location correlation of all the features through weighted sum. The weight multiplied by each feature location correlation will be determined based on its predictability of potential customers.

Definition 7: Feature-aware Location Correlation (FLC). Given a particular feature f_k, the *feature-aware location correlation* $flc_{f_k}(u(t_i), v(t_j))$ between nodes u and v, $(u, v) \in E$, in time periods t_i and t_j respectively can be derived from their feature difference $flc_{f_k}(u(t_i), v(t_j)) = \Delta f_k(u(t_i), v(t_j))$, where Δf_k is their feature difference, defined by $\Delta f_k = \|\mathbf{f_k}(u(t_i)) - \mathbf{f_k}(v(t_j))\|$.

Given a set of features $F = \{f_1, f_2, ..., f_m\}$, we combine *feature-aware location correlation* value $flc(u(t_i), v(t_j))$ between nodes u and v, $(u, v) \in E$, in time periods t_i and t_j, via the weighted sum of their correlation flc_{f_k}, given by:

$$flc(u(t_i), v(t_j)) = \exp(-\sum_{k=1}^{m} \pi_k \times flc_{f_k}(u(t_i), v(t_j))), \tag{1}$$

where π_k is the weight of feature f_k. The combined correlation is considered as the edge weight $w_{u(t_i), v(t_j)} = flc(u(t_i), v(t_j))$ between nodes u and v in CG.

Feature-based Top Store Detection. To determine feature weight π_k, we use the values of each feature f_k on existing stores to detecting stores with higher check-in numbers, and if f_k leads to higher precision scores, it will be assigned a higher feature weight. We use *Precision@X%* to evaluate the goodness of each feature. An instance is a store at a certain time period t_i, and there are $|S| \times |T|$ instances in total, where S is the set of all the stores. We denote the set of all instances to be ST. By setting $X\% = 10\%, 20\%, 30\%$ of stores with top/higher check-in numbers, we define the scores of *Precision@X%* as $|ST_{f_k, X\%} \cap ST_{X\%}| / |ST_{X\%}|$, where $ST_{X\%}$ is the set of stores with top $X\%$ check-in numbers, and $ST_{f_k, X\%}$ is the set of stores with top $X\%$ values of feature f_k. Features with higher precision scores provide more benefit on estimating potential customer numbers of stores. Therefore, we compute the weight π_k of each feature f_k by normalizing the average precision scores of f_k over all the features, $\pi_k = [0, 1]$.

5.4   Customer Inference Algorithm

We estimate the potential customer numbers of arbitrary locations over time $t_1, t_2, ..., t_n$ using the correlation graph. The idea is to iteratively update the number of potential customers $pc(v_o)$ of each unlabeled node v_o until the change of their potential customer numbers converges. Since the correlation of potential customer numbers among locations or stores is described by the correlation graph, we compute the potential customer number $pc(v_o)$ from its neighboring labeled or unlabeled nodes. This is fulfilled by averaging the potential customer numbers of v_o's neighbors, which are weighted by edge weights. Since the correlation graph provides benefits on modeling the temporal and spatial correlation

---

Algorithm 1. Potential Customer Estimation (PCE)

Input: (a) a set of existing store locations $V_\bullet$ with existing potential customer numbers $pc(v_\bullet)$ $(v_\bullet \in V_\bullet)$, (b) a set of query locations V_o, (c) the time periods to be observed $T = t_1, t_2, ..., t_n$

Output: $pc(v_o(t_i))$, where $v_o \in V_o$ and $t_i \in T$

1   $V \leftarrow V_\bullet \cup V_o$;

2   $\mathbf{f_k}(v) \leftarrow$ extracting feature f_k, $k = 1, 2, ..., m$, $v \in V$;

3   Construct CG from V and $\mathbf{f_k}(v)$, $v \in V$;

4   Compute feature weights π_k by *Precision@X%* with normalization;

5   $w_{uv} \leftarrow \exp(-\sum_{k=1}^{m} \pi_k \times flc_{f_k}(u(t_i), v(t_j)))$;

6   Initialize the potential customer number of each unlabeled node
$pc(v_o(t_i)) \leftarrow \sum_{u \in N(v_o(t_i)) \& u \in V_\bullet} w_{v_o(t_i), u} \times pc(u)$;

7   $\Delta avgPc \leftarrow \frac{1}{|V_o|} \times \sum_{v_o(t_i) \in V_o} pc(v_o(t_i))$;

8   **while** $\Delta avgPc > \epsilon$ **do**

9   **for** $v_o(t_i) \in V_o$ **do**

10        $pc(v_o(t_i)) \leftarrow \sum_{u \in N(v_o(t_i))} w_{v_o(t_i), u} \times pc(u)$;

11      $\Delta avgPc \leftarrow \frac{1}{|V_o|} \times \sum_{v_o(t_i) \in V_o} pc(v_o(t_i))$;

12   return $pc(v_o(t_i))$.

---

of potential customers, stores with higher correlation with v_o contribute more weights on the estimation of potential customer numbers for unlabeled nodes.

We give the complete algorithm of Potential Customer Estimator (PCE) in Algorithm 1. We first use both existing stores (i.e., labled nodes $V_\bullet$) and query locations (i.e., unlabeled nodes V_o) to construct the correlation graph based on the extracted features f_k for each node (line 1-3). With the *feature-based top store detection*, we can derive the weight of each feature π_k and use feature weight to initialize the edge weight w_{uv} in the correlation graph (line 4-5). Then we can further initialize the potential customer number of each unlabeled node $pc(v_o(t_i))$ from the set of $v_o(t_i)$'s neighboring labeled nodes $N(v_o(t_i)) \subset V_\bullet$ (line 6). We also initialize the difference of the average potential customer numbers between iterative rounds $\Delta avgPc$ by the sum of the initialized potential customer numbers of unlabeled nodes (line 7). In the iterative updating (line 8-11), we continue adjusting the potential customer numbers of unlabeled nodes based on those of its neighboring labeled and unlabeled nodes and the edge weights. This iterative process will terminate until $\Delta avgPc$ converges.

6   Experiments

We conduct experiments to exhibit the performance the proposed PCE model. The objective is three-fold. First, we aim to understand the effectiveness of PCE, comparing to a series of competitors. Second, we are eager to know whether or not PCE can successfully detect the locations with higher potential customer numbers. Three, we wonder how different combinations of features and different feature settings affect the performance of PCE.

6.1   Evaluation Plans

Competitive Methods. We compare PCE with a series of competitive methods, which are divided into four categories. The first is spatial k-nearest neighbors; the second category is two interpolation-based methods, i.e., Inverse Distance Weighting and Ordinary Kriging; the third is two conventional learning methods (i.e., Artificial Neural Network and Support Vector Regression); and the fourth is two state-of-the-art semi-supervised learning methods, i.e., Co-Training and Radial Basis Function-based SSL. Note that SVR is one of the methods that have the best performance popularity ranking on Geo-Spotting [8].

- **Spatial k-Nearest Neighbors (kNN)** considers the average potential customer number from the potential customer numbers of the k closet geographical neighboring locations as the estimated value.
- **Inverse Distance Weighting (IDW)** is a well-known interpolation method [4]. IDW assigns values of unlabeled locations by calculating the weighted averages of the values available on labeled locations. Locations lower geographical distances have higher weights.
- **Ordinary Kriging (OK)** [17] is a state-of-the-art method of spatial point interpolation. The prediction is calculated as weighted averages of geographical neighbors, in which the weights are determined by finding the *semi-variogram* values for instances between known locations and the semi-variogram values for instances between each unknown location and all known locations. Then a set of simultaneous equations are solved by minimizing the estimation error of each unknown location.
- **Artificial Neural Network (ANN)** with the commonly-used back propagation technique is used as another baseline. The constructed ANN contains one hidden layer in the experiments for the generality. We set a linear function for the neurons in the input layer and assign a sigmoid function for those in the hidden and output layers.
- **Support Vector Regression (SVR).** A version of SVM for regression is choose chosen to estimate the potential customer numbers. SVR utilizes the historical check-in data on locations as the training data and learn a cost function to build the predictive model.
- **Co-Training (CT)** is proposed by Nigam and Ghani [15] and serves as the state-of-the-art method for learning the correlation between real values. The co-training model consists of two separated classifiers. One is a spatial classifier based on artificial neural network to model the spatial correlation of labels. The other is a temporal classifier based on a *linear-chain conditional random field* (CRF) [10] to model the temporal dependency of labels.
- **Radial Basis Function-based Semi-supervised Learning (SSL)**, which is a state-of-the-art graph-based learning method [23], serves as a strong competitor. To apply RBF-SSL, the potential customer numbers of query locations within each time period are estimated separately, in which a graph is

constructed for each time period based on geographical distance. In addition, we quantize the potential customer numbers of locations as ten discrete labels, and consider the mean value of the predicted label to be the result.

Evaluation Metrics. We use two metrics in the experiments: *Hit Rate* and *Normalized Discounted Cumulative Gain* (NDCG). For a location v_o in the query set of locations V_o within time period $t_i \in T(T = t_1, t_2, ..., t_m)$, assume its estimated potential customer number is $pc(v_o(t_i))$ and the ground-truth potential customer number is $\tilde{pc}(v_o(t_i))$. Then the *hit rate* is defined as:

$$HitRate = \frac{\sum_{v_o \in V_o, t_i \in T} hit(pc(v_o(t_i)), \tilde{pc}(v_o(t_i)))}{|V_o| \cdot |T|}, \tag{2}$$

where $hit(pc(v_o(t_i)), \tilde{pc}(v_o(t_i))) = 1$ if $\tilde{pc}(v_o(t_i)) - \gamma \leq pc(v_o(t_i)) \leq \tilde{pc}(v_o(t_i)) + \gamma$, otherwise: $hit(pc(v_o(t_i)), \tilde{pc}(v_o(t_i))) = 0$, where the parameter γ determines the strictness of the evaluation through varying the *granularity* of the ground-truth potential customer numbers. A higher γ value indicates a loose generous evaluation and every methods would have higher accuracy in general; a lower γ value refers to a strict evaluation, and thus the accuracy tends to be lower for different methods. We choose to have a strict evaluation with $\gamma = 30$. The second evaluation metric is NDCG [7]. We use NDCG to estimate the ranking quality between the potential customer numbers estimated by a method and the ground-truth potential customer number. Higher scores of Hit Rate and NDCG mean better performance.

Basic Settings. To evaluate PCE, we use the potential customer numbers of stores of three retail chains, Starbucks (SB), McDonald's (MC) and Dunkin' Donuts (DD). We choose such three retail chains because their stores are three of the most popular and the most widely scattered in New York City. Such three retail chains are considered as three evaluation subsets. For each retail chain, we divide its stores into training and test parts. Assume there are n_S stores and n_T time periods, we randomly select 80% stores as training instances ($80\% \times n_S \times n_T$) and the other 20% stores are regarded as test instances, whose locations are used as the query and their potential customer numbers within each time period are removed and served as the ground truth. For the parameters used in the experiments, we have the following settings by default: (a) the geographical neighboring radius of feature extraction $r = 200$ meters, (c) two categories of time period granularity are considered: week and month, (d) all the three feature sets, geographical features (GF), temporal mobility features (TMF), cumulative mobility features (CMF), and social features (SF) are used together, and (e) the strictness parameter $\gamma = 30$ for the evaluation metric of accuracy.

Detailed Plans. To reach the three goals mentioned above, we have the following four detailed evaluation plans. The first is the *general evaluation*, in which the proposed PCE is compared to seven competitors. The general evaluation will be conducted under time periods of weeks and months for the three retail chains.

The second is the *top potential customers evaluation*, which is designed to understand whether or not the proposed PCE can successfully detect locations with higher potential customer numbers. The third is the *feature importance evaluation*. Through reporting the performance of PCE using different combinations of feature sets, including GF, TMF, CMF, and SF we can know which feature is more important in the estimation of potential customer numbers. The fourth is the *feature range evaluation*. Recall the feature values computed are constrained to a certain geographical radius r of neighborhood. We aim to present the performance by varying the radius r, to understand the predictability of geographical areas.

6.2 Experimental Results

General Evaluation. The results for the three retail chains under time periods of weeks and months are shown in Table 1. We can find that PCE significantly outperforms all of the competitors under all the cases. We think such promising results come from not only the investigation of location correlation as well as feature-aware location ranking, but also the simultaneous consideration of spatial and temporal dependency between locations and stores in the correlation graph. However, most of the competitors that purely learn the correlation between features and potential customer numbers. In more details, it can be observed that the accuracy is hard to exceed 0.8 under time periods of weeks, especially for the competitors. We think it is because we choose a strict evaluation with $\gamma = 30$ in the experiments. In addition, we can find that the performance of months is much better than that of weeks. It is due to the fact that the popularity value accumulated in each month is higher than that of each week. Therefore, the potential customer numbers of months tend to be a bit far apart from each other and make it a bit easier to be estimated.

Top Potential Customers Evaluation. We test if the proposed PCE is able to detect the locations with higher potential customer numbers. Following the settings described in the section of Feature-based Top Store Detection and using and the same evaluation metric $Precision@X\%$, we aim to present the estimated potential customer numbers by PCE by varying the percentage of locations with the highest potential customer numbers from 5% to 35%. We report the $Precision@X\%$ scores in Figure 4. We can find that the precision scores by PCE can have 0.8 precision scores for top 20% stores with the highest potential customer numbers. Such results exhibit the practical usages of PCE on estimating and finding hot zones in a city, and demonstrate the effectiveness of using stores as sensors to estimate the numbers of potential customers.

Feature Importance Evaluation. To understand which feature set is more important on potential customer estimation, we report the performance of different combinations of feature sets (i.e., GF, TMF, CMF, and SF) using PCE, as shown in Table 2. We can find that comparing to GF, CMF, and SF, TMF obtains the better results with higher scores of NDCG scores in general under

Table 1. General Evaluation Results on weeks and months.

| | Week | | | | | | Month | | | | | |
| | nDCG | | | HitRate | | | nDCG | | | HitRate | | |
	SB	MC	DD	SB	MC	DD	SB	MC	DD	SB	MC	DD
kNN	0.15	0.30	0.25	0.11	0.12	0.12	0.26	0.33	0.39	0.13	0.14	0.14
IDW	0.17	0.30	0.25	0.11	0.12	0.12	0.24	0.31	0.38	0.13	0.14	0.14
OK	0.18	0.35	0.28	0.19	0.24	0.23	0.29	0.34	0.39	0.16	0.16	0.15
ANN	0.53	0.58	0.60	0.52	0.54	0.53	0.57	0.61	0.64	0.69	0.67	0.69
SVR	0.58	0.61	0.62	0.58	0.60	0.56	0.62	0.64	0.65	0.72	0.73	0.75
CT	0.56	0.67	0.65	0.56	0.52	0.60	0.64	0.70	0.69	0.74	0.74	0.75
SSL	0.63	0.71	0.69	0.63	0.66	0.68	0.68	0.74	0.72	0.74	0.74	0.74
PCE	**0.71**	**0.79**	**0.78**	**0.79**	**0.84**	**0.81**	**0.76**	**0.82**	**0.80**	**0.83**	**0.88**	**0.88**

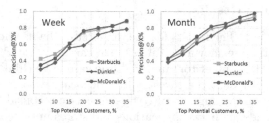

Fig. 4. Evaluation of Top Potential Customers using PCE, by varying the percentage of stores with the highest potential customer numbers.

both weeks and months and for all the three retail chains. We think the reason could be the TMF is capable of describe the neighboring human flows at the separate time periods while CMF can only capture the historical volume of human flow traveling in the neighborhood of a location, which reflects the total numbers of potential customers. As for GF and SF, what it captures is the properties and distributions of location categories and social activities in the neighborhood, and thus cannot directly exhibit the volume of potential customers. Therefore, GF and SF derives the worse estimation accuracy than CMF and TMF.

Feature Range Evaluation. Since the features extracted are constrained within a certain neighborhood via the radius r (in meters), we would like to which r is more effective and leads to better performance in PCE. The results are shown in Figure 5. We can find the performance of using $r = 300$ is the best for time periods of weeks while using $r = 200$ is the best for months. Too small or large radius values r leads to worse performance because features extracted constrain on a small area could not fully and precisely describe the neighborhood while constraining on a large area might include irrelevant features. The feature neighborhood radius $r = 300$ and $r = 200$ also is quite close to and responses to the optimal neighborhood radius 200 meters suggested by urban planning [14].

Table 2. Feature Importance Evaluation: the NDCG scores of different feature sets.

	Geographical Feat. (GF)						Temporal Mobility Feat. (TMF)					
	Week			Month			Week			Month		
	SB	DD	MC	SB	DD	MC	SB	DD	MC	SB	DD	MC
kNN	0.159	0.256	0.303	0.256	0.385	0.331	0.159	0.256	0.303	0.256	0.385	0.331
IDW	0.171	0.252	0.305	0.244	0.377	0.312	0.171	0.252	0.305	0.244	0.377	0.312
OK	0.189	0.284	0.352	0.289	0.393	0.342	0.189	0.284	0.352	0.289	0.393	0.342
SVR	0.322	0.332	0.455	0.355	0.514	0.648	0.541	0.582	0.656	0.589	0.635	0.731
ANN	0.329	0.342	0.452	0.368	0.522	0.651	0.517	0.568	0.632	0.552	0.622	0.722
CT	0.334	0.351	0.452	0.369	0.524	0.652	0.533	0.578	0.649	0.561	0.642	0.729
SSL	0.341	0.382	0.464	0.431	0.529	0.668	0.557	0.580	0.658	0.562	0.657	0.712
PCE	**0.349**	**0.401**	**0.481**	**0.462**	**0.552**	**0.681**	**0.582**	**0.619**	**0.713**	**0.663**	**0.685**	**0.756**
	Cumulative Mobility Feat. (CMF)						Social Feat. (SF)					
	Week			Month			Week			Month		
	SB	DD	MC	SB	DD	MC	SB	DD	MC	SB	DD	MC
kNN	0.159	0.256	0.303	0.256	0.385	0.331	0.132	0.247	0.280	0.194	0.375	0.316
IDW	0.171	0.252	0.305	0.244	0.377	0.312	0.168	0.255	0.271	0.195	0.379	0.284
OK	0.189	0.284	0.352	0.289	0.393	0.342	0.172	0.259	0.337	0.206	0.380	0.309
SVR	0.426	0.368	0.531	0.526	0.588	0.645	0.393	0.327	0.498	0.475	0.561	0.583
ANN	0.418	0.354	0.520	0.513	0.543	0.638	0.391	0.333	0.460	0.428	0.517	0.604
CT	0.426	0.391	0.536	0.535	0.596	0.659	0.432	0.369	0.555	0.396	0.520	0.637
SSL	0.435	0.408	0.519	0.529	0.585	0.673	0.471	0.388	0.526	0.466	0.594	0.650
PCE	**0.587**	**0.455**	**0.603**	**0.613**	**0.633**	**0.724**	**0.477**	**0.412**	**0.539**	**0.498**	**0.604**	**0.704**

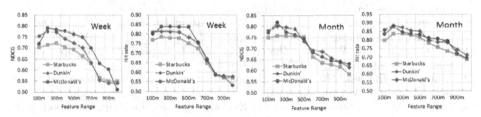

Fig. 5. Feature Range Evaluation, by varying the neighrborhood radius r using PCE.

7   Conclusion

Being able to acquire the knowledge about where and when the customers will show up can lead to many useful applications, including determining the locations of new business, choosing the right time and place to host campaign to maximize the advertise effect. This paper proposes a method to estimate the number of potential customers in an urban area. We leverage stores as a kind of sensors to estimate the potential customers of of any location during any given time span. A *PCE* model is developed and validated with promising performance. In the future, we aim to go beyond location-based services and further consider more heterogeneous urban information into the modeling of potential customers, such as traffic status, weather, and near-by activities.

References

1. Blondel, V.D., Guillaume, J.-L., Lambiotte, R., Lefebvre, E.: Fast unfolding of communities in large networks. Journal of Statistical Mechanics: Theory and Experiment (2008)
2. Chen, Z., Liu, Y., Wong, R.C.-W., Xiong, J., Mai, G., Long, C.: Efficient algorithms for optimal location queries in road networks. In: ACM SIGMOD (2014)
3. Cho, E., Myers, S.A., Leskovec, J.: Friendship and mobility: user movement in location-based social networks. In: ACM KDD (2011)
4. Donald, S.: A two-dimensional interpolation function for irregularly-spaced data. In: ACM National Conference (1968)
5. Fu, Y., Ge, Y., Zheng, Y., Yao, Z., Liu, Y., Xiong, H., Yuan, N.J.: Sparse real estate ranking with online user reviews and offline moving behaviors. In: IEEE ICDM (2014)
6. Hsieh, H.-P., Lin, S.-D., Zheng, Y.: Inferring air quality for station location recommendation based on urban big data. In: ACM KDD (2015)
7. Jarvelin, K., Kekalainen, J.: Cumulated gain-based evaluation of IR techniques. ACM TOIS (2002)
8. Karamshuk, D., Noulas, A., Scellato, S., Nicosia, V., Mascolo, C.: Geo-spotting: mining online location-based services for optimal retail store placement. In: ACM KDD (2013)
9. Kisilevich, S., Mansmann, F., Keim, D.: P-DBSCAN: a density based clustering algorithm for exploration and analysis of attractive areas using collections of geotagged photos. In: COM.Geo (2010)
10. Lafferty, J., McCallum, A., Pereira, F.: Conditional random fields: probabilistic models for segmenting and labeling sequence data. In: ICML (2001)
11. Li, Y., Steiner, M., Wang, L., Zhang, Z.-L., Bao, J.: Exploring venue popularity in four-square. In: IEEE INFOCOM (2013)
12. Lima, A., Musolesi, M.: Spatial dissemination metrics for location-based social networks. In: ACM UbiComp (2012)
13. Liu, Y., Wei, W., Sun, A., Miao, C.: Exploiting geographical neighborhood characteristics for location recommendation. In: ACM CIKM (2014)
14. Mehaffy, M., Porta, S., Rofe, Y., Salingaros, N.: Urban nuclei and the geometry of streets: The emergent neighborhoods' model. Urban Design International (2010)
15. Nigam, K., Ghani, R.: Analyzing the effectiveness and applicability of co-training. In: ACM CIKM (2000)
16. Monreale, A., Pinelli, F., Trasarti, R., Giannotti, F.: Where next: a location predictor on trajectory pattern mining. In: ACM KDD (2009)
17. Oliver, M.A., Webster, R.: Kriging: a method of interpolation for geographical information systems. IJGIS (1990)
18. Sadilek, A., Kautz, H., Bigham, J.P.: Finding your friends and following them to where you are. In: ACM WSDM (2012)
19. Scellato, S., Mascolo, C., Musolesi, M., Latora, V.: Distance matters: geo-social metrics for online social networks. In: WOSN (2010)
20. Tiwari, S., Kaushik, S.: User category based estimation of location popularity using the road GPS trajectory databases. Geoinformatica (2014)
21. Ying, J.-C., Lee, W.-C., Weng, T.-C., Tseng, V.S.: Semantic trajectory mining for location prediction. In: ACM SIGSPATIAL GIS (2011)
22. Zhang, C., Shou, L., Chen, K., Chen, G., Bei, Y.: Evaluating geo-social influence in location-based social networks. In: ACM CIKM (2012)
23. Zhu, X., Ghahramani, Z., Lafferty, J.: Semi-supervised learning using gaussian fields and harmonic functions. In: ICML (2003)

Exact Hybrid Covariance Thresholding for Joint Graphical Lasso

Qingming Tang[1], Chao Yang[1], Jian Peng[2], and Jinbo Xu[1]([✉])

[1] Toyota Technological Institute at Chicago, Chicago, USA
{qmtang,harryyang}@ttic.edu
[2] University of Illinois at Urbana-Champaign, Champaign, USA
jianpeng@illinois.edu

Abstract. This paper studies precision matrix estimation for multiple related Gaussian graphical models from a dataset consisting of different classes, based upon the formulation of this problem as group graphical lasso. In particular, this paper proposes a novel hybrid covariance thresholding algorithm that can effectively identify zero entries in the precision matrices and split a large joint graphical lasso problem into many small subproblems. Our hybrid covariance thresholding method is superior to existing uniform thresholding methods in that our method can split the precision matrix of each individual class using different partition schemes and thus, split group graphical lasso into much smaller subproblems, each of which can be solved very fast. This paper also establishes necessary and sufficient conditions for our hybrid covariance thresholding algorithm. Experimental results on both synthetic and real data validate the superior performance of our thresholding method over the others.

1 Introduction

Graphs have been widely used to describe the relationship between variables (or features). Estimating an undirected graphical model from a dataset has been extensively studied. When the dataset has a Gaussian distribution, the problem is equivalent to estimating a precision matrix from the empirical (or sample) covariance matrix. In many real-world applications, the precision matrix is sparse. This problem can be formulated as graphical lasso [1,22] and many algorithms [4,9,16,18,19] have been proposed to solve it. To take advantage of the sparsity of the precision matrix, some covariance thresholding (also called screening) methods are developed to detect zero entries in the matrix and then split the matrix into smaller submatrices, which can significantly speed up the process of estimating the entire precision matrix [12,19].

Recently, there are a few studies on how to jointly estimate multiple related graphical models from a dataset with a few distinct class labels [3,6–8,11,13, 14,20,23–25]. The underlying reason for joint estimation is that the graphs of these classes are similar to some degree, so it can increase statistical power and estimation accuracy by aggregating data of different classes. This joint graph

© Springer International Publishing Switzerland 2015
A. Appice et al. (Eds.): ECML PKDD 2015, Part II, LNAI 9285, pp. 593–607, 2015.
DOI: 10.1007/978-3-319-23525-7_36

estimation problem can be formulated as joint graphical lasso that makes use of similarity of the underlying graphs. In addition to group graphical lasso, Guo et al. used a non-convex hierarchical penalty to promote similar patterns among multiple graphical models [6] ; [3] introduced popular group and fused graphical lasso; and [20, 25] proposed efficient algorithms to solve fused graphical lasso. To model gene networks, [14] proposed a node-based penalty to promote hub structure in a graph.

Existing algorithms for solving joint graphical lasso do not scale well with respect to the number of classes, denoted as K, and the number of variables, denoted as p. Similar to covariance thresholding methods for graphical lasso, a couple of thresholding methods [20, 25] are developed to split a large joint graphical lasso problem into subproblems [3]. Nevertheless, these algorithms all use uniform thresholding to decompose the precision matrices of distinct classes in exactly the same way. As such, it may not split the precision matrices into small enough submatrices especially when there are a large number of classes and/or the precision matrices have different sparsity patterns. Therefore, the speedup effect of covariance thresholding may not be very significant.

In contrast to the above-mentioned uniform covariance thresholding, this paper presents a novel hybrid (or non-uniform) thresholding approach that can divide the precision matrix for each individual class into smaller submatrices without requiring that the resultant partition schemes be exactly the same across all the classes. Using this method, we can split a large joint graphical lasso problem into much smaller subproblems. Then we employ the popular ADMM (Alternating Direction Method of Multipliers [2, 5]) method to solve joint graphical lasso based upon this hybrid partition scheme. Experiments show that our method can solve group graphical lasso much more efficiently than uniform thresholding.

This hybrid thresholding approach is derived based upon group graphical lasso. The idea can also be generalized to other joint graphical lasso such as fused graphical lasso. Due to space limit, the proofs of some of the theorems in the paper are presented in supplementary material.

2   Notation and Definition

In this paper, we use a script letter, like $\mathcal{H}$, to denote a set or a set partition. When $\mathcal{H}$ is a set, we use $\mathcal{H}_i$ to denote the ith element. Similarly we use a bold letter, like $\boldsymbol{H}$ to denote a graph, a vector or a matrix. When $\boldsymbol{H}$ is a matrix we use $\boldsymbol{H}_{i,j}$ to denote its $(i, j)^{\text{th}}$ entry. We use $\{\mathcal{H}^{(1)}, \mathcal{H}^{(2)}, \ldots, \mathcal{H}^{(N)}\}$ and $\{\boldsymbol{H}^{(1)}, \boldsymbol{H}^{(2)} \ldots, \boldsymbol{H}^{(N)}\}$ to denote N objects of same category.

Let $\{\boldsymbol{X}^{(1)}, \boldsymbol{X}^{(2)}, \ldots, \boldsymbol{X}^{(K)}\}$ denote a sample dataset of K classes and the data in $\boldsymbol{X}^{(k)}$ ($1 \leq k \leq K$) are independently and identically drawn from a p-dimension normal distribution $N(\boldsymbol{\mu}^{(k)}, \boldsymbol{\Sigma}^{(k)})$. Let $\boldsymbol{S}^{(k)}$ and $\hat{\boldsymbol{\Theta}}^{(k)}$ denote the empirical covariance and (optimal) precision matrices of class k, respectively. By "optimal" we mean the precision matrices are obtained by exactly solving joint graphical lasso. Let a binary matrix $E^{(k)}$ denote the sparsity pattern of $\hat{\Theta}^{(k)}$, i.e., for any $i, j (1 \le i, j \le p), E_{i,j}^{(k)} = 1$ if and only if $\hat{\Theta}T_{i,j}^{(k)} \ne 0$.

Set Partition. A set $\mathcal{H}$ is a partition of a set $\mathcal{C}$ when the following conditions are satisfied: 1) any element in $\mathcal{H}$ is a subset of $\mathcal{C}$; 2) the union of all the elements in $\mathcal{H}$ is equal to $\mathcal{C}$; and 3) any two elements in $\mathcal{H}$ are disjoint. Given two partitions $\mathcal{H}$ and $\mathcal{F}$ of a set $\mathcal{C}$, we say that $\mathcal{H}$ is **finer** than $\mathcal{F}$ (or $\mathcal{H}$ is a **refinement** of $\mathcal{F}$), denoted as $\mathcal{H} \preceq \mathcal{F}$, if every element in $\mathcal{H}$ is a subset of some element in $\mathcal{F}$. If $\mathcal{H} \preceq \mathcal{F}$ and $\mathcal{H} \ne \mathcal{F}$, we say that $\mathcal{H}$ is strictly finer than $\mathcal{F}$ (or $\mathcal{H}$ is a strict refinement of $\mathcal{F}$), denoted as $\mathcal{H} \prec \mathcal{F}$.

Let Θ denote a matrix describing the pairwise relationship of elements in a set $\mathcal{C}$, where $\Theta_{i,j}$ corresponds to two elements $\mathcal{C}_i$ and $\mathcal{C}_j$. Given a partition $\mathcal{H}$ of $\mathcal{C}$, we define $\Theta_{\mathcal{H}_k}$ as a $|\mathcal{H}_k| \times |\mathcal{H}_k|$ submatrix of Θ where $\mathcal{H}_k$ is an element of $\mathcal{H}$ and $(\Theta_{\mathcal{H}_k})_{i,j} \cong \Theta_{(\mathcal{H}_k)_i(\mathcal{H}_k)_j}$ for any suitable (i, j).

Graph-based Partition. Let $\mathcal{V} = \{1, 2, \ldots, p\}$ denote the variable (or feature) set of the dataset. Let graph $G^{(k)} = (\mathcal{V}, E^{(k)})$ denote the kth estimated concentration graph $1 \le k \le K$. This graph defines a partition $\boxplus^{(k)}$ of $\mathcal{V}$, where an element in $\boxplus^{(k)}$ corresponds to a connected component in $G^{(k)}$. The matrix $\hat{\Theta}^{(k)}$ can be divided into disjoint submatrices based upon $\boxplus^{(k)}$. Let E denote the mix of $E^{(1)}, E^{(2)}, \ldots, E^{(K)}$, i.e., one entry $E_{i,j}$ is equal to 1 if there exists at least one k $(1 \le k \le K)$ such that $E_{i,j}^{(k)}$ is equal to 1. We can construct a partition $\boxplus$ of $\mathcal{V}$ from graph $G = \{\mathcal{V}, E\}$, where an element in $\boxplus$ corresponds to a connected component in G. Obviously, $\boxplus^{(k)} \preceq \boxplus$ holds since $E^{(k)}$ is a subset of E. This implies that for any k, the matrix $\hat{\Theta}^{(k)}$ can be divided into disjoint submatrices based upon $\boxplus$.

Feasible Partition. A partition $\mathcal{H}$ of $\mathcal{V}$ is feasible for class k or graph $G^{(k)}$ if $\boxplus^{(k)} \preceq \mathcal{H}$. This implies that 1) $\mathcal{H}$ can be obtained by merging some elements in $\boxplus^{(k)}$; 2) each element in $\mathcal{H}$ corresponds to a union of some connected components in graph $G^{(k)}$; and 3) we can divide the precision matrix $\hat{\Theta}^{(k)}$ into independent submatrices according to $\mathcal{H}$ and then separately estimate the submatrices without losing accuracy. $\mathcal{H}$ is uniformly feasible if for all k $(1 \le k \le K)$, $\boxplus^{(k)} \preceq \mathcal{H}$ holds.

Let $\mathcal{H}^{(1)}, \mathcal{H}^{(2)}, \ldots, \mathcal{H}^{(K)}$ denote K partitions of the variable set V. If for each k $(1 \le k \le K)$, $\boxplus^{(k)} \preceq \mathcal{H}^{(k)}$ holds, we say $\{\mathcal{H}^{(1)}, \mathcal{H}^{(2)}, \ldots, \mathcal{H}^{(K)}\}$ is a feasible partition of $\mathcal{V}$ for the K classes or graphs. When at least two of the K partitions are not same, we say $\{\mathcal{H}^{(1)}, \mathcal{H}^{(2)}, \ldots, \mathcal{H}^{(K)}\}$ is a non-uniform partition. Otherwise, $\{\mathcal{H}^{(1)}, \mathcal{H}^{(2)}, \ldots, \mathcal{H}^{(K)}\}$ is a class-independent or uniform partition and abbreviated as $\mathcal{H}$. That is, $\mathcal{H}$ is uniformly feasible if for all k $(1 \le k \le K)$, $\boxplus^{(k)} \preceq \mathcal{H}$ holds. Obviously, $\{\boxplus^{(1)}, \boxplus^{(2)}, \ldots, \boxplus^{(K)}\}$ is finer than any non-uniform feasible partition of the K classes. Based upon the above definitions, we have the following theorem, which is proved in supplementary material.

Theorem 1. *For any uniformly feasible partition $\mathcal{H}$ of the variable set $\mathcal{V}$, we have $\boxplus \preceq \mathcal{H}$. That is, $\mathcal{H}$ is feasible for graph G and $\boxplus$ is the finest uniform feasible partition.*

Proof. First, for any element $\mathcal{H}_j$ in $\mathcal{H}$, G does not contain edges between $\mathcal{H}_j$ and $\mathcal{H} - \mathcal{H}_j$. Otherwise, since G is the mixing (or union) of all $G^{(k)}$, there exists at least one graph $G^{(k)}$ such that it contains at least one edge between $\mathcal{H}_j$ and $\mathcal{H} - \mathcal{H}_j$. Since $\mathcal{H}_j$ is the union of some elements in $\boxplus^{(k)}$, this implies that there exist two different elements in $\boxplus^{(k)}$ such that $G^{(k)}$ contains edges between them, which contradicts with the fact that $G^{(k)}$ does not contain edges between any two elements in $\boxplus^{(k)}$. That is, $\mathcal{H}$ is feasible for graph G.

Second, if $\boxplus \preccurlyeq \mathcal{H}$ does not hold, then there is one element $\boxplus_i$ in $\boxplus$ and one element $\mathcal{H}_j$ in $\mathcal{H}$ such that $\boxplus_i \cap \mathcal{H}_j \neq \emptyset$ and $\boxplus_i - \mathcal{H}_j \neq \emptyset$. Based on the above paragraph, $\forall x \in \boxplus_i \cap \mathcal{H}_j$ and $\forall y \in \boxplus_i - \mathcal{H}_j = \boxplus_i \cap (\mathcal{H}_i - \mathcal{H}_j)$, we have $E_{x,y} = E_{y,x} = 0$. That is, $\boxplus_i$ can be split into at least two disjoint subsets such that G does not contain any edges between them. This contradicts with the fact that $\boxplus_i$ corresponds to a connected component in graph G.

3   Joint Graphical Lasso

To learn the underlying graph structure of multiple classes simultaneously, some penalty functions are used to promote similar structural patterns among different classes, including [3,6,7,13,14,16,20,21,25]. A typical joint graphical lasso is formulated as the following optimization problem:

$$\min \sum_{k=1}^{K} L(\Theta^{(k)}) + P(\Theta) \tag{1}$$

Where $\Theta^{(k)} \succ 0$ is the precision matrix ($k = 1, \ldots, K$) and Θ represents the set of $\Theta^{(k)}$. The negative log-likelihood $L(\Theta^{(k)})$ and the regularization $P(\Theta)$ are defined as follows.

$$L(\Theta^{(k)}) = -\log \det(\Theta^{(k)}) + \operatorname{tr}(S^{(k)} \Theta^{(k)}) \tag{2}$$

$$P(\Theta) = \lambda_1 \sum_{k=1}^{K} \|\Theta^{(k)}\|_1 + \lambda_2 J(\Theta) \tag{3}$$

Here $\lambda_1 > 0$ and $\lambda_2 > 0$ and $J(\Theta)$ is some penalty function used to encourage similarity (of the structural patterns) among the K classes. In this paper, we focus on group graphical lasso. That is,

$$J(\Theta) = 2 \sum_{1 \leq i < j \leq p} \sqrt{\sum_{k=1}^{K} (\Theta_{i,j}^{(k)})^2} \tag{4}$$

4   Uniform Thresholding

Covariance thresholding methods, which identify zero entries in a precision matrix before directly solving the optimization problem like Eq.(1), are widely used to accelerate solving graphical lasso. In particular, a screening method divides the variable set into some disjoint groups such that when two variables (or features) are not in the same group, their corresponding entry in the precision matrix is guaranteed to be 0. Using this method, the precision matrix can be split into some submatrices, each corresponding to one distinct group. To achieve the best computational efficiency, we shall divide the variable set into as small groups as possible subject to the constraint that two related variables shall be in the same group. Meanwhile, [3] described a screening method for group graphical lasso. This method uses a single thresholding criterion (i.e., uniform thresholding) for all the K classes, i.e., employs a uniformly feasible partition of the variable set across all the K classes. Existing methods such as those described in [3,20,25] for fused graphical lasso and that in [15] for node-based learning all employ uniform thresholding.

Uniform thresholding may not be able to divide the variable set into the finest feasible partition for each individual class when the K underlying concentration graphs are not exactly the same. For example, Figure 1(a) and (c) show two concentration graphs of two different classes. These two graphs differ in variables 1 and 6 and each graph can be split into two connected components. However, the mixing graph in (b) has only one connected component, so it cannot be split further. According to **Theorem 1**, no uniform feasible partition can divide the variable set into two disjoint groups without losing accuracy. It is expected that when the number of classes and variables increases, uniform thresholding may perform even worse.

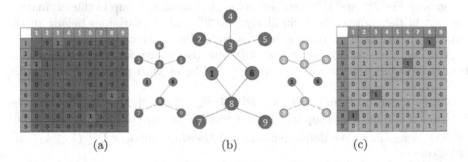

Fig. 1. Illustration of uniform thresholding impacted by minor structure difference between two classes. (a) and (c): the edge matrix and concentration graph for each of the two classes. (b): the concentration graph resulting from the mixing of two graphs in (a) and (c).

5   Non-uniform Thresholding

Non-uniform thresholding generates a non-uniform feasible partition by thresholding the K empirical covariance matrices separately. In a non-uniform partition, two variables of the same group in one class may belong to different

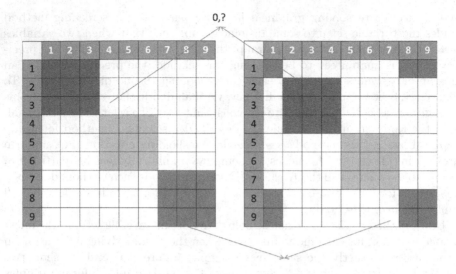

Fig. 2. Illustration of a non-uniform partition. White color indicates zero entries detected by covariance thresholding. Entries with the same color other than white belong to the same group.

groups in another class. Figure 2 shows an example of non-uniform partition. In this example, all the matrix elements in white color are set to 0 by non-uniform thresholding. Except the white color, each of the other colors indicates one group. The 7th and 9th variables belong to the same group in the left matrix, but not in the right matrix. Similarly, the 3rd and 4th variables belong to the same group in the right matrix, but not in the left matrix.

We now present necessary and sufficient conditions for identifying a non-uniform feasible partition for group graphical lasso, with penalty defined in Eq (3) and (4).

Given a non-uniform partition $\{\mathcal{P}^{(1)}, \mathcal{P}^{(2)}, \ldots, \mathcal{P}^{(K)}\}$ for the K classes, let $F^{(k)}(i) = t$ denote the group which the variable i belongs to in the kth class, i.e., $F^{(k)}(i) \Leftrightarrow i \in \mathrm{P}_t^{(k)}$. We define pairwise relationship matrices $\mathbf{I}^{(k)}$ $(1 \leq k \leq K)$ as follows:

$$\begin{cases} \mathbf{I}_{i,j}^{(k)} = \mathbf{I}_{j,i}^{(k)} = 0; \text{ if } F^{(k)}(i) \neq F^{(k)}(j) \\ \mathbf{I}_{i,j}^{(k)} = \mathbf{I}_{j,i}^{(k)} = 1; \text{ otherwise} \end{cases} \tag{5}$$

Also, we define $\boldsymbol{Z}^{(k)}(1 \leq k \leq K)$ as follows:

$$\boldsymbol{Z}_{i,j}^{(k)} = \boldsymbol{Z}_{j,i}^{(k)} = \lambda_1 + \lambda_2 \times \tau((\sum_{t \neq k} |\hat{\boldsymbol{\Theta}}_{i,j}^{(t)}|) = 0) \tag{6}$$

Here $\tau(b)$ is the indicator function.

The following two theorems state the necessary and sufficient conditions of a non-uniform feasible partition. See supplementary material for their proofs.

Algorithm 1 Hybrid Covariance Screening Algorithm

for $k = 1$ *to* K **do**
 Initialize $\mathbf{I}_{i,j}^{(k)} = \mathbf{I}_{j,i}^{(k)} = 1, \forall 1 \leq i < j \leq p$
 Set $\mathbf{I}_{i,j}^{(k)} = 0$, if $|\boldsymbol{S}_{i,j}^{(k)}| \leq \lambda_1$ and $i \neq j$
 Set $\mathbf{I}_{i,j}^{(k)} = 0$, if $\sum_{k=1}^{K}(|\boldsymbol{S}_{i,j}^{(k)}| - \lambda_1)_+^2 \leq \lambda_2^2$ and $i \neq j$
end for
for $k = 1$ *to* K **do**
 Construct a graph $\boldsymbol{G}^{(k)}$ for $\mathcal{V}$ from $\boldsymbol{I}^{(k)}$
 Find connected components of $G^{(k)}$
 for $\forall(i,j)$ *in the same component of* $\boldsymbol{G}^{(k)}$ **do**
  Set $\boldsymbol{I}_{i,j}^{(k)} = \boldsymbol{I}_{j,i}^{(k)} = 1$
 end for
end for
repeat
 Search for triple (x,i,j) satisfying the following condition:
 $\boldsymbol{I}_{i,j}^{(x)} = 0$, $|\boldsymbol{S}_{i,j}^{(x)}| > \lambda_1$ and $\exists s$, s.t. $\boldsymbol{I}_{i,j}^{(s)} = 1$
 if $\exists(x,i,j)$ satisfies the condition above **then**
  merge the two components of $\boldsymbol{G}^{(x)}$ that containing variable i and j into new
  component;
  for $\forall(m,n)$ in this new component **do**
   Set $\boldsymbol{I}_{m,n}^{(x)} = \boldsymbol{I}_{n,m}^{(x)} = 1$;
  end for
 end if
until No such kind of triple.
return the connected components of each graph which define the non-uniform feasible solution

Theorem 2. *If* $\{\mathcal{P}^{(1)}, \mathcal{P}^{(2)}, \ldots, \mathcal{P}^{(K)}\}$ *is a non-uniform feasible partition of the variable set* $\mathcal{V}$, *then for any pair* (i,j) $(1 \leq i \neq j \leq p)$ *the following conditions must be satisfied:*

$$\begin{cases} \sum_{k=1}^{K}(|\boldsymbol{S}_{i,j}^{(k)}| - \lambda_1)_+^2 \leq \lambda_2^2; & \text{if } \forall k \in 1, 2, \ldots, K, \boldsymbol{I}_{i,j}^{(k)} = 0 \\ |\boldsymbol{S}_{i,j}^{(k)}| \leq \boldsymbol{Z}_{i,j}^{(k)}; & \text{if } \boldsymbol{I}_{i,j}^{(k)} = 0 \text{ and } \exists t \neq k, \boldsymbol{I}_{i,j}^{(t)} = 1 \end{cases} \quad (7)$$

Here, each $\boldsymbol{S}^{(k)}$ *is a covariance matrix of the* k^{th} *class and* $x_+ = \max(0, x)$.

Theorem 3. *If for any pair* $(i,j)(1 \leq i \neq j \leq p)$ *the following conditions hold, then* $\{\mathcal{P}^{(1)}, \mathcal{P}^{(2)}, \ldots, \mathcal{P}^{(K)}\}$ *is a non-uniform feasible partition of the variable set* $\mathcal{V}$.

$$\begin{cases} \sum_{k=1}^{K}(|\boldsymbol{S}_{i,j}^{(k)}| - \lambda_1)_+^2 \leq \lambda_2^2; & \text{if } \forall k \in 1, 2, \ldots, K, \boldsymbol{I}_{i,j}^{(k)} = 0 \\ |\boldsymbol{S}_{i,j}^{(k)}| \leq \lambda_1; & \text{if } \boldsymbol{I}_{i,j}^{(k)} = 0 \text{ and } \exists t \neq k, \boldsymbol{I}_{i,j}^{(t)} = 1 \end{cases} \quad (8)$$

Algorithm 1 is a covariance thresholding algorithm that can identify a non-uniform feasible partition satisfying condition (8). We call **Algorithm 1** hybrid screening algorithm as it utilizes both class-specific thresholding (e.g. $|\mathbf{S}_{i,j}^{(k)}| \leq \lambda_1$)

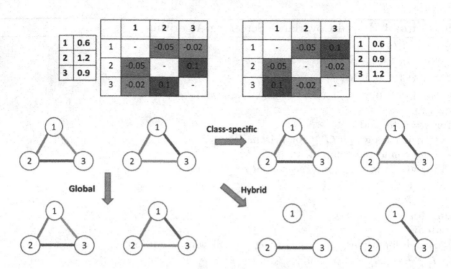

Fig. 3. Comparison of three thresholding strategies. The dataset contains 2 slightly different classes and 3 variables. The two sample covariance matrices are shown on the top of the figure. The parameters used are $\lambda_1 = 0.04$ and $\lambda_2 = 0.02$.

and global thresholding (e.g. $\sum_{k=1}^{K} (|\mathbf{S}_{i,j}^{(k)}| - \lambda_1)_+^2 \le \lambda_2^2$) to identify a non-uniform partition. This hybrid screening algorithm can terminate rapidly on a typical Linux machine, tested on the synthetic data described in section 7 with $K = 10$ and $p = 10000$.

We can generate a uniform feasible partition using only the global thresholding and generate a non-uniform feasible partition by using only the class-specific thresholding, but such a partition is not as good as using the hybrid thresholding algorithm. Let $\{\mathcal{H}^{(1)}, \mathcal{H}^{(2)}, \dots, \mathcal{H}^{(K)}\}$, $\{\mathcal{L}^{(1)}, \mathcal{L}^{(2)}, \dots, \mathcal{L}^{(K)}\}$ and $\mathcal{G}$ denote the partitions generated by hybrid, class-specific and global thresholding algorithms, respectively. It is obvious that $\mathcal{H}^{(k)} \preccurlyeq \mathcal{L}^{(k)}$ and $\mathcal{H}^{(k)} \preccurlyeq \mathcal{G}$ for $k = 1, 2, \dots, K$ since condition (8) is a combination of both global thresholding and class-specific thresholding.

Figure 3 shows a toy example comparing the three screening methods using a dataset of two classes and three variables. In this example, the class-specific or the global thresholding alone cannot divide the variable set into disjoint groups, but their combination can do so.

We have the following theorem regarding our hybrid thresholding algorithm, which will be proved in Supplemental File.

Theorem 4. *The hybrid screening algorithm yields the finest non-uniform feasible partition satisfying condition (8).*

6   Hybrid ADMM (HADMM)

In this section, we describe how to apply ADMM (Alternating Direction Method of Multipliers [2,5]) to solve joint graphical lasso based upon a non-uniform feasible partition of the variable set. According to [3], solving Eq.(1) by ADMM is equivalent to minimizing the following scaled augmented Lagrangian form:

$$\sum_{k=1}^{K} L(\Theta^{(k)}) + \frac{\rho}{2} \sum_{k=1}^{K} \|\Theta^{(k)} - Y^{(k)} + U^{(k)}\|_F^2 + P(Y) \tag{9}$$

where $Y = \{Y^{(1)}, Y^{(1)}, \ldots, Y^{(K)}\}$ and $U = \{U^{(1)}, U^{(1)}, \ldots, U^{(K)}\}$ are dual variables. We use the ADMM algorithm to solve Eq.(9) iteratively, which updates the three variables Θ, Y and U alternatively. The most computational-insensitive step is to update Θ given Y and U, which requires eigen-decomposition of K matrices. We can do this based upon a non-uniform feasible partition $\{\mathcal{H}^{(1)}, \mathcal{H}^{(2)}, \ldots, \mathcal{H}^{(K)}\}$. For each k, updating $\Theta^{(k)}$ given $Y^{(k)}$ and $U^{(k)}$ for Eq (9) is equivalent to solving in total $|\mathcal{H}^{(k)}|$ independent sub-problems. For each $\mathcal{H}_j^{(k)} \in \mathcal{H}^{(k)}$, its independent sub-problem solves the following equation:

$$(\Theta_{H_j^{(k)}}^{(k)})^{-1} = S_{\mathcal{H}_j^{(k)}}^{(k)} + \rho \times (\Theta_{\mathcal{H}_j^{(k)}}^{(k)} - Y_{\mathcal{H}_j^{(k)}}^{(k)} + U_{\mathcal{H}_j^{(k)}}^{(k)}) \tag{10}$$

Solving Eq.(10) requires eigen-decomposition of small submatrices, which shall be much faster than the eigen-decomposition of the original large matrices. Based upon our non-uniform partition, updating Y given Θ and U and updating U given Y and Θ are also faster than the corresponding components of the plain ADMM algorithm described in [3], since our non-uniform thresholding algorithm can detect many more zero entries before ADMM is applied.

7   Experimental Results

We tested our method, denoted as HADMM (i.e., hybrid covariance thresholding algorithm + ADMM), on both synthetic and real data and compared HADMM with two control methods: 1) GADMM: global covariance thresholding algorithm + ADMM; and 2) LADMM: class-specific covariance thresholding algorithm +ADMM. We implemented these methods with C++ and R, and tested them on a Linux machine with Intel Xeon E5-2670 2.6GHz.

To generate a dataset with K classes from Gaussian distribution, we first randomly generate K precision matrices and then use them to sample $5 \times p$ data points for each class. To make sure that the randomly-generated precision matrices are positive definite, we set all the diagonal entries to 5.0, and an off-diagonal entry to either 0 or $\pm r \times 5.0$. We generate three types of datasets as follows.

 - **Type A**: 97% of the entries in a precision matrix are 0.
 - **Type B**: the K precision matrices have same diagonal block structure.
 - **Type C**: the K precision matrices have slightly different diagonal block structures.

For **Type A**, r is set to be less than 0.0061. For **Type B** and **Type C**, r is smaller than 0.0067. For each type we generate 18 datasets by setting $K = 2, 3, \ldots, 10$, and $p = 1000, 10000$, respectively.

Table 1. Objective function values of HADMM and ADMM on the six classes type C data (first 4 iterations, $p = 1000$, $\lambda_1 = 0.0082$, $\lambda_2 = 0.0015$)

Iteration	1	2	3	4
ADMM	1713.66	-283.743	-1191.94	-1722.53
HADMM	1734.42	-265.073	-1183.73	-1719.78

7.1 Correctness of HADMM by Experimental Validation

We first show that HADMM can converge to the same solution obtained by the plain ADMM (i.e., ADMM without any covariance thresholding) through experiments.

To evaluate the correctness of our method HADMM, we compare the objective function value generated by HADMM to that by ADMM with respect to the number of iterations. We run the two methods for 500 iterations over the three types of data with $p = 1000$. As shown in Table 1, in the first 4 iterations, HADMM and ADMM yield slightly different objective function values. However, along with more iterations passed, both HADMM and ADMM converge to the same objective function value, as shown in Figure 4 and Supplementary Figures S3-5. This experimental result confirms that our hybrid covariance thresholding algorithm is correct. We tested several pairs of hyper-parameters (λ_1 and λ_2) in our experiment. Please refer to the supplementary material for model selection. Note that although in terms of the number of iterations HADMM and ADMM

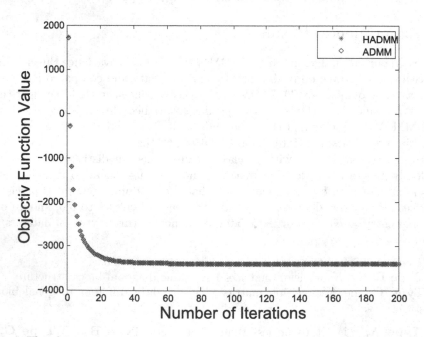

Fig. 4. The objective function value with respect to the number of iterations on a six classes type C data with $p = 1000$, $\lambda_1 = 0.0082$ and $\lambda_2 = 0.0015$.

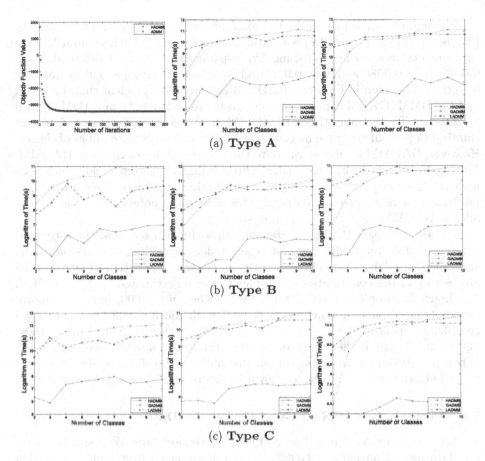

(a) **Type A**

(b) **Type B**

(c) **Type C**

Fig. 5. Logarithm of the running time (in seconds) of HADMM, LADMM and GADMM for $p = 1000$ on **Type A**, **Type B** and **Type C** data.

converge similarly, HADMM runs much faster than ADMM at each iteration, so HADMM converges in a much shorter time.

7.2   Performance on Synthetic Data

In previous section we have shown that our HADMM converges to the same solution as ADMM. Here we test the running times of HADMM, LADMM and GADMM needed to reach the following stop criteria for $p = 1000$: $\sum_{i=1}^{k} \|\boldsymbol{\Theta}^{(k)} - \boldsymbol{Y}^{(k)}\| < 10^{-6}$ and $\sum_{i=1}^{k} \|\boldsymbol{Y}^{(k+1)} - \boldsymbol{Y}^{(k)}\| < 10^{-6}$. For $p = 10000$, considering the large amount of running time needed for LADMM and GADMM, we run only 50 iterations for all the three methods and then compare the average running time for a single iteration.

We tested the running time of the three methods using different parameters λ_1 and λ_2 over the three types of data. See supplementary material for model selection. We show the result for $p = 1000$ in Figure 5 and that for $p = 10000$ in Figure S15-23 in supplementary material, respectively.

In Figure 5, each row shows the experimental results on one type of data (**Type A, Type B** and **Type C** from top to bottom). Each column has the experimental results for the same hyper-parameters ($\lambda_1 = 0.009$ and $\lambda_2 = 0.0005$, $\lambda_1 = 0.0086$ and $\lambda_2 = 0.001$, and $\lambda_1 = 0.0082$ and $\lambda_2 = 0.0015$ from left to right). As shown in Figure 5, HADMM is much more efficient than LADMM and GADMM. GADMM performs comparably to or better than LADMM when λ_2 is large. The running time of LADMM increases as λ_1 decreases. Also, the running time of all the three methods increases along with the number of classes. However, GADMM is more sensitive to the number of classes than our HADMM. Moreover, as our hybrid covariance thresholding algorithm yields finer non-uniform feasible partitions, the precision matrices are more likely to be split into many more smaller submatrices. This means it is potentially easier to parallelize HADMM to obtain even more speedup.

We also compare the three screening algorithms in terms of the estimated computational complexity for matrix eigen-decomposition, a time-consuming subroutine used by the ADMM algorithms. Given a partition $\mathcal{H}$ of the variable set of $\mathcal{V}$, the computational complexity can be estimated by $\sum_{\mathcal{H}_i \in \mathcal{H}} |\mathcal{H}_i|^3$. As shown in Supplementary Figures S6-14, when $p = 1000$, our non-uniform thresholding algorithm generates partitions with much smaller computational complexity, usually $\frac{1}{10} \sim \frac{1}{1000}$ of the other two methods. Note that in these figures the Y-axis is the logarithm of the estimated computational complexity. When $p = 10000$, the advantage of our non-uniform thresholding algorithm over the other two are even larger, as shown in Figure S24-32 in Supplemental File.

7.3   Performance on Real Gene Expression Data

We test our proposed method on real gene expression data. We use a lung cancer data (accession number GDS2771 [17]) downloaded from Gene Expression Omnibus and a mouse immune dataset described in [10]. The immune dataset consists of 214 observations. The lung cancer data is collected from 97 patients with lung cancer and 90 controls without lung cancer, so this lung cancer dataset consists of two different classes: patient and control. We treat the 214 observations from the immune dataset, the 97 lung cancer observations and the 90 controls as three classes of a compound dataset for our joint inference task. These three classes share 10726 common genes, so this dataset has 10726 features and 3 classes. As the absolute value of entries of covariance matrix of first class (corresponds to immune observations) are relatively larger, so we divide each entry of this covariance matrix by 2 to make the three covariance matrices with similar magnitude before performing joint analysis using unique λ_1 and λ_2.

The running time (first 10 iterations) of HADMM, LADMM and GADMM for this compound dataset under different settings are shown in Table 2 and the resultant gene networks with different sparsity are shown in Fig 6 and Supplemental File.

As shown in Table 2, HADMM (ADMM + our hybrid screening algorithm) is always more efficient than the other two methods in different settings. Typically, when λ_1 is small and λ_2 is large (**Setting 1**), our method is much faster than

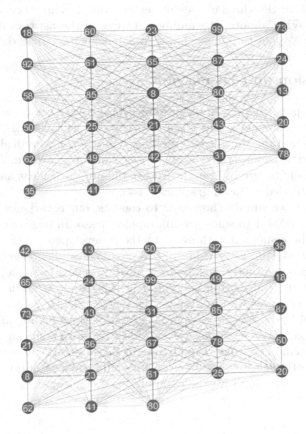

Fig. 6. Network of the first 100 genes of class one and class three for **Setting 1**.

Table 2. Running time (hours) of HADMM, LADMM and GADMM on real data. (**Setting 1**: $\lambda_1 = 0.1$ and $\lambda_2 = 0.5$; **Setting 2**: $\lambda_1 = 0.2$ and $\lambda_2 = 0.2$; **Setting 3**: $\lambda_1 = 0.3$ and $\lambda_2 = 0.1$; **Setting 4**: $\lambda_1 = 0.4$ and $\lambda_2 = 0.05$, and **Setting 5**: $\lambda_1 = 0.5$ and $\lambda_2 = 0.01$)

Method	Setting 1	Setting 2	Setting 3	Setting 4	Setting 5
HADMM	3.46	8.23	3.9	1.71	1.11
LADMM	> 20	> 20	13.6	3.72	1.98
GADMM	4.2	> 20	> 20	11.04	6.93

LADMM. In contrast, when λ_2 is small and λ_1 is large enough (**Setting 4** and **Setting 5**), our method is much faster than GADMM. What's more, when both λ_1 and λ_2 are with moderate values (**Setting 2** and **Setting 3**), HADMM is still much faster than both GADMM and LADMM.

As shown in Fig 6, the two resultant networks are with very similar topology structure. This is reasonable because we use large λ_2 in **Setting 1**. Actually, the networks of all the three classes under **Setting 1** share very similar topology structure. What's more, the number of edges in the network does decrease significantly as λ_1 goes to 0.5, as shown in Supplementary material.

8   Conclusion and Discussion

This paper has presented a non-uniform or hybrid covariance thresholding algorithm to speed up solving group graphical lasso. We have established necessary and sufficient conditions for this thresholding algorithm. Theoretical analysis and experimental tests demonstrate the effectiveness of our algorithm. Although this paper focuses only on group graphical lasso, the proposed ideas and techniques may also be extended to fused graphical lasso.

In the paper, we simply show how to combine our covariance thresholding algorithm with ADMM to solve group graphical lasso. In fact, our thresholding algorithm can be combined with other methods developed for (joint) graphical lasso such as the QUIC algorithm [9], the proximal gradient method [16], and even the quadratic method developed for fused graphical lasso [20].

The thresholding algorithm presented in this paper is static in the sense that it is applied as a pre-processing step before ADMM is applied to solve group graphical lasso. We can extend this "static" thresholding algorithm to a "dynamic" version. For example, we can identify zero entries in the precision matrix of a specific class based upon intermediate estimation of the precision matrices of the other classes. By doing so, we shall be able to obtain finer feasible partitions and further improve the computational efficiency.

References

1. Banerjee, O., El Ghaoui, L., d'Aspremont, A.: Model selection through sparse maximum likelihood estimation for multivariate gaussian or binary data. The Journal of Machine Learning Research **9**, 485–516 (2008)
2. Boyd, S., Parikh, N., Chu, E., Peleato, B., Eckstein, J.: Distributed optimization and statistical learning via the alternating direction method of multipliers. Foundations and Trends® in Machine Learning **3**(1), 1–122 (2011)
3. Danaher, P., Wang, P., Witten, D.M.: The joint graphical lasso for inverse covariance estimation across multiple classes. Journal of the Royal Statistical Society: Series B (Statistical Methodology) **76**(2), 373–397 (2014)
4. Friedman, J., Hastie, T., Tibshirani, R.: Sparse inverse covariance estimation with the graphical lasso. Biostatistics **9**(3), 432–441 (2008)
5. Gabay, D., Mercier, B.: A dual algorithm for the solution of nonlinear variational problems via finite element approximation. Computers & Mathematics with Applications **2**(1), 17–40 (1976)
6. Guo, J., Levina, E., Michailidis, G., Zhu, J.: Joint estimation of multiple graphical models. Biometrika, asq060 (2011)
7. Hara, S., Washio, T.: Common substructure learning of multiple graphical gaussian models. In: Gunopulos, D., Hofmann, T., Malerba, D., Vazirgiannis, M. (eds.) ECML PKDD 2011, Part II. LNCS, vol. 6912, pp. 1–16. Springer, Heidelberg (2011)

8. Honorio, J., Samaras, D.: Multi-task learning of gaussian graphical models. In: Proceedings of the 27th International Conference on Machine Learning, ICML 2010, pp. 447–454 (2010)
9. Hsieh, C.J., Dhillon, I.S., Ravikumar, P.K., Sustik, M.A.: Sparse inverse covariance matrix estimation using quadratic approximation. In: Advances in Neural Information Processing Systems, pp. 2330–2338 (2011)
10. Jojic, V., Shay, T., Sylvia, K., Zuk, O., Sun, X., Kang, J., Regev, A., Koller, D., Consortium, I.G.P., et al.: Identification of transcriptional regulators in the mouse immune system. Nature Immunology 14(6), 633–643 (2013)
11. Liu, J., Yuan, L., Ye, J.: An efficient algorithm for a class of fused lasso problems. In: Proceedings of the 16th ACM SIGKDD International Conference on Knowledge Discovery and Data Mining, pp. 323–332. ACM (2010)
12. Mazumder, R., Hastie, T.: Exact covariance thresholding into connected components for large-scale graphical lasso. The Journal of Machine Learning Research 13(1), 781–794 (2012)
13. Mohan, K., Chung, M., Han, S., Witten, D., Lee, S.I., Fazel, M.: Structured learning of gaussian graphical models. In: Advances in Neural Information Processing Systems, pp. 620–628 (2012)
14. Mohan, K., London, P., Fazel, M., Witten, D., Lee, S.I.: Node-based learning of multiple gaussian graphical models. The Journal of Machine Learning Research 15(1), 445–488 (2014)
15. Oztoprak, F., Nocedal, J., Rennie, S., Olsen, P.A.: Newton-like methods for sparse inverse covariance estimation. In: Advances in Neural Information Processing Systems, pp. 755–763 (2012)
16. Rolfs, B., Rajaratnam, B., Guillot, D., Wong, I., Maleki, A.: Iterative thresholding algorithm for sparse inverse covariance estimation. In: Advances in Neural Information Processing Systems, pp. 1574–1582 (2012)
17. Spira, A., Beane, J.E., Shah, V., Steiling, K., Liu, G., Schembri, F., Gilman, S., Dumas, Y.M., Calner, P., Sebastiani, P., et al.: Airway epithelial gene expression in the diagnostic evaluation of smokers with suspect lung cancer. Nature Medicine 13(3), 361–366 (2007)
18. Tseng, P., Yun, S.: Block-coordinate gradient descent method for linearly constrained nonsmooth separable optimization. Journal of Optimization Theory and Applications 140(3), 513–535 (2009)
19. Witten, D.M., Friedman, J.H., Simon, N.: New insights and faster computations for the graphical lasso. Journal of Computational and Graphical Statistics 20(4), 892–900 (2011)
20. Yang, S., Lu, Z., Shen, X., Wonka, P., Ye, J.: Fused multiple graphical lasso. arXiv preprint arXiv:1209.2139 (2012)
21. Yuan, M., Lin, Y.: Model selection and estimation in regression with grouped variables. Journal of the Royal Statistical Society: Series B (Statistical Methodology) 68(1), 49–67 (2006)
22. Yuan, M., Lin, Y.: Model selection and estimation in the gaussian graphical model. Biometrika 94(1), 19–35 (2007)
23. Yuan, X.: Alternating direction method for covariance selection models. Journal of Scientific Computing 51(2), 261–273 (2012)
24. Zhou, S., Lafferty, J., Wasserman, L.: Time varying undirected graphs. Machine Learning 80(2–3), 295–319 (2010)
25. Zhu, Y., Shen, X., Pan, W.: Structural pursuit over multiple undirected graphs. Journal of the American Statistical Association 109(508), 1683–1696 (2014)

Fast Inbound Top-K Query for Random Walk with Restart

Chao Zhang[✉], Shan Jiang, Yucheng Chen, Yidan Sun, and Jiawei Han

Department of Computer Science,
University of Illinois at Urbana-Champaign, Champaign, IL, USA
{czhang82,sjiang18,ychen233,ysun69,hanj}@illinois.edu

Abstract. Random walk with restart (RWR) is widely recognized as one of the most important node proximity measures for graphs, as it captures the holistic graph structure and is robust to noise in the graph. In this paper, we study a novel query based on the RWR measure, called the *inbound top-k* (Ink) query. Given a query node q and a number k, the Ink query aims at retrieving k nodes in the graph that have the largest weighted RWR scores *to* q. Ink queries can be highly useful for various applications such as traffic scheduling, disease treatment, and targeted advertising. Nevertheless, none of the existing RWR computation techniques can accurately and efficiently process the Ink query in large graphs. We propose two algorithms, namely SQUEEZE and RIPPLE, both of which can accurately answer the Ink query in a fast and incremental manner. To identify the top-k nodes, SQUEEZE iteratively performs matrix-vector multiplication and estimates the lower and upper bounds for all the nodes in the graph. RIPPLE employs a more aggressive strategy by only estimating the RWR scores for the nodes falling in the *vicinity* of q, the nodes outside the vicinity do not need to be evaluated because their RWR scores are propagated from the boundary of the vicinity and thus upper bounded. RIPPLE incrementally expands the vicinity until the top-k result set can be obtained. Our extensive experiments on real-life graph data sets show that Ink queries can retrieve interesting results, and the proposed algorithms are orders of magnitude faster than state-of-the-art method.

1 Introduction

Graphs have long been considered as one of the most important structures that can naturally model numerous real-life data objects (*e.g.*, the Web, social network, protein-protein interaction network). In most graph-related applications, it is fundamental to quantify node-to-node structural proximity. Among existing structural proximity measures, *random walk with restart* (RWR) is recognized as one of the most important, and has been widely adopted in Web search [15], item recommendation [12], link prediction [13], graph clustering [2], and many other tasks. Compared with other proximity measures like shortest path, RWR enjoys the nice property of capturing the holistic graph structure and being robust to noise in the graph.

© Springer International Publishing Switzerland 2015
A. Appice et al. (Eds.): ECML PKDD 2015, Part II, LNAI 9285, pp. 608–624, 2015.
DOI: 10.1007/978-3-319-23525-7_37

To date, much research effort has been devoted to RWR, including its efficient computation ([6], [16], [5], [7], [18], [8], [14]), top-k search ([7], [10], [3], [17]), and various mining tasks underpinned by RWR ([13], [2], [12]). However, insufficient attention has been paid to a fundamental task that arises in many graph-related applications, which is to determine the source nodes that *have a large amount of information flowing to* a given query node. To illustrate, consider a traffic flow network shown in Figure 1. Assume severe traffic congestion occurs at node q every day, then the following question is key to improving traffic scheduling and road network design: how do we find the nodes from which the traffic tends to flow into q and cause the congestion problem? Using the RWR measure, the node c is likely to be identified as a major source that causes congestion at q. Even though c is not the direct in-neighbor of q, there are many short paths from c to q. Given that c is a busy transportation hub, a large number of vehicles leaving from c tend to gather at q.

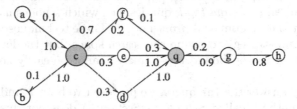

Fig. 1. An example traffic flow network. Each node is a road intersection, and the node size denotes the daily traffic volume at the intersection. Each edge is a road segment, and the attached number denotes the proportion of the traffic moving along that edge from a specific node.

We propose a novel query named the *inbound top-k* (Ink) query, which seeks to identify the nodes that have *a large amount of information flowing to* a query node based on RWR. Consider a query node q in a graph G. For any other node u in G, let $r_{u \rightsquigarrow q}$ be the RWR from u to q. Additionally, each node u has a nonnegative weight w_u.[1] Given G, q, and an integer k, the Ink query aims to find k nodes in G that have the largest scores in terms of $w_u \cdot r_{u \rightsquigarrow q}$.

The Ink query can be highly useful for a wide spectrum of applications besides traffic flow analysis. Think of a protein-protein interaction (PPI) network wherein each node is a protein, and a directed edge indicates one protein has a signal transduction to another protein to cause its formation or mutation. The signal transduction between proteins is essential to many biological processes and diseases (*e.g.*, Parkinson's disease, cancer). Querying by the characteristic protein of a disease, the Ink query can identify the top-k proteins that are most likely to cause the formation of the query protein. Another example application is *targeted advertising*. In an online social network like Facebook, suppose a company (*e.g.*, Walmart) wants to place advertisements on its Facebook page. With the Ink query, that company can easily identify the top-k Facebook users

---

[1] For instance, in our traffic flow example, w_u is the average daily traffic volume at node u.

that are most likely to visit its page. By statistically analyzing the profiles and preferences of these users, the company can adapt the advertising content to attract potential customers more effectively.

To the best of our knowledge, no existing methods can *accurately* and *efficiently* answer Ink queries. First, methods ([6], [16], [14], [18]) have been proposed to compute the approximate RWR between any two nodes with an error bound ϵ. However, it is hard to pre-specify a proper ϵ for an ad-hoc query node q, because a pre-specified ϵ may be either too coarse to generate the correct top-k results, or too fine to avoid unnecessary computation. Second, the k-dash method [7] can compute the exact RWR between any two nodes. However, it uses matrix LU decomposition as a pre-computation step, which has a time complexity of $O(n^3)$ and thus is prohibitively expensive for large graphs. Even assuming the LU decomposition is done, later we will see, it is costly and unnecessary to compute the RWR scores from all nodes to the query node q in order to answer Ink queries. Third, techniques ([16], [10], [5], [9], [8]) have recently been reported to process what we call *outbound top-k* queries, *i.e.*, which k nodes have the largest RWR if we start the random walk *from* node q? These techniques mostly use the branch-and-bound strategy to prune the search space, but the lower and upper bounds derived for the outbound top-k query cannot be easily adapted for the Ink query.

To efficiently answer the Ink query, we propose two branch-and-bound methods. Our first method, called SQUEEZE (Section 3) does not directly compute the exact RWR to q for each node in the graph, but maintains a lower bound and an upper bound. It then incrementally refines the bounds by performing matrix-vector multiplication. We prove that the error decreases exponentially as the iterative process continues, and thus the top-k results can be determined after a few number of iterations. Our second method, called RIPPLE (Section 4), is an even more efficient algorithm and thus suitable for extremely large graphs. Compared with SQUEEZE, RIPPLE leverages locality to gain significant performance improvement. The key observation is that the nodes falling in the vicinity of q tend to have large RWR scores to q. Hence, RIPPLE maintains a dynamic vicinity of q and estimates the RWR scores only for the nodes inside the vicinity. The outside nodes do not need to be evaluated because their scores are propagated from the boundary of the vicinity and thus upper bounded. RIPPLE progressively expands the vicinity, and refines the error bounds until the result set can be correctly identified.

Our theoretical analysis shows that both SQUEEZE and RIPPLE, without any pre-computation, can accurately answer the Ink queries in a fast and incremental manner. In addition, we have conducted extensive experiments on real-life graph data sets (Section 5). The results demonstrate the Ink query can retrieve interesting results. Meanwhile, SQUEEZE and RIPPLE outperform state-of-the-art method by orders of magnitude in efficiency.

Table 1. Notations used in the paper.

G	A graph $G = (V, E)$.	$\mathbf{P}$	The row normalized transition matrix for G.
n	The number of nodes in G.	p_{ij}	The transition probability from node i to j.
m	The number of edges in G.	c	The restart probability ($0 < c < 1$).
w_i	The weight of a node $i \in V$.	$\mathbf{e}_u$	$n \times 1$ vector, 1 for u's element and 0 for the others.
w_{ij}	The weight of an edge (i, j).	$\mathbf{r}_u$	The RWR score vector for the walk *from u*.

2 Preliminaries

In this section, we present some preliminaries for the Ink query. Table 1 lists the notations used throughout this paper.

2.1 Problem Description

Definition 1 (Transition Matrix). *For a node $i \in V$, let $d_i = \sum_{j=1}^{n} w_{ij}$ be the total out-degree of i. The transition matrix of G is an $n \times n$ matrix $\mathbf{P} = [p_{ij}]_{n \times n}$ where $p_{ij} = w_{ij}/d_i$ if $(i, j) \in E$ and 0 otherwise.*

Based on the definition of transition matrix, the *random walk with restart (RWR)* process is described as follows. Consider a surfer who starts RWR from the node $x_0 = u$. Suppose the surfer is at node $x_t = i$ at step t, she returns to u with probability c and continues surfing with probability $1 - c$. If she continues the surfing, she randomly moves to i's neighbor j with probability p_{ij}. The stationary distribution $\mathbf{r}_u$ of such a process, *i.e.*, the RWR scores of all the nodes in V, is the solution to the equation:

$$\mathbf{r}_u = (1 - c)\mathbf{P}^{\mathrm{T}}\mathbf{r}_u + c\mathbf{e}_u. \tag{1}$$

In $\mathbf{r}_u$, the element $\mathbf{r}_u(v)$ denotes the RWR score from u to v, namely $r_{u \rightsquigarrow v}$. Given a query node $q \subset V$, a restart probability c, and an integer k, the Ink query aims to find a set $S \subseteq V$ such that: (1) $|S| = k$; and (2) $\forall u \in S$, $\forall v \in V - S$, $w_u \mathbf{r}_u(q) \geq w_v \mathbf{r}_v(q)$.

2.2 Naïve Methods

In this subsection, we describe two naïve methods for answering Ink queries, and discuss why they are not satisfactory.

Power: Although directly solving Equation 1 costs $O(n^3)$, the power iteration can produce an approximate solution with time complexity $O(tm)$ where t is the number of iterations. Accordingly, a naïve solution, named Power, can answer the Ink query in two steps: (1) it computes $\mathbf{r}_u$ for every node $u \in V$ using the power iteration; and (2) it selects k nodes with the largest weighted RWRs to q.

LU: Another solution, named LU, is adapted from the *k-dash* method proposed by Fujiwara *et al.* [7]. As the solution of Equation 1 is $\mathbf{r}_u = c(\mathbf{I} - (1-c)\mathbf{P}^{\mathrm{T}})^{-1}\mathbf{e}_u$, they perform LU decomposition on the matrix $\mathbf{W} = \mathbf{I} - (1-c)\mathbf{P}^{\mathrm{T}} = \mathbf{LU}$ in an offline stage, and store $\mathbf{L}^{-1}$ and $\mathbf{U}^{-1}$ beforehand. Given that $\mathbf{r}_u = c\mathbf{W}^{-1}\mathbf{e}_u = c\mathbf{U}^{-1}\mathbf{L}^{-1}\mathbf{e}_u$, the exact RWR score between any two nodes can be computed in $O(n)$ time based on the matrices $\mathbf{L}^{-1}$ and $\mathbf{U}^{-1}$. Accordingly, LU does not need to compute the entire proximity matrix to answer the Ink query. Instead, it computes only one row that corresponds to the RWR scores from all the nodes to the query node q. Once the n RWR scores are obtained, the top-k nodes can be easily obtained.

Remark. For every $u \in V$, the Power method needs to compute u's RWR scores to all the nodes in V, leading to a time complexity $O(tmn)$, which is intolerable for large graphs. The LU method can directly compute the exact RWR scores from all nodes to the query node based on offline matrix decomposition. However, the time complexity of the on-line retrieval phase is $O(n^2 + n\log k)$, still time-consuming for large graphs. Later we will see, it is actually unnecessary and wasteful to compute the RWRs from all the nodes to q. Moreover, note that matrix LU decomposition has a time complexity of $O(n^3)$ and thus is prohibitively expensive for large graphs. Finally, the matrix $\mathbf{W}$ is dependent on the restart probability c. If the user launches an Ink query with a different c, the precomputed matrices $\mathbf{L}^{-1}$ and $\mathbf{U}^{-1}$ become useless and need to be recomputed.

2.3 Overview of Squeeze and Ripple

Before presenting the SQUEEZE and RIPPLE methods, we first analyze the relations of the RWR scores from all the nodes to the query node q based on the *Decomposition Theorem* proposed by Jeh and Windom [11].

Theorem 1. *Given a node u, and O_u, the set of u's out-neighbors, the RWR proximity vector from u satisfies $\mathbf{r}_u = (1-c) \sum\limits_{v \in O_u} p_{uv}\mathbf{r}_v + c\mathbf{e}_u$.*

Theorem 1 says that, the RWR vector of u can be derived by linearly combining the RWR vectors of u's out-neighbors, with extra emphasis on u itself. For any node $u \in V$, we have the RWR score from u to q computed as:

$$\mathbf{r}_u(q) = \begin{cases} (1-c) \sum\limits_{v \in O_u} p_{uv}\mathbf{r}_v(q) & \text{if } u \neq q \\ (1-c) \sum\limits_{v \in O_u} p_{uv}\mathbf{r}_v(q) + c & \text{if } u = q. \end{cases} \tag{2}$$

By writing down the decomposition for every node u in V according to Equation 2, we obtain a linear system $\mathbf{x}_q$ that consists of n variables. Specifically, letting $\mathbf{x}_q$ be an $n \times 1$ vector such that $\mathbf{x}_q(u) = \mathbf{r}_u(q)$ is the RWR score from u to q, and letting $\mathbf{A} = (1-c)\mathbf{P}$, then

$$\mathbf{x}_q = \mathbf{A}\mathbf{x}_q + c\mathbf{e}_q. \tag{3}$$

An intuitive idea to answer Ink query is to perform power iteration over Equation 3 and obtain a good "enough" approximation of $\mathbf{x}_q$. Based on this intuition, our first method SQUEEZE iteratively performs matrix-vector multiplication and provides an analytical error bound after each iteration. We prove that the error shrinks at an exponential rate as the iteration proceeds, hence SQUEEZE can prune the unqualified nodes quickly and retrieve the top-k results after a small number of iterations.

Though SQUEEZE is simple, performing matrix-vector multiplication over the whole graph can be costly if the graph is extremely large. Our second method RIPPLE addresses this problem by leveraging the locality of RWR. The key observation is that the nodes around q tend to have large RWR scores. As such, RIPPLE employs a local update strategy, which maintains a vicinity around q and evaluates RWR only for the nodes inside. According to Equation 3, the RWR scores of the nodes outside the vicinity are propagated from the boundary of the vicinity and thus upper bounded. By progressively pushing the boundary of the vicinity, the estimations for the inside nodes become more accurate, and the upper bound for the outside nodes becomes tighter. Finally, RIPPLE terminates the vicinity expansion once the top-k results can be correctly identified.

3   The Squeeze Algorithm

In this section, we describe the details of SQUEEZE. As aforementioned, the key idea of SQUEEZE is to iterate over Equation 3, and analyze the estimation errors on-the-fly. To begin with, we define the lower bound relation between two vectors.

Definition 2 (Lower Bound Vector). *Let* $\mathbf{x}$ *and* $\mathbf{y}$ *be two* $n \times 1$ *vectors.* $\mathbf{x}$ *is a lower bound vector of* $\mathbf{y}$ *if* $\forall 1 \leq i \leq n$, $\mathbf{x}(i) \leq \mathbf{y}(i)$, *denoted as* $\mathbf{x} \prec \mathbf{y}$.

SQUEEZE starts with the zero vector $\mathbf{x}_q^{(0)} = \mathbf{0}$, which serves as a lower bound vector for $\mathbf{x}_q$, the solution to Equation 3. Then it iteratively updates the lower bound vector according to the following equation:

$$\mathbf{x}_q^{(i+1)} = \mathbf{A}\mathbf{x}_q^{(i)} + c\mathbf{e}_q. \tag{4}$$

In the following, we prove: (1) each iteration produces a tighter lower bound vector, *i.e.*, $\mathbf{x}_q^{(i+1)}$ will be closer to $\mathbf{x}_q$, and (2) $\mathbf{x}_q^{(i)}$ finally converges to $\mathbf{x}_q$.

Theorem 2. *Let* $\mathbf{x}_q^{(0)} = \mathbf{0}$ *and* $\mathbf{x}_q^{(i+1)} = \mathbf{A}\mathbf{x}_q^{(i)} + c\mathbf{e}_q$. *It is ensured* $\forall i \geq 0, \mathbf{x}_q^{(i)} \prec \mathbf{x}_q^{(i+1)} \prec \mathbf{x}_q$; *and* $\mathbf{x}_q^{(i)} = \mathbf{x}_q$ *when* $i \to \infty$.

Proof. (1) Given $\mathbf{x}_q^{(0)} = \mathbf{0}$ and Equation 4, we have $\mathbf{x}_q^{(1)}(u) = c \cdot \mathbb{I}_{\{u=q\}}(u)$, where $\mathbb{I}$ is the indicator function. Clearly, $\mathbf{x}_q^{(0)} \prec \mathbf{x}_q^{(1)}$. Further, if $\mathbf{x}_q^{(i-1)} \prec \mathbf{x}_q^{(i)}$, then $\forall u$,

$$\mathbf{x}_q^{(i+1)}(u) - \mathbf{x}_q^{(i)}(u) = (1-c) \sum_{v \in O_u} p_{uv} \left[\mathbf{x}_q^{(i)}(u) - \mathbf{x}_q^{(i-1)}(u) \right] \geq 0.$$

(2) It is clear that $\mathbf{x}_q^{(0)} \prec \mathbf{x}_q$. Suppose $\mathbf{x}_q^{(i)} \prec \mathbf{x}_q$, then

$$\mathbf{x}_q^{(i+1)}(u) \le (1-c) \sum_{v \in O_u} p_{uv} \mathbf{x}_q(v) + c \cdot \mathbb{I}_{\{u=q\}}(u) = \mathbf{x}_q(u).$$

To prove $\lim_{i \to \infty} \mathbf{x}_q^{(i)} = \mathbf{x}_q^{(i)}$, note that the spectral radius of $\mathbf{A}$ satisfies $\rho(\mathbf{A}) \le 1 - c < 1$. Then $\lim_{i \to \infty} \mathbf{x}_q^{(i)} = \sum_{i=0}^{\infty} \mathbf{A}^i \mathbf{e}_q = c(\mathbf{I} - \mathbf{A})^{-1}\mathbf{e}_q$, which is the solution of Equation 4.   ∎

Theorem 2 tells us that the power iteration over Equation 4 produces a tighter lower bound after each iteration, and finally converges to the exact value of $\mathbf{x}_q$. Below, we proceed to analyze the RWR upper bound after each iteration.

Theorem 3. $\forall u \in V, \mathbf{x}_q(u) \le \mathbf{x}_q^{(i)}(u) + (1-c)^i$.

Proof. $\forall i > 0, \mathbf{x}_q^{(i+1)} - \mathbf{x}_q^{(i)} = \mathbf{A}(\mathbf{x}_q^{(i)} - \mathbf{x}_q^{(i-1)})$. Accordingly,

$$\|\mathbf{x}_q^{(i+1)} - \mathbf{x}_q^{(i)}\| \le \|\mathbf{A}\| \cdot \|\mathbf{x}_q^{(i)} - \mathbf{x}_q^{(i-1)}\| = (1-c) \cdot \|\mathbf{x}_q^{(i)} - \mathbf{x}_q^{(i-1)}\|.$$

Recursively applying the above inequality gives us $\|\mathbf{x}_q^{(i+1)} - \mathbf{x}_q^{(i)}\| \le (1-c)^i\|\mathbf{x}_q^{(1)} - \mathbf{x}_q^{(0)}\| = (1-c)^i c$. Moreover, $\forall m > i$,

$$\|\mathbf{x}_q^{(m)} - \mathbf{x}_q^{(i)}\| = \|\sum_{j=i}^{m-1} (\mathbf{x}_q^{(j+1)} - \mathbf{x}_q^{(j)})\| \le \sum_{j=i}^{m-1} \|\mathbf{x}_q^{(j+1)} - \mathbf{x}_q^{(j)}\|$$

$$\le (1-c)^i c \sum_{j=0}^{m-i-1} (1-c)^j = (1-c)^i c \frac{1 - (1-c)^{m-i}}{1 - (1-c)}.$$

By setting $m \to \infty$, we have $\|\mathbf{x}_q - \mathbf{x}_q^{(i)}\| \le (1-c)^i$.   ∎

Theorem 2 and 3 guarantee that after iteration i, $\forall u \in V$, $\mathbf{x}_q^{(i)}(u) \le \mathbf{x}_q(u) \le \mathbf{x}_q^{(i)}(u) + (1-c)^i$. Better still, the gap $(1-c)^i$ between the lower and upper bounds decreases exponentially as the iteration proceeds, which allows us to quickly identify the top-k results. Algorithm 1 sketches SQUEEZE. As shown, we first set $\mathbf{x}_q = \mathbf{0}$, and initialize a candidate set R that consists of all the nodes in V. Then we gradually refine $\mathbf{x}_q$ using power iteration (line 4), and select the k-th largest weighted RWR as the threshold τ (line 5). All the nodes whose weighted RWRs are smaller than τ can be safely pruned (lines 7-9). SQUEEZE terminates when the candidate set R contains only k nodes.

4   The Ripple Algorithm

In this section, we present the RIPPLE method, which employs a local update strategy to efficiently process Ink queries.

Algorithm 1. The SQUEEZE algorithm.

Input: query node q, number k, graph $G = (V, E)$, restart probability c

Output: the top-k result set R

1 $R \leftarrow V$, $i \leftarrow 0$, $\mathbf{x}_q \leftarrow \mathbf{0}$;

2 Construct the transition matrix $\mathbf{P}$;

3 **while** $|R| > k$ **do**

4 $\quad$ $\mathbf{x}_q \leftarrow \mathbf{A}\mathbf{x}_q + c\mathbf{e}_q$;

5 $\quad$ $\tau \leftarrow k$-th largest score in terms of $w_u \cdot \mathbf{x}_q(u)$;

6 $\quad$ $i \leftarrow i + 1$;

7 $\quad$ **foreach** $u \in R$ **do**

8 $\quad\quad$ **if** $w_u \cdot [\mathbf{x}_q + (1-c)^i] < \tau$ **then**

9 $\quad\quad\quad$ Remove u from R;

4.1   Algorithm Sketch

Given a query node q, we use N_q to denote a set of nodes falling in the vicinity of q, and F_q to denote the nodes falling outside, *i.e.*, $F_q = V - N_q$. Further, we call node $u \in N_q$ a *boundary node* if there exists a node $v \in F_q$ such that v is an in-neighbor of u, and use B_q to denote the set of boundary nodes. The key insight of RIPPLE is that the nodes close to q tend to have large RWR scores. Hence, starting from a small vicinity around q, RIPPLE estimates the RWRs for only the nodes in the vicinity. For the outside nodes, RIPPLE maintains one generic upper bound for them. As the vicinity is gradually expanded, the RWR estimations for the inside nodes as well as the upper bound for the outside nodes become more and more accurate. The expansion terminates when the estimations are accurate enough to produce the top-k results.

Algorithm 2 gives a sketch of RIPPLE. As shown, we initialize N_q and B_q to $\{q\}$, and iteratively select at most s boundary nodes with the largest RWRs (lines 3-6), where s is a pre-specified parameter of RIPPLE. Later we will see, the rationale of selecting the high-score boundary nodes is that such nodes determine the estimation error for the nodes in N_q, as well as the RWR upper bound for the nodes in F_q. We expand N_q by incorporating the in-neighbors of the selected nodes (lines 7-8). After each expansion, we iterate over the nodes in N_q for t times to refine their RWR estimations (lines 10-12). The RWR scores for all the nodes in F_q are set to 0 and do not need to be computed. Once the update operation is done, we select the k-th largest weighted RWR as the threshold τ, and prune the nodes whose upper bound scores are smaller than τ (lines 14-17). Such a process repeats until there are only k nodes left in R.

Figure 2 shows a concrete example of RIPPLE. Suppose $c = 0.2$, $s = 2$, and $t = 2$. First, the vicinity and boundary node sets are set to $N_q = B_q = \{q\}$. In the first round, q is the only node B_q. Hence, we expand q and obtain $N_q = \{2, 6, 7, 10\}$ and $B_q = \{2, 7, 10\}$. After the expansion, starting from $\mathbf{x}_q = \mathbf{0}$, RIPPLE updates the RWR for the nodes in N_q using 2 iterations, and obtains $\mathbf{x}_q(2) = 0.16, \mathbf{x}_q(6) = 0.2, \mathbf{x}_q(7) = 0.04, \mathbf{x}_q(10) = 0.16$. In the second round,

---

Algorithm 2. The RIPPLE algorithm.

Input: query node q, number k, graph $G = (V, E)$, restart probability c,
        number of to-expand nodes s, number of iterations t after expansion
Output: the top-k result set R

1   $R \leftarrow V$, $N_q \leftarrow \{q\}$, $B_q \leftarrow \{q\}$, $\underline{x}_q \leftarrow 0$;
2   **while** $|R| > k$ **do**
3   **if** $|B_q| \geq s$ **then**
4         $E \leftarrow s$ nodes in B_q with the largest RWRs;
5   **else**
6         $E \leftarrow B_q$;
7   **foreach** $u \in E$ **do**
8   Add u's in-neighbors into N_q;
9   Update B_q;
10   **for** $i = 1$ **to** t **do**
11   **foreach** $u \in N_q$ **do**
12           $\underline{x}_q(u) = (1 - c) \sum_{v \in O_u} p_{uv}\underline{x}_q(u) + c \cdot \mathbb{I}_{\{u=q\}}(u)$;
13     $\tau \leftarrow k$-th largest weighted RWR for the nodes in N_q;
14   **foreach** $u \in R$ **do**
15        $\overline{x}_q(u) \leftarrow$ the RWR upper bound;
16   **if** $w_u \cdot \overline{x}_q(u) < \tau$ **then**
17   Remove u from R;
18 **return** R;

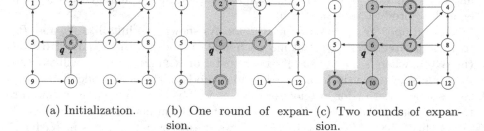

(a) Initialization.   (b) One round of expansion.   (c) Two rounds of expansion.

Fig. 2. Illustration of the RIPPLE algorithm. The nodes in the gray area are the vicinity nodes, and the double-ringed ones are the boundary nodes.

RIPPLE expands node 2 and 10, and derives $N_q = \{2, 3, 6, 7, 9, 10\}$ and $B_q = \{3, 7, 9\}$. With the previous $\mathbf{x}_q$, RIPPLE updates the new N_q using 2 iterations, and obtains $\mathbf{x}_q(2) = 0.16, \mathbf{x}_q(3) = 0.08, \mathbf{x}_q(6) = 0.2, \mathbf{x}_q(7) = 0.056, \mathbf{x}_q(9) = 0.128, \mathbf{x}_q(10) = 0.16$. The expansion process continues until the top-k nodes are obtained.

Several questions remain to be answered for Algorithm 2: (1) how do we compute the lower and upper bounds for the nodes in N_q and F_q? and (2) what is the reason of selecting high-score boundary nodes when expanding N_q? In what follows, we answer these questions in detail.

4.2 The Lower Bound

We first prove the RWR estimation is a lower bound when we set the RWR scores of the nodes in F_q to 0 and propagate the RWR scores only among the nodes in N_q.

Theorem 4. *Let $\underline{\mathbf{x}}_q$ be the solution to the equation $\underline{\mathbf{x}}_q = \mathbf{W}\underline{\mathbf{x}}_q + c\mathbf{e}_q$, where $\mathbf{W}$ is constructed from $\mathbf{A}$ by setting the rows of nodes in F_q to all zeros, then $\underline{\mathbf{x}}_q \prec \mathbf{x}_q$.*

Proof. The power method gives $\underline{\mathbf{x}}_q = \lim_{i \to \infty} c \sum_{j=1}^{i-1} \mathbf{W}^j \mathbf{e}_q$ and $\mathbf{x}_q = \lim_{i \to \infty} c \sum_{j=1}^{i-1} \mathbf{A}^j \mathbf{e}_q$. It suffices to prove $\forall i \geq 1, \mathbf{W}^i \mathbf{e}_q \prec \mathbf{A}^i \mathbf{e}_q$, which can be easily proved by induction. ∎

4.3 The Upper Bound

We proceed to analyze the RWR upper bounds. Let $M = \max_{u \in B_q} \mathbf{x}_q(u)$. Lemma 1 and Lemma 2 show that M determines the upper bound for the nodes in both F_q and N_q.

Lemma 1. $\forall u \in F_q, \mathbf{x}_q(u) \leq (1-c)M$.

Proof. When iterating over Equation 3 with $\mathbf{x}_q^{(0)} = \mathbf{0}$, Theorem 2 ensures $\forall u \in B_u, \forall i \geq 0, \mathbf{x}_q^{(i)}(u) \leq \mathbf{x}_q(u) \leq M$. Now consider any node $u \in F_q$: (1) when $i = 0, \mathbf{x}_q^{(0)}(u) = 0 \leq (1-c)M$ clearly holds. (2) $\forall i \geq 1$, assume $\forall v \in F_q, \mathbf{x}_q^{(i-1)}(v) \leq (1-c)M$. Since $\forall v \in B_q, \mathbf{x}_q^{(i-1)}(v) \leq M$, it is ensured $\mathbf{x}_q^{(i)}(u) = (1-c) \sum_{v \in B_q} p_{uv} \mathbf{x}_q^{(i-1)}(v) + (1-c) \sum_{v \in F_q} p_{uv}(1-c)\mathbf{x}_q^{(i-1)}(v) \leq (1-c)M$. ∎

Lemma 2. $\forall u \in N_q, \mathbf{x}_q(u) \leq \underline{\mathbf{x}}_q(u) + (1-c)^2 M$.

Proof. Let $\mathbf{d}_q = \mathbf{x}_q - \underline{\mathbf{x}}_q$. $\forall u \in N_q$, it suffices to prove $\mathbf{d}_q(u) \leq (1-c)^2 M$. Consider an $n \times 1$ vector $\mathbf{r}_F$, where the entries of the nodes in F_q are set to their accurate RWR scores, and the entries of the nodes in N_q are set to zeros. Then $\mathbf{d}_q = \mathbf{W}\mathbf{d}_q + \mathbf{r}_F$. By setting $\mathbf{d}_q^{(0)} = \mathbf{r}_F$ and using power iteration, we have $\mathbf{d}_q = \lim_{t \to \infty} \mathbf{d}_q^{(t)}$. Note that $\forall u \in V, \mathbf{d}_q^{(0)}(u) \leq (1-c)M$. The induction ensures $\lim_{t \to \infty} \mathbf{d}_q^{(t)}(u) \leq (1-c)^2 M$. ∎

Lemma 2 provides a generic upper bound for the vicinity nodes. In the following, we introduce the concept of *outward hop*, which allows us to derive tighter upper bounds for the vicinity nodes.

Definition 3 (Outward Hop). *For a node $u \in N_q$, the outward hop of u, denoted as $Hop(u)$, is the minimum number of steps that takes u to any node in F_q.*

The RWR estimation errors are propagated inwards layer by layer among the vicinity nodes. Hence, a larger outward hop implies a larger estimation error, as shown below.

Lemma 3. *Given a node $u \in N_q$, let $Hop(u) = h$, then $\mathbf{x}_q(u) - \underline{\mathbf{x}}_q(u) \leq (1 - c)^{h+1}M$.*

Proof. This lemma can be proved using induction, we omit the details to save space. ∎

Lemma 1 and 3 give us the RWR upper bounds for the nodes in F_q and N_q in terms of M, which is the maximum accurate RWR among the boundary nodes. Nevertheless, M is actually unknown to RIPPLE. Assume RIPPLE performs t power iterations over N_q to obtain an approximate vector $\mathbf{x}_q^{(t)}$, we now establish the connection between M and $\mathbf{x}_q^{(t)}$, and discuss how to compute the upper bounds based on $\mathbf{x}_q^{(t)}$.

Lemma 4. *Let $\Delta_N = \max\limits_{u \in V} \left[\mathbf{x}_q^{(1)}(u) - \mathbf{x}_q^{(0)}(u) \right]$, $\underline{M}^{(t)} = \max\limits_{u \in V} \mathbf{x}_q^{(t)}(u)$. We have*

$$M \leq \frac{1}{2c - c^2} \left[\underline{M}^{(t)} + (1 - c)^t \cdot \Delta_N/c \right].$$

Proof. Similar to Theorem 3, it can be shown $\forall u \in V$, $\mathbf{x}_q(u) - \mathbf{x}_q^{(t)}(u) \leq (1 - c)^t \cdot \Delta_N/c$. Let $\underline{M} = \max\limits_{u \in B_q} \mathbf{x}_q(u)$ and $w \in B_q$ be the node corresponding to $\underline{M}$. Then

$$\underline{M} - \underline{M}^{(t)} \leq \underline{M} - \mathbf{x}_q^{(t)}(w) \leq (1 - c)^t \cdot \Delta_N/c. \tag{5}$$

Further let $v \in B_q$ be the node corresponding to M. By Lemma 2, we know $M - \mathbf{x}_q(v) \leq (1 - c)^2M$, which gives $M \leq \mathbf{x}_q(v)/(2c - c^2)$. Combining this with Equation 5 completes the proof. ∎

Theorem 5. *$\forall u \in F_q$, we have*

$$\mathbf{x}_q(u) \leq \frac{1 - c}{2c - c^2} \left[\underline{M}^{(t)} + (1 - c)^t \cdot \Delta_N/c \right].$$

$\forall u \in N_q$, let $Hop(u) = h$, we have

$$\mathbf{x}_q(u) \leq \underline{\mathbf{x}}_q^{(t)}(u) + \frac{(1 - c)^{h+1}}{2c - c^2} \cdot \underline{M}^{(t)} + \frac{(1 - c)^{h+t+1} + (2c - c^2)(1 - c)^t}{2c^2 - c^3} \cdot \Delta_N$$

Proof. The first claim is obvious from Lemma 1 and 4, and the second claim is obvious from Lemma 3 and 4. ∎

5   Experiments

In this section, we evaluate the empirical performance of the proposed methods. All algorithms were implemented in JAVA and the experiments were conducted on a machine with Intel Xeon E5-2680 and 64GB memory.

5.1   Experimental Setup

Data Sets. Our experiments are based on two real graph data sets. Our first data set, referred to as 4SQ, is collected from Foursquare during a three-month period. The 4SQ data set consists of the check-in histories of 14,909 users living in New York. In 4SQ, each node is a place, and the node weight is set to the total number of visitors to reflect the popularity of the place. Meanwhile, there is a directed edge between two places if the check-ins at the two places occur within 3 hours. The weight of the edge is the number of users whose check-in history matches the transition. The 4SQ data set contains 48,564 nodes and 123,452 edges in total. The second data set, referred to as Wiki, is extracted from the Wikipedia graph. We have removed the non-English Wikipedia pages as well as the noisy pages that have less than 3 in-links. In the result Wiki data set, each node is a Wikipedia page, and the node weight is set to the number of in-links to reflect the importance of that page. Each edge is a directed link from one Wikipedia page to another, and all the edges have an equal weight. There are totally 4,382,715 nodes and 102,260,837 edges in the Wiki data set.

Compared Method. We described two naïve methods in Section 2.2, namely Power and LU. However, the time cost of Power is too expensive for our used data sets. Hence, we use the LU method for comparison in our experiments. LU involves an offline stage that performs matrix decomposition and an on-line stage that retrieves the top-k results, we only include the on-line retrieval time when measuring the performance of LU.

5.2   Illustrating Cases

In this subsection, we issue several test Ink queries on our data sets, and compare the results retrieved by the Ink query and those by the outbound top-k query [7].

Table 2 shows the inbound and outbound top-5 results for the queries "Yankee Stadium" and "Columbia University" on 4SQ,[2]. As shown, the results returned by the outbound top-k query are mostly famous places in New York, such as the Metropolitan Museum of Art and the Radio City Music Hall. As such places are structural hubs in the graph, the random walk from the query node is thus very likely to reach them, making their outbound RWR scores high. In contrast, the results returned by the Ink query are less famous places but have strong correlations with the query place. For example, the top result for the query "Yankee Stadium" is Yankee Tavern, which is a local pub close to the

---

[2] We do not include the query itself in the top-k results, same for the Wiki data set.

Table 2. A Comparison of the Inbound and Outbound Top-k Queries on 4SQ ($\mathbf{c}$ = 0.15 k = 5).

Query	Inbound Top-5 Results		Outbound Top-5 Results	
	Rank	Place Name	Rank	Place Name
Yankee Stadium	1	Yankee Tavern	1	The Metropolitan Museum of Art
	2	Stan's Sports Bar	2	Madison Square Garden
	3	Billy's Sports Bar	3	The Central Park
	4	New York Penn Station	4	Grand Central Terminal
	5	Grand Central Terminal	5	Brooklyn Museum
Columbia University	1	Morningside Park	1	Central Park
	2	Seeley Mudd Hall	2	116th St/Columbia University MTA Subway
	3	Whole Foods Grocery	3	Newark Liberty International Airport
	4	Dinosaur Bar-B-Que	4	Grand Central Terminal
	5	Starbucks	5	Lincoln Tunnel

Table 3. A Comparison of Inbound and Outbound Top-k Queries on Wiki ($\mathbf{c}$ = 0.15 k = 5).

Query	Inbound Top-5 Results		Outbound Top-5 Results	
	Rank	Page Title	Rank	Page Title
Information Retrieval	1	IDF	1	Computer Science
	2	Index Term	2	Information Science
	3	Keyword (Internet Search)	3	Linguistics
	4	Precision and Recall	4	Association for Computing Machinery
	5	Recall (Information Retrieval)	5	Mathematics
Microsoft Office	1	Microsoft Excel	1	2006
	2	Microsoft Word	2	2007
	3	Microsoft Windows	3	2008
	4	Microsoft Office 2007	4	Microsoft
	5	Microsoft FrontPage	5	Microsoft Office 2007

stadium. Yankee fans may love to gather together at the pub to have some beer and talk about their favorite players.

Table 3 shows the results for the queries "Information Retrieval" and "Microsoft Office" on Wiki. Again, we observe that the outbound top-k query tends to retrieve the pages that are popular, whereas the inbound top-k query obtains the pages that are more specific and strongly correlated to the query. For example, given the query "Information Retrieval", the results returned by Ink are all terminologies in the field of information retrieval, such as IDF and index term. In contrast, the outbound top-5 query returns general but more famous pages like Computer Science.

5.3   Efficiency Study

In this subsection, we study the efficiency of the proposed algorithms. An Ink query consists of two parameters: (1) the number k; and (2) the restart probability c. We set their default values as $k = 20$ and $c = 0.15$. We evaluate the effect of one parameter while the other is fixed at its default value, and run 1000 randomly generated queries with their average cost reported. LU and SQUEEZE are parameter-free, while RIPPLE has two parameters to tune: (1) s, the number of boundary nodes for expansion; and (2) t, the number of iterations after each expansion. We first fix $s = 30$ and $t = 1$ when comparing RIPPLE with the other two methods. Then we study the effect of s and t on the performance of RIPPLE.

Varying k. Figure 3 shows the running time of the three methods when k varies on 4SQ. As shown, k does not affect the running time of LU much, as the major cost of LU is the computation of the RWR scores of all the nodes in the graph. In contrast, the running time of SQUEEZE and RIPPLE increases with k, but at a quite slow rate. This phenomenon could be explained by the fact that, as k increases, the score gap between the k-th and the $(k + 1)$-th objects tends to become smaller. As a result, both SQUEEZE and RIPPLE need more iterations to retrieve the top-k results. Comparing the performance of the three methods, we find both SQUEEZE and RIPPLE outperform LU significantly even though the pre-computation time of LU has already been excluded. This fact suggests the branch-and-bound strategies used by SQUEEZE and RIPPLE are quite effective, they can largely prune the search space to avoid unnecessary RWR computations. Figure 3(b) shows the running time of SQUEEZE and RIPPLE on Wiki. We do not have the result of LU because the offline matrix decomposition stage fails to complete within one week on Wiki. Similarly, RIPPLE needs more iterations to produce the top-k results than SQUEEZE, but it takes much less time. Moreover, the performance gap between RIPPLE and SQUEEZE is even larger on Wiki than on 4SQ. This is explained by the fact that RIPPLE is a local search algorithm and is not so sensitive to the data set size. Therefore, RIPPLE is suitable for extremely large graphs.

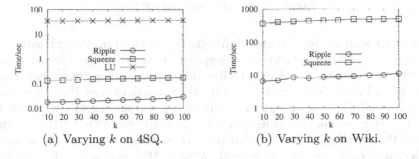

(a) Varying k on 4SQ.   (b) Varying k on Wiki.

Fig. 3. Running time v.s. k.

Varying c. Figure 4 shows the running time of the three methods on 4SQ when c varies from 0.05 to 0.2. The running time of SQUEEZE and RIPPLE decreases exponentially with c, which is in line with our expectation. As shown in Theorem 3, the error bound of SQUEEZE after i iterations is $(1 - c)^i$. For a larger c, SQUEEZE needs much less iterations to produce the top-k results. Similarly, for RIPPLE, Theorem 5 suggests that the error bounds for both inside and outside nodes are much tighter under a larger c, thus the number of iterations are fewer. To understand this phenomenon from another perspective, RIPPLE leverages RWR locality to answer Ink queries. When c is large, the random surfer has a higher probability to jump back to the start node, thus the nodes close to the query node are more likely to appear in the correct top-k results, making the vicinity-based estimation prune the search space more effectively.

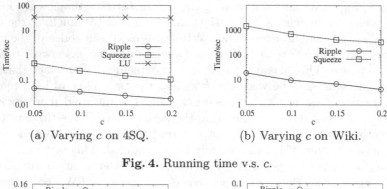

(a) Varying c on 4SQ.   (b) Varying c on Wiki.

Fig. 4. Running time v.s. c.

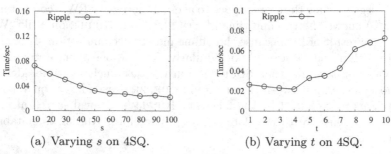

(a) Varying s on 4SQ.   (b) Varying t on 4SQ.

Fig. 5. Effects of s and t on RIPPLE.

Effects of s and t. Figure 5 shows the effects of s and t on RIPPLE using 4SQ (the results on Wiki are omitted to save space). As shown, when s increases from 10 to 100, the running time of RIPPLE first decreases and then gradually becomes stable. From Figure 5(b), we observe that the running time of RIPPLE is stable when t is small. However, the running time increases quite rapidly when t is too large. This suggests that iterating over a small vicinity for too many times cannot improve the efficiency of RIPPLE, but only incurs unnecessary computations. In practice, it is better to set t to a small value so that only a few iterations are performed before a new vicinity set is generated.

6   Related Work

The efficient computation of RWR has received a substantial amount of attention over the past decade. Though obtaining the closed-form solution of RWR requires the inversion of a matrix (Equation 1) and time-consuming, two popular strategies are widely adopted to address this problem: Monte Carlo sampling [3], [4] and power iteration [15]. Other techniques for efficiently approximating RWR have also been proposed. Tong et al. [16] introduced an efficient and novel algorithm for computing approximate RWR scores. Their method relies on a pre-processing step, which obtains the low-rank approximation of a large and sparse matrix. Zhu et al. [18] proposed to compute the approximate PPR vector using the inverse P-distance [18]. The key idea is to partition all random walk tours into different layers according to their contributions, and given priority to those important layers when computing the PPR vector. Methods [1] have also been proposed to compute the approximate RWR scores from all the nodes to a given query node. Unfortunately, these approximate algorithms cannot be easily applied to answer the Ink queries as it is hard to pre-specify the desired error bound for an ad-hoc query. Moreover, as suggested by RIPPLE, computing the RWR scores from all the nodes is actually unnecessary.

Along another line, much attention has been paid to the outbound top-k search problem. The goal is to retrieve the k nodes with the highest RWR/PPR scores from a query node. Most of the existing techniques for answering outbound top-k search resort to the branch-and-bound strategy to prune the search space. Specifically, Gupta et al. [10] proposed the Basic Push Algorithm, which computes PPR bounds based on bookmark coloring. Bahmani et al. [5] proposed a Monte Carlo based method for finding approximate top-k neighbors. Their results demonstrate that, by precomputing and storing a number of short random walk tours for all the nodes in the graph, the top-k neighbors can be fast approximated with satisfactory accuracy. Fujiwara et al. [7] proposed the k-dash algorithm to identify the top-k nearest neighbors of a query node based on matrix LU decomposition. They later proposed an method [8] that does not rely on offline pre-computation, but estimates the lower and upper bounds in an on-line manner. However, the lower and upper bounds derived for the outbound top-k query cannot be easily adapted for our Ink query.

7 Conclusions

We proposed the Ink query based on random walk with restart, which retrieves the top-k nodes that have high weighted RWR scores to a given query node. To efficiently process the Ink query, we designed the SQUEEZE and RIPPLE methods. SQUEEZE iteratively performs matrix vector multiplication and dynamically updates the lower and upper RWR bounds to generate the top-k result set. RIPPLE exploits RWR locality by maintaining a vicinity around the query node, and incrementally expands the vicinity to refine the RWR estimations. Our experimental results have demonstrated that both methods can answer Ink queries efficiently on large real-life graphs, while RIPPLE is especially suitable for extremely large graphs. Interesting future work includes investigating how the Ink query can benefit higher-level tasks such as link prediction, and how to adapt the RIPPLE method to a distributed version.

Acknowledgments. We thank the reviewers for their insightful comments. This work was sponsored in part by the U.S. Army Research Lab. under Cooperative Agreement No. W911NF-09-2-0053 (NSCTA), National Science Foundation IIS-1017362, IIS-1320617, and IIS-1354329, HDTRA1-10-1-0120, and grant 1U54GM114838 awarded by NIGMS through funds provided by the trans-NIH Big Data to Knowledge (BD2K) initiative, and MIAS, a DHS-IDS Center for Multimodal Information Access and Synthesis at UIUC.

References

1. Andersen, R., Borgs, C., Chayes, J.T., Hopcraft, J., Mirrokni, V.S., Teng, S.-H.: Local computation of pagerank contributions. In: Bonato, A., Chung, F.R.K. (eds.) WAW 2007. LNCS, vol. 4863, pp. 150–165. Springer, Heidelberg (2007)
2. Andersen, R., Chung, F.R.K., Lang, K.J.: Local graph partitioning using pagerank vectors. In: FOCS, pp. 475–486 (2006)
3. Avrachenkov, K., Litvak, N., Nemirovsky, D., Smirnova, E., Sokol, M.: Quick detection of top-k personalized pagerank lists. In: Frieze, A., Horn, P., Prałat, P. (eds.) WAW 2011. LNCS, vol. 6732, pp. 50–61. Springer, Heidelberg (2011)
4. Bahmani, B., Chakrabarti, K., Xin, D.: Fast personalized pagerank on mapreduce. In: SIGMOD Conference, pp. 973–984 (2011)
5. Bahmani, B., Chowdhury, A., Goel, A.: Fast incremental and personalized pagerank. PVLDB 4(3), 173–184 (2010)
6. Fogaras, D., Rácz, B., Csalogány, K., Sarlós, T.: Towards scaling fully personalized pagerank: Algorithms, lower bounds, and experiments. Internet Mathematics 2(3), 333–358 (2005)
7. Fujiwara, Y., Nakatsuji, M., Onizuka, M., Kitsuregawa, M.: Fast and exact top-k search for random walk with restart. PVLDB 5(5), 442–453 (2012)
8. Fujiwara, Y., Nakatsuji, M., Shiokawa, H., Mishima, T., Onizuka, M.: Efficient ad-hoc search for personalized pagerank. In: SIGMOD Conference, pp. 445–456 (2013)
9. Fujiwara, Y., Nakatsuji, M., Yamamuro, T., Shiokawa, H., Onizuka, M.: Efficient personalized pagerank with accuracy assurance. In: KDD, pp. 15–23 (2012)
10. Gupta, M.S., Pathak, A., Chakrabarti, S.: Fast algorithms for topk personalized pagerank queries. In: WWW, pp. 1225–1226 (2008)
11. Jeh, G., Widom, J.: Scaling personalized web search. In: WWW, pp. 271–279 (2003)
12. Konstas, I., Stathopoulos, V., Jose, J.M.: On social networks and collaborative recommendation. In: SIGIR, pp. 195–202 (2009)
13. Liben-Nowell, D., Kleinberg, J.M.: The link prediction problem for social networks. In: CIKM, pp. 556–559 (2003)
14. Lofgren, P., Banerjee, S., Goel, A., Comandur, S.: FAST-PPR: scaling personalized pagerank estimation for large graphs. In: KDD, pp. 1436–1445 (2014)
15. Page, L., Brin, S., Motwani, R., Winograd, T.: The pagerank citation ranking: Bringing order to the web. Technical report, Stanford University (1998)
16. Tong, H., Faloutsos, C., Pan, J.Y.: Fast random walk with restart and its applications. In: ICDM, pp. 613–622 (2006)
17. Yu, A.W., Mamoulis, N., Su, H.: Reverse top-k search using random walk with restart. PVLDB 7(5), 401–412 (2014)
18. Zhu, F., Fang, Y., Chang, K.C.C., Ying, J.: Incremental and accuracy-aware personalized pagerank through scheduled approximation. PVLDB 6(6), 481–492 (2013)

Finding Community Topics and Membership in Graphs

Matt Revelle[✉], Carlotta Domeniconi, Mack Sweeney, and Aditya Johri

George Mason University, Fairfax, VA 22030, USA
{revelle,carlotta}@cs.gmu.edu, {msweene2,ajohri3}@gmu.edu

Abstract. Community detection in networks is a broad problem with many proposed solutions. Existing methods frequently make use of edge density and node attributes; however, the methods ultimately have different definitions of community and build strong assumptions about community features into their models. We propose a new method for community detection, which estimates both per-community feature distributions (topics) and per-node community membership. Communities are modeled as connected subgraphs with nodes sharing similar attributes. Nodes may join multiple communities and share common attributes with each. Communities have an associated probability distribution over attributes and node attributes are modeled as draws from a mixture distribution. We make two basic assumptions about community structure: communities are densely connected and have a small network diameter. These assumptions inform the estimation of community topics and membership assignments without being too prescriptive. We present competitive results against state-of-the-art methods for finding communities in networks constructed from NSF awards, the DBLP repository, and the Scratch online community.

1 Introduction

Given a graph of self-organizing objects, we wish to estimate the latent topics around which the objects organize and discover community membership. We hypothesize groups with high edge density in graphs are evidence of communities whose members have similar attributes within a subset of the feature dimensions.

In this paper we present Seeded Estimation of Network Communities (SENC). SENC is a probabilistic method which uses both node attributes and graph structure to simultaneously estimate community feature distributions and members. We assume a community may exist around seed groups in the network. Many community detection methods build strong assumptions regarding community features into their models, which limits generalizability. SENC provides a flexible means of accounting for a variety of community structures through the use of configurable lower and upper bounds on discovered communities. The seed groups define the lower bounds, and they may in turn be defined by network structure or node and edge attributes. In the experiments presented in this paper, we consider every maximal k-clique in the network to be the core

© Springer International Publishing Switzerland 2015
A. Appice et al. (Eds.): ECML PKDD 2015, Part II, LNAI 9285, pp. 625–640, 2015.
DOI: 10.1007/978-3-319-23525-7_38

of a partially defined community. The upper bounds provide an intuitive way to incorporate knowledge about the degree of clustering in the network. Nodes may be members of multiple communities and communities may overlap.

Communities are defined by the associated distribution (topic) and a set of member nodes. Every seed group corresponds to a community, and the initial feature distributions are a weighted average of the seed members' attributes. We use the features of nodes in each group to compute initial estimates for the community feature distributions (topics). We then find initial estimates of the membership weights given these estimated per-community topics. After this initialization, membership weights and community feature distributions are iteratively updated. The feature distributions are updated by aggregating attributes of community members and finding the maximum likelihood of a mixture distribution where the parameters for all other communities are fixed.

The contributions of this paper are:

- A scalable probabilistic method for simultaneously finding highly interpretable community topics and node memberships (SENC).
- A flexible and intuitive method of influencing community estimation through the use of bounded seed groups.
- The introduction of several datasets with ground-truth communities used for comparative experiments with top-performing methods.

2   Related Work

There are many approaches to community detection and the state-of-the-art methods which use both network structure and node features are based on linking models [13,21], heuristic clustering [10,11,16], or topic models [14,15]. Previous work [17] has also considered initialization with candidate communities.

Linking models estimate the probability of links and node attributes. They are similar to block models [2,3] with link probabilities dependent on node attributes and community membership. Recent implementations are efficient and competitively find communities, but treat node and community features as binary values [21]. This results in a poor representation of the community's shared interest or topic.

There have been attempts at extending clustering methods to support network data, such as subspace clustering [10,11,16]. In contrast to linking models, these methods do not model edge probability and instead use observed edges and node attributes to identify dense, connected subgraphs with similar node attributes over a subset of the feature space. These methods are not probabilistic and rely on heuristics for detecting nodes with similar attributes. Further, they find many duplicates of a single detected community and require a distinct post-processing step to identify the optimal detected communities.

Topic model approaches extend basic models such as LDA [5] to estimate latent factors and introduce a dependence of edges on the latent factors. These models are generative and require a task-specific probabilistic graphical model. In the past they have been difficult to scale up for larger datasets due to the sampling methods on which they rely [9].

3   Background

A substantial proportion of community detection techniques do not use node attributes to detect communities or provide per-community feature distributions as output. Many solely rely on graph structure [8] or independently group objects by topics and structure [23]. The state-of-the-art methods for community detection have introduced linking models, subspace clustering, topic models, and heterogeneous networks to improve performance and simultaneously estimate topics and membership.

The intuition of our model is most similar to subspace clustering and topic models and both are further discussed. We assume community members are similar across a subset of the feature space and we consider node feature values to be drawn from per-community feature distributions.

The recent literature on linking models which incorporate node attributes [13,21] shows promising results. We aim to perform competitively with those methods by taking a different approach which is probabilistic but allows the use of heuristics to select seed groups.

Other literature [6] has focused on topic models for heterogeneous information networks. While our model is more general and does not require customization to support multiple types of nodes, we are still able to take advantage of the extra information provided by those networks by adding new features or edges.

3.1   Subspace Clustering

Subspace clustering is used to find clusters of objects that occur when the objects are embedded in a subset of the feature space dimensions. A survey of subspace clustering methods is provided in [12] which categorizes various approaches. Subspace clustering is frequently used on high-dimensional datasets and can be viewed as online feature selection for clustering [7].

A major challenge of subspace clustering is finding the optimal subspace clusters. A naive approach would exhaustively try every combination of features, but this is computationally infeasible for all but the smallest datasets. Our method is able to determine which features are relevant to each community by finding the maximal likelihood for the target community's feature distribution in the context of the mixture distribution which describes the node.

Our work extends research on subspace clustering in networks by introducing the use of probability distributions to describe the observed features and to estimate community topics and memberships. We view communities as having feature distributions which represent a common interest of all members.

3.2   Topic Models

Topic models are probabilistic models used to find the semantic structure of documents [4]. They are frequently generative and make assumptions about the relationships between topics, objects, and words. Some models support multiple topics per object or topic hierarchies, but the model is built with those assumptions. Topic models have been designed for networks which group related objects dependent on network structure [15].

The methods combining topic models with graph clustering tasks such as community detection are limiting. They either involve complex models which are only applicable to specific datasets or they independently find topics by treating vertices as documents and then attempt to fit the topics onto the graph to find clusters [23].

We represent node attributes as term-weight vectors and associate a topic with each community. Every cluster we find is a community, and each community has a single feature distribution or topic. We can then estimate a node's membership to a community by finding mixture weights which best explain the node's feature values through community topics.

4   Seeded Estimation of Network Communities

Network communities indicate interaction and attraction among members which is not shared by non-members. The nature of the interaction may be reflected in node attributes and we would expect for member nodes to be similar to one another. However, nodes may participate in multiple communities and the members of each community may be similar to each other in different ways.

To provide motivation for our method, let us discuss an example using an unspecified online social network. This social network allows users to join discussion areas for topics such as "computer science" or "coffee." Suppose a user is interested in both CS and coffee and participates in both communities. We expect the user's posts to the CS community will be different from her posts to the coffee community. We also expect the user's post in the CS community will be more similar in content to other posts in the CS community than to most posts in the coffee community.

Now assume we do not have access to individual user posts. Instead we have aggregated word counts for each user and we do not know which post contained which words. We can model a user's word frequencies as a random variable drawn from a multinomial distribution. Since each user may belong to different communities or have different levels of involvement then it's necessary to use a different multinomial distribution for each user. As previously hinted, we expect posts within a single community to have similar word frequencies. If we knew those per-community word distributions we could then represent each user's word distribution as a mixture distribution. This is akin to standard topic models such as LDA [4].

The Seeded Estimation of Network Communities (SENC) method described here has an advantage over state-of-the-art community detection methods in its exploitation of network structure to regularize and guide estimation. This is possible through the use of *seed groups*. A seed group is a subgraph with properties which indicate the nodes are a subset of a community.

Each seed group is considered to be a lower bound of a community and its members are representative of this corresponding community. The lower-bound

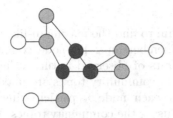

Fig. 1. Lower and upper bounds for a seed community. The lower-bound nodes are black, upper-bound nodes are grey, and excluded nodes are white.

members, or *seed members*, influence estimation; the members' attributes are used as the initial estimate for the corresponding community's word distribution. The community topic is updated as additional member nodes beyond the lower bound are found.

Along with a lower bound, each seed community has a corresponding upper bound. The definition of this upper bound can be dependent on the network and its selection guided by simple network statistics such as the clustering coefficient. Figure 1 depicts an example of the bound sets for a seed community where the distance of a node from lower-bound members is used to define the upper-bound set. The bounds serve as a gentle bias to flexibly model assumptions regarding the shape of communities in a network.

Table 1. Definition of notation.

N	number of nodes
C	number of communities
D	number of feature dimensions
Φ	community topics, $C \times D$ matrix
Θ	community memberships, $N \times C$ matrix
$G(V, E)$	graph defined by vertices and edges
$S_{c=1\ldots C} \subseteq V$	members of community c
$\boldsymbol{x}$	attributes of a node

4.1 Notation

Before continuing it is useful to introduce notation and additional terms for describing the proposed method. We use *topic* to refer to the characteristic features of a community as well as the associated probability distribution parameters for all C communities, Φ, where each row $\Phi_{c,*}$ is a parameter for the categorical distribution associated with community $c = 1, \ldots, C$ with length D, the number of feature dimensions.

The node attribute vector $\boldsymbol{x}$ is a D-length vector of node feature values. A *membership weight vector* or *membership vector* is denoted as $\Theta_{n,*}$ and refers to the probability weight vector associated with node n over all C communities. The individual membership vectors make up the N rows of the membership matrix Θ. The membership weights indicate the proportion of node features which are attributed to each community. For quick reference, basic notation used in the equations is available in Table 1.

4.2   Model

SENC uses an EM algorithm to find the maximum-likelihood estimates for community topics and node memberships. Per-node community memberships are estimated as weighted counts of observed feature values given the community topics in the E-step and per-community topics are maximized in the M-step.

Node memberships for each node n participating in a seed group, $n \in \bigcup_{c=1...C} S_c$, are estimated using the community topics. We represent the feature values of a node x as being drawn from a mixture distribution with per-node mixture weights $\Theta_{n,*}$ over all community distributions Φ using per-community topic distributions $\Phi_{c,*}$. A single term for a node n is drawn by first selecting a community c with probabilities $\Theta_{n,*}$ and then choosing a specific term with probabilities $\Phi_{c,*}$. For the data discussed in this paper, the community feature distributions are categorical distributions and node features x are generated by multiple trials of a mixture categorical distribution with proportions $\Theta_{n,*}\Phi$. A multinomial distribution is a categorical distribution with multiple, independent trials. We refer to the per-node feature distributions as multinomial distributions.

Nodes may be members of multiple communities and node features will then be characteristic of multiple community topics. In order to untangle the features characteristic of a community from those belonging to adjacent communities we define a mixture categorical likelihood function. This is the standard likelihood function but with the event probability vector p parameter computed as the matrix product of some $1 \times C$ mixture vector and $C \times D$ per-community topics matrix: $\Theta_{n,*}\Phi$.

We introduce γ as the sum of feature values from community members S_c to improve readability:

$$\gamma = \sum_{n \in S_c} x \tag{1}$$

When estimating $\Phi_{c,*}$ using community members S_c, the mixture vector θ is a weighted average of membership vectors $\{\Theta_{n,*} : n \in S_c\}$ weighted by the proportional number of observations contributed by each node $n \in S_c$:

$$\theta = \sum_{n \in S_c} \frac{(\sum_d x_d)\Theta_{n,*}}{\sum_d \gamma_d} \tag{2}$$

Using θ and γ we can now show how $\Phi_{c,*}$ may be updated. The event probability vector p is the parameter for a categorical distribution:

$$p = \theta\Phi \tag{3}$$

$$= \sum_{i=1}^{C} \theta_i \Phi_{i,*} \tag{4}$$

$$= \theta_c \Phi_{c,*} + \sum_{i=1, i \neq c}^{C} \theta_i \Phi_{i,*} \tag{5}$$

We can use the factoring of p in Equation (5) with the multinomial expected value to find the maximum-likelihood value of $\Phi_{c,*}$ given the community member observations γ from Equation (1).

The expected value for a single feature value i in random variable X drawn from $\text{Mult}(p, n)$ is $E\{X_i\} = np_i$, where n is the number of trials and p is the event probability vector. If we replace the expected value of each feature dimension with the summation of community members' S_c attributes γ then we can substitute the expected value with the observed value γ_i for feature i and define:

$$\gamma_i = (\sum_{d=1}^{D} \gamma_d)\theta\Phi_{*,i} \tag{6}$$

If we replace the expected value of each feature dimension with the summation of community members' S_c attributes γ then we can define the maximum likelihood of $\Phi_{c,*}$ as:

$$\gamma = (\sum_{d} \gamma_d)\theta\Phi \tag{7}$$

$$= (\sum_{d} \gamma_d)(\theta_c\Phi_{c,*} + \sum_{i=1,i\neq c}^{C} \theta_i\Phi_{i,*}) \tag{8}$$

$$\frac{\gamma}{\sum_{d}\gamma_d} = \theta_c\Phi_{c,*} + \sum_{i=1,i\neq c}^{C} \theta_i\Phi_{i,*} \tag{9}$$

$$\theta_c\Phi_{c,*} = \frac{\gamma}{\sum_{d}\gamma_d} - (\sum_{i=1,i\neq c}^{C} \theta_i\Phi_{i,*}) \tag{10}$$

$$\Phi_{c,*} = \frac{\frac{\gamma}{\sum_{d}\gamma_d} - (\sum_{i=1,i\neq c}^{C} \theta_i\Phi_{i,*})}{\theta_c} \tag{11}$$

Using Equation (11) we can easily estimate community topics using node attributes, per-node community membership weights, and the latest topic estimates for other communities.

We use Φ' to reference a modified version of Φ with normalized columns, each summing to 1. The per-node community memberships are found by performing a weighted count of node attributes over the communities, where α denotes a normalization scalar:

$$\Theta_{n,*} = \alpha\Phi'x^T \tag{12}$$

For each observed term, we assign a proportion of the count to each community according to the relative probability of that term occurring in each community. A community with a higher relative probability of a given term occurring will receive a larger proportion of the count than the others.

4.3   Algorithm

The SENC algorithm constructs per-community lower- and upper-bound matrices, initializes per-community topics Φ and per-node community memberships Θ, and then performs expectation-maximization iterations until estimates stop improving or the maximum number of iterations is reached. The algorithm requires the $N \times N$ graph adjacency matrix and the $N \times D$ node attribute matrix as input. The lower-bound matrix is a $C \times N$ binary matrix of the seed members where 1-values indicate node n belongs to community c. The upper-bound matrix is a binary $N \times C$ matrix where 1-values indicate node n may belong to community c. This prevents nodes from distant communities being assigned to communities with a similar topic. The construction of lower- and upper-bound sets for each community is dependent on the network being processed. Two matrices are produced as output: a $C \times D$ community topic matrix and an $N \times C$ community membership matrix. A goal of our method is to remove features representative of overlapping communities over EM iterations. The node membership vectors are estimated using the community topics to perform a weighted count over node attributes. These weighted counts are normalized to sum to one and used as membership weights.

Algorithm 1. Main Program: initialization, EM, termination.

Input: The *graph* and *node attributes*.
Output: The *community topics* and *membership*.
 1: Construct *lower-bound* and *upper-bound* matrices;
 2: Initialize community topics Φ and memberships Θ;
 3: **while** Not convergent or max iteration **do**
 4:     Call **E-step** to update membership Θ;
 5:     Call **M-step** to update community topics Φ;
 6:     Check for convergence;
 7: **end while**
 8: **return** Community topics Φ and membership Θ;

Algorithm 2. E-step: update per-node community memberships.

Input: The *community topics*, *upper-bound matrix*, and *node attributes*.
Output: The updated *membership*.
 1: **for** Each node n **do**
 2:     Identify which communities influence node n;
 3:     Select topics of influential communities;
 4:     Compute weighted counts from selected topics with Equation (12);
 5:     Assign normalized counts to membership vector $\Theta_{n,*}$;
 6: **end for**
 7: **return** Updated membership Θ;

Algorithm 3. M-step: update per-community topics.

Input: The *node attributes*, *membership*, *influence*, and *community topics* from the previous iteration.

Output: The updated *community topics*.

1: **for** Each seeded community c **do**
2:     Select all nodes with membership in c;
3:     Compute weighted average of selected nodes' membership by Equation (2);
4:     Estimate topic with Equation (11);
5:     Assign updated topic to parameter vector $\Phi_{c,*}$, if likelihood improves;
6: **end for**
7: **return** Updated topics Φ;

After initial estimates are calculated, the algorithm alternatively updates the node memberships and community topics. The per-node and per-community iterations within the E- and M-step are independent and computation may be distributed across multiple threads. The E-step in Algorithm 2 updates the per-node community memberships for all nodes given the community topics, influence matrix, and node attributes. This is done by computing the weighted counts of node attributes using the probability of each attribute for each community, as shown in Equation (12). The upper-bound complexity of the E-step is $O(\mathcal{NCD})$, where $\mathcal{C}$ and $\mathcal{D}$ are the number of communities to which a node may belong and the number of dimensions relevant to those communities. In practice, $\mathcal{C}$ and $\mathcal{D}$ will be much smaller than C and D.

The M-step, shown in Algorithm 3, updates the per-community feature distributions. We find a new estimate for $\Phi_{c,*}$ using Equation (11) and compare its log-likelihood to the previous iteration's estimate. The new estimate is used if it better explains the feature values of the member nodes. The M-step has computational complexity of $O(C(\mathcal{NC} + \mathcal{ND} + \mathcal{CD}))$, where $\mathcal{N}$ is the number of nodes in the upper-bound set of a community, $\mathcal{C}$ is the number of communities associated with the $\mathcal{N}$ nodes, and $\mathcal{D}$ is the number of feature dimensions relevant to all $\mathcal{C}$ communities and $\mathcal{N}$ nodes. Again, $\mathcal{C}$, $\mathcal{N}$, and $\mathcal{D}$ are usually much smaller than C, N, and D.

5 Experiments

We evaluate our proposed method on networks with varying structure to determine whether SENC's results are consistently competitive with state-of-the-art methods. The networks considered are: an NSF research collaboration network, several DBLP citation networks [19], and a Scratch project collaboration network. For comparison, we evaluate the performance of four state-of-the-art community detection methods: CESNA [21], CoDA [22], EDCAR [10], and Link Clustering [1]. CESNA and EDCAR use network structure and node attributes to detect communities; however, the current implementations struggled to process networks with a large number of features. In order to evaluate more methods we elected to use smaller datasets. CoDA and Link Clustering only use network structure.

An implementation of SENC and datasets used in experiments will be made available at the GMU DMML website[1].

5.1   Dataset Descriptions

We construct a research collaboration network from NSF awards granted by the Directorate for Computer and Information Science and Engineering (CISE) between January 1995 and August 2014. This is accomplished by forming undirected edges between the PI and co-PIs who received funding from the same award. The awards are associated with programs and we use the programs with at least three associated researchers as ground truth. We find 90% of researchers received funding from six or fewer programs; this suggests programs function well as ground-truth communities. There are a total of 768 programs in the CISE Directorate. NSF awards data is publicly available from the NSF website[2].

An online computer science bibliography, DBLP, contains entries for published papers with information about the authors, citations, and publication venues. The per-year DBLP citation networks were constructed from an existing citation dataset [19] by forming edges between authors who cited each other within that year. Papers are linked to a publication venue and these venues were used to define ground truth. Venues referenced only once were removed from our dataset. Venues with three or more associated authors were used as ground truth.

Scratch [18] is an online community where users may write and share projects (programs) with other users. One way in which Scratch users may interact is by remixing projects. Remixing allows a user to create a copy of any existing project which they may then modify. We created a co-remix affiliation network from the MIT Scratch Team's dataset containing users, projects, and remixes. An edge is formed when two users remix the same project. To reduce the total number of edges we used co-remix edges where users had three or more projects in common. Users may create project galleries which are curated collections of projects. Galleries corresponding to three or more users were used as ground truth. The Scratch dataset used to construct the network may obtained from the MIT Media Lab website[3].

Several of the methods make use of node attributes and these were provided as tf-idf weighted values for EDCAR and SENC and binary values for CESNA. For the NSF CISE network, terms associated with each researcher were taken from NSF award titles and abstracts. The DBLP author terms were taken from titles and abstracts of papers they wrote. Scratch user terms were extracted from titles, descriptions, and tags of their projects. The term features in all networks had stop words removed and terms stemmed.

---

[1] http://cs.gmu.edu/~dmml
[2] http://www.nsf.gov/awardsearch/download.jsp
[3] https://llk.media.mit.edu/scratch-data

Table 2. Network statistics. N: number of nodes, E: number of edges, D: number of node attributes, MC: number of maximal cliques with 3+ members, GCC: global clustering coefficient, LCC: average local clustering coefficient, G: number of ground-truth communities.

Dataset	N	E	D	MC	GCC	LCC	G
NSF	8,168	38,212	43,445	3,331	0.590	0.683	429
DBLP 2010	32,961	130,420	58,007	37,120	0.422	0.440	2,288
DBLP 2011	32,614	131,921	56,166	39,955	0.421	0.438	2,215
DBLP 2012	33,576	135,883	54,269	42,443	0.381	0.397	1,861
Scratch	1,714	17,824	36,494	7,705	0.584	0.704	718

Multiple connected components were found in all networks and the smaller components were removed as they may be trivially considered communities. Table 2 lists the network statistics for the largest component of each network used for experiments and analysis. All the networks used for experiments are undirected, but they vary in structure.

As shown in Table 2, the NSF and Scratch networks have higher clustering coefficients than the DBLP networks. This is unsurprising as the NSF and Scratch networks are affiliation networks (co-award and co-remix). Our experiments show that while SENC is able to perform competitively across all the networks other methods tend to either perform better on networks with higher or lower clustering coefficients.

5.2 Methods and Evaluation

The public implementations of CESNA, CoDA, EDCAR, and Link Clustering were used. CESNA and CoDA rely on an estimate of the number of communities. We provided the number of NSF programs, DBLP publication venues, and Scratch galleries as estimates. CoDA is designed for directed networks but can be used to find communities in undirected networks. It does this by processing the network twice, switching the direction of edges between runs. As a result, two sets of detected communities are generated. We combined both sets when evaluating the performance of CoDA. EDCAR requires 10 parameters and the suggested values from the implementation documentation were used. Link Clustering is parameter-less and only requires the edge list as input. Maximal cliques of size three and above were used as the lower-bound groups for SENC and the upper-bound groups were selected based on the clustering coefficient. The high clustering coefficients of the NSF and Scratch networks indicate tighter upper bounds should be used than with the DBLP networks. For the DBLP networks we extend the lower bounds by including all nodes adjacent to any lower-bound member. The upper bounds for the NSF and Scratch networks are simply the same maximal cliques.

Link Clustering and SENC require a post-processing step to define exact communities. The Link Clustering implementation includes a script to calculate the optimal dendrogram cut threshold and we use this to determine the communities for evaluation. SENC defines community membership with probabilities and does not perform a hard assignment of nodes to communities like the other evaluated methods. We account for this in our evaluation by filtering weaker memberships. For all nodes, we sort their memberships in descending order by weight and take all the assignments until the sum of weights reaches a minimum threshold value. An optimal threshold is used for each dataset.

We use the evaluation function described in [20,21] and recited in Equation (13) to compute the $F1$ score and Jaccard similarity of detected communities against ground-truth communities. This function is especially useful when the numbers of detected communities and ground-truth communities differ as occurs with several of the methods in our experiments. In Equation (13), C^* denotes a set of ground-truth communities, C a set of detected communities, and $\delta(\cdot)$ is a similarity metric.

$$\frac{1}{2|C^*|} \sum_{C_i^* \in C^*} \max_{C_j \in C} \delta(C_i^*, C_j) + \frac{1}{2|C|} \sum_{C_j \in C} \max_{C_i^* \in C^*} \delta(C_i^*, C_j) \tag{13}$$

5.3 Results

Using the evaluation function defined in Equation (13) we find the $F1$ score and Jaccard similarity between the detected communities from all methods and the ground-truth communities.

Table 3. $F1$ scores for all methods and datasets.

Method	Attr.	NSF	DBLP10	DBLP11	DBLP12	Scratch	Avg.
CoDA	No	0.216	0.278	0.273	0.263	0.283	0.263
Link Clust.	No	0.303	0.266	0.265	0.258	**0.399**	0.298
CESNA	Yes	0.228	0.272	0.263	0.255	0.356	0.275
EDCAR	Yes	0.164	N/A	N/A	N/A	N/A	N/A
SENC	Yes	**0.346**	**0.301**	**0.297**	**0.298**	0.365	**0.321**

Our results are provided in Tables 3 and 4 and show SENC outperforms most other methods over all datasets and achieves the highest average performance. Unfortunately, the current implementation of EDCAR was unable to process most of the networks. We believe this is partly due to the large number of features.

We note the relative difference in performance of CoDA and CESNA to Link Clustering flips between the networks with higher and lower clustering coefficients. In the NSF and Scratch networks, Link Clustering outperforms CoDA and CESNA but performs worse than CESNA on the DBLP10 network and worse than CoDA on every DBLP network. This may indicate these other methods include a biased definition of communities which is not found in all social networks. SENC performs well across all the networks and avoids this problem through the use of its configurable bounds chosen based on network statistics such as clustering coefficients.

Table 4. Jaccard index for all methods and datasets.

Method	Attr.	NSF	DBLP10	DBLP11	DBLP12	Scratch	Avg.
CoDA	No	0.132	0.172	0.168	0.162	0.174	0.162
Link Clust.	No	0.233	0.166	0.166	0.161	**0.265**	0.198
CESNA	Yes	0.139	0.167	0.161	0.156	0.228	0.170
EDCAR	Yes	0.112	N/A	N/A	N/A	N/A	N/A
SENC	Yes	**0.269**	**0.190**	**0.187**	**0.190**	0.235	**0.214**

5.4   Interpretation of Detected Communities

We also perform a qualitative analysis on communities discovered by SENC to illustrate the interpretability of its results. Several communities relating to data mining and machine learning were found in the NSF CISE network.

Table 5. Top-5 researchers of the AMPLab and Computational Learning communities with corresponding membership weights.

AMPLab		Comp. Learning	
Peter Bartlett	0.5084	Laurent El Ghaoui	0.5884
Laurent El Ghaoui	0.4116	Peter Bartlett	0.4916
Michael Franklin	0.1346	Jesse Snedeker	0.4647
Michael Jordan	0.1049	Federico Girosi	0.4134
Alexandre Bayen	0.0996	Robert Berwick	0.2830

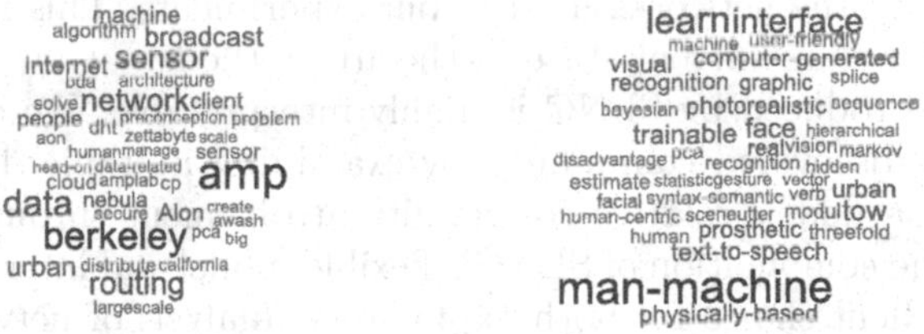

Fig. 2. Word clouds of the top-40 terms from the AMPLab community (left) and computational learning community (right).

We present the top-5 researchers and top-40 terms of two such groups in Table 5 and Figure 2. The first community is associated with Berkeley's AMPLab[4], which works on problems involving machine learning, cloud computing, and crowd-sourcing. The top-5 researchers are all EECS faculty at Berkeley and Michael Franklin and Michael Jordan are both directors of AMPLab. Recall membership weights are normalized per-researcher and a lower membership weight indicates the researcher's work is also captured by other community topics. Most of the terms are self-explanatory, but the term *Alon* refers to Alon Halevy of University of Washington whose name appears in several award abstracts and has collaborated with Michael Franklin.

---

[4] https://amplab.cs.berkeley.edu

We find another community with 12 members in common with the AMPLab community. Its topic may be described as computational learning and its applications to computer vision and natural-language processing. The AMPLab and computational learning communities have 41 and 34 members respectively, with roughly about one-third being shared. These common members include: Michael Jordan, Michael Franklin, Peter Bartlett, and Tomaso Poggio.

Although both communities are generally concerned with human-centric applications of machine learning, the AMPLab community is focused on computing architecture to solve such problems, while the computational learning community is focused on understanding human vision and motor control. This discovery of overlapping communities with shared general interests but distinct features exemplifies an advantage of SENC's initialization by seed groups.

6 Conclusion

We have introduced SENC — a probabilistic approach to community detection that outputs node memberships and community topics. Simple network statistics, such as the clustering coefficient, can be used to guide configuration of flexible bounds on seed groups. The bounded seed groups enable SENC to account for differences in underlying community structure across many networks. This contrasts with existing methods which build strong assumptions into their models. As a result, SENC is able to consistently outperform state-of-the-art community detection methods on a variety of networks. No other method performed consistently across all the networks used in our experiments. This indicates SENC generalizes better than current state-of-the-art methods.

The output produced by SENC is highly interpretable. We can understand the nature of a discovered community by examining its topic distribution. We can also review a node's relative community involvement through its membership weights. The combination of SENC's flexible model and interpretable results make it an excellent choice for both exploratory analysis of networks and community detection tasks.

Our experiments have raised several interesting questions for future work. We are interested in discovering how network characteristics affect assumptions made in community detection methods and how other approaches for defining bounded seed groups may further improve SENC's performance.

Acknowledgments. We appreciate the Lifelong Kindergarten group at MIT for publicly sharing the Scratch datasets. This work is partly based upon research supported by U.S. National Science Foundation (NSF) Awards DUE-1444277 and EEC-1408674. Any opinions, recommendations, findings, or conclusions expressed in this material are those of the authors and do not necessarily reflect the views of NSF.

References

1. Ahn, Y.-Y., Bagrow, J.P., Lehmann, S.: Link communities reveal multiscale complexity in networks. Nature **466**(7307), 761–764 (2010)
2. Airoldi, E.M., Blei, D.M., Fienberg, S.E., Xing, E.P.: Mixed membership stochastic blockmodels. Journal of Machine Learning Research **9**, 1981–2014 (2008)
3. Balasubramanyan, R., Cohen, W.W.: Block-LDA: Jointly modeling entity-annotated text and entity-entity links. In: Proceedings of the SIAM International Conference on Data Mining, vol. 11, pp. 450–461. SIAM (2011)
4. Blei, D.M.: Probabilistic topic models. Communications of the ACM **55**(4), 77–84 (2012)
5. Blei, D.M., Ng, A.Y., Jordan, M.I.: Latent Dirichlet allocation. Journal of Machine Learning Research **3**, 993–1022 (2003)
6. Deng, H., Han, J., Zhao, B., Yu, Y., Lin, C.X.: Probabilistic topic models with biased propagation on heterogeneous information networks. In: Proceedings of the ACM SIGKDD International Conference on Knowledge Discovery and Data Mining, pp. 1271–1279. ACM (2011)
7. Domeniconi, C., Papadopoulos, D., Gunopulos, D., Ma, S.: Subspace clustering of high dimensional data. In: Proceedings of the SIAM International Conference on Data Mining, pp. 517–521. SIAM (2004)
8. Fortunato, S.: Community detection in graphs. Physics Reports **486**(3), 75–174 (2010)
9. Geman, S., Geman, D.: Stochastic relaxation, Gibbs distributions, and the Bayesian restoration of images. IEEE Transactions on Pattern Analysis and Machine Intelligence **PAMI-6**(6), 721–741 (1984)
10. Günnemann, S., Boden, B., Färber, I., Seidl, T.: Efficient mining of combined subspace and subgraph clusters in graphs with feature vectors. In: Pei, J., Tseng, V.S., Cao, L., Motoda, H., Xu, G. (eds.) PAKDD 2013, Part I. LNCS, vol. 7818, pp. 261–275. Springer, Heidelberg (2013)
11. Günnemann, S., Färber, I., Boden, B., Seidl, T.: Subspace clustering meets dense subgraph mining: a synthesis of two paradigms. In: Proceedings of the IEEE International Conference on Data Mining, pp. 845–850. IEEE Computer Society (2010)
12. Kriegel, H.-P., Kröger, P., Zimek, A.: Subspace clustering. Wiley Interdisciplinary Reviews: Data Mining and Knowledge Discovery **2**(4), 351–364 (2012)
13. Leskovec, J., McAuley, J.: Learning to discover social circles in ego networks. In: Advances in Neural Information Processing Systems, pp. 539–547 (2012)
14. Liu, Y., Niculescu-Mizil, A., Gryc, W.: Topic-link LDA: Joint models of topic and author community. In: Proceedings of the International Conference on Machine Learning, pp. 665–672. ACM (2009)
15. McCallum, A., Wang, X., Mohanty, N.: Joint group and topic discovery from relations and text. In: Airoldi, E.M., Blei, D.M., Fienberg, S.E., Goldenberg, A., Xing, E.P., Zheng, A.X. (eds.) ICML 2006. LNCS, vol. 4503, pp. 28–44. Springer, Heidelberg (2007)
16. Moser, F., Colak, R., Rafiey, A., Ester, M.: Mining cohesive patterns from graphs with feature vectors. In: Proceedings of the SIAM International Conference on Data Mining, vol. 9, pp. 593–604. SIAM (2009)
17. Pool, S., Bonchi, F., Leeuwen, M.: Description-driven community detection. ACM Transactions on Intelligent Systems and Technology **5**(2), 28:1–28:28 (2014)
18. Resnick, M., Maloney, J., Monroy-Hernández, A., Rusk, N., Eastmond, E., Brennan, K., Millner, A., Rosenbaum, E., Silver, J., Silverman, B., et al.: Scratch: Programming for all. Communications of the ACM **52**(11), 60–67 (2009)

19. Tang, J., Zhang, J., Yao, L., Li, J., Zhang, L., Su, Z.: Arnetminer: Extraction and mining of academic social networks. In: Proceedings of the ACM SIGKDD International Conference on Knowledge Discovery and Data Mining, pp. 990–998 (2008)
20. Yang, J., Leskovec, J.: Overlapping community detection at scale: a nonnegative matrix factorization approach. In: Proceedings of the ACM International Conference on Web Search and Data Mining, pp. 587–596. ACM, New York (2013)
21. Yang, J., McAuley, J., Leskovec, J.: Community detection in networks with node attributes. In: IEEE 13th International Conference on Data Mining, pp. 1151–1156. IEEE (2013)
22. Yang, J., McAuley, J., Leskovec, J.: Detecting cohesive and 2-mode communities in directed and undirected networks. In: Proceedings of the ACM International Conference on Web Search and Data Mining, pp. 323–332. ACM (2014)
23. Zhao, Z., Feng, S., Wang, Q., Huang, J.Z., Williams, G.J., Fan, J.: Topic oriented community detection through social objects and link analysis in social networks. Knowledge-Based Systems **26**, 164–173 (2012)

Finding Dense Subgraphs in Relational Graphs

Vinay Jethava[(✉)] and Niko Beerenwinkel

Department of Biosystems Science and Engineering, ETH Zürich, Zurich, Switzerland
{vinay.jethava,niko.beerenwinkel}@bsse.ethz.ch

Abstract. This paper considers the problem of finding large dense subgraphs in relational graphs, i.e., a set of graphs which share a common vertex set. We present an approximation algorithm for finding the densest common subgraph in a relational graph set based on an extension of Charikar's method for finding the densest subgraph in a single graph. We also present a simple greedy heuristic which can be implemented efficiently for analysis of larger graphs. We give graph dependent bounds on the quality of the solutions returned by our methods. Lastly, we show by empirical evaluation on several benchmark datasets that our method out-performs existing approaches.

1 Introduction

Finding dense subgraphs is a key subtask in many applications [see 21, for a survey]. In many contexts, there exist several graphs encoding different relationships between the same set of actors. Then, a subset of actors having high degree of interconnections (dense) which recur in multiple graphs (frequent) often have a rich interpretation in the application domain. For example, dense recurrent subgraphs in multiple gene co-expression networks have been shown to correspond to known functional/transcriptional modules or protein complexes as well as phenotype-specific modules [12,24].

Several data-mining methods have addressed the problem of enumerating dense recurrent subgraphs [see,e.g., 6,12,13,24,26,31,33]. Most existing approaches enumerate all frequent quasicliques depending on parameters such as minimum support threshold and minimum relative density. This results in exponential growth of search space with increasing size of the returned subgraph, making the methods unsuitable for identifying large dense subgraphs in multiple graphs. The approach of [24] yields a non-convex cubic programming problem which is solved approximately using multi-stage convex relaxation [34] and used in the analysis of co-expression networks in order to identify small biologically relevant modules. [13] present a method to identify large dense subgraphs based on solving a Multiple Kernel Learning (MKL) problem [3,27] with precomputed kernels.

On the other hand, there is a rich body of work on approximation algorithms which addresses the problem of finding the densest subgraph (DS) [7,11] and its size-constrained variants (DkS, Dal$k$S, Dam$k$S) [1,5,20] including greedy approaches [2,9,28], truncated power method [32], linear programming (LP) based methods [7,20] and semidefinite programming (SDP) based methods

© Springer International Publishing Switzerland 2015
A. Appice et al. (Eds.): ECML PKDD 2015, Part II, LNAI 9285, pp. 641–654, 2015.
DOI: 10.1007/978-3-319-23525-7_39

[8,29]. We note that given a relational graph set [1], $\mathbb{G} := \{G^{(m)} = (V, E^{(m)})\}_{m=1}^{M}$, it is possible to construct integrated graphs, e.g., $G_{\cup}$, $G_{\cap}$ or $G_{\geq t}$ having edge sets respectively, $\bigcup E^{(m)}$, $\bigcap E^{(m)}$ or $\{e \mid \mathsf{supp}(e, \mathbb{G}) \geq t\}$ [2]; and identify dense subgraphs in the integrated graph using these methods. However, as noted by [16], a dense subgraph in the integrated graph may either not be dense in one or more of the original graphs, e.g., $G_{\cup}$ and $G_{\geq t}$; or, be too pessimistic in size of the returned subgraph, e.g., $G_{\cap}$.

In this paper, we formalize the notion of Densest Common Subgraph (i.e., a subset of nodes which maximizes the density of the induced subgraph in each of the graphs in the relational graph set) which was previously discussed in [13]. We present an approximation algorithm (DCS_LP) for finding the densest common subgraph in a relational graph set based on an extension of Charikar's LP based approach [7,20] for finding the densest subgraph (DS) in a single graph. We also present a simple greedy heuristic (DCS_GREEDY) which can be implemented efficiently for analysis of larger graphs. We give graph dependent bounds on the quality of the solutions returned by DCS_LP and DCS_GREEDY. Lastly, we show by empirical evaluation on several benchmark datasets and real-world datasets that our methods out-perform prior approaches.

Notation. We represent vectors using lower case bold letters $\mathbf{a}, \mathbf{b}, \ldots$, etc., and matrices using upper case bold letters $\mathbf{A}, \mathbf{B}, \ldots$ etc.; with a_i referring to i^{th} element of $\mathbf{a}$, and similarly A_{ij} referring to $(i, j)^{th}$ entry of matrix $\mathbf{A}$. We use notation $[n]$ to denote the set $\{1, 2, \ldots, n\}$. For a vector in $\mathbb{R}^d$, we denote the Euclidean norm by $\|.\|$ and the p-norm by $\|.\|_p$. The inequality $\mathbf{a} \geq 0$ is true if it holds element-wise.

Let $G = (V, E)$ be a simple undirected graph of order n with vertex set $V = [n]$ and edge set $E \subseteq V \times V$. Let $\mathbf{A} \in \mathbf{S}_n$ denote the adjacency matrix of G where $A_{ij} = 1$ if edge $(i, j) \in E$, and 0 otherwise. We use shorthand notation ij to mean (i, j) whenever clear from context. Let $\bar{G}$ denote the complement graph of G. The adjacency matrix of $\bar{G}$ is $\bar{\mathbf{A}} = \mathbf{e}\mathbf{e}^{\top} - \mathbf{I} - \mathbf{A}$, where $\mathbf{e} = [1, 1, \ldots, 1]^{\top}$ is a vector of length n containing all 1's, and $\mathbf{I}$ denotes the identity matrix. We denote the indicator vector for some set $S \subseteq V$ as $\mathbf{e}_S$ which is one for all $i \in S$ and zero in other co-ordinates.

We use notation $\deg_G(i)$ to denote the degree of node i in graph G, and $\deg_G(i, S)$ to denote the degree of node i in the subgraph $(S, E(S))$ induced by vertex set $S \subseteq V$ in graph G. The *density* $\delta_G(S)$ and *relative density* $\rho_G(S)$ of subgraph $(S, E(S))$ induced by vertex set $S \subseteq V$ in graph $G = (V, E)$ are given by $\delta_G(S) := \frac{|E(S)|}{|S|}$ and $\rho_G(S) := |E(S)|/\binom{|S|}{2}$ respectively. The induced subgraph $(S, E(S))$ is an α-*quasiclique* if $|E(S)| \geq \alpha\binom{|S|}{2}$, i.e., if the relative density $\rho_G(S)$ of the induced subgraph exceeds a threshold parameter $\alpha \in (0, 1)$. Let $\delta_G^* := \max_{S \subseteq V} \delta_G(S)$ denote the density of the densest subgraph (DS) in G.

---

[1] A relational graph set is defined as a set of simple undirected graphs which share a common vertex set.

[2] The support of an edge e in the relational graph set $\mathbb{G}$ denoted by $\mathsf{supp}(e, \mathbb{G}) := \#\{m : e \in E^{(m)}\}$ is the number of graphs $G^{(m)} \in \mathbb{G}$ which contain this edge.

Given a relational graph set $\mathbb{G} = \{G^{(m)} = (V, E^{(m)})\}_{m=1}^{M}$, we use short-hand notation $\deg_m(i)$, $\deg_m(i, S)$, $\delta_m(S)$, $\rho_m(S)$ and δ_m^* to denote $\deg_{G^{(m)}}(i)$, $\deg_{G^{(m)}}(i, S)$, $\delta_{G^{(m)}}(S)$, $\rho_{G^{(m)}}(S)$ and $\delta_{G^{(m)}}^*$, respectively, whenever clear from context. We use $n_{\mathbb{G}} := \sum_m |E^{(m)}|$ to denote the total number of edges in the relational graph set. The support of a subgraph $H = (V_H, E_H)$, $V_H \subseteq V, E_H \subseteq V \times V$ in $\mathbb{G}$ is given by $\text{supp}(H, \mathbb{G}) := \#\{m : E_H \subseteq E^{(m)}\}$.

2   Related Work

In this section, we review prior work on finding dense subgraphs and quasicliques. Section 2.1 discusses approximation algorithms (with known worst-case bounds) that find the subgraph with maximum density (DS) or its size-constrained variants (DkS, Dal$k$S, Dam$k$S). In Section 2.2 reviews methods for enumerating frequent dense subgraphs present in a relational graph set.

2.1   Finding a Dense Subgraph in a Single Graph

The problem of finding maximum density subgraph was studied by Goldberg [11] who introduced a maximum-flow algorithm for this problem (see also [10]). Kannan and Vinay [18] studied the problem for directed graphs and introduced an $O(\log n)$-approximation algorithm. Charikar [7] presented a linear programming relaxation from which the optimal solution can be recovered. They also showed that the related greedy algorithm of [2] yields 2-approximation for the problem. The work of Khuller and Saha [20] simplified the analysis in [7]. Bahmani et al. [4] obtained $O(2 + \epsilon)$-approximation algorithm for DS in the streaming model which makes $O(\frac{1}{\epsilon} \log n)$ passes over the data. Recently, Tsourakakis et al. [30] defined the notion of optimal α-quasiclique (OQC) which maximizes the edge surplus given by $f_\alpha(S) := (|E(S)| - \alpha\binom{|S|}{2})$; and, obtained a 2-approximation algorithm for finding optimal quasicliques analogous to Charikar's algorithm [7]. Comparing their approach ($\alpha = 1/3$) with Charikar's algorithm on several real-world graphs, the authors argue that OQC yields better results in terms of lower diameter, higher relative density and higher triangle density of the extracted subgraphs.

When there is a constraint on the size of the subgraph ($|S| = k$), the problem of finding the densest k subgraph (DkS) is NP-Hard [2,9]. Feige et al. [9] gave an $O(n^{1/3-\epsilon})$-approximation algorithm for DkS. The best known current bound for DkS is given by Bhaskara et al. [5] who obtain an $O(n^{1/4+\epsilon})$-approximation algorithm for DkS with any $\epsilon > 0$ having run time $n^{O(1/\epsilon)}$. Khot [19] showed that under reasonable complexity assumptions, DkS cannot be approximated within an arbitrary constant factor. Recently, Papailiopoulos et al. [25] obtained graph-dependent bounds for DkS based on low-rank approximation of the adjacency matrix, and experimentally showed that their bounds are tight for several large real-world graphs.

Two variants of the DkS problem where the size constraint is relaxed were introduced in [1], namely, densest at-most-k subgraph (Dam$k$S, $|S| \leq k$) and densest at-least-k subgraph (DalkS, $|S| \geq k$). Andersen and Chellapilla [1] gave an 2-approximation algorithm for DalkS which was subsequently improved in running time by [20]. Khuller and Saha [20] also showed that DamkS is as hard as DkS, specifically, an α-approximation for DamkS implies a 4α-approximation for DkS.

2.2   Finding Cross-Graph Quasicliques

The problem of identifying dense recurrent subgraphs has been studied in several guises with differing terminologies. Early works on frequent subgraph mining were based either on the apriori principle or pattern-growth approach [see 15, forarecentsurvey]. One key bottleneck in general frequent subgraph mining is handling graph (and subgraph) isomorphism; which is absent in relational graph sets since the graphs share a common vertex set.

Yan et al. [31] enumerated frequent *dense* subgraphs in a relational graph set where each edge has support greater than some threshold (frequent) and the subgraph has large minimum cut (dense). Zeng et al. [33] studied the problem of mining frequent quasicliques in a database of vertex labeled graphs by considering γ-quasicliques induced by node sets $S^{(a)}$ and $S^{(b)}$ in graphs $G^{(a)}$ and $G^{(b)}$ respectively "γ-isomorphic" if there is a one-to-one mapping between $S^{(a)}$ and $S^{(b)}$ which preserves their vertex labels. Boden et al. [6] present a method for enumerating frequent quasicliques (minimum support) in edge-labelled graphs.

Hu et al. [12] presented a method for identifying *coherent* dense subgraphs from gene microarray expression datasets by finding dense subgraphs in the *summary graph* $G_S = (V_S, E_S)$ where $V_S \subset V \times V$ and nodes (u, v) and (w, z) have an edge in E_S if their support sets in $\mathbb{G}$ have high Jaccard similarity. Pei et al. [26] presented an exhaustive approach for enumerating all cross-graph quasicliques defined as follows: given relational graph set $\mathbb{G}$, parameters $\gamma^{(1)}, \gamma^{(2)}, \ldots, \gamma^{(M)}$ and $min\_sup \in (0, 1]$; a vertex set S is a frequent cross-graph quasiclique (fCGQC) if it has relative density $\rho_m(S) \geq \gamma^{(m)}$ in at least $(M \cdot min\_sup)$ graphs.

Li et al. [24] present a method for identifying recurrent dense subgraphs in weighted graphs based on a non-convex optimization problem which is solved approximately using multi-state convex relaxation [34]. This is used to identify several recurrent heavy subgraphs in multiple co-expression networks.

Jethava et al. [13] present a parameter-less algorithm based on a multiple kernel learning (MKL) [3, 27] formulation for finding an ordering of vertices which maximizes the minimum relative density (across all graphs) of the induced subgraph. Their approach also provides weak graph-dependent bounds on the density of the induced subgraphs [see 14, Lemma 11 and Theorem 12]. However, their method has $O(n^3)$ complexity which cannot scale to large graphs.

3   Methods

We are interested in finding a set of nodes which induces a dense subgraph in each of the graphs in $\mathbb{G}$. We formalize this notion by defining the problem of densest common subgraph (DCS). Let $\delta_G(S) := \min_{G^{(m)} \in G} \delta_m(S)$ denote the density of the subgraph having minimum density among the subgraphs induced by S in graphs $G^{(m)} \in G$. In the sequel, we refer to $\delta_G(S)$ as the *common density* of vertex set $S \subseteq V$ in graph set G.

Definition 1 (Densest Common Subgraph). *Given relational graph set G, the densest common subgraph is given by:*

$$S_{DCS} := \arg\max_{S \subseteq V} \delta_G(S), \tag{1}$$

We use shorthand notation $\delta_G := \delta_G(S_{DCS})$ to denote the minimum density of any subgraph induced by S_{DCS} in the graph set G.

The DCS problem is related to the k-multicut (following Goldberg's construction [11]) which is known to be NP-Hard for $k \geq 3$, and therefore, DCS is suspected to be NP-Hard. However, a formal proof of hardness is non-trivial and it constitutes an interesting problem for future research.

We can define the LP relaxation (DCS_LP) of the DCS problem as:

$$\begin{aligned}
\max\ &t \\
\text{s.t.}\ &\sum_{i=1}^{n} y_i \leq 1 \\
&\sum_{ij \in E^{(m)}} x_{ij}^{(m)} \geq t \quad \forall\, E^{(m)},\, m \in [M] \\
&x_{ij}^{(m)} \leq y_i,\ x_{ij}^{(m)} \leq y_j \ \forall\, ij \in E^{(m)} \\
&x_{ij}^{(m)} \geq 0,\ y_i \geq 0 \qquad \forall\, ij \in E^{(m)},\, \forall\, i \subset [n],
\end{aligned} \tag{2}$$

where $(t, \{x_{ij}^{(m)}\}_{ij \in E^{(m)}}, \{y_i\}_{i \in [n]})$ denote the primal variables. We make the following observations:

Lemma 1. *For any $S \subseteq V$, the optimal value of DCS_LP in (2) is at least $\delta_G(S)$. In particular, the optimal value of DCS_LP is an upper bound on δ_G.*

Proof. Suppose $|S| = k$. We construct a feasible solution as follows:

$$y_i = \begin{cases} \frac{1}{k} & \text{if } i, j \in S \\ 0 & \text{otherwise} \end{cases}, \quad x_{ij}^{(m)} = \begin{cases} \frac{1}{k} & \text{if } i, j \in S \\ 0 & \text{otherwise} \end{cases}.$$

Then, $\sum_{ij \in E^{(m)}} x_{ij}^{(m)} = \frac{1}{k} \sum_{ij \in E^{(m)}(S)} 1 = \delta_{G^{(m)}}(S)$ and $t = \delta_G(S)$. □

We consider an algorithm which solves DCS_LP and returns $S(r) = \{i : y_i^* \geq r\}$ which maximizes $\delta_G(S(r))$. Note that there are at most n sets to consider corresponding to distinct values of y_i^*. The following theorem provides a lower bound on the quality of the returned subgraph.

Lemma 2. *Let $(t^*, x_{ij}^{(m)*}, y_i^*)$ denote optimal solution for DCS_LP (2) and let $S := \{i : y_i^* > 0\}$ denote the set of vertices with non-zeros y_i^*'s with $k := |S|$ and $y_{\min} = \min_{i \in S} y_i^*$. The following hold:*

(a) If $y_i^* = \frac{1}{k} \forall i \in S$, then S is a densest common subgraph, i.e., $\delta_G(S) = \delta_G$.

(b) For each graph $G^{(m)} \in \mathbb{G}$, there exists $S^{(m)} \subseteq S$ which induces a subgraph in $G^{(m)}$ with density at least t^*, i.e., $\delta_{G^{(m)}}(S^{(m)}) \geq t^*$.

(c) The density of the subgraph induced by S in any graph $G^{(m)}$ is at least $\frac{t^*\lceil 2t^*+1\rceil}{k}$, i.e., $\delta_{\mathbb{G}}(S) \geq \frac{t^*\lceil 2t^*+1\rceil}{k}$.

Proof (of Lemma 2). Without loss of generality, assume $x_{ij}^{(m)*} = \min(y_i{}^*, y_j{}^*)$ since for any feasible solution (x, y, t), we can construct another solution by setting $\bar{x}_{ij}^{(m)} := \min(y_i, y_j) \geq x_{ij}^{(m)}$ with $\bar{t} \geq t$.

(a) If $\mathbf{y}^* = \frac{1}{K}\mathbf{e}_S$, then $\bar{t} = \delta_{\mathbb{G}}(S)$. By Lemma 1, $t^* \geq \delta_{\mathbb{G}}$ and the result follows.

(b) Following the analysis of Charikar [7, Lemma 2], we define collection of sets indexed by a parameter $r \geq 0$. Let $S(r) := \{i : y_i{}^* \geq r\}$ and $E^{(m)}(r) := \{ij \in E^{(m)} : x_{ij}^{(m)*} \geq r\}$. Since $x_{ij}^{(m)*} = \min(y_i{}^*, y_j{}^*)$ by construction,

$$ij \in E^{(m)}(r) \Leftrightarrow i, j \in S(r).$$

Now, $\int_0^\infty |S(r)| dr = \sum_i y_i{}^* \leq 1$ and $\int_0^\infty |E^{(m)}(r)| dr = \sum_{ij \in E^{(m)}} x_{ij}^{(m)*} \geq t^*$. Then, for each $G^{(m)} \in \mathbb{G}$, there exists $r^{(m)}$ such that $\frac{|E^{(m)}(r^{(m)})|}{|S(r^{(m)})|} \geq t^*$. Otherwise, it leads to the following contradiction:

$$t^* \leq \int_0^\infty |E^{(m)}(r)| dr < t^* \int_0^\infty |S(r)| dr \leq t^*$$

We define $S^{(m)} := S(r^{(m)}) \subseteq S$ which induces a subgraph of density at least t^* in graph $G^{(m)} \in \mathbb{G}$.

(c) We observe $|S^{(m)}| \geq \lceil 2t^* + 1\rceil$ since $\delta_m(S^{(m)}) \geq t^*$. Consequently, for each graph $G^{(m)} \in \mathbb{G}$,

$$\delta_m(S) = \frac{|E^{(m)}(S)|}{k} \geq \frac{|E^{(m)}(S^{(m)})|}{k} = \frac{|S^{(m)}|}{k}\delta_m(S^{(m)}) \geq \frac{\lceil 2t^*+1\rceil}{k}t^* . \quad \square$$

The LP optimal t^* is at most $\delta_{\mathbb{G}}{}^{\min} := \min_{G^{(m)}} \delta_{G^{(m)}}$ where $\delta_{G^{(m)}}$ denotes the density of the densest subgraph in $G^{(m)}$. This yields an upper bound on the integrality gap of DCS_LP given by $\frac{\delta_{\mathbb{G}}{}^{\min}}{\delta_{\mathbb{G}}}$. The LP solution is particularly interesting whenever $\mathbf{y}^* = \frac{1}{|S|}\mathbf{e}_S$ since we get proof of optimality in that case. In the general case, DCS_LP can exhibit both integrality gap higher than 1 and an approximation ratio lower than 1 (obtained by $\delta_{\mathbb{G}}(S)$) in contrast to Charikar's LP in the densest subgraph problem (Section 4.1).

Furthermore, solving DCS_LP has running time polynomial in the number of edges in the graph set ($n_{\mathbb{G}}^{O(1)}$) which cannot be scaled to large graphs having more than a few hundred thousand edges. In the next section, we consider a simple greedy algorithm for finding common dense subgraphs in large graphs.

Algorithm 1 $(S, r) = \text{DCS\_GREEDY}(\mathbb{G})$

1: Initialize $V_1 := V$
2: **while** $\delta_{\mathbb{G}}(V_t) > 0$ **do**
3:     $m_t := \arg\min_m \delta_m(V_t)$ {$G^{(m)}$ with min. density induced subgraph}
4:     $V' := \{i \in V_t : \deg_{m_t}(i, V_t) > 0\}$ {Non-isolated nodes in $G^{(m_t)}(V_t)$}
5:     $i_t := \arg\min_{i \in V_t} \deg_{m_t}(i, V_t)$ {minimum degree node in $G^{(m_t)}(V_t)$}
6:     For all m, assign all edges $(i_t, j) \in E^{(m)}$ and $(j, i_t) \in E^{(m)}$ to node i_t.
7:     Set $\deg_t^+ := \max_m \deg_m(i_t, V_t)$ {Max. degree of i_t in any $G^{(m)}(V_t)$}
8:     Set $\deg_t^- := \min_{i \in V'} \deg_{m_t}(i, V')$ {Min. non-zero degree in $G^{(m_t)}(V_t)$}
9:     Set $k_t = \frac{\deg_t^+}{\deg_t^-}$ and $r_t = k_t \frac{|V_t|}{|V'|}$ {$\deg_t^+ \leq 2r_t \, \delta_{\mathbb{G}}(V_t)$}
10:    $V_{t+1} \leftarrow V_t \backslash i_t$
11:    $t \leftarrow t + 1$
12: **end while**
13: $S' := V_t$ {Clean-up phase}
14: **while** $|S'| > 0$ **do**
15:    Choose random $i' \in S'$
16:    $\forall m$, assign edges $(i', j) \in E^{(m)}(S')$ and $(j, i') \in E^{(m)}(S')$ to i.
17:    $S' \leftarrow S' \backslash i'$
18: **end while**
19: **return** $S := \arg\max_{V_t} \delta_{\mathbb{G}}(V_t)$ and $r := \max_t r_t$

3.1 A Greedy Algorithm for Densest Common Subgraph

In this section, we consider a peeling algorithm (DCS_GREEDY) for solving densest common subgraph problem. The algorithm iteratively constructs vertex set V_{t+1} at each time t by removing the node i_t from V_t where i_t is the minimum degree node in the subgraph $G^{(m_t)}(V_t)$ where $G^{(m_t)}(V_t)$ has the minimum density among the subgraphs induced by V_t in graph set $\mathbb{G}$. All edges (i_t, j) and (j, i_t) present in induced subgraphs $G^{(m)}(V_t)$ are assigned to node i_t. The set V_t which maximizes $\delta_{\mathbb{G}}(V_t)$ is returned as solution. Algorithm 1 gives the complete pseudocode for DCS_GREEDY.

We now consider an analysis of the above algorithm. We have the following invariant: after each iteration t, the set of edges $E^{(m)} \bigcap (V_t \times V_t)$ are unassigned while all other edges are assigned to a node in $V \backslash V_t$. At the termination of the algorithm, either all edges $(i, j) \in E^{(m)} \, \forall m$ are assigned to i, or all edges $(i, j) \in E^{(m)} \, \forall m$ are assigned to j.

Let $d_{max}^{(m)}$ denote the maximum number of edges (i, j) or (j, i) assigned to any node i in graph $G^{(m)}$ and let d_{minmax} denote the minimum value of $d_{max}^{(m)}$ among all graphs $G^{(m)} \in \mathbb{G}$. The following holds:

Lemma 3. *For any $S \subseteq V$, the value $\delta_{\mathbb{G}}(S)$ is at most d_{minmax}. In particular, d_{minmax} is an upper bound on $\delta_{\mathbb{G}}$.*

Proof (Sketch). *Each edge in $E^{(m)}(S)$ is assigned to a vertex in S, and therefore, $|E^{(m)}(S)| \leq d_{max}^{(m)} \cdot |S|$. Consequently, $\delta_{\mathbb{G}}(S) \leq d_{minmax}$.* □

The greedy algorithm of Charikar [7] yields a 2-approximation for the densest subgraph problem due to the following property: in the densest subgraph, the degree of the minimum degree subgraph is at least the average degree otherwise it can safely be removed to yield an even denser subgraph. This property is not true for DCS. The following result gives an lower bound on the quality of the solution returned by DCS_GREEDY.

Lemma 4. *The value d_{minmax} is at most $(2r \cdot \delta_{\mathbb{G}}(S))$ where (S, r) is the solution returned by the DCS_GREEDY algorithm. In particular, $\delta_{\mathbb{G}} \leq (2r \cdot \delta_{\mathbb{G}}(S))$.*

Proof. Let $m' = \arg\min d_{max}^{(m)}$ and i' denote the node in graph $G^{(m')}$ which has the maximum number of edges assigned to it. Let t' denote the iteration in which node i' is removed from $V_{t'}$ during the execution of DCS_GREEDY. By construction, edges are assigned to any node only when it is removed from the graph and therefore, $d_{minmax} = \deg_{t'}^{+}$. Further, by definition of $\deg_t^{-}$, we have $\deg_t^{-} \leq 2\frac{|E^{(m_t)}(V')|}{|V'|} = 2\,\delta_{\mathbb{G}}(V_t)\frac{|V_t|}{|V'|}$. Combining, we get

$$d_{minmax} = \deg_{t'}^{+} \leq 2r_{t'}\,\delta_{\mathbb{G}}(V_{t'}) \leq 2r\,\delta_{\mathbb{G}}(S). \qquad \square$$

The DCS_GREEDY algorithm can be efficiently implemented in $O(n + n_{\mathbb{G}})$ time by maintaining a list of node degrees for each graph $G^{(m)}$ and updating the neighbours of a node v in the degree list whenever node v is removed.

3.2 Densest Common at Least-k Subgraph (DCalkS)

We next consider the Densest Common at least-k Subgraph (DCalkS) problem which imposes constraint on the minimum size of the induced subgraphs.

Definition 2 (Densest Common at Least-k Subgraph). *Given relational graph set $\mathbb{G}$, the densest common at least-k subgraph (DCalkS) is given by,*

$$H := \arg\max_{S \subseteq V : |S| \geq k} \delta_{\mathbb{G}}(S). \tag{3}$$

In the sequel, we use the shorthand notation $n_h := |H|$ and $\delta_{\geq k} := \delta_{\mathbb{G}}(H)$ to denote the cardinality of the DCalkS subgraph and the minimum density of any subgraph by H induced in any graph $G^{(m)} \in \mathbb{G}$ respectively.

We note that DCalkS is NP-Hard (by restriction to DalkS whenever $|\mathbb{G}| = 1$) since DalkS is known to be NP-Hard [20, Theorem 3.1]. Consider the linear program P2(c) given by:

$$
\begin{aligned}
\max\ & t \\
\text{s.t.}\ & \textstyle\sum_{i=1}^{n} y_i = 1 \\
& \textstyle\sum_{ij \in E^{(m)}} x_{ij}^{(m)} \geq t && \forall\, E^{(m)},\, m \in [M] \\
& x_{ij}^{(m)} \leq y_i,\ x_{ij}^{(m)} \leq y_j && \forall\, ij \in E^{(m)} \\
& x_{ij}^{(m)} \geq 0,\ y_i \geq 0,\ y_i \leq \tfrac{1}{c}\, \forall\, ij \in E^{(m)},\, \forall\, i \in [n].
\end{aligned}
\tag{4}
$$

where $c \geq k$ is our guess for the size of the optimal DCalkS solution (n_h). The following result is a direct consequence of Lemma 1.

Lemma 5. *For any $S \subseteq V$ with $c := |S| \geq d$, the optimal value of P2(d) is at least $\delta_G(S)$. In particular, the optimal value P2(k) is an upper bound on $\delta_{\geq k}$.*

We consider an algorithm (DCalkS-LP) which solves P2(k) and returns $\{i : y_i^* > 0\}$. Then, the following holds:

Lemma 6. *Let $(t^*, x_{ij}^{(m)*}, y_i^*)$ denote optimal solution for P2(k) in (4) and let $S := \{i : y_i^* > 0\}$ denote the set of vertices with non-zeros y_i^*'s with $l := |S| \geq k$. Then,*

(a) *if $y_i^* = \frac{1}{l} \forall i \in S$ and zero otherwise, then $\delta_G(S) = \delta_{\geq k}$, i.e., S is a densest common at least-k subgraph; and,*

(b) *the density of the subgraph induced by S in any graph $G^{(m)}$ is at least $\frac{t^* \lceil 2t^* + 1 \rceil}{l}$, i.e., $\delta_G(S) \geq \frac{t^* \lceil 2t^* + 1 \rceil}{l}$.*

Proof (Sketch). The proof is analogous to the proof for Theorem 2. □

4 Experiments

We compare our approach (DCS_LP and DCS_GREEDY) with MKL-based formulation (MKL) in [13] and the greedy algorithm (CHARIKAR) of [7]. In order to compare with [13], we construct the sets $S(r) = \{i \in V : \alpha_i^* \geq r\}$ where α^* is the solution of the MKL optimization in [13] and choose the set which maximizes $\delta_G(S(r))$, i.e., $S_{MKL} := \arg\max_{S(r)} \delta_G(S(r))$. We run the greedy algorithm (CHARIKAR) to find a dense subgraph in the intersection graph $G_\cap$.

The tensor-based approach in [24] is not considered since it has significantly higher computational complexity (e.g., the authors report a running time of 200 hours on a high-performance computing node for analysis of co-expression networks [See 24, SupplementaryText,S6]).

All experiments were implemented in MATLAB R2014A and run on a laptop having 2.8GHz Intel Core i5 processor and 16GB RAM. The greedy algorithm was coded in MATLAB while the linear programs were solved using Gurobi 6.0 solver using its MATLAB API. Gurobi [3] is a state-of-the-art commercial solver for linear and non-linear optimization implemented in C programming language with APIs for several other programming languages.

4.1 Synthetic Dataset

We investigate the integrality gap and approximation ratio of DCS_LP by considering small graphs for which we can find the DCS solution by exhaustive search. Each graphset consists of m graphs each of which is generated independently as follows: We generate an Erdös-Renyi random graph $G(n, p)$ and then plant a clique of size k randomly. We generate a dataset having 100 such graphsets with the following parameters: $m = 3$, $n = 20$, $p \sim Uniform(0, 0.5)$ and $k \sim Uniform[1, \ldots, n/2]$.

---

[3] http://www.gurobi.com

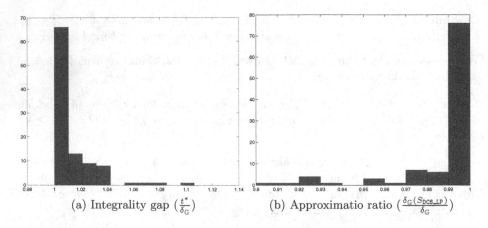

(a) Integrality gap ($\frac{t^*}{\delta_G}$)   (b) Approximatio ratio ($\frac{\delta_G(S_{DCS\_LP})}{\delta_G}$)

Fig. 1. Properties of DCS_LP

Figure 1 shows the histogram of (a) the LP objective value t^*, and, (b) the common density of subgraph obtained by rounding the LP solution $S = \{i : y_I^* > 0\}$ relative to the optimum solution δ_G, respectively. DCS_LP recovers the optimum solution in 70% of the cases but in general, it can and does exhibit both integrality gap $t^* > \delta_G$ and sub-optimal rounding $\delta_G(S) < \delta_G$.

4.2   Real-World Datasets

We consider the following datasets for evaluation of our methods:

- DIMACS. The DIMACS 1994 dataset [4] is a comprehensive benchmark for testing of clique finding and related algorithms. Each of the graph families in DIMACS (*brock, c-fat, p_hat, san, sanr*) is motivated by carefully selected real world problems e.g. fault diagnosis (*c-fat*), etc.; thus covering a wide range of practical scenarios [17]. This was used for experimental evaluation in [13] and we repeat their experimental setup.
- SNAP-AS. We consider the graph sets *Oregon-1* and *Oregon-2* from SNAP network database [5] [22,23] related to autonomous systems. Each dataset consists of $M = 9$ graphs having ~ 11000 nodes and $20000 - 30000$ edges.
- SNAP-AMAZON. This dataset consists of directed graphs *amazon0312, amazon0505* and *amazon0601* available from SNAP network database which consist of Amazon product co-purchasing networks on specific dates. We make the graphs undirected (by introducing edges (j, i) whenever (i, j) is present) and consider the nodes which are present in all three graphs. This yields a graph set having $n = 400727$ nodes and $n_G = 7157921$ edges in total.

---

[4] ftp://dimacs.rutgers.edu/pub/challenge/graph/benchmarks/clique/
[5] http://snap.stanford.edu/data/

Table 1. Comparison of DCS_LP and DCS_GREEDY with MKL [13] and CHARIKAR [7] for finding DCS in DIMACS and SNAP-AS datasets. M, n, n_G denote resp. the number of graphs, number of nodes in each graph, and the total number of edges summed over all the graphs in the graph set. $|S|$ and $\delta_G(S)$ denote the size and common density of the induced subgraph found by each method. $\delta_\cap(S)$ (in CHARIKAR) is the density of the subgraph in the intersection graph $G_\cap$ and t^* denotes the optimum of DCS LP. X marks the instances where the DCS LP failed to converge within 2 hours.

$\mathbb{G}$	M	n	n_G	MKL [13]		DCS_LP			DCS_GREEDY		CHARIKAR [7]		
				$|S|$	$\delta_G(S)$	$|S|$	$\delta_G(S)$	t^*	$|S|$	$\delta_G(S)$	$|S|$	$\delta_G(S)$	$\delta_\cap(S)$
c-fat200	3	200	13242	100	7.93	101	8.01	8.08	102	8.04	199	7.63	1.81
c-fat500	4	500	83416	140	9.65	140	9.65	9.65	140	9.65	140	9.65	9.65
brock200	4	200	29753	200	25.33	200	25.33	25.33	200	25.33	149	19.31	1.93
brock400	4	400	80245	400	50.04	400	50.04	50.04	400	50.04	110	14.08	1.20
brock800	4	800	447753	800	139.29	X	X	X	800	139.29	783	136.41	5.99
p_hat300	3	300	66251	300	36.44	286	36.65	36.65	286	36.65	286	36.65	36.65
p_hat500	3	500	188315	500	63.14	489	63.21	63.21	489	63.21	489	63.21	63.21
p_hat700	3	700	365737	700	87.14	X	X	X	679	87.27	679	87.27	87.27
p_hat1000	3	1000	738798	1000	122.25	X	X	X	973	122.41	973	122.41	122.41
p_hat1500	3	1500	1701127	1500	189.95	X	X	X	1478	190.05	1478	190.05	190.05
san200	5	200	17910	200	9.95	200	9.95	9.95	200	9.95	2	0.50	0.50
san400	3	400	119700	400	19.95	400	19.95	19.95	400	19.95	176	9.14	0.66
sanr200	2	200	8069	200	10.19	199	10.19	10.19	199	10.19	170	9.23	3.17
sanr400	2	400	63747	400	59.83	400	59.83	59.83	400	59.83	399	59.71	29.97
Oregon-1	9	11492	203127	54	11.37	76	12.03	12.03	76	12.03	64	11.28	10.09
Oregon 2	9	11806	284031	173	22.35	X	X	X	119	22.45	115	21.23	17.57

Table 1 shows the results for DIMACS and SNAP-AS datasets. In all instances (except *c-fat200*) where DCS_LP finishes within time, the DCS_LP solution is optimal as verified by $t^* = \delta_G(S)$. Further, DCS_GREEDY also finds the optimal solution in these instances. Both our methods out-perform MKL and CHARIKAR in all graph sets. This is especially striking in the case of *san* and *sanr* graph sets where the greedy algorithm (CHARIKAR) yields very poor results – highlighting the fact that taking the intersection of the graphs is unsuitable.

We note that the high computational complexity of DCS_LP ($n_G^{O(1)}$) and MKL ($O(n^3)$) prohibits their use for finding the densest common subgraph in the SNAP-AMAZON graph set. The DCS_GREEDY algorithm finds a subgraph with $\delta_G(S) = 5.90251$ while CHARIKAR finds a subgraph with $\delta_G(S) = 2.5$.

5   Discussion

This paper formalizes the Densest Common Subgraph (DCS) problem which extends the notion of densest subgraph to a relational graph set. We present an extension of Charikar's linear programming approach [7] to the problem of finding the densest common subgraph in a relational graph set.

The LP-based approach recovers the densest common subgraph in many cases (with proof of optimality). In other cases, it provides an upper bound on the common density (e.g. in the case of *c-fat200* graph set in the experiments) and a good starting point for further heuristic search approaches. We note that in the worst-case, the approximation guarantee is $O(\frac{n}{\delta_G})$ – which can be trivially obtained by taking original vertex set V as a solution. A tighter analysis of the LP relaxation can reveal more insight into the problem.

Our greedy algorithm DCS_GREEDY can be scaled to large graphs which is not possible with existing methods. Further, it substantially improves over the greedy approach (CHARIKAR) which only considers the integrated graph by taking intersection of the different edge sets. We note that the DCS_GREEDY algorithm is closely related to the dual LP. Designing a combinatorial primal-dual algorithm can lead to better results and will be addressed in future work.

Acknowledgments. We thank Chiranjib Bhattacharyya, Devdatt Dubhashi and Jack Kuipers for valuable comments and suggestions. Vinay Jethava is funded by the MERiC project as part of the European Research Council (ERC) synergy grant 2013.

References

1. Andersen, R., Chellapilla, K.: Finding dense subgraphs with size bounds. In: Avrachenkov, K., Donato, D., Litvak, N. (eds.) WAW 2009. LNCS, vol. 5427, pp. 25–37. Springer, Heidelberg (2009)
2. Asahiro, Y., Iwama, K., Tamaki, H., Tokuyama, T.: Greedily finding a dense subgraph. Journal of Algorithms **34**(2), 203–221 (2000)
3. Bach, F.R., Lanckriet, G.R.G., Jordan, M.I.: Multiple kernel learning, conic duality, and the SMO algorithm. In: Proceedings of the Twenty-First International Conference on Machine Learning, p. 6 (2004)
4. Bahmani, B., Kumar, R., Vassilvitskii, S.: Densest subgraph in streaming and mapreduce. Proceedings of the VLDB Endowment **5**(5), 454–465 (2012)
5. Bhaskara, A., Charikar, M., Chlamtac, E., Feige, U., Vijayaraghavan, A.: Detecting high log-densities: an o (n 1/4) approximation for densest k-subgraph. In: Proceedings of the Forty-Second ACM Symposium on Theory of Computing (STOC), pp. 201–210. ACM (2010)
6. Boden, B., Günnemann, S., Hoffmann, H., Seidl, T.: Mining coherent subgraphs in multi-layer graphs with edge labels. In: Proceedings of the 18th ACM SIGKDD International Conference on Knowledge Discovery and Data Mining, pp. 1258–1266. ACM (2012)
7. Charikar, M.: Greedy approximation algorithms for finding dense components in a graph. In: Jansen, K., Khuller, S. (eds.) APPROX 2000. LNCS, vol. 1913, pp. 84–95. Springer, Heidelberg (2000)
8. Feige, U., Langberg, M.: Approximation algorithms for maximization problems arising in graph partitioning. Journal of Algorithms **41**(2), 174–211 (2001)
9. Feige, U., Peleg, D., Kortsarz, G.: The dense k-subgraph problem. Algorithmica **29**(3), 410–421 (2001)
10. Gallo, G., Grigoriadis, M.D., Tarjan, R.E.: A fast parametric maximum flow algorithm and applications. SIAM Journal on Computing **18**(1), 30–55 (1989)

11. Goldberg, A.V.: Finding a maximum density subgraph. University of California at Berkeley, Berkeley (1984)

12. Hu, H., Yan, X., Huang, Y., Han, J., Zhou, X.J.: Mining coherent dense subgraphs across massive biological networks for functional discovery. Bioinformatics 21(suppl 1), i213–i221 (2005)

13. Jethava, V., Martinsson, A., Bhattacharyya, C., Dubhashi, D.: The lovasz θ function, svms and finding large dense subgraphs. In: Neural Information Processing Systems (NIPS), pp. 1169–1177 (2012)

14. Jethava, V., Martinsson, A., Bhattacharyya, C., Dubhashi, D.: Lovasz theta function, SVMs and finding dense subgraphs. Journal of Machine Learning Research 14, 3495–3536 (2014)

15. Jiang, C., Coenen, F., Zito, M.: A survey of frequent subgraph mining algorithms. The Knowledge Engineering Review 28(01), 75–105 (2013)

16. Jiang, D., Pei, J.: Mining frequent cross-graph quasi-cliques. ACM Transactions on Knowledge Discovery from Data (TKDD) 2(4), 16 (2009)

17. Johnson, D.S., Trick, M.A.: Cliques, coloring, and satisfiability: second DIMACS implementation challenge, October 11–13, 1993, vol. 26. Amer Mathematical Society (1996)

18. Kannan, R., Vinay, V.: Analyzing the structure of large graphs. Rheinische Friedrich-Wilhelms-Universität Bonn (1999)

19. Khot, S.: Ruling out ptas for graph min-bisection, dense k-subgraph, and bipartite clique. SIAM Journal on Computing 36(4), 1025–1071 (2006)

20. Khuller, S., Saha, B.: On finding dense subgraphs. In: Albers, S., Marchetti-Spaccamela, A., Matias, Y., Nikoletseas, S., Thomas, W. (eds.) ICALP 2009, Part I. LNCS, vol. 5555, pp. 597–608. Springer, Heidelberg (2009)

21. Lee, V.E., Ruan, N., Jin, R., Aggarwal, C.: A survey of algorithms for dense subgraph discovery. Managing and Mining Graph Data 303–336 (2010)

22. Leskovec, J., Krevl, A.: SNAP Datasets: Stanford large network dataset collection (June 2014). http://snap.stanford.edu/data

23. Leskovec, J., Kleinberg, J., Faloutsos, C.: Graphs over time: densification laws, shrinking diameters and possible explanations. In: Proceedings of the Eleventh ACM SIGKDD International Conference on Knowledge Discovery in Data Mining, pp. 177–187. ACM (2005)

24. Li, W., Liu, C.-C., Zhang, T., Li, H., Waterman, M.S., Zhou, X.J.: Integrative analysis of many weighted co-expression networks using tensor computation. PLoS computational biology 7(6), e1001106 (2011)

25. Papailiopoulos, D., Mitliagkas, I., Dimakis, A., Caramanis, C.: Finding dense subgraphs via low-rank bilinear optimization. In: Proceedings of the 31st International Conference on Machine Learning (ICML-14), pp. 1890–1898 (2014)

26. Pei, J., Jiang, D., Zhang, A.: Mining cross-graph quasi-cliques in gene expression and protein interaction data. In: Proceedings of the 21st International Conference on Data Engineering. ICDE 2005, pp. 353–356. IEEE (2005)

27. Rakotomamonjy, A., Bach, F., Canu, S., Grandvalet, Y.: Simplemkl. Journal of Machine Learning Research 9, 2491–2521 (2008)

28. Ravi, S.S., Rosenkrantz, D.J., Tayi, G.K.: Heuristic and special case algorithms for dispersion problems. Operations Research 42(2), 299–310 (1994)

29. Srivastav, A., Wolf, K.: Finding dense subgraphs with semidefinite programming. In: Jansen, K., Rolim, J.D.P. (eds.) APPROX 1998. LNCS, vol. 1444, pp. 181–191. Springer, Heidelberg (1998)

30. Tsourakakis, C., Bonchi, F., Gionis, A., Gullo, F., Tsiarli, M.: Denser than the densest subgraph: extracting optimal quasi-cliques with quality guarantees. In: Proceedings of the 19th ACM SIGKDD International Conference on Knowledge Discovery and Data Mining, pp. 104–112. ACM (2013)
31. Yan, X., Zhou, X., Han, J.: Mining closed relational graphs with connectivity constraints. In: Proceedings of the Eleventh ACM SIGKDD International Conference on Knowledge Discovery in Data Mining, pp. 324–333. ACM (2005)
32. Yuan, X.-T., Zhang, T.: Truncated power method for sparse eigenvalue problems. The Journal of Machine Learning Research 14(1), 899–925 (2013)
33. Zeng, Z., Wang, J., Zhou, L., Karypis, G.: Coherent closed quasi-clique discovery from large dense graph databases. In: Proceedings of the 12th ACM SIGKDD International Conference on Knowledge Discovery and Data Mining, pp. 797–802. ACM (2006)
34. Zhang, T.: Multi-stage convex relaxation for non-convex optimization. Technical report, Rutgers University (2009)

Generalized Modularity for Community Detection

Mohadeseh Ganji[1,3]([✉]), Abbas Seifi[1], Hosein Alizadeh[2], James Bailey[3], and Peter J. Stuckey[3]

[1] Amirkabir University of Technology, Tehran, Iran
sghasempour@student.unimelb.edu.au, aseifi@aut.ac.ir
[2] Iran University of Science and Technology, Tehran, Iran
halizadeh@iust.ac.ir
[3] NICTA, Victoria Laboratory, Department of Computing and Information Systems, University of Melbourne, Melbourne, Victoria
{baileyj,pstuckey}@unimelb.edu.au

Abstract. Detecting the underlying community structure of networks is an important problem in complex network analysis. Modularity is a well-known quality function introduced by Newman, that measures how vertices in a community share more edges than what would be expected in a randomized network. However, this limited view on vertex similarity leads to limits in what can be resolved by modularity. To overcome these limitations, we propose a generalized modularity measure called GM which has a more sophisticated interpretation of vertex similarity. In particular, GM also takes into account the number of longer paths between vertices, compared to what would be expected in a randomized network. We also introduce a unified version of GM which detects communities of unipartite and (near-)bipartite networks without knowing the structure type in advance. Experiments on different synthetic and real data sets, demonstrate GM performs strongly in comparison to several existing approaches, particularly for small-world networks.

Keywords: Community detection · Modularity · Generalized modularity · Vertex similarity · Resolution limit

1 Introduction

As many real-world systems can be represented by networks, much research has focused on analysing networks and finding underlying useful structural patterns. Examples include social and biological networks [1,2], in which vertices represent individuals or proteins and edges represent communications or interactions.

Among complex network analysis approaches, community detection is an important task which aims to find groups of vertices which could share common properties and/or have similar roles within the network [3]. This might reveal friendship communities in a social network or an unexpected hard-to-predict community structure in a biological dataset.

© Springer International Publishing Switzerland 2015
A. Appice et al. (Eds.): ECML PKDD 2015, Part II, LNAI 9285, pp. 655–670, 2015.
DOI: 10.1007/978-3-319-23525-7_40

Two important network structures covered in the literature are unipartite and bipartite networks. In unipartite networks like social networks [1], the assumption is connections within communities are dense and connections between communities are sparse. However, some real networks are bipartite which means they can be partitioned into two clusters such that no two vertices within the same cluster are adjacent [4]. People attending events [5] is one example of a bipartite network. In addition, there are some real networks with near-bipartite properties. In these networks, there are some connections inside the two communities but they are fewer than between-community connections. Networks of sexual relationships are an example of near-bipartite networks.

Among community detection criteria, *modularity* [6] is one of the most important because according to [7], "Modularity has the unique privilege of being at the same time a global criterion to define a community, a quality function and the key ingredient of the most popular method of graph clustering." After its introduction, modularity was rapidly adopted and physicists, computer scientists, and sociologists have all developed a variety of heuristic algorithms to optimize modularity. They are based on greedy algorithms [8] spectral methods [9], mathematical optimization [10] and other strategies [7,11].

Given an un-weighted undirected network $G(V, E)$, let d_i be degree of vertex i, m be total number of edges and A_{ij} be an element of the adjacency matrix which takes value 1 if vertices i and j are connected and 0 otherwise. Suppose the vertices are partitioned into communities such that vertex i belongs to community C_i. Then the modularity of the partition is defined by equation (1).

$$Q = \frac{1}{2m} \sum_{i,j} [A_{ij} - \frac{d_i d_j}{2m}] \delta(C_i, C_j) \qquad (1)$$

The matrix of elements $A_{ij} - \frac{d_i d_j}{2m}$ is called the modularity matrix which is denoted by W. The modularity matrix records the difference between the number of the edges connecting each pair of vertices and the expected number of edges in a randomly distributed network of the same size with the same vertex degree sequence (in the rest of this paper we call it a randomized network). If the number of edges between i and j is the same as what is expected in the randomized network, the corresponding element of the modularity matrix is zero. Hence, nonzero values of the modularity matrix represent deviation from randomness. The coefficient $\frac{1}{2m}$ normalizes modularity to the interval [-1,1].

For calculating the modularity of a network partition, one adds up the modularities between each pair of vertices that lie in the same community. In equation (1), $\delta(C_i, C_j)$, the Kronecker delta function performs this task by limiting the summation to just over vertex pairs of the same community. The Kronecker function has the value 1 if its arguments are equal and 0 otherwise.

Brandes *et al* [12] showed that finding a clustering with maximum modularity is an NP-hard problem. However, researchers have tackled community detection using exact and approximation methods for modularity maximization. Among exact methods, Aloise *et al* [10] introduced a column generation model which can find communities of optimal modularity value for problems of up to 512

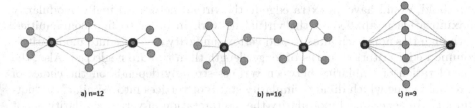

Fig. 1. Examples for neglecting neighbours by Modularity

vertices. Among the wide range of approximation algorithms for modularity maximization, the hierarchical iterative two phase method of Blondel *et al* [8] is one of the best (see [7]). In the first phase, communities are merged together only if this improves the modularity value of the partition, whilst the second phase reconstructs the network whose nodes are communities of the previous phase.

Limitations of Modularity. Although modularity performs effectively in many cases, some limitations have been noted about its performance [7,13]. First, modularity has a restricted interpretation of vertex similarity. Figure 1 illustrates this problem using hand-made examples. All structures shown in this figure have 17 edges but the number of vertices are different. Degrees of the bold vertices are the same and equal to 7 in all cases and they do not share any edges together. Hence, in all three cases, the modularity value between the two bold vertices is equal while one can clearly see that the structures are very different.

Figure 1 illustrates that modularity's interpretation of vertex similarity is limited to sharing common edges. While in reality, in addition to sharing an edge, pairs of vertices are more similar and more likely to lie in the same community when they have many neighbours in common. This is exactly one of the basic vertex similarity measures called common neighbour index [14]. There are also some other variations of vertex similarity measures based on number of common neighbours and paths of longer lengths [7,14–16].

We propose a new measure of community detection called generalized modularity (GM) which extends modularity's assumption about similar vertices. In addition to common edges, GM takes into account common neighbours and longer paths between vertices and compares the number of these paths to a randomly distributed network to achieve a more comprehensive interpretation of vertex similarity.

Although in the literature, some research has tried to detect communities based on a vertex similarity concept [1,17], these approaches have mostly failed to take advantage of modularity's strength in noticing common edges between vertices. The vertex similarity probability (VSP) model of Li and Pang [17] is one such approach which is just based on common neighbours of vertices but doesn't notice common edges or relations of longer lengths.

In other work, Alfalahi *et al* [1] proposed the concept of vertex similarity for modularity. They construct a virtual network which is initially the same as the original network. Then, vertices with higher Jaccard [15] similarity index (which is based on common edge and common neighbour concepts) than a pre-defined threshold, would have an extra edge in the virtual network. Finally, modularity maximization is applied to the virtual network in order to find communities. Although this approach aims to add vertex similarity concepts into modularity's common edge criterion, paths of longer length than two are neglected. Also, the consideration of similarity between vertices strongly depends on the choice of threshold value which divides similarity status of vertices into "similar" or "not-similar". In generalized modularity the interpretation of vertex similarity is not limited to 0 and 1. In addition, as opposed to Alfalahi's approach, GM benefits from the comparison to random graphs for measuring vertex similarity. In this sense, GM's interpretation of vertex similarity is close to Leicht et al [16] who proposed a vertex similarity index based on comparison to a randomized network, though there are basic differences in context and approach of comparison.

The second limitation of modularity, the resolution limit, arises from its null model. It causes the systematic merging of small communities into larger modules, even when the communities are well defined and loosely connected to each other [13]. Fortunato and Barthelemy in [13] and Fortunato in [7] discussed this issue in more detail. According to [7], in the modularity definition, the weak point of the null model is the implicit assumption that each vertex can communicate with every other vertex of the network. This is however questionable, and certainly wrong for large networks like the Web graph. To address the resolution limit problem, multiresolution versions of modularity have been introduced [18] which allow users to specify their target scale of communities. The choice of correct value for this scale parameter is still an issue with these approaches.

However, by considering longer paths, GM moderates the questionable assumption of modularity's null model. Because expecting network members to be able to share a neighbour with others is a more reasonable assumption. Even more realistic is the possibility of existence of paths with short lengths between members of a network, in particular, networks with the small-world property. According to Watts et al [19], small-world networks are those in which the typical distance L between two randomly chosen vertices grows proportionally to the logarithm of the size of the network. This means the transition from one vertex to any other vertex of the network requires just a few hops. It has been shown that a wide range of real-world complex networks like social networks, the connectivity of the Internet, wikis, collaboration networks and gene networks exhibit small-world network characteristics. In addition to small-world networks, Watts and Strogatz showed that in fact many real-world networks have a small average shortest path length between vertices [19]. Thus, although GM is a global criterion and considers the whole network for defining communities, the small-world property of real networks supports the assumption behind its null model.

Modularity maximization and most community detection criteria are designed for unipartite networks in which edges inside communities are more dense. In near-bipartite networks, however, connections between communities are denser than inside them and modularity maximization cannot find correct communities because it aims to minimize the number of edges between communities. Although there are community detection methods for bipartite networks like modularity minimization and some others [4], they require knowing the type of the data in advance. This problem is more important when it comes to near-bipartite networks, since the identification of such networks is more difficult. In this paper, we also propose a unified version of generalized modularity called UGM which can detect communities in unipartite, bipartite and near-bipartite networks without knowing the type of the network structure.

Briefly, the main contributions of this paper are:

- Extending the "modularity" community detection quality function and proposing a new criterion named Generalized Modularity (GM) which takes advantage of vertex similarity and longer paths between vertices.
- Proposing a more realistic null model in comparison to modularity, which enables generalized modularity to perform better than modularity in small-world data sets with communities of different scales.
- Introducing a unified version of the generalized modularity measure (UGM) which is able to detect communities in unipartite, bipartite and near-bipartite networks without any pre-knowledge about the structure of the data.
- Experimental comparison of the GM and UGM methods with some state of the art approaches and statistically demonstrating their high performance.

2   Generalized Modularity (GM)

The core concept of our proposed generalized modularity measure is to extend modularity to take advantage of indirect communications between vertices.

According to the definition of modularity, a pair of vertices is likely to be in the same community if they share more edges than what is expected from a randomly distributed network. Pairs of vertices can be also similar to each other based on the number of their shared neighbours [7,14]. In generalized modularity we believe that sharing more neighbours than what is expected (in a randomized network) also expresses how likely it is for the pair to lie in the same community. Likewise, two vertices are more likely to be in same community if they have more paths of length three or more, than the corresponding expected number in a randomize network. Hence, generalized modularity is inspired by the concept of vertex similarity while preserving the basic idea of modularity.

The general form of the proposed GM measure is presented in equation (2) which given a partition, adds up the elements of the W^{GM} matrix for pairs of the same communities. The generalized modularity matrix W^{GM} is the weighted summation of $W_{norm}^{(\ell)}$s (equation (3)) which are normalized generalized modularity matrices of level ℓ which means just relations with paths of length ℓ are considered. α_ℓ represents the weight of contribution of $W_{norm}^{(\ell)}$ in W^{GM}.

$$Q_{GM} = \sum_{i,j \in V} W_{i,j}^{GM} \delta(C_i, C_j) \qquad (2)$$

$$W^{GM} = \sum_{\ell=1}^{\infty} \alpha_\ell W_{norm}^{(\ell)} = \sum_{\ell=1}^{\infty} \alpha_\ell \frac{W^{(\ell)}}{||N^{(\ell)}||} = \sum_{\ell=1}^{\infty} \alpha_\ell \frac{[N^{(\ell)} - E^{(\ell)}]}{||N^{(\ell)}||} \qquad (3)$$

$N^{(\ell)}$ is the matrix representing the number of simple paths (paths containing no loops) of length ℓ between vertices. The matrix of $N^{(\ell)}$ is equal to the adjacency matrix power to ℓ, (A^ℓ), for $\ell = 1, 2$. $||N^{(\ell)}||$ is the entry-wise 1-norm of matrix $N^{(\ell)}$ which is summation of absolute values of the matrix elements. The matrix $E^{(\ell)}$ represents the expected number of paths of length ℓ between vertices in a randomized network. We can normalize each term by dividing it by the total number of paths of corresponding length which is denoted by $||N^{(\ell)}||$. According to equation (3), $W^{(1)}$ is exactly the same as the modularity matrix of Newman [6] while the matrix $W^{(2)}$ is the existing number of common neighbours (relations with paths of length 2) between vertices minus the expected number of such common neighbours in a corresponding randomized network. Other terms are also defined likewise.

The expected number of paths of length one between i and j is calculated by multiplying the number of edges connected to i (degree of vertex i) by the probability that an edge ends in j which is $d_j/2m$. By applying a similar approach, we calculate the expected terms in $W^{(2)}$ and $W^{(3)}$ for a pair of vertices in an un-weighted network. Note that the direct edges between two vertices cannot participate in any path of length 2 and 3 between them. So, in equation (4), apart from $d_i d_j/2m$ expected connections between i and j, we expect $(d_i - \frac{d_i d_j}{2m})$ remaining edges of i to contribute in simple paths of longer lengths. For these edges, the probability to be linked to the intermediate vertex k is $d_k/2m$ and then an edge from the set of $d_k - 1$ remaining edges of k must be linked to j with probability of $(d_j - \frac{d_i d_j}{2m})/2m$. Since the probability of existence of edges between vertices are independent to each other, the probability of existence of a path of length ℓ simply equals the multiplication of probabilities of each of its ℓ edges. Finally, as intermediate vertex k can be any vertex of the network except i and j, we have a summation over all possible ks.

$$E_{i,j}^{(2)} = \sum_{k \in V \setminus \{i,j\}} \left[\frac{(d_i - \frac{d_i d_j}{2m})d_k}{2m} \right] \left[\frac{(d_k - 1)(d_j - \frac{d_i d_j}{2m})}{2m} \right] \tag{4}$$

$$W_{ij}^{(2)} = N_{ij}^{(2)} - E_{i,j}^{(2)} = (A^2)_{ij} - \frac{(d_i - \frac{d_i d_j}{2m})(d_j - \frac{d_i d_j}{2m})}{(2m)^2} \sum_{k \in V \setminus \{i,j\}} d_k(d_k - 1) \tag{5}$$

Similarly, we can calculate the expected value for paths of length 3 which vertices i and j are connected through two intermediate vertices k and k'.

$$E_{i,j}^{(3)} = \sum_{k,k' \in V \setminus \{i,j\}} \left[\frac{(d_i - \frac{d_i d_j}{2m})d_k}{2m} \right] \left[\frac{(d_k - 1)d_{k'}}{2m} \right] \left[\frac{(d_{k'} - 1)(d_j - \frac{d_i d_j}{2m})}{2m} \right] \tag{6}$$

In the calculation of paths of length 3 as opposed to the two previous cases, there is a possibility for loops which are illustrated in Figure 2. Among these four topologies, just Figure 2-a is considered in the calculation of term $W_{ij}^{(3)}$, because the existence of the other three paths is dependent on the existence of a common

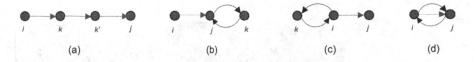

Fig. 2. Four different possible topologies for paths of length 3 between i and j

edge between i and j which we already considered in the calculation of $W_{ij}^{(1)}$. As matrix (A^3) counts all four topologies, we use matrix $(A^3)'$ of equation (7) which just represents the number of simple paths of length 3 between vertices. Hence, the third term of the generalized modularity is equal to the equation (8).

$$(A^3)'_{ij} = (A^3)_{ij} - A_{ij}(d_i + d_j - 1) \tag{7}$$

$$W_{ij}^{(3)} = N_{ij}^{(3)} - E_{i,j}^{(3)} = (A^3)'_{ij} - \left[\frac{(d_i - \frac{d_i d_j}{2m})(d_j - \frac{d_i d_j}{2m})}{(2m)^3}\left(\sum_{k \in V \setminus \{i,j\}} d_k(d_k - 1)\right)^2\right] \tag{8}$$

The number of terms in generalized modularity increases according to the path lengths considered, however, paths of length more than 3 are more complicated as the number of possible topologies and non simple paths rapidly increases. In addition, intuitively, it seems they would have smaller importance weight (α_l) than the first couple of terms. Therefore, in this paper, we limit generalized modularity to its first three terms which are related to paths of length ($\ell = 1, 2, 3$).

2.1 Comparison to Modularity

Although our GM quality function was initially inspired by modularity, extending the measure to consider neighborhoods with longer paths leads to improvements in several aspects.

First, GM is more comprehensive in its interpretation of similarity as it considers vertex similarity as well. Therefore, when edge related properties are still the same (as in Figure 1), GM can detect communities better than modularity since it uses common neighbours and the neighborhood of longer paths as well.

To illustrate how well generalized modularity can reveal the underlying community structures, we use visualization. The visual assessment of tendency (VAT) [20], is a tool for revealing the number of clusters. It uses the logic of Prim's algorithm and reorders the objects of symmetric square dissimilarity matrix R to show the number of clusters by squared shaped dark blocks along the diagonal in the VAT image. We scaled each element of modularity and GM matrices to $(-W_{ij} + 1)/2$ to ensure elements are in interval [0,1] and then we used them as dissimilarity matrix for VAT. Figure 3 presents VAT images of modularity and GM ($\alpha_1, \alpha_2, \alpha_3 = (0.25, 0.5, 0.25)$) for an LFR data set

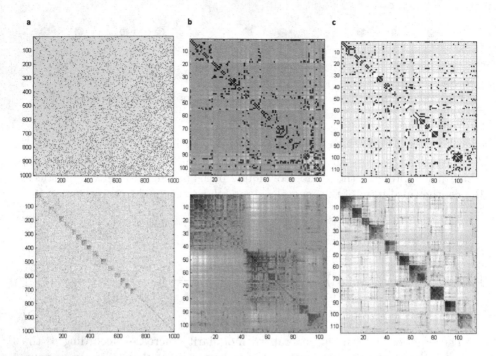

Fig. 3. VAT image of modularity matrix (top images) and generalized modularity matrix (bottom images) for three data sets (a) LFR, (b) Political Books, (c) American Football. Dark blocks in VAT images of GM correspond to communities in the data.

(which is a community detection synthetic benchmark proposed by Lancichinetti [21]) and also two real-world data sets. In this Figure, modularity's VAT image does not reveal the community structure of the data sets while dark blocks in GM's VAT image effectively distinguish community structures. A similar trend was also observed for the other real-world and artificial data sets used in our experiments.

The second advantage of GM is related to community detection in data sets of multi-scale communities. As explained about the resolution limit of modularity, it is related to the assumption/interpretation of vertex similarity in modularity. Modularity expects two similar vertices to share an edge while this is not reasonable, in the sense that, in large networks each vertex cannot know about all other vertices of the network. Although one cannot expect vertices to be able to directly communicate with all other members of the network, it is more sensible to expect them to be able to share a neighbour, or even more realistic, to expect them to have a longer path to other members of the network. This idea is powerful when it comes to small-world networks which are discussed in introduction and proved to have a small diameter [19]. Even in large networks with this property, although each vertex cannot communicate directly to all others, it is related to all other vertices with comparatively very short paths. This fact supports the more realistic underlying assumption in the definition of generalized modularity measure. So that in data sets with different community sizes in particular those of small-world networks, GM can achieve higher performance than modularity. However, GM is not completely free of resolution limit problems. Because GM is a global optimization criterion which considers the whole network for defining communities and resolution limit seems to be a general problem for all methods with a global optimization goal [7].

The third advantage of our generalized modularity is discussed in Section 2.2 which introduces a unified version of the generalized modularity measure.

2.2   Unified Generalized Modularity (UGM)

As explained in the introduction, modularity maximization cannot detect communities of bipartite networks. However, a specialised version of GM (in equation (9)) using the difference between the number of common neighbours and the expected such numbers in a randomized network, can detect communities in uni-partite, near-bipartite and bipartite networks without pre-knowledge of the network type. In equation (9), $W_{ij}^{(2)}$ is same as the term defined in equation (5).

$$Q_{UGM} = \frac{1}{||A^2||} \sum_{i,j \in V} W_{i,j}^{(2)} \delta(C_i, C_j) \tag{9}$$

In unipartite models, the basic community detection principle is "edges inside a community are dense and outside are sparse." Consider a partition of a unipartite network (Figure 4-a) detected by maximizing Q_{UGM}. As explained before, the elements of $W^{(2)}$ are higher for pairs of vertices who have more common neighbours than what is expected in a randomized network. As a general property of unipartite networks, vertices have neighbours of the same community. Hence, cluster members detected by Q_{UGM} have common neighbours which lie within the same community. This means density of connections inside communities is much more than the edge density between communities. Therefore, the Q_{UGM} criteria is completely aligned with properties of communities in unipartite networks.

However, in bipartite (near-bipartite) networks, all (most) common neighbours of members of the same cluster are definitely (probably) located in the opposite community. In these networks, the basic community detection principle is "edges inside communities are sparse and outside are dense."

Consider a partition of a bipartite or near-bipartite network (Figure 4-b) which is achieved by maximizing Q_{UGM} without any pre-knowledge about the type of the network. Q_{UGM} maximization, assigns vertices with more common neighbours— than the expected number in a randomized network— to the same community. This leads to high density of between-community edges because members of a community share neighbours which belong to the other community. Hence, Q_{UGM} maximization in bipartite and near-bipartite networks finds communities with sparse inter-community connections and dense between-community links. Therefore, the unified generalized modularity (UGM) measure is able to detect communities in unipartite, near-bipartite and bipartite networks without pre-knowledge of the network's structure.

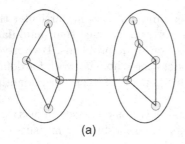 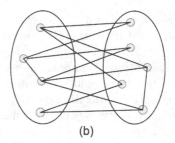

(a)   (b)

Fig. 4. Example of a unipartite network (left side) and a near-bipartite network (right side)

2.3   Finding Communities Based on the GM Quality Function

Similar to modularity maximisation, finding a partition with maximum generalized modularity is also an NP-hard problem. However, as GM can be represented as a matrix (similar to the modularity matrix), heuristic and exact algorithms of modularity maximization can be reused for partitioning data based on GM.

In this paper, we use an agglomerative community detection algorithm similar to one of Blondel *et al* in [8] which is also discussed in the introduction section. This algorithm considers each vertex as a community initially and then merges these small communities in a way that increases the GM value of the partition. It then updates the network information based on new communities and starts the next iteration and continues until no further improvement is possible.

3   Experiments

In this section, we present empirical analysis of generalized modularity and compare it with some state of the art approaches in the literature. All experiments are done on a PC with core i7 CPU 3.40 GHz and 16GB RAM.

Data Sets: In order to present a comprehensive comparison, we used four different categories of data sets which are common in the literature.

- We used LFR data sets proposed in [21]. In LFR data sets, degrees follow a power-law distribution $p(d) = d^{-\alpha}$ with parameter α and the community size a power-law distribution with parameter β. A mixing parameter, μ is the proportion of external degree for each vertex. Based on the original LFR data set in [21] we fixed α and β to be 2 and 1 respectively.
- To address the resolution limit of modularity, there are some structures of networks proposed in [13] where modularity fails to detect the underlying communities correctly. Similarly, we used four synthetic data sets with structures of Figure 5. In this figure, each circle represents a clique or complete graph which is denoted by k. For instance, k_{50} is a complete graph or clique of 50 vertices. Figure 5-d shows a circle of 30 cliques of size 5 [13].

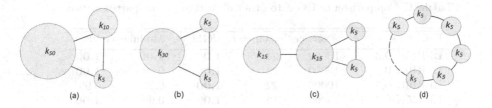

Fig. 5. Synthetic data sets for testing the resolution limit

- Four real-world data sets including Zachary Karate Club [22], Books about
  US Politics, American College Football [2] and Sampson's monastery data
  set [23] were selected. These data sets were chosen because their ground
  truth tags are known and we can measure performance by comparing the
  results to the ground truth.
- We also used the South Women data set [5] as a real bipartite network. We
  also generated random near-bipartite networks for further experiments.

Comparison Measure: Since we have the real ground truth of the data sets,
for evaluating quality of partitioning, we use the Normalized Mutual Information
of equation (10) which is proposed by Danon *et al* [24].

$$I_{norm}(A,B) = \frac{-2\sum_{i=1}^{CA}\sum_{j=1}^{CB} N_{ij}\log(N_{ij}N/N_{i.}N_{.j})}{\sum_{i=1}^{CA} N_{i.}\log(N_{i.}/N) + \sum_{j=1}^{CB} N_{.j}\log(N_{.j}/N)} \tag{10}$$

In equation (10), A represents the real communities and B represents the
detected communities while CA and CB are the number of communities in
A and B respectively. In this formula, N is the confusion matrix with rows
representing the original communities and columns representing the detected
communities. The value of N_{ij} is the number of common vertices that are in the
original community i but found in community j. The sum over the ith row is
denoted by $N_{i.}$ and the sum over the jth column is denoted by $N_{.j}$

In the rest of this section, first, we examine the unified version of generalized
modularity. Then we discuss the choice of model parameters α_l based on a set of
training experiments. Finally, we report the comparison with other approaches.

3.1 Testing Unified Generalized Modularity

We tested our unified generalized modularity of equation (9) on several real and
artificial unipartite, bipartite and near-bipartite networks to evaluate its performance. We compared the proposed UGM model with modularity and VSP
model of Li and Pang [17]. We chose the VSP model since it is one of the few
unified community detection algorithms and is expected to detect communities
without knowing the network structure type. For the sake of consistency, we
used the same algorithm (greedy algorithm of Blondel *et al* [8]) to maximize the
three examined measures. Information of the data sets are presented in three

Table 1. Comparison of UGM to other algorithms on unipartite networks

Data sets	#vertices	#cluster	UGM	Modularity	VSP
LFR10K-0.3	10000	24	**1.00**	**1.00**	**1.00**
LFR10K-0.4	10000	23	**1.00**	0.98	**1.00**
LFR10K-0.5	10000	22	**1.00**	0.97	0.99
LFR15K-0.3	15000	19	**1.00**	0.99	**1.00**
LFR15K-0.4	15000	20	**1.00**	0.99	**1.00**
LFR15K-0.5	15000	19	**1.00**	0.92	0.99
Karate Club	34	2	**0.83**	0.64	0.12
PolBooks	105	3	**0.54**	**0.54**	**0.54**
Football	115	12	0.17	**0.20**	0.16
Samson T4	18	4	**0.64**	0.59	**0.64**
Samson T1-T5	25	2	0.60	0.57	**0.62**
Figure 5-a	65	3	**1.00**	0.88	0.79
Figure 5-b	40	3	**1.00**	0.87	0.63
Figure 5-c	40	4	**0.93**	**0.93**	**0.93**
Figure 5-d	150	30	0.86	**0.89**	0.86
p-Value			baseline	0.0209	0.0588

first columns of Table 1. The normalized mutual information index achieved by UGM, modularity and VSP are shown in the remaining columns respectively. The real-world and artificial data sets were introduced earlier. We used mixing parameter 0.3–0.5 and generated large LFR data sets with average degree of 30 and maximum degree of 70. In Table 1, the LFR data sets are named based on their size and mixing parameter. The Samson data set represents affect relations among the novices in a New England monastery which were measured at five moments in time. The first Samson data set in Table 1 is just based on measurements on the fourth moment and the second data set is based on all measurements at five moments. Based on results of Table 1, UGM outperforms modularity and the VSP method in most data sets. Friedman statistical test results are also reported with null hypothesis of no difference in performance.

We also tested UGM on the bipartite network of Southern Women [5] who participated in social events. The proposed UGM measure 100% correctly detects the two groups in this bipartite network without knowing the structural type in advance.

For further comparison, we also generated near-bipartite networks. In randomly generated near-bipartite networks, each vertex shares an edge with a (randomly chosen) member of the same community with the probability P_{in} and the minimum degree of vertices is chosen uniformly from range of 1 and corresponding community size. In Figure 6, performance of GM, modularity and VSP are analysed over the change in P_{in} parameter. By increasing P_{in}, the data set becomes less and less bipartite as the percentage of inter-community edges increases. Figure 6 illustrates that the high performance of GM is maintained on near-bipartite networks up until they become close to unipartite.

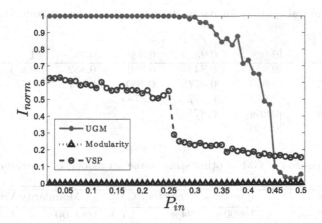

Fig. 6. Sensitivity analysis of methods on randomly generated near-bipartite network which has communities of size 500 and 300.

As explained earlier, when not knowing the type of data in advance, modularity maximization fails to detect communities and performs poorly on bipartite or near-bipartite data sets.

3.2 Training Parameters of Generalized Modularity

According to the definition of GM in equation (3), parameters α_1, α_2 and α_3 determine the importance of each term to the generalized modularity. We can tune these parameters based on use of training data.

For training the parameters, we chose the LFR benchmark data [21] because we can generate it in different sizes and features and it properly simulates real world [7]. Similar to Lancichinetti *et al* [21], we used LFR data sets of size 1000 while the size of communities is between 20 and 100 and the average degree is 20 and the maximum degree is set to be 50. The mixing parameter ranges from 0.1 to 0.5 and we also used a LFR with mixing parameter 0.7 in which communities are not well defined and community detection seems to be more challenging.

In the generalized modularity matrix of equation (3), without loss of generality, we assumed α_1, α_2 and α_3 to be between 0 and 1 and α_3 to be equal to $1 - \alpha_1 - \alpha_2$. We considered 5 levels for each parameter α_1 and α_2 and examined all 15 unique combinations of three parameters of our GM measure. Tabel 2 presents the average I_{norm} over all training data sets for each combination. Based on results reported in Table 2, all combinations of GM which include $W_{norm}^{(1)}$ ($\alpha_1 > 0$), on average, perform better than modularity. It shows that presence of $W_{norm}^{(1)}$ is essential but also using $W_{norm}^{(2)}$ and $W_{norm}^{(3)}$ improves the results. In Table 2 it is shown that the combination of $0.25W_{norm}^{(1)} + 0.5W_{norm}^{(2)} + 0.25W_{norm}^{(3)}$ has the best performance on average over our train data sets. Therefore, we use this combination in subsequent experiments for comparison with other methods.

Table 2. Empirical training for parameter configuration for generalized modularity

	α_2				
	0	0.25	0.5	0.75	1
0	0.831	0.844	0.850	0.852	0.861
0.25	0.866	0.877	**0.881**	0.877	
α_1 0.5	0.873	0.877	0.878		
0.75	0.869	0.874			
1	0.866				

Table 3. Comparison of GM to other state of the art models of community detection

Data sets	#vertices	#clusters	GM	Modularity	VSP
LFR10K-0.3	10000	24	**1.00**	**1.00**	**1.00**
LFR10K-0.4	10000	23	**1.00**	0.98	**1.00**
LFR10K-0.5	10000	22	**1.00**	0.97	0.99
LFR15K-0.3	15000	19	**1.00**	0.99	**1.00**
LFR15K-0.4	15000	20	**1.00**	0.99	**1.00**
LFR15K-0.5	15000	19	**1.00**	0.92	0.99
Karate Club	34	2	**1.00**	0.64	0.12
PolBooks	105	3	**0.56**	0.54	0.54
Football	115	12	**0.20**	**0.20**	0.16
Samson T4	18	4	**0.69**	0.59	0.64
Samson T1-T5	25	2	0.60	0.57	**0.62**
Figure 5-a	65	3	**1.00**	0.88	0.79
Figure 5-b	40	3	**1.00**	0.87	0.63
Figure 5-c	40	4	**0.93**	**0.93**	**0.93**
Figure 5-d	150	30	0.86	**0.89**	0.86
p-Value			baseline	0.0039	0.0196

3.3   Comparison with Other Methods

In this section, we compare our trained GM community detection model with some state of the art models in the literature. We compare GM with the modularity based algorithm of Blondel *et al* [8][1] and the vertex similarity probability (VSP) model of Li and Pang [17]. In order to be consistent in the experiments, we used the same algorithm of Blondel *et al* [8] for optimizing VSP and GM models. Table 3 reports average I_{norm} value of 10 independent runs.

Table 3 demonstrates that GM performs much better than the other methods over different data sets. It performs very strongly in large data sets. Besides, it detects communities of real world networks more precisely. Note that our results for the VSP model do not exactly match with experiments reported in [17], possibly due to differences in optimization procedure (which wasn't described in

---

[1] We also compared our method with modularity-based algorithms of Danon [24] and Newman [11] but as method of Blondel *et al* [8] outperforms the other two, we just report Blondel *et al* [8] here.

that paper). GM also detects small communities in data sets of Figures 5-a and 5-b. The reason that GM couldn't outperform modularity in the data set of Figure 5-d is because this data set has a large diameter in comparison to size of the network which shows its structure is very different from small-world networks. The result of a pairwise Friedman statistical test is also reported at the bottom of Table 3. The null hypothesis of this test is two algorithms have no significant difference in their performance. This hypothesis is rejected based on the very small p-values, indicating statistically significant differences in performance.

4 Conclusion

We have proposed a generalized modularity criterion (named GM) for community detection in complex networks. Generalized modularity extends the interpretation of modularity by taking into account paths between vertices rather than just common edges. The modelling of existence of paths between vertices enables GM to deliver better performance especially in small-world networks with different community sizes and it can also work on bipartite and near-bipartite networks. Although GM improves the resolution limit of modularity especially in small-world networks, still it can have this problem and approaches to solve it are a clear direction for future work.

Acknowledgments. NICTA is funded by the Australian Government through the Department of Communications and the Australian research Council through the ICT center of excellence program. James Baileys work is supported by an ARC Future Fellowship (FT110100112).

References

1. Alfalahi, K., Atif, Y., Harous, S.: Community detection in social networks through similarity virtual networks. In: Proceedings of the 2013 IEEE/ACM International Conference on Advances in Social Networks Analysis and Mining, pp. 1116–1123. ACM (2013)
2. Girvan, M., Newman, M.E.: Community structure in social and biological networks. Proceedings of the National Academy of Sciences 99(12), 7821–7826 (2002)
3. Adomavicius, G., Tuzhilin, A.: Toward the next generation of recommender systems: A survey of the state-of-the-art and possible extensions. IEEE Transactions on Knowledge and Data Engineering 17(6), 734–749 (2005)
4. Barber, M.J., Clark, J.W.: Detecting network communities by propagating labels under constraints. Physical Review E 80(2), 026129 (2009)
5. Davis, A., Gardner, B.B., Gardner, M.R.: Deep south. University of Chicago Press (1969)
6. Newman, M.E., Girvan, M.: Finding and evaluating community structure in networks. Physical Review E 69(2), 026113 (2004)
7. Fortunato, S.: Community detection in graphs. Physics Reports 486(3), 75–174 (2010)

8. Blondel, V.D., Guillaume, J.-L., Lambiotte, R., Lefebvre, E.: Fast unfolding of communities in large networks. Journal of Statistical Mechanics: Theory and Experiment **2008**(10), P10008 (2008)
9. Newman, M.E.: Finding community structure in networks using the eigenvectors of matrices. Physical Review E **74**(3), 036104 (2006)
10. Aloise, D., Cafieri, S., Caporossi, G., Hansen, P., Perron, S., Liberti, L.: Column generation algorithms for exact modularity maximization in networks. Physical Review E **82**(4), 046112 (2010)
11. Newman, M.E.: Fast algorithm for detecting community structure in networks. Physical Review E **69**(6), 066133 (2004)
12. Brandes, U., Delling, D., Gaertler, M., Gorke, R., Hoefer, M., Nikoloski, Z., Wagner, D.: On modularity clustering. IEEE Transactions on Knowledge and Data Engineering **20**(2), 172–188 (2008)
13. Fortunato, S., Barthelemy, M.: Resolution limit in community detection. Proceedings of the National Academy of Sciences **104**(1), 36–41 (2007)
14. Liben-Nowell, D., Kleinberg, J.: The link-prediction problem for social networks. Journal of the American Society for Information Science and Technology **58**(7), 1019–1031 (2007)
15. Blundo, C., De Cristofaro, E., Gasti, P.: EsPRESSo: efficient privacy-preserving evaluation of sample set similarity. In: Di Pietro, R., Herranz, J., Damiani, E., State, R. (eds.) DPM 2012 and SETOP 2012. LNCS, vol. 7731, pp. 89–103. Springer, Heidelberg (2013)
16. Leicht, E., Holme, P., Newman, M.E.: Vertex similarity in networks. Physical Review E **73**(2), 026120 (2006)
17. Li, K., Pang, Y.: A unified community detection algorithm in complex network. Neurocomputing **130**, 36–43 (2014)
18. Arenas, A., Fernandez, A., Gomez, S.: Analysis of the structure of complex networks at different resolution levels. New Journal of Physics **10**(5), 053039 (2008)
19. Watts, D.J., Strogatz, S.H.: Collective dynamics of small-world networks. Nature **393**(6684), 440–442 (1998)
20. Bezdek, J.C., Hathaway, R.J.: Vat: a tool for visual assessment of (cluster) tendency. In: Proceedings of the International Joint Conference on Neural Networks, vol. 3, pp. 2225–2230. IEEE (2002)
21. Lancichinetti, A., Fortunato, S., Radicchi, F.: Benchmark graphs for testing community detection algorithms. Physical Review E **78**(4) (2008)
22. Zachary, W.W.: An information flow model for conflict and fission in small groups. Journal of Anthropological Research 452–473 (1977)
23. Sampson, S.F.: A novitiate in a period of change: An experimental and case study of social relationships. Ph.D. dissertation, Cornell University (September 1968)
24. Leon Danon, A.D.-G., Arenas, A.: The effect of size heterogeneity on community identification in complex networks. Journal of Statistical Mechanics: Theory and Experiment P11010 (2006)

Handling Oversampling in Dynamic Networks Using Link Prediction

Benjamin Fish[1,2](✉) and Rajmonda S. Caceres[2]

[1] University of Illinois at Chicago, Chicago, IL, USA
bfish3@uic.edu
[2] MIT Lincoln Laboratory, Lexington, MA, USA

Abstract. Oversampling is a common characteristic of data representing dynamic networks. It introduces noise into representations of dynamic networks, but there has been little work so far to compensate for it. Oversampling can affect the quality of many important algorithmic problems on dynamic networks, including link prediction. Link prediction seeks to predict edges that will be added to the network given previous snapshots. We show that not only does oversampling affect the quality of link prediction, but that we can use link prediction to recover from the effects of oversampling. We also introduce a novel generative model of noise in dynamic networks that represents oversampling. We demonstrate the results of our approach on both synthetic and real-world data.

1 Introduction

Networks have become an indispensable data abstraction that captures the nature of a diverse list of complex systems, such as online social interactions and protein interactions. All these systems are inherently dynamic and change over time. A common abstraction for incorporating time has been the "dynamic network," a time series of graphs, each graph representing an aggregation of a discrete time interval of the observed interactions. While in many cases the system under observation naturally suggests the size of such a time interval, it is more often the case that the aggregation is arbitrary and is done for the convenience of the data representation and analysis. However, an abundance of literature has demonstrated that the choice of the time interval at which the network is aggregated has great implications on the structures observed and inferences made [7,14,21,24].

1.1 Oversampling

We view the system through the filter of the data we collect. This data is typically collected opportunistically, with the temporal rate of data not always matching

---

This work is sponsored by the Assistant Secretary of Defense for Research & Engineering under Air Force Contract FA8721-05-C-0002. Opinions, interpretations, conclusions and recommendations are those of the authors and are not necessarily endorsed by the United States Government.

© Springer International Publishing Switzerland 2015
A. Appice et al. (Eds.): ECML PKDD 2015, Part II, LNAI 9285, pp. 671–686, 2015.
DOI: 10.1007/978-3-319-23525-7_41

that of the system. With the advent of microelectronic data collection systems such as GPS and RFID sensors, it is often the case that data is sampled at orders of magnitude more frequently than the temporal scale of the underlying system. Therefore, it is important that the aggregation process that transforms the collected data into a dynamic network representation correctly accounts for the oversampling effects.

Oversampling is an aspect of the data collection process that can help with the issue of representing continuous time discretely. It helps reduce the number of missing interactions and allows us to better identify persistent interactions. On the other hand, oversampling affects our ability to distinguish between noisy local temporal orderings and critical temporal orderings. For example, when analyzing email communication networks, the data is collected at a resolution of seconds, but the causality of email interactions and the emergence of complex structures such as communities is often more accurately represented and detectable at much coarser scales.

The concept of oversampling has been studied extensively in the signal processing community. However, to the best of our knowledge, there is no natural, well-defined notion of oversampling that translates directly to the domain of graph sequences. So here we use oversampling to mean a *distribution over graph sequences* that displays the typical effects of sampling too frequently: noisy local temporal orderings, spurious interactions, etc.

In this paper, we assume we are given a noisy dynamic network that has been observed at a fixed oversampled rate. Our goal is to recover from this oversampling by aggregating the network together in such a way as to remove any artifactual temporal orderings, while preserving any real temporal information such as edge co-occurrences and critical temporal orderings. We represent a dynamic network as a sequence of discrete snapshots, that is, as a sequence of graphs. Given a window size w, we bin together into a graph all edges that occur within each length-w time span. This results in a new dynamic network. The goal becomes to find the window size w that best recovers from the noise. At one extreme, if this window size w is the entire length of the original dynamic network, then this means that there is no temporal information and the network is static. At the other extreme, if this window size w lasts for a single snapshot of the original dynamic network, then this means there is no oversampling at all and the network is observed at the right temporal scale. Thus, the window size can be seen as a proxy for the amount of temporal information stored in the dynamic network, and finding a good window size can be seen as finding the temporal scale at which the network is evolving.

1.2   Link Prediction

In this work, we take the approach that the inference task should inform how to recover from oversampling, whether the task is link prediction, community detection, or something else. Link prediction, in particular, is an important inference task with many applications. It has beeen used in the analysis of the internet [1,27], social networks [2], and biological networks [3,11]. It has also been used for designing recommendation systems [16,18] and classification systems [12]. See the survey of Al Hasan and Zaki [4] for more applications and an introduction to the various techniques used in link prediction.

Given the importance of link prediction, we investigate whether this task can serve as a good driver for recovering the correct temporal scale of an oversampled dynamic network. Our approach is simple: we use the quality of a link prediction algorithm as a score for the window size. If we can predict links better, it means that we have found a good window size, not only because by definition we have improved the performance of our inference task, but because at this scale we can better capture the evolution of the network.

To our knowledge, we are the first to formally study the relationship between temporal oversampling in networks and link prediction. As such, we need to introduce a model of noise for dynamic networks that captures effects of oversampling. We define such a model by leveraging the following observation: when a dynamic network is oversampled, edge occurrences are recorded near the "natural" occurrence time, but are spread out around that time. We present a model that generates noisy, oversampled dynamic networks by distributing the times of edges over a Gaussian distribution. This model can be extended to capture unique characteristics of a given data collection process. It also is applied independently of the underlying process generating the network, allowing for a wide variety of phenomena to be modeled.

In this paper, we show that oversampling affects the quality of link prediction algorithms. Furthermore, using synthetic data, we show that link prediction performs better on graph sequences aggregated near the ground-truth window size than on other window sizes. In this light, we can recover from the effects of oversampling by using link prediction to find a good window size. Finally, we show that our method results in robust results on real-world networks.

In Section 2, we give a more formal description of the oversampling problem for dynamic networks and present our link prediction based approach. We define a generative model for Gaussian-type noise representing oversampling in Section 3, as well as a generative model for dynamic networks for testing link prediction algorithms. In Section 4, we use our synthetic generative model to test our approach. We demonstrate that our method yields reasonable results and show the impact of a variety of different parameters on our results. We then show the results of our approach on two real-world networks in Section 5. We end with a few concluding remarks in Section 6.

1.3   Related Work

Extensive literature has demonstrated that the choice of aggregation window greatly impacts the quality of the corresponding dynamic network [7,21,24]. Some work has been done in developing heuristics for identifying the "right" window size or temporal partitioning, especially for numerical time series [15,26]. In dynamic networks, though, most work does so only in limited or slightly different contexts. Peel and Clauset, for example, consider the problem of finding change points, points at which the generative process of a dynamic network is itself changing [19]. Sun et al. also considers the problem of finding change points - and more generally considers the problem of finding a partition of the network - this time in the context of community detection [25]. In [9] and [7], they analyze the discrete Fourier transform of time series of different graph metrics to identify important frequencies in a dynamic network. Similarly, Sulo et al. [24] use information-theoretic tools to analyze time series of a variety of different graph metrics for graph sequences that have been aggregated at different temporal scales.

Prediction as a tool to inform model selection is not new, especially in the literature of time series analysis. Central in this literature is the principle of minimum description length, which says that if data displays any regularity (i.e. is predictable), then that regularity can be used to shorten an encoding of the data. Therefore finding a short encoding is finding a good predictor, and vice versa. See [13] for a survey on the subject. In the context of temporal networks, this approach is not as common, but link prediction has been used to infer useful static networks from data [8].

There are a number of models in the literature for dynamic networks, including the GHRG model of Clauset and Peel [19], the activity-driven model of Perra et al. [20], and the dynamic latent-space model of Sarkar and Moore [22]. Some models for static networks can also naturally be seen as a model for dynamic networks in which the number of nodes is growing over time, such as in the preferential attachment model. However, none of these models are well-suited for modeling oversampling or for controlling the quality of a link prediction algorithm as we will need, so we use a novel model, which we detail in Section 3.

2   Problem Formulation and Methodology

To formalize the problem, we assume that a noisy dynamic network is a sequence of discrete graphs (each graph representing one time step) on a fixed set of n nodes. Furthermore, this graph sequence is the output of some noise process being applied to a (possibly shorter) sequence of graphs - the 'ground truth' - on the same set of n nodes. Specifically, we assume that the noisy sequence can be partitioned into 'windows' of fixed size, and that any temporal orderings between edges in a single window are spurious. Thus recovery consists of the two step process of first choosing a window size w (which in this paper we will also refer to as a *temporal scale* of the graph sequence) and then aggregating all edges within each w consecutive graphs into a single graph. This results in a new graph sequence that is a factor of w shorter. Figure 1 gives an illustration of this process.

In general, the goal is to recover the ground truth from the noisy sequence. Of course, this is not always possible. In the degenerate case, the ground-truth

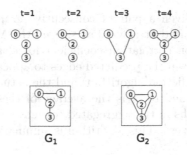

Fig. 1. Example of aggregating a time series of graphs into a coarser window $w = 2$.

sequence is just noise and this task is impossible. In this paper, we assume the ground-truth sequence is sufficiently non-noisy, so that a link prediction algorithm will perform well on it. Link prediction is the classification task of finding those pairs of vertices (where there is not already an edge) which are most likely to form an edge in the next time step. Any similarity score on pairs of vertices can naturally be considered a link prediction algorithm: two vertices with a high similarity score are assumed to be more likely to have a link in the future. Liben-Nowell and Kleinberg [17] give a detailed list of such similarity scores. We will also need similarity scores to create ground-truth synthetic data, as detailed in Section 3. Let $score(x, y)$ denote the similarity score between two vertices and $\Gamma(x)$ the neighborhood of x. With this notation, we use the following four scores, as given in [17]:

1. Adamic-Adar: $score(x, y) := \sum_{z \in \Gamma(x) \cap \Gamma(y)} \frac{1}{\log |\Gamma(z)|}$
2. Katz$_\beta$: $score(x, y) := \sum_{\ell=1}^{\infty} \beta^\ell \cdot |\{\text{paths of exactly length } \ell \text{ from } x \text{ to } y\}|$
3. Graph distance: $score(x, y) := -d(x, y)$, where $d(x, y)$ is the distance between x and y.
4. Rooted PageRank$_\alpha$: Consider a random walk on the graph that resets to vertex u with probability α and moves to a random adjacent vertex with probability $1 - \alpha$. Then $score(x, y)$ is the stationary probability of y when resetting to x plus the stationary probability of x when resetting to y.

To test the quality of a window size w, we analyze the performance of a link prediction algorithm on the graph sequence aggregated at w. The link prediction algorithm gets as input the graph at time t and predicts new edges in the graph at time $t+1$. We predict the edges with the top k scores, where k is the number of edges that actually are created in the next time step in the sequence. As is common, we view this as a binary classification task, where pairs of vertices are either in the category of new edges or not. When viewed in this way, it is natural to score the algorithm as the correlation between the algorithm's predicted edges and those edges that actually appear, which is called the Matthews correlation coefficient[1](MCC) [5]. Given a pair of consecutive graphs G_i and G_{i+1} from a graph sequence aggregated at a window size w, the Matthews correlation is the $[-1, 1]$-valued Pearson correlation coefficient for classification defined as a normed χ^2 statistic between the predicted edges to appear in G_{i+1} (using G_i as the input to the link prediction algorithm) and the actual new edges appearing in G_{i+1}. The score assigned to w is the average of these correlations over all pairs of consecutive graphs in the aggregated sequence.

We test all window sizes[2]. This algorithm is summarized in Algorithm 1.

Algorithm 1. Assign scores to each window size

$\mathcal{G} = G_1, \ldots, G_n$
for $w = 1$ to $\lfloor n/3 \rfloor$ **do**
    Let $\mathcal{G}'$ be $\mathcal{G}$ aggregated at size w, so $\mathcal{G}' = G'_1, \ldots, G'_{\lfloor n/w \rfloor}$.
    for $i = 0$ to $|\mathcal{G}'| - 1$ **do**
        new links $= E(G'_{i+1}) \setminus E(G'_i)$
        predicted links $= LP(G'_i, G'_{i+1})$
        $score_i = MCC(\text{new links}, \text{predicted links})$
    end for
    $score_w = \frac{\sum_{i=0}^{|\mathcal{G}'|-1} score_i}{|\mathcal{G}'|}$
end for
return all pairs $w, score_w$

Since this is a computationally-expensive task, for sufficiently long sequences, a random ten percent of the consecutive pairs of aggregated graphs are tested instead of all of the pairs. Based on our empirical analysis, considering only a subsample of the windows does not significantly affect the scores.

Depending on the application, it may be desirable to use more than one different window size - more than one time scale may be interesting or perform highly. In addition, for a given time scale, there may be a range of window sizes centering around that time scale that is of interest. Moreover, we leave for future work determining how significant differences in quality scores are, given that we make no assumptions on how many different time scales are of interest for the application. In this light we do not return just the top-quality window size, as seen in Algorithm 1.

---

[1] The MCC measure is used, rather than accuracy or precision, because MCC skirts the issue of bias that accuracy and precision have: the number of edges appearing is often a very small fraction of the total number of possible edges. We use MCC over other measures also resistant to unequally sized categories, such as AUC, because MCC is computationally very fast. In addition, it seems to emphasize differences in scores better than other measures for our link prediction task. Regardless, other measures give very similar scores as AUC - different measures seem to preserve the order of the qualities of the window sizes.

[2] To be more precise, we only test window sizes up to a third of the length of the input. We assume that if the actual window size is any bigger, then there is no temporal information that we can utilize and the underlying network is really a static network.

3   Generative Models for Graph Sequences

To create synthetic data, we will use two generative models, one for the ground-truth sequence representing the noiseless dynamic network, and the other, a noise model that takes as input a ground-truth sequence (which we sometimes refer to as the underlying sequence) and outputs a noisy oversampled sequence.

3.1   Generative Model for the Ground-Truth Graph Sequence

The ground-truth graph sequence can be formed from any existing model of a dynamic network (such as the latent-space model or the activity-driven model mentioned in the introduction), but to test the performance of a link-prediction-based approach, we instead use a novel and simple generative process that allows us to test our approach, which has the advantage that the quality of the link prediction algorithm is a parameter of the generative process. This generative process starts with an initial graph G and adds a fixed but parameterized number of edges δ for every subsequent graph. The edges added are the non-edges with the top scores as rated by a given similarity score. This model can easily be extended as needed, for example by deleting the edges with the lowest similarity score every time step as well, or still further to a probabilistic edge creation and deletion process.

For the initial graph G, in this paper we consider both the Erdős-Rényi model $G(n, p)$ [10] and the preferential attachment model $BA(n, m)$ [6]. For the similarity scores, we use Adamic-Adar and Katz. The use of different similarity scores allows us to test our approach both when the quality of the link prediction algorithm does well and when it does not do well. For example, if we use Adamic-Adar to both create the graph sequence and to do link prediction (assuming no noise) the link prediction algorithm will perform perfectly. However if the sequence is instead made with the Katz similarity score, the link prediction algorithm will not perform as well. Since we can't guarantee the quality of the link prediction algorithm on non-synthetic data, this model allows us to see how link prediction performance affects our approach.

3.2   Generative Model for Oversampling

We now need to model the sampling process by which non-synthetic data would be gathered. Our primary approach to modeling noise, specifically oversampling, is to assume that for a given time step, the edges that occur are measured to be near that time step, but not necessarily at that time step. Furthermore, we assume the distribution of these edges in time is Gaussian. Given an input graph sequence of length t and parameters $\mu \in \mathbb{N}, \sigma^2 \in \mathbb{R}_{\geq 0}$, this model outputs a graph sequence of length approximately $\mu \cdot t$ that represents the sequence being oversampled at a constant rate μ, with σ^2 controlling how concentrated the edges occur around the "true" times. Specifically, given these two parameters μ and σ, for an edge that occurs in the ground-truth graph at time step i, the edge will occur in the new noisy graph sequence at time step $j \sim \mathcal{N}(\mu i, \sigma^2)$, where j is rounded to the nearest integer.

If σ is sufficiently small, then there is likely to be intermediate graphs where there are no edges. If windows start and end within these gaps, we can recover fully from this noisy process. However, as σ gets larger, it becomes more and more difficult, as more and more edges from distinct graphs in the original network start getting added to the same graphs in the new noisy network. In the limit, all temporal information is destroyed. Figure 3 gives two examples of the number of edges in each time step for two different settings of parameters. It's worth noting that the oversampling noise model presented here can be extended so that the oversampling rate is non-constant, the distribution used is non-Gaussian, etc.

4   Results for Synthetic Data

In this section, we analyze a variety of oversampled, dynamic networks generated by models detailed in Section 3 and aggregated at different window sizes. We investigate the effect of window size on the performance of link prediction. We show that link prediction performs better on sequences aggregated at window sizes close to ground-truth than on other window sizes. Furthermore, this holds when the ground-truth sequence uses a different similarity score than the one used to perform link prediction. We also show that larger values of μ, while fixing σ (increasing the separation between means), smaller values of σ, while fixing μ (higher degree of concentration around each mean), larger values of δ (more edges appearing each time step), and higher quality of link prediction all make the task of recovering the ground-truth window size easier. Yet even at higher levels of noise we show that this approach recovers reasonable results and gives some evidence that it outperforms the simpler approaches that rely solely on extracted time series information, such as the number of edges per time step.

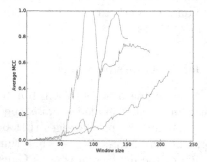

Fig. 2. MCC scores as a function of window size for three different noisy sequences. Parameters used were $\mu = 100$, $\delta = 50$, and from shortest to longest, $\sigma = 8, 20, 40$. The underlying graph sequence is generated by starting graph $G(n, p)$ and Adamic-Adar as the similarity score.

For the sake of brevity, given the number of parameters, including μ, σ, δ, the number of link prediction algorithms both for creating the sequence and for finding the quality scores, we do not show results for all possible combinations, but instead show a representative sample.

In the remainder of this section, we fix the number of vertices n to be 250 and fix the edge probability p for the Erdős-Rényi model at 0.05. Figure 2 shows the drastic improvement of link prediction performance at larger windows of aggregation. For the sequence generated using $\sigma = 8$, as shown in this figure, performance at window size of 1 (no aggregation) is essentially random (average MCC $\approx -10^{-4}$), but when aggregated at a window size of 95, performance is perfect (average MCC $= 1.0$). Note that the best performing window size is very close to the mean separation $\mu = 100$. When the standard deviation is comparatively higher, as when $\sigma = 40$, Figure 2 shows that windowing has a comparatively smaller effect. Here the link prediction is not able to separate consecutive graphs and therefore recommends to aggregate all graphs in the sequence together.

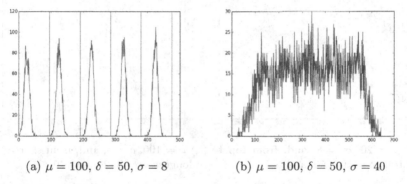

(a) $\mu = 100$, $\delta = 50$, $\sigma = 8$ (b) $\mu = 100$, $\delta = 50$, $\sigma = 40$

Fig. 3. Number of edges in two noisy sequences over time with the given parameters. The underlying sequences were created with $G(n, p)$ as the starting graph and Adamic-Adar as the similarity score. The vertical lines indicate the borders of windows when the window size is 95.

Figure 3 gives further evidence for why this is the case. When $\sigma = 40$, mixing (of edges) between graphs is much higher than in the case when $\sigma = 8$. In this latter case, there is in fact no mixing at all, which is why there are many window sizes where link prediction performs perfectly. Since link prediction here is so good - it will be perfect in the absence of noise - it represents an easier case. However, this case still demonstrates a convenient benefit to our approach: we can still perform well even when the target windowing is not uniform over the length of the input noisy sequence. That is, the noisy sequence may have shifted windows and as a result, one window could be smaller than the others. In our case, window size 95 performs better than a window size 100, indicating we can still find a *uniform* window size that performs well, even if it is not the same window size as the mean separation window.

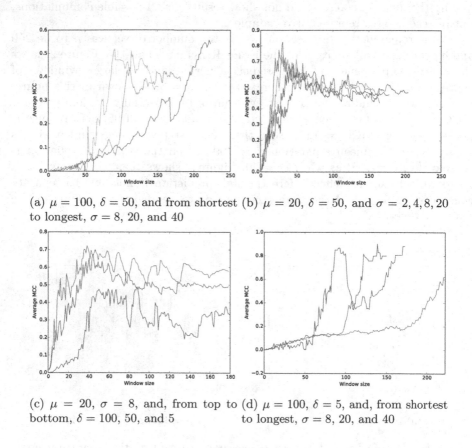

(a) $\mu = 100$, $\delta = 50$, and from shortest to longest, $\sigma = 8$, 20, and 40

(b) $\mu = 20$, $\delta = 50$, and $\sigma = 2, 4, 8, 20$

(c) $\mu = 20$, $\sigma = 8$, and, from top to bottom, $\delta = 100$, 50, and 5

(d) $\mu = 100$, $\delta = 5$, and, from shortest to longest, $\sigma = 8$, 20, and 40

Fig. 4. MCC scores as a function of window size for different noisy sequences where the Katz similarity score was used to generate the underlying sequence. In Figure 4d, the starting graph is the preferential attachment graph $BA(n, 5)$ instead of $G(n, p)$.

Figure 4a illustrates a similar behavior when the ground truth sequence was instead created with the Katz similarity score (testing of each window size is still done using the Adamic-Adar score as the link prediction algorithm). While overall quality scores are lower, as the link prediction algorithm itself is worse, the same pattern holds: near a window size of 100, the link prediction algorithm does significantly better.

Our approach is more resilient to a higher value of σ, as seen in Figure 4b. However, even when edge mixing is very high ($\sigma = 20$ and $\mu = 20$), link prediction is still helpful in identifying a good window size (near $\mu = 20$). This gives significant evidence that our approach outperforms simpler approaches that rely solely on extracted time series information such as the number of edges per time step. Finally, link prediction does better at recovering a good window size when δ is higher - it retains its resiliency even for very sparse cases, as seen in Figure 4c.

As mentioned above, our generative model easily extends to more general models. Again, for the sake of brevity, we will not discuss the performance of our approach on every possible variation of our model, but we do want to note that our results do extend. For example, Figure 4 shows our results when the following two changes have been made: The initial graph is instead of $G(n,p)$ an instance of the preferential attachment model. In addition, instead of just adding edges, edges are deleted as well. Namely, δ existing edges that have the lowest Katz score are deleted. These changes show very little impact on our results, as seen by comparing Figures 2 and 4a.

5   Results for Real-World Data

The validation process on real-world data is difficult, in general, for inference tasks on networks, but especially for the problem of temporal scale identification. A significant barrier is the lack of ground truth and/or formal notions of what should be considered a good temporal scale. Real-world dynamic networks often exhibit multiple critical temporal scales corresponding to the evolution of different important features (e.g. communities) and processes (e.g random walks) [21,24], but the relationship between these important features and processes and their corresponding temporal scales is not well understood.

Our expectation is that there can be multiple peaks in the quality of a link prediction algorithm as window size increases, each corresponding to different important temporal scales. We validate our results by making sure that different link prediction algorithms behave similarly as a function of window size, despite following different mechanisms for scoring future edges. When two different link prediction algorithms show peaks at the same time scale despite not necessarily predicting the same edges, this gives evidence that the peaks are inherit to the sequence and not a function of the particular algorithm. We now show how our approach performs on real-world data sets, namely the Haggle Infocom network [23] and the MIT Reality Mining network [9].

5.1   Haggle Infocom

The Haggle Infocom dataset is the result of 41 users equipped with Bluetooth phones at the Infocom 2005 conference, over the course of four days. There is an undirected edge between two users at time t if they were in proximity at that time. The data we used was initially binned at 10 minute time intervals. The results are shown in Figure 5. The four link prediction algorithms behave consistently with clear peaks at approximately $w = 75$ (≈ 12.5 hours), $w = 110$ (≈ 18 hours), and $w = 130$ (≈ 22 hours). The interactions present in the Haggle network have a periodic nature imposed by the regular conference structure and our algorithm seems to identify such periodicities, in particular the half and one-day periodicities.

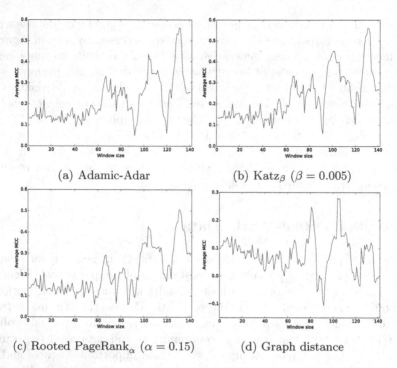

(a) Adamic-Adar

(b) Katz$_\beta$ ($\beta = 0.005$)

(c) Rooted PageRank$_\alpha$ ($\alpha = 0.15$)

(d) Graph distance

Fig. 5. MCC scores for four link prediction algorithms as a function of window size for the Haggle sequence.

5.2   MIT Reality Mining

The MIT Reality Mining dataset consists of 90 grad students and professors' data from their cell phones in the 2004-2005 academic year. Timestamps were kept for three types of data: Bluetooth proximity, cell tower proximity and phone call communications. This naturally yields three different dynamic networks that we extracted from the raw data. The first network has an undirected edge at time t whenever Bluetooth recorded two cell phones as close at time t, the second has an undirected edge between two participants at time t if they were recorded near the same cell tower, and finally the third has an undirected edge between two participants at time t whenever one participant called the other. We will refer to these as the Bluetooth sequence, the cell tower sequence, and the call sequence respectively. Each network is initially windowed at a size of one day.

The results are shown in Figure 6. Of interest is that the three different networks perform - in terms of the quality of the link prediction - consistently differently from each other, implying that they really are different types of networks. The Bluetooth sequence shows clear signs of oversampling; a window size of $w = 1$ has smaller quality than larger window sizes. The performance of link prediction on this sequence stabilizes after $w = 14$ (roughly 2 weeks), while in

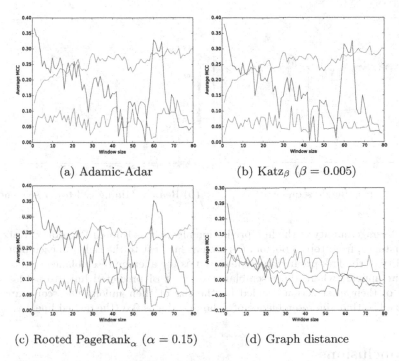

(a) Adamic-Adar   (b) Katz$_\beta$ ($\beta = 0.005$)

(c) Rooted PageRank$_\alpha$ ($\alpha = 0.15$)   (d) Graph distance

Fig. 6. MCC scores for four link prediction algorithms as a function of window size for the Reality Mining sequence. At the far left of each plot, from top to bottom is the cell tower sequence, the Bluetooth sequence, and the call sequence respectively.

the case of the cell tower sequence, the performance drops and then prominently improves at $w = 65$ (roughly 2 months). Considering the weekly, monthly, and semester structure of academic activities, these windows of aggregations appear reasonable in capturing the underlying dynamics of the Reality Mining networks.

These results are given weight by the agreement between three link prediction algorithms, Adamic-Adar, Katz, and Rooted PageRank, as seen in Figure 6. The exception is the graph distance algorithm, whose performance is significantly worse than the other three algorithms, ultimately making it difficult to discern the quality of different window sizes. The graph distance algorithm has been studied in the literature before and is identified as a very low performing link prediction algorithm [17]. Figure 7b shows the relative performance of the four link prediction algorithms for the Haggle and Reality Mining datasets and re-emphasizes the conclusions in [17]. In picking the graph distance algorithm for our problem, we wanted to investigate whether the performance of a bad prediction algorithm can be substantially improved by a better window of aggregation. As the analysis of the Haggle and Reality Mining datasets shows, while some loss in the quality of prediction can be overcome (as in the case of the Haggle sequence), sufficiently bad prediction cannot (as in the case of Reality Mining sequence), as should be expected.

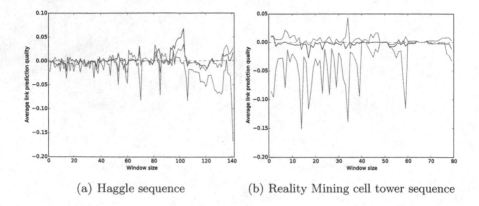

(a) Haggle sequence   (b) Reality Mining cell tower sequence

Fig. 7. Average quality of the link prediction algorithms Rooted PageRank, Katz, and graph distance, from top to bottom (as seen at furthest to the right) in each plot. Here the quality of the link prediction is measured as the average resemblance to the actual links that appear, where the resemblance of a set of predicted links to actual set is the size of their intersection divided by the size of their union. The scores are then normalized so that the resemblance of Adamic-Adar is 0.0 (the dotted line).

6   Conclusions

In this work, we treat link prediction as a signal that helps us understand the temporal dynamics of a network. Making the connection between the ability to predict new edges and the appropriate temporal scale that captures the evolution of the network, we present a novel, task-driven algorithm for de-noising oversampled dynamic network into more robust representations. We formally define a model of oversampling in dynamic networks that captures general properties of the noise it induces and allows for a more rigorous analysis of our algorithm. Across a variety of synthetic instances of noisy, oversampled dynamic networks, and two real dynamic networks, we show that the performance of a link prediction algorithm can serve as a good quality score for identifying the appropriate window of aggregation.

Our work opens up several potential avenues for further investigation. As mentioned above, we do not discuss in this paper how to select which windows are outliers in terms of link prediction quality. We would also be interested in extending the framework presented here to non-uniform partitions of the timeline of the network, instead of an inform window of aggregation.

We only investigate link prediction but other tasks, such as community detection or other inference tasks, may be used to drive the choice of window size. One direction for further work is to investigate how the choice of inference task affects the choice of window size and to what degree do other inference tasks agree with link prediction.

Finally, we leave issues of scalability - both in terms of properties of link prediction as the size of the graphs gets very large and in terms of how to make our algorithmic approach suitably fast for such large graphs - for future work.

Acknowledgments. The authors would like to thank Tanya Berger-Wolf for helpful discussions and her generous support.

References

1. Adafre, S.F., de Rijke, M.: Discovering missing links in Wikipedia. In: Proc. of the 3rd Int. Workshop on Link Discovery, pp. 90–97. ACM (2005)
2. Adamic, L.A., Adar, E.: Friends and neighbors on the web. Social Networks 25(3), 211–230 (2003)
3. Airoldi, E.M., Blei, D.M., Xing, E., Fienberg, S.: Mixed membership stochastic block models for relational data, with applications to protein-protein interactions. In: Proc. of Int. Biometric Society - ENAR Annual Meetings, vol. 5 (2006)
4. Hasan, M.A., Zaki, M.J.: A survey of link prediction in social networks. In: Social Network Data Analytics, pp. 243–275. Springer (2011)
5. Baldi, P., Brunak, S., Chauvin, Y., Andersen, C.A.F., Nielsen, H.: Assessing the accuracy of prediction algorithms for classification: an overview. Bioinformatics 16(5), 412–424 (2000)
6. Barabási, A.-L., Albert, R.: Emergence of scaling in random networks. Science 286(5439), 509–512 (1999)
7. Clauset, A., Eagle, N.: Persistence and periodicity in a dynamic proximity network. DIMACS Workshop on Computational Methods for Dynamic Interaction Networks (2007)
8. De Choudhury, M., Mason, W.A., Hofman, J.M., Watts, D.J.: Inferring relevant social networks from interpersonal communication. In: Proc. of the 19th Int. Conf. on World Wide Web, pp. 301–310. ACM (2010)
9. Eagle, N., Pentland, A.: Reality Mining: sensing complex social systems. Personal and Ubiquitous Computing 10(4), 255–268 (2006)
10. Erdős, P., Rényi, A.: On random graphs I. Publ. Math. Debrecen 6, 290–297 (1959)
11. Freschi, V.: A graph-based semi-supervised algorithm for protein function prediction from interaction maps. In: Stützle, T. (ed.) LION 3. LNCS, vol. 5851, pp. 249–258. Springer, Heidelberg (2009)
12. Gallagher, B., Tong, H., Eliassi-Rad, T., Faloutsos, C.: Using ghost edges for classification in sparsely labeled networks. In: Proc. of the 14th ACM SIGKDD Int. Conf. on Knowledge Discovery and Data Mining, pp. 256–264. ACM (2008)
13. Hansen, M.H., Bin, Y.: Model selection and the principle of minimum description length. Journal of the American Statistical Association 96(454), 746–774 (2001)
14. Holme, P., Saramäki, J.: Temporal networks. Physics Reports 519(3), 97–125 (2012)
15. Hu, B., Rakthanmanon, T., Hao, Y., Evans, S., Lonardi, S., Keogh, E.: Discovering the intrinsic cardinality and dimensionality of time series using MDL. In: IEEE 11th Int. Conf. on Data Mining (ICDM) 2011, pp. 1086–1091, December 2011
16. Huang, Z., Li, X., Chen, H.: Link prediction approach to collaborative filtering. In: Proc. of the 5th ACM/IEEE-CS Joint Conf. on Digital Libraries, pp. 141–142. ACM (2005)

17. Liben-Nowell, D., Kleinberg, J.: The link prediction problem for social networks. In: Proc. of the 12th Int. Conf. on Information and Knowledge Management. CIKM 2003, pp. 556–559. ACM, New York (2003)
18. Liu, Y., Kou, Z.: Predicting who rated what in large-scale datasets. ACM SIGKDD Explorations Newsletter 9(2), 62–65 (2007)
19. Peel, L., Clauset, A.: Detecting change points in the large-scale structure of evolving networks (2014). CoRR, abs/1403.0989. Pre-print
20. Perra, N., Gonçalves, B., Pastor-Satorras, R., Vespignani, A.: Activity driven modeling of time varying networks. Scientific Reports 2 (2012)
21. Ribeiro, B., Perra, N., Baronchelli, A.: Quantifying the effect of temporal resolution on time-varying networks. Scientific Reports 3 (2013)
22. Sarkar, P., Moore, A.W.: Dynamic social network analysis using latent space models. ACM SIGKDD Explorations Newsletter 7(2), 31–40 (2005)
23. Scott, J., Gass, R., Crowcroft, J., Hui, P., Diot, C., Chaintreau, A.: CRAW-DAD data set cambridge/haggle (v. 2006-01-31) (January 2006). Downloaded from http://crawdad.org/cambridge/haggle/
24. Sulo, R., Berger-Wolf, T., Grossman, R.: Meaningful selection of temporal resolution for dynamic networks. In: Proc. of the 8th Workshop on Mining and Learning with Graphs, pp. 127–136. ACM (2010)
25. Sun, J., Faloutsos, C., Papadimitriou, S., Yu, P.S.: Graphscope: Parameter-free mining of large time-evolving graphs. In: Proc. of the 13th ACM SIGKDD Int. Conf. on Knowledge Discovery and Data Mining. KDD 2007, pp. 687–696. ACM, New York (2007)
26. Wagner, N., Michalewicz, Z.: An analysis of adaptive windowing for time series forecasting in dynamic environments: Further tests of the DyFor GP Model. In: Proc. of the 10th Conf. on Genetic and Evolutionary Computation. GECCO 2008, pp. 1657–1664. ACM, New York (2008)
27. Zhu, J., Hong, J., Hughes, J.G.: Using Markov chains for link prediction in adaptive web sites. In: Bustard, D.W., Liu, W., Sterritt, R. (eds.) Soft-Ware 2002. LNCS, vol. 2311, pp. 60–73. Springer, Heidelberg (2002)

Hierarchical Sparse Dictionary Learning

Xiao Bian[1], Xia Ning[2(✉)], and Geoff Jiang[3]

[1] Electrical and Computer Engineering Department,
North Carolina State University, Raleigh, NC 27695, USA
xbian@ncsu.edu
[2] Department of Computer and Information Science,
IUPUI, Indianapolis, IN 46202, USA
xning@cs.iupui.edu
[3] Autonomic Management Department, NEC Labs America,
Princeton, NJ 45237, USA
gfj@neclabs.com

Abstract. Sparse coding plays a key role in high dimensional data analysis. One critical challenge of sparse coding is to design a dictionary that is both adaptive to the training data and generalizable to unseen data of same type. In this paper, we propose a novel dictionary learning method to build an adaptive dictionary regularized by an a-priori over-completed dictionary. This leads to a sparse structure of the learned dictionary over the a-priori dictionary, and a sparse structure of the data over the learned dictionary. We apply the hierarchical sparse dictionary learning approach on both synthetic data and real-world high-dimensional time series data. The experimental results demonstrate that the hierarchical sparse dictionary learning approach reduces overfitting and enhances the generalizability of the learned dictionary. Moreover, the learned dictionary is optimized to adapt to the given data and result in a more compact dictionary and a more robust sparse representation. The experimental results on real datasets demonstrate that the proposed approach can successfully characterize the heterogeneity of the given data, and leads to a better and more robust dictionary.

1 Introduction

Sparse representation has been demonstrated as very powerful in analyzing high dimensional data [1–3], where each data point can be typically represented as a linear combination of a few atoms in an over-complete dictionary. Assume $\mathbf{x} \in R^d$ is a data vector and $\mathbf{D}$ is the dictionary, and then the sparse representation of $\mathbf{x}$ can be formulated as to find the sparse code $\mathbf{w}$ over $\mathbf{D}$ by solving the following optimization problem,

$$\min_{\mathbf{w}} \quad \|\mathbf{w}\|_0$$
$$\text{s.t.} \quad \|\mathbf{Dw} - \mathbf{x}\| \leq \sigma,$$

where σ is a pre-defined threshold. The pursued sparse code $\mathbf{w}$ can been considered as a robust representation of $\mathbf{x}$, and can be used for clustering [4,5], classification [6] and denoising [2,7].

© Springer International Publishing Switzerland 2015
A. Appice et al. (Eds.): ECML PKDD 2015, Part II, LNAI 9285, pp. 687–700, 2015.
DOI: 10.1007/978-3-319-23525-7_42

One key question is how to construct such an over-complete dictionary that is suitable for sparse representation. There are two major approaches for constructing such dictionaries: analytic approaches and learning-based approaches [8]. In an analytic approach, the dictionary is carefully designed a priori, e.g. with atoms such as wavelets [9], curvelets [10] and shearlets [11,12]. One advantage of the analytic approaches is that the dictionary can be designed as well-conditioned for stable representation, for example, to have a better incoherence condition or restricted isometry property [13,14].

In learning-based approaches, the dictionaries are learned from the given data [2,15,16]. Compared to the designed dictionaries in analytic approaches, the learned dictionaries are usually more adaptive to the given data, and therefore lead to more robust representations. Therefore, the learning-based approaches outperform analytic approaches in many tasks such as denoising and classification, etc [1,17]. The dictionary learning problem in the learning-based approaches is typically formulated as the following optimization problem,

$$\min_{D \in \mathcal{C}, \mathbf{W}} \|\mathbf{X} - \mathbf{DW}\|_F^2$$
$$\text{s.t.} \quad \|\mathbf{W}\|_0 \leq k, \tag{1}$$

where $\mathbf{X}$, $\mathbf{W}$ and $\mathbf{D}$ represent the data, their sparse codes and the dictionary, respectively, and $\mathcal{C}$ is a pre-specified feasible region for $\mathbf{D}$. However, (1) is non-convex and thus it is very difficult to find the global optimal solution or even a good local optimum.

In this paper, we propose to integrate both analytic approaches and learning-based approaches and learn from data a dictionary that is also built upon and regularized by an a-priori dictionary. The learned dictionary will be adaptive to the training data and its size will be determined by the intrinsic complexity of the training data. Meanwhile, due to the regularization from the a-priori dictionary, the non-convex optimization problem will have a more stable and better local minimum solution, and requires fewer training data. We compare the new method with the state-of-the-art methods on various aspects and our experimental results demonstrate superior performance of the new method.

2   Hierarchical Sparse Structures on Dictionaries

In the dictionary learning problem in (1), the constraint $\mathbf{D} \in \mathcal{C}$ is critical to regularize $\mathbf{D}$. In the state of the art, $\mathcal{C}$ is typically specified as $\mathcal{C} = \{\mathbf{D} : \forall \mathbf{d}_i \in \mathbf{D}, \|\mathbf{d}_i\|_2 \leq c\}$ [2,15] or $\mathcal{C} = \{\mathbf{D} : \|\mathbf{D}\|_F \leq c\}$ [16]. Intuitively, in both cases, $\mathcal{C}$ tames the amplitude of $\mathbf{D}$. However, these constraints do not consider any prior knowledge on $\mathbf{D}$, if available. Prior knowledge is valuable to learn a dictionary that is more powerful to characterize the data. For example, a dictionary for image patches is expected to have finer structures that might be further represented using DCT or wavelets. Incorporating such knowledge into dictionary learning can result in superior results [8,18].

Given an a-priori over-complete dictionary $\mathbf{\Phi}$ for data $\mathbf{X}$ based on some prior knowledge about $\mathbf{X}$, we aim to learn a dictionary $\mathbf{D}$ based on $\mathbf{\Phi}$ so that $\mathbf{D}$ is more adaptive to $\mathbf{X}$. In specific, we propose hierarchical sparse structures among $\mathbf{\Phi}$, $\mathbf{D}$ and $\mathbf{X}$, that is, $\mathbf{D}$ is constructed from $\mathbf{\Phi}$ via sparse combination of $\mathbf{\Phi}$'s atoms, and $\mathbf{X}$ is constructed from $\mathbf{D}$ via sparse combination of $\mathbf{D}$'s atoms. Mathematically, the hierarchical sparse structure of $\mathbf{D}$ over $\mathbf{\Phi}$ can be specified using the feasible region $\mathcal{C}$ as follows,

$$\mathcal{C} = \{\mathbf{D} : \mathbf{D} = \mathbf{\Phi U}, \|\mathbf{u}_i\|_0 \leq l, \forall i\}, \tag{2}$$

where $\mathbf{U}$ is the sparse coefficients for $\mathbf{D}$ over $\mathbf{\Phi}$. Given the dictionary $\mathbf{D} = \mathbf{\Phi U}$, data $\mathbf{X}$ can then be represented as

$$\mathbf{X} = \mathbf{DW} = \mathbf{\Phi UW}, \tag{3}$$

where $\mathbf{W}$ is the sparse coefficients over $\mathbf{D}$.

The hierarchical structures in (3) share some properties with deep architectures of learning models. Deep architectures have been empirically demonstrated as very effective for many complicated AI tasks [19]. Compared to a shallow model, a deep architecture is able to characterize complex data with alleviated overfitting. Our experimental results demonstrate that the hierarchical structures among dictionaries can also reduce overfitting and improve generalizability of the model.

In this paper, we propose a learning framework to learn the dictionary $\mathbf{D}$ and the sparse codes $\mathbf{W}$ in (3). The primary contributions of this paper include

- the proposed hierarchical sparse structures among an a-priori over-complete dictionary $\mathbf{\Phi}$, the pursued dictionary $\mathbf{D}$ and the given training data $\mathbf{X}$ as in (3);
- the formulation of a hierarchical sparse dictionary learning problem to learn $\mathbf{D}$ and $\mathbf{W}$ in Sect. 3; and
- the solution algorithm for the problem in Sect. 3.

3   Hierarchical Sparse Dictionary Learning

We formulate the problem of learning a dictionary $\mathbf{D}$ from data $\mathbf{X}$ and $\mathbf{X}$'s sparse representation $\mathbf{W}$ over $\mathbf{D}$, where $\mathbf{D}$ is built upon an a-priori over-complete dictionary $\mathbf{\Phi}$, as in the following optimization problem.

$$\min_{\mathbf{D},\mathbf{W}} \quad \|\mathbf{X} - \mathbf{DW}\|_F^2$$

$$\text{s.t.} \quad \|\mathbf{W}\|_0 \leq k \tag{4}$$

$$\mathbf{D} \in \mathcal{C} = \{\mathbf{D} : \mathbf{D} = \mathbf{\Phi U}, \|\mathbf{u}_i\|_0 \leq l, \forall i\}.$$

We denote the learning problem in (4) as Hierarchical Sparse Dictionary Learning (HiSDL). The major difficulty in HiSDL is that the feasible region $\mathcal{C}$ is non-convex and even not path-connected, and thus optimization over $\mathcal{C}$ is very challenging. We solve the problem by first giving an approximated sparsity of $\mathbf{D}$ on $\mathbf{\Phi}$ in Sect. 3.1, and then a corresponding optimization algorithm in Sect. 3.2.

3.1 Approximated sparsity of D on Φ

We first reformulate the feasible region constraint in (4) as a regularizer in the objective function, and then consider its convex approximation. Specifically, using the ℓ_1 convex relaxation of $\|\cdot\|_0$, we define an $\mathcal{C}$-function of $\mathbf{D}$ as follows,

$$
\begin{aligned}
\mathcal{C}(\mathbf{D}) \quad &= \sum_i \min_{\mathbf{d}_i} \|\mathbf{u}_i\|_1 \quad \text{s.t.} \quad \mathbf{D} = \mathbf{\Phi U} \\
&= \min_{\mathbf{D}} \|\mathbf{U}\|_1 \qquad \text{s.t.} \quad \mathbf{D} = \mathbf{\Phi U}.
\end{aligned}
$$

Thus, the dictionary learning problem in (4) can be reformulated as

$$
\begin{aligned}
\min_{\mathbf{D},\mathbf{W}} \quad & \frac{1}{2}\|\mathbf{X} - \mathbf{DW}\|_F^2 + \gamma\|\mathbf{U}\|_1 \\
\text{s.t.} \quad & \|\mathbf{W}\|_0 \le k, \\
& \mathbf{D} = \mathbf{\Phi U}.
\end{aligned} \tag{5}
$$

Then we consider a convex approximation of $\mathcal{C}(\mathbf{D})$ based on the following theorem.

Theorem 1. *Assume a $d \times p$ dictionary $\mathbf{\Phi}$ with incoherence μ, and $\mathbf{D} = \mathbf{\Phi U}$ with all $\mathbf{u}_i$ k-sparse and $k < 1 + 1/\mu$, then*

$$
\alpha\|\mathbf{\Phi}^\mathsf{T}\mathbf{D}\|_1 \le \|\mathbf{U}\|_1 \le \beta\|\mathbf{\Phi}^\mathsf{T}\mathbf{D}\|_1,
$$

where $\alpha = \frac{1}{1+(p-1)\mu}$, $\beta = \frac{1}{1-(k-1)\mu}$. In particular, if $\mathbf{\Phi}$ is an orthonormal basis, then $\|\mathbf{U}\|_1 = \|\mathbf{\Phi}^\mathsf{T}\mathbf{D}\|_1$.

The proof of Theorem 1 is presented in the Appendix section.

Since $\mathbf{\Phi}$ is a pre-designed dictionary with a well-constrained incoherence, based on Theorem 1, we choose $\|\mathbf{\Phi}^\mathsf{T}\mathbf{D}\|_1$ to approximate $\mathcal{C}(\mathbf{D})$ and thus to regularize the sparsity of $\mathbf{D}$ on $\mathbf{\Phi}$. Furthermore, we relax and reformulate the sparse constraint of $\mathbf{W}$ as an ℓ_1-norm regularizer in the objective function. The resulting dictionary learning problem is thus as follows.

$$
\min_{\mathbf{D},\mathbf{W}} \frac{1}{2}\|\mathbf{X} - \mathbf{DW}\|_F^2 + \lambda\|\mathbf{W}\|_1 + \gamma\|\mathbf{\Phi}^\mathsf{T}\mathbf{D}\|_1. \tag{6}
$$

Due to the convexity of $\|\mathbf{\Phi}^\mathsf{T}\mathbf{D}\|_1$, the objective function in (6) is convex with respect to $\mathbf{D}$.

3.2 Optimization Algorithm

There are two key steps in a typical dictionary learning algorithm: sparse coding and dictionary update. In the sparse coding step, the goal is to find the sparse coefficients $\mathbf{W}$ with a fixed dictionary $\mathbf{D}$ from the last iteration. In the dictionary update step, $\mathbf{D}$ is further optimized with respect to the pursued $\mathbf{W}$. The objective function is therefore minimized in an alternating fashion.

For the objective function as in (6), the sparse coding step is similar to that in [15], that is, it is to find $\mathbf{W}$ by solving the following problem after fixing $\mathbf{D}$.

$$\min_{\mathbf{W}} \frac{1}{2}\|\mathbf{X} - \mathbf{D}\mathbf{W}\|_F^2 + \lambda\|\mathbf{W}\|_1. \tag{7}$$

It is a classical linear inverse problem with l_1 regularization. We utilize the FISTA algorithm [20], due to its efficiency and robustness, to solve (7).

During the dictionary update step, the objective is to pursue the dictionary $\mathbf{D}$ by solving the following problem after fixing $\mathbf{W}$.

$$\min_{\mathbf{D}} \frac{1}{2}\|\mathbf{X} - \mathbf{D}\mathbf{W}\|_F^2 + \gamma\|\mathbf{\Phi}^{\mathsf{T}}\mathbf{D}\|_1. \tag{8}$$

To solve the above problem, we introduce an auxiliary variable $\mathbf{H} = \mathbf{\Phi}^{\mathsf{T}}\mathbf{D}$. Thus, the problem in (8) can be reformulated as follows,

$$\hat{\mathbf{H}} = \arg\min_{\mathbf{H}} \frac{1}{2}\|\mathbf{X} - \mathbf{\Phi}^{\dagger}\mathbf{H}\mathbf{W}\|_F^2 + \gamma\|\mathbf{H}\|_1, \tag{9}$$

$$\hat{\mathbf{D}} = \mathbf{\Phi}^{\dagger}\hat{\mathbf{H}}, \tag{10}$$

where $\mathbf{\Phi}^{\dagger} = (\mathbf{\Phi}\mathbf{\Phi}^{\mathsf{T}})^{-1}\mathbf{\Phi}$.[1] The problem in (9) is again a linear inverse problem with ℓ_1 regularization. We can solve it similarly as for (7) in the sparse coding step. Thus, the entire procedure for solving (6) is presented in Algorithm 1.

Algorithm 1. Hierarchical Sparse Dictionary Learning (HiSDL)

Input: Data matrix $\mathbf{X} \in R^{m \times n}$, dictionary $\mathbf{\Phi}$
Initialize: λ, γ, $\mathbf{D}_0$
for $t = 1, 2, \ldots, T$ do
    // Sparse coding: solve (7)
    $\mathbf{W}_t = \arg\min_{\mathbf{W}} \frac{1}{2}\|\mathbf{X} - \mathbf{D}_{t-1}\mathbf{W}\|_F^2 + \lambda\|\mathbf{W}\|_1$
    // Dictionary update: solve (8)
    $\mathbf{H}_t = \arg\min_{\mathbf{H}} \frac{1}{2}\|\mathbf{X} - \mathbf{\Phi}^{\dagger}\mathbf{H}\mathbf{W}_t\|_F^2 + \gamma\|\mathbf{H}\|_1$
    $\mathbf{D}_t = \mathbf{\Phi}^{\dagger}\mathbf{H}_t$
end for
return $\mathbf{D}_T$

3.3 Analysis of HiSDL Algorithm

Atom selection in HiSDL Generally, the number of atoms in $\mathbf{D}$ is largely determined by the complexity of the given data, and is therefore difficult to determine a priori. Moreover, the non-convex nature of the objective function in (6) inevitably leads to non-global optima. Therefore, it is very challenging to

---

[1] $\mathbf{\Phi}\mathbf{\Phi}^{\mathsf{T}}$ is invertible since $\mathbf{\Phi}$ is an over-complete frame [9].

find the correct size of a dictionary $\mathbf{D}$ and its associated atoms that result in a good local minimum [2,21]. Interestingly, HiSDL as in Algorithm 1 has an "atom selection" property. In particular, the obsolete atoms in $\mathbf{D}$ will be automatically eliminated, and thereby the size of $\mathbf{D}$ is well-controlled. To verify this property of HiSDL, we first have the following lemma.

Lemma 1. *For any atom* $\mathbf{d}_i \in \mathbf{D}$, *if* $\mathbf{d}_i^t = \mathbf{0}$, *then* $\mathbf{d}_i^{t+1} = \mathbf{0}$, *where* $\mathbf{d}_i^t$ *is the i-th atom of* $\mathbf{D}$ *at the t-th iteration as in Algorithm 1.*

The proof of Lemma 1 is presented in the Appendix section.

Different from other state-of-the-art approaches as in [2,15], Lemma 1 states that if one atom degenerates to $\mathbf{0}$, then it will stay as $\mathbf{0}$ since then. This essentially addresses the dictionary pruning problem, i.e. the unused atoms are automatically set to zero. Indeed, if one atom dose not contribute much to the reduction of the empirical error $\|\mathbf{X} - \mathbf{DW}\|_F$, then it will be set to zero in the dictionary update step based on the following theorem.

Theorem 2. *At iteration* t_0, *if* $\|\mathbf{\Psi}^\mathsf{T}\mathbf{R}_i\mathbf{W}^\mathsf{T}\|_\infty < \gamma$, *where* $\mathbf{\Psi} = \mathbf{\Phi}^\dagger$, *and* $\mathbf{R}_i = \mathbf{X} - \mathbf{D}_{-i}\mathbf{W}$ *is the empirical error without using* $\mathbf{d}_i$ *in* $\mathbf{D}$, *then* $\mathbf{d}_i = 0$ *for* $t > t_0$.

The proof of Theorem 2 is presented in the Appendix section.

Theorem 2 ensures that the unnecessary atoms will degenerate to $\mathbf{0}$ as the empirical error reduces during the learning process. We are therefore able to maintain a compact dictionary in an on-line fashion.

Computational Complexity of HiSDL The sparse coding step (7) and the dictionary update step (8) dominate the computational complexity of HiSDL. In particular, the sparse coding step and the dictionary update step are essentially the same constrained ℓ_1-minimization problem, of which the computational complexity is mainly from matrix multiplication when using soft-thresholding methods such as FISTA [20]. Specifically, if $\mathbf{X}$ is of dimension $d \times m$, $\mathbf{D}$ is of dimension $d \times n$, and $\mathbf{\Phi}$ is of dimension $d \times p$, where typically $m > p > n$ and $m > d, p > d^2$, then the computational complexity of each soft-thresholding iteration in sparse coding is $O(mnd)$, and similarly $O(pdn)$ for dictionary update.

4 Related Work

Structured dictionary learning has been explored in previous works [22–24] from different perspectives. For example, in [22], a tree-like hierarchical structure is learned among the atoms in a dictionary, instead of treating each atom independently. Group sparsity among atoms is also considered in [23] and is applied to model spatial relations between atoms. In [24], a smooth prior on the sparse

---

[2] The number of samples m in $\mathbf{X}$ should be larger than p, the number of atoms in $\mathbf{\Phi}$, and $p > n$, the number of atoms in a more compact dictionary D. However, n is determined by the richness of $\mathbf{X}$, and may therefore be larger or smaller than d.

coefficients W is used in order to get a more stable representation. In contrast to these methods, we introduce a known dictionary representing the prior knowledge of the given data, and the hierarchical structure is imposed on the known dictionary and the learned dictionary rather than among the atoms in the learned dictionary. In addition, in our model, the known dictionary is used directly to regularize dictionary learning rather than to enforce structures in W as in [22–24]. As shown in this paper later, the use of the known dictionary and the hierarchical structures among the known dictionary and the learned dictionary enable a sparser representation with lower empirical errors.

5   Experimental Results

In this section, we present experimental results on synthetic data to empirically evaluate HiSDL. We also demonstrate the applications of HiSDL using real-world data. In particular, we test HiSDL on the following two datasets:

1. Synthetic data: we synthesize 200 time series of length 100 using DCT and Haar wavelets to simulate the real-world time series. DCT and Haar wavelets are composed into the a-priori over-complete dictionary Φ. Then a few atoms from Φ are randomly selected and combined with amplitudes following a uniform distribution in $[-1, 1]$ into an atom in a dictionary D, and in the end D has 100 atoms. A random sparse matrix W is then generated and used so as to generate the synthetic time series from D.
2. Chemical plant time series (CPT): This dataset includes 1625 time series from various sensors monitoring an entire manufacture process of a chemical plant. Every time series is the output of one sensor, and each sensor collects one observation every minute. The data exhibit high heterogeneity in nature, e.g., there are both continuous and discrete time series, smooth and non-smooth time series, etc.

5.1   Evaluation on Empirical Errors

Fig. 1 shows the empirical errors of HiSDL and of the state-of-the-art method [15], denoted as BatchDL, during learning iterations with different parameter λ values on the synthetic data ($\gamma = 0.05\lambda$; other λ values give similar trends; the optimal λ and γ combinations are from grid search). For each of the λ values, the sparsity of the learned W is relatively similar from both HiSDL and BatchDL. However, HiSDL consistently achieves smaller empirical errors than BatchDL after each learning iteration. This demonstrates that by introducing a regularization of D with respect to an a-priori over-complete Φ, the optimization process in (4) may have a better chance to end up at a better local minimum within the reduced (and better) search space. In addition, HiSDL achieves smaller empirical errors faster than BatchDL. This implies that HiSDL can quickly find a more accurate sparse representation than BatchDL.

We further compare the performance of HiSDL and BatchDL on the CPT dataset. We randomly pick one-day data in the dataset for dictionary learning,

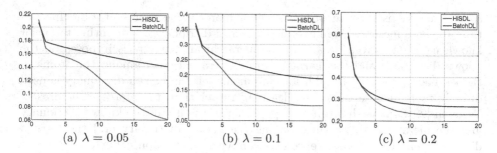

(a) $\lambda = 0.05$ (b) $\lambda = 0.1$ (c) $\lambda = 0.2$

Fig. 1. Empirical errors vs learning iterations on synthetic data

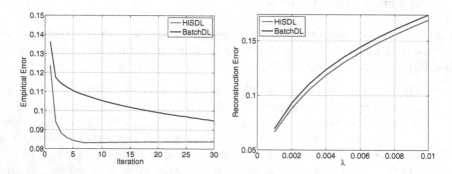

Fig. 2. Empirical errors vs learning **Fig. 3.** Reconstruction errors on CPT
iterations on CPT   testing data

and the data from a later day for testing. For CPT, Φ is constructed as a combination of DCT and Haar wavelets, of which the number of atoms is twice as the length of time series. However, the learned dictionary is composed of only 120 atoms. Fig. 2 shows the empirical errors during learning iterations with $\lambda = 0.001$ and $\gamma = 0.02\lambda$ (the λ and γ values and combinations are optimized from grid search). Again, on the real dataset, HiSDL achieves smaller empirical errors faster than BatchDL. Fig. 3 shows the reconstruction errors of HiSDL and BatchDL on CPT testing data with different λ values. In Fig. 3, HiSDL consistently achieves smaller reconstruction errors than BatchDL, which implies that HiSDL is able to find more robust and generalizable dictionaries than BatchDL.

5.2   Evaluation on Atom Recovery

Fig. 4 presents some sample atoms learned from HiSDL on the synthetic data. These atoms exhibit finer structures as a linear combination of DCT and Haar wavelets, which demonstrates the capability of HiSDL recovering the building structures of the data. However, as shown in Fig. 5, the learned atoms by BatchDL on the synthetic data appear less structured, more homogeneous and do not conform to the true structures underlying the data. This is due to that fact that

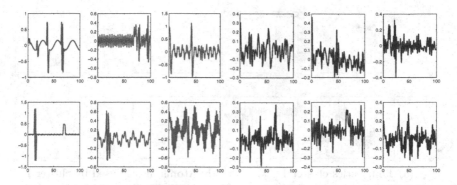

Fig. 4. Sample atoms from HiSDL on synthetic data   **Fig. 5.** Sample atoms from `BatchDL` on synthetic data

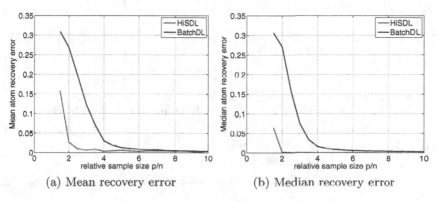

(a) Mean recovery error   (b) Median recovery error

Fig. 6. Recovery errors vs sample size

`BatchDL` constraints the norm of each atom and thus biases the search of the atoms towards a bad local minimum.

To further test the performance of the methods on the discovery of latent atoms, we evaluate HiSDL and `BatchDL` on the blind source separation problem [25,26] on a set of synthetic datasets. These synthetic datasets have the same a-priori over-complete dictionary Φ and dictionary $\mathbf{D}$ as generated as before, but different number of time series (150, 200, 250 up to 1000). The success of the recovery of latent atoms relies on the ratio of the given sample size to the number of latent atoms. Intuitively, we can only expect to recover all latent atoms when every atom has been sufficiently used in the given sample set. Naturally, this recovery goal is more likely to be achieved when we have a large dataset.

Denote the learned dictionary as $\hat{\mathbf{D}}$, and the relative recovery error of each atom $\mathbf{d}_i \in \mathbf{D}$ is then defined as follows,

$$r_i = \min_{\hat{\mathbf{d}}_j \in \hat{\mathbf{D}}} \{1 - \cos \Theta(\hat{\mathbf{d}}_j, \mathbf{d}_i)\}, \tag{11}$$

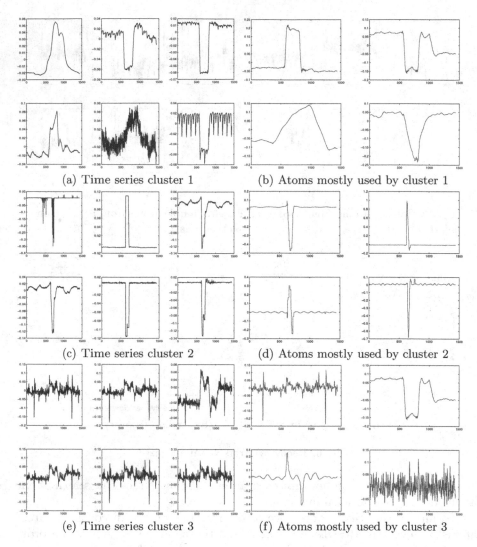

(a) Time series cluster 1   (b) Atoms mostly used by cluster 1

(c) Time series cluster 2   (d) Atoms mostly used by cluster 2

(e) Time series cluster 3   (f) Atoms mostly used by cluster 3

Fig. 7. Sample time series clusters in CPT and the corresponding mostly used atoms

where $\Theta(\hat{\mathbf{d}}_j, \mathbf{d}_i)$ is the angle between $\hat{\mathbf{d}}_j$ and $\mathbf{d}_i$. In specific, $r_i \in [0, 1]$, and if there exists $\hat{\mathbf{d}}_j \in \hat{\mathbf{D}}$ that satisfies $\mathbf{d}_i = \hat{\mathbf{d}}_j$, then $r_i = 0$. In Fig. 6, we show the mean/median atom recovery error vs the relative sample size (i.e., p/n in Fig. 6, the ratio of sample sizes over the dictionary size) on the synthetic datasets.

In this set of experiments, the initial dictionary size for both algorithms is 150, and the dictionary of ground truth is of size 100. Each point in Fig. 6 is the mean of 5 experiments under the same training set, but with different initialization. We can see that when the sample size is large, i.e. sufficient information is provided, both HiSDL and BatchDL work well. However, when the sample size is small, such as two to four times the number of latent atoms, HiSDL shows a substantially superior performance. It further demonstrates that by integrating a priori knowledge of the given dataset, HiSDL achieves better generalizability.

5.3   Evaluation on Sparse Codes

We also explore the ability of HiSDL to process heterogeneous time series data by studying the clustering results using sparse codes. Intuitively, if the learned dictionary successfully characterizes the given data, the clustering generated from their associated sparse codes should exhibit good structures. Fig. 7 shows some clusters of time series from CPT data and their frequently used dictionary atoms. The clustering is done by using spectral clustering algorithm [27] on the sparse codes. As Fig. 7 shows, the CPT data are heterogeneous including step signals, piece-wise linear signals, periodical signals and even brownian motion-like signals. However, after representing the time series over the learned hierarchical dictionaries, the clustering over their sparse codes shows high homogeneity within each cluster, and the frequently used atoms for each cluster represent dominant features of the cluster. This demonstrates that HiSDL has the capability of capturing the most representative features from even highly heterogeneous time series.

6   Conclusion

In this paper, we introduce a novel dictionary learning framework HiSDL which utilizes a hierarchical sparse structure to characterize observed data. The experiments demonstrate that the hierarchical sparse structure within the model regularizes potential solutions, and enables smaller empirical errors. In addition, HiSDL is able to identify the most representative latent atoms from a few training samples, and thus well characterizes the training data. Future work may include constructing nonlinear and deep structures on dictionary learning models. Also it would be interesting to see a more thorough evaluation of HiSDL on other types of data, such as videos and images.

Appendix

Proof of Theorem 1

Proof. We first show the right part of the inequality, $\beta\|\boldsymbol{\Phi}^{\mathsf{T}}\mathbf{D}\|_1 \geq \|\mathbf{U}\|_1$.
    since $\mathbf{D} = \boldsymbol{\Phi}\mathbf{U}, \|\mathbf{u}_i\|_0 \leq k, \forall \mathbf{u}_i \in \mathbf{U}$, it follows that

$$\|\boldsymbol{\Phi}^{\mathsf{T}}\mathbf{D}\|_1 = \|\boldsymbol{\Phi}^{\mathsf{T}}\boldsymbol{\Phi}\mathbf{U}\|_1$$
$$= \sum_i \|\boldsymbol{\Phi}^{\mathsf{T}}\boldsymbol{\Phi}\mathbf{u}_i\|_1. \tag{12}$$

Moreover, according to the definition of incoherence,

$$\|\boldsymbol{\Phi}^\mathsf{T}\boldsymbol{\Phi}\mathbf{u}_i\|_1 \geq \sum_j (1 - (k-1)\mu)|u_{ij}|_1$$
$$= (1 - (k-1)\mu)\|\mathbf{u}_i\|_1. \tag{13}$$

Consequently, we have

$$\|\boldsymbol{\Phi}^\mathsf{T}\mathbf{D}\|_1 \geq \sum_i \|\mathbf{u}_i\|_1 (1 - (k-1)\mu)$$
$$= (1 - (k-1)\mu)\|\mathbf{U}\|_1, \tag{14}$$

and let $\beta = \frac{1}{1-(k-1)\mu}$, it follows that $\beta\|\boldsymbol{\Phi}^\mathsf{T}\mathbf{D}\|_1 \geq \|\mathbf{U}\|_1$.

We next prove the left part of the inequality, $\alpha\|\boldsymbol{\Phi}^\mathsf{T}\mathbf{D}\|_1 \leq \|\mathbf{U}\|_1$. Proceeding from (12), we further have

$$\|\boldsymbol{\Phi}^\mathsf{T}\boldsymbol{\Phi}\mathbf{u}_i\|_1 \leq \sum_j (1 - (p-1)\mu)|u_{ij}|_1$$
$$= (1 - (p-1)\mu)\|\mathbf{u}_i\|_1. \tag{15}$$

and as a result,

$$\|\boldsymbol{\Phi}^\mathsf{T}\mathbf{D}\|_1 = \sum_i \|\boldsymbol{\Phi}^\mathsf{T}\boldsymbol{\Phi}\mathbf{u}_i\|_1$$
$$\leq (1 - (p-1)\mu)\|\mathbf{U}\|_1. \tag{16}$$

Since $\alpha = \frac{1}{1+(p-1)\mu}$, we therefore have $\alpha\|\boldsymbol{\Phi}^\mathsf{T}\mathbf{D}\|_1 \leq \|\mathbf{U}\|_1$.   ∎

Proofs of Lemma 1 and Theorem 2

We first show the proof of Lemma 1.

Proof. If $\mathbf{d}_i^t = \mathbf{0}$, then

$$\mathbf{W}_{t+1} = \arg\min_\mathbf{W} \frac{1}{2}\|\mathbf{X} - \mathbf{D}_t\mathbf{W}\|_F^2 + \lambda\|\mathbf{W}\|_1 \tag{17}$$

implies that the ith row of $\mathbf{W}_{t+1}$, $\mathbf{w}_{t+1}^i$, is also $\mathbf{0}$.

Now consider

$$\mathbf{D}_{t+1} = \arg\min_\mathbf{D} = \frac{1}{2}\|\mathbf{X} - \mathbf{D}\mathbf{W}_{t+1}\|_F^2 + \lambda\|\boldsymbol{\Phi}^\mathsf{T}\mathbf{D}\|_1, \tag{18}$$

since $\mathbf{w}_{t+1}^i = \mathbf{0}$, we therefore have $\mathbf{d}_i^{t+1} = \mathbf{0}$.   ∎

Having Lemma 1 proved, we can then proceed to prove Theorem 2.

Proof. At iteration t_0, let $g(\mathbf{H}) = \gamma\|\mathbf{H}\|_1 + \frac{1}{2}\|\mathbf{X} - \boldsymbol{\Psi}\mathbf{H}\mathbf{W}\|_F^2$, and assume $\hat{\mathbf{H}} = \arg\min_H g(\mathbf{H})$, we then have

$$\partial g(\hat{\mathbf{H}}) = \gamma\partial\|\hat{\mathbf{H}}\|_1 + \boldsymbol{\Psi}^{\mathsf{T}}(\mathbf{X} - \boldsymbol{\Psi}\hat{\mathbf{H}}\mathbf{W})\mathbf{W}^{\mathsf{T}} \ni \mathbf{0}. \tag{19}$$

Rewrite $\mathbf{H}$ as $\mathbf{H} = \mathbf{H}_i + \mathbf{H}_{-i}$, where $\mathbf{H}_i = [\mathbf{0}, \dots, \mathbf{0}, \mathbf{h}_i, \mathbf{0}, \dots, \mathbf{0}]$ and $\mathbf{H}_{-i} = [\mathbf{h}_1, \dots, \mathbf{h}_{i-1}, \mathbf{0}, \mathbf{h}_{i+1}, \dots, \mathbf{h}_n]$, then we have

$$\partial g(\hat{\mathbf{H}}_i) \ni \mathbf{0}, \partial g(\hat{\mathbf{H}}_{-i}) \ni \mathbf{0}. \tag{20}$$

Note that

$$\partial g(\hat{\mathbf{H}}_i) = \gamma\partial\|\hat{\mathbf{H}}_i\|_1 + \boldsymbol{\Psi}^{\mathsf{T}}(\mathbf{R}_i - \boldsymbol{\Psi}\hat{\mathbf{H}}_i\mathbf{W})\mathbf{W}^{\mathsf{T}}, \tag{21}$$

where $\mathbf{R}_i = \mathbf{X} - \boldsymbol{\Psi}\hat{\mathbf{H}}_{-i}\mathbf{W} = \mathbf{X} - \mathbf{D}_{-i}\mathbf{W}$.

Consider the condition $\|\boldsymbol{\Psi}^{\mathsf{T}}\mathbf{R}_i\mathbf{W}^{\mathsf{T}}\|_\infty < \gamma$, combined with the subgradient of ℓ_1-norm that $\partial\|x\|_1 = (-1, 1)$, we have

$$\hat{\mathbf{H}}_i = \mathbf{0} \Leftrightarrow \partial g(\hat{\mathbf{H}}_i) \ni \mathbf{0}. \tag{22}$$

When $\mathbf{h}_i = \mathbf{0}$, since $\mathbf{D} = \boldsymbol{\Psi}\mathbf{H}$, it follows that $\mathbf{d}_i = \mathbf{0}$ at $t_0 + 1$. According to Lemma 1, we consequently have $\mathbf{d}_i = \mathbf{0}$ for $t > t_0$. ∎

References

1. Mairal, J., Bach, F., Ponce, J., Sapiro, G., Zisserman, A.: Supervised dictionary learning. In: Koller, D., Schuurmans, D., Bengio, Y., Bottou, L. (eds.) NIPS, Curran Associates Inc., pp. 1033–1040 (2008)
2. Aharon, M., Elad, M., Bruckstein, A.: K-svd: An algorithm for designing overcomplete dictionaries for sparse representation. IEEE Transactions on Signal Processing **54**, 4311–4322 (2006)
3. Elad, M.: Sparse and redundant representation modeling: What next? IEEE Signal Processing Letters **19**, 922–928 (2012)
4. Bian, X., Krim, H.: Robust Subspace Recovery via Bi-Sparsity Pursuit (2014). ArXiv e-prints
5. Soltanolkotabi, M., Elhamifar, E., Candes, E.: Robust subspace clustering (2013). arXiv preprint arXiv:1301.2603
6. Huang, K., Aviyente, S.: Sparse representation for signal classification. In: Advances in neural information processing systems, pp. 609–616 (2006)
7. Elad, M., Aharon, M.: Image denoising via sparse and redundant representations over learned dictionaries. IEEE Transactions on Image Processing **15**, 3736–3745 (2006)
8. Rubinstein, R., Zibulevsky, M., Elad, M.: Double sparsity: Learning sparse dictionaries for sparse signal approximation. IEEE Transactions on Signal Processing **58**, 1553–1564 (2010)
9. Mallat, S.: A wavelet tour of signal processing. Academic press (1999)
10. Candes, E.J., Donoho, D.L.: Curvelets: A surprisingly effective nonadaptive representation for objects with edges. Technical report, DTIC Document (2000)

11. Yi, S., Labate, D., Easley, G.R., Krim, H.: A shearlet approach to edge analysis and detection. IEEE Transactions on Image Processing **18**, 929–941 (2009)
12. Labate, D., Lim, W.Q., Kutyniok, G., Weiss, G.: Sparse multidimensional representation using shearlets. In: Optics & Photonics 2005, International Society for Optics and Photonics, pp. 59140U–59140U (2005)
13. Eldar, Y.C., Kutyniok, G.: Compressed sensing: theory and applications. Cambridge University Press (2012)
14. Candès, E.J., Romberg, J.K., Tao, T.: Stable signal recovery from incomplete and inaccurate measurements. Communications on pure and applied mathematics **59**, 1207–1223 (2006)
15. Mairal, J., Bach, F., Ponce, J., Sapiro, G.: Online dictionary learning for sparse coding. In: Proceedings of the 26th Annual International Conference on Machine Learning, 689–696. ACM (2009)
16. Kreutz-Delgado, K., Murray, J.F., Rao, B.D., Engan, K., Lee, T.W., Sejnowski, T.J.: Dictionary learning algorithms for sparse representation. Neural computation **15**, 349–396 (2003)
17. Elad, M., Figueiredo, M.A., Ma, Y.: On the role of sparse and redundant representations in image processing. Proceedings of the IEEE **98**, 972–982 (2010)
18. Lee, H., Ekanadham, C., Ng, A.: Sparse deep belief net model for visual area v2. In: Advances in neural information processing systems, pp. 873–880 (2007)
19. Bengio, Y.: Learning deep architectures for AI. Foundations and Trends in Machine Learning **2**, 1–127 (2009)
20. Beck, A., Teboulle, M.: A fast iterative shrinkage-thresholding algorithm for linear inverse problems. SIAM Journal on Imaging Sciences **2**, 183–202 (2009)
21. Rubinstein, R., Faktor, T., Elad, M.: K-svd dictionary-learning for the analysis sparse model. In: 2012 IEEE International Conference on Acoustics, Speech and Signal Processing (ICASSP), pp. 5405–5408. IEEE (2012)
22. Jenatton, R., Mairal, J., Obozinski, G., Bach, F.: Proximal methods for hierarchical sparse coding. The Journal of Machine Learning Research **12**, 2297–2334 (2011)
23. Jenatton, R., Audibert, J.Y., Bach, F.: Structured variable selection with sparsity-inducing norms. The Journal of Machine Learning Research **12**, 2777–2824 (2011)
24. Bradley, D.M., Bagnell, J.A.: Differential sparse coding (2008)
25. Zibulevsky, M., Pearlmutter, B.: Blind source separation by sparse decomposition in a signal dictionary. Neural Computation **13**, 863–882 (2001)
26. Li, Y., Cichocki, A., Amari, S.: Analysis of sparse representation and blind source separation. Neural Computation **16**, 1193–1234 (2004)
27. Ng, A.Y., Jordan, M.I., Weiss, Y.: On spectral clustering: Analysis and an algorithm. In: Advances in Neural Information Processing Systems, pp. 849–856. MIT Press (2001)

Latent Factors Meet Homophily
in Diffusion Modelling

Minh-Duc Luu[✉] and Ee-Peng Lim

School of Information Systems, Singapore Management University,
80 Stamford Road, Singapore 178902, Singapore
{mdluu.2011,eplim}@smu.edu.sg

Abstract. Diffusion is an important dynamics that helps spreading
information within an online social network. While there are already
numerous models for single item diffusion, few have studied diffusion of
multiple items, especially when items can *interact* with one another due
to their inter-similarity. Moreover, the well-known homophily effect is
rarely considered explicitly in the existing diffusion models. This work
therefore fills this gap by proposing a novel model called *Topic level
Interaction Homophily Aware Diffusion* (TIHAD) to include both latent
factor level interaction among items and homophily factor in diffusion.
The model determines item interaction based on latent factors and edge
strengths based on homophily factor in the computation of social influence. An algorithm for training TIHAD model is also proposed. Our
experiments on synthetic and real datasets show that: (a) homophily
increases diffusion significantly, and (b) item interaction at topic level
boosts diffusion among similar items. A case study on hashtag diffusion
in Twitter also shows that TIHAD outperforms the baseline model in
the hashtag adoption prediction task.

1 Introduction

Ubiquitous presence of online social networks (OSN) has made information diffusion an important topic that attracts much research interests. While many items
may diffuse in a social network simultaneously, most existing models of diffusion
are built upon *independent* contagion assumption whereby the diffusion of each
item is assumed (at least implicitly) to happen independent of other items. The
interaction among items during diffusion is thus left out of the picture. This is
obviously not true in the complex dynamics of diffusion process. For instance,
the diffusion of iPhones in the Facebook friendship network may interact favorably with that of iPad; and the diffusion of a catchy phrase on Twitter also aids
the diffusion of its variants.

Interaction Among Items. Modeling these interactions is crucial in both
theory and practice since it helps us understand the detailed dynamics of multiple
item diffusion. It is also valuable for business to develop suitable strategies to
promote diffusion of their own items considering the other items that have been
diffused recently or are being diffused. It may be good to time the diffusion of

© Springer International Publishing Switzerland 2015
A. Appice et al. (Eds.): ECML PKDD 2015, Part II, LNAI 9285, pp. 701–718, 2015.
DOI: 10.1007/978-3-319-23525-7_43

a new item with the diffusion of other similar items (possibly by the business or other businesses) to achieve a larger reach. This idea of diffusion with item interaction can be further illustrated in the following motivating example.

Example. *A user may be inspired to watch the movie version of "Hunger Games" after observing some neighbors already read the book. Moreover, if both the book and the movie versions were adopted by a neighbor, the user will even be more likely to adopt the movie than if only one of them was adopted by the neighbor (as he may be more convinced that the movie is good in the former case).*

The example not only highlights that diffusion of an item can support that of another *similar* item but suggests other deeper ideas which will distinguish our work from the rest. These ideas are:

1. *The more similar items are, the more interaction will happen between them in diffusion.* In other words, item similarity can be used as a proxy for *item interaction*. This idea will be formulated in Section 3.2 where we propose a general diffusion framework for modeling item interaction when there is more than one item diffusing.
2. *Whether or not a user adopts an item i is affected not only by neighbors who adopted exactly the item but also by those who adopted other items.* A neighbor who already adopted another item i' can still influence the decision (see Example) as i' may be very similar to i.
3. *Each neighbor's social influence on a user's adoption decision should include all contributions from a set of items adopted by the neighbors*, not just limited to one item as in the existing models.

Homophily Factor. Another important aspect which also has great impact on item diffusion, is the well-known *homophily* phenomenon. Homophily refers to the tendency of individuals to associate and bond with similar others. It is well known that homophily affects the mechanisms in which item diffusion happens, be it innovation [14], information [3] or behavior [2]. Thus, it is important to integrate homophily into diffusion models so that we can better quantify its effect on diffusion. In this work, we assume a global homophily level of the network and learn it from the diffusion cascade data. Given that networks with homophily involves more similar users connecting with one another, it also plays a role in determining if an item can more smoothly diffuse across the network links.

Research Objectives. In this paper, we therefore propose to consider the above two factors in the design of a new diffusion model. To involve both item similarity (which helps to estimate item interaction) and user similarity (due to homophily), our modeling approach employs latent factors (LF) to represent both items and users (e.g. [13], [10]) where each user or item is represented as a vector in a common feature space with dimension much smaller than that of items and users. The similarity between two items (or users) can then be defined by the cosine similarity of the respective item (or user) vectors. Unlike the collaborative filtering approach taken by recommender systems, our diffusion modeling work also consider social influence among users. Although there are recently hybrid models ([9], [12]) which combine latent factor approach with social networks, they still do not model a user adopting an item influenced by the neighbors' past adoption of similar items and the strength of relationships with these neighbors. Based on our proposed model, we seek to answer some interesting research questions related to multiple-item diffusion in homophily networks.

Summary of Contributions. In summary, our work makes the following contributions.

- We develop an extended diffusion framework which incorporates both item interaction and homophily into modeling diffusion. To the best of our knowledge, this is the first attempt to combine the two factors. The framework is flexible and can offer useful insights to multiple item diffusion.
- We propose a specific diffusion model based upon the new framework. This model, known as TIHAD, utilizes latent factors to capture item interaction and homophily effect for effective modeling diffusion processes of multiple items.
- We formulate the parameter learning of model as a constrained optimization problem, and devise an effective learning algorithm using Projected Gradient Descent.
- We conduct experiments on both synthetic and real datasets to show that: (a) homophily increases diffusion significantly, and (b) item interaction at topic level boosts diffusion among similar items. We also shows that TIHAD outperforms the baseline model in the hashtag adoption prediction task.

Paper Outline. We will next give an overview of the related works. In Section 3, we present our proposed diffusion model known as TIHAD. The learning of this model is given in Section 4. Section 5 describes experiments that evaluate the TIHAD using both synthetic and real datasets. We finally conclude the paper in Section 6.

2   Related Works

Our work is closely related to very well studied adoption and diffusion modeling research: (i) Latent Factor models and (ii) Social Influence models. In the following sections, we briefly review these research works and relate them with our work.

2.1   Latent Factor Models

These models ([16], [13], [10]) take a user-item adoption matrix and factorize it into a set of user and item vectors with f dimensions where f is much smaller than the number of users or items. For each item i, a latent factor vector $q_i \in \mathbb{R}^f$

is derived and it contains the relevance weights of the latent factors for the item i. Similarly, a latent factor vector $\boldsymbol{p}_u \in \mathbb{R}^f$ is derived for each user u to represent the weights u has for the latent factors. Thus, the amount of interest u has towards item i can be defined as the inner product $\boldsymbol{p}_u^T \boldsymbol{q}_i$. Unlike latent factor models which focus on user-item interactions only, our work considers both user-item and item-item interactions in the diffusion setting. We are therefore also interested in the effect of item similarity. We exploit the latent factor space by defining the similarity between two items i and j as the inner product $\boldsymbol{q}_i^T \boldsymbol{q}_j$.

For better interpretability, many Latent Factor (LF) models (see [13], [15]) require latent factor vectors to have positive elements. We also follow this practice and consider only positive latent factor vectors. Although LF models enjoy the benefit of dimension reduction by matrix factorization, they do not consider the underlying social network which forms the substrate over which diffusion occurs. To address this shortcoming, recent research proposed to exploit social influence in the modeling of user-item adoptions (or ratings).

2.2 Social Influence and Diffusion Models

Social influence modeling works takes into account social interest and social trust as additional input to achieve better accuracy for recommendation ([9], [12], [18], [17], [4]). These works proposed various ways of modeling the social dimension such as factorizing the social network graph ([12]) or modeling social factors of users as another set of latent factors ([17], [4]). While these works focus on recommendation tasks, they are similar to diffusion models in that both estimate social influence on user-item adoptions. Social diffusion models on the other hand consider only influence from a subset of neighbors, called the set of *active* neighbors $\mathcal{A}_u$, who adopt exactly the target item ([6], [8], [11]). For example, Linear Threshold (LT) model is a social diffusion model which estimates social influence by the sum of weights of active neighbors. Thus, its standard form is

$$social\ influence = \sum_{v \in \mathcal{A}_u} w_{v,u} \tag{1}$$

As pointed out in our motivating example, items similar to the item being diffused i can affect diffusion. Even though a neighbor has not yet adopted item i, he can still affect the target user's decision on adopting i, when the neighbor adopted item(s) similar to i. Such a diffusion scenario has been largely overlooked in the existing social diffusion models.

3 Proposed Framework and Model

Before we present our proposed modeling framework and the TIHAD model, we first introduce the notations used in the problem formulation.

3.1 Basic Notations

We represent a social networks as a (directed), weighted graph $G = (U, E)$ whose nodes represent users and edges represent links among the users. For each edge (u, v), the edge weight $w_{v,u}$ represents the social influence that v exerts on u. To model diffusion over the network during a time period, we bin the continuous time into discrete time steps $\{1, 2, \cdots, T\}$ and consider adoptions in each step.

Denote adoption decision of a user u on item i at time step t as $a_{u,i,t}$. At first sight, it seems that $a_{u,i,t}$ is simply a binary label which is 1 when u adopt i and 0 otherwise. However, it is often that a user does not adopt an item because he has not been exposed to the item. It is thus incorrect to assume that he rejects the item, and underestimate his preference for the item. We can avoid this by considering, at each time step, only items which are exposed to the user. When the user did not adopt an item he has exposed to, we say that the case is a non-adoption. We call these user-exposed items as the candidate items in Definition 1, which in turn help us to define adoption labels properly in Definition 2.

Definition 1 (Candidate item). *At a given time step t, a candidate item for a user is an item that: (i) he has not yet adopted before t; and (ii) he is exposed to it through some source (e.g., recent adoptions by his neighbors). The set of candidate items for a user u at time t is denoted by $C_{u,t}$.*

Definition 2 (Adoption label). *Given an item $i \in C_{u,t}$, adoption label $a_{u,i,t}$ is a binary variable which is 1 if u adopts i at time t and 0 otherwise.*

3.2 Framework

Our proposed framework extends the latent factor model framework by considering both personal interest and social influence in the modeling of user-item adoption at different time steps. Personal interest is estimated by user-item similarity in a latent space and social influence is an aggregation of individual influences from neighbors. However, that influence from a neighbor $v \in N_u$ (the set of neighbors of u) now depends on: (i) the link weight $w_{v,u}$, and (ii) the interaction level between item i and a certain set of items adopted by v. We also follow common practice (e.g. [9]) by including in the framework global bias μ, user bias b_u and item bias b_i.

For easy reading, we first state the core formula of the framework in Eqn. (2) and provide the reasoning behind the formulae subsequently. By denoting personal interest and social influence as $\phi(u, i)$ and $\sigma(u, i, t)$ respectively, we can express the framework as follows (the logic behind will be explained soon).

$$\widehat{a}_{u,i,t} := \mu + b_u + b_i + \overbrace{\boldsymbol{p}_u^T \boldsymbol{q}_i}^{\phi(u,i)} + \overbrace{\sum_{v \in N_u} w_{v,u} \cdot \lambda\left(v, t, i\right)}^{\sigma(u,i,t)} \tag{2}$$

where we introduce the following

1. $w_{v,u}$: link weight, which will later be estimated by a function of user similarity parameterized by the so-called *homophily level*, which will be denoted as h
2. $\lambda(v,t,i)$: the *interaction level* (will be defined formally later) between the items adopted by v and item i at time step t.

Our framework adapts the general formula by proposing in Eqn. (2) a novel estimation of social influence term $\sigma(u,i,t)$ and a homophily derived link weight $w_{v,u}$. As can be seen from the definitions, the estimation will incorporate both *item interaction* and *homophily factor*. To keep the framework tractable, we assume the latent factors are static. Given this framework, we can now apply it for modeling *interacting diffusion* processes of items over a social network as follows.

Framework (Interacting Diffusion of Items). *Consider a set of items I and a social network G. For each such candidate item i, its adoption label $\widehat{a}_{u,i,t}$ can be estimated by Eqn. (2). Candidate i will be adopted by u if the estimation is close enough to 1 (i.e., $\widehat{a}_{u,i,t} \geq 1 - \theta$). Thus, at each time step, a user can adopt several candidate items which satisfy this criterion. The process continues until no more adoption can happen.*

We proceed by providing the logic behind Eqn. (2) of our framework. The logic includes two parts: how to define item interaction and how to incorporate homophily.

Item Interaction. The interaction level depends on a certain set of v's adopted items which can actually affect u's decision. This leads us to the concept of *effective item set* defined as follows.

Definition 3 (Effective item set). *For a given neighbor v of user u, the set of items adopted by v which can influence adoption decision $a_{u,i,t}$ is called* effective item set *from the neighbor at time step t and denoted as $I_{eff}(v,t)$.*

Given effective item set $I_{eff}(v,t)$, we now need to estimate the interaction level $\lambda(v,t,i)$ between the adopted items of v and candidate item i and time step t. We now provide a general estimation of $\lambda(v,t,i)$ in Definition 4.

Definition 4 (Interaction level). *The interaction level $\lambda(v,t,i)$ is defined as the sum of interactions (i.e. similarities) between the effective item set of v and i at time step t.*

$$\lambda(v,t,i) := \sum_{j \in I_{eff}(v,t)} q_j^T q_i \tag{3}$$

The social influence from neighbor v will then be $w_{v,u} \times \lambda(v,t,i)$. In total, social influence on u will be estimated by

$$\sigma(u,i,t) := \sum_{v \in N_u} w_{v,u} \times \lambda(v,t,i) = \left(\sum_{v \in N_u} \sum_{j \in I_{eff}(v,t)} w_{v,u} q_j \right)^T q_i \tag{4}$$

Note that for directed networks, N_u will be replaced by the followee set of u.

Replace (4) into (2), we obtain our novel estimation for adoption label

$$\hat{a}_{u,i,t} := \mu + b_u + b_i + \boldsymbol{p}_u^T \boldsymbol{q}_i + \left(\sum_{v \in N_u} \sum_{j \in I_{eff}(v,t)} w_{v,u} \boldsymbol{q}_j \right)^T \boldsymbol{q}_i \qquad (5)$$

This new estimation allows our framework to capture item interaction. Thus, in the context of interacting diffusion, we expect it to provide a better model than existing models (e.g. [11]). This will be realized later in our experiments on synthetic data.

Incorporating Homophily. Eqn. (5) involves link weight $w_{v,u}$ which is determined by homophily factor. Due to homophily effect, more similar individuals tend to be connected. We therefore propose to estimate $w_{v,u}$ as an *increasing* function of the similarity between u and v. In other words, for a social network with an underlying homophily level $h \in [0,1]$ (smaller h implies low homophily), we propose to define $w_{v,u}$ as:

$$w_{v,u} := g(\boldsymbol{p}_u^T \boldsymbol{p}_v | h) \qquad (6)$$

where $g(.)$ is an increasing function parameterized by h. Since weights are in $[0,1]$, we also choose functions g with range in $[0,1]$.

Finally, by replacing Eqn. (6) in (5) and using estimation of $\lambda(v,t,i)$, we obtain Eqn. (7), the main estimation of our framework.

$$\hat{a}_{u,i,t} := \mu + b_u + b_i + \boldsymbol{p}_u^T \boldsymbol{q}_i + \sum_{v \in N_u} g(\boldsymbol{p}_u^T \boldsymbol{p}_v | h) \cdot \sum_{j \in I_{eff}(v,t)} \boldsymbol{q}_j^T \boldsymbol{q}_i \qquad (7)$$

3.3 Topic Interaction and Homophily Aware Diffusion (TIHAD) Model

To apply our general framework, we need to give specific definitions for $g(\boldsymbol{p}_u^T \boldsymbol{p}_v | h)$, and $I_{eff}(v,t)$. This leads to our proposed Topic Interaction and Homophily Aware Diffusion (TIHAD) Model.

In TIHAD, we define the function $g(.)$ as a linear function of user similarity $\boldsymbol{p}_u^T \boldsymbol{p}_v$ as follow.

$$g(\boldsymbol{p}_u^T \boldsymbol{p}_v | h) := h \cdot (\boldsymbol{p}_u^T \boldsymbol{p}_v), \ \forall(u,v), \ v \in N_u \qquad (8)$$

There are other interesting forms of function $g(.)$ including $(\boldsymbol{p}_u^T \boldsymbol{p}_v)^h$. In this work, we focus on the linear form due to its tractability and leave other forms for future research.

For $I_{eff}(v,t)$, we choose the set of items *adopted recently* by neighbor v. This is based on the common intuition that a user usually pays attention only to those recent items (e.g., Twitter users only focus on recent hashtags from their followees [19]). Thus, for each time t, we choose the effective set as the set of k items which neighbor v adopted most recently with respect to time step t, which we denote as $r_v^{k,t}$. Hence, $I_{eff}(v,t) = r_v^{k,t}$.

The TIHAD model is therefore expressed as Eqn. (9).

$$\hat{a}_{u,i,t}^{\text{tihad}} = \mu + b_u + b_i + p_u^T q_i + h p_u^T \left[S_t(u) \right] q_i \tag{9}$$

where the matrix $S_t(u)$ is

$$S_t(u) := \sum_{v \in N_u} p_v \left(\sum_{j \in r_v^{k,t}} q_j \right)^T \tag{10}$$

$S_t(u)$ can be interpreted as the matrix characterizing the social influence from u's neighbors recent adoption events.

3.4 Linear Threshold with Latent Factors (LTLF)

In the special case when $I_{eff}(v,t) = \{i\}$, Eqn. (4) becomes

$$\tilde{\sigma}(u,i,t) = \left(\sum_{v \in \mathcal{A}_{u,t}^i} w_{v,u} \right) \|q_i\|^2$$

where v now is not an arbitrary neighbor of u but instead an *active* neighbor i.e. one who actually adopted i at time t. This estimation of social influence is obviously an extension of Eqn. (1) commonly used in Linear Threshold models ([6], [1]). Thus, by substituting it into Eqn. (5), we obtain the following model, called Linear Threshold with Latent Factors (LTLF)

$$\hat{a}_{u,i,t}^{\text{ltlf}} := \mu + b_u + b_i + p_u^T q_i + \left(\sum_{v \in \mathcal{A}_{u,t}^i} w_{v,u} \right) \|q_i\|^2 \tag{11}$$

4   Learning of TIHAD Model

We formulate the learning of TIHAD model parameters as a constrained optimization problem, which can be solved by Projected Gradient Descent (PGD). We provide the detailed formula to solve the problem and a pseudocode for model learning. For brevity, we use P and Q matrices to denote user and item latent factors respectively. All parameters of TIHAD then can be compactly represented by $\Pi = (h, \mu, \{b_u\}_{u \in U}, \{b_i\}_{i \in I}, P, Q)$. We also use $\hat{a}_{u,i,t}$ in place of $\hat{a}_{u,i,t}^{\text{tihad}}$ for brevity.

4.1 Optimization Formulation

Let A_1^T denote the set of all adoption labels in a diffusion cascade during the time span $[1, T]$.

$$A_1^T := \{a_{u,i,t} : t \in [1,T], \ u \in U \text{ and } i \in C_{u,t}\} \tag{12}$$

Diffusion data is then represented by a tuple of item set, the social network and the adoption labels as $\mathcal{D} = \left(I, G, A_1^T\right)$. Given $\mathcal{D}$, we formulate the model learning problem as finding the optimal parameters $\boldsymbol{\Pi}^*$ that minimize squared error upon generating the adoption labels.

For a given $\boldsymbol{\Pi}$, the squared error at time step t is the sum

$$SE_t(\boldsymbol{\Pi}|\mathcal{D}) = \sum_{u \in U} \sum_{i \in C_{u,t}} \left[\hat{a}_{u,i,t}(\boldsymbol{\Pi}) - a_{u,i,t}\right]^2 \tag{13}$$

Hence, over the whole time span $[1, T]$, the total error is

$$\mathcal{E}(\boldsymbol{\Pi}|\mathcal{D}) = \sum_{t=1}^{T} SE_t(\boldsymbol{\Pi}|\mathcal{D}) = \sum_{t=1}^{T} \sum_{u \in U} \sum_{i \in C_{u,t}} \left[\hat{a}_{u,i,t}(\boldsymbol{\Pi}) - a_{u,i,t}\right]^2 \tag{14}$$

To avoid over-fitting, we also define a regularizer as

$$\mathcal{R}(\boldsymbol{\Pi}) := h^2 + \sum_u b_u^2 + \sum_i b_i^2 + \|P\|_F^2 + \|Q\|_F^2 \tag{15}$$

where $\|.\|_F$ denotes the usual Frobenius norm. Hence, the objective function is

$$J(\boldsymbol{\Pi}|\mathcal{D}) = \frac{1}{2}\left[(\mathcal{E}(\boldsymbol{\Pi}|\mathcal{D}) + \delta\mathcal{R}(\boldsymbol{\Pi})\right]$$

We now can formulate the learning as the following constrained optimization problem.

Problem 1. *Given diffusion data set* $\mathcal{D} = \left(I, G, A_1^T\right)$. *We learn parameters* $\boldsymbol{\Pi}$ *by solving for optimal parameters which minimize the objective function*

$$\boldsymbol{\Pi}^* = \underset{\boldsymbol{\Pi}}{\operatorname{argmin}} \, J(\boldsymbol{\Pi}|\mathcal{D}) = \underset{\boldsymbol{\Pi}}{\operatorname{argmin}} \, \frac{1}{2}\left[\mathcal{E}(\boldsymbol{\Pi}|\mathcal{D}) + \delta\mathcal{R}(\boldsymbol{\Pi})\right] \tag{16}$$

subject to constraints

$$\boldsymbol{p}_u \geq 0, \ \forall u \in U, \quad \boldsymbol{q}_i \geq 0, \ \forall i \in I \quad and \ 0 \leq h \leq 1 \tag{17}$$

4.2 Optimization Solution

In general, the above problem is not convex. Thus, we resort to a solver which uses grid search and Projected Gradient Descent (PGD). For that, we provide formulae of gradients in the following sections. Due to space constraints, proofs of these formulae are not provided, interested readers can find it in the technical note.[1]

---

[1] http://goo.gl/2ltY9I

Derivatives for Bias Variables

$$\frac{\partial}{\partial \mu}J = \sum_t \sum_{u \in U} \sum_{i \in C_{u,t}} \overbrace{(\widehat{a}_{u,i,t}(\boldsymbol{\Pi}) - a_{u,i,t})}^{e_{u,i,t}} \tag{18a}$$

$$\forall u \in U, \ \frac{\partial}{\partial b_u}J = \delta b_u + \sum_t \sum_{i \in C_{u,t}} e_{u,i,t} \tag{18b}$$

$$\forall i \in I, \ \frac{\partial}{\partial b_i}J = \delta b_i + \sum_t \sum_{u \in U : i \in C_{u,t}} e_{u,i,t} \tag{18c}$$

Derivative for Homophily Variable

$$\frac{\partial}{\partial h}J = \delta h + \sum_t \sum_u \boldsymbol{p}_u^T [\boldsymbol{S}_t(u)] \, \boldsymbol{q}_t^{err}(u) \tag{19}$$

where $\boldsymbol{S}_t(u)$ is defined in Eqn. (10) and $\boldsymbol{q}_t^{err}(u) := \sum_{i \in C_{u,t}} e_{u,i,t} \cdot \boldsymbol{q}_i$.

Derivatives for User and Item Factors

1. (Gradient w.r.t user factor $\boldsymbol{p}_u$) For each given user u, we have

$$\nabla_{\boldsymbol{p}_u} J = \delta \boldsymbol{p}_u + \sum_t \left[\boldsymbol{M}_t(u) \boldsymbol{q}_t^{err}(u) + h \eta_t(u) \boldsymbol{q}_t^k(u)\right] \tag{20}$$

where matrix $\boldsymbol{M}_t(u)$ and scalar $\eta_t(u)$ are defined as

$$\boldsymbol{M}_t(u) := \boldsymbol{Id} + h \boldsymbol{S}_t(u) \text{ and } \eta_t(u) := \sum_{v \in N_u} \boldsymbol{p}_v^T \boldsymbol{q}_t^{err}(v) \tag{21}$$

where $\boldsymbol{Id}$ denotes the identity matrix.

2. (Gradient w.r.t. item factor $\boldsymbol{q}_i$) For each given item i, we have

$$\nabla_{\boldsymbol{q}_i} J = \delta \boldsymbol{q}_i + \sum_t \left(h \sum_{u \in U} \left[\boldsymbol{q}_t^{err}(u) \boldsymbol{\varphi}_{u,i,t}^T\right] \boldsymbol{p}_u + \sum_{u : C_{u,t} \ni i} e_{u,i,t} \left[\boldsymbol{M}_t(u)\right]^T \boldsymbol{p}_u \right) \tag{22}$$

where vector $\boldsymbol{\varphi}_{u,i,t} := \sum_{recent\ adopters} \boldsymbol{p}_v$ is the sum of factors of neighbors who adopted i recently.

Now that all derivatives are available, we can use them in Projected Gradient Descent (PGD) with grid search to update the corresponding parameters. Thus, we repeat Algorithm 1 with different initial parameter values to learn the parameters of TIHAD model. All the derivatives in the algorithm are computed using Eqns. (18a) – (18c) and (19) – (22).

Algorithm 1. PGD for TIHAD model using an initial guess $\boldsymbol{\Pi}_0$

1: **procedure** TRAIN($\mathcal{D}$, $\boldsymbol{\Pi}_0$, ε)
2:     Initialize $\boldsymbol{\Pi}_c \leftarrow \boldsymbol{\Pi}_0$
3:     **while** (!*converge*) **do**
4:         Compute objective value: $j_c \leftarrow J(\boldsymbol{\Pi}_c | \mathcal{D})$ ▷ use Eqns. (14) − (16)
5:         Compute gradients: $\boldsymbol{g}_c \leftarrow \nabla J(\boldsymbol{\Pi}_c | \mathcal{D})$ ▷ use Eqns. (18a) − (22)
6:         Descend & project: $\boldsymbol{\Pi}_n \leftarrow$ GRADPROJ($\boldsymbol{\Pi}_c, j_c, \boldsymbol{g}_c$)   ▷ see gradproj() in [7]
7:         Check convergence: *converge* $\leftarrow$ ($|\boldsymbol{\Pi}_n - \boldsymbol{\Pi}_c| < \varepsilon$)
8:         $\boldsymbol{\Pi}_c \leftarrow \boldsymbol{\Pi}_n$
9:     **end while**
10:     **return** $\boldsymbol{\Pi}_n$
11: **end procedure**

---

5   Experiments

In this study, we want to be able to evaluate TIHAD model with some parameter settings that control the item interaction and homophily factor during the diffusion process. Hence, we need a synthetic diffusion data generation method with the following input parameters: (a) M items, (b) N users, (c) N_e relationships among the users, (d) f latent factors, (e) homophily value h for the social network, (f) T number of time steps, and (g) k recently adopted items. The generation steps are described below:

1. (Generation of M items and N users in latent space) We generate M items and N users as f-dimensional vectors $\boldsymbol{q}_i$'s and $\boldsymbol{p}_u$'s respectively. The item and user vectors are generated such that each of them has a dominant factor. The set of users and items are denoted by U and I respectively.
2. (Generation of a social network with homophily value h) We generate N_e edges among the users using Algorithm 2. The resultant network, $G_h = (U, E_h)$ where E_h denotes the set of N_e edges, satisfies the required homophily level h.
3. (Generation of an initial adoption state) We want to ensure that every user in the network initially has adopted at least k items. We assign k items to each user based on his latent factor interests.
4. (Generation of a diffusion cascade) We randomly assign a user as the single seed of diffusion. The seed user will adopt all M items initially. We then employ TIHAD model to start generating a data set of simultaneous diffusion of the items over the network G_h within the time interval $[1, T]$. The details of this step are given in Algorithm 3.

We generate N diffusion cascades by performing steps 3 and 4 with a different initial adoption state and different user as the seed each time. Hence every diffusion cascade share the same network with identical user and item latent factor vectors. We finally generate N different data sets so that we can get empirical distribution of cascade sizes.

Algorithm 2. Generation of a network with a given homophily level

---
1: **procedure** BUILDNETWORK(U, N_e, h)
2:     $Pairs \leftarrow \{(u, v) : u \neq v \in U\}$
3:     **for** each user pair $(u, v) \in Pairs$ **do**
4:         Compute user-item similarity: $sim(u, v) \leftarrow \boldsymbol{p}_u^T \boldsymbol{p}_v$
5:         Compute edge weight: $\rho(u, v) \sim \exp(h \cdot sim(u, v))$
6:     **end for**
7:     Normalize: $p(u, v) \leftarrow \frac{\rho(u,v)}{\sum \rho(u',v')}$, $\forall (u, v) \in Pairs$
8:     Collect probabilities: $\boldsymbol{probs} \leftarrow (p(u, v) : (u, v) \in Pairs)$
9:     Sample N_e edges based on the probabilities: $E_h \leftarrow sample(Pairs, N_e, \boldsymbol{probs})$
10: **return** Network $G_h = (U, E_h)$
11: **end procedure**

Algorithm 3. Generation of diffusion data

1: **procedure** CREATEDIFFUSION(I, G_h, θ, T, u_s)   ▷ $G_h = (U, E_h)$: network in
        Algo. 2
2:     **for** $t \in [1, T]$ **do**
3:         Initialize $A_t \leftarrow \emptyset$ ▷ Set of adoption records at time t
4:         **for** $u \in U$ **do**
5:             Derive $C_{u,t}$ by Definition 1   ▷ use seed u_s to get $C_{u,1}$, $\forall u$
6:             **for** $i \in C_{u,t}$ **do**
7:                 Compute adoption label approximation $\hat{a}_{u,i,t}$ by Eqn. (9)
8:             **end for**
9:             Pick adoptions $I_t(u) \leftarrow \{i \in C_{u,t} : \hat{a}_{u,i,t} \geq 1 - \theta\}$ ▷ approx. is close to 1
10:             $A_t \leftarrow A_t \cup \{(u, i, t) : i \in I_t(u)\}$ ▷ Add to adoption records at time t
11:         **end for**
12:     **end for**
13:     Collect all adoption records: $\boldsymbol{A}_1^T \leftarrow \bigcup_{t=1}^T A_t$
14:     **return** $\mathcal{D} = (I, G_h, \boldsymbol{A}_1^T)$
15: **end procedure**

5.1   Impact of Homophily on Diffusion

Experiment Setup: We study how the size of diffusion cascade is affected by different degrees of homophily h. Thus, we generate items and users by setting $f = 10$. We then generate diffusion in five different networks G_h's each with a different h value, $h \in \{0, 0.2, 0.4, 0.6, 0.8\}$. These networks however share the same set of users and same number of edges to minimize the effect of choices of users and number of relationships among them. For each such network, we generate N diffusion cascades of M items using TIHAD and study distribution of the average cascade size over the M items. Detailed statistics of this experiment is provided in Table 1.

Result: As the homophily level increases, the diffusion cascade also becomes larger (see Figure 1a). This trend is observed for all items. To evaluate the robustness of the result, we repeat the experiment for $f = 15$ and $f = 20$.

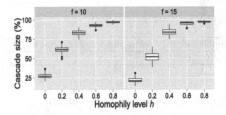 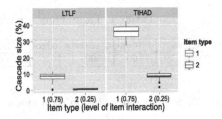

(a) Impact of homophily (cascades generated by TIHAD under different settings for number of factors f)

(b) Impact of item interaction (cascades generated by both models, for TIHAD we set parameters $h = 0.1$ and $f = 10$)

Fig. 1. Impact of homophily and item interaction on diffusion.

We report here results for $f = 10$ and $f = 15$. This result is expected as homophily facilitates diffusion ([5], [3]). It also shows that our model has incorporated homophily effect properly.

5.2   Impact of Item Interaction on Diffusion

Experiment Setup: In this experiment, we change our focus to study how item interaction (i.e. support among items) affects diffusion. We now generate diffusion cascades on the same network with a fixed homophily level $h = 0.1$. The item set is however generated differently. We partition the item set I into the *majority* set I_1 (occupy 75% of I) and the *minority* set I_2. In each subset, items are generated such that they are similar to each other. Thus, items in I_1 receive more interaction than items in I_2 and we can study difference in cascade sizes of items in two sets. Other statistics of this experiment is the same as in Table 1.

Under this setting, we use TIHAD model to simulate diffusion as done in the previous experiments. We then compare cascade size distribution of items in I_1 against that of items in I_2. We also want to see if cascades generated by TIHAD are significantly different from those generated by a baseline diffusion model that does not consider item interaction. Hence, we generate another set of cascades following the same process using the LTLF model. The cascade size distributions of the two models are then compared.

Result: Figure 1b shows several interesting insights. First, it provides strong evidence that TIHAD model can capture the item interaction effect (among similar items) currently ignored by the existing models including LTLF. The figure shows that the cascade size of an item diffused with TIHAD is much larger than that of the item when it is diffused using LTLF. Moreover, the more similar an item with previous items, the larger cascade size it can reach. This makes sense since an item will receive more support in diffusion if it is more similar to other previously adopted items.

Table 1. Parameters used in synthetic data generation

# factors	# items	# users	# edges	Homophily level	# recent items	# time steps
$f \in \{10, 15, 20\}$	100	500	$70K$	$h \in \{0, 0.2, 0.4, 0.6, 0.8\}$	$k = 5$	$T = 20$

Table 2. Statistics of diffusion data among Singapore Twitter users in Valentine Day

Data set	# hashtags	# users	# follow links	# adoptions	# time steps	# adoption labels
Training	4002	1000	9935	$11,565$	12	$60,875$
Test	1219	884	8754	9390	12	$39,375$
Total	4002	1000	9935	$20,955$	24	$100,250$

5.3   Hashtag Diffusion Prediction Evaluation

This experiment aims to evaluate TIHAD using real dataset and compare it with the baseline LTLF model which does not consider item interaction.

Data Set: We first collected the diffusion of hashtags in the Twitter network among Singapore users during on 14 February 2014, the Valentine Day. We expected that there should be some interesting diffusion cascades on this special day. We extracted the tweets of about 150,000 Singapore users from 3 to 16 February and sampled 1000 active users who adopted at least 3 hashtags per day. These users are connected by a social network with 9935 follow links.

We next wanted to determine the time step when each user first adopted a hashtag during the Valentine Day. Each time step duration is set as one hour. We confined ourselves to *fresh* hashtags which only appeared during Valentine Day but not the days during [3 Feb, 13 Feb]. We then identified the time step a user adopted a hashtag as the first time step in 14 February he used the hashtag. We obtained $20,847$ hashtags which the active users adopted from 00:00am to 11:59pm on the Valentine day. By filtering away unpopular hashtags, i.e., those with less than 5 active users adopting them, we were left with 4002 hashtags and $20,955$ adoptions. Based on Definition 2, we derived $100,250$ adoption labels (both adoption and non-adoption) associated with these 24 hours. Adoptions of the users on previous day (13 Feb) were used as their initial adoption histories. The hashtag diffusion data on 14 February from 0:00am to 11:59am is then used as the training data, while the remaining data on 14 February is used as the test data. The statistics of combined training and test datasets is summarized in Table 2.

Training Process: We trained both TIHAD and LTLF using the diffusion training dataset on February 14. We tried different values for the regularization constant and observed that $\delta = 0.1$ gives the best result in terms of minimizing RMSE. We also tried different values for the number of recent items $k \in \{1, \ldots, 10\}$ and found that $k \in \{3, 4\}$ yield the best RMSE result for this

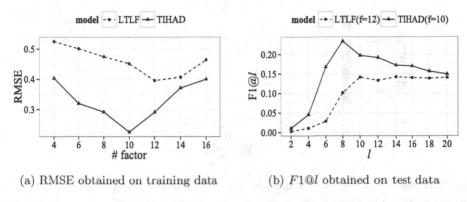

(a) RMSE obtained on training data   (b) $F1@l$ obtained on test data

Fig. 2. Comparing TIHAD against baseline LTLF. Both models were trained with regularization coefficient $\delta = 0.1$; for TIHAD, the number of recent items k is set as 3.

training dataset. In the learning process, we observed that both models can achieve smallest RMSE for the training data.

Evaluation Metrics: For evaluations, we used two accuracy metrics: (i) RMSE for measuring the model performance during training, and (ii) $F1@l$ when using the trained models for the *hashtag adoption prediction task* on the test data. To compute $F1@l$, we use the trained models to predict hashtag adoptions (based on estimated adoption labels) from 12:00 noon to 11:59pm of 14 Feb 2014. We selected those users who appear in both the training and test datasets and extracted from their tweets generated during the test period the hashtags that already appeared in the training set. The resultant test set had 884 users and 1219 hashtags which were actually adopted during the test period (detailed statistics of the test set can be found in Table 2).

Results: We first focus on the accuracy of trained models using RMSE defined on the training data. As shown in Figure 2a, the RMSE obtained by TIHAD is much smaller than that of LTLF when they are trained using the same dataset for different latent factor settings (i.e., $4 \leq f \leq 16$). TIHAD achieves the best RMSE when $f = 10$, while LTLF achieves best RMSE at $f = 12$.

In the prediction task, TIHAD shows a huge improvement over LTLF as shown in Figure 2b. Other than $l = 2$, TIHAD outperforms LTLF for all other i values. The highest $F1$ achieved by TIHAD ($F1@8$) is more than 150% that of LTLF ($F1@10$).

As TIHAD performs best for 10 factors, we would like to know what are the 10 factors. We manually check the top hashtags of each latent factor. We discover that the latent factors are topical and manually assign them topical labels. Table 3 shows the latent factors and their top 3 hashtags (due to limited space). Most of the latent factors (e.g., Music tour, Valentine, Electronics, Self-Improve) are self explanatory based on hashtags. The "Music bands/Singers" latent factor covers names of singers (e.g., Siti Nurhaliza and Eminem) and music concert (e.g., SUL14). The "Local movies/actors" latent factor covers popular

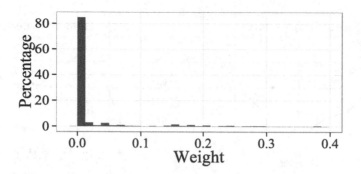

Fig. 3. Histogram of influence weights $w_{v,u}$ which TIHAD learned for the network of Twitter users in our experiment.

Table 3. Latent factors and their top-3 hashtags

Latent Factors	Hashtags
Music bands/Singers	eminemftw, DatoSitiNurhaliza, SUL14
Local movies/actors	YouWhoCameFromTheStars, BrothersKeeper, GongLi
International movies/actors	frozen, jimmyfallon, KristenWiig
Music tour	RedAsiaTour, TheScriptUSTour, BANGERZTour2014
Sport	ICC2014, F1NightRace, LFCfacebook
Beauty	ILoveWTF, Dior, maybellinesg
Valentine	happyvalentine, firstvalentine, TweetforLove
Scandal/Controversy	AsylumSeekers, bigimmigrationrow, LittleIndiaRiot
Electronics	Xiaomi, ipadmini, Logitech
Self-improve	limitless, nickvijucic, empoweryourself

movies (e.g., "You Who Came From the Stars", "Brothers Keeper") and actor (e.g., Gong Li). The other latent factors can be interpreted in a similar manner.

Finally we would like to see what TIHAD can tell us about the network based on the homophily level and influence weights it learned. The homophily level learned by TIHAD is $h = 0.08$. This value is quite small and can be explained due to the sparseness of the network under study. Moreover, the histogram of influence weights $w_{v,u}$ in Figure 3 shows that most weights are very small (80% of them are close to 0), which matches the nature of weak links among most Twitter users.

6   Conclusion

This work deals with the challenging problem of modeling multiple simultaneous diffusion processes where topic level interaction exists among items being diffused in a social network with homophily. We successfully incorporate item interaction and homophily by proposing a novel way to model social influence from recent adoptions of user's neighbors. Behavior of the model under different settings and parameters have been investigated. Results on synthetic data show that both homophily and interaction at topic level can increase diffusion remarkably. Experiment on hashtag diffusion on Twitter shows that TIHAD can model interacting diffusion effectively and give better prediction as well.

Since training TIHAD is not a convex problem, we are currently using grid search to deal with the non-convexity. However, the problem is still convex for each set of parameters if others are kept fixed. Thus, we plan to use Alternating Descent to develop a more rigorous algorithm.

Acknowledgement. This research is supported by the Singapore National Research Foundation under its International Research Centre @ Singapore Funding Initiative and administered by the IDM Programme Office, Media Development Authority (MDA).

References

1. Chen, W., Yuan, Y., Zhang, L.: Scalable influence maximization in social networks under the linear threshold model. In: Proceedings of the 10th ICDM, pp. 88–97. IEEE (2010)
2. Christakis, N.A., Fowler, J.H.: The spread of obesity in a large social network over 32 years. New England Journal of Medicine **357**(4), 370–379 (2007)
3. De Choudhury, M., Sundaram, H., John, A., Seligmann, D.D., Kelliher, A.: birds of a feather: Does user homophily impact information diffusion in social media? (2010). arXiv:1006.1702
4. Delporte, J., Karatzoglou, A., Matuszczyk, T., Canu, S.: Socially enabled preference learning from implicit feedback data. In: Blockeel, H., Kersting, K., Nijssen, S., Železný, F. (eds.) ECML PKDD 2013, Part II. LNCS, vol. 8189, pp. 145–160. Springer, Heidelberg (2013)
5. Golub, B., Jackson, M.O.: How homophily affects the speed of learning and best-response dynamics. The Quarterly Journal of Economics **127**(3), 1287–1338 (2012)
6. Granovetter, M.: Threshold models of collective behavior. American Journal of Sociology, 1420–1443 (1978)
7. Kelley, C.T.: Iterative methods for optimization, vol. 18. SIAM (1999)
8. Kempe, D., Kleinberg, J., Tardos, É.: Maximizing the spread of influence through a social network. In: Proceedings of the 9th KDD, pp. 137–146. ACM (2003)
9. Koren, Y.: Factorization meets the neighborhood: a multifaceted collaborative filtering model. In: Proceedings of the 14th KDD, pp. 426–434 (2008)
10. Koren, Y., Bell, R., Volinsky, C.: Matrix factorization techniques for recommender systems. Computer **42**(8), 30–37 (2009)
11. Lin, S., Hu, Q., Wang, F., Yu, P.S.: Steering information diffusion dynamically against user attention limitation. In: Proceedings of the 14th ICDM (2014)
12. Ma, H., Yang, H., Lyu, M.R., King, I.: Sorec: social recommendation using probabilistic matrix factorization. In: Proceedings of the 17th CIKM, pp. 931–940. ACM (2008)
13. Mnih, A., Salakhutdinov, R.: Probabilistic matrix factorization. In: Proceedings of the 20th NIPS, pp. 1257–1264 (2007)
14. Rogers, E.M.: Diffusion of innovations. Free Press, New York (1983)

15. Salakhutdinov, R., Mnih, A.: Bayesian probabilistic matrix factorization using markov chain monte carlo. In: Proceedings of the 25th ICML, pp. 880–887. ACM (2008)
16. Sarwar, B., Karypis, G., Konstan, J., Riedl, J.: Item-based collaborative filtering recommendation algorithms. In: Proceedings of the 10th WWW (2001)
17. Shen, Y., Jin, R.: Learning personal+social latent factor model for social recommendation. In: Proceedings of the 18th KDD, pp. 1303–1311. ACM (2012)
18. Su, X., Khoshgoftaar, T.M.: A survey of collaborative filtering techniques. Advances in Artificial Intelligence **2009**:4 (2009)
19. Weng, L., Flammini, A., Vespignani, A., Menczer, F.: Competition among memes in a world with limited attention. Scientific Reports (2012)

Maintaining Sliding-Window Neighborhood Profiles in Interaction Networks

Rohit Kumar[1]([⊠]), Toon Calders[1], Aristides Gionis[2], and Nikolaj Tatti[2]

[1] Department of Computer and Decision Engineering,
Université Libre de Bruxelles, Brussels, Belgium
r.kumar@ulb.ac.be
[2] Helsinki Institute for Information Technology and Department of Computer
Science, Aalto University, Espoo, Finland

Abstract. Large networks are being generated by applications that keep track of relationships between different data entities. Examples include online social networks recording interactions between individuals, sensor networks logging information exchanges between sensors, and more. There is a large body of literature on computing exact or approximate properties on large networks, although most methods assume static networks. On the other hand, in most modern real-world applications, networks are highly dynamic and continuous interactions along existing connections are generated. Furthermore, it is desirable to consider that old edges become less important, and their contribution to the current view of the network diminishes over time.

We study the problem of maintaining the neighborhood profile of each node in an *interaction network*. Maintaining such a profile has applications in modeling network evolution and monitoring the importance of the nodes of the network over time. We present an online streaming algorithm to maintain neighborhood profiles in the sliding-window model. The algorithm is highly scalable as it permits parallel processing and the computation is node centric, hence it scales easily to very large networks on a distributed system, like Apache Giraph. We present results from both serial and parallel implementations of the algorithm for different social networks. The summary of the graph is maintained such that query of any window length can be performed.

1 Introduction

Modern big-data systems are confronted with scenarios in which data are gathered in exceedingly large volumes. In many cases, the system entities are modeled as graphs, and the recorded data represent fine-grained activity among the graph entities. Traditionally, graph mining has focused on studying static graphs. However, as the emergence of new technologies makes it possible to gather detailed information about the behavior of the graph entities over time, a growing body of literature is devoted to the analysis of dynamic graphs.

In this paper we focus on a dynamic-graph model suitable for recording interactions between the graph entities over time. We refer to this model as *interaction*

A. Appice et al. (Eds.): ECML PKDD 2015, Part II, LNAI 9285, pp. 719–735, 2015.
DOI: 10.1007/978-3-319-23525-7_44

networks [26], while it is also known in the literature as *temporal networks* [21] or *temporal graphs* [23]. An interaction network is defined as a sequence of time-stamped interactions $\mathcal{E}$ over edges of a static graph $G = (V, E)$. In this way, many interactions may occur between two nodes at different time points. Interaction networks can be used to model the following modern application scenarios:

1. the set of nodes V represents the users of a social network or a communication network, and each interaction over an edge represents an interaction between two users, e.g., emailing, making a call, re-tweeting, etc.;
2. the set of nodes V represents autonomous agents, and each edge represents an interaction between two agents, e.g., exchanging data, being in the physical proximity of each other, etc.

We study the problem of maintaining the *neighborhood profile* of each node of a interaction network. In particular, we are interested in maintaining a data structure that allows to answer efficiently queries of the type *"how many nodes are within distance r from node v at time t?"* Graph neighborhood profiles have been studied extensively for static graphs [6,25]. They provide a fundamental primitive for mining large graphs, either for characterizing the global graph structure, or for discovering important and central nodes in the graph. In this work, we extend the concept of neighborhood profiles for interaction networks, and we develop algorithms for computing neighborhood profiles efficiently in large and rapidly-evolving interaction networks. Our methods can be used for network monitoring, and allow detecting changes in the graph structure, as well as keeping track of the evolution of node centrality and importance.

To make our methods scalable to large and fast-evolving networks, we design our algorithms under the *data-stream model* [18,24]. This model requires to process the interactions in an online fashion, and perform fast memory updates for each interaction processed. To make our model adaptable to changes and allow concept drifts we focus on the sliding-window model [14], a data-stream model that incorporates a forgetting mechanism, by considering, at any time point, only the most recent items up to that point. One uncommon benefit of our algorithm is that because of the data structure we incrementally maintain, the user can decide about the exact window length at query time.

Concretely, in this paper we make the following contributions: *(i)* we introduce a new problem of efficiently querying neighborhood profiles on interaction networks in Section 3; *(ii)* we develop and analyze an exact but memory-inefficient (Section 4) and an inexact but more efficient streaming algorithm for the sliding-window model (Section 5); *(iii)* we provide experimental validation of the algorithms in Section 7.

2   Preliminaries

We consider a static underlying graph $G = (V, E)$. An *interaction* over G is a time-stamped edge $(\{v, w\}, t)$ indicating an interaction between nodes v and w. An *interaction network* over G is now defined as a pair $(G, \mathcal{E})$, where G is a static graph and $\mathcal{E}$ is a set of interactions. We should point out that we do not need to know E beforehand.

If the set of interactions $\mathcal{E} = \{(\{u,v\},t)\}$ is ordered by time, it can be seen as a *stream of edges*, and written as $\mathcal{E} = \langle (e_1,t_1), (e_2,t_2), \ldots \rangle$, with $t_1 \leq t_2 \leq \ldots$. Note that two fixed nodes may interact multiple times in $\mathcal{E}$.

In our model we are only interested in recent events, and hence queries over our interaction network will always include a window length w — recall that the summary will be maintained in such a way that all window lengths are possible, i.e., every query can use a different window length. The *snapshot graph at time t for window w*, denoted $G(t,w)$, is the triplet $(V, E(t,w), recent)$ in which $E(t,w) = \{e \mid (e,t') \in \mathcal{E} \text{ with } t - w < t' \leq t\}$, and *recent* is a function mapping an edge $e \in E(t,w)$ to the most recent time stamp that an interaction between the endpoints of e occurred, that is, $recent(e) = \max\{t' \mid (e,t') \in \mathcal{E} \text{ such that } t - w < t' \leq t\}$.

Furthermore, for the graph G we have the usual definitions; a *path* of length k between two nodes $u, v \in V$ is a sequence of nodes $u = w_0, \ldots, w_k = v$ such that $\{w_{i-1}, w_i\} \in E$, for all $i = 1, \ldots, k$, and all w_i are different. The *distance* between u and v in the graph G is defined as the length of the shortest path between u and v, if such a path exists, otherwise it is infinity. The *distance* between nodes u and v in the graph G is denoted by $d_G(v,w)$, or simply $d(v,w)$, if G is known from the context.

3   Problem Statement

The central notion we are computing in this paper is the *neighborhood profile*:

Definition 1. *Let $G = (V, E)$ be a graph and let $u \in V$ be a node. The r-neighborhood of u in G, denoted $N_G(u,r)$, is is the set of all nodes that are at distance r from node u, i.e., $N_G(u,r) = \{v \mid d_G(u,v) = r\}$. We write $n_G(u,r) = |N_G(u,r)|$ to denote the cardinality of the r-neighborhood. We will call the sequence $p_G(u,r) = \langle n_G(u,1), n_G(u,2), \ldots, n_G(u,r) \rangle$ the r-neighborhood profile of the node u in graph G.*

In this paper we study the problem of *maintaining* the neighborhood profile $p_{G(t,w)}(u,r)$, for all nodes $u \in V$, as new interactions arrive in $\mathcal{E}$. Our solution allows w to vary; hence, at a time point t, we should be able to query for the neighborhood profile $p_{G(t,w)}(u,r)$ for any w. If there is an upper bound given for w, say w_{max}, then we can use this information to improve memory consumption. However, this is optional, and we can set $w_{max} = \infty$. On the other hand, r is given and fixed. Obviously, by computing $p_{G(t,w)}(u,r)$ we also compute $p_{G(t,w)}(u,r')$ for $r' < r$.

Let $H = G(t,w)$. To simplify the notation we will denote $N_H(u,r)$, $n_H(u,r)$, $p_H(u,r)$ by $N_{t,w}(u,r)$, $n_{t,w}(u,r)$, $p_{t,w}(u,r)$, respectively. Moreover, if $w = w_{max}$, then we will use $N_t(u,r)$, $n_t(u,r)$, $p_t(u,r)$, respectively. We will also write $G(t) = G(t, w_{max})$ and $E(t) = E(t, w_{max})$.

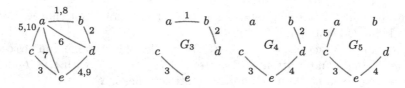

Fig. 1. A toy interaction network, and three snapshot graphs with a window size of 3.

Example 1. Consider the illustration given in Figure 1 of an edge stream over the set of nodes $V = \{a, b, c, d, e\}$. The numbers on the edges denote the time of interactions over the edges. Let the window length be 3. The snapshot graphs $G(t)$ at times $t = 3, 4, 5$ are also depicted in Figure 1. The 3-neighborhood profiles of node c in these graphs are respectively $(1, 0, 0)$, $(1, 1, 1)$, and $(2, 1, 0)$.

To accomplish our goal we maintain a summary S_t of the snapshot graph $G(t, w_{max})$, from which we can efficiently compute the neighborhood profiles $p_{t,w}(u, r)$, for every node u in the graph G. More concretely, we require that the summary S_t has the following properties:

1. The summary S_t of $G(t, w_{max})$ should require limited storage space.
2. The size of the r-neighborhood $n_{t,w}(u, r)$ should be easy to compute from S_t. The time to compute $n_{t,w}(u, r)$ from S_t will be called *query time*.
3. There should be an efficient update procedure to compute S_{t_i} from $S_{t_{i-1}}$ and the edge e_{t_i} on which the interaction at time-stamp t_i is taking place.

4   Maintaining the Exact Neighborhood Profile

We first introduce an *exact*, yet memory-inefficient solution. This exact solution will form the basis of a memory-efficient and faster *approximate* solution based on the well-known *hyperloglog sketches*.

4.1   Summary for Neighborhood Functions

An essential notion in our solution is the *horizon of a path*, which expresses the latest time that needs to be included in the sliding window in order for the path to exist; i.e., if the sliding window starts after the horizon the path will not exist in it anymore.

Definition 2. *Let* $G(t) = (V, E, recent)$ *be a snapshot graph and* $p = \langle v_0, \ldots, v_k \rangle$ *a path in it. The edge horizon of* p *in* $G(t)$, *denoted by* $h_t(p)$, *is the time stamp of the oldest edge on that path:* $h_t(p) = \min\{recent((v_{i-1}, v_i)) \mid i = 1, \ldots, k\}$.

We will next define the horizon between two nodes u and v. Let $\mathcal{P}_H(u, v)$ be all the paths from u to v in a graph H. If $H = G(t)$, then we will write $\mathcal{P}_t(u, v)$.

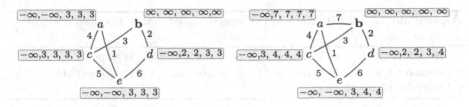

Fig. 2. Two toy snapshot graphs along with $h(u, b, i)$ for $i = 0, \ldots, 4$.

Definition 3. *The horizon for length i between two different nodes u and v is the maximum horizon of any path of at most length i between them; that is, $h_t(u, v, i) = \max\{h_t(p) \mid p \in \mathcal{P}_t(u, v), |p| + 1 \le i\}$. We set $h_t(u, v, i) = -\infty$ if no such path exists. For any node u, $h_t(u, u, i)$ is defined to be ∞.*

Example 2. Consider the leftmost graph given in Figure 2, along with, for every node $u \in \{a, b, c, d, e\}$, the list of horizons $h(u, b, 0), \ldots, h(u, b, 4)$. In this graph $h(d, b, 1) = h(d, b, 2) = 2$, as there is an edge with a time stamp of 2. However, $h(d, b, 3) = 3$ as there is a path $\langle d, e, c, b \rangle$ with a horizon of 3.

The horizon between two nodes u and v for a length i is very important for our algorithm as it expresses in which windows u and v are at a distance i or less. Windows that include the horizon will have the nodes at distance i, shorter windows will not. Hence, if for a node u we know all horizons $h_t(u, v, i)$, for all distances i and all other nodes v, we can give the complete neighborhood profile for u for any window length. Hence, the summary S_t of the snapshot graph $G(t)$ will be the combination, for all nodes u and distances $i = 0, \ldots, r$, of the summaries S_t^u for $N_t(u, i)$. In other words, for every node u, we will be maintaining the summary $S_t^u = (S_t^u[0], \ldots, S_t^u[r])$, where $S_t^u[i] = \{(v, h_t(u, v, i)) \mid h_t(u, v, i) > -\infty\}$.

Example 3. For the snapshot graph given in Fig. 2, the summary S_t consists of $S^u[i]$, $i = 0, \ldots, r$. Assuming $r = 3$, the summaries for a and b are as follows:

S^a						S^b				
distance a	b	c	d	e		distance a	b	c	d	e
0 ∞						0 ∞				
1	∞	3	4	1		1	∞	3	2	
2	∞	3	4	1	4	2 3	∞	3	2	3
3	∞	3	4	4	4	3 3	∞	3	3	3

4.2   Updating Summaries

We describe how to update the summary S_t as new edges arrive in the stream $\mathcal{E}$ or old edges expire. The latter event happens for edges whose time-stamp

---

Algorithm 1. ADDEDGE($\{a, b\}, t$), updates a summary upon addition of $\{a, b\}$ at time t

---
1 **foreach** $i = 0, \ldots, r - 1$ and $(x, t') \in S^a[i]$ **do** $g(b, x, i + 1) \leftarrow \min(t', t)$;
2 **foreach** $i = 0, \ldots, r - 1$ and $(x, t') \in S^b[i]$ **do** $g(a, x, i + 1) \leftarrow \min(t', t)$;
3 PROPAGATE($\{g(v)\}_{v \in V}$)

Algorithm 2. PROPAGATE($\{g(v)\}_{v \in V}$), Processes all propagations that are in the general register g.

---
1 **foreach** $i = 1, \ldots, r$ **do**
2   **foreach** $v, x \in V$ *such that* $g(v, x, i)$ *is set* **do**
3   **if** MERGE($x, v, g(v, x, i), i$) **then**
4   **foreach** $(v, u) \in E_t \setminus \{a, b\}$ **do**
5             $horizon \leftarrow \min(g(u, x, i), recent(v, u))$;
6   **if** $g(u, x, i + 1)$ *not set* **or** $horizon > g(u, x, i + 1)$ **then**
7                $g(u, x, i + 1) \leftarrow horizon$;

---

becomes smaller than $t - w_{max}$. Removing an edge is easy enough; we need to remove all pairs (x, t') from summaries $S_t^u[i]$, for all $u, x \in V$, $i = 1, \ldots, r$, and $t' \leq t - w_{max}$. This operation could also be postponed and executed in batch. Updating the summary S_t to reflect the addition of a new-coming edge e_t, however, is much more challenging. Let us first look at an example.

Example 4. Consider the horizons of the two graphs given in Figure 2. Notice that adding an edge $\{a, b\}$ changed $h(d, b, 4)$ from 3 to 4 because we introduced a path $\langle d, e, c, a, b \rangle$. However, the key observation is that we also changed $h(e, b, 3)$ to 4 due to the path $\langle e, c, a, b \rangle$, $h(c, b, 2)$ to 4 due to the path $\langle c, a, b \rangle$, and $h(a, b, 1)$ to 6 due to the path $\langle a, b \rangle$.

As can be seen in the example, the addition of an edge may result in a considerable number of non-trivial changes. However, the example also hints that we can propagate the summary updates.

Assume that we are adding an edge $\{a, b\}$, and this results in change of $h(u, v, i)$. This change is only possible if there is a path $p = \langle u = v_0, \ldots, v_k = v \rangle$ through $\{a, b\}$. Moreover, we will also change $h(u, v_{k-1}, i - 1)$. By continuing in this logic, it is easy to see that all the updates can be processed via a *breadth-first search* from node b. Furthermore, whenever we can conclude that $h(u, v, i)$ does not need to be updated, we can stop exploring this branch since we know that no extensions of this path will result in updates. The pseudo-code for this procedure is given in Algorithms 1–3.

In the algorithm we update the summaries, distance by distance, and we set new (earlier) horizons that have possibly appeared due to the newly added edge. To maintain the updates we use a function g; $g(u, x, i) = h$ indicates that there is

---

Algorithm 3. MERGE(x, v, t, i), adds x to a summary of v with a distance of i and edge horizon t. If false is returned, then the branch can be pruned.

---

1 **if** $(x, t') \in S^v[i]$ *for some* $t' \geq t$ **then** **return false**;
2 remove all (x, t') from $S^v[i]$ for which $t' < t$;
3 add x, t to $S^v[i]$;
4 **return true**;

---

a new path between u and x of length i and horizon h. As not every new path of length i will lead to an improved horizon, we do not propagate this information immediately to the summary of the neighboring nodes, but rather wait until we have processed all paths of length $i - 1$. For those new paths that improve the summary of a node u, we will then propagate this information further on in the graph. For every distance i, when we process an update to a summary we will record potential updates to horizons of length $i + 1$ as follows: if $g(u, x, i)$ leads to a better horizon of length i between u and x; that is, either there is not yet an entry (x, h) in $S^u[i]$, or $h < g(u, x, i)$, then we will propagate this information to its neighbors u. Let $t = \min(recent(u, v), g(v, x, i))$, then we will propagate $g(u, x, i + 1) = t$, if $t > g(u, x, i + 1)$, that is, we were able to improve our potential update.

Example 5. We will continue our running example given in Figure 2. Let us demonstrate how the horizons of $h(u, b, i)$, $u \in \{a, b, c, d, e\}$ are updated once we introduce the edge $\{a, b\}$. In Figure 3 we illustrate how the propagation is done. At the beginning of each round we compare the current summary $S^u[i]$ against the new candidate horizon $g(u, b, i)$. If the latter is larger, then we update the summary as well as propagate new candidate horizons to the neighboring nodes. In the subsequent figures it is indicated what are the changes with respect to the distances to node b. In the first step, due to the addition of edge $\{a, b\}$ at time 7, for distance 1 the update $g(a, b, 1) = 7$ is propagated. When processing this update indeed it is seen that the summary $S^a[1]$ is updated. Therefore, this update is further propagated to the neighbors, leading to the following

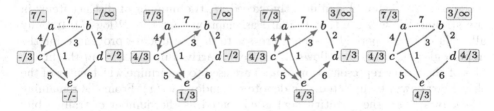

Fig. 3. Propagation of updates for the vertex b when adding (a, b) for the rounds $i = 1, \ldots, 4$. The format of boxes is y/z, where y is the time of b in $S^v[i]$ and $z = g(v, b, i)$ at the beginning of ith round. The edges used for propagation during ith round are marked in red. We do not show propagation during the last round as it is not needed.

updates: $\{g(c, b, 2) = 4, g(e, b, 2) = 1\}$. As only the first update changes the summary $S^c[2]$, only this update will be further propagated. Furthermore, for a there is the update $g(a, b, 2) = 7$ that needs to be processed. Propagation leads to the following new updates (first three for $g(c, b, 2)$, last two for $g(a, b, 2)$): $\{g(a, b, 3) = 4, g(b, b, 3) = 3, g(e, b, 3) = 4, g(c, b, 3) = 4, g(e, b, 3) = 1\}$. The last update $g(e, b, 3) = 1$ will never be considered as it is dominated by the update $g(d, b, 3) = 4$. These updates are then processed and those implying changes in the summary are again propagated.

The proofs of the following proposition is omitted due to space constraints.

Proposition 1. ADDEDGE *updates the summary correctly. Let* $n = |V|$, $m = |E|$, *and* r *be the upper bound on the distances we are maintaining. The time complexity of* ADDEDGE *is* $\mathcal{O}(rmn \log(n))$. *The space complexity is* $\mathcal{O}(rn^2)$.

5   Approximating Neighborhood Function

The algorithm presented in the previous section computes the neighborhood profiles exactly, albeit, it has high space complexity and update time. In this section we describe an approximate algorithm, which is much more efficient in terms of memory requirement and update time.

The approximate algorithm is based on an adaptation of the hyperloglog sketch [17] to the sliding-window context, similar to the adaptation by Chabchoub and Hébrail [9]. The resulting sliding hyperloglog sketch has the following properties: (i) it provides a compact summary of a stream of items, and (ii) it allows to answer the following question: *"How many different items have appeared in the stream since a given time point t?"* Subsequently, this sketch can replace the neighbor sets that need to be maintained by the exact algorithm.

5.1   Hyperloglog and Sliding-Window Hyperloglog Sketches

The hyperloglog sketch [17] consists of an array of numbers, whose size is 2^k, and a hash function η that assigns each item of the stream in a uniformly-random number in the range $[0, 2^n - 1]$. The value of n should be sufficiently large in the sense that 2^{n-k} should significantly exceed M, the number of distinct items in the stream. We will use the standard assumption that $n \in \mathcal{O}(\log M)$. Initially all cells of the hyperloglog sketch are set to 0. The update procedure for the hyperloglog sketch is as follows: if an item x arrives in the stream, the first k bits of the binary representation of $\eta(x)$ are used to determine which entry of the sketch array will be updated. We denote this index by $\iota(x)$. From the remaining $n - k$ bits $\eta'(x)$, the quantity $\rho(x)$ is computed as the number of trailing bits in the binary representation of $\eta'(x)$ that are equal to 0, plus 1. If the current value at the entry $\iota(x)$ of the sketch is smaller than $\rho(x)$, we update the value of that entry. Clearly, the more different items in the stream, the more likely it is to observe large tails of 0's and the higher the numbers in the hyperloglog sketch will become.

In order to make the hyperloglog sketch working in the sliding-window setting, we need to store multiple values per entry. Initially the sliding-HLL sketch will start with an *empty set* for each entry. The process a new item x arriving in the stream at time t, we first need to retrieve the set of time-value pairs associated with the index $\iota(x)$. We then need to add the pair $(t, \rho(x))$ to that set and remove all entries (t', β) for which $\beta \leq \rho(x)$ (as t is the most recent time-stamp, it is also $t' < t$). We denote the sliding-HLL sketch after processing the stream of events $S = \langle \sigma_1, \ldots, \sigma_n \rangle$ by $sHLL(S)$. More formally:

Definition 4. *Let $S = \{(t_1, \beta_1), \ldots, (t_n, \beta_n)\}$ be a set of time-value pairs. Define the subset of time-decreasing values of S as*

$$dec(S) = \{(t_i, \beta_i) \mid \beta_i > \beta_j \text{ for all } (t_j, \beta_j) \in S \text{ with } t_i \leq t_j\}.$$

A sliding hyperloglog sketch sHLL of dimension k is an array of length 2^k in which every entry contains a set of time-value pairs. For a stream S, $sHLL(S)$ is recursively defined as follows:

- *If $S = \langle \rangle$, then $sHLL(S)[i] = \{\}$, for all indices $i = 1 \ldots 2^k$.*
- *Otherwise, if $S = \langle S', (x, t) \rangle$ then $sHLL(S)[i] = dec(sHLL(S')[i] \cup \{(t, \rho(x))\})$ for $i = \iota(x)$; while $sHLL(S)[i] = sHLL(S')[i]$ for all other $i = 1 \ldots 2^k$.*

Example 6. Suppose that the hash η, ι, and ρ are as follows (recall that η determines the other two quantities):

item	a	b	c	d	e
η	100 01	101 11	010 11	010 10	001 10
ι	1	3	3	2	2
ρ	3	1	2	2	1

For the stream of items a, b, a, c, d, e, the resulting sliding HLL sketches are respectively the following:

ι	0	1	2	3
ρ	{}	{}	{}	{}

$\xrightarrow{a}$

ι	0	1	2	3
ρ	{}	(1,3)	{}	{}

$\xrightarrow{b}$

ι	0	1	2	3
ρ	{}	(1,3)	{}	(2,1)

$\xrightarrow{a}$

ι	0	1	2	3
ρ	{}	(3,3)	{}	(2,1)

$\xrightarrow{c}$

ι	0	1	2	3
ρ	{}	(3,3)	{}	(4,2)

$\xrightarrow{d}$

ι	0	1	2	3
ρ	{}	(3,3)	(5,2)	(4,2)

$\xrightarrow{e}$

ι	0	1	2	3
ρ	{}	(3,3),(6,1)	(5,2)	(4,2)

When b arrives, cell 3 gets value 1, which is updated later on when c arrives, since c has the same index, but a higher value. For d and e the situation is opposite; first d arrives giving a value of 2 in cell 2. Later on, when e arrives this value is not updated even though e has the same index because its value is lower.

The next proposition shows that with the sliding HLL sketch we can indeed obtain an approximate answer regarding the number of different items since time s, for any s specified at query time. We omit the proof as it follows immediately from the definition.

Proposition 2. *Let $\mathcal{S} = \langle \sigma_1, \ldots, \sigma_n \rangle$ be a stream of events in which event σ_t arrives at time t. Then for every index $1 \leq s \leq n$, it holds that for every entry $i = 1, \ldots, 2^k$, it is $HLL(\sigma_s, \ldots, \sigma_n)[i] = \max\{r \mid (t, r) \in sHLL(\mathcal{S})[i] \text{ and } t \geq s\}$, where $\max(\{\}) = 0$.*

5.2   Computation of Neighborhood Profiles Based on Sliding HLL

We are now ready to describe our technique for computing the approximate neighborhood profiles. Recall that we are working over a streaming graph with nodes from a set V and a stream of edges $\mathcal{E} = \{(e_1, t_1), (e_2, t_2), \ldots\}$. We have used E_t to denote the set of edges arrived until time t, i.e., $E_t = \{(e, t') \in \mathcal{E} \mid t' \leq t\}$. The approximate sketch is very similar to the exact sketch, with the exception that all sets of (node,time)-pairs are replaced by the much more compact sliding HLL sketch. Furthermore, in order to be able to propagate the updates to its neighbors, for every node we should know its neighbors. Hence, at time t, the summary consists, for every node u, of the following components:

$$N_t^u = \{(v, recent(u, v)) \mid (u, v) \in E_t\} \quad \text{and} \quad C_t^u = \langle C_t^u[1], C_t^u[2], \ldots, C_t^u[r] \rangle,$$

where $C_t^u[i] = sHLL(\{(v, h_t(p)) \mid p \in \mathcal{P}_t(u, v), |p| \leq i\})$.

The set N_t^u specifies the neighbors of node u in the graph $G_t = (V, E_t)$. Note that in the set N_t^u we keep pairs (v, t) such that v is a neighbor of u and t is the most recent time-stamp that an interaction between u and v took place. This time-stamp is needed to decide whether the neighbor v is active for a given window length that is specified at query time.

To update the summary C_t from the summary at the previous time instance, after the addition of an edge (a, b) at time t, we follow the almost exact same propagation method as the exact algorithm. The only difference is that instead of keeping all pairs $(v, h_t(p))$, we now keep a sliding HLL sketch over those pairs, as specified in the previous section. Updating a sliding HLL sketch is slightly more involved than updating the exact summary since we need to keep the sketch as a time-decreasing sequence. The pseudo-code for this is given in Algorithm 4.

Finally, to update the sketch, we use Algorithms 1 and 2, with the exception that the summary $S^u[\cdot]$ is replaced with the sketch $C^u[\cdot][j]$ for a fixed bucket j. We then execute 2^k copies of the algorithm, each handling its own bucket. As these algorithms are syntactically the same to the ones of the exact algorithm, we omit them. The proof of the following proposition is omitted due to space constraints.

Proposition 3. *The sketch version of* ADDEDGE *performs correctly. Let $n = |V|$, $m = |E|$, and r be the upper bound on the distances we are maintaining. The time complexity of the sketch version of* ADDEDGE *is $\mathcal{O}(2^k rm \log^2(n))$. The space complexity is $\mathcal{O}(2^k nr \log^2 n)$.*

Note that a naïve way to maintain approximate neighborhood profiles is to execute the sketching algorithm from scratch after each newly-arriving interaction. In the worst case, this brute-force method has roughly the same space and

---

Algorithm 4. SKETCHMERGE(x, v, t, i), adds x to a summary of v with a distance of i and edge horizon t.

---

1 **if** $(y, t') \in C^v[i]$ *for some* $t' \geq t$, $y \geq x$ **then** **return false**;
2 remove all (y, t') from $C^v[i]$ for which $t' \leq t$ and $y \leq x$;
3 add (x, t) to $C^v[i]$;
4 **return true**;

---

time complexity as our incremental algorithm. However, the brute-force method is expected to require as much space and time as indicated by the worst-case bound, while for our method the worst-case analysis is very pessimistic: most of the times the summaries will not by propagated at the whole network and updates will be very fast. This is demonstrated in our experimental evaluation.

6 Related Work

During the last two decades, a large body of work has been devoted to developing algorithms for mining data streams. Interestingly, the area started with processing *graph streams* [20], but a lot of emphasis was put on computing statistics over streams of items [12,18], and many fundamental techniques have been developed for that setting. Many different models have been studied in the context of data-stream algorithms, including the *sliding-window* model [14], which incorporates a forgetting mechanism where data items expires after W time units from the moment they occur. Existing work has considered estimating various statistics in this model [2,3].

The concept of *sketching* is closely related to data streams, as efficient streaming algorithms operate by maintaining compact sketches, which provide approximate statistics and summaries of the data stream seen so far. Popular data-stream sketches include the *min-hash sketch* [10], the *LogLog sketch* [15], and its improvement, the *hyperloglog sketch* [17], all of which have been used to approximate distinct counts. *Distance distribution sketches* [6,11] are built on top of the distinct-count sketches, and provide a powerful technique to approximate the number of neighbors of a node in a graph within a certain distance. Such sketches have been used extensively in graph-mining applications [6,25].

As graphs provide a powerful abstraction to model a wide variety of real-world datasets, and as the amount of data collected gives rise to massive graphs, there is growing interest on algorithms for processing *dynamic graphs* and *graph streams*. This includes work on data structures that allow to perform efficient queries under structural changes of the graph [16,19], as well as the design of algorithms for computing graph primitives under data-stream models. Work in the latest category includes algorithms for counting triangles [4,5,27] and other motifs [7,8], computing graph sparsifiers [1], and so on. Most of the above papers consider the standard data stream model, although Crouch et al. [13] study many graph algorithms on the sliding-window model.

Table 1. Characteristics of interaction networks.

Dataset	Nodes	Distinct edges	Total edges	Clustering coefficient	Diameter	Effective diameter
Facebook	4 039	88 234	88 234	0.60	8	4.7
Cit-HepTh	27 771	352 801	352 801	0.31	13	5.3
Higgs	166 840	249 030	500 000	0.19	10	4.7
DBLP	192 357	400 000	800 000	0.63	21	8.0

Table 2. Average relative error as a function of ℓ.

ℓ	Facebook	Cit-HepTh	Higgs	DBLP
16	0.28	0.23	0.22	0.22
32	0.13	0.16	0.19	0.15
64	0.10	0.12	0.16	0.12
128	0.08	0.10	0.14	0.09

7   Experimental Evaluation

We provide an empirical evaluation of the approximate algorithm presented in Section 5. We evaluate the space requirements, time, and accuracy. We compare the approximate algorithm with the exact algorithm presented in Section 4 and the *off-line* HyperANF algorithm [6]. Since our implementations have not been optimized, we compare to a HyperANF version developed under the same conditions and without low-level optimizations such as broad-word computing.

Datasets and Setup: We use four real-world datasets obtained from SNAP repository [22]. We take snapshots of the largest datasets Cit-HepTh and DBLP of 500 000 and 400 000 edges, respectively. Three of the data sets, Facebook, DBLP, and Cit-HepTh, have unique edges and do not contain any time information. To create an interaction network out of these static graphs, we order the edges randomly. In the case of DBLP we allow edges to repeat until we have 800 000 edges. Statistics of these datasets are reported in Table 1.

As a maximum window size we use $w_{max} = \infty$, that is, we do not delete any previous edges. We also set $r = 3$, except for one experiment where we vary r.

Accuracy of the Sketch: In order to test the accuracy of the sketch algorithm, we compare the algorithm with the exact version, and we compute the average relative error as a function of number of buckets ($\ell = 2^k$). Running the exact algorithm is infeasible for the large datasets due the memory requirements, and hence we use only a subset of the large datasets to measure accuracy. The results are given in Table 2. As expected from previous studies, the accuracy increases with ℓ.

Running Time for Updating Summaries: Our next goal is to study the running time needed to update the summary upon adding an edge. The average

Table 3. Average time in seconds needed to process 1 000 edges as a function of ℓ

ℓ	Facebook	Cit-HepTh	Higgs	DBLP
16	0.06	7.20	3.92	0.80
32	0.08	12.57	6.84	1.31
64	0.12	28.64	12.12	2.10
128	0.17	50.74	21.38	3.45

running time for every 1 000 edges is reported in Table 3.[1] Detailed time measurements are shown in Figure 4. We took average run time by running 3 iteration of Facebook and Cit-HepTh and 2 iteration of Higgs and DBLP datasets.

The time needed to process an edge depends on two factors. First, as we increase the number of buckets ℓ, the processing time increases. Second, a single edge may cause a significant number of updates if it connects two previously disconnected components. We see the fluctuating nature and peaks in the processing time in Figure 4 as some edge-addition updates require more time than others whenever an edge between two disjoint cluster of nodes comes close the propagation list grows and hence the time taken increases. Interestingly enough, for large datasets, DBLP and Higgs, the time taken to process a new edge becomes almost constant after the snapshot graphs stabilize.

The average processing time depends greatly on the characteristics of the dataset. For example, we can process DBLP quickly despite its size. We suspect that this is due to high diameter and high clustering coefficient.

We parallelize the algorithm to measure the speed up. In Figure 5 we see that by using 4 threads we are able to process the edges 4 times faster.

We also study the processing time as a function of the maximum distance r. Here we use Facebook and DBLP, and vary $r = 2, \ldots, 5$. The results are given in Figure 6. We see that the processing time increases exponentially as a function of r. This is expected as the neighborhood sizes also increase at a similar rate.

Space Complexity: We also evaluate the memory usage of our method. The results are shown in Figure 7. Initially, the need for space increases rapidly as new nodes are added with every edge. Once all the nodes are seen the memory increase drops as only the sketches of the nodes are increasing. Note that we are not pruning any edges. As expected, the memory requirement increases linearly with ℓ.

Comparison with Off-line Method: Finally, for reference, we compare with a non-streaming algorithm that uses the same hyperloglog technology, the Hyper-ANF algorithm of Boldi et al. [6]. To support querying of any window length as supported by our algorithm we modified the HyperANF algorithm to a Sliding-HyperANF algorithm by replacing the HyperLogLog sketch with Sliding HyperLogLog sketch. Running the Sliding-HyperANF algorithm in DBLP takes 3.6 seconds per sliding window. In contrast, for the same data-set, our streaming algorithm gives a rate of 0.003 seconds per sliding window.

---

[1] We measure the time for batches to get a more accurate reading.

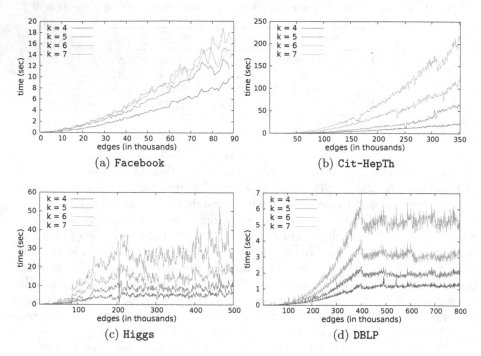

Fig. 4. Time needed to process 1 000 edges for different ℓ

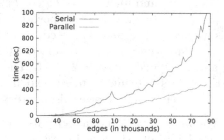

Fig. 5. Running times for DBLP with parallelized version of the algorithm.

8   Concluding Remarks

We studied the problem of maintaining the neighborhood profile of the nodes of an interaction network—a graph with a sequence of interactions, in the form of a stream of time-stamped edges. The model is appropriate for many modern graph datasets, like social networks where interaction between users is one of the most important aspects. We focused on the sliding-window data-stream model, which allows to forget past interactions and adapt to new drifts in the data. Thus, the proposed problem and approach can be applied to monitoring large

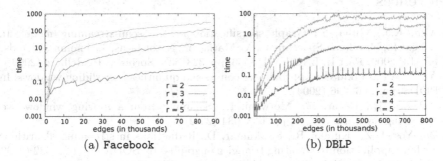

Fig. 6. Time needed to process 1 000 edges as a function of distance r

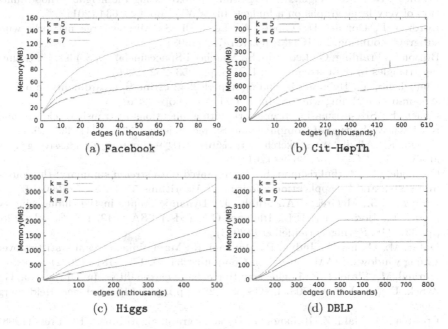

Fig. 7. Memory utilization as a function of ℓ

networks with fast-evolving interactions, and used to reason how the network structure and the centrality of the important nodes change over time.

We presented an exact algorithm, which is memory inefficient, but it set the stage for our main technique, an approximate algorithm based on sliding-window hyperloglog sketches, which requires logarithmic memory per network node, and has fast update time, in practice. The algorithm is also naturally parallelizable, which is exploited in our experimental evaluation to further improve its performance. One desirable property of our algorithm is that the sketch we maintain does not depend on the length of the sliding window, but the length can be specified at query time.

References

1. Ahn, K.J., Guha, S.: Graph sparsification in the semi-streaming model. In: Albers, S., Marchetti-Spaccamela, A., Matias, Y., Nikoletseas, S., Thomas, W. (eds.) ICALP 2009, Part II. LNCS, vol. 5556, pp. 328–338. Springer, Heidelberg (2009)
2. Arasu, A., Manku, G.: Approximate counts and quantiles over sliding windows. In: PODS, pp. 286–296 (2004)
3. Babcock, B., Datar, M., Motwani, R.: Sampling from a moving window over streaming data. In: SODA, pp. 633–634 (2002)
4. Bar-Yossef, Z., Kumar, R., Sivakumar, D.: Reductions in streaming algorithms, with an application to counting triangles in graphs. In: SODA, pp. 623–632 (2002)
5. Becchetti, L., Boldi, P., Castillo, C., Gionis, A.: Efficient semi-streaming algorithms for local triangle counting in massive graphs. In: KDD (2008)
6. Boldi, P., Rosa, M., Vigna, S.: Hyperanf: approximating the neighbourhood function of very large graphs on a budget. In: WWW, pp. 625–634 (2011)
7. Bordino, I., Donato, D., Gionis, A., Leonardi, S.: Mining large networks with subgraph counting. In: ICDM, pp. 737–742 (2008)
8. Buriol, L., Frahling, G., Leonardi, S., Marchetti-Spaccamela, A., Sohler, C.: Counting triangles in data streams. In: PODS, pp. 253–262 (2006)
9. Chabchoub, Y., Hébrail, G.: Sliding hyperloglog: estimating cardinality in a data stream over a sliding window. In: ICDM Workshops (2010)
10. Cohen, E.: Size-estimation framework with applications to transitive closure and reachability. Journal of Computer and System Sciences 55(3), 441–453 (1997)
11. Cohen, E.: All-distances sketches, revisited: HIP estimators for massive graphs analysis. In: PODS, pp. 88–99 (2014)
12. Cormode, G., Muthukrishnan, S.: An improved data stream summary: the count-min sketch and its applications. Journal of Algorithms 55(1), 58–75 (2005)
13. Crouch, M.S., McGregor, A., Stubbs, D.: Dynamic graphs in the sliding-window model. In: Bodlaender, H.L., Italiano, G.F. (eds.) ESA 2013. LNCS, vol. 8125, pp. 337–348. Springer, Heidelberg (2013)
14. Datar, M., Gionis, A., Indyk, P., Motwani, R.: Maintaining stream statistics over sliding windows. SIAM Journal on Computing 31(6), 1794–1813 (2002)
15. Durand, M., Flajolet, P.: Loglog counting of large cardinalities. In: Di Battista, G., Zwick, U. (eds.) ESA 2003. LNCS, vol. 2832, pp. 605–617. Springer, Heidelberg (2003)
16. Eppstein, D., Galil, Z., Italiano, G.: Dynamic graph algorithms. CRC Press (1998)
17. Flajolet, P., Fusy, É., Gandouet, O., Meunier, F.: Hyperloglog: the analysis of a near-optimal cardinality estimation algorithm. In: Proceedings of the DMTCS (2008)
18. Gama, J.: Knowledge discovery from data streams. CRC Press (2010)
19. Henzinger, M., King, V.: Randomized fully dynamic graph algorithms with poly-logarithmic time per operation. Journal of the ACM 46(4), 502–516 (1999)
20. Henzinger, M., Raghavan, P., Rajagopalan, S.: Computing on data streams. In: DIMACS Workshop External Memory and Visualization, vol. 50 (1999)
21. Holme, P., Saramäki, J.: Temporal networks. Physics Reports 519(3), 97–125 (2012)
22. Leskovec, J., Krevl, A.: SNAP Datasets: Stanford large network dataset collection, June 2014. http://snap.stanford.edu/data
23. Michail, O.: An introduction to temporal graphs: An algorithmic perspective (2015). arXiv:1503.00278

24. Muthukrishnan, S.: Data streams: Algorithms and applications (2005)
25. Palmer, C., Gibbons, P., Faloutsos, C.: ANF: A fast and scalable tool for data mining in massive graphs. In: KDD, pp. 81–90 (2002)
26. Rozenshtein, P., Tatti, N., Gionis, A.: Discovering dynamic communities in interaction networks. In: Calders, T., Esposito, F., Hüllermeier, E., Meo, R. (eds.) ECML PKDD 2014, Part II. LNCS, vol. 8725, pp. 678–693. Springer, Heidelberg (2014)
27. Tsourakakis, C., Kang, U., Miller, G., Faloutsos, C.: Doulion: counting triangles in massive graphs with a coin. In: KDD, pp. 837–846 (2009)

Response-Guided Community Detection: Application to Climate Index Discovery

Gonzalo A. Bello[1], Michael Angus[1], Navya Pedemane[1], Jitendra K. Harlalka[1],
Fredrick H.M. Semazzi[1], Vipin Kumar[2], and Nagiza F. Samatova[1,3]([✉])

[1] North Carolina State University, Raleigh, NC, USA
samatova@csc.ncsu.edu
[2] University of Minnesota, Minneapolis, MN, USA
[3] Oak Ridge National Laboratory, Oak Ridge, TN, USA

Abstract. Discovering climate indices–time series that summarize spatiotemporal climate patterns–is a key task in the climate science domain. In this work, we approach this task as a problem of *response-guided community detection*; that is, identifying communities in a graph associated with a response variable of interest. To this end, we propose a general strategy for response-guided community detection that explicitly incorporates information of the response variable during the community detection process, and introduce a graph representation of spatiotemporal data that leverages information from multiple variables.

We apply our proposed methodology to the discovery of climate indices associated with seasonal rainfall variability. Our results suggest that our methodology is able to capture the underlying patterns known to be associated with the response variable of interest and to improve its predictability compared to existing methodologies for data-driven climate index discovery and official forecasts.

Keywords: Community detection · Spatiotemporal data · Climate index discovery · Seasonal rainfall prediction

1 Introduction

Detecting communities in real-world networks is a key task in many scientific domains. Oftentimes, domain scientists are particularly concerned with finding communities associated with a response variable of interest that can be used to analyze or predict this response variable. For example, in climate science, such communities may represent spatiotemporal climate patterns associated with a particular weather event [24], while in biology, they may represent groups of functionally associated genes associated with a particular phenotype [12].

However, community detection techniques are traditionally unsupervised learning methods, and thus do not take into account the variability of the response variable of interest. Therefore, the communities identified may not necessarily be associated with this response variable. Furthermore, even though semi-supervised methods have been proposed to incorporate prior knowledge

© Springer International Publishing Switzerland 2015
A. Appice et al. (Eds.): ECML PKDD 2015, Part II, LNAI 9285, pp. 736–751, 2015.
DOI: 10.1007/978-3-319-23525-7_45

to the community detection process, these methods do not consider a response variable either and require partial information about the community memberships, which may not be available [6]. For this reason, we introduce the problem of *response-guided community detection*–that is, identifying communities in a graph associated with a response variable of interest–and study its application to the discovery of *climate indices*, an important task in the climate science domain.

Climate indices are time series that summarize spatiotemporal patterns in the global climate system. These patterns are often associated with temperature, pressure, and wind anomalies, which can have a significant impact on regional climate. Consequently, climate indices are frequently used to analyze and predict regional weather events. For example, climate indices defined for El Niño Southern Oscillation (ENSO) are used to forecast Atlantic hurricane activity [9].

Climate indices were traditionally the product of hypothesis-driven research. However, the increasing amount of climate data available has led to the adoption of data-driven approaches to guide and accelerate climate index discovery, most commonly by using Principal Component Analysis (PCA) to identify major modes of variability in the data. Nonetheless, the use of PCA has important limitations in regards to the physical interpretability of the climate indices obtained and its ability to detect weaker patterns [22].

As an alternative, the application of clustering techniques, such as Shared Nearest Neighbor (SNN) clustering, to identify regions of homogeneous long-term variability in climate data has been proposed [22]. More recently, a network representation of the data has been adopted to better capture the dynamics of the global climate system [23–25]. Then, the climate index discovery task has been approached as a community detection problem [24]. The validity of the clusters or communities identified as climate indices has been evaluated in terms of their ability to predict a response variable of interest [22,24]. However, since these are unsupervised learning methodologies, the climate indices discovered may not necessarily be good predictors.

Therefore, to discover climate indices associated with a response variable of interest, we propose a methodology that explicitly incorporates information of this response variable during the discovery process by using response-guided community detection. We apply this methodology to the discovery of climate indices associated with seasonal rainfall variability in the Greater Horn of Africa, and validate the climate indices discovered in terms of their predictive power and climatological relevance. Discovering climate indices associated with a response variable of interest allows us to identify its sources of variability. Moreover, using these climate indices as predictors allows us to improve forecasts of this response variable, which is one of the major current challenges in climate science [20].

The main contributions of this paper are as follows. First, we formulate the problem of response-guided community detection (Section 2.1) and propose a general strategy to identify communities in a graph associated with a response variable of interest by explicitly incorporating information of this response variable during the community detection process (Section 2.2).

And second, we propose a methodology to discover climate indices associated with a response variable of interest from multivariate spatiotemporal data by using response-guided community detection (Section 3). As part of this methodology, we introduce a network representation of multivariate spatiotemporal data that, unlike existing network construction methodologies [23–25], builds the network in a response-guided manner, while also incorporating multiple covariates, spatial neighborhood information, and multiple related response variables to the network construction process (Section 3.1).

Finally, we should note that in this paper we only demonstrate the value of response-guided community detection in the context of climate index discovery. Its application to other problems and domains is the subject of future work.

2   Response-Guided Community Detection

In this section, we formally define the problem of response-guided community detection (Section 2.1), describe a general strategy for response-guided community detection, and present two examples of community detection algorithms that can be adapted to identify communities highly associated with a response variable of interest (Section 2.2).

2.1   Problem Statement

Let $X = \{x_{t,d,f} \in \mathbb{R} \mid t \in T, d \in D, f \in F\}$ be a multivariate spatiotemporal data set and $Y = \{y_t \in \mathbb{R} \mid t \in T\}$ be a response variable, where T is a set of time steps, D is a set of spatial points, and F is a set of covariates. For our motivating application of climate index discovery, X may be a global climate data set for a given month, Y may be the total rainfall at a target region for a given season, T may be a set of years, D may be a set of global coordinates, and F may be a set of climate variables (e.g., temperature, pressure, humidity).

Let data set X be represented as a graph $G = (V, E)$, where $V \subseteq D$ is the set of vertices, E is the set of edges, and each edge $(d_1, d_2) \in E$ is defined based on a domain-specific relationship between the data at spatial points d_1 and d_2 for all covariates $f \in F$ and over all time steps $t \in T$. For our motivating application of climate index discovery, an edge (d_1, d_2) may represent a statistically significant correlation between the data at spatial points d_1 and d_2.

Informally, we define *response-guided community detection* as the task of partitioning graph G into a set of communities C, such that every community $c_i \in C$ is highly *associated* with the response variable Y. To quantify this association, we construct an *index* for each community.

Definition 1. Given a community c_i, the *index* constructed for c_i using covariate $f \in F$, $I_{i,f}$, is defined as

$$I_{i,f}(t) = \frac{1}{|c_i|} \sum_{d \in c_i} x_{t,d,f} \quad \forall t \in T \tag{1}$$

Definition 2. Given a community c_i, the *association* of c_i with the response variable Y, ϕ_{c_i}, is defined as

$$\phi_{c_i} = \max_{f \in F} |r_{I_{i,f},Y}| \tag{2}$$

where $r_{I_{i,f},Y}$ is the Pearson's linear correlation coefficient between index $I_{i,f}$ and the response variable Y over all time steps $t \in T$.

Finally, we formally define the problem of *response-guided community detection*: Given a graph $G = (V, E)$ and a response variable Y, partition G into a set of communities $C = \{c_1, c_2, ..., c_{|C|}\}$, where $c_i \subseteq V$ for all $c_i \in C$, $c_i \cap c_j = \emptyset$ for all $c_i, c_j \in C$ with $i \neq j$, and $\bigcup_{i=1}^{|C|} c_i = V$, such that the average association with the response variable Y over all $c_i \in C$, $\bar{\phi}_C$, is maximized.

2.2 Algorithms for Response-Guided Community Detection

Community detection is one of the most widely studied topics in graph data analytics and, as a result, numerous methods have been proposed for this problem [8,11]. A common approach to community detection is to find the set of communities that maximizes a given quality function that measures the "goodness" of the partition of the graph. For traditional community detection, a "good" partition of the graph is generally such that there are many edges within the communities but few edges among them. However, for response-guided community detection, our goal is to identify communities highly associated with a response variable of interest. Therefore, we must maximize not only the "goodness" of the partition of the graph, but also the association of the communities in the partition with this variable.

To this end, we introduce a joint optimization criterion, $\mathcal{F}$, given by

$$\mathcal{F} = \alpha \cdot q(C) + (1 - \alpha) \cdot \bar{\phi}_C \tag{3}$$

where C is a set of communities, $q(C)$ is a function of the "goodness" of C, $\bar{\phi}_C$ is the average association of the communities in C with the response variable of interest (see Definition 2), and α is a tuning parameter to balance the trade-off between the "goodness" of C and the association of the communities with the response variable.

The "goodness" function is typically a metric that quantifies some structural properties of the partition of the graph. In this paper, we choose modularity– "by far the most used and best known quality function" for community detection [8]–as the "goodness" function. The modularity of a given partition of a graph is defined as the difference between the number of edges within the communities and the expected number of such edges in a random graph with the same degree distribution [17]. For a simple graph $G = (V, E)$ which vertices are partitioned into communities, the modularity Q [16] of the partition is given by

$$Q = \frac{1}{2m} \Sigma_{vw} \left[A_{vw} - \frac{k_v k_w}{2m} \right] \delta(v, w) \tag{4}$$

where A is the adjacency matrix of the graph (that is, A_{vw} is 1 if vertices v and w are connected and 0 otherwise), $m = \frac{1}{2}\Sigma_{vw}A_{vw}$ is the number of edges in the graph, $k_v = \Sigma_w A_{vw}$ is the degree of vertex v, and $\delta(i,j)$ is the Kronecker delta function (that is, $\delta(i,j)$ is 1 if i and j belong to the same community and 0 otherwise). Modularity optimization is an NP-complete problem [3], but many heuristic algorithms have been proposed [8].

A general strategy for response-guided community detection is to adapt modularity optimization algorithms by replacing modularity with the joint optimization criterion $\mathscr{F}$ defined in Equation 3 as the objective function. To illustrate this strategy, we next present two algorithms that can be adapted in this way to identify communities highly associated with a response variable of interest: the Louvain method, a very efficient greedy algorithm for modularity optimization, and simulated annealing, a computationally demanding but potentially more accurate optimization technique.

Greedy Algorithms for Response-Guided Community Detection. In general, greedy algorithms for modularity optimization identify communities by iteratively merging vertices or communities that result in the largest increase in the modularity of the graph partition [2,4].

In this paper, we focus on the Louvain method [2], a well-known greedy algorithm that has been shown to outperform other community detection algorithms in empirical comparative studies [15]. The Louvain method is adapted for response-guided community detection by using the joint optimization criterion $\mathscr{F}$ as the objective function.

Initially, each vertex is assigned to a different community. In the first phase of the algorithm, each vertex is iteratively and sequentially assigned to the community that yields the highest positive gain in the joint optimization criterion, $\Delta\mathscr{F}$, given by

$$\Delta\mathscr{F} = \alpha \cdot \Delta Q + (1 - \alpha) \cdot \Delta\bar{\phi} \tag{5}$$

where ΔQ and $\Delta\bar{\phi}$ are the gain in modularity and the gain in average association with the response variable of interest over all communities resulting from the change in the communities, respectively.

In the second phase of the algorithm, a new graph is constructed by aggregating the vertices in each community into a single meta-vertex. These two phases are repeated iteratively until no further improvement of the joint optimization criterion $\mathscr{F}$ can be achieved.

Simulated Annealing for Response-Guided Community Detection. Another strategy that has been employed for modularity optimization is simulated annealing [14], an optimization technique that avoids local optima by incorporating stochastic noise into the search procedure. The level of noise is defined by a computational temperature $\mathcal{T}$, which decreases after each iteration.

In this paper, the simulated annealing algorithm proposed by Guimerà et al. [10] is adapted for response-guided community detection by using the joint optimization criterion $\mathscr{F}$ as the objective function.

Initially, each vertex is assigned to a different community. At each temperature T, the algorithm performs (typically) n^2 random local movements (i.e., moving a vertex to another community) and n random global movements (i.e., merging two communities and splitting a community in two). Each of these local and global movements is accepted with probability

$$p = \begin{cases} 1, & \text{if } \Delta\mathscr{F} \geq 0 \\ \exp\left(\dfrac{\Delta\mathscr{F}}{T}\right), & \text{if } \Delta\mathscr{F} < 0 \end{cases} \tag{6}$$

where $\Delta\mathscr{F}$ is the gain in the joint optimization criterion resulting from the change in the communities, as defined in Equation 5.

After all local and global moves have been evaluated, the current temperature T is decreased to $T' = c \cdot T$, where $c \in (0,1)$ is a cooling parameter (typically between 0.990 and 0.999). The algorithm stops when a minimum temperature is reached or when there is no change in the joint optimization criterion $\mathscr{F}$ for a given number of consecutive iterations.

3 Climate Index Discovery

In this section, we describe our proposed methodology for the discovery of climate indices associated with a response variable of interest from multivariate spatiotemporal data by using response-guided community detection.

Our proposed methodology is comprised of two main steps. First, we represent the multivariate spatiotemporal data as a graph using our proposed network construction methodology (Section 3.1). Second, we identify communities in this graph using one of our adapted algorithms for response-guided community detection (see Section 2.2). For each community c_i identified, we construct an index I_{i,f_i^*} (see Definition 1) potentially associated with the response variable, where f_i^* is the *representative covariate* of the community, defined as

$$f_i^* = \arg\max_{f \in F} |r_{I_{i,f},Y}| \tag{7}$$

3.1 Network Construction Methodology

Spatiotemporal data can be represented as a graph, where each vertex is a spatial point and each edge indicates a significant relationship between a pair of spatial points. This type of representation has been adopted to model climate data, because it captures the dynamical behavior of the data's underlying system [23–25]. Furthermore, communities in these networks often have a higher association with the response variable of interest than clusters obtained using traditional clustering techniques, such as spectral clustering and the k-means clustering algorithm [24].

In this paper, we propose a methodology for the construction of climate networks associated with a response variable of interest. The key features of this methodology are as follows.

First, we construct the network in a response-guided manner. Existing methodologies for climate network construction consider all the spatial points in the data set as vertices and build the network by computing the correlation between every pair of vertices [24, 25], which can be computationally expensive. In contrast, we only consider as vertices the spatial points associated with the response variable.

Second, we incorporate multiple covariates to the network construction process. Some existing methodologies have incorporated multiple covariates by defining a cross correlation function to weight the edges of the network [23]. Here, instead, we leverage the information of multiple covariates to assess the statistical significance of each edge in the network.

And third, we incorporate spatial neighborhood information and multiple related response variables to the network construction process, to increase its robustness in the case of data sets with small sample size.

Selecting the Set of Vertices. The set of vertices V of the network is selected based on the statistical significance of the relationship between each spatial point in the data set and the response variable of interest for multiple covariates. To assess this statistical significance, we first calculate the Spearman's rank correlation coefficients between the time series for each covariate at each spatial point and the response variable. Spearman's rank correlation is used to capture nonlinear relationships known to exist in climate data.

For each spatial point d, the p-values of the Spearman's rank correlation coefficients computed for each covariate are combined using Fisher's $\mathcal{X}^2$ test [7]; that is, by calculating the p-value of the test statistic given by

$$-2 \sum_{f \in F} ln(p_{X_{d,f},Y}) \tag{8}$$

where $p_{X_{d,f},Y}$ is the p-value of the Spearman's rank correlation coefficient between the time series for covariate f at spatial point d, $X_{d,f}$, and the response variable Y, over all time steps $t \in T$. The use of this combined probability test allows us to capture relationships between multiple covariates and the response variable. Finally, the set S of spatial points with a statistically significant combined p-value ($p < 0.01$) is selected as the set of vertices V of the network (i.e, spatial points potentially associated with the response variable of interest).

Defining the Set of Edges. The set of edges E of the network is defined based on the statistical significance of the relationship between each pair of spatial points in V for multiple covariates. To assess this statistical significance, we first calculate the Pearson's linear correlation coefficients between the time series for each covariate at each pair of spatial points. Climate networks constructed using Pearson's linear correlation coefficient have been shown to be highly similar to those constructed using nonlinear measures, such as mutual information [5].

For each pair of spatial points $d_1, d_2 \in V$, the p-values of the Pearson's linear correlation coefficients computed for each covariate are combined using Fisher's $\mathcal{X}^2$ test [7]; that is, by calculating the p-value of the test statistic given by

$$-2 \sum_{f \in F} ln(p_{X_{d_1,f},X_{d_2,f}}) \qquad (9)$$

where $p_{X_{d_1,f},X_{d_2,f}}$ is the p-value of the Pearson's linear correlation coefficient between the time series for covariate f at spatial point d_1, $X_{d_1,f}$, and at spatial point d_2, $X_{d_2,f}$, over all time steps $t \in T$. Finally, an edge $(d_1, d_2) \in E$ is defined for every pair of spatial points $d_1, d_2 \in V$ with a statistically significant combined p-value ($p < 10^{-10}$, as defined in previous studies [24]).

Incorporating Spatial Neighborhood Information and Multiple Response Variables. Data sets with small sample size, such as the ones used in this study, can often lead to the selection of spatial points with spurious associations with the response variable of interest as vertices. To increase the robustness of the vertex selection in these cases, we leverage the spatial structure of the data and the information of multiple related (i.e., highly correlated) response variables (e.g., seasonal rainfall at multiple stations in the same region) by finding a consensus set of spatial points, S^*, given by

$$S^* = \bigcap_{j=1}^{h} S_j \cup \{N(d) \mid d \in S_j\} \qquad (10)$$

where h is the number of response variables, S_j is the set of spatial points potentially associated with the jth response variable and $N(d)$ indicates the spatial points spatially adjacent to spatial point d. We incorporate spatial neighborhood information because, given the strong spatial autocorrelations present in spatiotemporal data, it is likely that if a spatial point is associated with the response variable of interest, then its spatially adjacent points will also be associated with the response variable.

We then construct a climate network for the multiple related response variables using the previously described methodology with the consensus set of spatial points S^* as the set of vertices V of the network. Note that the rest of our proposed methodology for climate index discovery, including the response-guided community detection algorithms, can also be extended to incorporate multiple related response variables. In this case, the association of a community c_i, ϕ_{c_i} (see Definition 2), is redefined as the average association of c_i over all response variables Y_j for $j = 1, 2, ..., h$.

4   Experimental Evaluation

In this section, we describe the experimental evaluation of our proposed methodology for climate index discovery and report the results obtained. We applied our proposed methodology to the discovery of climate indices associated with October to December (OND) rainfall variability in the Greater Horn of Africa (GHA), using data from four (4) stations with highly correlated rainfall patterns located in the North Eastern Highlands of Tanzania (Arusha, Kilimanjaro, Moshi, and Same).

4.1   Data Description

We used monthly gridded ocean data for the following climate variables: Sea Surface Temperature (SST), obtained from the NOAA Extended Reconstructed Sea Surface Temperature version 3 (ERSST V3) data set (data available from 1854 to present at 2° latitude-longitude resolution) [21], and Sea Level Pressure (SLP), Geopotential Height at 500 mb (GH), Relative Humidity at 850 mb (RH) and Precipitable Water (PW), obtained from the NCEP/NCAR Reanalysis 1 data set (data available from 1948 to present at 2.5° latitude-longitude resolution) [13]. SST, SLP, and GH are the most frequently used variables in identifying global climate patterns. We also include RH and PW as secondary variables for the temperature and water vapor content of the atmosphere.

Monthly rainfall data (52 years, from 1960 to 2011) and seasonal rainfall forecasts (14 years, from 1998 to 2011) for stations in Tanzania were provided by the Tanzania Meteorological Agency (TMA). Data was divided into a training set (38 years, from 1960 to 1997) and a test set (14 years, from 1998 to 2011). Note that only the training set was used to construct the climate networks and discover the climate indices presented in Section 4.3 and Section 4.4, respectively.

4.2   Data Preprocessing

Climate data exhibits complex characteristics, such as seasonal trends and strong spatial and temporal autocorrelations, that may hinder the performance of data mining techniques. To remove seasonality and minimize autocorrelations, we normalized the data using monthly z-scores transformations by subtracting the mean and dividing by the standard deviation of the data over the training set [24]. Since the focus of this study is on interannual variability, we also linearly detrended the data. Furthermore, all experiments were performed using a spatial resolution of 10° latitude-longitude for the gridded ocean data.

4.3   Climate Networks Constructed

Climate networks were constructed using our proposed network construction methodology with OND rainfall variability in the GHA as the response variable of interest (see Section 3.1). To capture time-lagged relationships, which are often present in climate data, five (5) climate networks were constructed, one for each month, starting four (4) months before the season (June) until the first month of the season (October). It is worth noting that when constructing a climate network for the month of May, no spatial points were selected as potentially associated with the response variable, suggesting that this month may be too early before the season to yield significant climate indices.

Each climate network was constructed by leveraging the information of four (4) related stations in the North Eastern Highlands of Tanzania. Since these stations are located in the same climatological region and exhibit highly correlated rainfall patterns, they are expected to be associated with the same global climate patterns. Hence, the use of the consensus set allows us to filter out spatial points with potentially spurious associations with the response variable. Interstation variability is due to local factors, which are out of the scope of this paper.

4.4 Climate Indices Discovered

Communities associated with OND rainfall variability in the GHA were identified in the climate networks constructed using both the Louvain method and the simulated annealing algorithm adapted for response-guided community detection (see Section 2.2). As previously explained, we use a tuning parameter α to balance the trade-off between the modularity of the network partition and the association of the communities with the response variable of interest. For this experimental evaluation, we set the value of α to the multiple of 0.05 in the interval $[0.75, 1]$ that yields the set of communities with the highest average association with the response variable over the training set. Lower values of α were not considered to ensure a good modularity value. For each community identified, a climate index was constructed by computing the spatial average over the community of its representative climate variable (see Figure 1).

We compare our climate indices with those discovered using a baseline methodology and the state of the art [24]. For the baseline methodology, communities were identified in multivariate climate networks (i.e., one network was constructed for all covariates via a combined probability test, as described in Section 3.1) using both the original Louvain method [2] and the original simulated annealing algorithm for community detection [10]. For the state of the art [24], communities were identified in univariate climate networks (i.e., one network was constructed for each covariate) using Walktrap, a community detection algorithm based on random walks [19]. In both cases, the community detection and the network construction were performed in an unsupervised manner.

Table 1 summarizes the properties of the climate networks constructed and the climate indices discovered using each methodology. Given that our response-guided community detection algorithms do not exclusively optimize the "goodness" of the network partitions, our climate networks exhibit a lower modularity than those constructed using unsupervised methodologies (0.34 vs. 0.74, 0.75, and 0.59). However, our communities have a higher internal density (0.62 vs. 0.29, 0.28, and 0.47) and a lower internal variability (0.63 and 0.62 vs. 0.77, 0.78, and 0.74), indicating a well-defined structure.

We also observe that, unlike most of the climate indices discovered using the baseline and the state of the art, the majority of our climate indices (66.67%) have a statistically significant linear correlation ($p < 0.01$) with the response variable of interest over the training set. Moreover, our proposed methodology

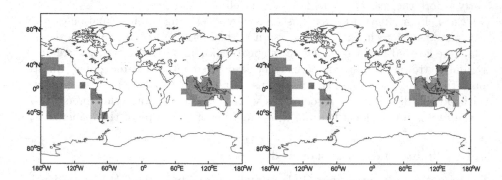

Fig. 1. Climate indices discovered using our proposed methodology with the response-guided community detection algorithm based on the Louvain method (left) and simulated annealing (right), respectively, and with OND rainfall variability in the GHA as the response variable of interest. Each color represents a different index, and diamonds indicate overlaps between indices. To improve visualization, only the top 10 indices with the highest association with the response variable over the training set are shown in each figure. Best viewed in color.

Table 1. Properties of networks constructed and climate indices discovered for OND rainfall variability in the GHA, using the proposed, baseline, and state-of-the-art (SOTA) [24] methodologies with the Louvain method (LM), simulated annealing (SA) and Walktrap as the community detection algorithms: number of networks (Num Nets), average number of vertices and edges per network (Avg Vtxs, Avg Edges), average modularity (Avg Mod), number of indices (Num Idxs), average number of vertices, standard deviation, and internal density per index (Avg Vtxs, Avg Std, Avg Dens), and percentage of indices with a statistically significant ($p < 0.01$) linear correlation with the response variable of interest (% Idxs). Best values are highlighted in bold.

Method	Algorithm	Networks				Indices				Significant Indices	
		Num Nets	Avg Vtxs	Avg Edges	Avg Mod	Num Idxs	Avg Vtxs	Avg Std	Avg Dens	Num Idxs	% Idxs
Proposed	Adapted LM	5	40.80	169.20	0.34	18	11.33	0.63	**0.62**	12	**66.67**
	Adapted SA	5	40.80	169.20	0.34	18	11.33	**0.62**	**0.62**	12	**66.67**
Baseline	Original LM	5	446.00	2614.60	0.74	49	45.51	0.77	0.29	6	12.24
	Original SA	5	446.00	2614.60	**0.75**	50	44.60	0.78	0.28	4	8.00
SOTA	Walktrap	25	444.80	7493.80	0.59	265	41.96	0.74	0.47	6	2.26

Table 2. Average linear correlation with OND rainfall at each station and at the GHA region, over the training set and the test set, of climate indices discovered for OND rainfall variability in the GHA using the proposed, baseline, and state-of-the-art (SOTA) [24] methodologies with the Louvain method (LM), simulated annealing (SA) and Walktrap as the community detection algorithms. Check marks ($\checkmark$) indicate that our proposed methodology performs significantly better according to a two-way ANOVA at the 95% confidence level. Best values are highlighted in bold.

Station	Proposed				Baseline				SOTA	
	Adapted LM		Adapted SA		Original LM		Original SA		Walktrap	
	Train	Test	Train	Test	Train	Test	Train	Test	Train	Test
Arusha	**0.4436**	**0.2999**	0.4431	0.2848	0.2496	0.2495	0.2489	0.2639	0.1481	0.2261
Kilimanjaro	0.4103	**0.3752**	**0.4300**	0.3583	0.2586	0.2437	0.2629	0.2525	0.1567	0.2230
Moshi	0.3629	**0.2980**	**0.3764**	0.2791	0.2404	0.2552	0.2317	0.2501	0.1393	0.2481
Same	0.4292	**0.3403**	**0.4341**	0.3119	0.2574	0.2111	0.2572	0.2429	0.1589	0.2148
GHA	0.4502	**0.3478**	**0.4614**	0.3272	0.2763	0.2356	0.2749	0.2497	0.1558	0.2219
Two-way ANOVA ($\alpha = 0.05$)		$\checkmark$	$\checkmark$	$\checkmark$	$\checkmark$	$\checkmark$	$\checkmark$		

performs significantly ($p < 0.05$) better than the baseline and the state of the art across all stations in terms of the average linear correlation between the climate indices and the response variable of interest over the training set and the test set (see Table 2). This shows that, as expected, our proposed methodology is able to discover climate indices more highly associated with the response variable of interest than those discovered using unsupervised methodologies.

4.5   Seasonal Rainfall Prediction

We validate the climate indices discovered with our proposed methodology by assessing their predictive power for OND rainfall in the GHA. To this end, we trained linear regression models to predict rainfall at each station, and average rainfall at the region, using our climate indices as predictors. As specified in Section 4.1, data from 1960 to 1997 was used for training and data from 1998 to 2011 was used for testing. For comparison, linear regression models were also built using the climate indices discovered with the baseline and state-of-the-art [24] methodologies introduced in Section 4.4.

In order to avoid overfitting given the small sample size of the data sets, only the top six (6) climate indices with the highest average correlation with OND rainfall in the GHA over the training set were used to build the models. This number of predictors was selected because it yielded relatively stable performance over the training set across all methodologies (see Figure 2). Furthermore, to evaluate the ability of the models to make predictions before the start of the OND rainfall season, all experiments were preformed using data up to the month of August (one-month lead time). Climate indices discovered for the months of September and October were reconstructed using August data.

The correlations between predicted and true rainfall and the root mean squared errors (RMSE) obtained for each methodology are shown in Table 3.

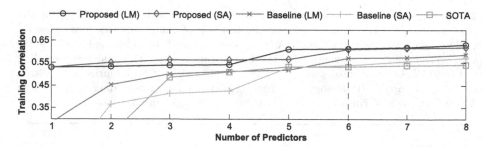

Fig. 2. Average linear correlation between true and predicted rainfall for predictions of OND rainfall at each station in the GHA region over the training set using the proposed, baseline, and state-of-the-art (SOTA) [24] methodologies with the Louvain method (LM), simulated annealing (SA) and Walktrap as the community detection algorithms vs. the number of predictors used to build the regression models. The dashed line indicates the number of predictors selected for further analysis.

Table 3. Linear correlation between true and predicted rainfall (Corr) and RMSE scores for predictions of OND rainfall at each station and at the GHA region from 1998 to 2011 obtained using the proposed, baseline, and state-of-the-art (SOTA) [24] methodologies with the Louvain method (LM), simulated annealing (SA) and Walktrap as the community detection algorithms. Check marks ($\checkmark$) indicate that our proposed methodology performs significantly better according to a two-way ANOVA at the 95% confidence level. Best values are highlighted in bold.

| Station | Proposed | | | | Baseline | | | | SOTA | |
| | Adapted LM | | Adapted SA | | Original LM | | Original SA | | Walktrap | |
	Corr	RMSE	Corr	RMSE	Corr	RMSE	Corr	RMSE	Corr	RMSE
Arusha	**0.7143**	**0.5017**	0.5869	0.5215	0.2462	0.5023	0.3432	0.6853	0.2034	0.5779
Kilimanjaro	**0.7629**	**0.5034**	0.6736	0.5619	0.1844	1.0432	0.2053	0.7477	0.2940	0.7874
Moshi	0.6561	0.4719	**0.6564**	**0.4664**	-0.0319	0.5937	0.1059	0.7055	0.3088	0.6231
Same	**0.7237**	0.4779	0.6896	**0.4749**	0.1470	0.6796	0.1806	0.6929	0.2575	0.7121
GHA	**0.7722**	0.4133	0.7425	**0.4007**	0.1501	0.6316	0.2135	0.6390	0.2665	0.6053
Two-way ANOVA ($\alpha = 0.05$)					$\checkmark$	$\checkmark$	$\checkmark$	$\checkmark$	$\checkmark$	$\checkmark$

We observe that the models built using our climate indices yield a significantly ($p < 0.05$) higher correlation and lower RMSE than those built using climate indices discovered using unsupervised methodologies. This suggests that climate indices more highly associated with the response variable of interest, as the ones discovered using our proposed methodology, have greater predictive power.

We further assess the predictive power of our climate indices by comparing our predictions with the official forecasts of the OND rainfall season issued by the TMA every year on September. To this end, the rainfall season for each year was categorized according to the guidelines of the TMA as *below normal, normal,* or *above normal* (rainfall below 75%, between 75% and 125%, or above 125% of long-term averages, respectively). Long-term averages were computed using the training set. Similarly to the regression models, decision trees to classify the

OND rainfall season at each station were trained using data up to the month of August and considering only the top six (6) climate indices discovered with our proposed methodology as predictors. The decision trees were built using the Gini index as the split criterion and pruning to avoid overfitting.

The classification accuracies obtained are shown in Figure 3. We observe that the accuracy of the decision trees built using our climate indices is higher than that of the official forecasts for three (3) out of four (4) stations. This suggests that the use of the climate indices discovered using our proposed methodology can potentially improve forecasts of the response variable of interest.

4.6 Physical Interpretation of Climate Indices Discovered

Finally, we discuss the climate indices discovered in terms of their climatological relevance. Rainfall variability in the GHA is known to be mainly associated with

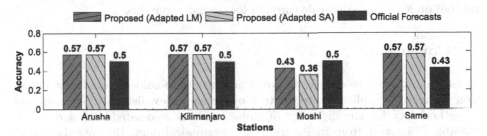

Fig. 3. Classification accuracy of the prediction of the OND rainfall season at each station in the GHA region from 1998 to 2011 obtained using the proposed methodology with the Louvain method (LM) and simulated annealing (SA) as the community detection algorithms, as well as official forecasts issued by the TMA.

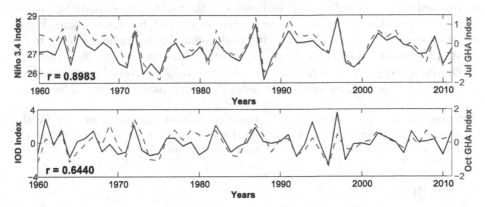

Fig. 4. Time series of the Niño 3.4 index (upper, solid line) and the IOD index (lower, solid line) with climate indices discovered in July (upper, dashed line) and October (lower, dashed line) using our proposed methodology with the adapted Louvain method as the community detection algorithm and OND rainfall variability in the GHA as the response variable of interest. The linear correlation between the time series is shown in the lower left corner of each figure.

ENSO in the equatorial Pacific Ocean [18] and the Indian Ocean Dipole (IOD) in the tropical Indo-Pacific Ocean [1].

Climate indices significantly correlated ($p < 0.01$) with ENSO, in particular with the Niño 3.4 index, were discovered in June, July, August, September, and October using both adapted community detection algorithms (for example, see Figure 4). The representative climate variable selected for these climate indices is mostly either SST or PW, a close proxy of SST in the equatorial Pacific Ocean in the NCAR/NCEP Reanalysis 1 data set. Higher SSTs in the equatorial Pacific Ocean are associated with a suppression of East African rainfall, by modulating the strength of the global upper level wind flow [18].

Climate indices significantly correlated ($p < 0.01$) with the IOD were discovered in July, August, September and October using both adapted community detection algorithms (for example, see Figure 4). These climate indices were generally discovered closer to the onset of the OND rainfall season than the ones in the equatorial Pacific Ocean, as the IOD exerts its influence on East African rainfall on a shorter timescale through local wind anomalies [1].

5   Conclusions

In this paper, we introduced the problem of response-guided community detection through its application to the task of climate index discovery. We proposed a methodology for the discovery of climate indices associated with a response variable of interest from multivariate spatiotemporal data, the contribution of which is twofold. First, we proposed a general strategy for response-guided community detection, and second, we introduced a network representation of the data that incorporates information from multiple variables.

We applied our proposed methodology to the discovery of climate indices associated with seasonal rainfall variability in the GHA. The climatological relevance of the climate indices discovered is supported by domain knowledge, as evidenced by their association with traditional climate indices known to be related to seasonal rainfall in the region. Furthermore, our results show that our methodology improves the forecast skill for this response variable with respect to existing methodologies for climate index discovery, as well as official forecasts.

Acknowledgments. This material is based upon work supported in part by the Laboratory for Analytic Sciences, the U.S. Department of Energy, Office of Science, Advanced Scientific Computing Research, and NSF grant 1029711.

References

1. Black, E., Slingo, J., Sperber, K.R.: An observational study of the relationship between excessively strong short rains in coastal East Africa and Indian Ocean SST. Mon. Weather Rev. **131**(1), 74–94 (2003)
2. Blondel, V.D., Guillaume, J.L., Lambiotte, R., Lefebvre, E.: Fast unfolding of communities in large networks. J. Stat. Mech. Theor. Exp. **2008**(10), P10008 (2008)

3. Brandes, U., Delling, D., Gaertler, M., Görke, R., Hoefer, M., Nikoloski, Z., Wagner, D.: On finding graph clusterings with maximum modularity. In: Brandstädt, A., Kratsch, D., Müller, H. (eds.) WG 2007. LNCS, vol. 4769, pp. 121–132. Springer, Heidelberg (2007)

4. Clauset, A., Newman, M.E.J., Moore, C.: Finding community structure in very large networks. Phys. Rev. E **70**(6), 066111 (2004)

5. Donges, J.F., Zou, Y., Marwan, N., Kurths, J.: Complex networks in climate dynamics. The European Physical Journal-Special Topics **174**(1), 157–179 (2009)

6. Eaton, E., Mansbach, R.: A spin-glass model for semi-supervised community detection. In: Proc. of the 26th AAAI Conference on Artificial Intelligence, pp. 900–906. AAAI (2012)

7. Fisher, R.A.: Statistical methods for research workers. Edinburgh (1934)

8. Fortunato, S.: Community detection in graphs. Phys. Rep. **486**(3), 75–174 (2010)

9. Gray, W.M.: Atlantic seasonal hurricane frequency. Part I: El Niño and 30 mb quasi-biennial oscillation influences. Mon. Weather Rev. **112**(9), 1649–1668 (1984)

10. Guimerà, R., Amaral, L.A.N.: Functional cartography of complex metabolic networks. Nature **433**(7028), 895–900 (2005)

11. Harenberg, S., Bello, G.A., Gjeltema, L., et al.: Community detection in large-scale networks: a survey and empirical evaluation. WIREs Comput. Stat. (1939-0068) (2014)

12. Harenberg, S., Seay, R.G., Ranshous, S., et al.: Memory-efficient query-driven community detection with application to complex disease associations. In: Proc. of the 2014 SIAM Int. Conf. on Data Mining, pp. 1010–1018. SIAM (2014)

13. Kalnay, E., Kanamitsu, M., Kistler, R., et al.: The NCEP/NCAR 40-year reanalysis project. Bull. Amer. Meteor. Soc. **77**(3), 437–471 (1996)

14. Kirkpatrick, S., Gelatt, C.D., Vecchi, M.P.: Optimization by simmulated annealing. Science **220**(4598), 671–680 (1983)

15. Lancichinetti, A., Fortunato, S.: Community detection algorithms: a comparative analysis. Phys. Rev. E **80**(5), 056117 (2009)

16. Newman, M.E.J.: Analysis of weighted networks. Phys. Rev. E **70**(5), 056131 (2004)

17. Newman, M.E.J., Girvan, M.: Finding and evaluating community structure in networks. Phys. Rev. E **69**(2), 026113 (2004)

18. Omondi, P., Ogallo, L.A., Anyah, R., et al.: Linkages between global sea surface temperatures and decadal rainfall variability over Eastern Africa region. Int. J. of Climatol. **33**(8), 2082–2104 (2013)

19. Pons, P., Latapy, M.: Computing communities in large networks using random walks. J. Graph Algorithms Appl. **10**(2), 191–218 (2006)

20. Schiermeier, Q.: The real holes in climate science. Nature **463**(7279), 284–287 (2010)

21. Smith, T.M., Reynolds, R.W., Peterson, T.C., Lawrimore, J.: Improvements to NOAA's historical merged land-ocean surface temperature analysis (1880–2006). J. Climate **21**(10), 2283–2296 (2008)

22. Steinbach, M., Tan, P.N., Kumar, V., et al.: Discovery of climate indices using clustering. In: Proc. of the 9th ACM SIGKDD Int. Conf. on Knowledge Discovery and Data Mining, pp. 446–455. ACM (2003)

23. Steinhaeuser, K., Chawla, N.V., Ganguly, A.R.: An exploration of climate data using complex networks. ACM SIGKDD Explor. Newsl. **12**(1), 25–32 (2010)

24. Steinhaeuser, K., Chawla, N.V., Ganguly, A.R.: Complex networks as a unified framework for descriptive analysis and predictive modeling in climate science. Statistical Analysis and Data Mining 4(5), 497–511 (2011)

25. Tsonis, A.A., Roebber, P.J.: The architecture of the climate network. Phys. A **333**, 497–504 (2004)

Robust Classification of Information Networks by Consistent Graph Learning

Shi Zhi[1]([⊠]), Jiawei Han[1], and Quanquan Gu[2]

[1] Department of Computer Science,
University of Illinois at Urbana-Champaign, Champaign, IL, USA
{shizhi2,hanj}@illinois.edu
[2] Department of Systems and Information Engineering,
University of Virginia, Charlottesville, VA, USA
qg5w@virginia.edu

Abstract. Graph regularization-based methods have achieved great success for network classification by making the label-link consistency assumption, i.e., if two nodes are linked together, they are likely to belong to the same class. However, in a real-world network, there exist links that connect nodes of different classes. These inconsistent links raise a big challenge for graph regularization and deteriorate the classification performance significantly. To address this problem, we propose a novel algorithm, namely *Consistent Graph Learning*, which is robust to the inconsistent links of a network. In particular, given a network and a small number of labeled nodes, we aim at learning a consistent network with more consistent and fewer inconsistent links than the original network. Since the link information of a network is naturally represented by a set of relation matrices, the learning of a consistent network is reduced to learning consistent relation matrices under some constraints. More specifically, we achieve it by joint graph regularization on the nuclear norm minimization of consistent relation matrices together with ℓ_1-norm minimization on the difference matrices between the original relation matrices and the learned consistent ones subject to certain constraints. Experiments on both homogeneous and heterogeneous network datasets show that the proposed method outperforms the state-of-the-art methods.

Keywords: Robust classification · Information network · Consistent link · Consistent network · Consistent Graph Learning

1 Introduction

Information networks have been found to play increasingly important role in real-life applications. Generally speaking, information networks can be categorized into two families: (1) homogeneous information networks where there is only one type of nodes and links. Examples include friendship network in Facebook[1], co-author and citation network in DBLP[2], and the World Wide Web;

---

[1] http://www.facebook.com
[2] http://www.informatik.uni-trier.de/~ley/db/

© Springer International Publishing Switzerland 2015
A. Appice et al. (Eds.): ECML PKDD 2015, Part II, LNAI 9285, pp. 752–767, 2015.
DOI: 10.1007/978-3-319-23525-7_46

and (2) heterogeneous information networks where there exist multiple types of nodes and links. A bibliographic information network is an example of heterogeneous information network, which contains four types of objects: papers, authors, conferences and terms. Papers and authors are linked by the relation of "written by" and "write". Papers and conferences are linked by "published in" and "publish". Papers and terms are linked by "contain" and "contained in".

In the past decade, many methods have been proposed for classification of both homogeneous information networks [5,9,11,13,14,18–21] and heterogeneous information networks [7,8], which are based on the link structure and the node content of networks. Among these methods, graph regularization-based methods [5,8,9,18,19,21] have achieved superior performance over other methods. These methods assume that if two nodes are linked in a network, their labels are likely to be the same. Start from a small number of labeled nodes, labels are propagated along linking nodes to preserve the local consistency. Therefore, they heavily depend on the link structure of a network and implicitly require the links of the network to be consistent with node labels. However, in many cases, this requirement is not satisfied. For example, in Figure 1(a), there are two classes of nodes denoted by different colors. The black edge links two nodes of the same class, while the yellow edge links node from different classes. We define black link as *Consistent Link*, and yellow link as *Inconsistent Link*. Due to the existence of inconsistent links, graph regularization-based methods may fail to correctly classify the nodes residing on both sides of the inconsistent edges. In our study, we call the network with inconsistent links as *Inconsistent Network*. Since inconsistent links are prevalent in real-world networks, it is of central importance to develop learning models for classification of inconsistent networks.

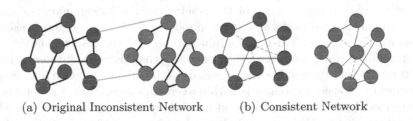

(a) Original Inconsistent Network   (b) Consistent Network

Fig. 1. An example of (a) Inconsistent Network, and (b) Consistent Network. There are two classes of nodes denoted by red and green. The black links are consistent with the labels, while the nodes linked by yellow links have different labels. The blue dashed links are added consistent links. The goal is to remove red links and add blue links.

Intuitively, if there are no inconsistent links, e.g., the yellow edges in Figure 1(a), graph regularization-based classification methods [18] can achieve good results. This motivates us to handle an inconsistent network in the following two ways. First, if we can detect which links are inconsistent, we can delete these inconsistent links. Second, if we can add more consistent links between the nodes of the same classes, we can compensate the effect of the inconsistent links as well. It is desirable to get a network as shown in Figure 1 (b), where we remove the inconsistent yellow edges and add the dashed blue edges, i.e., consistent links. Based on this modified network, graph regularization-based methods may work better than using original relation matrices. In our study, we will show that we can learn an approximately consistent network using a small number of labeled data under certain constraints.

In this paper, based on the above discussion, we propose a novel regularization technique, namely *Consistent Graph Learning*, which is robust to those inconsistent links of a network. Our goal is to learn an approximately consistent network based on a small number of labeled data. Since the link information of a network can be naturally represented by a set of relation matrices, the learning of a consistent network can be transformed into learning consistent relation matrices. More specifically, we assume that each original relation matrix can be decomposed into a consistent relation matrix and a residue matrix. In a fully consistent network, each pair of nodes of the same class are linked while those of different classes are not linked. Though the real-world network is usually sparse, nodes in a consistent network connect much more to the nodes of the same class rather than a different class. Thus, consistent relation matrix intrinsically has the low-rank property. We can achieve this low-rank characteristics by applying nuclear-norm minimization on consistent relation matrix. By doing this, more consistent links are added to the original inconsistent network. On the other hand, since in real-world network nodes of the same class tends to have much more links than those of different classes do, the consistent network should be similar to the original network and the norm of the residue matrix should be small. To remove inconsistent links and keep consistent links, we aim to have a sparse residue matrix with non-zero elements as fewer as possible instead of changing the value of every element in the original relation matrix. It can be achieved by minimizing ℓ_1-norm of the residue matrix. In summary, to satisfy both requirements, we perform a joint graph regularization on the consistent relation matrix with nuclear-norm minimization, and the residue matrix with ℓ_1-norm minimization, subject to the constraint that the sum of the consistent relation matrix and the residue matrix equals to the original relation matrix, and each element of consistent matrix is within a certain range. Given a set of labeled data, our model can learn the consistent network by alternating direction method of multipliers [2] (ADMM) method that solves a convex optimization problem by breaking it into smaller pieces, each of which is easier to handle. We can use the consistent network to classify all the other nodes by any network classification method. Experiments on both homogeneous and heterogeneous network datasets show that the proposed method outperforms the state-of-art methods.

The main contributions of this paper are as follows: (1) We raise and analyze the inconsistency of real-world networks; (2) we propose a consistent graph learning technique which is able to learn an approximately consistent network given a small number of labeled data; and (3) we validate the effectiveness of the proposed method on both homogeneous and heterogeneous networks. The remainder of this paper is organized as follows. In Section 2 we present a model for classification of the information networks with inconsistent links. In Section 3, we discuss several related work to our method. The experiments on Cora and DBLP datasets are demonstrated in Section 4. Finally, we draw a conclusion and point out the future work in Section 5.

2   The Proposed Method

In this section, we present Consistent Graph Learning for semi-supervised classification of information networks. Before going deep into the proposed method, we first present some preliminary definitions of information network.

2.1   Preliminary Definitions

Definition 1. *An information network consists of m types of objects $\mathcal{X}^{kl} = \{\mathcal{X}^k\}_{k=1}^m$, where $\mathcal{X}^k$ is a set of objects belonging to the k-th type. A weighted graph $\mathcal{G} = (\mathcal{V}, \mathcal{E}, R)$ is called an **information network** on objects $\mathcal{X}$, if $\mathcal{V} = \mathcal{X}$, $\mathcal{E}$ is a binary relation on $\mathcal{V}$, and $R : \mathcal{E} \to \mathbb{R}$ is a weight function mapping from an edge $e \in \mathcal{E}$ to a real number $w \in \mathbb{R}$. Specially, we call such an information network **heterogeneous network** when $m \geq 2$; and **homogeneous network** when $m = 1$.*

We can treat homogeneous information network as a special case of heterogeneous information network. The crucial difference of using heterogeneous network is that we work on each relation matrix between two types of nodes instead of working on a large relation matrix between all nodes of different types. Therefore, we will introduce the proposed method in the context of heterogeneous network, which is more general. Now we present the formal definitions of *Consistent Link* and *Consistent Network*.

Definition 2. *A link is **consistent** if the nodes it connects belong to the same class. An information network is **consistent** if and only if all of its links are consistent.*

The definitions of *Inconsistent Link* and *Inconsistent Network* can be deduced analogously, hence we omit them. Note that the definitions in this paper are specific to our problem, i.e., classification of networks. There may exist other definitions of *Consistent Link* and *Consistent Network* in the literature.

2.2   Notation

A heterogeneous network can be represented by a collection of relation matrices, each of which models the pairwise relation between a node in one type and another node in a different type. Mathematically speaking, in a heterogeneous information network, suppose there are m types of entities, i.e., $\mathcal{X}^k, 1 \leq k \leq m$, where $\mathcal{X}^k = \{x_1^k, \ldots, x_n^k\}$. A relation graph $\mathcal{G}^{kl}$ can be built corresponding to each type of link relationships between two types of data entities $\mathcal{X}^k$ and $\mathcal{X}^l$, $1 \leq k \leq m$. Let $\mathbf{R}^{kl}$ be an $n_k \times n_l$ relation matrix corresponding to graph $\mathcal{G}$, in which R_{ij}^{kl} denotes the weight on link from x_i^k to x_j^l. Note that $\mathbf{R}^{kl}$ is not symmetric. One possible definition of $\mathbf{R}^{kl}$ is as follows.

$$R_{ij}^{kl} = \begin{cases} 1 & \text{if there is a link from } x_i^k \text{ to } x_j^l \\ 0 & \text{otherwise} \end{cases} \tag{1}$$

If we consider a weighted graph, the definition of $\mathbf{R}^{kl}$ can be extended to

$$R_{ij}^{kl} = \begin{cases} m & \text{if there are } m \text{ links from } x_i^k \text{ to } x_j^l \\ 0 & \text{otherwise} \end{cases} \tag{2}$$

Suppose there are c classes, in order to encode label information of each type, we basically define a label matrix for each type, i.e., $\mathbf{Y}^k \in \mathbb{R}^{n_k \times c}$, such that

$$Y_{il}^k = \begin{cases} 1 & \text{if } x_i^k \text{ is labeled to the } l\text{-th class} \\ 0 & \text{otherwise} \end{cases} \tag{3}$$

Note that if x_i^k is unlabeled, then $Y_{il}^k = 0$ for $\forall l$.

For each type of objects, we are going to learn a class assignment matrix $\mathbf{F}^k \in \mathbb{R}^{n_k \times c}$, whose definition is similar to $\mathbf{Y}^k$. We denote the i-th row of $\mathbf{Y}^k$ by $\mathbf{Y}_{i\cdot}^k$, and the i-th row of $\mathbf{F}^k$ by $\mathbf{F}_{i\cdot}^k$. For a matrix $\mathbf{E}^{kl}$, its ℓ_1-norm is defined as $||\mathbf{E}||_1 = \sum_{ij} |E_{ij}|$. For a $m \times n$ matrix $\mathbf{W}$, its nuclear norm is defined as $||\mathbf{W}||_* = \sum_i^{\min\{m,n\}} \sigma_i$, where $\mathbf{W} = \mathbf{U}\boldsymbol{\Sigma}\mathbf{V}^T$ is the Singular Value Decomposition (SVD) of $\mathbf{W}$, $(\boldsymbol{\Sigma})_{ii} = \sigma_i$. For a matrix $\mathbf{D}$, its Frobenius norm is defined as $||\mathbf{D}||_F = \sqrt{\sum_{ij} D_{ij}^2}$. Notation $\circ$ is used to get the entry-wise product of two matrices, e.g., $\mathbf{D} \circ \mathbf{E}$ is a matrix whose each element equals to $D_{ij}E_{ij}$. Matrix $\mathbf{0}$ is a matrix of all zeros, and matrix $\mathbf{1}$ is a matrix of all ones.

2.3   Standard Graph Regularization

The basic assumption of graph regularization is that if two objects x_i^k and x_j^l are linked together, then their labels F_{ip}^k and F_{jp}^l are likely to be the same. It can be mathematically formulated as [16],

$$\min_{\mathbf{F}^k, \mathbf{F}^l} \frac{1}{2} \sum_{i=1}^{n_k} \sum_{j=1}^{n_l} ||\mathbf{F}_{i\cdot}^k - \mathbf{F}_{j\cdot}^l||_2^2 R_{ij}^{kl}, \tag{4}$$

where $\mathbf{R}^{kl}$ could either be the original relation matrix or the normalized one. As we can see, if $R_{ij}^{kl} > 0$, Eq. (4) will push the label of x_i^k and the label of x_j^l close. If $R_{ij}^{kl} = 0$, the labels are determined by other terms of the objective function. This is the rationale of graph regularization. To give an example, in a homogeneous citation network ($k = l = 1$), if the i-th paper cites the j-th paper, standard graph regularization tends to classify these two papers into the same class. However, we will show that it is not true in real world citation dataset when there are inconsistent links, as we will show in the experiments. In this case, graph regularization would fail. This shows the drawback of standard graph regularization technique for network classification.

2.4 Consistent Graph Learning

The basic idea of our method is to learn an approximately consistent network, based on which we apply graph regularization and learn a classifier. Since a heterogeneous information network can be represented by a set of relation matrices, i.e., $\{\mathbf{R}^{kl}\}$, the learning of a consistent network can be transformed into learning a set of consistent relation matrices, i.e., $\{\mathbf{W}^{kl}\}$. Here we say a relation matrix is consistent if and only if its corresponding network is consistent. For each relation matrix $\mathbf{R}^{kl}$ where there may exist some entries which are inconsistent with the labels, we decompose it into a consistent relation matrix $\mathbf{W}^{kl}$ whose entries are consistent with the node labels, and a residue matrix $\mathbf{E}^{kl}$ whose entries are inconsistent with the labels. Hereafter, we call $\mathbf{W}^{kl}$ as *Consistent Relation Matrix* and $\mathbf{E}^{kl}$ as *Residue Matrix*. It is mathematically described as

$$\mathbf{R}^{kl} = \mathbf{W}^{kl} + \mathbf{E}^{kl}, 0 \leq \mathbf{W}^{kl} \leq \max(1, \mathbf{R}^{kl}), \tag{5}$$

where the function max takes the larger value of 1 and each element in $\mathbf{R}^{kl}$. We add a box constraint on $\mathbf{W}^{kl}$ to make it both lower and upper-bounded. The reason is that the original relation matrix $\mathbf{R}^{kl}$ is bounded. We hope that the learned consistent relation matrix $\mathbf{W}^{kl}$ is also bounded.

In order to make the learned relation matrix $\mathbf{W}^{kl}$ consistent, we need to specify additional constraints as follows:

1. $\mathbf{W}^{kl}$ should be consistent with the labeled data $\mathbf{Y}$. It can be achieved by standard graph regularization on $\mathbf{W}^{kl}$ with respect to $\mathbf{Y}$.
2. We assume that the number of inconsistent links is only a portion of the links in $\mathbf{R}^{kl}$. Hence, we require the residue matrix to be sparse. To obtain this goal, we apply ℓ_1-norm minimization to $\mathbf{E}^{kl}$.
3. In principal, we prefer not to remove links that connect nodes of large degree because removing such links may take a risk to disconnect more unlabeled nodes with labeled ones such that some labels cannot be propagated through the consistent links. To handle this issue, we take entry-wise product of $\mathbf{E}^{kl}$ in the ℓ_1-norm term and $\mathbf{D}^{kl}$, where $D_{ij}^{kl} = \sqrt{d_i d_j}$, $d_i = \sum_i R_{ij}^{kl}$ is the out-degree of node x_i^k and $d_j = \sum_j R_{ij}^{kl}$ is the in-degree of node x_j^k.
4. As we mentioned before, nodes in a consistent relation matrix connect much more to the nodes of the same class rather than different class. To pursue the low-rank property of consistent relation matrix $\mathbf{W}^{kl}$, we apply nuclear norm minimization to $\mathbf{W}^{kl}$. Note that there may be some extreme cases when a low-rank matrix is not necessarily a consistent matrix (e.g., an all-ones matrix). However, we can prevent our method from converging to these cases by balancing different regularizations.

We bring in an auxiliary $\mathbf{Q}^{kl}$ and let it be equal to $\mathbf{W}^{kl}$. The advantage of it is to allows us to solve nuclear norm minimization and box constraint in separate steps, which is easy to solve. Putting all the above constraints together, we obtain

$$\min_{\{\mathbf{Q}^{kl}, \mathbf{W}^{kl}, \mathbf{E}^{kl}\}} \sum_{i=1}^{n_k} \sum_{j=1}^{n_l} ||\mathbf{Y}_{i\cdot}^k - \mathbf{Y}_{j\cdot}^l||_2^2 W_{ij}^{kl} + \gamma_{kl} ||\mathbf{D}^{kl} \circ \mathbf{E}^{kl}||_1 + \beta_{kl} ||\mathbf{W}^{kl}||_*$$

$$\text{subject to} \quad \mathbf{R}^{kl} = \mathbf{W}^{kl} + \mathbf{E}^{kl}, \mathbf{W}^{kl} = \mathbf{Q}^{kl}, 0 \le \mathbf{Q}^{kl} \le \max(1, \mathbf{R}^{kl}) \quad (6)$$

where $\gamma_{kl} > 0$ and $\beta_{kl} > 0$ are regularization parameters that controls the sparsity of $\mathbf{E}^{kl}$ and the low-rank property of $\mathbf{W}^{kl}$, respectively. These two parameters essentially control the balance among the label-link consistency, sparsity and low-rank property. The larger γ_{kl} is, the sparser $\mathbf{E}^{kl}$ will be. Larger β_{kl} forces the nuclear norm of $\mathbf{W}^{kl}$ to be smaller. We call the model in Eq. (6) as *Consistent Graph Learning*. Note that if we set $\gamma_{kl} = \infty$ and $\beta_{kl} = 0$, $\mathbf{W}^{kl}$ will be exactly equal to $\mathbf{R}^{kl}$. If we have prior knowledge indicating some of the relation matrices $\mathbf{R}^{kl}$ are consistent, we can set the corresponding γ_{kl} to ∞ and $\beta_{kl} = 0$. Note that we cannot guarantee that learned $\mathbf{W}^{kl}$ is totally consistent because we only have partial labels of the nodes, but the learned relation matrix has fewer inconsistent links and more consistent links than the original one. In the following, we will introduce how to solve it.

2.5   Optimization

Due to the decomposition equality constraint in Eq. (6), we use the alternating direction method of multipliers [2] (ADMM). We will derive an algorithm based on ADMM for solving Eq. (6). Before that, we first briefly introduce augmented Lagrangian multiplier [3] method. Augmented Lagrangian [3] (ALM) is a method for solving equality constrained optimization problem. It reformulates the problem into an unconstrained one by adding Lagrangian multipliers and an extra quadratic penalty term for each equality constraint.

As to our method, the augmented Lagrangian function is as follows by ignoring the inequality constraints on $\mathbf{E}^{kl}$

$$L(\mathbf{Q}^{kl}, \mathbf{W}^{kl}, \mathbf{E}^{kl}, \mathbf{Z}^{kl}) = \sum_{i=1}^{n_k} \sum_{j=1}^{n_l} ||\mathbf{Y}_{i\cdot}^k - \mathbf{Y}_{j\cdot}^l||_2^2 W_{ij}^{kl} + \beta_{kl} ||\mathbf{W}^{kl}||_* + \gamma_{kl} ||\mathbf{D}^{kl} \circ \mathbf{E}^{kl}||_1$$

$$+ \text{tr}((\mathbf{Z}^{kl})^T (\mathbf{W}^{kl} + \mathbf{E}^{kl} - \mathbf{R}^{kl})) + \frac{\mu}{2} ||\mathbf{W}^{kl} + \mathbf{E}^{kl} - \mathbf{R}^{kl}||_F^2$$

$$+ \text{tr}((\mathbf{X}^{kl})^T (\mathbf{W}^{kl} - \mathbf{Q}^{kl})) + \frac{\zeta}{2} ||\mathbf{W}^{kl} - \mathbf{Q}^{kl}||_F^2, \quad (7)$$

where $\mathbf{Z}^{kl}$ and $\mathbf{X}^{kl}$ are Lagrangian multipliers, μ and ζ are penalty parameters.

In the following, we will derive the updating formula for each variable. In other words, we solve each variable when fixing the other variables. This is also known as alternating direction method of multipliers [2] (ADMM).

Computation of $\mathbf{W}^{kl}$. Given other variables fixed, the optimization of Eq. (7) with respect to $\mathbf{W}^{kl}$ is reduced to

$$\min_{\mathbf{W}^{kl}} \quad \mathrm{tr}((\mathbf{S}^{kl})^T \mathbf{W}^{kl}) + \beta_{kl}||\mathbf{W}^{kl}||_*$$

$$+\mathrm{tr}((\mathbf{Z}^{kl})^T(\mathbf{W}^{kl} + \mathbf{E}^{kl} - \mathbf{R}^{kl})) + \frac{\mu}{2}||\mathbf{W}^{kl} + \mathbf{E}^{kl} - \mathbf{R}^{kl}||_F^2$$

$$+\mathrm{tr}((\mathbf{X}^{kl})^T(\mathbf{W}^{kl} - \mathbf{Q}^{kl})) + \frac{\zeta}{2}||\mathbf{W}^{kl} - \mathbf{Q}^{kl}||_F^2, \tag{8}$$

where the matrix $\mathbf{S}^{kl}$ is defined as $S_{ij}^{kl} = ||\mathbf{Y}_{i\cdot}^k - \mathbf{Y}_{j\cdot}^l||_2^2$. Eq. (8) is equivalent to

$$\min_{\mathbf{W}^{kl}} \quad \frac{\beta_{kl}}{\mu + \zeta}||\mathbf{W}^{kl}||_* + \frac{1}{2}||\mathbf{W}^{kl} - \mathbf{A}^{kl}||_F^2, \tag{9}$$

where

$$\mathbf{A}^{kl} = \frac{\zeta \mathbf{Q}^{kl} - \mu(\mathbf{E}^{kl} - \mathbf{R}^{kl}) - \mathbf{S}^{kl} - \mathbf{Z}^{kl} - \mathbf{X}^{kl}}{\mu + \zeta}. \tag{10}$$

Eq. (9) has a closed-form solution

$$\mathbf{W}^{kl} = \mathbf{U}\boldsymbol{\Sigma}_*\mathbf{V}^T, \tag{11}$$

where $\mathbf{A}^{kl} = \mathbf{U}\boldsymbol{\Sigma}\mathbf{V}^T$ is the SVD of $\mathbf{A}^{kl}$, and $\boldsymbol{\Sigma}_*$ is the diagonal with $(\boldsymbol{\Sigma}_*)_{ii} = \max\{0, (\boldsymbol{\Sigma})_{ii} - \beta_{kl}/(\mu + \zeta)\}$. By setting small singular values to zero, the nuclear norm of $\mathbf{W}^{kl}$ is reduced.

Computation of $\mathbf{E}^{kl}$. Given other variables fixed, the optimization of Eq. (7) with respect to $\mathbf{E}^{kl}$ boils down to

$$\min_{\mathbf{E}^{kl}} \quad \gamma_{kl}||\mathbf{D}^{kl} \circ \mathbf{E}^{kl}||_1$$

$$+\mathrm{tr}((\mathbf{Z}^{kl})^T(\mathbf{W}^{kl} + \mathbf{E}^{kl} - \mathbf{R}^{kl})) + \frac{\mu}{2}||\mathbf{W}^{kl} + \mathbf{E}^{kl} - \mathbf{R}^{kl}||_F^2, \tag{12}$$

which is equivalent to

$$\min_{\mathbf{E}^{kl}} \quad \frac{\gamma_{kl}}{\mu}||\mathbf{D}^{kl} \circ \mathbf{E}^{kl}||_1 + \frac{1}{2}||\mathbf{E}^{kl} + \mathbf{W}^{kl} - \mathbf{R}^{kl} + \frac{1}{\mu}\mathbf{Z}^{kl}||_F^2. \tag{13}$$

Eq. (13) has a closed-form solution as follows,

$$E_{ij}^{kl} = \begin{cases} B_{ij}^{kl} - \frac{\gamma_{kl}D_{ij}^{kl}}{\mu} & \text{if } B_{ij}^{kl} \geq \frac{\gamma_{kl}D_{ij}^{kl}}{\mu} \\ 0 & \text{if } -\frac{\gamma_{kl}D_{ij}^{kl}}{\mu} < B_{ij}^{kl} < \frac{\gamma_{kl}D_{ij}^{kl}}{\mu} \\ B_{ij}^{kl} + \frac{\gamma_{kl}D_{ij}^{kl}}{\mu} & \text{if } B_{ij}^{kl} \leq -\frac{\gamma_{kl}D_{ij}^{kl}}{\mu} \end{cases}, \tag{14}$$

where $\mathbf{B}^{kl} = -\mathbf{W}^{kl} + \mathbf{R}^{kl} - \frac{1}{\mu}\mathbf{Z}^{kl}$. This step essentially only allows E_{ij}^{kl} to be non-zero when falling out of a certain range. We can see that by introducing matrix D_{ij}^{kl}, E_{ij}^{kl} is more likely to be zero if D_{ij}^{kl} is larger. Thus, the link between nodes of large in-degree and out-degree will be less likely to be removed.

Computation of $\mathbf{Q}^{kl}$. Given other variables fixed, the optimization of Eq. (7) with respect to $\mathbf{Q}^{kl}$ boils down to

$$\min_{\mathbf{Q}^{kl}} \quad \mathrm{tr}((\mathbf{X}^{kl})^T(\mathbf{W}^{kl} - \mathbf{Q}^{kl})) + \frac{\zeta}{2}||\mathbf{W}^{kl} - \mathbf{Q}^{kl}||_F^2$$
$$\text{subject to} \quad \mathbf{0} \leq \mathbf{Q}^{kl} \leq \max(\mathbf{1}, \mathbf{R}^{kl}), \tag{15}$$

which has a closed-form solution

$$Q_{ij}^{kl} = \begin{cases} R_{ij}^{kl} & Q_{ij}^{kl} \geq \max(1, R_{ij}^{kl}) \\ W_{ij}^{kl} + \frac{1}{\zeta}X_{ij}^{kl} & 0 < Q_{ij}^{kl} < \max(1, R_{ij}^{kl}) \\ 0 & Q_{ij}^{kl} \leq 0 \end{cases}. \tag{16}$$

By making $\mathbf{Q}^{kl} = \mathbf{W}^{kl}$ and adding a box constraint on $\mathbf{Q}^{kl}$, $\mathbf{W}^{kl}$ is essentially upper and lower-bounded.

Computation of $\mathbf{Z}^{kl}$ and $\mathbf{X}^{kl}$. Taking the derivative of L with respect to $\mathbf{Z}^{kl}$ and $\mathbf{X}^{kl}$, we obtain

$$\frac{\partial L}{\partial \mathbf{Z}^{kl}} = \mathbf{W}^{kl} + \mathbf{E}^{kl} - \mathbf{R}^{kl} \quad \text{and} \quad \frac{\partial L}{\partial \mathbf{X}^{kl}} = \mathbf{W}^{kl} - \mathbf{Q}^{kl}, \tag{17}$$

which leads to the following updating formula for Lagrangian multiplier $\mathbf{Z}^{kl}$,

$$\mathbf{Z}^{kl} = \mathbf{Z}^{kl} + \mu(\mathbf{W}^{kl} + \mathbf{E}^{kl} - \mathbf{R}^{kl}). \tag{18}$$

Similarly, the updating formula for Lagrangian multiplier is $\mathbf{X}^{kl}$,

$$\mathbf{X}^{kl} = \mathbf{X}^{kl} + \zeta(\mathbf{W}^{kl} - \mathbf{Q}^{kl}). \tag{19}$$

In summary, we present the algorithm in Algorithm 1. In our experiments, we set $\mu = 10$ and $\zeta = 10$, which leads to fast convergence. In addition, we initialize $\mathbf{W}^{kl}$ as the original relation matrix $\mathbf{R}^{kl}$ with a small perturbation by adding a random matrix $\mathbf{M}^{kl}$ to $\mathbf{W}^{kl}$. Note that the random perturbation matrix helps the convergence of the ADMM algorithm [17]. We can see that in each outer iteration of Algorithm 1, it learns the underlying consistent network between $\mathcal{X}^k$ and $\mathcal{X}^l$, i.e., $\mathbf{W}^{kl}$ by ADMM. Note that k and l can be either the same or different. We can see later the learned consistent matrices can improve the accuracy of classification in the next step.

2.6 Estimation of Unlabeled Data

After we compute the consistent relation matrix $\mathbf{W}^{kl}$, we can apply existing semi-supervised classification algorithms to estimate the unlabeled data. In the experiment, we use LLGC [18] for homogeneous network classification and GNet-Mine [8] for heterogeneous network classification. In the experiments, for the citation and co-author sub-networks, we transform the learned consistent relation matrix into a symmetric one by setting W_{ij}^{kl} to the larger element between W_{ij}^{kl} and W_{ji}^{kl}. We do the same symmetrization on original relation matrix for input of LLGC and GNetMine.

Algorithm 1 Robust Classification of Network by *Consistent Graph Learning* (CGL)

   Input: $\mathbf{R}^{kl}$, $\beta_{kl} > 0$, $\gamma_{kl} > 0$, $\mathbf{Y}$, μ, ζ;
   Output: $\mathbf{W}^{kl}$, $\mathbf{E}^{kl}$, $k, l = 1, \ldots, m$;
   for $k, l = 1 \rightarrow m$ **do**
      Initialize $\mathbf{W}^{kl} = \mathbf{R}^{kl} + \mathbf{M}^{kl}$, $\mathbf{Z}^{kl}$, $\mathbf{X}^{kl}$
      repeat
         Compute $\mathbf{W}^{kl}$ as in Eq. (11)
         Compute $\mathbf{E}^{kl}$ as in Eq. (14)
         Compute $\mathbf{Q}^{kl}$ as in Eq. (16)
         Compute $\mathbf{Z}^{kl}$ as in Eq. (18)
         Compute $\mathbf{X}^{kl}$ as in Eq. (19)
      until Convergence
   end for

2.7 Analysis

The convergence of Algorithm 1 is stated in the following theorem.

Theorem 1. *Algorithm 1 is theoretically guaranteed to converge to the global minima of the problem in Eq. (6).*

Proof sketch: The global convergence of the algorithm can be proved using the technique in [10] [6].

Now we analyze the time complexity of Algorithm 1. Let c be the number of classes, $|V|$ denote the total number of objects, and $|E|$ denote the total number of links in the information network. In each inner iteration of Algorithm 1, it takes $O(n^2 c)$ to update $\mathbf{W}^{kl}$, $O(|E|)$ to update $\mathbf{Q}^{kl}$, $O(|E|)$ to update $\mathbf{E}^{kl}$, and $O(|E|)$ to update $\mathbf{X}^{kl}$ and $\mathbf{Z}^{kl}$. Hence the total time complexity of Algorithm 1 is $O\big(T(|E| + n^2 c)\big)$, where T is the average number of inner iterations. In our empirical study, we found that algorithm usually converges within 30 iterations.

3 Related Work

In this section, we review some work which are closely related to our study.

Classification of information networks has been extensively studied in the past decade. Earlier studies mainly focus on the homogeneous network. For example, [18, 21] studied classification of undirected networks while [19] studied classification of directed networks. [20] proposed link-content matrix factorization (LCMF) method, which integrates content and link information into a joint matrix factorization framework. Sen et al. [14] studied collective classification of networked data. Li and Yeung [9] proposed probabilistic relational principal component analysis (PRPCA), which is the state-of-the-art subspace learning method for networks. More recently, [1,15] suggested active learning for networked data, whose goal is to minimize the labeling effort while maximize the classification accuracy. Gu and Han [5] proposed a feature selection approach for homogeneous networked data, which selects a subset of features, such that they are consistent with the link structure of the network. Recently, classification of heterogeneous information networks received increasing attention. For instance, as a natural generalization of [18], Ji et al. [8] proposed a model for classification of heterogeneous networks. Later, Ji et al. [7] proposed to integrate ranking and classification for heterogeneous networks, where they pay more attention to the nodes whose ranking scores are higher. All the methods mentioned above are heavily depending on the link structure of the network. They should perform well if we remove inconsistent links and add consistent links. However, their classification performance is limited when the networks are inconsistent. This motivates us to develop a new model which is robust to the inconsistent links and performs well on inconsistent networks.

We notice that Chen et. al [4] proposed a similar technique for sparse graph clustering. However, their method does not take into account the label information and heavily relies on the planted partition model assumption. Luo et. al [12] proposed a similar method namely forging the graph, while their method does not leverage label information either.

4   Experiments

In this section, we empirically evaluate the effectiveness of the proposed method. All the experiments are performed on a PC with Intel Core i5 3.20G CPU and 48GB RAM.

4.1   Data Sets

In our experiments, we use two benchmark datasets: one is a homogeneous citation network, the other is a heterogeneous bibliographic network.

Cora: It contains the abstracts and references of about 34,000 research papers from the computer science community. The task is to classify each paper into one of the subfields of data structure (DS), hardware and architecture (HA), machine learning (ML), and programming language (PL), based on the citation relation between the papers. We only use the link information of this dataset. The statistics about the Cora data set are summarized in Table 1. Before we run all the baselines and our algorithm, we first make adjacent matrices symmetric, i.e. set $r'_{ij} = \max(r_{ij}, r_{ji})$.

Table 1. Description of the Cora dataset

Data Sets	#samples	#links	#classes
DS	751	1283	9
HA	400	793	7
ML	1617	4046	7
PL	1575	4918	9

DBLP: We extract a sub-network of the DBLP data set on four areas: database, data mining, information retrieval and artificial intelligence, which naturally form four classes. By selecting five representative conferences in each area, papers published in these conferences, the authors of these papers and the terms that appeared in the titles of these papers, we obtain a heterogeneous information network that consists of four types of objects: paper, conference, author and term. Within that heterogeneous information network, we have four types of link relationships: paper-conference, paper-author, paper-term and author-term. The data set we used contains 14376 papers, 20 conferences, 6401 authors and 4483 terms, with a total number of 192003 links. For evaluation, we use a labeled data set of 2876 authors, 100 papers and all 20 conferences. The statistics about the DBLP data set are summarized in Table 2.

Table 2. The statistics of the DBLP dataset

#paper	14376	#paper-author	33720
#author	6401	#paper-conference	14376
#conference	20	#paper-term	110187
#term	4483	#author-term	33720

4.2 Baselines and Parameter Settings

We compare the proposed method with the state-of-the-art network classification algorithms. The methods and their parameter settings are summarized as follows.

Network-only Link-based Classification (nLB). [13] We use network-only derivative of nLB because local features are not available in our problem. We use the implementation from NetKit-SRL[3].

Weighted-vote Relational Neighbor Classifier (wvRN). [13] We only create a feature vector for each node based on the structure information and use the implementation from NetKit-SRL.

Learning with Local and Global Consistency (LLGC). [18] LLGC is a graph-based transductive classification algorithm. The regularization parameter is tuned by searching the grid $\{0.01, 0.1, 1, 10, 100\}$.

GNetMine. [8] GNetMine is a heterogeneous generalization of LLGC. According to [8], we set the regularization parameters λ to be the same for every pair of k, l, and tune it by searching the grid $\{0.01, 0.1, 1, 10, 100\}$. It uses three relation matrices: paper-author, paper-conference and paper-term.

Consistent Graph Learning (CGL). The regularization parameters β_{kl} are set to be the same for all k, l, and tuned by searching the grid $\{0.1 : 0.1 : 1\}$ and $\{1 : 1 : 20\}$. Similarly, we turn regularization parameters γ_{kl} by searching the grid $\{1 : 1 : 10\}$ and $\{10 : 10 : 100\}$. After learning the consistent relation matrices,

---

[3] http://netkit-srl.sourceforge.net

Table 3. Classification Accuracy (%) on the Cora dataset

subset	DS			HA			ML			PL		
#labeled node	20%	50%	80%	20%	50%	80%	20%	50%	80%	20%	50%	80%
nLB	39.49	41.08	57.87	38.50	48.27	64.13	45.14	55.32	60.09	46.82	58.75	61.40
wvRN	45.56	58.97	64.93	42.56	57.24	67.38	49.97	60.77	63.69	50.90	59.76	64.84
LLGC	62.77	73.39	77.32	70.38	81.71	83.50	73.54	81.05	82.79	66.66	75.33	76.16
CGL	**64.39**	**77.02**	**81.45**	**79.72**	**87.31**	**88.38**	**76.26**	**82.68**	**84.94**	**69.54**	**76.90**	**80.68**

we use LLGC for Cora and GNetMine for DBLP to estimate the unlabeled data. Parameters are tuned in the same way as LLGC and GNetMine.

The searching grids are set based on heuristics. We found for CGL, the balance of different regularization terms is more important than the absolute values. Note that we do not compare our method with [1,5,7,9,20] because they either need the content information of the nodes or come from a different line of ideas.

4.3 Classification Results on Cora

For each subset of Cora dataset, we randomly choose 20%, 50%, 80% objects as labeled samples, and the rest as test samples. We repeat the selection 10 times and report the average result.

The semi-supervised classification results on the Cora data are shown in Table 3. We can see that the proposed method outperforms the other methods significantly on all the subsets with different proportion of labeled samples. More specifically, considering that LLGC is the special case of our method without relation matrix learning, it indicates that finding the consistent network is of essential importance for classification of network.

4.4 Classification Results on DBLP

For DBLP dataset, according to [8], we randomly choose 0.1%, 0.2%, 0.3%, 0.4%, 0.5% of authors and papers, and use their label information in the classification task. When applying LLGC to heterogeneous network, we tried different settings. In detail, when classifying authors and papers, we tried constructing homogeneous author-author (A-A) and paper-paper (P-P) subnetworks in various ways, where the best results reported for author are given by the co-author network, and the best results for papers are generated by linking two papers if the are published in the same conference. The above two approaches are referred as LLGC (A-A) and LLGC (P-P). Note that we do not use labels of conferences in training, so we cannot build a conference-conference (C-C) sub-network for classification. We also try to apply LLGC on all the objects without considering their different types. It is denoted by LLGC (A-C-P-T). The key difference between LLGC (A-C-P-T) and GNetMine (A-C-P-T) is the normalization of the relation matrix: the former one normalizes the whole big relation matrix, while the latter one normalizes the small relation matrices respectively. The semi-supervised classification results of paper, author and conference on DBLP

Table 4. Classification Accuracy (%) of Paper on the DBLP dataset

% of authors and papers labeled	LLGC (P-P)	CGL (P-P)	LLGC (A-C-P-T)	GNetMine (A-C-P-T)	CGL (A-C-P-T)
0.1%	63.26±2.81	67.02±2.97	58.49±2.23	71.74±2.93	**74.42±2.50**
0.2%	69.58±2.49	74.15±1.98	61.27±2.41	80.01±2.67	**82.94±2.53**
0.3%	80.70±2.68	82.47±1.73	69.82±2.78	84.91±2.31	**87.75±2.23**
0.4%	79.76±2.29	82.29±2.39	67.38±1.90	83.81±1.86	**87.62±2.29**
0.5%	79.64±1.76	82.57±2.37	74.64±2.15	83.57±2.18	**87.00±2.75**

Table 5. Classification Accuracy (%) of Author on DBLP dataset

% of authors and papers labeled	LLGC (A-A)	CGL (A-A)	LLGC (A-C-P-T)	GNetMine (A-C-P-T)	CGL (A-C-P-T)
0.1%	41.31±3.04	45.39±3.18	58.68±2.89	80.42±2.62	**83.77±2.55**
0.2%	45.51±2.83	51.47±2.93	60.86±3.41	81.24±3.46	**84.71±3.58**
0.3%	47.72±2.72	55.53±2.75	66.39±3.08	84.50±2.55	**87.31±2.44**
0.4%	47.47±3.81	55.50±3.41	70.32±2.19	85.14±2.05	**88.97±1.94**
0.5%	50.64±2.16	59.11±2.31	71.17±1.72	86.84±1.55	**89.93±1.67**

Table 6. Classification Accuracy (%) on Conference

% of authors and papers labeled	LLGC (A-C-P-T)	GNetMine (A-C-P-T)	CGL (A-C-P-T)
0.1%	73.50±4.84	80.50±3.25	**82.00±3.94**
0.2%	77.50±2.35	83.50±2.30	**86.50±2.12**
0.3%	82.00±3.37	87.00±2.89	**90.00±2.82**
0.4%	78.00±2.83	88.00±2.22	**92.00±2.59**
0.5%	82.50±2.17	90.00±2.77	**94.50±2.80**

dataset are shown in Tables 4, 5 and 6 respectively. Since wvRN and nLB perform much worse than the other methods on this dataset, we omit their results due to space limit. Similar observations are reported in [8].

We can observe that:

1. CGL (A-C-P-T) outperforms the state-of-the-art method, i.e., GNetMine, significantly on all the types of objects. The reason is that CGL is able to learn an approximately consistent heterogeneous network. This indicates the effectiveness of CGL on heterogeneous information networks.
2. CGL (P-P) is better than LLGC (P-P) and CGL (A-A) is better than LLGC (A-A). This strengthens the effectiveness of CGL on homogeneous networks.

5   Conclusions and Future Work

In this paper, we proposed a *Consistent Graph Learning*, which is robust to inconsistent links in networks. Experiments on both homogeneous and heterogeneous network datasets show that the proposed method outperforms the state-of-the-art methods. In the future, we plan to develop theoretical analysis on the conditions under which the relation matrices can be recovered. Also, it is interesting to analyze how the percentage of inconsistent links in a network affect the classification performance, and test the algorithm in data set with large number of classes.

Acknowledgments. Research was sponsored in part by the U.S. Army Research Lab. under Cooperative Agreement No. W911NF-09-2-0053 (NSCTA), National Science Foundation IIS-1017362, IIS-1320617, and IIS-1354329, HDTRA1-10-1-0120, and grant 1U54GM114838 awarded by NIGMS through funds provided by the trans-NIH Big Data to Knowledge (BD2K) initiative, and MIAS, a DHS-IDS Center for Multimodal Information Access and Synthesis at UIUC.

References

1. Bilgic, M., Mihalkova, L., Getoor, L.: Active learning for networked data. In: ICML, pp. 79–86 (2010)
2. Boyd, S., Parikh, N., Chu, E., Peleato, B., Eckstein, J.: Distributed optimization and statistical learning via the alternating direction method of multipliers. Foundations and Trends® in Machine Learning **3**(1), 1–122 (2011)
3. Boyd, S., Vandenberghe, L.: Convex optimization. Cambridge University Press, Cambridge (2004)
4. Chen, Y., Sanghavi, S., Xu, H.: Clustering sparse graphs. In: Advances in Neural Information Processing Systems, pp. 2204–2212 (2012)
5. Gu, Q., Han, J.: Towards feature selection in network. In: CIKM, pp. 1175–1184 (2011)
6. Hong, M., Luo, Z.-Q.: On the linear convergence of the alternating direction method of multipliers. arXiv preprint arXiv:1208.3922 (2012)
7. Ji, M., Han, J., Danilevsky, M.: Ranking-based classification of heterogeneous information networks. In: KDD, pp. 1298–1306 (2011)
8. Ji, M., Sun, Y., Danilevsky, M., Han, J., Gao, J.: Graph regularized transductive classification on heterogeneous information networks. In: Balcázar, J.L., Bonchi, F., Gionis, A., Sebag, M. (eds.) ECML PKDD 2010, Part I. LNCS, vol. 6321, pp. 570–586. Springer, Heidelberg (2010)
9. Li, W.-J., Yeung, D.-Y., Zhang, Z.: Probabilistic relational pca. In: NIPS, pp. 1123–1131 (2009)
10. Lin, Z., Chen, M., Wu, L.: The augmented lagrange multiplier method for exact recovery of corrupted low-rank matrices. Analysis, math.OC:-09-2215 (2010)
11. Lu, Q., Getoor, L.: Link-based classification. In: ICML, pp. 496–503 (2003)
12. Luo, D., Huang, H., Nie, F., Ding, C.H.: Forging the graphs: a low rank and positive semidefinite graph learning approach. In: Advances in Neural Information Processing Systems, pp. 2960–2968 (2012)
13. Macskassy, S.A., Provost, F.J.: Classification in networked data: A toolkit and a univariate case study. Journal of Machine Learning Research **8**, 935–983 (2007)
14. Sen, P., Namata, G., Bilgic, M., Getoor, L., Gallagher, B., Eliassi-Rad, T.: Collective classification in network data. AI Magazine **29**(3), 93–106 (2008)
15. Shi, L., Zhao, Y., Tang, J.: Combining link and content for collective active learning. In: CIKM, pp. 1829–1832 (2010)
16. Smola, A.J., Kondor, R.: Kernels and regularization on graphs. In: Schölkopf, B., Warmuth, M.K. (eds.) COLT/Kernel 2003. LNCS (LNAI), vol. 2777, pp. 144–158. Springer, Heidelberg (2003)

17. Sun, R., Luo, Z.-Q., Ye, Y.: On the expected convergence of randomly permuted admm. arXiv preprint arXiv:1503.06387 (2015)
18. Zhou, D., Bousquet, O., Lal, T.N., Weston, J., Schölkopf, B.: Learning with local and global consistency. In: NIPS (2003)
19. Zhou, D., Huang, J., Schölkopf, B.: Learning from labeled and unlabeled data on a directed graph. In: ICML, pp. 1036–1043 (2005)
20. Zhu, S., Yu, K., Chi, Y., Gong, Y.: Combining content and link for classification using matrix factorization. In: SIGIR, pp. 487–494 (2007)
21. Zhu, X., Ghahramani, Z., Lafferty, J.D.: Semi-supervised learning using gaussian fields and harmonic functions. In: ICML, pp. 912–919 (2003)

Author Index